U0942843

孜孜以求，耕耘攀登，
硕果累累，巾帼强人。

祝贺李琦学友文集出版

洪绂曾

己丑二月

章琦近照

恩师林世成教授

1989年林世成等在南宁

1983年林世成出席国际水稻研究所（IRRI）理事会

1985年林世成会见IRRI的苗东华博士夫妇

党和国家领导人接见二〇〇〇年度国家科学技术奖获奖代表 2001年2月19日于人民大会堂

“我国抗稻白叶枯病粳稻近等基因系的培育及应用”荣获国家科学技术发明二等奖

1994年，课题组成员杨文才、施爱农、章琦、王春莲

1997年，课题组成员邢全党、周永力、章琦、王春莲、赵炳宇

1999年，“九五”攻关中期总结会议合影

1977年，沈阳市农业科学研究所杂交稻南繁人员

1977年，章琦、阙更生在海南南繁

1988年，陈成斌、李道远、章琦等
在南宁野生稻田进行抗性遗传调查

网室中携有*Xa23*基因的NIL

携有*Xa23*基因的抗病水稻品系

2008年，携有*Xa23*基因的抗病水稻品系（安徽省南繁）

2001年，在海南繁殖携有*Xa23*基因的水稻品系

2005年，安徽省海南育种基地

携*Xa23*的品系中野5114与JG30

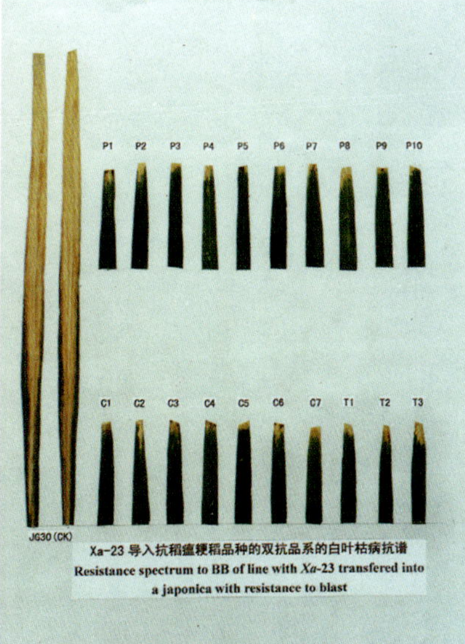

携有*Xa23*基因的广抗谱水稻叶片

*Xa23*对28个白叶枯病代表菌株的抗性反应

2007年，广东省农业科学院导入*Xa23*基因的育种田

2003年，安徽桐庐，携有*Xa23*的品系中野5114与JG30抗、感比较

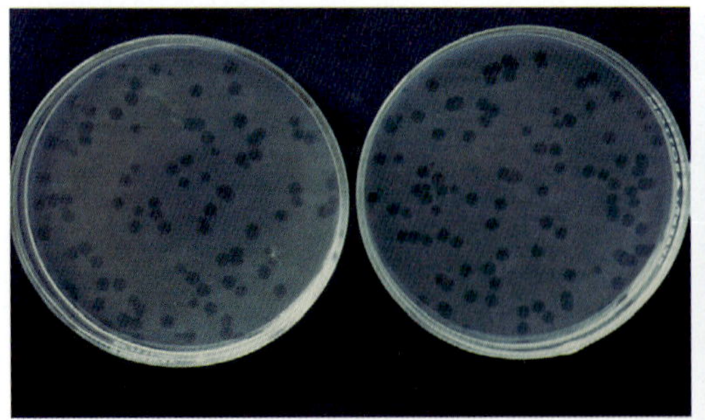

水稻白叶枯病菌的噬菌斑

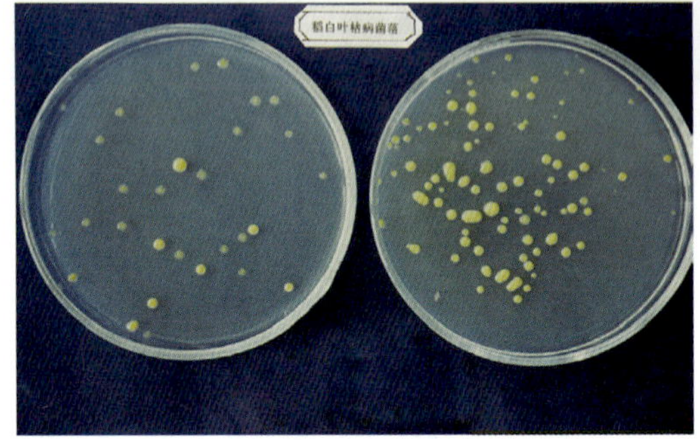

*Xoo*菌单菌落

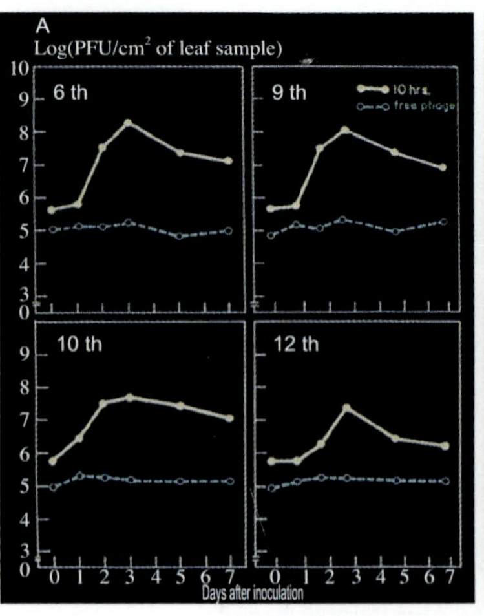

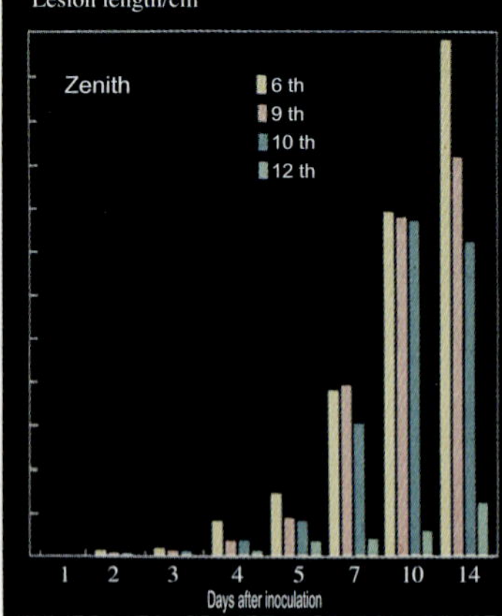

用噬菌体法测定成株抗性品种Zenith

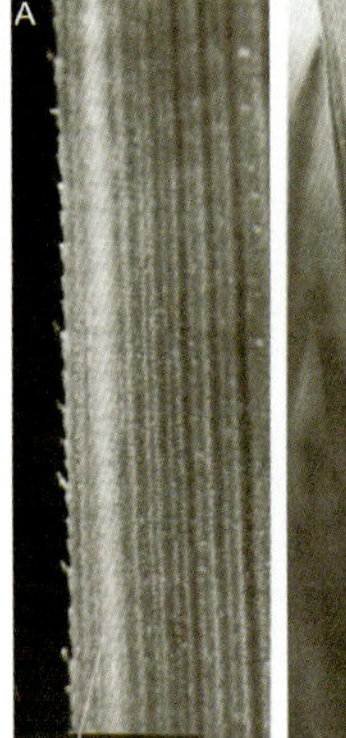

水稻白叶枯病症状

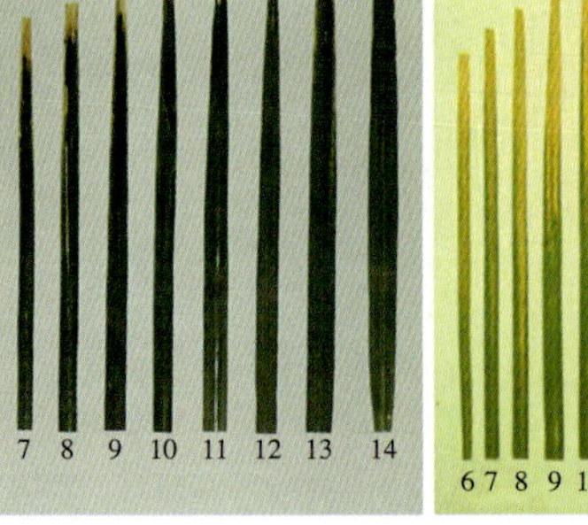

水稻对白叶枯病不同生育期抗性类型品种的病斑反应

A.全生育期抗性　B.成株抗性　C.感病对照

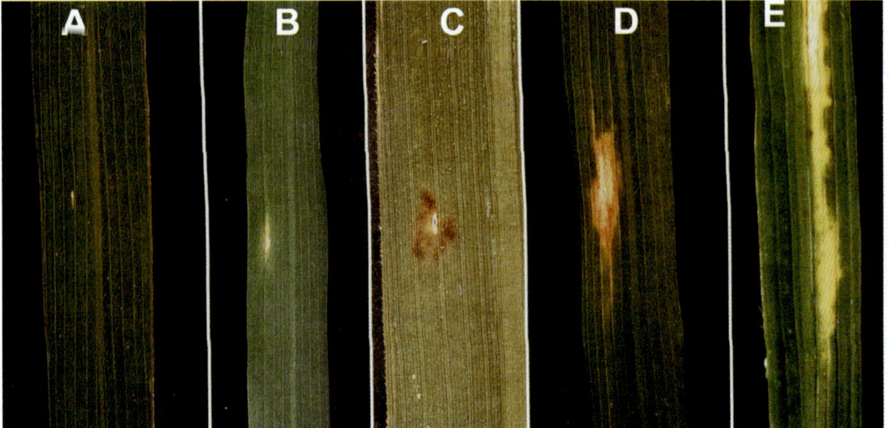

水稻和白叶枯病不亲和互作

水稻白叶枯病双针接种法

中国粳稻抗白叶枯病近等基因系图A

中国粳稻抗白叶枯病近等基因系图B

江苏连云港东海安峰乡白叶枯病严重发生的稻田（2005年）

1991年，中日稻病抗性遗传育种学术研讨会与会代表合影

1993年，分子生物技术讲习班成员

2007年，第二届国际水稻白叶枯病大会合影

2009年，章琦在国际杂交稻会议上作报告

2007年，南京，在国际水稻白叶枯病第二届会议上章琦与合作者在一起

2007年，南京，在国际水稻白叶枯病第二届会议上章琦与林汉明、王世平博士在一起

2010年，章琦与林汉明博士讨论本书

1989年，卢良恕（左四）、闵绍楷（左五）等访问国际水稻研究所（IRRI）［左二为张德慈博士，左三为IRRI副所长和韩国专家（右二）］

1985年，国际水稻研究所，国际水稻白叶枯病研讨会合影

1984年，国际水稻研究所病理系专家，Mew，Lee，章琦，Hibino，Bonman（从左到右）

1982年，章琦在温室调查

1989年，章琦在网室调查野栽杂交F_1

1987年，章琦和Vera Cruz在IRRI

1989年，章琦在实验室

1983年，章琦在IRRI的“中国之夜”

1985年，在稻麦抗性研讨会上章琦（后排右一）与吴友三（前排右三）、朱有钰（前排右二）、韦石泉（前排右四）等合影

1980年，欧世璜访问南京农业大学（自左至右，过崇俭、方中达、梅籍芳、欧世璜、章琦等）

1989年，李竞雄与先锋公司专家和章琦在一起

1997—1999年洛克菲勒基金和国际植物病理学会议　上左，章琦、Leach（杭州）；上右，章琦（泰国）下左，章琦、朱立煌、洪孟民、朱玉贤；下右，辛志勇、何礼远、章琦、成卓敏等

1993年，Leach 实验室人员合影，后排右二为Leach，前排右二为章琦

1993年，章琦在堪萨斯州立大学（KSU）植物病理系与系主任、Leach合影

1993年，KSU实验展廊

1994年，在Leach家过圣诞节

1995年，在Conlorado Leach家

1958年，大学时代，沈阳农学院篮球队队员

1958年，沈阳农学院篮球队

1998年秋游颐和园

2009年在云谷山庄

2009年在云谷山庄

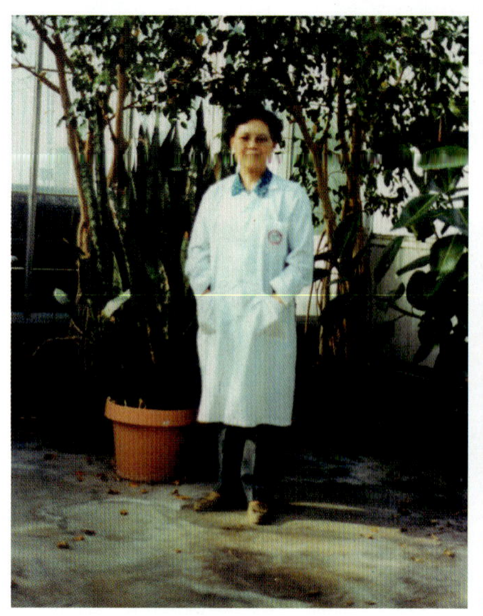

1993年在KSU温室

2010年6月在中国农业科学院研究生院作专题讲座

阙更生、章琦的大学时代

1967年去“五七”干校前

1968年，在沈阳居所门前

1968年，两个孩子在“文革”期间

2008年在水立方残奥会上

2009年老两口在鸟巢

祖孙乐

开心2008

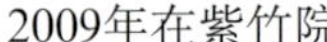

2009年在紫竹院

2009年在颐和园

2009年合家欢

2010年金婚

耕耘——章琦稻病抗性研究选集

TILLAGE–Selected Works of ZHANG Qi on Rice Disease Resistance

章　琦　林汉明　编

Edited by

Qi ZHANG Hon–Ming LAM

中　国　农　业　出　版　社

China　Agriculture　Press

序言一

初识章琦老师是在1989年秋，我正准备考研。扬州大学原校长顾铭洪教授建议我攻读章琦先生的硕士研究生，我们便开始有书信往来。1990年我如愿以偿地成了她的弟子，毕业后留在她的课题组参加抗性遗传研究和粳稻抗白叶枯病近等基因系的培育，直到1997年3月赴美。虽然我现在没能从事水稻病害方面的工作，但仍在进行蔬菜抗病性遗传的应用基础研究，与章老师一直保持着密切的联系。章老师既是一位著名的植物病理学家，又在抗性遗传研究方面有着很深的造诣，这些都源于她孜孜不倦的学习精神和严谨踏实的治学风格。在从事科学研究的几十年里，章老师一直保持敏捷的才思，始终把握住科学研究的前沿，不断拓深自己的研究内容，在科研和生产一线勤勤恳恳地工作。2007年，章老师出版了学术专著《水稻白叶枯病抗性的遗传及改良》之后，又将发表的论文整理成集，期望能再为年青一代做点什么，这又给我们增添了一笔宝贵的财富。

文集收录了章老师已发表的主要论文和科研成果，虽然这不是她的全部著作，但体现了她在水稻病害方面研究的系统性。从中可以看到，随着时间的推移，章老师的研究步步深入。从20世纪70年代开始，她就对水稻纹枯病的发生规律、品种抗性差异、农药筛选和栽培措施等综合防治进行了研究，较早地开始了水稻对稻瘟病和白叶枯病抗性遗传行为的分析。随后，她从病原菌和寄主两个层面，聚焦于白叶枯病，开展病原菌分化研究。90年代初，她开始采用分子技术探索我国白叶枯病菌的群体遗传结构及其与寄主抗性基因利用的关系，又同时进行品种对白叶枯病的抗性遗传研究，在评价了大量栽培品种的抗性后，系统地发掘利用野生资源抗源。通过从经典遗传学分析到现代的分子基因定位、从传统育种到标记辅助选择等工作，她不仅成功地培育了国际首套粳稻近等基因系，并发掘鉴定了深受国内外同行瞩目的优异抗

白叶枯病新基因 *Xa23*，育成了几个携带该基因的水稻高抗品系。文集已经展示了章琦老师的学术成就和深厚的造诣，这里不再赘述。

白叶枯病严重威胁水稻的生产。早在 20 世纪 50 年代我国著名植物病理学家方中达先生就开始对我国水稻白叶枯病的基础生物学问题进行了探讨，为我国的白叶枯病研究打下了坚实的基础。然而，80 年代以前我国一直缺乏从抗源筛选到抗性遗传表现以及利用等方面的连续性研究。要开展这种系统性的研究，研究者不仅自身要有渊博的知识，更需要有精诚合作的态度。章老师就是一位这样的科学家，她充分发挥了植物病理学方面的特长，勤奋学习遗传育种和分子生物学方面的知识，并且与国内外植物病理、遗传育种和分子生物学家开展了广泛而紧密的合作，取得了一系列成果，也赢得了国内外同行的尊重。这本文集既是章老师几十年不倦工作的真实写照，也是我国科学家对水稻白叶枯病的研究发展史。

众所周知，在某一个领域开展应用基础研究都需要长期的积累，期间可能会遇到种种困难和挫折。章老师几十年的科研生涯并不是一帆风顺的，但是她带领课题组中的年轻人执著地追求真知，求实进取，踏踏实实地埋头苦干，硬是凭借着那种“永不言弃”的精神克服了经费不足、人手少、科研条件有限等困难，朝着自己的人生目标迈进。她的这种精神永远值得我们年青一代学习。

我很荣幸为此文选写序，借此表达对恩师的万分敬意。

中国农业大学教授 杨文才

2010 年 1 月 20 日

Preface 1

I first became acquainted with Professor Qi Zhang through Professor Minghong Gu, former president of Yangzhou University, in the fall of 1989 when I was preparing for the entrance exam for graduate study. Professor Gu suggested me to pursue my master degree under Professor Zhang's supervision, so I started communication with Professor Zhang via mail. In 1990, I became her student as wished. After graduation, I joined her research group to work on resistance genetics and develop near isogenic *Japonica* rice lines for resistance to bacterial blight till I left for the US in March 1997. Although I am not engaged in studies on rice diseases any more, I am still conducting fundamental studies on genetics and breeding for resistance to vegetable diseases. Thus, I have kept in touch with her closely. Professor Zhang is a famous plant pathologist. She also had great achievements in genetics of disease resistance. These should be attributed to her tireless and rigorous scholarship. For decades in scientific research, Professor Zhang has kept an agile mind, always grasping the front edge of science and constantly expanding her research areas. She is diligent and conscientious in scientific research and production. In 2007, she published her academic monograph, *Genetics and Improvement of Resistance to Bacterial Blight in Rice*. Nevertheless, Professor Zhang wishes she can do more things for the younger generation. So she compiled her publications into this volume and gave us another valuable asset.

This volume includes Professor Zhang's main publications and academic achievements. Although it is not a complete collection, it demonstrates her systematic research in rice diseases and how her work has progressed in greater depth over time. She started in the 1970s by working on integrated control of rice sheath blight, including the pattern of occurrence, difference of varietal resistance, chemical screening and cultivation measure. She is also one of the pioneers to analyze the inheritance of resistance to rice blast and rice bacterial blight. Her work focused on pathogen differentiation of bacterial blight on the levels of both pathogen and host. In early 1990s, she investigated the genetic structure of *Xanthomonas oryzae* pv. *oryzae* population in China and the relationship between the pathogen and the utilization of resistance genes using molecular technology. Meanwhile, she studied the inheritance of resistance to bacterial blight. After evaluating resistance in many cultivars, she explored and utilized the resistance source in wild relatives. Her work varied from classical genetic analysis to molecular mapping and from traditional breeding to marker-assisted selection. She not only developed the first set of near isogenic lines in *Japonica* rice, but also discovered a novel gene, *Xa23*, with strong resistance to rice bacterial blight. This gene caught the attention of scholars worldwide and was later developed into several highly resistant lines. As Professor Zhang's academic achievements and attainments have already been well demonstrated in this book, I shall not go into the details here.

Bacterial blight puts rice production under great threat. As early as in the 1950s, Professor Zhongda Fang, a famous Chinese plant pathologist, began investigating the basic biology of bacterial blight and built a solid foundation for the study of bacterial blight in China. However, there had been a lack of a systematic study from source screening to the genetics of resistance and its utilization before the 1980s. The establishment of a systematic research requires a researcher with broad knowledge and a sincere attitude for collaboration. Professor Zhang was just the right person. Giving full play to her professional knowledge in plant pathology and learning assiduously on genetics, breeding and molecular biology, she broadly and closely collaborated with national and international plant pathologists, geneticists, breeders, and molecular biologists. Her achievements won her respects from colleagues here and abroad. The book is a genuine portrayal of Professor Zhang's hard work in the past decades. It also describes the development of the study on bacterial blight of rice by Chinese scientists.

It is well known that establishing basic applied research requires persistent efforts and would probably face various difficulties and setbacks. Professor Zhang's scientific career has not been all plain sailing. However, she led the young ones in her group to relentlessly seek knowledge through rigorous work. With the belief of "never give up", she overcame difficulties including lack of funds, inadequate manpower and limited support in scientific research to continue striving for her goal. Her spirit is always worthy of study by the younger generation.

Honored to be given the chance to write this preface, I would like to take this opportunity to express my deepest gratitude to my mentor.

Professor, China Agricultural University **Wencai Yang, Ph. D.**

January 2010

序言二

生平第一次写序，就是为我的恩师章琦老师撰写的《耕耘——章琦稻病抗性研究选集》作序，难免诚惶诚恐。章琦老师治学严谨，精益求精。拜读了文选初稿，我再一次为章老师凡事追求完美的风格所感动。

章琦老师致力于研究水稻病害，特别是在白叶枯病的防治和抗性方面。40余载，她的著述丰富，研究成果涵盖了从水稻生产实践、病虫害综合防治、病原菌分化、水稻抗病性遗传，到现代分子生物学技术的应用等广泛内容。这些文章本身就见证了老一代科学家承前启后、奋斗终生、呕心沥血、教书育人，培养新一代农业科研工作者的历程。这些论文也见证了我国农业科研人员脚踏实地的从植物病害的发生发展规律观察到深入至寄主抗病性遗传分子机理研究的发展过程。

我有幸在章琦老师门下读书、工作了将近6年，参加了她主持的课题组有关发掘利用野生稻抗白叶枯病资源的研究。章老师言传身教、耳提面命、严格要求，使我在科学研究、做学问和做人等基本素质方面得到了很好的启蒙教育，令我终生受益。章老师的名言是“做农业研究一定要有头啃地的精神，要既能动手又能动脑，有想象力，有真诚合作精神，只有这样才能真正从事研究工作，也才能领会到科学研究的真正意义：从实践到理论研究，再从理论研究回到农业实践”。

我想作为年轻的农业科研工作者对文选内容的阅读，最要紧的是从中学习体会章琦老师的治学方式和她勤奋耕耘、锲而不舍、高度负责的精神。她满腔热情的对事业的执著追求和献身精神也是农业科研工作者学习的榜样。文选有关白叶枯病抗性研究的内容已在章琦老师近年发表的专著《水稻白叶枯病抗性的遗传及改良》中有较系统的论述。本文选是章琦老师专著的姊妹篇，在研究方法、结果和论述方面更为具体。相互参阅，定会受益匪浅。目前，水稻白叶枯病的研究已经成为探索植物与微生物分子互作的模式系统，大量文献纷纭繁复，新成果、新发现日新月异。为了尽快理清稻白叶枯病研究的历史进程，明白其间的来龙去脉，章琦老师的文选是不可不读的文献。

学生 赵炳宇

（弗吉尼亚理工大学，助理教授）

Preface 2

With reverence, I wrote my first preface for this book, *Tillage*, written by my mentor, Professor Qi Zhang. Professor Zhang is rigorous in scholarship and persistent in constant improvement. Reading the manuscript, I am once again impressed by Professor Zhang's striving for perfection.

Professor Zhang devoted herself in research on rice diseases, particularly on bacteria blight and resistance. For over 40 years, she wrote plenty of research papers covering topics from applications on rice production, integrated control, pathogenic differentiation, resistance inheritance in rice to the applications of modern molecular biology technology. These articles are testimonies of how preceding scientists battled painstakingly to educate and foster the new generation of researchers in agricultural sciences. The papers also demonstrated the development of study of diseases, from observation of the patterns of occurrence to in-depth investigation on the mechanism of resistance inheritance in hosts, by the solid work of Chinese researchers in agriculture.

I was fortunate to learn from and work with Professor Zhang for almost 6 years, as well as participate in her research group investigating the resistance of bacterial blight by the use of wild rice resources. She walked the talk and led by example. By growing a seed on the importance of discipline in scientific research and scholarship, her strict teaching will benefit all my life. Her famous remark was "doing agricultural research, one needs to work like a Trojan" . One has to use the hands as well as the brain, be creative and genuinely cooperative. Only by these he/she can work in research and understand the meaning of scientific research to develop theories based on practices and to bring theories back to agricultural applications.

When reading the collection, it is most important for young agricultural scientists to learn from Professor Zhang's scholarship and her diligent, persistent and responsible personality. Her passion, dedication and devotion on research are also exemplary for researchers. While Professor Zhang's previous book, *Genetics and Improvement of Resistance to Bacterial Blight in Rice*, provided a systematic description on the research of resistance to bacterial blight, this volume is more specific on research methods, results and analyses. One shall benefit most from cross-referencing the two books. Research on bacterial blight in rice has now been a model for research on the interaction of plants and biological molecule, leading to a lot of articles and new discoveries. This volume is not a book to be overlooked for people who want to understand the development of research in bacterial blight.

Bingyu Zhao,
Assistant Professor,
Program of Molecular Plant Sciences,
Department of Horticulture,
Virginia Polytechnic Institute and State University

前言

1960年我毕业于沈阳农学院植物保护系，原想当一名熟练的植物医生，努力掌握作物病虫害防治的基础知识与实用技术。然而，面对当时学习、生活境遇的变化不定和本职工作的需求，单凭一点肤浅的书面专业知识是远远不够的，必须虚心向有经验的人请教，善于不断在实践中学习，才能逐渐拓宽知识、积攒经验。勤奋加上机遇，让我有缘在水稻抗病遗传改良研究这块土地上耕耘了大半生，留下些点滴痕迹，或许可供年青一代同行借鉴。这就是我虽年逾古稀还要出版这本书的初衷。

大学毕业后至退休前，除在“文革”期间当了3年农民外，先后在基层单位从事作物病虫害防治工作，在科研机构进行水稻抗病育种的应用研究，然后逐渐向相关领域延伸。在近半个世纪的科研生涯中，得到诸多良师益友的热忱无私提携和帮助，其中许多细节在林汉明教授为我撰写的传略中已有提及。饮水思源，我在此再谈些个人感受。1979年国际水稻研究所（IRRI）理事会中国首任理事，已故林世成先生的提携，是我人生的重大转折点，通过他的竭力引荐和一系列艰难繁杂的运作，将我全家从东北调入北京，并安排我在中国农业科学院作物研究所水稻育种室“水稻遗传评价和利用”研究组工作。他是国内知名的水稻科学家，谦虚平和，乐于助人，学风严谨朴实，业务基础扎实，工作认真负责，深受同仁推崇。我耳濡目染，潜移默化，被引领入系统研究抗性遗传育种应用基础领域。1982—1984年，林先生又推荐我去IRRI深造，合作研究白叶枯病抗性及其遗传，获益良多。工作告一段落，对方希望我能留下继续协作研究，而当时国内则更为急需进修人员回国服务。不料一向宽厚的林先生却让我自己决定是否回归课题组，这一慷慨举动深深触动了我，我就义无反顾地决定回国工作。

在IRRI学习、交流合作期间，我首先应该感谢博学、热情的植物病理学家苗东华博士，接受我作为访问学者与他协同开展研究。他真诚、毫无保留地以自己亲身体会与我讨论研究立题、试验设计及其他细节，使我很快明确、熟悉和进入工作状态，因而得以在两年内就做完了本该4年完成的研究工作。1988年，是他在日本大阪国际植物病理大会的细菌病害分会上，推荐我入选为国际水稻白叶枯病研究工作委员会5个委员之一。此后，我们之间建立了长期密切的互动合作关系。在IRRI的头两年，我还有幸认识国际著名的水稻资深遗传学家张德慈博士。他知识渊博、真诚坦率、助人为乐，曾多次抽暇讲解遗传学基础知识，利用他编写的经典遗传学简明教材，帮助分析我在IRRI初期获取的抗病性遗传研究数据结果，并惠赠

Present knowledge of Rice Genetics and Cytogenetics 、*The Conservation and Use of Rice Genetic Resources* 等专著，为以后进一步自学经典遗传学打下重要基础。

20 世纪 90 年代初，应用分子生物学技术研究水稻病原菌群体多样性已是国际研究热点之一。为了不落后于国际发展，1992 年我应美国堪萨斯州立大学分子微生物学家 Jen E. Leach 教授邀请，前往她的实验室进行合作研究，第一次运用 RFLP、Rep - PCR 等分子技术分析中国水稻白叶枯病菌群体遗传结构，这是一次新的挑战。Leach 博士曾多次热情、耐心而又扼要地讲解这些分子技术的基本原理，并亲自操作示范，使我能够尽快地熟练掌握这类基本操作。在先后两次赴美与她合作研究中，我们有过多次深入交流、共同切磋，从不同相关学科交叉融合分析所得研究结果，深化诠释病原菌 DNA 单元型（haplotype）与其毒性群的对应性；以及寄主基因型分布对病原菌群体遗传多样性的影响。她的睿智、宽厚与真诚，深受学生的爱戴。我两次访美都由她亲自接送，并在生活上给予无微不至的照顾，让我难以忘怀。1992 年，她访问中国，在全国分子生物技术应用讲习班上承担主讲半个月，促进了我国这一研究的发展，我们之间也建立了更加密切的联系。1994 年，承蒙她的推荐，我的研究项目获得洛克菲勒基金的资助。

历经 10 年，我和所内外的同事们从中国普通野生稻（*Oryza rufipogon*）中发掘鉴定出白叶枯病抗性新基因 *Xa23* 并初步应用。这项工作包括广西农业科学院以李道远研究员为首的课题组人员的材料接种、花培提纯；中国科学院遗传研究所朱立煌研究员的技术指导和提供基因分子定位实验条件；农民育种家江祺祥协助育成中野 5112 等双抗稻瘟病和白叶枯病品系，以及由广东省农业科学院、安徽省农业科学院、中国水稻研究所的周少川、李成荃、杨长登等育种家成功地应用于水稻抗病育种，这确是一项集体协作的结晶。

1991—1994 年，为了进一步促进我国水稻抗病性遗传研究，两位水稻遗传学家朱立宏教授、谢岳峰教授和我，本着学科互补、集思广益的精神，联合提出“我国水稻白叶枯病抗性遗传研究标准化商榷”的建议，由国家自然科学基金立项，在各单位不计较年均不到7 000多元经费的艰难条件下，紧密协作，取得了预期效果。

2001 年，我参与香港中文大学分子生物学家林汉明教授主持的有关“水稻抗病信号传导研究”。基于学术上的共识和学科的互补，以及林汉明教授的敬业精神、严谨治学和诚挚待人，使彼此成为忘年之交。我们曾共同在国际刊物发表 3 篇论文，并合写了《水稻白叶枯病抗性的遗传及改良》中的两章书稿。然后他又欣然承担共同编辑这本书的任务，并在百忙中抽暇审阅了全书。

1998 年 12 月我退休了，2003 年在国家自然科学基金和洛克菲勒基金的资助下，我又在编写一本题为“水稻白叶枯病抗性的遗传及改良”的专著，2007 年由科学出版社出版。

以上不厌啰嗦地对自己一生科研历程的回顾，其目的是想阐明：不论是个人知识学问的成长，还是一个领域或相关领域以致不同领域的学术思想的发展与完善，都是在各抒己见、自由讨论的基础上通过相互沟通、渗透、融合而各自去粗存精、集思广益后逐步形成的。我国科学研究的力量是雄厚的，改革开放，鼓励协作兼善天下。从大处着眼，小处下手，完全可以在现有的人力、物力条件下实现事半功倍的高效快速发展。

2008 年我在整理发表过的论文中，选出 60 篇编成这本选集“耕耘”，分为六个部分，其要点如下。

Ⅰ　水稻抗病育种研究

选入“水稻品种对白叶枯病和稻瘟病的抗性研究”等11篇。

包括水稻纹枯病综合防治研究，揭示水稻对白叶枯病和稻瘟病的抗性遗传有显性、不完全显性和隐性之分，根据不同稻瘟病抗性基因品种的抗谱与抗病型的相关性，提出抗病育种材料取舍的依据，通过我国主栽品种白叶枯病抗性系谱渊源分析，阐明了不同基因型水稻白叶枯病成株抗性品种的抗性机理等。

Ⅱ　水稻白叶枯病抗性遗传研究

选入“我国水稻白叶枯病抗性遗传研究标准化的几点商榷”等15篇。

提出培育近等基因系的关键技术和方法，抗性遗传研究标准化要点，包括抗感亲本和所用菌株的选择、接种和调查的要求；遗传研究方法：配组、群体容量、分析、基因鉴定、定位和命名的要求。

Ⅲ　水稻白叶枯病抗性新基因的鉴定与应用

选入“普通野生稻抗水稻白叶枯病新基因 *Xa23* 的鉴定和分子标记定位”等10篇。

阐述从普通野生稻中发掘白叶枯病抗性新基因 *Xa23* 的关键技术。包括抗谱比较、野栽杂交、花培、构建近等基因系、遗传分析和分子定位等。

Ⅳ　水稻白叶枯病病理、群体遗传结构研究

选入“中国水稻白叶枯病菌致病型的研究”等9篇。

包括水稻白叶枯病致病性合作研究，将中国水稻白叶枯病菌分为7个致病型；90年代，IRRI、美国和亚洲国家首次协作，揭示单元型多元演化的异质性小种能孕育新毒性型而分离出新小种；寄主基因型是引起病菌遗传适应性演变的主因，通过病菌分子型遗传多样性差异图像，推演出 *Xoo* 菌群系统发育及其历史渊源，初步对品种—病菌—环境间的互动关系做出阐述。

Ⅴ　学术报告与述评

选入在国际会议上的“Characterizing adult plant resistance to BB”（1984），和“Genetics and improvement of resistance to BB in hybrid rice in China”（2008）等4个报告，“水稻白叶枯病抗性基因鉴定进展及其利用”（2005）等3篇论述。

2004年和2007年，分别在日本和中国举行第1和第2届国际水稻白叶枯病研讨会，提出野生稻基因利用及中国科学家们半个世纪以来在水稻白叶枯病抗性遗传研究上的贡献。2008年，在第5届国际杂交水稻学术研讨会上，介绍中国杂交水稻白叶枯病抗性遗传改良的进展及面临的挑战。

在三篇论述中，概述20世纪90年代后水稻抗性遗传改良状况，以及半个多世纪以来水稻白叶枯病抗性基因鉴定史。

Ⅵ　专著

选自两本书：《Bacterial Blight in Rice》（IRRI，1989）中的“Types of resistance in rice to bacterial blight”和《水稻白叶枯病抗性的遗传及改良》（章琦编，2007）中的“水稻白叶枯病质量抗性遗传和抗病主基因鉴定”、“水稻白叶枯病抗性的机理”等7篇。

专著是集一个多世纪以来，水稻白叶枯病抗性及其遗传改良的显著进展和国内外学者对抗性及遗传改良的理解与认识。它综合经典遗传和分子遗传学的知识，对质量抗性遗传的一般规律与主基因系统鉴定进行论述。透过探讨病原菌群体结构遗传多样性与寄主抗性基因互作的协同演变，以及白叶枯病抗性机理等方面的研究历史、现状、近期进展和今后值得探索的问题，可以了解我国现代水稻抗病育种研究和生产发展的总体情况。

最后，愿借此机会感谢院、所领导和有关职能部门的长期支持，特别要感谢草地科学专家、原农业部副部长洪绂曾教授拨冗为本书题词，赵炳宇博士和杨文才博士分别为本书作序，万建民教授除一贯支持本选集的编写工作外，还为其出版筹措资助。我主持的课题组全体成员与我长期共同作战，他们在上述研究成果中都做出了重要贡献，顺致诚挚的感谢。

本书的出版得到中国农业出版社的大力支持，责任编辑张洪光编审做了大量细致的校阅、编辑等工作，部分插图和文献汇总由郑天清博士完成，在此一并鸣谢。

伴我走过半个世纪的先生阙更生研究员，在水稻育种上给予我很大帮助，始终默默地支持我的工作，风雨同舟，无言感激。

章琦

2010年1月20日

Foreword

In 1960, I graduated from the Department of Plant Protection, Shenyang Agricultural College. With a career goal to become a skilled plant doctor, I paid great effort to learn technologies of plant protection. Facing the changing circumstances in my life and the job's need of continuous in-depth investigation, using what I learnt from my education and books was far from adequate. I must humbly seek advice from the experienced and improve myself through practices. To accumulate knowledge and experience, devotion and diligence were needed. Striving hard and given chances, more than half of my lifetime was tightly tied with the study on genetics and improvement of rice resistance to diseases, leaving traces of footprints through my "tillage" in this research field. Should these traces I left inspire the young researchers, it would serve the objective of my publishing this book at the age of seventy.

Except the three years working as a farmer during the "Cultural Revolution", since graduation until retirement I worked on crop pest protection and application research on rice breeding for diseases resistance. Sincere helps and encouragements from mentors and friends provided great support throughout my near half century of academic life. These memories will always linger in the depth of my heart. Many details were given in the biography written by Professor H. M. Lam. Here, I express my heartfelt thanks to those named and unnamed who had accompanied me in the path that I had been through.

Professor S. C. Lin, who became the first Chinese member in the International Rice Research Institute (IRRI) Board of Trustee in 1979, was critical to my career. Sweeping away all obstacles, he helped to relocate my family to Beijing to join the research group "Genetic Evaluation and Utilization" in the Rice Breeding Unit of the Crop Breeding and Cultivation Institute at the Chinese Academy of Agricultural Sciences. Professor Lin was a scholar with profound knowledge in genetics. He was a model of serious and dedicated scientist. Subtly, his charisma had influenced the development of my scientific mind and value, and led me into the research field of genetic improvement of disease resistance in rice breeding. From 1982 to 1984, Professor Lin gave me an opportunity to further my study on the inheritance of bacterial blight (BB) resistance at the IRRI. At the completion of this collaborated research, he generously allowed me to choose to stay in IRRI despite a tight demand of manpower in his laboratory. His kindness was most touching, so I unfalteringly decided to return to China.

I must thank Prof. T. W. Mew, a knowledgeable and enthusiastic plant pathologist, for his kindness during my exchange to IRRI. He wholeheartedly discussed and planned the scope and content of my research. This help brought me a quick start and allowed me to complete my project in two years instead of a normal four-year period. In the 1988 meeting of the Bacterial Pathogen Chapter of the International Plant Pathology Association, Prof. Mew recommended me to become one of the five committee members of the International Rice Bacterial Blight Research Consortium. Since then, we have established a close and long-lasting collaboration relationship. In the beginning of my two-year stay at IRRI, I also met the renowned rice geneticist Prof. T. T. Zhang. He gave personal tutorials to enrich my fundamental knowledge of genetics. Using the "Summary of Common Genetic Ratios Formed by Indistinguishable or Absent Classes" complied by him, he helped to analyze the initial data of my project on disease resistance genetics. He also kindly shared with me his publications such as *Present Knowledge of Rice Genetics and Cytogenesis* and *The Conservation and Use of Rice Genetic Resources*. These materials formed an integral part of my foundation in classical genetics.

In the early 1990s, the application of molecular biology technologies to study the population diversity of rice pathogens had caused great interest in the international scientific area. To catch up with this new development, I became a visiting scientist to the laboratory of Prof. J. E. Leach, a molecular microbiologist at Kansas State University, in 1992. For the first time, I adopted unfamiliar technologies such as RFLP and Rep-PCR to analyze the population structure of the rice bacterial blight pathogen. Prof. Leach patiently and concisely explained and demonstrated to me the technologies involved, allowing me to learn the basic operation speedily. During my two visits to her laboratory, we had plenty of opportunities to discuss science in depth. Prof. Leach's wisdom and compassion earned her great respect from students. She was truthful and helpful. She took care of every detail of my stay in Kansas, even including rides to the airport. In 1992, Prof. Leach paid a return visit to China and hosted a 2-week national workshop on molecular technologies. This effort had helped the development of similar research in China as well as a closer relationship between us. In 1994, with her strong recommendation, I received supports from the Rockefeller Foundation to my further research works.

For ten years, my colleagues and I had been dedicated to the identification and application of the new bacterial blight resistant gene *Xa23* from wild rice (*Oryza rufipogon*). This achievement was a product of an extensive collaboration. Professor D. Y. Li and others from Guangxi Academy of Agricultural Sciences contributed to the large scale pathogen inoculation and anther cultured. Prof. L. H. Zhu provided technical advices and facilities for molecular mapping. The breeder Q. X. Jiang helped to generate lines with resistance to BB and rice blast such as Zhongye 5112, which were later employed in rice breeding programs led by the breeders S. C. Zhou, C. Q. Li, and C. D. Yang from Guangdong and Anhui Academy of Agricultural Sciences and China Rice Research Institute.

To advance the national standard on inheritance of rice disease resistance, rice geneticist professors L. H. Zhu, Y. F. Xie and I proposed a National Science Foundation project during 1991 to 1994. With an annual budget of less than RMB 7 000, our complementary expertise and collected efforts resulted in a proposal on the standardization of the national genetic study for BB resistance in rice. This is another successful example of deep collaboration across different disciplines.

In 2001, I participated in the research program on "Studies on signal transduction of rice defense responses" led by plant molecular biologist Prof. Hon-Ming Lam from the Chinese University of Hong Kong. He is a rigorous and dedicated scientist. His upright and passionate personality built the foundation of our deep friendship. We jointly published three papers in international journals and wrote two chapters in my edited book *Genetics and Improvement of Resistance to Bacterial Blight in Rice*. Without hesitation, he agreed to co-edit this book despite his tight working schedule.

I retired and stepped down from the research frontier in December 1998. Supported by both the National Natural Science Foundation of China and the Rockefeller Foundation, a monograph *Genetics and Improvement of Resistance to Bacterial Blight in Rice* was published by Science Press (China) in 2007.

Through my wordy reiteration of my path in scientific research I hope to enunciate that the growth of knowledge, no matter on a personal level, in a discipline or across disciplines, can only happen in the foundation of free expressions and discussions. Through communication and cooperation, knowledge is condensed and consolidated. Being open-minded and collaborating, China has great strength in scientific research. Its development can certainly be accelerated with the talents and resources we have, as long as we focus on large issues and tackle them bit by bit.

Since 2008, I have begun editing papers for the publication of *Tillage*. Sixty papers were selected. They were divided into six sections with the following themes.

1 Studies on breeding rice for resistance to diseases

There were eleven papers in this part, including "Studies on resistance to bacterial blight and blast in rice" (1978), etc.

The content includes integrated control of rice sheath blight, genetic behaviors of resistance to blast and BB in rice (dominant, incomplete dominance and recessive), and the correlation between the resistance spectrum and the strength of blast resistance. Based on the analysis of the pedigree of cultivars, the genetic base of BB was found to be critically narrow in major indica rice in China. These papers clarified the BB adult plant resistance mechanism of different genotypes in rice.

2 Studies on genetics of resistance to bacterial blight

Fifteen papers were chosen for this section, including "Genetics of resistance to BB in two rice cultivars" (1986), etc. The content includes the techniques and methods for studies on the inheritance of BB resistance and development of near-isogenic lines (NILs). The key issues related to the standardization of the genetic study of BB resistance are: 1) selection of susceptible parents and the resistant genetic materials, scientific basis for strains adoption, time of inoculation and identification, and standard operational procedures of investigation; and 2) methods including parental combinations and the population size of their progenies, parameters for analysis, identification, molecular mapping, and nomenclature of new genes.

3 Identification and application of new BB resistance gene

Ten papers including "Identifying and mapping a new gene *Xa23* for resistance to BB from *O. rufipogon*" (2000), etc. were chosen for this section.

This section expatiated on the crucial factors and methods leading to the discovery of new BB resistance genes from *O. rufipogon*. The novel *Xa23* gene conferring broad spectrum, dominant, strong, and all-growth-stage resistance toward *Xoo* was identified by comparison of the resistance spectrum using international evaluation BB races, hybridization between wild and cultivated rice, anther culture, and construction of NILs.

4 Studies on phytopathology and population genetic structure of *Xoo* pathogen

The nine papers in this part include "Studies on pathotypes of *Xoo* in China" (1990), etc.

From 1988—1990, a collaborative study in China classified *Xoo* into seven pathotypes. A joint effort was launched in 1990s which included scientists from USA, the Philippines, China, Korea, Nepal, India, Indonesia, and Malaysia to perform combined molecular and pathogenic studies related to the evolution genetics of *Xoo* population. With this effort, it was discovered that the evolution of polyphyletic molecular types in a heterogeneous population led to new pathotypes, and may result in the rising of new races. The result also concluded that the genotype of host plants was the main factor inducing an adaptive evolution in *Xoo* genetics. Through investigating the diversity of molecular types, the phylogenetics of *Xoo* was revealed and the dynamic relationship among host genotypes, *Xoo* pathotypes and environmental factors was deduced.

5 Scientific presentation and review

Four research reports to international meetings are included in this section, including "Characterizing adult plant resistance to BB" (1984), "Genetics and improvement of resistance to BB in hybrid rice in China" (2009), etc.

There are also three overviews such as "Highlights in identification and application of resistance genes to BB" (2005), etc. They were presentations in the 1st and 2nd International Conference on Bacterial Blight of Rice (ICBB) held in Japan (2004), in Nanjing, China (2007), and in the 5th International Hybrid Rice Symposium in Changsha, Hunan (2008).

6 Edited Books

The content of this section was chosen from two books: "Types of resistance in rice to bacterial blight" in *Bacterial Blight in Rice* (IRRI, 1989); and "The genetics of quality resistance and identification of major resistance genes" and "The mechanism of resistance to BB" etc., in *Genetics and Improvement of Resistance to Bacterial Blight in Rice* (edited by Q. Zhang, 2007).

Contents in the books covered the accumulated effort of scientists and researchers who strived to broaden and deepen our understandings on the genetic improvement for BB resistance and its basic principles in the past 100 years. The authors summarized and integrated the knowledge on phytopathology, plant genetics, breeding, and molecular biology of BB by a brief historical account and a detailed discussion on the course of advancement and the present status of various BB studies with respect to their actual or potential applications in rice production.

I also acknowledge the continuous support from the leadership and administration of the Chinese Academy of Agricultural Sciences. Special thanks go to Prof. F. Z. Hong, a grassland expert and the former Deputy Minister of Agriculture for his epigraph for this book. Prof. W. C. Yang and Dr. B. Y. Zhao kindly wrote the preface. Prof. J. M. Wan gave me strong encouragement and kindly arranged financial support for the publication of this book. Members of my research group had fully participated and made significant contributions to all scientific achievements. I salute these comrades.

The kind assistance of the Chinese Agricultural Press and the detailed and dedicated works of their staff in publishing this book are gratefully appreciated. Dr. T. Q. Zheng assisted in finishing part of the illustrations and bibliography.

Last but the least, my unfailing life-long companion and husband, Professor G. S. Que provided tremendous support in rice breeding. Without grievances, he has stayed by my side through all rainy and sunny days. My gratitude to him goes beyond words can ever express.

Zhang Qi

January 20, 2010

传　略

章琦，中国农业科学院作物科学研究所研究员。1934 年 7 月出生于江苏武进县厚余镇一个书香之家，小学至中学就读于常州私立恺乐教会学校，在其叔父著名小麦遗传育种学家章锡昌教授的影响下，萌生学习农业科学之意愿，50 多年来孜孜不倦地进行农业研究至今。

农业是民生之根本。对于拥有超过 13 亿人口，9 亿农民的中国来说，农业更是一个关乎温饱，关乎生存，关乎经济和国家发展的命脉。正如许多农业科研人员一样，章琦正是一个怀着这个信念，在大地上留下她的汗水和足印。

献身农业　苦乐兼程

原沈阳农学院是章琦农业研究历程的开始。1956 年，章琦入读该院的植物保护系。那里云集了由复旦大学农学院迁移来的陈恩凤、曹诚英、徐天锡、赵仁镕、任明道、吴友三、杨守仁等众多治学严谨、载誉海外的著名学者，他们继承了复旦大学“博学而笃志，切问而近思”的治学精神，重视基础理论和基本知识教学，强调扩大学生的知识面，为业者在“博大”的基础上走向“精深”创造条件。

1960 年，章琦大学毕业后，分配在沈阳市农业局植保植检站工作，负责全市农作物病虫害防治。那时，郊区农业生产还比较落后，各种农作物的常见病虫害每年都有发生。她以满腔热情投入工作，长年累月同郊区农技站技术人员密切配合，与当地农民齐心协力控制病虫害。风天一身土，雨天一身泥。这位带着眼镜、看上去有点娇气的女技术员，终于赢得了农民和农技站员工的欢迎和赞赏。

在基层工作，经常会在农业生产上碰到与自己专业有关的各种各样的实际问题，这对初出茅庐的章琦来说，既是能力的检验，又是意志的锻炼。她在应对各种困难的实践中摸索出一个好办法，即虚心向人求教，拜能者为师。章琦曾与辽宁省农业科学院忻亦芬研究员合作试验防治玉米螟和大豆食心虫；与母校李进教授合作调查全市稻田杂草种类；与沈阳化工研究院王桂云技师共同测试水田除莠剂（weedicide)，与沈阳市生产资料公司的司东钧技术员试用新农药等等，她在相互合作中不断增强自己的业务能力。

短短几年内，她就掌握了高粱蚜、黏虫、玉米螟、大豆食心虫，稻瘟病、纹枯病、白叶枯病以及单、双子叶杂草等病、虫、草害的发生发展规律和综合防治措施。这一段与能者的默契而愉快的合作共事，对她后来科研事业的发展很有裨益。

1966 年夏，风华正茂的章琦及其丈夫阙更生却在“文革”中遭到冲击。1969 年，他们同两个孩子在险被遣送回原籍时，又改送去农村插队落户，开始了晨曦出、星夜归的“战天斗地”式生活。但是，章琦并没有因此而意志消沉。到辽中县老观坨大队，她就注意学习、熟悉农业生产实际操作，播种、间苗、铲地、施肥、砍高粱秆、掰玉米棒 、挑泥积肥、沤麻、砍柴等，什么农活都干，每天总是湿透全身，但从不怨天尤人。东北地区大田生产中铲高粱地是一项十分辛苦的劳动，要求间苗、松土、除草三道工序一气呵成。对此，尽管她使出浑身力气，还是远远落在后面。可是老乡们谁先铲到地头，谁就转过身来默默地接着铲章琦那条垄，当她瞥见迎面伸过来的锄头，一抬头遇到那敦厚、亲切的目光，就鼻子酸了，泪水同汗水一起溢满面颊，这段艰苦时期的经历始终让章琦的内心充满着对这些普通劳动者的无限感激和敬仰。一年后，刚刚适应了大灶坑烧柴、大铁锅炖菜、贴玉米面大饼子的东北农村生活，县里又将其全家搬到水稻单产极低、全县最贫困的潘家堡张家村大队。在这里，全队种植的水稻都是成片叶色灰紫、叶片发蔫、不能分蘖的。其丈夫（水稻育种家）阙更生意识到这是明显的严重缺磷症状，她就配合其进行土壤改良，用自行车搬移、采集全大队所有低产田的土样，进行磷肥效益试验。试验取得成效，凡是施用磷肥的小区都显著增产，这令全县掀起了低产稻田改良土壤的高潮。章琦也在这段时间学会了水稻育秧、栽培管理、收获等整套稻作技术。3 年中，章琦与纯朴、善良的乡民们为邻，朝夕相处，共同劳动。乡民们所表现出的宽容包纳和通情达理，使她深深地感受到人间真情的可贵。同时，作为一个农业科技工作者，章琦也深刻地体会到国家的农业生产还是如此落后，这里的农民大众仍过着吃玉米和稗子混合面还得不到温饱的那种日子，她决心要为改变这种状况而努力。

1972 年，章琦被调到沈阳市农科所工作。1973—1974 年，因为水稻纹枯病严重，章琦到浑河站公社朝鲜族满融大队试验站驻点，研究如何进行综合防治。两年间，她与生产经验丰富、有多年稻田试验基础的农民技术员们紧密合作，进行了越冬菌源、病情发生规律、农药筛选、品种抗性、稻田深浅水层等设计处理的调查、取样、测产、考种一系列工作。在掌握沈阳地区水稻纹枯病发生发展规律的基础上，章琦初步提出了切实可行的综合防治措施，1976 年在《沈阳市农科所科学研究资料通讯》上发表了关于“沈阳地区水稻纹枯病的发生规律及综合防治研究”的论文。

1975 年，沈阳市农科所开始进行粳稻杂种优势利用研究，章琦承担了其中稻种资源及对稻瘟病、白叶枯病的抗性研究部分。面对这个全新的课题，章琦反复阅读杂交水稻遗传育种学原理，密密麻麻地记下几大本笔记，翻阅了大量资料，并将国际水稻研究所（IRRI）著名育种家 G S Khush 编著的《Disease and Insect Resistance in Rice》译成中文（经章锡昌教授修改），便于人们更好地了解当时国际上水稻抗病（虫）性遗传育种的研究动态。在这段时间里，章琦将两个孩子送到上海姐姐家借读，而自己则随所内育种队去海南，进行了大量抗、感病品种间杂交组合，观察上百个抗、感正、反交组合 F_1 的抗性，分析这些 F_1 对白叶枯病和稻瘟病的抗性遗传行为。1977 年，章琦在全国稻瘟病抗性协作研究会上报告了这些结果，揭示水稻对白叶枯病和稻瘟病的抗性都有显性、不完全显性和隐性遗传三种表现，抗病性主要受核基因控制，少数表现为核质互作。此发现立即引起了中国农业科学院林世成先生和与会者的注意。1978 年，她的论文“水稻对白叶枯病和稻瘟病的抗性研究”发表在《沈阳市农科所科学研究资料通讯》上，这些研究结果在当时还鲜有报道。

鉴于她的表现，1978 年章琦被选为沈阳市第八届科技界人大代表，这是对她踏实工作、潜心研究的最好肯定。

幸遇伯乐　走出国门

1977 年，IRRI 理事会中国首任理事林世成先生正在为组建中国水稻多学科协作组物色人选。他在全国稻瘟病协作研究大会上发现了章琦的科研潜力，经过一番艰难周折，终于在 1979 年将章琦调入中国农业科学院作物育种栽培研究所。从此，她的科研生涯又迈向一个新的里程：以水稻主要病害的抗性及其遗传研究为中心，为育种服务为最终方向。

1982 年，为了进一步历练章琦，林世成先生推荐她赴菲律宾 IRRI 与著名植物病理学家 Mew T W 合作，探索水稻对白叶枯病抗性类型和抗性遗传行为，重点研究“成株抗性”。她借鉴水稻高产栽培中根据叶龄指数来调节水肥管理的调查方法，对上百个水稻品种的植株从 3 叶期开始，逐叶标记叶龄直至剑叶，配以分段接种白叶枯病菌，比较不同水稻基因型在各生育期对此病的抗性差异，测定成株抗性品种在由感“转抗”前后，抗、感叶片组织中细菌增殖的动态，以及对不同病菌小种的反应，探索病原菌与寄主间的互作关系。

因为水稻白叶枯病菌（*Xanthomonas oryzae* pv. *oryzae*，*Xoo*）的生长十分缓慢，极易污染，每张叶片组织内成活细菌的增殖及其速率，只能间接地从白叶枯病菌噬菌体（*Xoo* phage）的增值率来获取佐证。这个做法，当时在国内外皆不多见。章琦做的上百次测试都失败了，但她锲而不舍，认真查阅文献资料，冷静、细致地检查每个步骤，分析失败原因，终于发现了快速提取菌液、减少叶片污染、筛选两对最适白叶枯菌及其专化噬菌体菌株（specific phage strain）、确定最适临界浓度等一系列关键点，于是制定出一个最佳方案，并运用熟练的操作技巧，制服了那变幻莫测的噬菌体快速复制的特性。因为在测试中，要面对需进行多次稀释用的数百个指形管，在片刻之间必须完成这套操作，才能获得准确数据，那就需要思想高度集中、凭记忆来节省在指形管上做注释的分分秒秒。尤其在这项合作研究的最后 4 个多月，她已记不清楚是怎样度过的，一走进试验室后便不分昼夜，只得用周六下午来弥补睡眠的不足。经过不懈的努力，章琦终于在掌握噬菌体的增殖规律之后，获得了预期结果。

1984 年 4 月 28 日，在 IRRI 例行的星期六报告会（Saturday Seminar）的大报告厅（Chandler Hall），章琦做了“水稻对白叶枯病的成株抗性特性”报告。她逻辑清晰、精彩独到的论述，得到了与会科学家们热烈的掌声。会议主席，IRRI 副所长、昆虫学家帕撒克博士连声说：“excellent”。Mew T W 博士更是兴高采烈，他满意地看到合作者经过两年的努力，便较系统地揭示了当时国际上还不很明确的水稻对白叶枯病成株抗性的机理、基因表现特性以及中国主体抗源的抗性遗传基础。章琦的论文“Adult-Plant Resistance of Rice Cultivars to Bacterial Blight”被选入 IRRI 1984 年度的“Research Highlight（研究精粹）”，其他相关研究发表于美国《Phytopathology》（植物病理学报），1984 和《Plant Disease》（植物病害），1985。

章琦在 IRRI 看到那国际一流的配套实验设备和高效率的科学管理体系，犹如久旱逢甘霖，便下决心一定要争分夺秒，找回“文革”期间失去多年的珍贵岁月。她充分利用自控生长箱、人控暗处理室、网室和温室，在两年内种植了 7 个水稻生长季节，完成了加倍的工作量。此外，她坚持参加IRRI为非英语国家人员设立的英文课程，努力提高英语水平。在基础理论上，经著名遗传学家张德慈博士多次面授机宜，章琦获益良多。与她合作研究多年、亦师亦友的Mew T W博士由衷地说："章琦是我见到过少有的如此专注于研究的人，她具有极优秀的研究素质，别人至少要三至四年才能完成的工作，她只用了两年"。对Mew博士，章琦始终心存感激，因为她所取得的科研进展，是基于Mew博士的启迪和鼓励，以及众多学者的热忱帮助。

看到章琦的出色表现，IRRI病理系研究稻瘟病的J M Bonman博士很希望留她再工作几年，同时闵绍楷研究员则动员她去杭州刚刚成立的中国水稻研究所工作，这些都是很诱人的机会。林世成先生告诉章琦说："不能影响你的前途，由你决定去、留吧！"章琦感动极了，她感念林先生当年经过千艰万难将自己全家调入北京，而且在人手短缺、许多工作急需开展的情况下，还送自己到IRRI合作进修两年，于是断然谢绝了各方盛情，于1984年6月按时回国，开始探索水稻抗病性应用基础的研究。

开拓创新　世代传承

水稻抗病性遗传改良研究的起步

回到中国农业科学院水稻育种室后，章琦承担了水稻对稻瘟病和白叶枯病的抗性遗传改良工作。以鉴定育种材料对两病的抗性和筛选广谱抗瘟材料为基础，比较了22个携有不同抗稻瘟病基因的主要供体品种以及丰产性较好的品系对中国A～G7个小种15个代表菌株的抗谱，证明拓宽抗谱在抗稻瘟的育种计划中具有重要的实践意义。论文"水稻抗病育种研究Ⅱ.品种（系）对稻瘟菌的抗谱鉴定初报"发表在中国农业科学，1981（3）：1～9。

章琦此后集中进行水稻对白叶枯病的抗性研究。以追溯"水稻品种对白叶枯病抗性的系谱渊源"为切入点，首先测定和分析我国近百个水稻品种对白叶枯病6个代表菌株的抗性及其系谱，进而研究78个品种和抗源的抗性遗传行为和基因推导，发现籼稻的3个主要抗源所携有的抗白叶枯病基因都没有越出*Xa4*的范围，明确了我国籼稻品种的白叶枯病抗性主要由*Xa4*所控制，而粳稻品种的白叶枯病抗性都来源于成株抗性基因*Xa3*。显然，我国水稻品种对白叶枯病抗性的遗传基础是脆弱的。章琦连续发表了"水稻抗病育种研究Ⅲ.我国主栽品种对若干白叶枯病菌株的抗性分析"（中国农业科学，1988）；"我国水稻白叶枯病抗性遗传的评价和利用"（中国农业科学，1991）；"我国水稻抗白叶枯病基因的应用及策略"（植物保护学报，1995）等论文，及时向育种界发出了抗性将继续下降或全面丧失的报道。不久，携带*Xa4*和*Xa3*的水稻品种都在我国南方部分稻区遇到新毒性菌株的侵染。

显然，进一步的研究目标，应该是从我国蕴藏着的丰富的稻种资源中发掘新的优异抗性基因，拓宽品种抗性的遗传基础。

发掘优异抗病基因

1986年，林世成研究员主持国家自然科学基金项目"水稻野栽杂交抗病育种"，章琦与广西农业科学院品种资源研究所李道远、陈成斌、黄勇等合作，连续三年，在南宁对该所保存的包括13个野生稻种近3 000份编号材料进行白叶枯病抗性初测，再用3个毒性菌株分别接种其中871个编号材料各稻株的不同分蘖，完成了很大的工作量，再经复测，筛选出22份新抗源，发现其中一份普通野生稻RBB16表现抗强毒菌株而且抗性稳定。经与感病籼稻品种金刚30杂交，获得抗性疯狂分离的野栽杂交F_1，再经花培，获得抗性纯合系。再以金刚30为轮回亲本，育成一对近等基因系CBB23。通过比较抗谱、抗性遗传分析、基因等位性测验、分子检测等系统研究，确认了CBB23携有一对新的抗性基因。后通过洛克菲勒基金资助，派其优秀研究生赵炳宇到IRRI，在H. Leung博士指导下鉴定另一个野生稻新抗源WBB2的抗性基因，学习分子标记定位技术。赵炳宇回国后，在著名分子遗传学家朱立煌教授的指导下，将CBB23的新基因定位于第11染色体上。2001年在国际水稻新基因命名委员会注册为*Xa23*。1998—2001年间，出于改良杂交稻抗性的需要，又育成以携有恢复基因IR24为轮回亲本的*Xa23*近等基因系CBB23（B）。截至2006年，在国内外鉴定的30个白叶枯病抗性基因中，*Xa23*以其抗谱最广、呈完全显性遗传、全生育期高抗、抗性传递性能强等特点，受到国内外瞩目，已被70多个育种单位广泛引用。中国水稻研究所及广东、安徽、湖北等省农业科学院已育成一批优良的抗病品系。面对东南亚国家白叶枯病严重肆虐的状况，以“同一个世界，同一个梦想”的精神，她认为应该像为全球提供携带*Xa21*、*Xa7*、*xa13*、*xa5*、*Xa27*等抗性基因的载体材料那样，携*Xa23*的品系材料也应尽快在国际上发挥其应有的作用。

当然，章琦不会忘记，*Xa23*基因的发掘是基于广西农业科学院提供的种质资源，他们的有关专家也参与了大量筛选和花培等工作；中国科学院遗传研究所朱立煌分子实验室提供了基因定位的条件和关键技术指导，应该说，这个成果是大协作的结晶。

在此期间，南京农业大学朱立宏教授和华中农业大学谢岳峰教授分别从太湖流域和云南稻种资源中评价出大量抗白叶枯病的材料。在深入开发新基因时，大家一致认为我国缺乏遗传分析的统一标准和急待查清我国种质资源的抗性遗传基础。1991—1994年，朱、谢两位资深遗传育种学家和章琦在国家自然科学基金资助下，共同研讨，制定了统一研究方案协作研究抗病种质的抗性遗传问题，初步提出水稻品种对白叶枯病主基因抗性遗传研究标准化程序的商榷，希望改变因研究单位分散和所用菌株、方法及调查标准不一，对相同品种却得出不同遗传解释，导致结果难以比较的尴尬局面。本着学科互补、诚挚交流、相互尊重的协作精神，圆满完成上述目标。此后他们在国家攻关项目、农业部专项中一直保持密切的合作。

论文“我国水稻白叶枯病（*Xanthomonas oryzae* pv. *oryzae*）抗性遗传研究标准化的几点商榷”发表于中国农业科学（1996）。

培育粳稻白叶枯病近等基因系

20世纪80年代，日本热带农业研究中心和IRRI合作培育了具IR24（籼）、密阳23（籼/粳交）和丰锦（粳）三套不同遗传背景、携有10个抗水稻白叶枯病基因的近等基因系(NILs)。章琦在IRRI曾用菲律宾鉴别小种多次测定它们的抗性。以IR24为轮回亲本的NILs携带的各基因之专化抗性清晰、稳定，农艺性状一致，但是，其他两套的专化抗性不清晰、中抗偏感、不稳定，并且保留了供体的某些表型性状。因此，在1986—1996年，章琦请教了经常在同一个网室工作的日本遗传学家Ogawa博士，并在其方法的基础上，严格选择轮回亲本、认真进行对目标基因转入后的验证，全课题组历经10年，成功地育成了一套抗白叶枯病粳稻近等基因系。各系目标基因的专化抗性清晰稳定、遗传背景一致、农艺性状优良，在基因克隆、寄主—病菌互作机制研究和育种应用等方面，均取得了良好结果。当时在浙江农业科学院工作的徐建龙博士曾利用 CBB7（G38）育成了 3 个全生育期抗性、抗谱广的高产晚粳品系 D601、D602 和 D603，成为周边地区育种单位的优良抗源；中国农业科学院作物科学研究所花培组李梅芳研究员用 CBB4 为抗病供体，育成了抗病、高产粳稻品种中花 14 和中花 15。这套近等基因系为基因分子定位、克隆和植物抗病性、病原菌致病性机理研究也提供了理想工具。中国科学院朱立煌研究员实验室用沈农 1033/CBB7 的 F_2 植株为作图群体，将 CBB7 携有的基因定位于第 11 染色体，鉴定出在该基因两侧定位的分子标记 G1465 和 RG1109 与 G1465 的图距为 2.0cm，通过染色体步移起点，已调出 2 个候选基因。

2002 年，笔者有幸与章琦教授探讨了植物—病原菌互作在水稻对白叶枯病原菌方面的分子传导表现。她对 BB 抗、感叶片组织中 *Xoo* 增殖动态、寄主—病菌小种的专化互作等方面非常熟悉，自此，我们在采用 CBB 系统 NILs 利用 DNA 基因芯片技术等方面一直进行紧密的合作，从分子水平上寻找水稻防御信号传导的新基因，论文发表在《New Phytologist》(2008)，《Plant Cell & Environment》(2009) 及《Journal of Experimental Botany》(2007) 上。章琦对科研的执著和她那独到的见解、活跃的思维、对人毫无保留的真诚始终令我敬佩。

开拓新研究领域——病菌群体遗传结构

寄主丧失抗性是育种家关注的焦点。水稻白叶枯病原菌具有独特的适应寄主抗性的遗传潜能，因而其群体结构表现区域差异，有些地区出现新的小种或小种的毒性增强了，寄主对病菌群体的变化也有较强的选择作用，品种基因型的分布反过来影响菌株“亚群”的分布。因此，要成功地利用寄主的抗病性，需要选择适宜的寄主基因型来控制地区优势菌群。系统观察病菌群体的时空变化是进行抗性基因部署的依据，这两者都涉及了分子生物学原理。

20 世纪 90 年代初，章琦用世界银行贷款赴美国探究中国 *Xoo* 群体遗传结构时空变迁与寄主抗性基因利用的关系，先后采用 RFLP、Rep-PCR 等分子标记技术分析了 1984—1994 年中国 20 个省、市 272 个 *Xoo* 菌株遗传多样性，初步揭示了：1）中国 *Xoo* 菌群体遗传结构简单，明显受制于寄主抗病基因利用集中，遗传基础狭窄。侵袭 *Xa4* 基因的广东Ⅴ型遗传多样性值上升快，传递了品种丧失抗性的信息；2）菌株 DNA 分子型与致病型有一定对应关系，多数致病型的分子型具遗传异质性，广东省Ⅴ型菌及云南菌群特异，后者含有除中国 7 群菌之外的多个新毒性型，具备发现鉴别品种和新小种的条件；3）不同稻区病菌分子型有一定差异，提供了因地制宜利用抗病基因的信息。

这又是一段难忘的经历，1992 年冬，章琦已年近花甲，第一次在美国 Kansas 州立大学 Jan E Leach 教授实验室开始了上述研究。她同青年学子一道，每天工作至深夜，并很快熟悉和掌握了分子标记技术的基本要领，用 4 个月的时间完成了一年的任务。其勤奋和高效着实令 Leach 吃惊，她推荐章琦申请洛克菲勒基金，Leach 写道：“我所以愿与章琦合作，不仅是她的聪明才智和丰富的知识，还有她那对事业的献身精神。”1994 年秋，章琦获洛克菲勒基金，1997 年再次获得资助，1999 年，在多数水稻项目被缩减的时候，却保留了她的项目。1994 年冬，Leach 邀请她再次赴美参加亚洲 7 个国家 *Xoo* 群体遗传多样性协作研究。章琦只有 1989 年在 IRRI 短期培训 RFLP 技术的初步知识，如果没有 Leach 在分子学上的启蒙、耐心指导和热忱支持，是不可能取得这些成果的。

章琦从科研一线退下来后，回顾已有100多年研究历史的水稻白叶枯病，至今在深度和广度上都发生了巨大变化。一个世纪以来，不论国外或国内，人们对水稻白叶枯病抗性及其遗传改良的理解与认识，从基础理论到对应策略都有显著发展，积累了大量的论文和资料，但是缺乏一部比较系统、全面论述有关这一专门领域的专著。她就组织本领域的一些专家教授，将国内外数量庞大的文献综合起来，特别是中国培育的和各省种植的数百个抗白叶枯病品种以及数千份有关分子遗传学、抗性遗传等新领域研究的文献，编写成一部题为《水稻白叶枯病的抗性遗传及其改良》的专门论著，便于年轻科学家参考。此书于2007年由科学出版社出版。

回首自己所走过的历程时，章琦感慨万千，她说："自己仅有植物保护专业基础，因为工作的需要，只有虚心学习、刻苦勤奋、精诚协作、取长补短，才有可能不断开拓；也许能给年轻科研人员有所借鉴；事实上所取得的点滴成绩，也都与课题组、众人给予自己无私的帮助、支持、鼓励，以及诚挚的友谊是密不可分的。"

对于章琦的努力和成果，与章琦多年合作研究的Mew T W博士说："1982年章琦第一次到IRRI两年。她对科学研究的全身心投入和刻苦钻研的精神，成为许多同她在一起工作的科学家的楷模。她付出了20多年的心血，潜心研究白叶枯病的抗性和遗传。她对该研究的见解对于我们对寄主抗性遗传学的理解以及对新基因*Xa23*的发现都做出了重要贡献。通过她本身的工作以及与国内外科学家的共同努力，已经使水稻—白叶枯病系统成为植物病理学中寄主—病原菌互作系统的模式。回顾起来，在水稻白叶枯病的无数研究工作中都留下了章琦教授的足迹。"德高望重的著名小麦遗传育种家庄巧生院士评价："章琦一直从事水稻病害的抗性遗传育种应用基础研究，在前人的基础上有所创新，成果显著，不仅拓宽了这一领域的研究基础，而且明显促进了我国水稻抗病育种、抗性研究水平的提高；从深化遗传育种研究的角度，常规育种怎样与高科技相结合是目前农业科技界讨论的热点。章琦的工作处于这个交叉的前沿…，她工作勤奋、治学严谨、基础扎实、重视知识更新、善于开拓和钻研、善于合作、热诚认真指导学生和科技人员，是我所屈指可数的优秀科研人员。"

章琦所追求的不是大量的研究经费和高影响因子的国际期刊文章，她是致力解决农业的重要问题；她获得的荣誉和认可不是来自刻意的人事及组织关系钻营，而是源于那份洋溢的工作热情和无私的奉献精神。

◆ 研究简历

1960年　毕业于沈阳农学院植物保护系

1960—1969年　任沈阳市农业局植保植检站技术员

1969—1972年　沈阳市辽中县老观坨、张家村大队插队落户

1972—1979年　任沈阳市农业科学研究所助理研究员

1978年　当选沈阳市第八届人民代表大会代表

1979—1981年　任中国农业科学院助理研究员

1982—1984、1987、1989年　国际水稻研究所访问学者

1986年10月　任中国农业科学院副研究员

1988—2010年　中国农业科学院研究生院高级育种学专题讲授兼职教授

1992年3月　任中国农业科学院研究员

1993年、1995年（各4个月）　美国堪萨斯州立大学客座教授

1998年12月　退休

1988年　日本第五届植物病理大会当选为国际水稻白叶枯病研究委员会理事

1992年10月　农业科技事业做出突出贡献，获国务院颁发的政府特殊津贴和证书

2004—2010年　任《作物学报》编委会委员、编委会常务编委

◆研究和成果简介

自20世纪80年代初开始，章琦获历届国家、农业部遗传育种攻关资助，5项“国家自然科学基金”、洛克菲勒基金、CAAS-IRRI合作、“863”、“948”、“973”项目。

章琦主要从事农作物病虫害防治及水稻纹枯病、稻曲病、稻瘟病和白叶枯病的抗性研究。她集中探索了水稻白叶枯病（BB）抗性遗传育种理论与应用结合的途径；明确了我国20世纪70年代以来水稻主栽品种的抗性状况及其系谱渊源，揭示了我国BB抗性基因应用单一的问题；与广西农业科学院合作，与课题组共同从普通野生稻中发掘一个抗谱广、完全显性遗传、全生育期高抗、抗性传递性能高的BB新基因*Xa23*，育成携有*Xa23*的两个品系中野5112和5114、两对近等基因系CBB23和CBB23（B），已有70多个育种单位引用。与IRRI合作，明确了BB成株抗性特点及病原菌与品种互作机理。与美国堪萨斯州立大学合作，最早揭示了中国水稻白叶枯病原菌群体遗传结构与寄主基因型布局互作的动态，为科学育种合理应用抗性基因提供依据。主持育成国际上水稻白叶枯病6个抗性单基因粳稻近等基因系CBB2、CBB3、CBB4、CBB7、CBB12、CBB14。

获奖

1981年　我国水稻稻瘟病抗源的筛选　农牧渔业部技术改进一等奖（参加）

1989年　水稻白叶枯病的抗性和遗传研究及应用效果　农业部科学技术进步三等奖（第1完成人）

1991年　稻、麦、玉米抗病性遗传分析　农业部科学技术进步三等奖（第4完成人）

1991年　中国稻白叶枯菌致病型研究　国家科学技术进步三等奖（第6完成人）

2000年　我国抗稻白叶枯粳稻近等基因系的培育及应用　国家技术发明二等奖（第1完成人）

在国内外核心刊物和国际会议上发表论文84篇，编写专著《水稻白叶枯病抗性遗传及其改良》(2007)，参与撰写5本著作。

林汉明

香港中文大学　农业生物技术国家重点实验室及生物系

2010年1月20日

Biography

Qi Zhang, a researcher in Institute of Crop Sciences, the Chinese Academy of Agricultural Sciences (ICS-CAAS), was born in a literary family in Houyu, Wujin County, Jiangsu in July 1934 and received her elementary and secondary education in Changzhou Kaile Private Mission School. Her interest in the study on agriculture was developed early by her uncle, Prof. Xi-Chang Zhang, a prominent scientist in wheat genetics. In the past 50 years, she has been dedicated unfailingly to research in agriculture.

Agriculture builds the foundation of our living. For China, a country with a population of 1. 3 billion and 0. 9 billion peasants, agriculture is vital to the survival of people, the economy and the development of the country. Zhang carries the same belief shared by many researchers in agriculture, leaving behind her sweat and footprints in the field.

Devoted to Agriculture: A Journey with Joy and Bitterness

Zhang started her research journey in agriculture at Shenyang Agricultural College. In 1956, Zhang was enrolled in the Department of Plant Protection. The institute gathered numerous rigorous and internationally renowned scholars from Agriculture College, Fudan University, including En-Feng Chen, Cheng-Ying Cao, Tian-Xi Xu, Ren-Rong Zhao, You-San Wu and Shou-Ren Yang. Inherited the motto of Fudan University, "rich in knowledge and tenacious of purpose; inquiring with earnestness and reflecting with self-practice", these scholars valued training in fundamental theories and basic knowledge, emphasized the broadening of students' perspective and created the condition which enabled scientists to progress.

In 1960, Zhang was assigned to work at the Plant Protection Station of the Shenyang Agriculture Bureau and was responsible for crop pest control in the whole city. Agricultural production in the rural region at the time was relatively primitive, with yearly common crop pest infestation. Zhang was passionate about her work, working closely with technicians in Agriculture Technology Station and peasants over the years to implement pest control. Covered with sand at windy days and mud in rainy seasons, this eyeglass-wearing seemingly pampered female technician finally won the accolade from the peasants and staff in the stations.

Working at the grassroots level, Zhang often faced practical problems related to her profession during the process of agricultural production. For the fledging scientist, it was a test of her ability and training of her volition. When faced with different challenges, she had learnt a good way to response: enquire with modesty and learn from people who are able. Zhang co-operated with Professor Yi-Fen Xin, researcher in Liaoning Academy of Agricultural Sciences, to experiment on the control of corn borer and soybean pod borer, investigated with Professor Jin Li from her alma mater on weed diversity in rice field, tested herbicide on paddy field with Ms. Gui-Yun Wang, assistant engineer in Shenyang Academy of Chemical Industry, and tried new pesticides with Mr. Dong-Jun Si, technician of Shenyang Production Goods Company. Through co-operation with different parties she enhanced her professional ability.

In few years, she mastered the cause, pattern of development and methods of integrated control of multiple diseases, pests and weeds, including broomcorn aphid, armyworm, cornborer, soybean pod borer, rice blast, sheath blight, bacterial blight, as well as monocotyledonous and dicotyledonous weeds. This period of working with able people benefited the growth of her career in scientific research in the future.

In the summer of 1966, supposedly Zhang's heyday, she and her husband, Geng-Sheng Que were slashed by the "Cultural Revolution" . Right before both of them and their children being sent back to Zhejiang, her husband's domicile of birth, they were rearranged to work in a production team in the rural area and started battling with work from dawn to dusk.

However, Zhang was not discouraged by the political turmoil. Since her arrival at the production brigade at Laoguantuo, Liaozhong County, she paid attention to learn and get familiarized with the actual operation of agricultural production. She had involved herself with all kinds of farm work, from sowing, thinning, ploughing and weeding, fertilization, slashing sorghum stalks, harvesting corn spikes, tilting, manure composting, flax retting to mowing and chopping firewood. Every day, she worked so hard that she was all covered with sweat; yet she had never moaned or blamed others.

Shoveling sorghum field in the northeast region was laborious, requiring the completion of thinning, soil breaking and weeding all in a row. Although she exercised all her strength, she was always way behind other fellow townsmen. Whoever finished the ridge first, he would turned around and quietly continued on Zhang's ridge. When Zhang raised her head to see the approaching hoe and the caressing and good-natured eyes, her nose got sore and tears and sweat flew down her face. The tough experience filled Zhang's heart with boundless gratitude and respect to the most ordinary laborers.

One year later, when Zhang slowly got adapted to the lifestyle of the northeast village, like the firewood burning in bolar cooking range, braised vegetables in big wok and broiled corn flour bread, the county moved her family to the Zhangjiacun production brigade, Panjiabao Commune , one of the poorest villages in the county with extreme low rice yield. The team there planted in a large piece of paddy field of no tillering ability, resulting in a sea of withering or grizzled-purple leaves. Zhang's husband, Geng-Sheng Que, an expert in rice breeding, realized that it was a symptom of phosphorus shortage. Zhang assisted in the implementation of a soil improvement scheme, biking around to collect and transport soil samples of all low-yield fields for a phosphorous fertilizers effect trial.

The trial was successful. All fields using phosphate fertilizers recorded significant growth in yields, leading the whole county to a wave of soil improvement projects in low-yield paddy fields. Zhang, at the same time, had learnt the whole set of skills in rice cultivation, including rice breed seedling, growth management and harvest. She worked day and night with the kind and pure-minded villagers. Their tolerance and understanding had impressed her of the treasure of love and respect among people. At the same time, Zhang, as a worker of agricultural science, deeply felt that agricultural production in the country had been too primitive, resulting in the mass population living barely on "cake" made of corn and grass-seed mixed . She vowed to alter the devastating situation.

In 1972, Zhang was assigned to work in the Shenyang Agricultural Research Institute. During 1973—1974, due to serious sheath blight, Zhang, stationing at Hunhe Commune's Chaoxian nation Manrong Experimentation Station, researched on the implementation of integrated control. In these two years, she worked with farmer technicians, who had rich production knowledge and multiple years of experience in basic paddy field research, to carry out a series of experiments including growth investigation, sampling, production estimation and rice spike examining in order to study the sources of over-wintering fungi, occurrences of diseases, fungicide selections, varietal resistance, fertilization and watering management, etc. After mastering the basic of the development and growth of sheath blight in Shenyang, Zhang suggested measures of integrated control and published a paper, "Studies on development status and integrated control of sheath blight in Shenyang" in the *Newsletter of Research Information*, *Shenyang Agricultural Research Institute* (1976) .

In 1975, Shenyang Agricultural Research Institute started researching on japonica hybrid vigor utilization. Zhang was in charge of the research on rice resources and the resistance of rice fungal blast and bacterial blight (BB) . Facing a new research topic, Zhang repetitively studied the principles of hybrid rice genetics and breeding. Not only did she reviewed a large amount of data and made volumes of notes, she also translated *Disease and Insect Resistance in Rice* by Dr. G. S. Khush, a famous breeder of International Rice Research Institute (IRRI), into Chinese (edited by Prof. Xi-Chang Zhang), in order to help others to understand the trend of contemporary international research on the inheritance and breeding of resistance in rice. Zhang also sent her two children to live with her sister in Shanghai to free herself to follow the breeding team of the institute to Hainan Island , in which she made a lot of hybridization combinations between susceptible and resistant rice varieties, observed over a hundred of reciprocal crosses of R/S F_1, and analyzed how these F_1 affect the genetic behavior of rice bacterial blight and fungal blast.

In 1977, Zhang presented in the National Collaborative Research Conference on Rice Fungal Blast Resistance, revealing that all three relationships of dominance, incomplete dominance and recessiveness could be found among rice and the resistance of rice fungal blast and bacterial blight. She concluded that resistance was determined not by cytoplasmic inheritance, but mainly by resistance genes in nucleus and partly by nucleo-cytoplasmic interaction. Such findings immediately raised the attention of many participants, among them professor Shi-Cheng Lin of Chinese Academy of Agricultural Sciences. In the following year, Zhang's paper titled "Studies on Resistance of Bacterial Blight and Fungal Blast in Rice" was published in *Newsletter of Research Information, Shenyang Agricultural Research Institute* (1978). These research results were rarely found at that time.

Due to her outstanding performance, Zhang was elected a Shenyang deputy in science and technology of the 8th Shenyang People's Congress in 1978, which was the best recognition for her hard work and devotion in research.

Talent Spotted, Going International

In 1977, professor Shi-Cheng Lin, the first Chinese member in the IRRI Board of Trustee, was looking for candidates to form a multidiscipline group for the Genetic Evaluation and Utilization Program in China. Seeing Zhang's potential in the National Collaborative Research Conference on Rice Blast Resistance, although having to go through series of difficulties, he finally recruited Zhang to the Chinese Academy of Agricultural Sciences (CAAS). That brought Zhang's career on scientific research to a new journey: focusing on rice resistances to common diseases and their inheritance to ultimately advance breeding.

In 1982, Lin recommended Zhang to work with Mew T W, a famous plant pathologist in IRRI in the Philippines, for training and experience. Targeting adult plant resistance, they investigated on the types of resistance on BB and their inheritance. Borrowing the research method of adjusting fertilization and watering management in accordance to leaf age index in high yield managing cultivation, Zhang investigated over a hundred rice varieties. Since the third-leaf stage, she marked each leaf of its age until flagged and inoculated each of them with *Xanthomonas oryzae* pv. *oryzae* (*Xoo*) in phases to compare the differences in the bacterial resistance among different rice genotypes at each stage of growth, tested the dynamic of bacteria multiplication in leaf tissues from stages of susceptibleness to resistance in varieties with adult-plant resistance as well as their reaction to different strains of bacteria, in order to explore the interaction between pathogenic bacteria and their host.

Because the growth of *Xoo* was slow and easily hampered with contamination, the estimation and enumeration of the bacteria in plant tissues can only be indirectly assessed by the bacteriophage technique which evaluated the growth of *Xoo* phage with efficiency and reliability. It was not a common technique and Zhang failed more than a hundred times. With persistence, she studied related archives and examined each step in detail, analyzed the reasons of her previous failures and finally developed the best plan as she discovered a series of key points on how to quickly extract bacteria, minimize leaf contamination, select the most suitable two pairs of *Xoo* and its specific phage strain as well as decide on the fittest critical concentration. In combination with her adept operational skills, she overcame the unpredictable characteristics of the quickly replicating phage.

In order to collect accurate data, the operation of experiment must be completed in one set. Therefore, Zhang needed to be highly concentrated and utilized her memory to save time from making marks on the hundreds of test-tube which were used multiple times for dilution. Especially in the last four months of this collaborating research, she could hardly recall how time passed, as she worked day and night in the laboratory. She could only make up for lost sleep on saturday afternoons. Through her unremitting strife, Zhang at last mastered the regularity of phage's propagation and got the result as expected.

On 28th April, 1984, Zhang presented on the Saturday Seminar of IRRI in Chandler Hall on "Adult-Plant Resistance to Bacterial Blight" . Her clear logic and brilliant discursion won her a round of applause from the attending scientists, including entomologist Dr. M D Pathak, the deputy director general of IRRI and chairman of the conference, repeatedly exclaiming her excellence. Dr. Mew T W was overjoyed and contended to see that, after two years of collaborating hard work, Zhang systematically revealed that the unclear mechanism of adult-plant resistance of rice to bacterial blight, characteristics of gene expression and the hereditary basis of resistance in the main antigens in China.

Her paper of the same title was selected as one of the research highlights in IRRI in 1984 and related research were published in American journals *Phytopathology* (1984) and *Plant Disease* (1985) .

In IRRI, Zhang was determined to make up for the lost time during years of "Cultural Revolution" with their top-ranked experiment facilities and efficient scientific management system. Utilizing the automatic incubators, manual dark rooms, screen houses and greenhouses, she grew seven rice cropping in two years and finished the doubled workload. In addition, she insisted in participating in English courses provided by IRRI for non-native English speakers to improve her English. She also learnt much on the principle theories from famous geneticist Dr. T T Chang. Dr. Mew T W, who was to her a friend, a mentor and a colleague, commented, "Zhang is very focused on research and has great potential. She can finish what others may need three to four years to do in two years." To Dr. Mew T W, Zhang was thankful, since she understood deep in her heart that her progress in scientific research was built upon his inspiration and encouragement, as well as the enthusiasm of many scholars.

Witnessing Zhang's outstanding performance, Dr. J M Bonman, an expert in rice blast in the Department of Plant Pathology in IRRI wanted her to stay for some extra years.

Professor Shao-Kai Min also persuaded her to work for the newly established China National Rice Research Institute in Hangzhou. These were all alluring opportunities. Professor Shi-Cheng Lin tried not to influence her future and told her to decide on her own. She was moved. As she thought of how Lin brought her to Beijing despite many difficulties and sent her to be trained in IRRI for two years albeit the shortage of manpower and urgent needs to start new projects, she decided to decline other offers and return to China in June 1984 to start research on the basic application of resistance of rice.

Invention and Inheritance

The Beginning of Research on Improvement of Rice Resistance Inheritance

Since her return to the rice breeding room in the CAAS, Zhang was responsible for the work to improve the inheritance of resistance to rice fungal blast and BB. To examine the resistance to the two diseases in breeding materials and select materials with broad-spectrum resistance, she compared 22 main species donors with different blast-resisting genes and the spectrum resistance in 15 signature lines from 7 (A to G) races on high-yield varieties, in order to prove the practical significance in broadening spectrum resistance in the blast-resisting breeding plan. A paper, "Studies on Disease Resistance in Rice Breeding. Ⅱ. A Preliminary Report on the Identification of Spectrum of Blast Resistance in Rice Varieties and Hybrid Lines" was published in *Scientia Agricultura Sinica* (1981).

Since then, Zhang focused on research of BB resistance. Using "the origin of pedigree for resistance to bacterial blight in rice" as an entry point, she first examined and analyzed nearly a hundred rice varieties and their resistance and pedigree to 6 representative strains of *Xoo*, then proceeded to study the resistance inheritance and gene postulation of 78 varieties and resistant sources, discovering that the BB resistance gene of the three main resistant sources in indica rice did not exceed the range of *Xa4*. It demonstrated that in China the resistance of BB is mainly controlled by *Xa4* in indica rice and *Xa3* in japonica rice. It clearly showed that the inheritance base of BB resistance in rice in China was vulnerable. Zhang consecutively published papers, including "Studies on Disease Resistance in Rice Breeding: Ⅲ. Resistant- Susceptible Reaction of Main Rice Cultivars to Six Bacterial Blight Strains of China" in *Scientia Agricultura Sinica* (1988), "Genetic Evaluation and Utilization of Resistance to Bacterial Blight in China" in *Scientia Agricultura Sinica* (1991), "Utilization and Strategy of Gene for Resistance to Bacterial Blight in China" in *Acta Phytophylacica Sineca* (1995), to warn breeders of the continuous decline or even a complete disappearance of BB resistance. Soon after, rice varieties carrying *Xa4* and *Xa3* were infested by a new BB race in the southern part of China.

Obviously, the next research goal would be to broaden the inheritance base of resistance by discovering new meritorious resistance gene from the affluent crop resources in China.

Discovering Meritorious Resistance Gene

In 1986, professor Shih-Cheng Lin hosted the project of "Breeding Rice for Resistance to Bacterial Blight by Hybridization between Wild *Oryzae* and Cultivation Rice", under the National Natural Science Foundation of China. Cooperating with Dao-Yuan Li and Cheng-Bin Chen in Guangxi Academy of Agricultural Sciences in Nanning for three consecutive years, Zhang carried out preliminary experiments on the BB resistance of nearly 3000 accessions from 13 wild *oryzae* specieses and inoculated with 3 *Xoo* races on different tillers of the 871 accessions. Despite heavy workload, she carried out re-tests and selected 22 new resistance sources, discovering that acc. RBB16 was highly and steadily resistant to the strain PXO99 with strong virulence. The RBB16 was crossed with a susceptible variety JG30 and a wild/cultivated hybridization F_1 with mad segregation of resistance was obtained. After anther culture, homologous resistance was obtained. A pair of near-isogenic line, CBB23, was developed by using JG30 as recurrent parent. Through systematic research including comparing resistance spectrum, analyzing resistance inheritance, gene allelism test and molecular detection, CBB23 was confirmed to carry a new resistance gene.

With the support of Rockefeller Foundation, her research student, Bing-Yu Zhao, was sent to IRRI for further study under Dr. H Leung's guidance, learning the techniques of molecular marking while identifying the resistance gene in WBB2, a resistance source in another common wild rice. When Zhao returned to China, he marked the new gene from CBB23 on the 11th chromosome under the teaching of professor Li-Huang Zhu, a famous molecular geneticist. In 2001, the new gene symbol was registered as *Xa23* at the Committee on Rice Gene Symbolization. In 1998—2001, due to a need to improve the resistance of hybrid rice, CBB23 (B), an NIL of *Xa23*, was bred by using the restore-gene-carrying IR24 as recurrent parent. By 2006, among the 30 BB resistence genes identified worldwide, *Xa23* was the most eye-catching one and was widely adopted by over 70 breeding units due to its characteristics, including the broadest resistance spectrum, complete dominance inheritance, high resistance during the complete development period and high resistance transferring property. A series of resistance strains were bred by the China National Rice Research Institute and academies of agricultural sciences in Guangdong, Anhui and Hubei. With the spirit of "One World, One Dream", Zhang believed that, considering the rage of bacterial blight in Southeast Asian countries, strain materials with *Xa23* should soon be able to bring its assumed effects to play, just as other globally supported resistance gene carriers like *Xa21*, *Xa7*, *xa13*, *xa5*, *Xa27* did.

Zhang would never forget *Xa23* was found upon the germplasm resources provided by the Guangxi Academy of Agricultural Sciences and the support of their experts in the work of selection and anther culture, as well as the guidance on gene mapping techniques and conditions by Li-Huang Zhu's Laboratory of Molecular Biology in the Institute of Genetics, Chinese Academy of Sciences. In other words, the achievement was a result of collaborative effort.

At that time, Prof. Li-Hong Zhu, Nanjing University and Prof. Yue-Feng Xie , Huazhong Agricultural University evaluated a large sum of BB resistant materials from rice germplasms in Taihu Basin and Yunnan respectively. Before developing the new gene, it was generally agreed that the country lacked a unified standard for genetic analysis and had an urgent need to investigate the hereditary basis of resistance in germplasm resource. In 1991—1994, with the support of National Natural Science Foundation, Zhang, Zhu and Xie jointly studied and drafted a centralized research protocol and collaboratively researched on the issue of resistance inheritance bacterial blight. They initiated to standardize the procedures in research of the inheritance of key BB resistence gene in rice, in the hope of solving the embarrassing situation in which comparison was impossible because different explanations of inheritance were given on the same variety as a result of differences in research units, uses of bacterial strain, methods and investigating standards. In the spirit of support, sincere communication and mutual respect in the discipline, they satisfactorily reached the goal. Since then, they maintained a close working relationship in programs under the Ministry of Agriculture.

Breeding Near-isogenic Japonica Rice Lines for Resistance to Bacterial Blight

In the 1980s, Center for Tropical Agriculture in Japan and IRRI co-operated in the development of NILs which had 10 BB resistance genes from three different genetic backgrounds of IR24 (indica), Milyang 23 (hybrid of indica and japonica) and Toyonishike (japonica) . Zhang had multiply tested their resistance in IRRI with the philippine races of *Xoo*. Specific resistance in the NILs using IR24 as recurrent parents was clear and stable with consistent agronomic characters, yet the specific resistance in the other two was unclear, medium resistant inclining to susceptibility, unstable and inherited certain phenotypic characters from the donor. Therefore, in 1986—1989, Zhang consulted Dr. Ogawa, a famous rice geneticist from Japan, in the same screen house. Building on the foundation of his method, she strictly selected recurrent parents and conscientiously inspected the targeted gene after its transfer. The project team worked over 10 years and a set of near-isogenic japonica rice lines with six major genes for BB resistance was developed. Each line clearly and stably exhibits specific resistance as same as the donor and genetic background similarly with common agronomic characters. They also had good results in cloning, research on interaction mechanism between host and pathogen as well as application in breeding.

At the time, Dr. Jian-Long Xu of Zhejiang Academy of Agricultural Sciences used CBB7 (G38) to breed D601, D602 and D603, three high-yield late-season japonica varieties with BB resistance in whole growth period and broad resistance spectrum, making them good resistance sources for the breeding units in nearby regions. Rice breeder, Mei-Fang Li, researcher in the Institute of Crop Breeding and Cultivation, CAAS, used CBB4 as resistant donor and bred Zhonghua 14 and 15, japonica rice varieties with high yield and resistance. This set of NIL was also a genetic tool for molecular mapping, gene cloning, comparing resistance genes in terms of their similarities and differences, and resistance mechanism and other properties. Researchers in the laboratory of Dr. Li-Huang Zhu in Chinese Academy of Agricultural Sciences, using F_2 plants of Shennong 1033/CBB7 as linkage group, located gene of CBB7 on the 11th Chromosome and identified the molecular marking of G1465 and the map distance of G1465 and RG1109 was 2.0cM. Through the start site of chromosome walking, two candidate genes were elected.

In 2002, I had the opportunity to discuss molecular signal transduction of plant-pathogen interaction with Zhang. She possesses deep and profound knowledge on BB resistance, dynamics of *Xoo*/ multiplication in susceptible leaf tissues, and specific interaction between the plant host and bacteria strains. Since then, we have collaborated on applying gene chip technology on the to the NILs of the CBB system to search for new signaling components in rice defense response. The results were published in /*New Phytologist* (2008), /*Plant Cell & Environment* (2009), and /*Journal of Experimental Botany* (2007). Her persistence in scientific research, original insights, brisk thought and unreserved genuineness were always admired.

Opening new research territory: pathogen population genetic structure of Xoo

The breaking down of host resistance has been a central issue for breeders. The *Xoo* pathogen possesses unique genetic potential adaptability of host resistance, leading to regional differences in its group structure. Some regions develop new race or strengthened virulence. Host selection on changes of bacterial group is strong. Distribution of genotype reversely affects the subgroup of bacterial strains. Therefore, choosing the suitable host genotype to control regional dominant bacteria is needed for the successful utilization of resistance in host. Systematic observation of the space-time variety of bacterial group is the basis of the deployment of resistance gene, and both are related to the principle theories in molecular biology.

In the early 1990s, Zhang used loans from World Bank to go to the USA for the investigation in the relationship of the time-space variety in *Xoo* population genetic structure and the utilization of resistance gene in host. Using molecular marking skills like RFLP and Rep-PCR, Zhang analyzed the genetic diversity of 272 *Xoo* strains collected in 20 provinces in China from 1984 to 1994. Analysis of *Xoo* population genetic structure in China led to the following findings. First, the simple population genetic structure was due to the cultivation of host plants carrying a single R (*Xa4*) gene. The rapid increase in genetic diversity of the *Xa4* insensitive Guangdong V indicated the breakdown of host resistance. Second, some correlations were found between molecular types and pathotypes. Most pathotypes were heterogeneous populations. The Chinese *Xoo* pathotype V's molecular type and *Xoo* population from Yunnan were distinct from those which from other regions. The *Xoo* population from Yunnan included new pathotypes differed from the 7 known *Xoo* pathotypes in China. New and unique races and pathotypes may be identified therein. It may help in refining the national *Xoo* evaluation system. Third, different molecular types were found in different rice cultivation areas. This provides information for the utilization of disease resistance genes in different environments.

That was another unforgettable experience. In the winter of 1992, Zhang was almost 60 years old when she started the research above in the laboratory under Prof. Jan E. Leach in Kansas State University. She worked with young scholars till late and soon mastered the basics of the technology in molecular marking, completing one year's task in 4 months. Her efficiency amazed Leach, who recommended her to apply for funding in Rockefeller Foundation. Leach wrote that she was willing to work with Zhang not only because of her intelligence and knowledge, but also her devotion to the career. In the autumn of 1994 and again in 1997, Zhang got funding from Rockefeller Foundation. When many programs on rice were cut in 1999, Zhang's project was kept. In the winter of 1994, Leach invited her to go to the USA again to participate in a collaborative research in the genetic diversity of *Xoo* population in 7 Asian countries. Zhang, back in 1989, only received short-term training in IRRI on the basic knowledge of the RFLP technique. Without the enlightenment, patient instruction and enthusiastic support of Dr. Leach in the study of molecules, it was impossible to accomplish such achievements.

After stepping down from the frontier of research, Zhang reflected on the tremendous changes in the depth and the width of the study of BB over the past century. Although papers and data on BB resistance and its genetic improvement from theoretic level to corresponding strategies were building up, there had been a lack of a systematic and comprehensive publication in BB. She wanted to synthesize the voluminous literature in China, specifically, document the hundreds of rice varieties developed and grown in different provinces in China, as well as the tremendous new frontier research on molecular genetics, the resistance and the genetics in one volume as easy reference to young scientists of the future. She felt that it is important for young scientists to understand the historical development of our science in relation to BB of rice. With her background and wide experience, this would be an excellent contribution to scientists of current and future generation interested in rice and BB . Coordinating some experts and professors, she edited a book, "*Genetics and Improvement of Resistance to Bacterial Blight in Rice*" and had it published by Science Press in 2007. Looking back, Zhang's feeling is mixed. She said, "I received my rudimentary training in plant protection. Due to the need at work, I kept exploring, through learning humbly, hard work, faithful cooperation and overcoming my own weaknesses with the strengths of others. " Maybe young researcher can learn from it. As a fact, even the slightest achievement is closely related to the altruistic help, support, encouragement and sincere friendship from the team.

On Zhang's effort and achievement, Dr. Mew T W, who worked with Zhang for years, said, "In 1982, Zhang came to IRRI. Her dedication and assiduousness in scientific research was a role model for many scientists with whom she was working. She spent 20 years on the research of bacterial blight resistance and its inheritance. Her ideas greatly contributed to our understanding of host resistance genetics and the discovery of a new gene *Xa23*. Looking back, Prof. Zhang left footprints on countless research on bacterial blight in rice." The renowned wheat breeder Qiao-Sheng Zhuang appraised, "Zhang building on the foundation of her predecessors, she broadened and deepened the research base and significantly accelerated the level of resistance breeding and research in resistance. The application of high technology with common breeding is a hot issue in agricultural science. Zhang's research was at the tip of such crossing.... She worked hard, studied rigorously, developed a sound basis as well as paid attention to updating one's knowledge. She was also good at exploring new areas and digging into a topic, good to work with and sincerely and conscientiously instructing students and technicians. Zhang is an outstanding scientist that only few can be compared to."

Zhang did not look to a large sum of research fund or influential publication in international journals; her focus was to solve an important problem in agriculture. The honor and recognition she received did not come from intentional affiliations to other people or organizations, but from her passion at work and unreserved altruism.

◆ Profile

10th July, 1934 Born in Houyu, Wujin County, Jiangsu

1956—1960 Plant Protection/Pathology, Agricultural College, Shenyang

1960—1969 Technician, Plant Protection Station, Agricultural Bureau, Shenyang

1969—1972 Farmer, Panjiabao Commune, Zhangjia Village, Liaozhong County, Shenyang

1972—1979 Technician, Assistant Professor, Shenyang Agricultural Research Institute

1978 Shenyang deputy in science and technology of the 8th Shenyang City-Liaoning province People's Congress

1979—1986 Assistant Professor, Institute of Crop Breeding and Cultivation, CAAS, Beijing.

1982—1984, 1987, 1990 Research Fellow/ Visiting Scientist, IRRI

1986—1992 Associate Professor, Institute of Crop Breeding and Cultivation, CAAS, Beijing.

1988—2010 Lecturer on special topic in advanced breeding for disease resistance, Graduate School of Chinese Academy of Agricultural Sciences

1992 Professor, Institute of Crop Breeding and Cultivation, CAAS, Beijing

1992 Awarded certificate and stipend by the State Council for her contribution in agricultural sciences

1993, 1995 Visiting Professor, Kansas State University

1998 Retired from CAAS.

2004—2010 Member of the standing committee, the editorial board of *Acta*

Agronomica Sinica

Research Accomplishments

Since the 1980s, Zhang was given financial support from the country and Ministry of Agriculture, five National Natural Science Funds and funding from Rockefeller Foundation. She also cooperated with CAAS-IRRI and participated in projects numbered “863”, “948” and “973” .

Zhang researched mainly on disease control in crops and resistance in false smut, blast, BB and sheath blight in rice. She focused on the integration of the theory on the inheritance of resistance in BB and its application, clarified the condition of resistance in rice and its spectrum resistance's origin and exposed the problem of dependency on a single resistance gene. Worked with Guangxi Academy of Agricultural Sciences, she and her research group identified *Xa23*, a resistance gene of BB in wild rice with broad resistance spectrum, complete dominance inheritance, complete development cycle resistance and good resistance inheritance, as well as bred the popular adopted Zhongye 5112 and 5114 with two pairs of near-isogenic line CBB23 and CBB23 (B) now used in over 70 breeding units. She cooperated with IRRI and clarified the characteristics of adult plant resistance and the interaction of pathogens and varieties. With Kansas State University, she exposed the interaction of the pathogenic race of BB in China and the genotype of its host, providing scientific basis on the application of resistance gene in breeding. Zhang also managed the breeding of 6 near-isogenic lines of japonica rice with BB resistance gene, which are CBB2, CBB3, CBB4, CBB7, CBB12 and CBB14.

Professional Society Memberships

Board Member, International Committee of Research on Bacterial Blight in Rice, 5th International Congress of Plant Pathology (1988)

Chinese Society of Genetics

Chinese Society of Plant Pathology

Chinese Society of Biotechnology

Chinese Society of Crop

Honors, Awards

1989, 1991 Awarded two Science and Technology Awards from the Ministry of Agriculture

1991 Awarded State Science and Technology Awards respectively in 1991

1992 Outstanding contributor for Agricultural Science Technology with awards of specific allowance and certificate from the government

2000 Awarded the national Science Technology Invention Award (second class, first awardee)

Research Publications

Zhang published 84 papers in 1978-2009.

Book Chapters：

Genetics and Improvement of Resistance to Bacterial Blight in Rice. The monograph Edited by Zhang Q. 2007. Beijing, Sciences Press. Supported by National Natural Science Foundation of China and the Rockefeller Foundation.

Advances in Research on Resistance to Diseases in Major Crops. Edited by Zhu L. H., Lu W. Z. and Xie Y. F. Jiangsu Science Technology Publishing House, 1990, pp. 1 - 14; 42 - 63.

Zhang Qi, T. W. Mew, "Types of resistance in rice to bacterial blight", in: *Bacterial Blight of Rice*, Proceedings of the International Workshop on Bacterial Blight of Rice, pp. 125 - 133, The International Rice Research Institute, Manila, Philippines.

Basic Research on Breeding Technique of Rice. Edited by Xiong Z. M. and Liang C. B.. Beijing, Sciences Press. 1991. pp. 15 - 23; 152 - 153.

Hon-Ming Lam
The Chinese University of Hong Kong
January 20, 2010

目 录

Contents

I 水稻抗病育种研究

Part 1 Studies on Rice Breeding for Resistance to Diseases

沈阳地区水稻纹枯病的发生规律及综合防治研究

章 琦[1] 张龙吉[2] 朴健勇[2] 智益浩[2]

（1. 沈阳市农业科学研究所；2. 沈阳市浑河站公社满融试验站）

Studies on Development Status and Integrated Control of Rice Sheath Blight in Shenyang

水稻纹枯病属土壤传播的真菌病害。在我国江、浙两湖等南方稻区发生比较普遍（中国农林科学院，1974）。随着全国水稻生产的发展，施肥水平的提高、种植密度加大等因素，此病迅速向北方稻区扩展。1956 年，沈阳市最早在水稻高产区浑河农场大面积发现纹枯病。此后，邻近的浑河站公社、八一公社等新水田也发生了。1960 年以后，水稻纹枯病已经成为沈阳市的一个重要病害。1974—1975 年，本所与浑河站公社满融屯试验站合作，共同对水稻纹枯病的消长、发病因子、损失率以及综合防治方法等方面进行了观察和试验，了解水稻纹枯病在沈阳稻区的发生发展动态，提出控制此病的建议。

1 材料和方法

1.1 田间设计和品种

1.1.1 病情消长观察圃 1974—1975 年，选择一块地力均匀稻田（约 660m^2）作为病情消长观察田，种植粳稻品种京引 3 号，按常规种植密度（行株距，10 寸* ×3 寸），每穴 3～4

* 寸、亩、斤、两均为非法定计量单位，因该文章为 20 世纪 70 年代已发表文章；故寸在本书中延用，1 寸≈3.3cm、1 亩≈667m^2、1 斤＝0.5kg、1 两＝0.05kg。

株苗。

1.1.2 品种抗病性及其生育期与发病的关系 观察、比较试验站13个参加全省水稻产量区试的粳稻品种：笹稔、越实、丰锦、γ-18-8、γ-25-3、263、深育1号、3071、403、401、253、6908-7-4、清杂45。按生产田的种植密度插秧，田间水肥常规管理。

1.1.3 水层管理与发病关系 设浅水灌溉（保持1寸）和深水灌溉（长期3寸）两个处理。

1.1.4 施肥与发病的关系 品种京引3号，设置3个因子7个处理：不同基肥量加相同化肥量；等量基肥60斤、分别单施氮、磷、钾；早施（5月5日）和晚施（6月10日）两个不同追肥时期。

1.1.5 种植密度与发病的关系 种植粳稻品种丰锦，设置行株距（寸）4×3、6×3、8×3、10×3、11×3、12×3共6个处理。

1.1.6 农药防治试验 通过预备试验，从多菌灵、井冈霉素、托布津、稻脚青（甲基砷酸锌）、稻宁（甲基砷酸钙）、田安（甲基砷酸铁铵）、纹枯利、退菌特、稻瘟净、克瘟散共10种农药中，选出药效较好、对植株和人畜安全的多菌灵、井冈霉素、托布津，进行大面积防治示范。

1.1.2～1.1.6 均设3次重复。

1.1.7 越冬菌核处理 插秧前打捞菌核与不打捞菌核两种处理。面积各为1亩*。

1.2 调查方法

1.2.1 病情消长观察 在所选田块中，从东、南、西、北四面，共选择4个调查点，分别离田边2m为起点，各两行计40穴。从稻苗分蘖初期开始逐日调查，记载发病株率，当病斑始发期后，隔3～5d调查一次，直至抽穗后病情稳定时结束。

1.2.2 调查记载发病、计算发病率 分级标准：

0级 全株无病；

1级 顶叶以下第三张叶片、叶鞘发病；

2级 顶叶以下第二张叶片、叶鞘发病；

3级 剑叶及其以下叶片、叶鞘都发病；

4级 全株发病。

1.2.3 产量影响调查 选择生长均匀的高产田（京引3号），按分级标准，每级随机取样100穴，计算每穴平均有效穗数，考查不实率、千粒重、折算损失率。

2 结果和分析

2.1 纹枯病的发生和消长

沈阳地区在水稻苗期时气温较低，不利发病。从分蘖期初期才出现病斑，分蘖末期到抽穗期病情迅速发展。发病初起时，在稻苗齐水面的叶鞘上出现水渍状、边缘不甚清晰的小病斑，随后向上伸展到叶片，严重时几个病斑相连，形成云纹状大斑。病菌主要侵染叶鞘和叶片，重病株的病斑扩展到剑叶和穗部。7月末进入雨季后，空气相对湿度提高，从病部伸出浅色菌丝体，菌丝体攀缘邻近植株、扩展蔓延，随后结成疏松菌丝球，进而形成褐色菌核，菌核干燥后掉落灌溉水，成为再侵染源。菌丝穿透茎秆内部而引起植株倒伏，瘪粒增多、甚至白穗，结实率和千粒重明显降低而减产。

1974—1975 年，从育苗到收获的全程观察表明，纹枯病的发展消长趋势主要与当年气候条件和水稻生长发育有关。1974 年前期低温，水稻生长滞缓，分蘖期推迟了，7 月上旬才开始发病，比常年延晚 1 周。随着气温从日平均温度 22℃提高到 25℃以后，株发病率便从 7 月 6 日的 0.3% 迅速上升到 7 月 27 日的 14.1%，8 月份进入雨季，相对湿度高达 92%～96%，加速了病情的扩展蔓延，在 8 月中旬出现高峰，8 月 14 日株发病率达 84.1%，发病指数 43.6。9 月中旬以后，病情基本稳定。从开始见病到出现高峰约 40d。

1975 年的气温高于常年，水稻长势迅猛，纹枯病发病提前，6 月 20 日即开始出现病斑。较 1974 年早了 10d。7 月 22 日的株发病率为 14.1%；8 月中旬以后，连续阴雨、相对湿度达 89%～97%，8 月 12 日进入发病高峰，株发病率 78.6%，发病指数 35.3。高峰期延续较长，至 9 月中旬病情才开始稳定，从开始发病到高峰期约 60d（表 1）。

表 1 沈阳稻区水稻纹枯病发生消长观察

（1974—1975 年）*

1974 年（月/日）	发病率（%）	发病指数	日均温（℃）	相对湿度（%）	1975 年（月/日）	发病率（%）	发病指数	日均温（℃）	相对湿度（%）
7/6	0.30	0.04	22.3	76	6/25	0.20	0.05	21.7	81
7/7	0.90	0.13	23.7	87	6/28	1.20	0.25	24.6	76
7/8	1.00	0.18	23.4	88	7/1	1.80	0.45	24.0	79
7/9	1.10	0.30	23.8	84	7/4	2.60	0.65	24.9	84
7/10	1.10	0.30	22.3	97（雨）	7/7	7.50	1.40	27.2	73
7/11	1.70	0.40	26.2	75	7/15	17.60	4.60	25.8	缺
7/12	2.10	0.50	25.6	76	7/22	49.00	14.10	25.1	92（雨）
7/16	2.30	0.60	24.7	81	8/5	66.40	25.90	28.2	80
7/21	14.70	4.70	24.9	91（雨）	8/12	78.60	35.30	23.8	91[b]（小雨）
7/27	42.90	14.10	25.6	84	8/21	61.20	24.40	25.9	76
8/2	69.20	31.50	26.3	87	8/28	52.20	30.40	22.5	78
8/8	69.90	28.50	26.5	80[a]	9/12	50.10	26.70	18.0	80
8/14	84.10	43.60	22.6	92（雨）					
8/21	54.10	36.30	27.6	67					
9/12	36.60	13.10	22.9	79					

[a] 1974 年 8 月 9 日至 11 日连续降雨，9 日至 13 日相对湿度分别为 96%、96%、92%、87%、83%；

[b] 1975 年 8 月 11 日至 13 日连雨，其相对湿度分别为 89%、91%、97%；

* 根据当地气象站资料整理。

水稻纹枯病是个喜高温和高湿的病害。两年的初步观察，温度和湿度综合地影响其发生消长。始病期和终止期的推迟和提前主要取决于温度。1974 年的 6 月份有效积温仅 605.9℃，比历年低 37.5℃；常年开始发病的 6 月中、下旬日平均气温都低于 20℃，推迟了始发病期，到 7 月上旬才开始出现病斑，比常年延后 1 周。1975 年高温，6 月份有效积温高达 669.7℃，比 1974 年和常年分别高 63.8℃和 26.3℃，6 月 20 日便出现病斑，6 月中旬以后的日平均气温都在23℃～25℃以上，纹枯病在沈阳地区气温达23℃、相对湿度80%时就能发病。此后，日均气温高于23℃时，田间相对湿度便成为病情发展速度的主导条件。1974和1975年的发病高峰期虽然相差1周，但高峰来临前都遇到连续阴雨，1974年8月9～11日3d连降中雨，相对湿度高达92%～96%，病株率由8月8日的69.9%急剧上升到8月14日的84.1%，发病指数由28.5上升到43.6。1975年的情况类似，8月11～13日连雨，相对湿度89%～97%，病株率由8月5日的66.4%上升到8月12日的78.6%，发病指数由25.9上升到35.3（表1）。9月中旬以后，虽然阴雨连绵，相对湿度不低，但是气温均大大低于20℃，抑制了病害的发展。

2.2 品种及生育期与发病的关系

水稻品种之间对纹枯病的抵抗力差异虽不像对稻瘟病那样明显，但也有差异。在相同水肥管理条件下，品种植株的形态与发病有密切关系。1975年对参加省水稻品种区域试验的13个品种的定期、定点观察和调查，一般矮秆多穗型重于高秆大穗型，但早熟高秆型品种在抽穗后期的发病也很重，如清杂45在前期发病较轻，后期发展很快，发病指数由前期的10.8上升为30.5，病斑延伸到剑叶；强分蘖力的品种重于分蘖力弱的品种。在13个品种中，发病较重的有笹稔、越实、γ-18-8三个；中度发病的有403、6908-7-4、401、253、丰锦和清杂45六个；263、γ-25-3、深育1号、3071四个品种发病较轻。

沈阳稻区在分蘖盛期开始发病，拔节到抽穗时，病情激增，一方面由于此时温、湿度较高，同时也因稻株发根机能减退，植株养分大量转运于生殖生长，机体抗病力减弱，发病最烈；后期稻株基部叶老枯黄，茎秆老化，株间湿度下降，病情发展终止。

2.3 水肥与发病的关系

1974年，进行了浅水、深水灌溉两个水层处理，3次重复。浅水区保持1寸，发病晚而轻，长期深灌处理田的发病早且重，8月13日调查浅水、深水的发病率分别为34.1 %和54.7 %，发病指数为14.0和26.4（表2）。

表2 稻田水层管理与纹枯病的发病关系

（1974年）

调查时间（月/日）	浅灌		深灌	
	发病率（%）	发病指数	发病率（%）	发病指数
7/12	0	0	0.20	0.05
7/15	0.18	0.05	0.61	0.15
7/21	3.06	0.97	9.64	2.85
7/26	4.41	1.09	17.85	4.65
8/1	14.30	3.99	39.30	10.80
8/7	23.30	9.01	44.80	20.14
8/13	34.10	17.60	54.90	28.40
8/19	28.30	14.00	47.40	26.40

1975 年高温，稻苗长势迅猛，全市稻田及时进行烤田，防止水稻封行过早，降低了田间相对湿度，直至将浓绿披叶烤得挺直发黄，有效地抑制了病菌气生菌丝的滋生蔓延。发病后整个 1 个月的病情发展缓慢，对控制病害发展的作用非常明显。

两年来进行了肥料的不同配比和不同追肥时期、不同底肥量对发病的影响。结果如下：

1）相同基肥量（60 斤），单施氮、磷、钾的 3 个处理，在病情高峰期的发病率分别为 32.2%、16.6%、6.6%，发病指数为 13.9、6.6 和 2.8。

2）不同基肥量，40 斤*、60 斤、80 斤的发病率分别为 39.9%、38.5%、14.5%，发病指数为 20.1、19.9 和 8.4。

3）不同追肥期，5 月 5 日、6 月 10 日追肥的发病率分别为 12.7%、35.6%，发病指数为 7.5、19.5。

上述结果表明，单施氮肥、追肥晚、基肥量少而追肥多的处理发病重。施足基肥、早施追肥，增施磷、钾肥处理，促使稻株增多纤维素、木质素，茎秆健壮，抗病力有明显的提高。

2.4 不同密度与发病的关系

1974 年设置了 6 个插秧密度，行株距（寸为单位）为 4×3、6×3、8×3、10×3、11×3、12×3，3 次重复，定点、定期调查 11 次。结果表明，上述密度对发病的影响有 3 个幅度，以发病高峰期调查为例，4×3、6×3 最重，8×3、10×3、11×3 的病情相近，以 12×3 的病情最轻。表明稻株种植密度过大，相对湿度提高、增加了病株接触传播的几率，有利于病害扩展。12×3 的行间宽，遮光小、通风好，不利发病，但在生产中必须具有足够的单位面积有效穗数才能保证增产的群体，必须考虑在丰产的前提下采用合理的种植密度。

2.5 不同病情严重度对产量重要组分的影响

1974 年调查了一块生长均衡、种植品种京引 3 号的高产田。按 0～4 级 5 个病情分级标准随机取样。每级取样 100 穗，考察不实率、千粒重，并根据试验站作定期生育调查的每穗平均粒数、每穴有效分蘖以及 10×3 的密度，计算理论产量，折算损失率。结果表明，病情 2 级以上严重度的千粒重明显下降，4 级的千粒重比 0 级降低 2.4 克；不实率比 0 级增加 2.5 倍，高达 34.3%；病情 3、4 级的产量损失达 2～3 成（表 3）。

表 3 水稻纹枯病不同严重度对主要产量组分的影响

（1974 年）

病情级别	总粒数	秕粒数	秕粒率（%）	饱满粒数	饱满粒率（%）	千粒重	比 0 级减重	减产率（%）
0 级	10 201	1 325	13.0	8 876	87.0	27.0		
1 级	9 931	1 722	17.3	8 209	82.7	26.2	0.8 克	7.85
2 级	10 203	2 260	22.2	9 743	77.8	25.6	1.4 克	15.2
3 级	8 691	2 333	26.9	6 358	73.1	25.1	1.9 克	23.4
4 级	7 072	2 422	34.3	4 650	65.7	24.1	2.4 克	32.5

2.6 打捞菌核减少菌源的效果

1974 年设置打捞与不打捞对照两个处理，各 1 亩地。两块地在 7 月 6 日都开始发病，但发病穴数不同，打捞田的病穴率为 1.6%，对照田为 5.8%。随后的调查结果显示，打捞田发病速度缓慢，对照发展快，并影响病斑伸展；打捞区的病斑多数在剑叶下第 2～3 张叶片（1～2 级），对照区病株的病斑伸展到剑叶下第 1 张叶片（病情 3 级）。发病高峰期的发病穴率分别为 59%和 85%。

1975 年进行了大面积试验，在 1974 年的一块不捞菌核小区进行了连续打捞，在发病高峰期显示，发病率由 1974 年的 85%下降为 29%，重复 2 处理小区的结果相似，发病率由 1974 年的 82%下降为 37%。

打捞菌核减少菌源具有抑制发病效果的推测，从实践中得到了证实，这确是个有效措施。1974 年在满融试验站所在的大队在发病高峰期后，在未打捞菌核田内，进行了四角交叉定点调查稻株上的菌核量，每平方米平均为 386 粒，1975 年对这块田进行了打捞菌核处理，同期调查的结果是每平方米平均为 112 粒。尽管 1975 年气温高、水稻长势好、适于病害流行，但是由于该队连续两年全面、彻底地打捞菌核，减少了菌源，明显地减轻了病害。

在生产中，稻田在插秧前都要进行翻耕、泡田、耙地、耘田等操作，大部分越冬菌核都浮在水面泡沫中，插秧前也必须捞出浮渣，其中有部分菌核，问题是没有妥善处理，一般都将浮渣撂在田埂上面，随着灌水等劳作，捞出的菌核又回到田里。只有将捞出的菌核转移到远离稻田之处，才能奏效。

2.7 药剂防治试验

经两年小试、中试、从 10 种农药中筛选出了多菌灵、井冈霉素、托布津 3 种农药，进行了大面积试用。结果表明，多菌灵、井冈霉素使用安全，50%多菌灵 500～1 000 倍，1%～1.5%井冈霉素（50～80mg/kg）的效药效都能达到 95%左右，防病效果明显。多菌灵的药效期长达 2 个月，井冈霉素的药效期较短，约半个月，但其治疗效果是明显的，在病情盛发期使用的效果最佳。50%托布津可湿性粉剂 500～1 000 倍喷施的效果达 90%，药效较慢。稻脚青、稻宁、田安等有机砷制剂农药虽然药效好、残效期长，但容易发生药害，在水稻孕穗期更不宜使用。

2.8 药剂防治策略

由于大部分纹枯病菌核在插秧前都分布在稻田四周，沈阳地区春季都刮南风，使得混杂菌核的浮渣漂向北面，因此北面的发病早并且重。在稻田中部基本不见病株或发病轻。两年来我们连续普查满融大队 5 种类型计 50 个地块的病情分布，每块地从四边分别向田中央 3m、6m、10m 长，调查发病穴率，两年的结果是类似的：北重南轻，发病率和严重度依次为，3m>6m>10m。10m 以远的植株很少发病。在采取打捞菌核、合理水肥管理的基础上，如果需要辅以药剂防治的田块，则可进行挑治，从南、北两面 10m 之内、东西各打药 12 行（约 10m），经济而有效。

3 生产综合防治措施

根据水稻纹枯病在沈阳地区的消长情况，影响发病因子的观察调查及生产实践结果，初步认为防治水稻纹枯病应采用预防为主，综合防治的策略，包括减少菌源、栽培措施和辅以药剂防治。

3.1 打捞菌核减少菌源

在插秧前、后进行两次打捞和深埋菌核工作。严重地区坚持连年、全面处理菌核，达到逐渐减少菌量，满融大队的实践有力地说明这是一个事半功倍的措施。

3.2 合理水层管理

促进植株根系生长，防止植株过早封行，注意水稻各生育阶段的群体与个体之间的协调关系。做到浅水发棵、撤水促根、薄水养胎、润湿长穗。进入分蘖末期做到及时烤田，改善土壤通透条件，降低湿度，对控制病害的效果十分明显。烤田要看地看苗，肥田重烤，直至叶片挺直，绿中带黄，漏风土、沙性土轻烤。孕穗期适当加深水层，破肚轻烤，足水抽穗，达到防早衰，防倒伏，增强抗病力。

3.3 合理施肥

重施基肥、早施追肥、巧施穗肥。保证前期早生、早发、早分蘖，中期不徒长、后期不贪青。收割时青秆黄熟。施肥种类的配合，土杂肥与化肥、长效肥与速效肥相结合、氮肥配合磷肥和钾肥。增强抗病、抗倒伏。

3.4 药剂防治

1）50%多菌灵500～1 000倍，在病重田施药两次，第一次在分蘖末期至孕穗期，烤田灌水后施药。第二次在抽穗前进行。中等病情田，在打捞菌核的基础上，整个生育期于孕穗期到抽穗前用500倍液打1次。

2）1.5%井冈霉素需要打两遍，第1次在发病穴率达30%左右进行，第二次在抽穗期进行，每亩用1.5斤药兑100斤水。

3）50%托布津在发病始期应用，每亩1两兑50～100斤水。

以上的用药量均根据菌核分布田边的特点，只打10m以内的药液量。

本试验在沈阳地区水稻试验基础较好、研究规范的满融试验站所在地满融大队进行了两年，全部试验与生产计划紧密结合，在水稻高产栽培措施、品种产量试验等方面，均有多年实践经验的技术人员规划、指导。在满融大队采取综合防治方法对控制水稻纹枯病的试验、示范，获得了增产效益，被当地农民接受，是值得推广的。

4 存在的问题

这项试验至少应该进行3年，却在两年中匆忙结束。工作做得较粗，未来可进一步研究的内容：

4.1 在沈阳冬季严寒下越冬菌核的成活力测定；

4.2 病情消长变化仅有低温的1974年和高温的1975年的资料，缺少全面、系统观察；

4.3 药剂残效只有田间调查，缺少接种试验。

原载：沈阳市农业科学研究所，科学研究资料通讯，1976年。

水稻品种对白叶枯病和稻瘟病的抗性研究

章　琦

（沈阳市农业科学研究所）

Study on Resistance to Bacterial Blight and Rice Blast in Rice Cultivars

摘要　进行品种资源对稻瘟病和白叶枯病的抗性遗传分析，为杂交水稻抗病育种提供抗性显性遗传的供体亲本，采用两个毒力强的白叶枯病菌株崖凋 1－1 和E 2－2，3 个稻瘟病小种中 F1、中 D1、中 D5 接种，分析双抗两病或高抗一个病害的品种与感病亲本杂交子一代的抗性反应。观察品种不同生育期的抗性表达差异。结果表明，参试的抗病品种对所用的白叶枯病菌株的抗性有显性、部分显性或隐性遗传。抗稻瘟病的材料对小种中 F1、中 D1、中 D5 的抗性多数为显性、部分显性、并有少数为隐性遗传。抗、感品种间杂交 F_1 的抗性有互作现象。分别对两个病害的抗、感品种正反交组合 F_1 的测定表明，其抗性反应基本一致，初步表明水稻对稻瘟病和白叶枯病的抗性主要受细胞核所控制，也存在微效基因修饰作用。在参试品种中，约 90％的籼稻、76％的粳稻品种在不同生育阶段对白叶枯病的抗、感反应是一致的。约 91％的籼稻、76％的粳稻在不同生育阶段对稻瘟病的抗、感反应一致。水稻品种对两种病害都有全生育期抗性和成株期抗性两种类型。

从近代科学育种工作开始以来，改良品种的抗病性就成为育种的一个重要目标。我国从 70 年代开始，抗稻瘟病和白叶枯病被列为育种主要目标之一。

1975 年，我们开始参加全国水稻对稻瘟病和白叶枯病抗性鉴定协作项目。用本地和两病的优势菌株鉴定全国种质交换材料的抗性，在此基础上，进行本所杂交水稻育种材料的双抗及显性遗传抗源的鉴定和分析，并观察抗源在不同生育阶段的抗性表现。

1　材料和方法

1.1　材料

1.1.1　配置抗感杂交组合获 F_1　用 12 个双抗（中抗）稻瘟病和白叶枯病的品种，东海 2 号、IR28、IR30、早金凤五号、早生爱国 3 号、矢租、特大米，城堡一号、京引 139、岩锦、京引 83 分别同感（中感）、中抗品种配置正反交组合，获得 73 抗白叶枯病和 67 个抗稻

瘟病 F_1 种子。

1.1.2 配置抗性遗传改良的杂交组合 1）用抗两病的品种与感两病恢复系或不育系配组杂交；2）转育籼稻抗瘟性。

1.1.3 分析不同生育阶段抗性品种 抗病鉴定材料 611 份，其中粳稻 458 份、籼稻 153 份。

1.2 所用菌株

1.2.1 白叶枯病 两个强毒力菌株，崖凋 1－1 和沈阳优势菌 E2－2。经 3 年 6 次重复测试均保持稳定的毒性。

1.2.2 稻瘟病 全国生理小种沈阳地区菌株中 F1、中 D1、中 D5。

1.3 试验方法

1.3.1 种植 4 月 18 日将试材播于塑料大棚抗病鉴定圃。各 F_1 群体与其双亲以单株相邻种植，按生产田正常管理稻苗。

1.3.2 稻瘟病接种 用中 F1、中 D1、中 D5 三个小种混合菌液接种。在高粱培养基上培养病菌孢子，菌龄 12～14d，孢子浓度在 100 倍显微镜下平均每视野约 50 个。当大部分稻苗在 7～8 叶期，于日落后喷雾接种，连续 3d。接种后，用草帘（每条间隔 50 cm）覆盖于大棚顶部遮阴、降温，保持相对湿度 95 %以上，温度控制在 33℃以下。10d 后即普遍发病，当病情稳定后，按全国稻瘟病协作试验方案标准调查，叶瘟以病斑反应型分级：抗（R）、中抗（M）、感（S）。穗瘟在最高分蘖期喷雾接种两次，在水稻黄熟初期调查分为 R、M、S 型。

1.3.3 白叶枯病接种 用崖凋 1－1 和 E2－2 两个菌株接种。在 PSA 培养基上培养 3d，以 1∶1混合，接种液浓度 3×10^9/ml 细菌，于苗期 8～9 叶期、剑叶平展时各剪叶接种一次。每株剪顶叶和顶叶下一张叶片，各材料接种 40～50 张。根据病情发展，在接种后 20、25、30、35d 直至稳定后，共调查 4～5 次。按全国白叶枯病协作标准记载，分为 0～5 级：高抗（HR≤1.5 级），抗（R 1.6～2.0 级），中抗（MR 2.1～2.9 级），中感（MS 3.0～3.9 级），感（S 4.0～5.0 级）。F_1 群体单株调查，抗和中抗归为抗病级，感和中感归为感病级。

2 结果

2.1 杂交组合 F_1 的抗性

2.1.1 抗白叶枯病的反应 在品种间杂交的 73 个 F_1 中，20 个抗×抗组合有 19 个抗、1 个中感；在 37 个抗×感组合中，15 个抗、13 个中抗、9 个感；16 个感×感组合中，15 个感，1 个中抗（表 1）。

表 1 杂交组合 F_1 对白叶枯病的抗性反应

组合方式[a]	组合数	F_1 抗性反应			
		R	MR	MS	S
R×R	6	5	1		
R×MR	6	5		1	
MR×R	8	5	3		
小计	20	15	4	1	

（续）

组合方式[a]	组合数	F_1 抗性反应			
		R	MR	MS	S
R×S	7	5	1	1	
S×R	12	8	4		
MR×S	8	1	3	4	
S×MR	10	1	5	4	
小计	37	15	13	9	
S×S	16		1		15
合计	73	30	18	10	15

[a] R=抗，MR=中抗，S=感。

在粳稻抗病品种中，高抗品种矢租（1.2级）分别与5个感病品种和4个抗病品种的正交或反交组合 F_1 对所用的菌株全部抗病；成株期表现广谱抗性的日本品种早生爱国3号（2.1级）分别与抗或感病品种杂交 F_1 一般都达到中抗（2.2～2.5级），与高感品种几内亚杂交的 F_1 因生育期太长，当年未能反映其成株期抗性，表明成株抗性品种的杂种后代也要到成株期才能呈现。高抗品种中新120（1.5级）与几内亚杂交的 F_1 也因为未到成株期，只表现中感（3.0级）。东海2号（1.7级）与抗、感品种杂交的 F_1，除（B9×丰锦）品系外，都为高抗至中抗（1.0～2.9级）。

在籼稻抗病品种中，IR28和IR30与抗病品种东海2号、早生爱国3号或中抗的京引83和岩锦杂交组合的 F_1 都提高了抗性水平，前者可达到高抗（1.0～1.5级），后者达到抗～中抗，但与感病品种杂交的 F_1，一般都在中抗与中感之间（2.5～3.5级）。高抗品种早金凤5号（1.4级）分别与抗、感品种杂交的组合之 F_1 的抗性水平也都有所提高（1.0～2.2级）。

以上结果初步表明，参试品种对于所采用的白叶枯病菌株的抗性，分别表现完全显性、部分显性遗传（表2）。

表2 抗×抗、抗×感品种间杂交 F_1 对白叶枯病的抗性反应[a]

组　合	亲本抗性反应		F_1 抗性反应
	母本	父本	
矢租×不落籼43	1.2	5.0	1.5
矢租×二九龙	1.2	5.0	2.0
松粳2号×矢租	5.0	1.2	1.5
（IR24×矮科情×京引83）×矢租	4.2	1.2	1.5～2.0
（B9×丰锦-1）×矢租	4.2	1.2	2.0
矢租×城堡2号	1.2	2.3	1.5
矢租×IR28	1.2	2.0	1.0
矢租×（珍汕98×意大利B）	1.2	2.6	2.0
早生爱国3号×矢租	2.1	1.2	1.2

（续）

组　　合	亲本抗性反应		F_1 抗性反应
	母本	父本	
早生爱国 3 号×松粳 2 号	2.1	5.0	2.5
早生爱国 3 号×BL7	2.1	3.5	2.2
早生爱国 3 号×几内亚	2.1	4.5	4.5
早生爱国 3 号×东海 2 号	2.1	1.7	2.0～2.3
中新 120×几内亚	1.5	4.5	3.0
中新 120×东海 2 号	1.5	1.7	2.0
东海 2 号×松粳 2 号	1.7	5.0	2.0～2.9
东海 2 号×74-206	1.7	5.0	1.8～2.4
东海 2 号×几内亚	1.7	5.0	1.0～1.4
东海 2 号×（B9×丰锦）	1.7	5.0	3.1～5.0
东海 2 号×早生爱国 3 号	1.7	2.1	2.0
东海 2 号×中新 120	1.7	1.5	1.5
东海 2 号×IR28	1.7	2.0	1.0
IR28×京引 83	2.0	2.3	2.0
IR28×426	2.0	4.5	3.1～3.5
IR28×74-206	2.0	5.0	3.5
（B9×丰锦-1）×IR28	4.2	2.0	2.5～3.0
松粳 2 号×IR28	5.0（S）	2.0	2.2～3.4
京引 139×IR28	2.0	2.0	2.5
岩锦×IR28	2.5	2.0	2.0
早生爱国 3 号×IR28	2.1	2.0	1.5
早生爱国 3 号×IR30	2.1	1.5	2.0
松粳 2 号×IR30	5.0	1.5	1.4～1.8
74-206×早金凤 5 号	4.3	1.4	2.0
426×早金凤 5 号	4.5	1.4	2.2
松粳 2 号×早金凤 5 号	5.0	1.4	2.0
早生爱国 3 号×早金凤 5 号	2.1	1.4	1.8
东海 2 号×早金凤 5 号	1.7	1.4	1.0

[a] HR（高抗）≤1.5 级，R（抗）＝1.6～2.0 级，MR（中抗）＝2.1～2.9 级，MS（中感）＝3.0～3.9 级，S（感）＝4.0～5.0 级。

2.1.2　抗稻瘟病的反应　在品种间杂交的 69 个组合中，全部 31 个抗×抗组合 F_1 都抗；在 36 个抗×感组合中，26 个抗、1 个中抗、9 个感。两个感×感组合 F_1 全感（表 3）。

表 3 杂交组合 F_1 对稻瘟病的抗性反应

组合方式	组合数	F_1 抗性反应		
		R	M	S
R×R	7	7		
R×MR	8	8		
MR×R	13	12	1	
MR×MR	3	3		
小计	31	30	1	
R×S	4	4		
S×R	23	15		8
MR×S	4	4		
S×MR	5	3	1	1
小计	36	26	1	9
S×S	2			2
合计	69	56	2	11

高抗稻瘟病的籼粳交粳型品种城堡 1 号对稻瘟病的抗性受一对显性基因 *pi-zt* 控制，有高抗品种 TKM1 号的亲缘，它同中抗、高感品种杂交 F_1 全部高抗，在由中国农科院选育的 85 个品系中，多数在全生育期高抗稻瘟病的品系也携有城堡 1 号的亲缘。籼稻 IR28、IR30、IR2071、75 p12、几内亚、早金凤 5 号和矮脚窄叶青等抗瘟品种与感病品种杂交组合的 F_1 也都是抗病的。对叶瘟和穗瘟都抗的籼稻品种不落籼 421 和浙红与感病品种杂交的 F_1 是感的，虽然鲜有隐性抗瘟基因的报道，但本试验的 23 个感病（S）母本×抗病（R）父本组合中有 8 个表现隐性遗传（表 3）。抗白叶枯病、中抗稻瘟病品种东海 2 号与不同感病品种杂交组合的 F_1 提高了抗性水平，都有超亲现象。上述结果表明，参试品种对所用的稻瘟病菌株的抗性，表现完全显性、部分显性和隐性抗性遗传（表 4）。

表 4 抗×抗、抗×感品种间杂交 F_1 对稻瘟病的抗性反应

组　　合	母本反应	父本反应	F_1 抗性反应
城堡 1 号×松粳 2 号	R	S	R
松粳 2 号×城堡 1 号	S	R	R
C57×城堡 1 号	S	R	R
（Pi5×喜峰）×IR410－15	R	S	R
京引 131×IR28	S	R	R
松粳 2 号×IR28	S	R	R
（B9×丰巾）×IR28	S	R	R
IR28×74－206	R	S	R
γ72－25－8×γ72－21－3）×IR2071	S	R	R

（续）

组　　合	母本反应	父本反应	F_1 抗性反应
74-206×（初锦×IR30）	S	R	R
松粳2号×IR30	S	R	R
（姬穗波×IR26×丰锦）×75p12	S	R	R
γ72-25-8×γ72-21-3）×75p12	S	R	R
426×早金凤5号	S	R	R
G99×内福17-4	S	R	R
古76-2118×矮脚窄叶青	S	R	R
松粳2号×几内亚	S	R	R
中新120×几内亚	S	R	R
松粳2号×不落粃421	S	R	S
清锦×浙红	S	R	S
东海2号×74-206	M	S	R
东海2号×中新120	M	S	R
中新120×东海2号	S	M	R
松粳2号×东海2号	S	M	R
东海2号×松粳2号	M	S	R
松粳2号×早生爱国3号	S	M	R
74-206×东海2号	S	M	M
早生爱国3号×松粳2号	M	S	M
岩锦×IR28	R	R	R
IR28×京引83	R	R	R
京引139×IR28	R	R	R
初锦×（IR30×不落粃43）	R	R	R
岩锦×青秆黄	R	R	R
东海2号×IR28	M	R	R
特大米×东海2号	R	M	R
东海2号×早生爱国3号	M	M	R
东海2号×（B9×丰锦）	M	R	R
（B9×丰锦）×东海2号	R	M	R
东海2号×早金凤5号	M	R	R
东海2号×几内亚	M	R	R
城堡1号×早生爱国3号	R	M	R
BL7×早生爱国3号	R	M	R
早生爱国3号×BL7	M	R	R
IR30×早生爱国3号	R	M	R

（续）

组　　合	母本反应	父本反应	F_1 抗性反应
（IR24×矮科情×京引 83）×早生爱国 3 号	R	M	R
早生爱国 3 号×IR30	M	R	R
早生爱国 3 号×早金凤 5 号	M	R	R
早生爱国 3 号×几内亚	M	R	R
早生爱国 3 号×（初锦×IR30）	M	R	R
早生爱国 3 号×IR28	M	R	R
早生爱国 3 号×东海 2 号	M	M	R
早生爱国 3 号×矢租	M	M	R
矢租×IR28	M	R	R
IR28×矢租	R	M	R
矢租×二九龙	M	R	M
二九龙×矢租	R	M	R
矢租×城堡 1 号	M	R	R
城堡 1 号×矢租	R	M	R
（IR24×矮科情×京引 83）×矢租	R	M	R

2.1.3 双抗品种的抗性遗传 几个双抗品种分别与感病品种杂交组合 F_1 对稻瘟病和白叶枯病的抗性反应是不同的，对两病的抗性遗传相互独立。如东海 2 号抗白叶枯是 1.7～2.0 级，在分别与 4 个高感品种杂交组合的 F_1 中，有 3 个抗、1 个感；该品种中抗稻瘟病，分别与 3 个高感品种杂交的 F_1 却都是抗的，抗性水平有所提高，有超亲现象。

2.1.4 抗病、感病品种正反交组合 F_1 的抗性反应 在 14 对抗、感白叶枯病品种正反交组合的 F_1 中，有 13 对的抗性是一致的，只有东海 2 号（R，1.7 级）与（B9×丰锦）（S，4.2 级）的正反交略有不同，其 F_1 的抗性都降低为中感水平；在 8 对抗、感稻瘟病品种间正反交组合的 F_1 中，7 对的抗性一致，也有一对（74－206 与 IR 28 正反交）组合 F_1 的抗性相左，以 74－206 为母本的杂交组合 F_1 感，IR28 为母本组合的 F_1 抗；该组合 F_1 对白叶枯病的抗性效应也不佳，介于双亲之间且偏感（3.5 级）（表 5）。

表 5 正反交组合 F_1 对白叶枯病、稻瘟病抗性反应

组　　合	双亲和 F_1 抗白叶枯病反应			双亲和 F_1 抗稻瘟病反应		
	母本	父本	F_1	母本	父本	F_1
松粳 2 号×东海 2 号	S（5.0）	R（1.7）	R～MR（2.0～2.2）	S	R	R/M
反交	R		R～MR（2.0～2.5）	R	S	R/M
松粳 2 号×早生爱国 3 号	S（5.0）	MR（2.1）	MR（2.5～2.8）	S	M	M
反交	MR	S	MR（2.2～2.5）	M	S	M
松粳 2 号×城堡 1 号	S（5.0）	MR（2.3）	MR（2.5～2.9）	S	M	M
反交	MR	S	MR（2.5～2.6）	M	S	M

（续）

组　合	双亲和 F_1 抗白叶枯病反应			双亲和 F_1 抗稻瘟病反应		
	母本	父本	F_1	母本	父本	F_1
东海 2 号× IR28	R (1.7)	R (2.0)	HR (1.0～1.2)	R	R	R
反交	R	R	HR (1.0)	R	R	R
东海 2 号×早生爱国 3 号	R (1.7)	MR (2.1)	R (2.0)	R	M	R
反交	MR	R	R～MR (1.7～2.5)	M	R	R/M
东海 2 号×74 - 206	R (1.7)	S (4.3)	R～MR (1.8～2.8)	R	S	R/M
反交	S	R	R～MR (2.0～2.8)	S	R	R/M
东海 2 号×中新 120	R (1.7)	HR (1.5)	HR (1.5)	R	R	R
反交	HR	R	R (2.0)	R	R	R
东海 2 号× (B9×丰锦)	R (1.7)	S (4.2)	MS～S (3.0～5.0)	R	S	S
反交	S	R	MR～MS (2.1～3.0)	S	R	M/S
早生爱国 3 号×BL7	MR (2.1)	S (3.5)	R (2.0)	M	S	R
反交	MS	MR	R (2.0)	S	M	R
矢租×城堡 1 号	HR (1.2)	MR (2.3)	HR (1.5)	R	M	R
反交	MR	HR	HR (1.5)	M	R	R
矢租×IR28	HR (1.2)	R (2.0)	HR (1.0)	M	R	R
反交	R	HR	HR (1.0～1.5)	R	M	R
426× IR28	S (4.5)	R (2.0)	MS (3.2)	S	R	S
反交	R	S	MS (3.2)	R	S	S
74 - 206× IR28	S (5.0)	R (2.0)	MS (3.2)	S	R	S
反交	R	S	MS (3.5)	R	S	S
矢租×二九龙	HR (1.2)	S (5.0)	R (2.0)	R	S	R
反交	S	HR	MR (2.7)	S	R	MR

2.2 水稻品种不同生育阶段与抗病性的关系

2.2.1 品种不同生育阶段对白叶枯病的抗性 多数品种在苗期和孕穗期对白叶枯病的抗、感反应相同。一些品种前期感后期抗，但是也有极少数品种表现前抗后感。在参试的 379 个粳稻品种中，有 101 个在全生育期都抗（多数中抗），其中矢租表现高抗；有 60 个品种在孕穗期之后转抗，如中新 120、早生爱国 3 号、秋光、邳早 15、南粳 15、贵州麻谷、丸 10、筑紫晴、东海 2 号、兵系酒、秀峰等。少数前期抗，后期感的品种有时中 8 号和中系 7602 - 17 等。在 133 个参试的籼稻中，15 个抗病，其中 11 个全生育期抗病，如 IR28、IR30、早金凤 5 号、印度糯、宁晚 1 号等，4 个在成株期抗病。

2.2.2 品种不同生育阶段对稻瘟病的抗性 在参试的 458 个粳稻中，有 102 个品种全生育期中抗-抗，值得推荐的有城堡 1 号、城堡 2 号、南 56、南 65 及一些中系品种。多数籼稻品种在苗期和孕穗期对稻瘟病的抗、感反应相同。在 153 个籼稻中，130 个品种全生育期抗，

9 个成株期抗。有些抗穗瘟品种对叶瘟的感染也较轻，虽然是 S 型病斑，但病斑极少，为 S^{+}，或 $S^{\pm}$。在全生育期感的品种中，大多数在抽出剑叶以后又发生严重的节稻瘟，这种植株多数到后期便高感穗颈瘟。

水稻品种在不同生育阶段表现抗瘟性不同的原因，主要取决于其所含有的抗性基因，还可能与不同季节流行不同的病菌小种有关。我们每年抗性鉴定所采用的菌株有来自二季作和单季稻田里不同品种上的混合菌株，在不同时期也可能分布有不同的生理小种，这也是形成品种在不同生育阶段抗性差异的另一个原因。

3 讨论

3.1 杂交亲本的选配

抗病亲本的抗性越强，对后代的影响越大。本试验中采用的 IR30、矢租、城堡 1 号、早金凤 5 号、东海 2 号等都高抗一种病、中抗另一种病害，对后代的影响较大，是比较理想的抗源。有些高抗品种的抗性传递效应还受另一个亲本的影响，其抗病水平受到不同品种背景的牵制。如高抗白叶枯病的东海 2 号分别与几个感病品种杂交的子一代多数抗，但也有高感的 F_1；早生爱国 3 号和 IR28 与不同感病品种杂交 F_1 的抗性水平不同，在（抗×中感）时，F_1 表现中抗到中感，抗×高感时，则表现中感到高感。表明抗源的抗性传递状况，还与农艺亲本的感病程度有关。

对两个病害的抗、感亲本正反交组合的绝大多数 F_1 抗病性的差异很小。因此，一般可按育种要求，以农艺性状好的品种为母本，抗源作父本。籼粳稻杂交后代多数为不育株，将籼稻抗瘟性转入粳稻时，尤其应以粳稻为母本。

3.2 抗病性鉴定安排

3.2.1 抗瘟鉴定 应选用全生育阶段都抗的材料。随着抗病育种研究的深入，大量原始材料、杂交及回交后代的鉴定量越来越大。东北地区除一些夏秋阴雨寡照日数较多的稻区，像丹东市东沟县这样理想的自然发病鉴定场所外，多数地区要借温室、塑料薄膜大棚或其他保湿条件好的设施。需要考虑用地、人工以及争取时间等问题。我们认为，原始材料可以先经过用地较少的苗瘟鉴定，多设重复、淘汰大批高感材料，将多次表现轻感、抗病的材料再作穗瘟鉴定，可以获得品种前后期抗性一致的结果，经过初筛、复筛后，也避免遗漏优良材料；杂交或回交后代的 F_1 和 F_1B_1 的种子量不多，可直播在穗瘟鉴定圃；一般 F_2 群体大，首先经苗瘟鉴定，汰除大量病苗后，待抗病株到孕穗期再鉴定穗瘟。有些高代材料虽然含有抗病亲本，如果在加代过程中不进行抗病鉴定，很可能在选择农艺性状的同时丢弃了抗病株，因此，在抗病育种计划中，应紧密配合抗病鉴定，才能培育出优良、抗病的品种。

3.2.2 抗白叶枯病鉴定 我们鉴定的粳稻品种，多数在苗期抗性弱，到成株期抗性增强。因此，重点在成株期鉴定白叶枯病。但是，如果只在成株期鉴定一次，可能漏选一些全生育期抗病材料。因此最好进行苗期和成株期两次鉴定。在苗期鉴定时，单株相隔 1 寸直播，省地也便于操作。

国外较早就比较深入而系统的进行了抗病性遗传研究，促进了抗病育种工作（Ou，

1972)。我们目前还只能对杂种后代进行简单的抗性鉴定，比较粗浅。我国目前的病原菌毒性分化的研究状况，还缺乏有鉴别性的专化菌株，无法进一步确定抗性基因，也不具备恒温条件，对于病原菌的发生及侵染力不无影响，因此，不同地区、季节的接种试验，都有可能影响其结果的准确性。未来的抗病育种研究，多学科联合进行，选育出抗性的品种，将会成为客观发展的必然趋势。

原载：沈阳市农业科学研究所，科学研究资料通讯，1978 年。

水稻抗病育种研究Ⅱ. 品种（系）对稻瘟病的抗谱鉴定初报

林世成　章　琦　冯瑞英　阙更生　袁　珏

（中国农业科学院作物育种栽培研究所）

Studies on Disease Resistance in Rice Breeding Ⅱ. A Preliminary Report on the Identification of Spectrum of Blast Resistance in Rice Cultivars and Hybrid Lines

Abstract　Spectra of resistance to 15 strains of 13 races representation 7 different groups（A，B，C，… G）of blast were identified of 21 rice cultivars and 22 hybrid lines in 1980. Among the japonica cultivars，Toride No. 1 and No. 2（carrying $Pi-z^t\ Pi-a$）and some of their hybrid lines had the broadest resistance spectrum covering all the pathogenic strains tested，and had higher percentage of R-type reaction within their spectra. Pi 5（carrying $Pi-ta^z$）and some of its hybrid lines and two indica cultivars Tetep and Tadukan had also broader spectrum of resistance. Among the BL group cultivars（carrying $Pi-b$），BL 7 and BL 2 exhibited better blast resistance then others. Most of the above mentioned cultivars were resistant to both blast and bacterial blight.

Among those hybrid lines，Zhong-Xi 7602－5，7604，7702，7759，and 7850 had not only broad spectra of blast resistance but promising agronomic characters also. It is suggested that they are to be introduced to commercial production or as parents in rice breeding programs.

Most cultivars and hybrid lines are resistant to some strains and susceptible to others，but their resistance spectra are different. Therefore，it is of practical value that a newly released with broad spectrum，and especially to local prevailing blast races should be provided.

It was found that the correlation between the broadness of spectrum and the strength of blast resistance was highly significant，that is，the broader the spectrum the stronger the resistance and the higher the percentage of R-type reaction within the spectrum（R＋M）. Moreover, the unanimity of resistance to blast among rice cultivars and hybrid lines was mainly in case of R-type reaction. So, in the breeding program for blast resistance the unanimity of quantitative and qualitative effects of resistance should be emphasized. The breeder should pay more attention to R-type reaction in the breeding materials.

Using two strains of bacterial blight for inoculation it was found that strain L2－2 had a stronger virulence than the strain OS－118, and cultivars Shi-Zu, Tetep and Toride No. 2 were superior in resistance to bacterial blight. The correlation between virulence of the two strains was highly significant, while only negligible negative correlations between the spectrum of blast resistance and the virulence of either of those two bacterial strains have been observed.

在作物抗病育种工作中，育种材料的抗性鉴定是一个重要环节。水稻对叶瘟的抗性一般根据病斑和病斑数量，以“高抗”到“高感”的级别来表示，并以此作为筛选的依据。Flor（1971）“基因对基因”学说认为，相当于病原菌的非致病性或致病性基因，在寄主植物上也相应的有抗病或感病基因。一旦病原菌的非致病基因转变为致病基因，则寄主的抗病性开始丧失。Ou（1972）的大量研究证明，稻瘟菌的生理分化是十分普遍的，在一个病斑上产生的孢子，可分化出许多小种。清泽等（1980）报道，稻瘟病的“自发突变”（spontaneous mutation）和致病性变异的频率是很高的。因此水稻品种丧失抗性的频率也很高。实践证明，许多抗病品种在生产上种植2～10年即丧失抗病性，但Ou（1979）从国际稻瘟病圃的大量资料看出，具有广抗谱的水稻品种则很少发生病斑。从广抗谱品种上的病斑分离的新菌株，经过回接，不容易再侵染原品种，但却能使某些感病品种严重发病。因此，在利用垂直抗性作为抗源的情况下，培育对稻瘟病有广谱抗性的携有多基因或单基因水稻品种，可能比抗谱较窄的“高抗”品种，具有更重要的利用价值。

本文报道水稻品种（系）对稻瘟病菌的抗谱鉴定研究的初步结果。

1 材料和方法

1.1 供试品种和品系

供试材料选用了具不同抗病基因和还不了解抗病基因的常用抗稻瘟病亲本品种21个，和几年来选育的抗瘟性较好的品系22个。以高度感病品种丽江新团黑谷作对照。具体的品种（系）见表1、表2。

试材于1980年5月23日播于大磁盘内，每盘各播45份材料，其中有2份为对照，每格63 cm^2 内播15～20粒种子，重复2次，共播30盘。播前播后喷洒1 000倍液敌克松防治立枯病。播种前后各施1次氮肥，稻苗4～6叶期接种。

表 1 品种对中国 7 群稻瘟病小种及 2 个白叶枯病菌株的抗性反应

品种名称	抗稻瘟基因型[a]	抗小种数	测试的稻瘟病菌小种[b]															白叶枯病级	
			A_1	A_{13}	A_{61}	B_1	B_{13}	B_{29}	C_1	C_{13}	D_1	E_1 晋	E_1 吉	E_3	F_1 黔	F_1 吉	G_1	L_{2-2}	OS-118
城堡 1 号	*Pi-z*^{t}	15	M	R	R	M	R	R	R	M	R	R	M	R	R	R	R	3.0	3.1
城堡 2 号	*Pi-z*^{t}、*Pi-a*	15	R	R	R	R	R	R	R	R	M	R	R	R	R	R	R	2.8	2.0
BL7	*P** *-b*	14	R	R	R	M	M	R	R	R	R	R	S^+	R	R	R	R	2.5	3.7
Pi5	*Pi-ta*^{2}	14	R	S^+	R	M	M	M	R	M	M	R	M	R	R	M	R	2.8	2.8
Tetep	*Pi-k*^{n} 及其他	13	R	S^+	S^+	R	R	R	R	R	R	R	R	R	R	R	R	1.8	2.6
Tadukan	*Pi-ta*	13	S^+	R	R	M	R	M	R	R	R	R	R	R	S^+	R	R	2.0	2.9
BL2	*Pi-b*	13	R	M	M	M	S^+	R	R	R	R	S^+	R	M	R	R	R	3.0	2.6
BL4	*Pi-b*	13	M	R	R	S^{++}	M	M	R	S^+	R	R	R	R	R	R	R	3.0	2.9
BL6	*Pi-b*	13	R	R	R	S^+	R	M	R	S^+	M	R	R	R	M	R	R	4.7	4.0
BL3	*Pi-b*	12	R	R	S^+	S^{++}	M	M	R	S^+	R	R	M	M	R	R	R	3.1	3.0
Pi1	*Pi-ta*、*Pi-a*	11	R	S^+	R	R	S^+	R	R	S^{++}	R	M	R	R	S^+	R	R	3.2	2.9
BL1	*Pi-b*	10	M	R	S^+	S^{+++}	R	M	S^+	S^+	R	M	M	S^+	R	M	R	3.4	2.7
抗一		10	S^+	S^+	R	R	R	R	R	R	R	S^+	S^+	M	S^+	R	R	2.6	3.0
抗二		10	S^+	R	S^+	R	M	M	R	R	R	S^+	S^+	M	S^+	R	R	2.9	3.0
矢租		9	S^+	S^+	S^+	R	S^+	S^+	R	S^+	R	R	M	M	M	M	R	1.4	1.8
岩粳早		9	R	S^+	S^+	R	R	R	R	R	R	S^+	S^{+++}	S^{+++}	R	S^+	R	3.2	3.7
下北	*Pi-ta*、*Pi-a*	8	R	S^+	S^+	M	R	M	S^+	S^+	R	M	S^+	S^+	S^+	R	R	3.5	3.0
阿苏实		7	M	S^+	R	S^{++}	S^+	M	M	S^+	S^{+++}	S^+	S^{++}	M	R	S^+	R	2.8	2.3
关东 51	*Pi-k*	6	S^+	R	R	S^+	S^+	M	S^+	R	S^+	S^{++}	S^{+++}	S^{+++}	R	S^+	R	4.1	3.3
福锦	*Pi-z*	5	M	S^{++}	S^+	S^+	S^+	M	M	M	S^+	S^+	S^{+++}	S^{+++}	S^+	S^{++}	R	3.2	3.3
BL5	*Pi-b*	5	S^+	S^+	S^+	S^{++}	S^+	M	R	S^+	S^+	S^+	S^{+++}	R	S^+	M	R	2.8	3.0
丽江新团黑谷（对照）		0	S^{+++}	S^{+++}	S^{+++}	S^{+++}	S^{+++}	S^+	S^{+++}	S^{+++}	S^{+++}	S^{+++}	S^{+++}	S^{+++}	S^{++}	S^{+++}	S^+	3.5	3.0

[a] 清泽测定或推定；

[b] R=抗，M=中抗，S=感（S^+=轻感，S^{++}=中感，S^{+++}=重感）。

表 2　品系对中国 7 群稻瘟病小种及 2 个白叶枯菌株的抗性反应

品种名称	杂交组合	抗小种数	测试的稻瘟病菌小种															白叶枯病级	
			A_1	A_{13}	A_{61}	B_1	B_{13}	B_{29}	C_1	C_{13}	D_1	E_1 晋	E_1 吉	E_3	F_1 黔	F_1 吉	G_1	L_{2-2}	OS-118
中系 7602-5	城堡 1 号/喜峰	15	R	R	R	R	R	R	R	R	R	R	R	R	R	R	R	4.1	3.6
中系 7604	城堡 1 号/咸南 10 号	15	R	M	R	R	R	R	R	R	M	R	M	R	R	R	R	3.7	2.7
中系 7709	红旗 12/70-523//下 北///城堡 1 号	15	R	M	R	R	R	R	R	R	R	R	R	R	M	R	R	3.0	3.3
中系 7722	抗 二/513//中系 7609	13	R	M	R	M	R	R	S^+	R	R	R	M	S^+	M	R	R	3.5	3.3
中系 7850	抗 二/513//中系 7609	1[illegible]	R	M	R	M	R	R	R	R	R	S^+	S^{+++}	R	R	R	R	3.8	3.1
中系 7603	大和晴/喜峰	12	R	R	R	S^+	R	R	R	R	R	S^+	S^+	R	R	R	R	3.7	3.3
中系 7647	大和晴/喜峰	12	R	R	R	S^+	R	R	R	R	R	S^+	S^+	R	R	R	R	3.2	3.2
中系 7648	津辐 1 号/福锦	12	R	R	R	S^+	R	R	R	R	R	S^+	R	S^+	R	R	R	4.5	3.3
中系 7652	70-001/长元 16	12	R	R	R	R	R	M	R	R	R	S^+	S^{++}	R	R	S^+	R	2.7	2.6
中系 7702	城堡 2 号/70-001	12	R	M	R	M	R	S^+	R	R	R	R	M	S^+	S^+	R	R	2.7	2.4
中系 7705	黎明/福锦	12	S^+	R	R	S^+	R	R	R	R	R	R	R	R	S^+	M	R	3.6	3.3
中系 7814	城堡 1 号/喜峰	12	R	R	M	M	S^+	M	M	R	S^+	R	M	M	M	S^+	R	3.7	4.3
中系 7844	福锦/矮秆稻	12	S^+	M	M	S^+	R	M	M	M	R	R	M	S^+	M	R	R	3.4	2.5

（续）

品种名称	杂交组合	抗小种数	测试的稻瘟病菌小种															白叶枯病级	
			A_1	A_{13}	A_{61}	B_1	B_{13}	B_{29}	C_1	C_{13}	D_1	E_1 晋	E_1 吉	E_3	F_1 黔	F_1 吉	G_1	L_{2-2}	OS-118
中系 7825	抗二/513/中系 7609	11	R	R	S^+	S^+	R	R	R	M	R	S^+	S^+	R	R	M	R	3.0	3.2
中系 7836	城堡 2 号/72048//垦丰 5 号	11	M	R	R	R	R	M	M	S^+	S^+	S^+	S^+	M	M	M	R	2.7	2.3
中系 7839	梯特普/京引47//5001	11	R	R	S^+	S^{+++}	S^+	R	R	S^+	R	M	M	M	R	M	R	3.4	2.7
中系 7703	城堡 1 号/雨良//大和晴/喜峰	10	M	M	R	M	R	S^{++}	M	R	R	S^+	S^+	S^{+++}	S^+	R	R	3.3	3.3
中系 7823	福锦/矮秆稻	10	M	S^+	S^+	S^{++}	R	S^+	M	S^{++}	R	R	M	M	M	M	R	3.2	3.3
中系 7824	抗二/513//中系 7609	10	R	R	R	S^+	R	S^+	M	M	R	S^+	S^{++}	S^+	R	R	R	4.5	3.0
中系 7720	抗二/513//中系 7609	9	R	S^+	R	S^+	M	S^+	R	R	R	S^+	S^{++}	M	M	S^+	R	3.6	2.9
中系 7872	Pi5/喜峰//中新 120	7	S^+	R	R	S^+	R	R	S^+	M	S^+	S^+	S^{++}	S^+	R	S^+	R	3.8	2.9
中系 7710	水原三百粒/城堡 1 号//黎明	6	M	S^+	S^+	S^+	S^+	M	R	R	S^+	S^+	S^{+++}	S^{+++}	M	S^+	R	3.5	3.5
丽江新团黑谷（对照）		0	S^{++}	S^{+++}	S^{+++}	S^{+++}	S^{+++}	S^{++}	S^{+++}	S^{+++}	S^{+++}	S^{+++}	S^{+++}	S^{+++}	S^{++}	S^{+++}	S^+	3.5	3.3

1.2 菌株和接种

由浙江省农科院植保所提供稻瘟病菌株，另由吉林省农科院植保所提供 Po78－30 和 Po79－103 两个菌株，共有代表中国稻瘟菌生理小种的 7 个群、13 个小种、15 个菌株。接种前 1 个月将试管斜面菌种移植至高粱培养基，在低温下保存。接种前 2d 取出，冲洗净菌丝体，摊开并覆盖 2 层纱布保湿，在 24℃下培养 2d，长满孢子后，制成孢子悬浮液。浓度在 120 倍显微镜下每视野约 40 个孢子。各菌株在有严格隔离的特制塑料分隔小室中接种，每分室喷孢子液 50ml。接种后封闭保湿 24h 后移出，每天辅以清水喷雾。接种后 12 d，当对照区充分发病、叶面布满病斑时进行第 1 次调查，病情稳定后进行第 2 次调查。

病情按全国稻瘟病协作研究的标准记载。

1.3 白叶枯病抗性鉴定

为了分析参试材料对白叶枯病的抗性与稻瘟病抗谱的关系，本研究选用了两个白叶枯病菌株，沈阳市致病菌 L2－2 和江苏省 OS－118。以牛肉汁蛋白胨培养基扩大繁殖，菌龄 3d，浓度为 3×10^9/ml。在 8～9 叶期，用单菌株分别剪叶接种，重复 3 次，接种后 25 d 调查，以后在同株的剑叶上接种，20 d 后调查其成株抗性。按 1978 年全国白叶枯病协作会制订的标准记载。

2 结果与讨论

供试品种和品系对 15 个稻瘟病菌株的抗性鉴定结果分别列于表 1 和表 2。由表所列资料，就以下问题进行分析和探讨。

2.1 关于抗病品种和感病品种的概念

国内外稻瘟病圃多点多年试验结果一致表明，在供试的大量品种中，没有发现一个品种在各病圃都表现抗病，即能抗所有的小种。但某些品种却具有抗多种小种的广谱抗性（全国稻瘟病科研协作组，1980；Ou，1972）。我们用代表 7 个群、13 个小种的 15 个菌株进行抗谱鉴定的结果也表明，不论是品种或品系均有 86.4%的试材既抗一些菌株，也感另一些菌株，只有感病对照品种丽江新团黑谷有 0～4.5%不抗任何菌株。在 1976—1977 年全国稻瘟病圃的 20 个点上种植，表现 R 型的 9 个点，M 型的 8 个点，S 型的 3 个点，R+M 的点占 85%。可见，所谓抗病或感病品种，如不联系具体的小种来考虑，是不确切的。因此，介绍一个抗稻瘟的新品种，应该指出其所抗的小种，如果只笼统地说“抗”或“感”，那是不全面的。清泽（1980）为了研究品种的抗瘟性遗传，选定了能鉴别日本全部抗病基因，并在培养基上能形成多量孢子、致病性稳定的 7 个菌株。后来，凡是对 7 个菌株中的任何一个表现抗性反应的品种，被称为抗病品种。我国稻区辽阔，稻瘟病生理小种相当复杂，要在全国范围内提出统一的抗病品种的标准是不现实的。但是在一个省或地区内根据优势小种的构成，提出抗这些优势小种的品种为抗病品种，则是切实可行的。一旦优势小种发生变化，原先的抗病品种也可能变成感病品种。

2.2 不同基因型的抗谱比较

抗稻瘟病不同基因型的抗谱很不相同。由表3所示，携有 $Pi-z^{t}$ 基因的“城堡1号”、“城堡2号”以及由“城堡1号”与其他品种杂交的抗性较强的后代“中系7602－5”、“中系7604”和“中系7709”，都抗所有的15个菌株，其中“中系7602－5”且全部为R型，抗性超亲。1979年的鉴定结果也是如此。看来这些品种（系）受小种威胁的机会较少。携有 $Pi-ta$ 或 $Pi-ta^{2}$ 基因的品种，以Pi 5的抗谱最广（14/15），但4个品种的抗性首先受A群小种威胁，其次受F群和C群的小种威胁。抗一和抗二的抗性基因不明，其抗谱均为10/15，同样受A群、E群、F群小种威胁。“中系7722”等5个姊妹系，抗谱为9/15－13/15，其抗性则受E群小种和B群小种的威胁。具有 $Pi-b$ 基因的BL1～BL7其抗谱为5/15－14/15，幅度很大，其抗性则受B群、C群和E群小种的威胁。以上的初步结果可作为进一步增广稻瘟抗谱的参考。至于具有 $Pi-k$ 和 $Pi-z$ 的品种，由于参试品种数太少，而且其抗性表现不很规律，则难以分析。

表3 不同基因型水稻品种和杂交后代的抗谱比较

品种（系）	杂交组合	基因型	抗谱（R+M）/（R+M+S）	感病的菌株
城堡1号	TKM1/农林8号[4]	$Pi-z^{t}$	11+4/15	0
城堡2号	CO25/农林8号[4]	$Pi-z^{t}$ $Pi-a$	14+1/15	0
中系7602－5	城堡1号/喜峰	$Pi-z^{t}$	15+0/15	0
中系7604	城堡1号/咸南10号	$Pi-z^{t}$	13+2/15	0
中系7709	红旗12/70523//下北///城堡1号	$Pi-z^{t}$	13+2/15	0
Tadukan	菲律宾品种	$Pi-a$	11+2/15	A_{1}、F_{1}
Pi 5	Tadukan /东山38[3]	$Pi-ta^{z}$	7+7/15	A_{13}
Pi 1	Tadukan /千本旭[3]	$Pi-ta$ $Pi-a$	10+1/15	A_{13}、B_{13}、C_{13}、F_{1}
下 北	八甲田//Pi/农林29	$Pi-ta$ $Pi-a$	5+3/15	A_{13}、A_{61}、C_{1}、C_{13}、E_{1}、E_{3}、F_{1}
抗 一	矮落200/垦跃209		9+1/15	A_{1}、A_{13}、E_{1}、E_{1}、F_{1}
抗 二	矮落200/垦跃209		6+4/15	A_{1}、A_{61}、E_{1}、E_{1}、F_{1}
中系7722	抗二/513//中系7609		9+4/15	C_{1}、E_{1}
中系7850	抗二/513//中系7609		11+2/15	E_{1}、E_{1}
中系7825	抗二/513//中系7609		9+2/15	A_{61}、B_{1}、E_{1}、E_{1}
中系7824	抗二/513//中系7609		8+2/15	B_{1}、B_{29}、E_{1}、E_{1}、E_{3}
中系7720	抗二/513//中系7609		6+3/15	A_{13}、B_{1}、B_{29}、E_{1}、E_{1}、F_{1}
BL1		$Pi-b$	5+5/15	A_{61}、B_{1}、C_{1}、C_{13}、E_{3}
BL2		$Pi-b$	9+4/15	B_{13}、E_{1}
BL3		$Pi-b$	8+4/15	A_{61}、B_{1}、C_{13}
BL4		$Pi-b$	10+3/15	B_{1}、C_{13}
BL5		$Pi-b$	2+3/15	A_{1}、A_{13}、A_{61}、B_{1}、B_{13}、C_{13}、D_{1}、E_{1}、E_{1}、F_{1}

（续）

品种（系）	杂交组合	基因型	抗谱（R+M）/（R+M+S）	感病的菌株
BL6		*Pi-b*	10+3/15	B_1、C_{13}
BL7		*Pi-b*	12+2/15	E_1
关东 51	杜稻/中生银坊主	*Pi-k*	5+1/15	A_1、B_1、B_{13}、C_1、D_1、E_1、E_1、E_3、F_1
Tetep	越南当地品种	*Pi-k^n*	13+0/15	A_{13}、A_{61}
福 锦	54BC-68/初锦	*Pi-z*	1+4/15	A_{13}、A_{61}、B_1、B_{13}、D_1、E_1、E_1、E_3、F_1、F_1

2.3 抗谱（R+M）与病斑型 R 的相关

水稻品种（系）对稻瘟病的抗谱存在很大差异，但抗谱的广度（数量）与抗性的强度（质量）之间是否有相关？我们曾进行了分析。结果 r＝0.470 2** （n＝43），相关极显著（图 1），说明抗谱愈广，抗性愈强，即 R 在 R+M 中所占的比例愈大，反之，抗病谱愈窄，抗病性愈弱，即中抗型（M）占的比例愈大。抗性的量与质的一致性，说明增广抗谱在抗稻瘟的育种计划中具有重要的实践意义。

由于自变数是由 R－M 组成，那么抗谱与中抗型 M 必然形成同样强度的负相关。水稻的抗瘟性易于丧失，因此经常有人提出应该注意选中抗品种，那么，抗谱与所谓“理想”的病斑型之间便存在矛盾。关于这些问题，今后还要反复地加以验证。

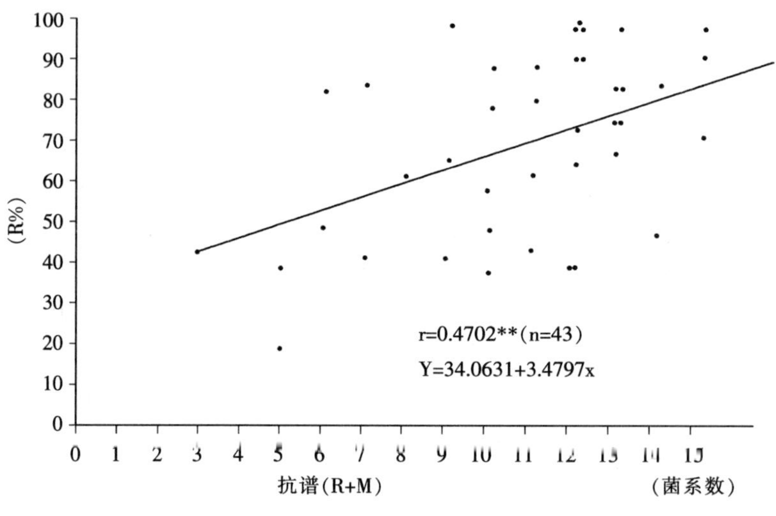

图 1　抗谱（R+M）与 R 型病斑（%）的相关

2.4 水稻品种（系）对稻瘟菌株抗、感反应的相关性

本研究所用的水稻品种和品系之间虽然不都是直接的亲缘关系，但从抗瘟性的角度看，它们之间的关系还是比较密切的。我们将品种和品系对 15 个菌株的抗感关系，进行了相关

分析，结果见图 2。(1) 从抗谱（R＋M）的资料进行分析（图 2 上），品种与品系的相关不显著，r＝0.389 3。(2) 从抗病型 R 的资料进行分析（图 2 中），品种与品系的相关极显著，r＝0.677 0**。(3) 用中抗型 M 的资料分析（图 2 下），二者的相关极不明显，r＝0.156 8。这些结果表明，供试品种和品系之间，抗、感关系的一致性，主要表现在 R 型的抗性方面，而 M 型抗性关系不大。

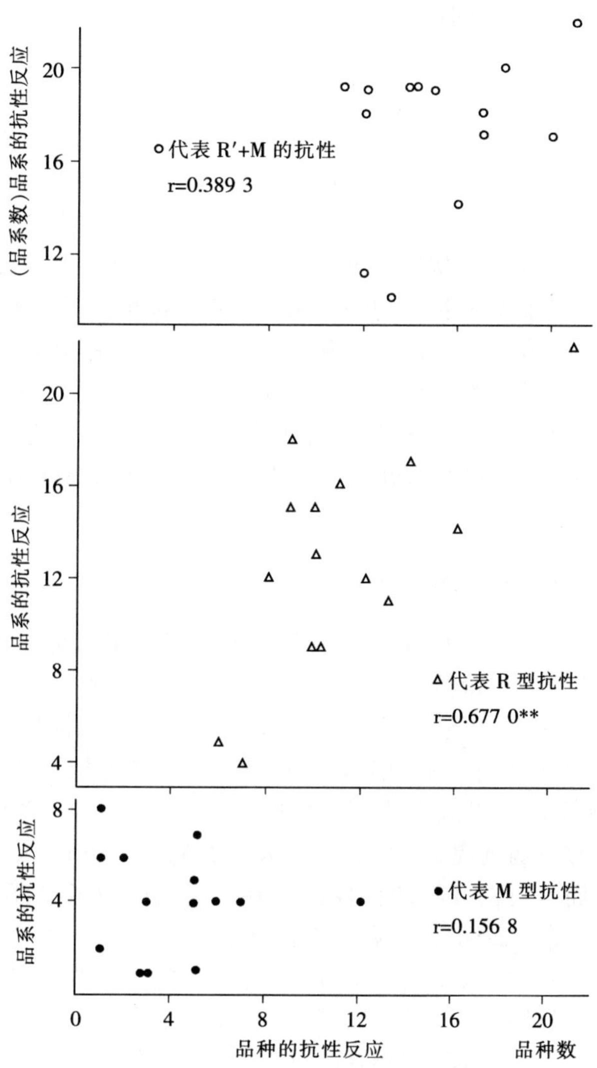

图 2　水稻品种与品系对稻瘟菌株抗、感反应

上　品种与品系对 15 个菌系抗谱（R＋M）的相关

中　品种与品系对 15 个菌系 R 型抗性的相关

下　品种与品系对 15 个菌系 M 型抗性的相关

2.5 品种（系）对白叶枯的抗性及其稻瘟抗谱的关系

为了研究品种（系）对稻瘟和白叶枯抗性的关系，我们除鉴定试材的稻瘟病抗谱外，并测定它们对两个白叶枯病菌株 L2－2 和 OS－118 的抗性。由表 1、2 所列资料看，品种（系）对白叶枯的成株抗性反应，由抗性最强的品种矢租对 L2－2 的 1.4 级到抗性最弱的“BL6”对同一菌株的 4.7 级，以3.0～3.5级的次数最多，约占 46.7%，可见这些材料对白叶枯的抗性是偏低的。

变量分析结果表明，菌株间的差异显著，品种（系）间的差异极显著。菌株间平均病级差异显著标准为 2.018×0.086 7＝0.175 0（P＝0.05）和 2.700×0.086 7＝0.234 1（P＝0.01）。L2－2 与 OS－118 的致病力也存在极显著的正相关，r＝0.539 8**。至于品种（系）间平均病级差异显著标准为 2.018×0.406 7＝0.82（P＝0.05）和 2.70×0.406 7＝1.10（P＝0.01）。即在供试品种（系）间，白叶枯病级相差达到 0.82 级，差异显著；达到 1.1 级，差异极显著。“矢租”的抗性最强（平均为 1.6 级），与“Tetep”（2.2 级）和“城堡 2 号”（ 2.4 级）差异不显著，而与其他品种、品系的抗性比较均达到显著或极显著标准。对于稻瘟病抗谱与白叶枯病抗性之间的关系，我们也进行了相关分析。结果稻瘟病抗谱与 L2－2 和 OS－118 两个菌株的致病力都没有相关关系，前者 r＝－0.079，后者 r＝－0.043。这显然不能反映稻瘟病与白叶枯病抗性的全部关系。

根据基因分析，现已发现 13 个稻瘟的抗性基因，分布于 7 个连锁群，已发现 9 个白叶枯的抗性基因，其连锁关系还没有研究清楚。所以稻瘟抗性与白叶枯抗性的关系如何，还有待深入系统地研究。

2.6 对部分供试品种和品系的评价

本研究选用的 22 个水稻品种都是抗病育种工作中常用的授体亲本，经过抗谱和抗性鉴定，其中有些具有较高的应用价值。以 $Pi-z^t$ 抗病基因为主体的“城堡 1 号”、“城堡 2 号”对 15 个稻瘟菌株全抗，但“城堡 2 号”又比“城堡 1 号”更为优越：(1)“城堡 2 号”的 R 型病斑占 14/15，(2) 对白叶枯的抗性好，与抗性最好的“矢租”差异不显著。其杂交后代“中系 7702”和“中系 7836”也都具有双抗特性。

在由 Tadukan 转移抗病基因 $P-ta$ 或 $P-ta^2$ 的品种（“Pi 1”、“Pi 5”、“下北”）中，“Pi 5”不仅抗谱广，而且对白叶枯的抗性也较好，用它作为亲本育成的中丹 1 号、中丹 2 号和中丹 3 号（林世成等，1980），都属于双抗类型。

具 $Pi-b$ 抗病基因的“BL 1”至“BL 7”品种中，以“BL7”（抗谱为 14/15）和“BL2”（双抗）比其他品种稍佳，“BL7”还具有早熟特点。

籼稻品种“Tetep”具有对稻瘟菌的抗谱广、R 型反应多和抗白叶枯病等特点，也是较好的授体亲本。该品种株型欠佳，但与株型较好的粳稻品种杂交后，其后代的株型不难改进。“Tadukan”也具有类似特点。

在供试的 22 个品系中，抗性和其他经济性状比较好的品系列于表 4，总的特点是对稻瘟菌株的抗谱广（有 3 个品系全抗），早熟，分蘖力强，株、穗形较好，糙米率高，根据历年试验丰产性较好。其中“中系 7602－5”、“中系 7604”、“中系 7709”、“中系 7702”、“中系 7850”等品系有可能在北方稻区生产上直接利用或作为育种亲本材料。

表 4 部分供试品系的主要性状和特性

品系	杂交组合	分蘖数	单株穗数	株高(cm)	穗长(cm)	主穗粒数	空秕率(%)	糙米率(%)	千粒重(g)	全生育期(d)
中系7602-5	城堡1号/喜峰	18.9	11.6	96	19.1	122	5.7	83.7	26.2	139
中系7603	大和晴/喜峰	15.8	7.9	107	19.0	120	5.4	81.8	25.8	136
中系7604	城堡1号/咸南10号	14.8	6.6	108	19.9	131	5.7	83.0	27.4	159
中系7702	城堡2号/70-001	12.5	9.9	104	20.1	177	6.0	82.5	26.5	145
中系7703	城堡1号/雨良//大和晴/喜峰	14.3	7.7	105	22.2	156	13.3	82.2	25.0	141
中系7705	黎明/福锦	16.9	11.0	88	18.6	97	2.4	82.2	27.0	143
中系7709	红旗12/70-523//下北///城堡1号	16.6	10.9	107	19.0	128	5.6	83.9	25.5	144
中系7850	抗2/513//中系7609	17.9	17.9	104	—	107	14.0	83.0	26.9	139

原载：中国农业科学，1981（3）：1～9.

Adult-plant Resistance of Rice Cultivars to Bacterial Blight

ZHANG Qi, Research Fellow, and
T. W. Mew, Plant Pathologist,

(International Rice Research Institute,
IRRI, Los baños, Laguna, Philippines)

Abstract Selected rice cultivars expressed more resistance to bacterial blight (*Xanthomonas campestris* pv. *oryzae*) as plants developed. This resistance was comparable to adult-plant resistance to diseases of other crops. Some cultivars were resistant to four races, and others, to only one or two races. The disease severity on some cultivars gradually decreased from seedling to flag leaf, whereas in others, it showed a distinct change to a resistant reaction on a certain leaf. The leaf showing resistance varied with cultivars. Wase Aikoku 3 had adult plant resistance to three races but seedling resistance to another. The adult plant resistance of rice to bacterial blight appears to race-specific.

Adult-plant resistance is defined as the plant resistance that is expressed as the plant matures (Mew et al., 1982). This phenomenon has been observed and studied with various plant diseases. Cases of adult-plant resistance of wheat to stem and leaf rust and that of oats to crown rust have been documented (Heagle et al., 1972; Peterson, 1944; Simons, 1955).

For rice bacterial blight (*X. c.* pv. *oryzae*), Wakimoto and Yoshii (1954) noted that disease reactions varied when rice cultivars were tested at different growth stages. Horino and Ezuka (1973) devised a spray-inoculation method to test rice cultivars for bacterial blight resistance. When the plants were inoculated at the two or the four to five leaf stage, the reaction to the strains was not always the same, and some seedlings of Wase Aikoku 3 showed reactions different from those of adult plants. They speculated that the resistance was not fully expressed in young seedlings. Ezuka et al (1974). found that the resistance of Wase Aikoku 3 increased as the plants advanced in age. They concluded that the resistance of Wase Aikoku 3 and other similar cultivars was not fully expressed at seedling stage, postulating a form of adult resistance.

At the International Rice Research Institute (IRRI), resistance of rice cultivars to bacterial blight was distinguished into two general groups: seedling resistance and adult-plant resistance (Mewet al., 1981). The growth stage when the transition from susceptible to resistant responses occurred varied with cultivars (Zhang et al., 1984). Because the adult-plant resistance of rice to bacterial blight has still not been well confirmed and characterized, 13 cultivars were selected for further evaluation on the basis of preliminary tests. This paper reports our results.

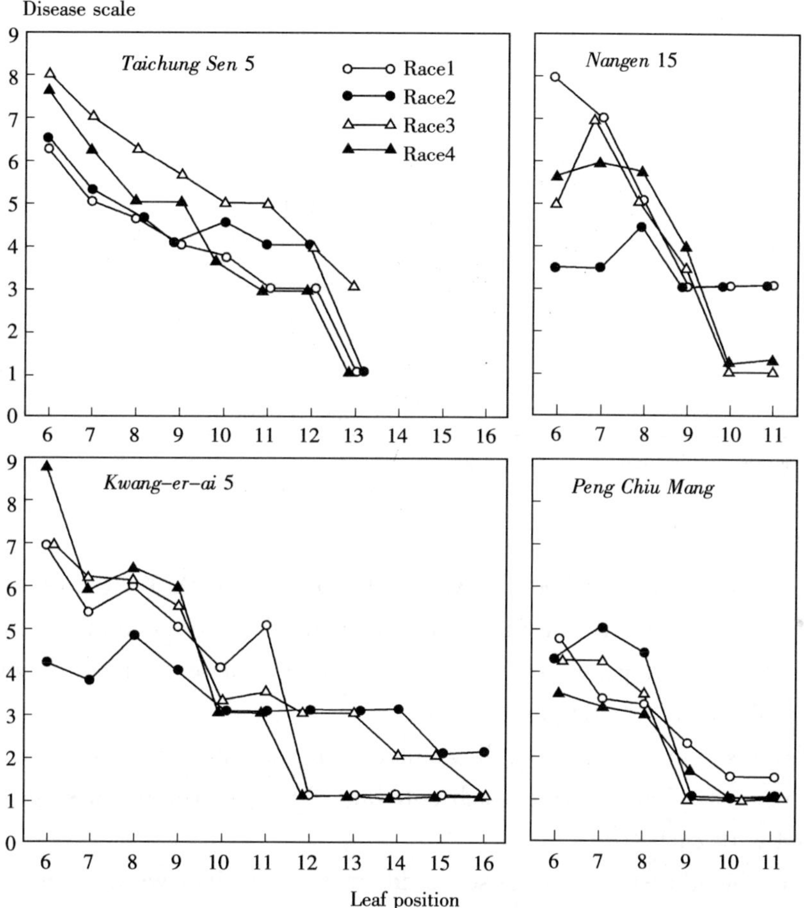

Fig. 1 Gradual increase in resistance from low to high leaf positions of four rice cultivars in response to infection by the four races of *Xanthomonas c.* pv. *oryzae*.

Materials and Methods

Thirteen cultivars were studied: nine were grown commercially and seven were used as resistance sources for bacterial blight in China. Wase Aikoku 3, which in Japan showed resistance as plants matured, was also included. Taichung Native 1 (TNI) was used as a susceptible check, and IR 1545 - 339 (IR 1545) was included as a check for differential resistance to races 1 , 2, 3, and 4 at both tillering and flowering stages. Shi-Zu identified earlier at 1RRI as resistant check (Table 1).

Table 1 Cultivar identification and sources

Cultivar	IRRI Acc. No.	Origin	Gene for resistance
Guei Zhou Magu[a]		China [b]	Unknown
Nangeng 11[a]		China [b]	*Xa3*
3330[a]		China [b]	Unknown
Peng Chu Mang[a]	7287	China [c]	*Xa3*
Nangeng 15[a]		China [c]	*Xa3*
Kwang er-ai *	32551	China [c]	Unknown
Taizhong Sen 5	38893	China [c]	Unknown
Wan Mi Hsiang	1149	China [c]	Unknown
Chin Kong Tao	7303	China [c]	Unknown
Wase Aikoku 3		Japan	*Xa3*
IR1545	32624	IRRI	*xa5*
TN1	105	China[c]	Unknown
Shi-zu		North Korea[b]	Unknown

[a] Resistant sources used in China.

[b] From China.

[c] From IRRI germ plasm bank.

X. c. pv. *oryzae* strains PXO61, PXO86, PXO79, and PXO71 representative races 1, 2, 3, and 4, respectively , in the Philippines were used. The inoculum of each strain was prepared from 72 h culture grown on peptone-sucrose agar slants. Inoculum was adjusted to a concentration of about 10^9 cells per milliliter.

To test for resistance at the seedling stage, the 13 cultivars were sown plastic trays with 15 seeds per row. Four rows of each cultivar, 20 days (three-leaf-stage) after sowing were inoculated with each strain.

To evaluate the adult-plant response, seeds of each cultivar were sown in seed boxes in the greenhouse. Seedlings were transplanted to 15-cm clay pots, four plants per pot, 14 days after seeding. The soil was fertilized with PNK (9 - 6 - 6 - g/ m^2) in the forms of $(NH_4)_2SO_2$, P_2O_5, and KO_2, before translating, and additional nitrogen at the rate of 3 g/m^2 was applied one week before inoculation. The positions of leaves on the main culm were marked in order to emergence, starting from the second leaf to the flag leaf. The disease scores of different leaf positions in a cultivar were therefore based only on data from the culm.

Fully expended leaves were inoculated in the following sequence: first at 2 - to 3 - leaf stage, then at the 6 - to 8 - , 9 - to 10 -, 11 - to 12 -, and 13 - to 14 - leaf stage, and finally on the flag leaf of the same plant. Usually, 2 - 3 (leaf stage) referred to seedling stage, 6～12 to tellring to maximum tillring stage, and 13－15 to booting stage, depending on the cultivar. The leaf-clipping method was used for inoculation (Kauffman et al. 1973). A split-plot design for two-factor experiments with three replications was used (Gomes et al., 1976). Three leaves of each leaf position per replicate were inoculated with each strain. Disease was measured according to the Standard Evaluation System for Rice (IRRI, 1980) 2weeksafter inoculation.

Results

All cultivars were evaluated and 45 days after sowing and at the booting stage. Most cultivars were susceptible at the seedling stage but became resistant to one or all of the four races as the plants grew older (Table 2). Disease scores were significantly different between the three growth stages of the cultivars. Disease scores to race 1 of Guai Zhou Mage decreased from 8.5 to 5.5 to 1.0 at the seedling, maximum tillering, and booting stages, respectively. A similar trend was observed for the other races. With the four races, the disease score in Kwang-er-ai 5 decreased from a mean of 8.3 at seedling to 6.3 at maximum tillering, to 2.4 at booting. With Shi-zu, however, disease score were not affected by plant grown stage. TN1 was susceptible to moderately susceptible to the four races at all growth stages. The lesions, however, were longer on seedlings than on mature plants. IR 1545 was resistant to races 1, 2, and 3, but susceptible to race 4, regardless of growth stage (Table 2).

The cultivars that became more resistant as the plants matured showed two patterns of disease development from seedling to flag leaf. Resistant cultivars such as Taizhong Sen 5, Nangeng 15, Kwang-er-ai 5, and Peng Zhu-Mang gradually increased in resistance with each new leaf, and resistance was fully expressed by the flag leaf (Fig. 1). With Guai Zhou Magu, plants were susceptible until the nine leaf stage and then became distinctly resistance from the 10th leaf to flag leaf (Fig. 2).

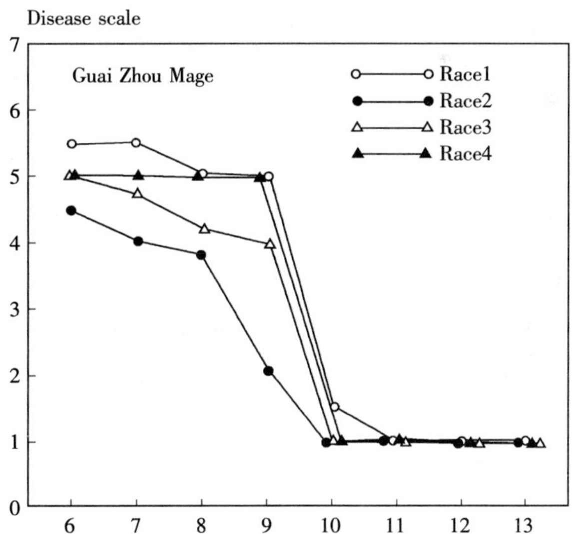

Fig. 2 Distinct resistance response of leaf position 10 of Guai Zhou Magu to infection by races 1, 2, 3, and 4 of *X. c.* pv. *oryzae*. The 13th leaf was the flag leaf

Leaves of cultivars responding to infection with *X. c.* pv. *oryzae* were at different leaf positions. Some cultivars showed resistance earlier than others. Guai Zhou Magu, Nangeng 15, Nangeng 11, and Peng Zhu-Mang were resistant from the ninth leaf position, whereas, Taizhong Sen 5 was resistant from the 11th or 12th leaf positions. The flag leaves, however, were all very resistant to bacterial blight.

Guai Zhou Magu, Nangeng 11, Nangeng 15, Peng Zhu Mang, Kwang-er-ai 5 and Taizhong Sen 5 were resistant to these races at booting stage, whereas, Wang Mi Hsiang, 3330, and Chin Kong Tao 3 were resistant to one or two races at booting. The main effects of cultivars, and of races were highly significant as was the interaction effect. Wase Aikoku 3 was susceptible to races 1, 2, and 4 at seedling and resistant at booting, but it was resistant to race 3 at all growth stages (Fig. 3).

Table 2 Adult -plant resistance of rice caltivars at seedling. maximum tiliering, and adult growth stages

Cultlvar	Disease scores'to suain (race)											
	PX061 (race 1)			PX086 (race 2)			PX079 (race 3)			PX071 (race 4)		
	SS[z]	MTS	BS	SS	MTS	BS	SS	MTS	BS	SS	MTS	BS
Shi-zu	1.0 a	1.3 a	1.0 a	3.0 a	1.0 a	1.0 a	1.0 a	2.5 a	1.3 a	1.0 a	1.2 a	1.0 a
IR 1545－339	1.8 a	3.0 b	1.0 a	3.5 a	3.0 b	1.0 a	3.2 a	2.0 a	1.0 a	6.8 c	5.0 d	6.0 d
Peng Chiu Mang	6.5 b	3.1 b	1.2 a	6.1 b	1.3 a	1.0 a	6.0 b	3.0 b	1.0 a	6.2 b	3.2 b	1.0 a
Nangen 11	6.5 b	4.0 c	1.0 a	6.3 bc	3.0 b	1.0 a	6.5 bc	3.3 b	1.0 a	6.0 b	3.5 b	1.0 a
Nangcn 15	6.7 b	5.5 e	2.0 b	6.5 bc	3.0 b	1.7 bc	6.5 bc	5.6 d	2.8 b	7.0 c	5.1 d	1.0 a
Chin Kong Tao 3	7.0 bc	4.3 cd	3.5 d	6.7 bc	3.7 c	1.7 bc	6.7 bc	4.8 c	1.2 a	7.3 d	6.6 c	6.8 f
Taichung Sen 5	7.0 bc	4.7 d	3.0 cd	7.5 d	4.7 d	2.0 c	9.0 f	5.4 cd	3.0 b	8.0 d	4.3 c	3.0 b
Wase Aikoku 3	7.0 bc	6.5	3.0 cd	7.0 cd	6.8 e	2.0 c	1.0 a	2.5 b	2.0 b	5.8 b	5.5 de	2.5 c
3330	7.5 c	5.7 ef	3.5 d	6.7 cd	3.3 bc	1.6 abc	7.2 d	5.4 cd	2.7 b	7.2 b	5.5 de	4.0 c
Guai Zhou Magu	8.5 d	5.5 e	1.0 a	8.5 e	3.1 b	1.1 a	8.0 e	5.2 cd	1.0 a	7.0 c	5.0 d	1.4 a
Kwang-er-ai 5	8.5 d	6.2 f	2.7 c	6.5 cd	5.0 d	1.5 bcd	9.0 f	7.0 e	2.7 b	9.0 e	7.0 fg	2.8
Wan Mi Hsiang	9.0 d	5.4 e	4.7 e	7.0 cd	3.0 b	1.3 bc	8.5 f	5.0 c	5.0 c	9.0 e	6.0 e	5.0 d
TNI	9.0 d	8.0 g	6.8 e	9.0 a	6.7 c	4.3 d	9.0 f	7.3 e	6.3 c	9.0 e	7.3 g	7.0 e

[y] Based on disease scale of 1~9, where 1=1% of leaf area affected and 9=50% of leaf area infected. Means in a column followed by a common letter are not significantly different (p=0.05) according to Duncan's multiple range test.

[z] SS=seedling stage, MTS=maximum tillering stage, and BS = booting stage.

Discussion

Adult-plant resistance to plant disease has been observed in various crops. In the wheat cultivar Thatcher, Nazareno and Roeifs (1981) found no relationship between the independent *Sr* genes for seedling resistance and adult-plant resistance to rust. Heagle and

Moore (1970) indicated that less crown rust developed on adult than seedling oat plants. These resistances were largely non-race-specific. Bartos et al (1969) analyzed the adult-plant resistance to leaf rust in wheat cultivars Thatcher and Red Bobs and indicated that as seedlings, both were susceptible to races 9 and 161. By plant maturity, however, Thatcher became resistant to race 9 but not to 161, whereas Red Bobs remained susceptible to both races.

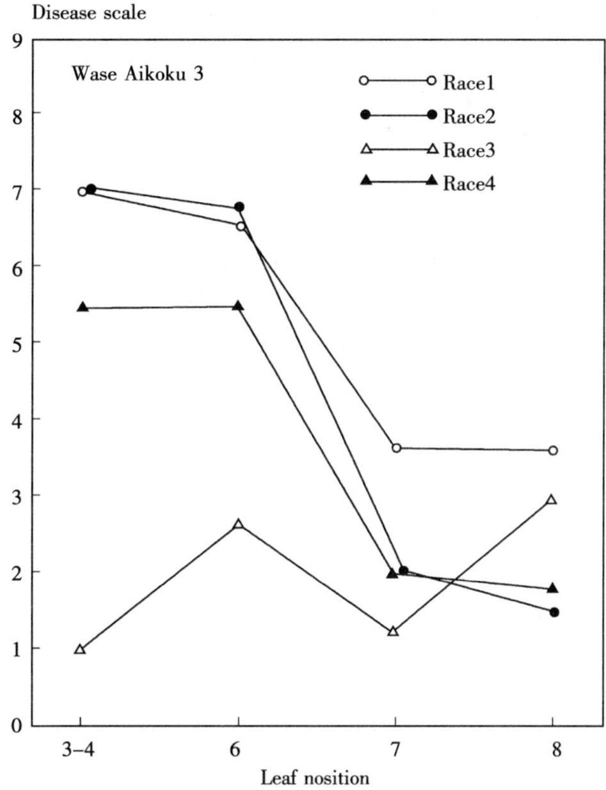

Fig. 3 Responseof Wase Aikoku 3 from the third leaf position to flag leaf (no. 8) to four races of *X. campestris* pv. *oryzae*

With rice bacterial blight, Wakimoto and Yoshii (1954) noted the variation in disease reactions among rice cultivars when plants were tested at different growth stages. When evaluation was done at the seedling stage, results were more variable. Horino and Ezuka (1973) and Ezuka et al. (1974) showed the resistance expression of Wase Aikoku 3, and a group of rice-derived cultivars had a high level of resistance. The disease reactions to the bacterial strains were more stable on the flag leaves than on seedlings and were quite consistent in all strains in Japan. The *Xa6* gene, conferring bacterial blight resistance in Zenith and Malagkit Sungsong, was expressed at the booting stage (Sidhu et al., 1978; Zhang et al., 1984). In our study, the resistance of Wase Aikok 3 was expressed at the flag leaf stage, but the resistance to race 3 was also expressed in the seedling stage. The adult-plant resistance of rice to bacterial blight appears quite stable. Our results indicated that resistance effective at the seedling stage, as in Shi-Zu, was not affected by growth stage or races. The susceptibility of TN1 to the four races was not altered, but the disease score decreased as plants grew older. Similarly, the different resistance of IR 1545 was independent of growth stage, although the disease score tended to decrease at later growth stages. Adult-plant resistance was expressed consistently on the flag leaves of the cultivars evaluated (Table 2). The actual leaf position when resistance was shown varied according to cultivars-earlier in some, such as Guai Zhou Mag, Nangen 15, Peng Chu Mang, etc ., and later in others such as Taichung Sen 5. The adult-plant resistance was also race-specific, as in Wan Mi Hsiang, which was resistant to one race, and 3330 which was resistant to two races.

Published in: America, Plant Disease, 1985, 69: 896 - 898.

水稻对白叶枯病的成株抗性

章　琦[1]　T. W. Mew[2]

（1. 中国农业科学院；2. 国际水稻研究所）

Characterization of Adult Plant Resistance of Rice to Bacterial Blight

Abstract　Some cultivars showed adult plant resistance to bacterial blight (*Xanthomonas campestris* pv. *oryzae*). They were susceptible at seedling stage but became resistant as plants mature.

Among thirteen cultivars analyzed possessing adult plant resistance to bacterial blight, some were resistant to all the four Philippine races tested while some others to one or two races only. The adult plant resistance of rice to bacterial blight seemed to be also race-specific. The disease score in some cultivars decreased gradually from seedling to booting stage, while in others there was a distinct shift from susceptible to resistant reaction at a definite leaf position. The leaf position showed a shift of resistance varied with different cultivars.

Temperatures appeared to stress no effect on the adult plant resistance conveyed by the *Xa6* gene. It was quiet stable in high temperature. On the contrary, TN1, the susceptible check and those cultivars whose adult plant resistance had not reached the stage of transition, the lesion progressed faster at higher temperature (33℃) than at lower (25℃), while IR1545 - 339 carrying gene *xa5* for differential resistance to the 4 races tested, gave a similar expression as TN 1 did when matched with a pathogenic race.

Key words　rice; bacterial blight ; adult plant resistance

摘要　在分析的 13 个成株抗性品种中，有些能抗供试的 4 个菲律宾小种，而另一些只抗 1 个或 2 个小种。菌株与品种间有较强的互作反应，表明有些水稻品种对白叶枯病的成株抗性似乎属“小种专化”。抗性转变的形式是多种多样的，有些品种从苗期到孕穗期逐渐由感转为抗，另一些品种从感转抗的界限却很明显。不同品种对同一个菌株的转抗叶位不同，同一品种对不同菌株的转抗的叶位也不一样。携有 *Xa6* 基因品种的成株抗性仍表现稳定。高温对感病品种和成株抗性品种在转抗前的感病阶段能促进病情发展。携有 *xa5* 基因的 IR1545－339 在高温下仍然保持其鉴别抗性反应，对其有毒力的小种 4 的病情发展与 TN I 是类似的。

关键词 水稻，白叶枯病，成株抗性

水稻白叶枯病的抗性研究始于 50 年代。日本 Wakimoto 等（1954）最早注意到有些水稻品种对白叶枯病的抗性在不同生育期表现不同。Sidhu 等（1978）将水稻对白叶枯病的抗性分为苗期抗性和成株期抗性，由 *Xa3*、$Xa4^b$、*Xa6*、*xa8* 和 *xa9* 抗性基因控制的品种对某些菌株在苗期感病，直到孕穗期或花期才转为抗病。作者在 1975—1977 年进行水稻品种对白叶枯和稻瘟病的抗性研究，在 243 个抗白叶枯品种中，观察到它们的抗性反应类型是不同的，其中有 64 个品种表现为成株抗性（章琦，1978）。张端品、谢岳峰等（1984）在研究成株抗性中，也探讨了一些国外品种与病原菌互作及抗性变化动态的问题。

一般认为外界环境条件对白叶枯病的病情发展有密切关系。Horino 等（1982）报道一些携有 *Xa4* 基因的品种在高温下的病斑发展迅速甚至丧失抗性，但 DV85 和 CAS 209 等品种却不受温度影响。

本研究较系统地研究水稻对白叶枯病成株抗性的类型、特性、品种与菌株相互作用的反应以及温度对携有 *Xa6* 基因品种的影响，为抗性遗传研究和抗病育种提供参考。

1 材料和方法

1.1 品种来源、携带的抗病基因、抗病反应等见表 1，供试菌株为 PXO61、PXO86、PXO79 和 PXO71，分别代表菲律宾小种 1、2、3 和 4。

表 1 供试水稻品种（系）

Table 1 Rice varieties (lines) tested

(IRRI) 登记号	品种（系）	来源	抗病基因	抗性类型
	贵州麻谷	中国	未知	成抗
	南粳 11	中国	未知	成抗
	3330	中国	未知	成抗
	南粳 15	中国	未知	成抗
7298	蚌珠芒	中国	未知	成抗
32551	广二矮 5 号	中国	未知	成抗
1149	台中籼 5 号	中国	未知	成抗
7303	黄米香	中国	未知	成抗
755	陈光稻 3 号	中国	未知	成抗
4038	Zenith	美国	$Xa6^c$	成抗
756	Malagkit Sungsong	菲律宾	$Xa6^c$	成抗
	IR944	IRRI	$Xa6^c$	成抗
	IR1696	IRRI	$Xa6^c$	成抗
	BG 90－2	斯里兰卡	$Xa4^d$	全抗

（续）

(IRRI) 登记号	品种（系）	来源	抗病基因	抗性类型
	Wase Aikoku 3	日本	*Xa3*	成株期抗日本小种 1～3
	矢租（对照）	朝鲜	*xa5*	全生育期抗菲小种 1～4
	IR1545－339（鉴别品种对照）	IRRI	*xa5*	全生育期抗菲小种 1～3，感小种 4
	TN1（对照）	中国	未知	感菲小种 1～4

1.2 种植和管理

在温室浸种催芽后播于育苗箱，12d 后移栽于瓦钵，每盆 4 株，每立方米加 9－6－6g 氮、磷、钾混合肥，接种前一周追加 3g 氮/m^3。

1.3 接种

用菲律宾白叶枯病 4 个小种的代表菌株，在 PSA 斜面培养基上于 28℃恒温箱中培养 72h，菌液浓度约 10^9/mL。用剪叶法接种。

1.3.1 成株抗性品种转抗的叶位差异 用两种方式接种，1）在同一植株上分期接种，分别在叶片生长到第 3～4、第 6～7、第 8～9、10～11、12～13、14～15 叶、最后至剑叶时，共接种 6～7 次。2）同期在不同植株上接种。因品种生育期不同，采用分期播种，生育期长的品种早播、生育期短的晚播，随后，各品种分别每隔两周播种一次，共播 3 次。当早播的植株抽出剑叶、最后一期播的展出第六、七片叶时，就可同时接种各个叶位。

1.3.2 温度对成株抗性的影响 在人工气候生长室测定。共测试 4 个携有 *Xa6* 基因的品种（系）和 IRl545－339 以及 TN1，同样进行分期播种。秧苗按每盆 4 株单本移栽至直径为 20cm 的陶瓷盆。先在温室中生长，在首次接种前 8d，将植株分别移至昼夜温度为 33/25℃、29/21℃、25/20℃三种设置的人工气候生长室，相对湿度 70%。在同一植株上分期接种，从第 6～8 个叶位开始，依次按序接种直至剑叶，共接种 4～5 次，每次接种后两周按 IRRI 标准进行调查。但测定温度对携带 *Xa6* 基因品种的影响时，分别在接种后第 5、10、15 和 20d 进行成株期病情调查，以观察这些品种成株抗性的稳定性。

本试验同时记载病级及病斑占叶面积百分率。按以下标准调查

病　级	1～2	3～4	5～6	7～8	9
病斑占叶面积百分率（%）	<5	5～24	25～<50	50～<75	>75

试验设计　双因子和三因子分别采用 2 级或 3 级裂区设计，3 次重复。

以上试验于 1982—1983 年于菲律宾的两个湿季一个旱季反复测试 2～3 次。

2 结果

2.1 成株抗性品种在不同生育期的表现及其与菌株的相互作用

表 2 列出 18 个品种在不同生育期对 4 个菲律宾小种代表菌株的抗性表现。对照品种矢

租在全生育期对 4 个小种表现全抗。IR1545－339 在全生育期抗小种 1、2 和 3，感小种 4，仍然显示其鉴别抗性。TNI 感染所有 4 个小种。供试的 15 个成株抗性品种在不同生育期对白叶枯病的抗性表现大致相同。它们都在苗期感病，到分蘖期开始表现中感到中抗，孕穗期在剑叶和剑叶下第 1～2 张叶位则明显转抗。但是这些品种与 4 个小种间的相互作用可分为 3 种类型，1）抗谱较广。如携有 *Xa6* 基因的 4 个品种（品系）Zenith、Malagkit Sungsong、IR1695、IR944 和贵州麻谷、广二矮 5 号、南粳 11、南粳 15、蚌珠芒以及台中籼 5 号共 10 个品种对菲律宾的 4 个小种都表现典型的成株抗性，如图 1 ，M. Sungsong 所示；2）抗谱较窄，品种与小种间有较强互作。如黄米香、陈光稻 3 号、BG90－2 和 3330，只对 4 个小种中的 1 到 2 个有成株抗性而对其他小种在全生育期感病。其中有些品种与小种之间表现

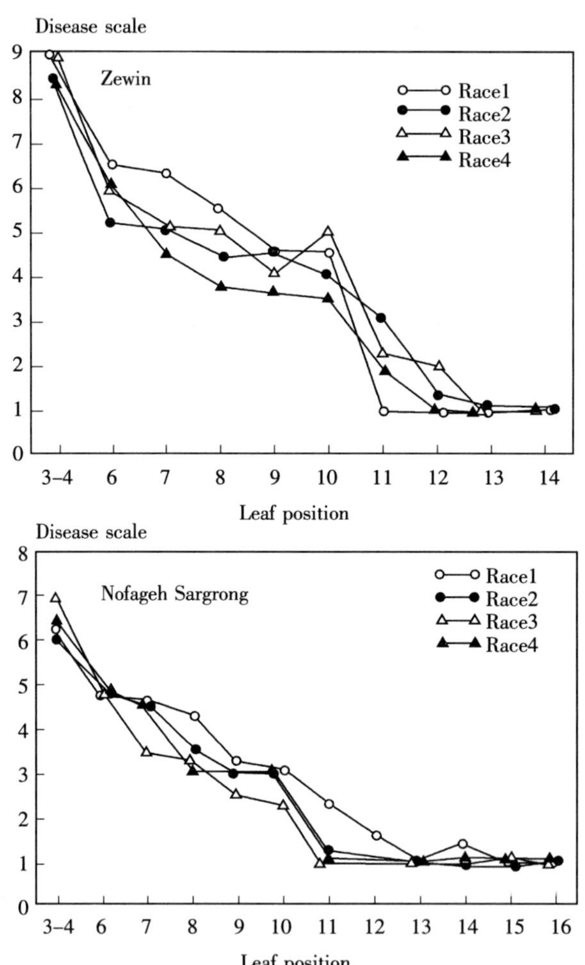

图 1　Zenith 和 Malagkit Sungsong 在温室中对菲律宾 *Xoo* 小种的病情反应

Fig. 1　Reaction of two cultivars inoculated at the same time to 4 races *of X. c.* pv. *oryzae* from 3thd leaf to flag leaf

“小种专化抗性”的互作模式。如陈光稻3号、BG90－2和黄米香这3个品种在苗期对供试的4个小种都感染，随着植株成长，陈光稻3号对小种2和小种3转抗，但不抗小种1和4；而BG90－2则相反，它对小种1和4转抗，而不抗小种2和3（表2）。黄米香仅对小种2转抗，对其余3个小种都不抗；3）类似日本的携有*Xa3*基因的代表品种早生爱国3号的表现，对菲律宾小种1、2和4表现成株抗性、而对小种3为全生育期抗性。这些结果表明水稻品种对白叶枯病的成株抗性是与特定菌株相互作用的结果，似乎也有“小种专化”的特性。

2.2 成株抗性品种的抗性转化特点

在温室中用两种方式接种的结果基本一致。这些品种一般都在某一叶位开始转抗，但是，不同品种对相同菌株由感转抗的叶位有差异。一些品种较早，如 Malagkit Sungsong、贵州麻谷、南粳15等对有的菌株从第八或第九张叶位就开始转抗，而 Zenith、台中籼5号则在第11～12叶表达抗（图1）。品种的成株抗性不一定都到生殖生长阶段才表达，抗性转化的迟早与品种生育期的长短没有联系。如 Malagkit Sungsong 的生育期为135d，约有16～17张叶片，在第八张叶位上就呈现对小种2的抗性。而携有相同基因*Xa6*的 Zenith 有100d生育期，约13～14张叶片，但它对这4个小种的成株抗性直至第12张叶位才表现出来。这种转抗叶位不同的现象还存在于一个品种与不同小种互作的反应中。如南粳15对小种1、2在第9叶位转抗，而对小种3、4则在第10叶位。

表2 供试水稻品种在不同生育期对4个菲律宾*Xoo*小种的抗性反应*

Table 2 Resistance reaction of rice cultivars to the Philippine 4 races of *Xoo* at different growth stages

品种（系）	PXO61			PXO86			PXO79			PXO71		
	SS	MTS	BS	SS	MTS	BS	SS	MTS	BS	SS	MTS	BS
贵州麻谷	8.5	5.5	1.0	8.5	3.1	1.1	8.0	5.2	1.0	7.0	5.4	1.4
南粳11	6.5	4.0	1.0	6.3	3.0	1.0	6.5	3.3	1.0	6.0	3.5	1.0
蚌珠芒	6.5	3.1	1.2	6.1	1.3	1.0	6.0	3.0	1.0	6.2	3.2	1.0
南粳15	6.7	5.5	2.0	6.5	3.0	1.7	6.5	5.6	2.8	7.0	5.1	1.0
广二矮5号	8.5	6.2	2.7	6.5	5.0	1.5	9.0	7.0	2.7	9.0	7.0	2.8
台中籼5号	7.0	4.7	3.2	7.5	4.7	2.0	9.0	5.4	3.0	8.0	4.4	3.0
3330	7.5	5.7	3.5	6.7	2.3	1.6	7.5	5.4	2.7	7.2	5.5	4.0
陈光稻3号	7.0	4.3	3.5	6.7	3.7	1.7	6.7	4.8	1.2	7.3	6.6	6.8
黄米香	9.0	5.4	4.7	7.0	3.0	1.3	8.5	5.0	5.0	9.0	6.0	5.0
BG90－2	7.0	3.6	2.0	7.0	4.2	4.0	8.6	6.5	5.0	7.0	4.4	3.0
Zenith	9.0	5.0	1.0	8.5	5.0	1.0	9.0	4.6	1.0	8.0	4.7	2.0
IR1695	9.0	6.0	1.5	9.0	5.3	1.0	9.0	5.4	1.0	8.5	5.0	1.5
Malagkit Sungsong	6.3	4.0	1.0	6.0	2.5	1.0	7.0	3.7	1.0	6.5	3.0	1.0
IR944	9.0	5.3	3.2	9.0	5.0	2.0	8.5	5.4	3.0	8.5	5.0	2.0

（续）

品种（系）	PXO61			PXO86			PXO79			PXO71		
	SS	MTS	BS	SS	MTS	BS	SS	MTS	BS	SS	MTS	BS
Wase Aikoku 3	7.0	6.5	3.0	7.0	6.7	1.5	7.0	2.7	3.0	5.5	5.5	1.8
IR1545 - 339	1.8	3.0	1.0	3.5	3.0	1.0	3.2	2.0	1.0	6.8	5.0	6.0
矢租（对照）	1.0	1.3	1.0	3.0	1.0	1.0	2.5	1.3	1.0	1.0	1.2	1.0
TN1	9.0	8.0	6.8	9.0	6.7	4.3	9.0	7.3	6.3	9.0	7.3	7.0

* SS=苗期（seedling stage），MTS=最高分蘖期（Maximum tillering stage），BS=孕穗期（Booting stage）。

2.3 温度对成株抗性的影响

Malagkit Sungsong、Zenith 苗期在 33/25℃的高温生长室中，对 4 个小种表现高感，其病情比在 29/21℃和 25/20℃中的发展快且重，但是其抗性仍随着植株的成长而增强，与生长在温室里的植株相同，一进入成株期，病斑发展转缓，出现褐色病痕后，扩展即止，即使在接种后 20d 调查，也稳定地保持高抗（图 2）另见彩图。

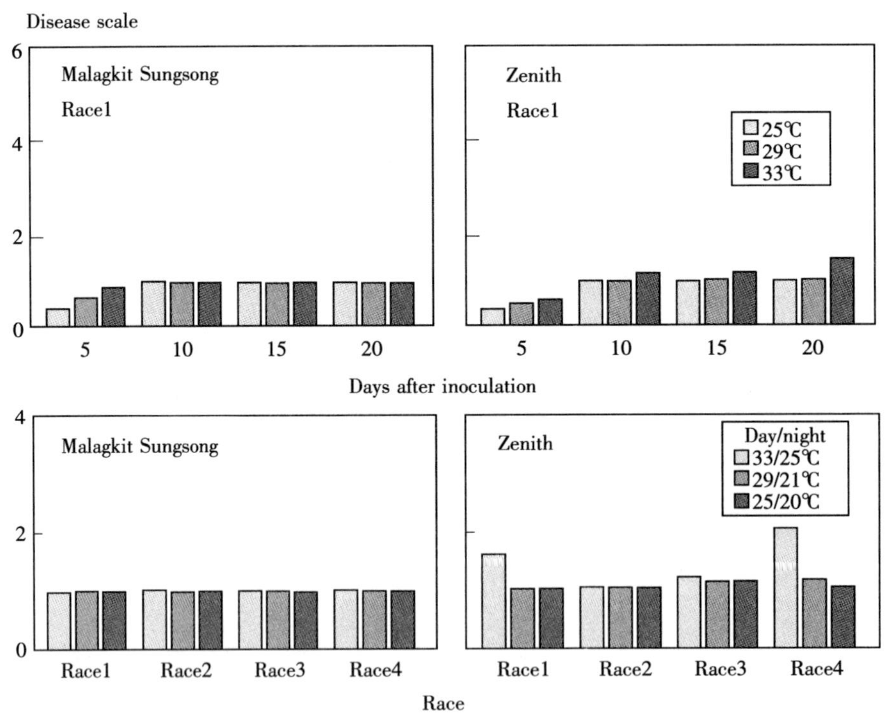

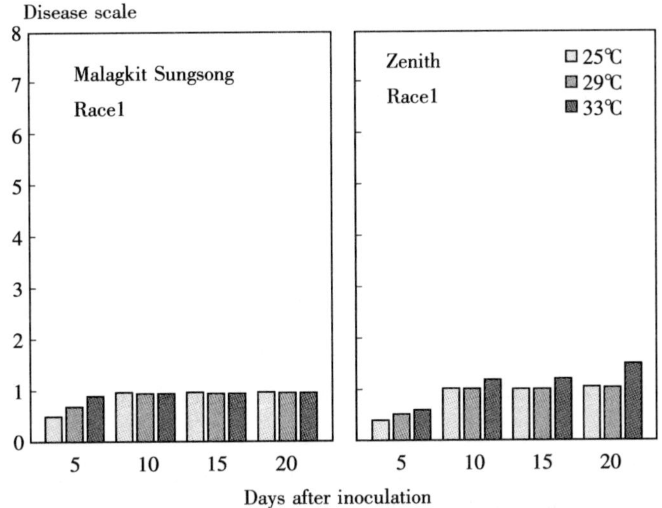

图 2 不同温度对 Malagkit Sungsong 和 Zenith 在孕穗期接种后 20d 的病情

Fig. 2 Effect of temperature on disease reaction caused by 4 races of *Xoo* on Malagkit Sungsong and Zenith

IR944 和 IR1695 的成株抗性分别来自 Malagkit Sungsong 和 Zenith，但是在高温下，其转抗时期比亲本延后，抗性水平也有下降，保持在中抗水平，可能有遗传背景的影响。

高温有利于病斑发展的现象，也表现在专化抗性品系 IR1545 上，它对亲和小种 4 随着温度的增高而加重，而对其不亲和的小种 1，却不受温度的影响（图 3）另见彩图。

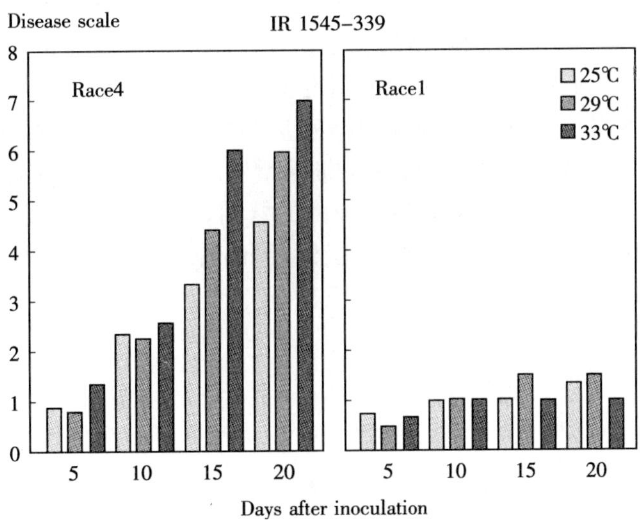

图 3 IR1545 对菲律宾白叶枯病小种 1 和 4 在 3 种不同温度条件下的抗性反应

3 讨论

3.1 本试验证实了水稻品种对白叶枯病也存在典型的成株抗性。对于控制像白叶枯病这种一般在水稻生长后期暴发的细菌性病害，成株期抗性也具有实用价值。贵州麻谷、蚌珠芒、南粳15、南粳11、广二矮5号、台中籼5号及Zenith、Malagkit Sungsong等成株抗性品种的抗谱都比较广，是育种中的理想抗源。

3.2 水稻对白叶枯病的成株抗性有明显的寄主一病原菌的相互作用，类似"小种专化"现象。在寄主与病原菌互作下，不同品种对相同菌株或同一品种对不同菌株的成株抗性表现，在由感转抗的方式、时期、抗谱等方面都有差别。国内外均报道了地区间的白叶枯病菌株分布不同。因此，针对性地采用当地菌株研究品种的抗性可为本地区育种计划提供更有应用价值的资料。

3.3 品种对白叶枯病的抗性反应是多种多样的。初次评价抗性的品种资源最好在苗期和孕穗期都进行接种鉴定。可以筛选出全生育期和成株期抗性的品种。在外界条件有利于发病的情况下，苗期可在5叶左右、成株期在剑叶和剑叶下第1～2张叶片接种。对于育种材料的抗性筛选，如果仅作一次接种，则在成株期进行。

3.4 Horino等（1982）报道高温影响携有*Xa4*基因的品种的抗性，IR20和IR26在高温下由抗病变成中度感病，在其同一个试验中，DV85（*xa5*、*Xa7*）和CAS 209（*Xa10*）对某些菌株的抗性在高温下并未受到影响。本试验结果表明，携有*Xa6*基因的品种Zenith和M. Sungsong的成株抗性在高温下十分稳定。IR1545－339（*xa5*）的抗性在不同温度下也没有什么变化。由此可见，高温促进白叶枯病发展的现象是指一般情况的感病品种和某些抗病基因而言的。因此对携有不同抗性基因的品种做出准确的结论需分别进行测试才能确定。

3.5 关于细菌在叶片中的增殖测定早有报道。Chand和Walker（1964）用*Pseudmonas lachrymans*接种于一株黄瓜的老叶和嫩叶，发现老、嫩叶中的细菌增殖明显不同，老叶的细菌量较嫩叶中的少20～40倍。白叶枯病原细菌在水稻不同叶片中的增殖与成株抗性品种的转抗是否也有关系？我们在另一个试验中测定了成株抗性品种Zenith的抗和感叶位中的*Xoo*增殖情况。结果表明抗病叶片中的细菌增殖量明显低于感病叶片，增殖的速度也缓慢（Zhang & Mew，1984）那么影响抗病叶片中细菌增殖的机理或成株抗性品种由感转抗的内在因素是什么？假如是由于有关物质抑制了细菌的繁殖、生长，那么这又是一种什么物质？这种物质是否可能被用以进行抗性筛选或将其转移到品种中，成为品种遗传性的一个组成部分？这将是未来值得进一步探讨的有趣课题。

原载：作物学报，1986，12（4）：241～247.
本研究为CAAS－IRRI合作项目，1982—1984年在菲律宾国际水稻研究所进行。

水稻抗病育种研究Ⅲ．我国主栽品种对若干白叶枯病菌株的抗性分析

章　琦　王春莲　林世成

（中国农业科学院作物育种栽培研究所）

Studies on Disease Resistance in Rice Breeding Ⅲ. Resistant-Susceptible Reaction of Main Rice Cultivars to Six Bacterial Blight Strains of China

Abstract Thirty two indica and 45 japonica rice cultivars including resistant and susceptible checks of each group were tested for resistance to six bacterial blight (*Xanthomonas oryzae* pv. *oryzae*) strains from the southern, central, and northeast parts of China for three seasons. It indicated that 78% indica cultivars are susceptible to all six strains. Cultivars Qing-Hua-Ai, Er-Jiu-Feng, 910, 929, hybrid rice Wei - You No. 6 and the resistant check IR20 are highly resistant to all 6 strains. For the japonica cultivars, 40% are resistant to the six strains and about 56% resistant to 3 to 5 strains. Among the resistant cultivars, Xiu-shui 48, Cheng-Te 232, Pi-Zao 15, Ken-Xi 2, Wase-Aikoku 3, and the resistant check Ginga are the most promising. By comparative analysis of the pedigree constitution of these cultivars, it is obvious that the difference in the reaction of these groups is due to the limited existence and utilization of resistant gene (s) in indica rice breeding, and the extensive utilization consciously or unconsciously the resistante gene (s) of the parental types in japonica rice breeding programs.

Key words commercial cultivars, strains; bacterial blight of rice; resistance

摘要　用华南、长江流域和华北的6个白叶枯病菌株测试32个籼稻和45个粳稻主栽品种的抗性。籼稻中有78%全感6个菌株。其中青华矮、二九丰、910、929、威优6号及对照IR20高抗全部菌株。粳稻中，有40%抗6个菌株，约占56%抗3～5个菌株。其中秀水48、城特232、邳早15、垦系2号、早生爱国3号及对照品种银河表现突出。本文试分析其系谱渊源，上述籼稻品种的亲缘大多不抗白叶枯病，而粳稻品种培育中，有意或无意地广泛采用了抗

病亲本。

关键词 主栽品种；菌株；水稻白叶枯病；抗性

近代水稻育种目标，随着生产发展的需要，从过去以高产为主，逐步向优质、丰产、多抗（病、虫、不良环境）综合目标发展（中国稻作学，1986）。抗病育种也由初期的单抗转向抗多种病害。病理工作者对大量品种资源进行了主要病害的抗性鉴定，越来越多广谱抗性或多抗性的宝贵材料用于育种计划。抗性研究和筛选抗源是抗病育种的重要环节，同时结合不同专业进行探索，能达到相互促进共同提高的效果。本研究就我国主要稻区部分主栽品种对南北若干有代表性的白叶枯病菌株的侵染反应进行测试，并初步分析其系谱，为进一步利用抗源培育新品种提供资料。

1 材料及方法

1.1 供试品种

由粤、桂、滇、蜀、湘、鄂、苏、浙、辽、吉、冀、陕和北京等 24 个省市（区）农科院提供了 77 个当地水稻主栽品种和若干抗源，其中籼稻 32 个，粳稻 45 个。IR20 和 TN1、银河和沈农 1033 分别作为籼稻和粳稻的抗、感对照品种。

1.2 菌株

由江苏、广东省农科院提供及本所采集的 6 个菌株 KS6－6、KS1－20、浙 173、064－2 和 HN01、HB29，分别来自江苏、浙江、广东、河南及河北。菌株都经回接到感病品种上进行复壮，分离病叶、挑取单菌落，各菌株繁殖 10 支斜面培养菌种，保存于低温下备用。

1.3 种植和接种调查

分 3 期播种育苗，每期种子经浸种消毒催芽，播于育秧盘。秧苗 4 叶期按每盆 6 株单本移栽。为测试各品种整个生育期的抗、感反应，在秧苗的第 2 张叶片展开时，开始标记叶龄，直至剑叶。不同植株同期接种，即当早播植株的剑叶平展、最后一期播种的植株展开第 6、7 张叶片时，同时接种各个叶位，使病情发展处于一致的外界条件。对部分生育期长的品种作了短日照处理。接种以前，每盆 6 株稻苗分别用 6 种不同色泽塑带标示 6 个菌株接种。

病　级	1	3	5	7	9
病情类型	高抗	抗（含中抗）	中感	感	高感
病斑占叶片面积百分率（%）	<5	5～<20	20～<50	50～<75	>75

试验设计：分区组随机排列，3 次重复。

全程试验从 1984—1985 年的两夏一冬共 3 季，包括预试、正式试验和复测。

各菌株在肋本哲氏改良培养基上于28℃培养72h，以灭菌蒸馏水稀释菌液浓度约10^9/mL。接种后两周按以下标准进行调查：

2 结果与讨论

供试品种和品系用6个白叶枯病菌株接种后的病情反应列于表1、2。就以下问题进行分析和探讨。

2.1 籼稻品种对白叶枯病菌株的抗性及其主要亲缘分析

在参试的32个籼稻中，包括感病对照TNl在内的25个品种对6个菌株全部感染，约占78%。这些品种的病斑面积虽然随生长发育而有所减小，但至孕穗期的病斑面积仍>25%，呈典型水渍状病斑。青华矮、二九丰、910、929和杂交稻组合威优6号以及抗病对照IR20高抗所有菌株，其中多次测试的青华矮和IR20在整个生育期始终表现高抗。包选矮和寒露早在成株期中抗南方稻区的4个菌株，但受河南、河北菌株的侵染。这个结果极易令人认为籼稻不抗白叶枯病，然而国际水稻所（IRRI）育成的大量籼稻品系表明，到1974年的产量试验登记材料中，抗白叶枯病的品系达到98%（Khush，1977；Mew et al.，1981）。上述我国所育成的高抗品种也都说明籼稻中不乏抗白叶枯的品种，但是目前的数量确实不多。形成这种状况的主要原因是自20世纪开展杂交育种以来，相当长的一段时期都在探索耐肥抗倒、实现高产目标，最后确定了矮化育种方向。从50年代以来，发现和广泛利用了矮子占、南特号，胜利籼、低脚乌尖及其衍生种矮脚南特、广场矮、珍珠矮等矮源，并选育出大批适应各地区的矮秆高产优良籼稻品种（中国稻作学，1986）。例如桂朝2号、广陆矮4号、沪双1011、二九青等在我国生产中有很大贡献，但多数品种的矮源均不具有抗白叶枯病基因，因此多数品种也是高度感病的（表1）。20世纪70年代以来，水稻育种目标有所改进，育种家们在选育优良农艺性状的同时，重视利用抗源。如高抗品种青华矮的主要抗源供体是华竹矮，华竹矮由lR20/竹印2的后代杂交育成。青华矮的抗性可能渊源于抗菲律宾当年的优势小种1的具有*Xa4*基因之IR20。910和929也均源于具有抗菲律宾小种1、2、3、4的斯里兰卡品种BG-90-2（章琦等，1986）。育成上述品种的单位也体现了学科间的密切配合。

表1 我国籼稻主栽品种对6个水稻白叶枯病菌株的抗性反应

Table1 Reactions of indica commercial cultivars from main rice growing regions to 6 strains of *X. o.* pv. *oryzae* in China

品种 Cultivar	组合 Cross	Strain 菌株[b]						所属衍生系统 Origin
		KS6-6	ZHE173	064-2	KS1-20	HN01	HB29	
桂潮2号	桂阳矮49/朝阳早18	S	S	S	S	S	S	矮子占
竹科2号	竹莲矮/科矮13	S	S	S	S	S	S	矮子占
双桂1号	桂阳矮C17/桂潮2号	S	S	MS	S	S	MS	矮子占
湘矮早7号	朝阳1号选	S	S	S	S	S	S	矮子占、南特号
湘矮早9号	IR8/湘矮早4号	S	S	S	S	S	S	矮子占、南特号

（续）

品种 Cultivar	组合 Cross	Strain 菌株[b]						所属衍生系统 Origin
		KS6-6	ZHE173	064-2	KS1-20	HN01	HB29	
广陆矮 4 号	广场矮 3784/陆财矮	S	S	S	S	S	S	矮子占、南特号
广二 104	选自广二矮 5 号	S	S	S	S	S	S	矮子占、南特号
泸南早 1 号	二九矮/矮南早	S	S	S	S	S	S	矮子占、南特号
泸南早 2 号	南洋密种/东南亚五谷种	S	S	S	S	S	S	矮子占、南特号
内中 152	八茂 22/科六	S	S	S	S	S	S	矮子占、南特号
二九青	二九矮 7 号/青小金早	S	S	S	S	S	S	矮子占、南特号
青秆黄	广陆矮 4 号/龙革 16 号	S	S	S	S	S	S	矮子占、南特号
早籼 141	桐乡不脱早/温革	S	S	S	S	S	S	矮子占、珍珠矮
竹系 26	选自竹莲矮	S	S	S	S	S	S	矮子占、南特号
南京 11	南京 6 号/二九矮 4 号	S	S	S	S	S	S	矮子占、胜利籼
楚籼 1 号	南特号 14/秕五升	S	S	S	S	S		
泸双 1011	矮脚南特/马边双须谷	S	S	S	S	S		
泸科 3 号	科揭选/泸双 1011	S	S	S	S	S		
湘矮早 10 号	湘矮早 8 号/湘矮早 9 号	S	S	S	S	S		
原丰早	科 6 辐	S	S	S	S	S		
80-133	IR24/原丰早	S	S	S	S	S		
双竹占	中竹 2915/塘竹 2913	S	S	S	S	S		
二白矮 1 号	秋二矮/秋白早	S	S	S	S	S		
青华矮 6 号	（晚青/青兰）F_4/华竹矮	*R[a]	*R[a]	*R[a]	*R[a]	*R[a]		
二九丰		*R[a]	*R[a]	*R[a]	*R	*R[a]		
910	选自 BG90-2	*R[a]	R[a]	*R[a]	R	*R[a]		
929	选自 BG90-2	*R[a]	*R[a]	*R[a]	R	R		
威优 6 号	V20A//IR8/IR127	R	*R[a]	R[a]	R	R		
包选 2 号	选自包胎白	MR	MR	MR	MR	S		
寒露早	选自广二矮	MR	MR	MR	—	S		
IR20	抗病对照 check (R)	*R[a]	*R[a]	*R[a]	*R[a]	*R[a]		
TN1	感病对照 check (S)	S	S	S	S	S		

[a] 全生育期抗性（Over-all resistance）；其余为成株抗性（The others are adult-plant resistance）.

[b] R=抗（Resistant），MR=中抗（Moderatly resistant），S=感（Susceptible）

2.2 粳稻品种对供试菌株的抗性特点及其主要亲缘分析

在 45 个粳稻品种中，除抗、感对照品种银河和沈农 1033 分别抗和感所有 6 个菌株外，其中 18 个约占 40%的品种抗所有 6 个菌株，尤以秀水 48、城特 232、邳早 15、垦系 2 号、早生爱国 3 号和抗病对照银河表现突出。其余 25 个品种抗 3～5 个菌株（表 2）。表明我国粳稻主栽品种对白叶枯病的抗性较好。追溯其渊源，我国粳稻育种中主要引进日本品种。推广面积较大的这些当家粳稻品种大多数与农垦 58、爱国、坊主等日本品种具有亲缘关系。农垦 58（原名世界），经多次接种测试，对本试验中所用的菌株除浙 173 外，都有中抗到抗的水平。由于该品种表现矮秆、早熟、丰产，1957 年在太湖地区种植以来，普遍推广到苏、沪、浙、皖、鄂、湘、蜀等省，并被大量用作亲本，培育出农虎系统、秀水系统以及浙粳 66、矮粳 23、沪晚 4 号、鄂晚 5 号、紫金糯等抗病丰产良种。另一个日本品种金南风（农垦 57）对日本或菲律宾的白叶枯菌株均高度感染（章琦等，1986），但是由它育成的盐粳 2 号、南粳 11、南粳 15、邳早 15、南粳 35、中百 4 号等品种，却都具有广谱而稳定的抗性，其原因还不清楚。此外，云南省主要推广品种及育种亲本西南 175 对南方菌株具有抗性，但受北方菌株侵染。这种专化抗性的特点同样传递到其衍生品种云粳 136 中。

本试验中测试的粳稻品种，无论是全抗 6 个菌株或抗部分菌株的品种，绝大多数到成株期才呈现其抗性。这种成株抗性同其抗源品种相似。此外，在 23 个抗部分菌株的品种中，与有些菌株间的互作中，表现出某种程度的交叉反应（表 2），如鄂晚 5 号感浙 173、抗 HN 01，而紫金糯则抗浙 173、感 HN 01；滇渝 1 号和中丹 2 号对 064－2 和 HN 01 两个菌株以及中百 4 号和黎明对浙 173 和 HN 01 均呈现交叉反应，但不很强，这种近似品种与菌株间的特异性反应尚需进一步测试。但是启示人们在选择抗源亲本时，需要考虑本地区的菌株分布或选择具有广谱抗性的品种。

表 2　我国粳稻主栽品种对 6 个水稻白叶枯病菌株的抗性反应

Table 2　Reactions of japonica commercial cultivars from main rice growing regions to 6 strains of *X. o.* pv. *oryzae* in China

品种 Cultivar	组合 Cross	Strain 菌株						所属衍生系统 Origin
		KS6－6	ZHE173	064－2	KS1－20	HN01	HB29	
浙粳 66	矮粳 23/百哥	R	MR	R	MR	R	R	矮子占
矮粳 23	矮粳 22/桂花黄//农红 73	R	R	R	R	R	MR	矮子占
秀水 48	辐农 709/京引 154//辐农 709	R	R	R	R	R	R	矮子占
城特 232	矮粳 23/矮城//农特	R	R	R	R	R	R	矮子占、南特号
矮城 804		R	R	MR	R	MR	R	矮子占、南特号
嘉湖 5 号	58 糯虎/BL7	R	MR	MR	MR	MR	R	矮子占、南特号
农虎 6 号	农垦 58/老虎稻	R	R	R	R	MR	MR	矮子占、南特号
更新农虎		MR	MR	MR	MR	MR	MR	矮子占、南特号
丰锦	屉锦/奥羽 239 Sasanishiki / Ou239	R	MR	R	MR	R	R	矮子占、南特号
邳早 15	南粳 15 号辐	R	R	R	*R	R	R	矮子占、南特号
京越 1 号	水沅 300 粒/农垦 21	R	MR	R	R	MR	MR	矮子占、南特号
731		R	MR	R	R	MR	R	
垦系 2 号		*R	R	R	R	R	R	矮子占、南特号
秋光[a]Kihikari		R	R	R	R	MR	R	矮子占、珍珠矮
喜峰[a]Kiho	银河/藤坂 5 号 Ginga / Fujisaka5	R	MR	R	R	MR	R	矮子占、南特号

（续）

品种 Cultivar	组合 Cross	Strain 菌株 KS6－6	 ZHE173	 064－2	 KS1－20	 HN01	 HB29	所属衍生系统 Origin
台 24		R	MR	R		R	R	矮子占、胜利籼
早生爱国 3 号 Wase Aikoku3		R	R	R		R	R	爱国 Aikoku
南粳 15	农垦 57/农垦 51	R	MS	R		R	MR	旭、滋贺中生旭 Asahi，Shiga-chuseishin asahi
南粳 34	早丰（辐）	MR	MR	R	MR	S	S	爱国、旭 Aikoku，Asahi
南粳 35	665/IR26//501－1	R	MR	R	S	MR	R	金南风（旭） Kinmaze
盐粳 2 号	选自南粳 11	R	S	R	R	MR	R	金南风（旭） Kinmaze（Asahi）
盐粳 902－20－1		R	MS	MR	R	S	MR	旭
农垦 58 （世界稻） Nong-ken58		R	S	R	MS	MR	MR－MS	农垦 58
鄂晚 5 号	鄂晚 3 号//四上裕/IPR	R	bS	R	R	bR	R	农垦 58
泸晚 11 号		R	MR	R	S	MR	R	农垦 58
泸晚 4 号	选自苏州青 2	MR	MR	R	S		S	农垦 58
紫金糯	复虹糯 3 号// （辛尼斯/南粳 15）F1	R	bR	R	MS	bS	MR	农垦 58
秀水 27		R	MR	R	R	S	R	农垦 58
南粳 33	桂花黄/农垦 46	R	S	R	S	MR	S	巴利拉 Balila
西南 175		R	MR	MR	R	S	S	
云粳 136		*R	MR	*MR	R	S	MR－MS	西南 175
云玉 1 号		R	R	R	R	S	R	
临伦科砂 1 号		MR	R	MR	R	S	R	
滇渝 1 号	紫米/科情 3 号	R	MS	bS	R	bR	S	爱国、坊主、龟尾 Aikoku，Bozu，Kame-no-o
中丹 2 号	Pi5/喜峰 Pi5/Kiho	R	MR	bR	R	bS	R	爱国、旭 Aikoku，Asahi
中花 9 号	砦 2 号/京系 17//京系 17 Toride2/Jingxi17//Jingxi17	R	MR	R	R	S	MR	旭 Asahi

（续）

品种 Cultivar	组合 Cross	Strain 菌株						所属衍生系统 Origin
		KS6－6	ZHE173	064－2	KS1－20	HN01	HB29	
中百4号	喜峰/南粳15 Kiho/Nangeng15	R	[b]S	R		[b]R	R	旭
中作75	京系9号//合江18/窄叶青8号	R	MR	R	R	S	S	坊主
黎明 Reimei	藤稔用r射线辐射 Radiated from Fujiminori	R	MR	R	R	S	S	爱国、旭
晋红1号	晋宁2号/矮棵红	R	[b]R	R	MR	[b]S	R	矮棵红
晋稻1号		R	MR	R	R	S	MR	
奥羽244 Ou 244		R	S	R	MS	S	R	
辽粳5号		R	S	R	MR	S	MR	
银河（对照）Ginga（check）R	秀峰/陆羽132	R	R	R	R	*R	R	
沈农1033（对照）（check）S		S	S	S	S	S	S	

[a] 秋光和喜峰，作者于1977—1978年，用东北沈阳地区菌株E2－2测试3次，表现在苗期感病，剑叶期中抗（Akihikari and Kiho were tested with isolate E2－2 collected from Northeast area for three times by Zhang Qi in 1977—1978，they were susceptible at seedling stage，but became resistant at the adult-plant stage）。

[b] 交叉反应（Reversal reaction between two cultivars and two isolates）。

2.3 对水稻抗病育种问题的几点商榷

本试验77个品种中的绝大多数是我国水稻主栽品种，少数是育种计划中多年来采用的抗源亲本。用南北方6个菌株进行了3季测定，初步了解我国当前生产中的主栽品种对白叶枯病的抗性状况。华南、华中双季和单、双季稻作带，大面积种植的多数籼稻品种的白叶枯抗性较差。问题在于抗病基因源的挖掘和采用。IRRI自确定发展多抗性改良株型品种的育种目标后，从1965年始，开展抗稻瘟、白叶枯病育种，以后逐年增加抗病虫种类。1969年，只有分别为12％和17％的产量试验登记材料抗稻瘟和白叶枯病，1974年的这一比例就分别增加到60％和98％（Khush 1977；Mew 1981）。该所主要的经验是筛选抗源、累积抗性基因，进行同步育种。就白叶枯病而言，采用得较多的是对菲律宾当年的优势小种1有抗性的基因 *Xa4* 的代表品种，来自印尼的Sigadis和印度的TKM6，以及携有 *xa5* 基因的品种。在我国已经引进的广谱抗性的IR26、1R20、1R1545－339、DV85等均是较好的抗源。

我国粳稻主栽品种基本上对白叶枯病有不同程度的抗性，但大部分源出于日本品种。所培育的矮粳23、南粳15、邳早15等都是较好的抗源。喜峰有银河的亲缘，具有抗稻瘟和白叶枯病的优点。在引进国外抗源计划中，除日本品种外，也可以考虑籼稻中的抗源，将籼稻的抗性导入适应于我国生态地理条件的地方品种中，培育一批中间材料，以扩大抗性遗传基础。

随着经济发展，在抗病育种计划中发展我国的抗性基因库，将本国的地方抗性基因、国外引进或导入的抗性资源进行逐一的测试、整理，将促进我国的水稻育种工作。

原载：中国农业科学，1988，21（1）：41～49.

用 PCR 技术诊断水稻的白叶枯病抗性

陆朝福[1]　李晓兵[1]　朱立煌[1]　章　琦[2]
杨文才[2]　赵炳宇[2]　王春莲[2]

（1. 中国科学院遗传研究所；2. 中国农业科学院作物育种栽培研究所）

Diagnosis for Resistance to Bacterial Blight in Rice by PCR Technology

摘要　植物育种中应用分子标记辅助选择要求分子标记与目的基因紧密连锁，而且分析手段经济简便、重复性好。*Xa21* 是最近发现的一个具有广谱抗性的水稻叶枯病抗性基因，利用一个携有 *Xa21* 基因的品系 IRBB21，分别与 2 个感病品种杂交获得两个 F_2 群体。用 4 对引物分别对 3 个亲本进行 PCR 分析，其中一对引物（PB78）的 PCR 产物在抗、感品种之间可揭示多态性。对两个 F_2 群体进一步的遗传分析表明，PCR 标记和 *Xa*21 的白叶枯病抗性紧密连锁，其重组率仅为 2.48%。根据该标记选择在基因型纯合的抗性植株，其准确率可达 100%。本文还就植物育种中分子标记的检测途径进行了评价。

关键词　水稻白叶枯病，抗病育种，PCR，分子标记

植物杂交育种中根据植株的表现型进行选择通常会受到环境条件的影响，难以在早代对其农艺性状的表现做出准确的判断。利用遗传标记对目标性状进行选择可以大大提高育种效率，但是直到近年，由于分子标记的出现以及分子连锁图谱的建立，分子标记辅助的育种手段才受到育种家们的关注（Tanksley et al.，1992；Tanksley et al.，1989）。然而 DNA 诊断真正用于育种实践还受到许多条件的限制。首先必须找到和目标基因紧密连锁的 DNA 标记，更重要的是为使育种工作者易于接受，检测分子标记的方法必须简便、重复性好、并且经济可靠。

水稻是重要的粮食作物，其基因组较小（C=450Mb），已建立了高密度的 RFLP 分子遗传图谱（徐吉成等，1994；McCouch et al.，1988；Tanksley et al.，1992），并在分子图谱上定位了稻瘟病抗性基因（*Pi-zh*、*Pi-2*（*t*）、*Pi-4*（*t*）、*Pi-5*（*t*）、*Pi-7*（*t*）（徐吉成等，1994；Wang et al.，1994；Yu et al.，1991），半矮秆基因（*sdg*）（Liang et al.，1994）；香味基因（*fgr*）（Ahn et al.，1992），与胚乳蒸煮延展性相关的 QTL 基因（Ahn et al.，1993），稻瘿螨抗性基因（*Gm2*）（Mohan et al.，1994），白叶枯病抗性基因 *Xa*21（Rafalski et al.，1993）等。因而很适合于进行分子标记辅助的品种改良研究。

水稻白叶枯病是由 *Xanthomonas oryzae* pv. *oryzae* 引起的细菌性病害。我国大多数水

稻品种对白叶枯病抗源的抗性基因单一，潜伏着抗性丧失的危险，因而迫切需要拓宽和利用新的抗病基因（章琦，1990）。目前已知的白叶枯病抗性基因有 21 个之多。其中 *Xa21* 是最近从长药野生稻（*Oryza longistaminata*）中发现的 1 个显性抗病基因。由于其表现广谱抗性，因而具有重要的育种利用价值。Xa21 已被成功地转入栽培稻中（Khush et al.，1990），并已被定位在水稻第 11 染色体上，该基因与 3 个分子标记（RG103、RAPD 248 和 RAPD 818）紧密连锁，它们之间的遗传距离仅为 1.2cM（Ronald et al.，1992）。本文即在这些研究的基础上利用 PCR 的方法对 *Xa21* 基因进行分子遗传诊断，结果证明分子标记在水稻抗白叶枯病育种中有十分重要的应用价值。

1 材料和方法

1.1 供试品种和菌株

供试材料包括 IRBB21、珍珠矮和金刚 30 三个籼稻品种（系）。其中 IRBB21 携有抗性基因 *Xa*21，作为为抗性供体导入轮回亲本、感病籼稻品种 IR24 后所转育的抗病品系；珍珠矮和金刚 30 均为矮秆、农艺性状较好的籼稻感病品种。将这 2 个品种分别与 IRBB21 配制杂交组合，获得各组合 F_1 和 F_2 代群体。采用与抗性基因 *Xa21* 相对应的菌株 PXO99（P6）。

1.2 白叶枯病抗性鉴定

将 3 个亲本、2 个杂交组合的 F_1、F_2代种植在中国农业科学院隔离网室。在植株分蘖盛期用剪叶法接种。白叶枯病菌种 P6 保存于低温冻干状态，用前 1 个月进行复壮、毒性测定。接种物的菌龄 2d，接种浓度为 10^9/mL。接种后 15d，发病情况趋于稳定时调查各亲本、F_1 和 F_2 的感病性，以 F_2 群体抗性反应参数分布的峰槽趋势作为抗、感划分依据。本研究将病斑面积＜20％的为抗病（包括中抗），＞20％的为感病（S）。

1.3 DNA 提取和 PCR 反应

分别剪取 3 个亲本、2 个 F_1 及 F_2 群体各单株的新鲜叶片，按徐吉成等（1994）所述方法提取 DNA。PCR 反应在 Perkin Elmer Cetus 480 型 DNA 扩增仪上进行，25μl 反应液中含 10 mmol/L Tris-HCl（pH 9.0）、50 mmol/L KCl、1.8 mmol/L $MgCl_2$、0.1％ Triton-X 100。反应条件为 94℃，1min；60℃，1min；72℃，1.5min，循环 35 次；72℃，10min。4 对 PCR 引物分别是 PB78（B7，B8）；PRG103（RGl03f，RG103r）；pTA818（pTA818f，pTA8l8r）和 pTA560（pTA560f，pTA560 r）。这些引物均由 Ronald 博士（Univ. of California，Davis）根据与 *Xa21* 紧密连锁的标记即 RG103、RAPD818、RAPD560 设计序列，在本实验室合成。其中可揭示亲本多态性并进一步用于 F_2 代植株分析的引物 PB78 的核苷酸序是：B75′d［AGACGCGGAAGGGTGGTTCCCGGA］3′ B8 5′ d［AGACGCGGTA-ATCGAAAGATGAAA］3′。反应产物用 1.5％琼脂糖凝胶、3V/cm 电压电泳检测。

2 结果与分析

2.1 亲本、F_1 和 F_2 群体的白叶枯病抗性分析

用菌株 P6 分别接种亲本珍珠矮、金刚 30 和 IRBB21 各 30 株。结果显示，IRBB21 高

抗，其余 2 个品种均高度感病。由于只携有 *Xa21* 基因的植株才抗 P6，因此用 P6 来检测 *Xa21* 基因。在珍珠矮/IRBB21 和金刚 30/IRBB21 的 2 个杂交组合中各调查的 30 株 F_1，均表现抗病，证实了 *Xa21* 为显性抗性。2 个 F_2 群体出现了抗、感分离（表 1），表明 *Xa21* 已被转入 2 个感病品种。在金刚 30/IRBB21 的 165F_2 植株中，119 株表现抗病、46 株感病；在珍珠矮/IRBB21 的 145 个 F_2 植株中，99 株抗病，46 株感病，两个群体的抗、感植株的分离均符合 1 对基因的孟德尔公式 3∶1 分离比。

表 1 两个 F_2 群体的抗性及 PCR 分析

Table 1 Analysis of segrigation for resistance to bacterial blight and PCR genotype in F_2 population of two crosses

组合 Cross	F_2 分离[a] Segregation of F_2		χ^2 3∶1	PCR 基因型[b] PCR genotype			χ^2 1∶2∶1	重组率 Recombination freq.（%）
	R	S		S	R/S	R		
珍珠矮/ IRBB21	99	46	3.15	52	75	19	14.10*	
金刚 30/ IRBB21	119	46	0.59	50	77	34	3.06	2.48

[a] R=抗病株数（Resistant），S=感病株数（Suceptible）。

[b] R=抗病亲本基因型（Resistant parent genotype），R/S=杂合型（Heterozygous），S=感病亲本基因型（Susceptible parent genotype）。

2.2 亲本和 F_2 的 PCR 分析

用 4 对 PCR 引物分别对感病品种珍珠矮、金刚 30 和抗病品系 IRBB21 进行分析。4 个引物对 3 个亲本都能扩扩增出 1 条带，但只有 PB78 可在抗病和感病品种间揭示多态性。抗病品系 IRBB21、感病品种珍珠矮和金刚 30 的 PCR 产物片段大小分别约 900bp、700bp 和 600bp（图 1）。F_1 有 2 条带，说明该分子标记为共显性的。遂以 PB78 为引物进行 F_2 的 PCR 分析。

F_2 群体的 PGR 分析结果如图 2。在 F_2 中出现了 3 种带型，即与抗性亲本 IRBB21 的纯合基因型（第 22 道）、感病亲本珍珠矮或金刚 30 的纯合型 11 相同的带型，及同时呈现双亲的 2 条带型（杂合型 12）。在金刚 30/IRBB21 的 F_1 中，这 3 种带型的分离比例符合 1∶2∶1的理论值，在珍珠矮/IRBB21 的 F_2 中，与感病亲本一致的带型显著偏多，表现为偏分离。

2.3 抗病性和分子标记的共分离及分子标记在育种上的应用

用遗传标记对单基因控制的性状进行选择时，该标记必须与目的性状紧密连锁。本研究的两个 F_2 群体之绝大多数植株的 PB78 标记和植株的抗病性表现为共分离。由于作物育种中希望选择带有纯合抗病基因的单株，我们根据分子标记 PB78 表现的抗病亲本带型（22）所入选的植株，全部表现抗病（表 2）。结果还表明，绝大多数 PB 标记（97%以上）为杂合带型（12）的植株也表现抗病，但部分 PB78 基因型为纯合感病带型（11）的植株却表现抗病，这是导致 PCR 诊断结果与植株抗病反应的符合率下降的主要原因。PCR 是在分子水平上对基因型进行分析，其结果不受环境条件和其他基因作用的影响，因此在珍珠矮/IRBB21 的 F_2 中有 6 株、金刚 30/IRBB21 的 F_2 中有 4 株 PCR 为感病亲本

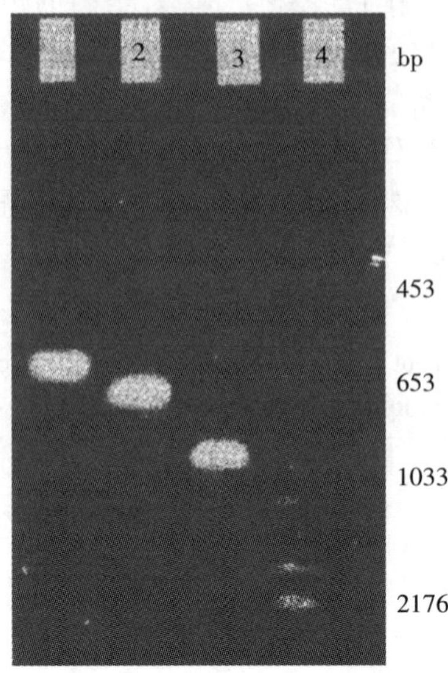

图 1 PCR 标记 PB78 在亲本间的多态性

1. 珍珠矮，2. 金刚 30，3. IRBB21，4. DNA 标记

Fig. 1 The polymorphism of PCR marker PB78 among 3 varieties (line).

1. Zhenzhuai，2. Jingang 30，3. IRBB21，4. DNA marker

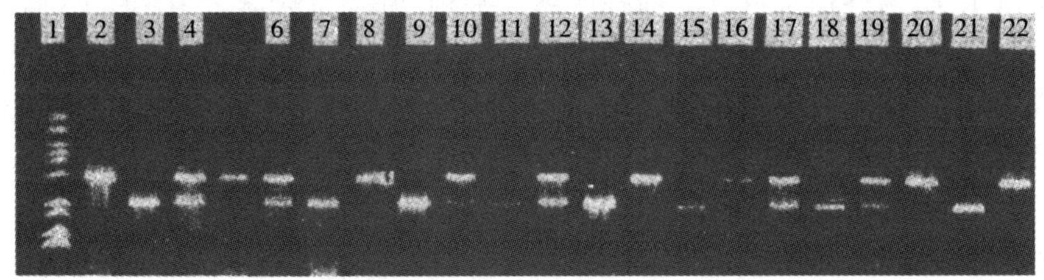

图 2 PB78 在金刚 30/IRBB21F_2 群体中的分离

1. DNA 标记，2. 金刚 30，3. IRBB21，4～22. F_2 植株

Fig. 2 Segregation of PB78 in F_2 of Jingang 30/ IRBB21

1. DNA MARKER，2. Jingang 30，3. IRBB21，4～22. F_2 plants

带型。而植株 PCR 为感病亲本带型、植株却表现抗病的原因，除了由于 PB78 与 *Xa21* 之间可能产生的重组外，还应考虑到白叶枯病抗性调查标准的人为因素。至于国内外的 PCR 分析显示杂合抗病带型，而植株却表现感病的情况，可能由于 *Xa*21 基因受到不同的遗传背景的影响，这些分析还需要对水稻白叶枯病抗性基因抵御病菌机制作进一步研究。尽管这样，我们分析的金刚 30/IRBB21 的 F_2 的结果表明，分子标记 PB78 和 *Xa21* 之间的重组率仅为 2.48%；另一组合的 F_2 群体虽受偏分离的影响不能准确估算重组率，但根据 PB78 的纯合基因型（22）选择抗病株，其准确率也达到 100%。总之，从本实验看，分子标记与抗病基因之间的紧密连锁和分子标记与植株抗病表现的很高的符合度，可以满足其在抗病育种中的应用。

表 2 用 PCR 标记诊断水稻白叶枯病抗性的符合率

Table 2 Consistency between the PCR genotype and BB resistance in two F_2 populations

组合 Cross	分子诊断 Molecular diagnosis	PCR 带型 PCR genotype	F_2 群体（F_2 population）		
			抗病株 R No.	感病株 S No.	符合率 Consistency（%）
珍珠矮/IRBB21	R	22	19	0	100
		12	73	2	97.3
	S	11	7	44	86.3
金刚 30/IRBB21	R	22	34	0	100
		12	77	0	100
	S	11	5	45	90.0

3 讨论

目前，鉴定的植物抗病（虫）性等，主要受单基因或少数主效基因控制。育种选择的目的是获得目标基因型纯合的个体。对目标性状利用植株的表现型进行直接选择，并要求对杂交后代进行遗传分析以区分纯合型和杂合型。其结果往往由于其他基因的互作或外界环境的影响而不够准确，并且需要花费大量的人力和土地。对于植物抗病育种来说，传统的抗病性鉴定不仅时间长，而且受到许多条件的限制（Meichinge，1990）。比如，田间环境条件的变化很难取得多年一致的结果、有些检疫性病害的测定必须严格控制在一定的范围内、大多数病害的发展与植物的发育阶段有关等。而RFLP、RAPD 等分子标记属于中性的遗传标记，它们的表现不受环境条件的影响，分析手段快速简便，因而寻找与抗性基因紧密连锁的分子标记以辅助转育抗性品种可以减轻上述的繁琐工作，从而大大提高育种效率。此外，许多隐性抗病基因如白叶枯病抗性基因 *xa5*，*xa8* 等在常规育种中很难利用。而用分子标记辅助选择可在 F_2 中迅速筛选到含纯合目的基因的抗性植株。

近年来分子遗传标记的出现使植物育种摆脱了选择标记十分贫乏的限制（Tanksley et al.，1989。分子标记几乎可以在任何亲本之间揭示多态性，利用分子连锁图谱可以很快找到与目标性状相应的分子标记。对于那些没有建立分子图谱的植物来说，也有比较方便的方法可以迅速找到与目标性状连锁的分子标记。常用的有效手段是获得携带该基因的近等基因系（NIL）(Martin et al.，1991)。Michelmore 等（1991）提出了一种对目的性状分池（bulked segregant analysis）的分析方法，可以利用任何一个单交后代的分离群体进行RFLP 或 RAPD 分析，快速检测到与目的性状连锁的分子标记。

植物育种中可以利用的分子标记有许多种（Rafalski et al.，1993），常用的是 RFLP 和RAPD 等。RFLP 是共显性的，只要选择含有与目的基因紧密连锁标记的植株，即可区分纯合型和杂合型。但 RFLP 的分析程序比较复杂，而且十分昂贵。RAPD 标记是显性的，只要找到与目的基因连锁的相斥型（repulsion-phase）的 RAPD 标记，仍可在 F_2 代进行有效的选择（Haley et al.，1994）。RAPD 的分析程序简便，但重复性较差，因此在实际应用中也存在一定的困难。本研究中使用的是专一扩增的 PCR 标记，它兼具 RFLP 和 RAPD 的优点，像前者一样是一种共显性的标记，能区分纯合型和杂合型，又像后者一样基于 PCR 反应，测定时所需的 DNA 量很少，反应程序简单迅速，测定结果重复性好，只需要一套 PCR 反应仪和电泳设备，在一般实验室即可进行分析，因而很适合于作为育种工作中的一种有力的辅助手段。

原载：遗传学报，1996，23（2）：110～116.

稻曲病的发生趋势初析——一个值得警惕的病害

章　琦　周永力　王春莲　施爱农　杨文才　赵炳宇

（中国农业科学院作物育种栽培研究所）

Preliminary Analysis of Bevelopment Trend of Rice False smut

随着农作物单位面积产量的提高，生态环境的变迁和农业集约措施的加强，植物病害的群体结构也发生了变化。20 世纪 50 年代以来我国水稻生产的两次突破性飞跃——杂交水稻和和籼粳交育种，使一些病害从次要地位上升为主要病害，如稻曲病。几年来我们每年进行全国性的水稻生产主栽品种的利用和病虫害发生的调查中发现，除稻瘟、白叶枯和纹枯病三大病害之外，稻曲病已逐渐遍及全国，所造成的产量损失已不容忽视。

1　稻曲病的分布及危害

稻曲病是由子囊菌纲麦角菌科的 *Ustilagonoidea virens*（Cooke）Tak.（Takatlashi，1896）引起的真菌病害。我国从 20 世纪 80 年代以来，随着杂交水稻和籼粳杂交育成的高产品种的推广和采用相应的增产措施，稻曲病在南北稻区的分布面积和严重度都逐年上升。1982 年南方双季晚稻普遍严重发生，湖南省严重发病的面积高达 74.5%（谭荫初，1983）；江西省有 30%的水稻种植面积严重发生，大面积种植的籼稻品种“桂潮 2 号”的穗发病率高达 30.14%（李火苟等，1984）；贵州省 1982—1983 年发病面积均超达 40%（付淑云等，1987）；江苏省 1991—1992 年调查，新发展的杂交粳稻发病也非常严重，每穗病粒竟高达 72～120 粒，一般减产 1～2 成（江苏省农业科学院等，1993）；1988 年，湖北省鹤峰县因稻曲病造成的损失占水稻总产的 5.4%；翌年，洪湖市杂交稻汕优 63 的病穗率达 30%～40%。我国北方单季粳稻的稻曲病尤其严重；1987 年山东滨海地区的发病面积为 36%，重病田块的损失产量在 10%以上；辽宁省 1984—1985 年连续大流行，有 43%的稻田面积严重罹病，1987 年该省新金县严重田病穴率达 81.5%（俞孕珍等，1994）。我们 1996 年调查辽宁省种植面积较大的品种辽盐 241 和辽粳 454 的稻田块块发病，有些田块的病穴率高达 90%以上，病穗率最高的约 40%，辽粳 454 的每穗稻曲数平均 5 粒，最高 20 粒，沈阳市苏家屯区公路沿途的稻田的稻穗布满病菌，已成一片墨绿。农民反应非常强烈，他们在脱粒重病稻谷时，全身布满飞扬的病菌厚垣孢子，令人触目惊心。研究表明，此病的危害不仅限于病粒，并能通过病粒影响养分的输送，扩展到染病的枝梗，致使结实率和千粒重降低而减产。受稻曲病菌污染的谷粒不仅降低了米质，并且危害人畜健康，家畜或家禽食用了混有病粒的谷糠，都能引起慢性中毒，内脏发生病变，导致肉猪生长减慢、仔猪存活率下降，家禽产蛋率降低（高骏，1987）。综上所述，稻曲病已从微不足道的地位上升为一个不容忽视的主要病害。我们1996年在云南省调查时，该省植保工作者认为，云南省90年代的重点病害是稻曲病，这也代表了我国水稻生产上面临一个新的挑战。

2　水稻对稻曲病的抗性

许多报道指出，水稻品种之间对稻曲病的抗性有明显差异。了解品种与稻曲病的互作关系，是进行抗病育种的重要参考资料。我们于1986—1988年连续在沈阳于洪区的重病区，对200份育种亲本和高代材料进行了设有3次重复小区的自然发病鉴定。连续3年调查各小区抗病、感病品种的病穴和病穗率，根据其平均值比较，两者均为0的抗病品种有，DL-10、矢租和Zenith，在感病品种中的“中作8524”的病穴和病穗率分别为100%和78.2%。1996年在沈阳市汪家乡调查，辽粳454的病穴率、病穗率分别为90%和40%，其每穗的病曲数竟有10～20粒；在同一块地里的感病品种辽盐241的病穴、病穗率分别有90%和35%以上；抗病品种辽盐2号的病穴率低于10%，稻曲粒仅有1～2个。沈阳农业大学的调查认为，灰杂、辽优1号、秋杂和黎杂等杂交水稻组合都比较感病；籼粳杂交高产品种辽粳5号也是高度感病的（俞孕珍等，1994）。南方大面积种植的品种桂潮2号、紫金糯、矮粳23、秀水48、嘉17等品种及其品系发病都很严重；汕优6号也很感病。表1所列部分品种的抗性状况表明，品种对稻曲病的抗性具有明显差异，初步可以认为，品种对此病的抗性由主效基因控制，自然条件、气候因子和农业措施对稻曲病的发生也有一定影响，利用寄主的抗性可以发挥控制作用，是一项经济有效的措施。

表1　水稻品种（系）种植在稻曲病自然发病区的观察

品种（系）/杂交稻	抗感反应	穴发病率（%）	病穗率（%）	病粒率（%）	参考文献
DL10	R	0	0	0	章琦、施爱农等
矢租	R	0	0	0	（1986—1988）
Zenith	R	0	0	0	
H12	MR	0.6			
喜峰	MR	0.6			
早生爱国3号	MR	0.6			
银河	MR	0.6			
中作8524	S	100	78.2		
中作8751	S	72			
辽粳5号（CK）	S	41.2			
黄金光	R	0	0	0	俞孕珍等，1994
202	R	0	0	0	
108	R	0	0	0	
18	R	0	0	0	

（续）

品种（系）/杂交稻	抗感反应	穴发病率（%）	病穗率（%）	病粒率（%）	参考文献
灰杂	S	47.0	41.0	1.2	
辽优1号	S	26.2	12.7	3.14	
秋杂	S	28.8	1.89	0.16	
黎杂	S	18.8	9.41	0.23	
辽粳5号	S	39.3	9.0	0.46	
双糯4号	R	0	0	0	林元札，1983
汕优6号	S		56.0	92.0	
测24	S		32.0	38.0	
测48	S		11.0	11.0	
矮粳23	S		5.0	6.0	
紫金糯	S		10.1		黄正兴等，1984
苏粳2号	S				
桂潮2号	S		30.14		李火苟，1985
科印糯	S		14.4		
洞庭晚籼	S		12.6		
广二104	R	0	0	0	黎毓干，1993
沂优63	R				
桂潮	S				
威优35	R	0	0	0.72	谭荫初，1983
台中青AI测296	S			6.94	
虎21糯	R	0	0	0	金敏忠等，1987
宁119/103	R	0	0	0	
H129	R	0	0	0	
嘉11	R	0	0	0	
T8368	R	0	0	0	
秀水48	S	57.5	6.68		
嘉17	S	42.5	5.98		
052	S	30.0	4.05		

3　病情发展趋势的预测及控制策略

20世纪80年代以来我国大面积种植的大多数品种对稻曲病的抗病性不佳，以致南北稻区普遍发生且有严重趋势，病田中必然积累大量病菌，稻曲病在今后的发展形势将是严峻的，将可能出现大流行的局面。虽然此病已开始引起人们的关注，但与稻瘟病、白叶枯病相比，有关稻曲病的病理学、生物间遗传学、遗传育种以及综合防治等方面的研究不多。在病原菌分化、致病性分类以及病理生理等方面还都是空白，目前难以进行品种抗病性评价、基因鉴定等方面的系统研究，影响了抗病育种的开展，据此，提出控制稻曲病的初步建议。

3.1 综合防治

3.1.1 在有代表性的重病区建立品种抗性自然评定，筛选种植适应性强的高产抗病品种，压缩严重感病品种的种植面积。

3.1.2 合理施肥，基肥重施，采用氮、磷、钾复合肥，避免偏施氮肥。

3.1.3 水稻孕穗末期，辅以药剂防治，6%多菌酮粉，每公顷喷施 37.5 kg 或 3 kg 4%井冈霉素。

3.2 设立稻曲病应急防治关键技术研究项目

3.2.1 病原菌分化研究 明确中国稻曲病原菌的致病性分化（是否存在生理小种或病原型），为建立鉴别系统提供基本依据（包括寄主一病原菌互作明显的鉴别菌株及鉴别品种）。

3.2.2 全国水稻品种、种质的抗性评价、分群及抗病性遗传基础的研究。

3.2.3 组建抗病和农艺性状优良的育种材料库，纳入育种计划。

原载：中国科学技术协会专家论述论文集，1997，84～87.

稻曲病菌分离技术的初探

周永力　章　琦

（中国农业科学院作物育种栽培研究所）

Preliminary Report on the Isolation Technique of *Ustilagonoidea Virens*（Cooke）Tak.

Abstract Effect of the kinds of media，isolation methods and the stored periods of false smut ball on the isolation of *Ustilagonoidea virens* were primarily studied to seek ideal isolation technique. The results showed that Wakimoto Toceshi + chloromycetin (50 mg/mL) is an ideal medium for the pathogen isolation，and rice grain is the best medium for the sporulation of chlamydospores studied so far. The methods of chlamydosporal suspension can be used for rapid isolation of the pathogen，and can be isolated successfully the pathogen from false smut balls if the inner tissue，which has been stored for a long time (10 months) already.

Key words *Ustilagonoidea virens*；culture media；colony；isolation technique；spore

摘要　用7种培养基和2种分离方法对稻曲病菌的培养及分离技术进行了研究。结果表明，胁本哲氏培养＋抑制细菌生长的氯霉素（50μg/mL）是分离稻曲病菌最佳培养基。稻谷也是产生厚垣孢子的理想培养基。厚垣孢子悬液法可用于稻曲病菌的快速分离，而采用稻曲球不同层次组织块分离法，可从已经保存了10个月这样长时间的稻曲球中分离到稻曲菌。

关键词　稻曲病菌；培养基；菌落；分离技术；孢子

稻曲病是由子囊菌纲麦角菌科 *Ustilagonoidea virens*（Cooke）Tak. 引起的真菌病害。20世纪80年代以来，随着杂交水稻的推广种植，稻曲病的危害面积逐渐扩大到全国稻区，上升为重要的水稻病害。章琦等（1997）观察到水稻品种间对稻曲病的抗性有差异，表明利用寄主的抗性应该是控制此病的有效方法，病原菌的分离是进行抗性评价的关键技术，但是稻曲病菌的分离培养难度较大，人工接种的发病率很低，在病原菌的生理分化和品种抗性研究等方面鲜有报道。我们采用不同培养基和分离方法，连续对来自不同地区的115份标样进行分离，以寻找快速、有效的分离技术，为抗病品种鉴定和筛选提供基本条件。

1 材料与方法

1.1 供试材料

1996 年，在稻曲病发病高峰期，在辽宁、吉林、湖北、云南、河北、广西等地的不同水稻品种上采集了 115 份稻曲病菌标样，保存于 4℃冰箱中。

1.2 试验方法

1.2.1 培养基

分离培养基：1）胁本哲氏培养基[马铃薯 300g、蛋白胨 5g、蔗糖 15g、$Ca(NO_3)_2 \cdot 4H_2O$ 0.5g、$Na_2HPO_4 \cdot 12H_2O$ 2.0g、琼脂 16g]，2）大米粒培养基，3）米粉培养基，4）玉米粉培养基，5）燕麦片培养基，6）PSA 培养基，7）PDA 培养基。根据这 7 种培养基的分离效果，选出最佳者。

将分离到的菌种转接到上述培养基和胁本哲氏、燕麦片、PS 和 PD 液体培养基上观察菌落的培养性状和产孢情况。

1.2.2 分离方法

厚垣孢子悬液法：用 75%乙醇将稻曲球消毒 5～10s，再用无菌水冲洗 3 次，置于 Eppendorf 管中振荡，配制成厚垣孢子悬浮液，取 2～3 滴，滴加在培养皿中，待培养基冷却至 45℃时，倒入培养皿中，轻摇，使之与孢子悬浮液均匀混合，待培养基凝固后置于 28℃恒温箱，在黑暗条件下培养，逐日观察分离结果。

稻曲球不同层次组织块分离法：将在 75%乙醇中消毒 5～10s 的稻曲球再用 0.1%升汞消毒 5～8min 后，用无菌水冲洗 3 次，将稻曲球的外、中和内层分别切成 2mm×3mm 的小块，置于培养基表面，每皿 5 块，培养条件与厚垣孢子悬浮液法相同。

1.2.3 分离时期

采用厚垣子悬浮液法和组织块法，每隔 7d 将保存于冰箱中的稻曲病标样分离一次，连续分离，直至分离不到稻曲病菌为止。

2 结果和分析

2.1 不同培养基的分离效果

稻曲菌球营养丰富，常伴生大量细菌和真菌，加之此菌生长缓慢，以致分离的成功率很低。本试验的结果表明，稻曲菌在不同培养基上的生长速度不同。其速度依次为，胁本哲氏>PSA>PDA>大米粒>米粉>燕麦片>玉米粉。培养至 1 个月时，病菌在大米粒培养基上的菌落面积最大（图 1），缺点是其初期生长较慢，易于污染；在胁本哲氏培养基上的生长速度最快，菌丝旺盛，因此胁本哲氏培养基+抑制细菌生伏的氯霉素（50μg/mL）是分离稻曲病菌最适宜的培养基，其次为 PSA。

2.2 不同培养基对稻曲病菌培养性状和产量的影响

2.2.1 固体培养基

在所用的7种固体培养基上，稻曲菌初期菌落的性状基本相同，但至1个月时，菌落大小和厚垣孢子的产孢量出现明显差异。图1显示，以大米粒培养基上的产孢量最高，其他培养基的产孢量依次为，胁本哲氏＞米粉＞PDA，在PSA、燕麦片和玉米粉培养基上基本不产生厚垣孢子。

在同一种培养基上，各菌株培养初期的菌落形态和色泽相近，培养至1个月时，显现不同特征。如在胁本哲氏固体培养基上有以下类型，1）菌落中央凸起呈鲜黄色，边缘黄绿色；2）菌落中央黄绿色，边缘凸起为黄色；3）菌落黄绿色，局部形成黄色凸起；4）菌落黄绿色，周围长出一至多个白色小菌落；5）菌落中央灰绿色、边缘白色；6）菌落白色。此外，产生厚垣孢子的能力差异很大，前3种类型的菌落容易产生厚垣孢子堆，后3种菌落基本不产生厚垣孢子。

2.2.2 液体培养基

在4种液体培养基中，燕麦片最利于产生厚垣孢子堆，在燕麦片液体培养基中，各菌株的培养性状基本相同，培养至1周时，形成片状白色菌落，1个月时，菌落呈黄绿色或橄榄绿色，在液面形成黄色厚垣孢子堆。各菌株所形成的厚垣孢子堆之数量不同。在PS中，供试菌株的培养性状有两类：1）一类菌株在培养7d左右时即形成大量的分生孢子，10d后，孢子开始萌发产生白色菌落，但始终不形成厚垣孢子；2）另一类菌株的菌丝生长旺盛，但不产生分生孢子，培养至1个月时，在培养基表面形成黄色厚垣孢子堆。

2.3 不同分离方法的比较

2.3.1 厚垣孢子悬液法

其优点是可快速分离出稻曲病菌。在4℃冰箱中不超过1个月的稻曲球的厚垣孢子悬液在胁本哲氏培养基上，约1周即长出白色、馒头状单孢菌落，但分离成功率随着标样保存时间的延长而降低，如果标样保存超过了2个月，则不宜用此方法。

2.3.2 组织块分离法

成熟稻曲球的切面可分为3个层次，绿色的最外层，橙黄色的中层和白色的内层。采用不同层次的组织块分离的成功率明显不同，外层组织伴生大量快速生长的细菌和真菌，很难分离到稻曲病菌。以稻曲球内层组织的分离成功率最高，15～20d时从组织块的周围长出白色、致密的菌丝。稻曲球内层主要是菌丝。在4℃条件下，菌丝可存活1年以上。因此，对长时间保存的标样，可采用内层组织块法进行分离。我们保存了3个月的稻曲球的分离成功率（成功的培养皿数/所用的培养皿总数）仍然达到100%，用该法还可以从保存了10个月的稻曲球分离到稻曲病菌。

我们发现随着继代培养次数的增加，稻曲病菌的产孢能力急剧下降，转接到第4代时，所有菌株就都不产生厚垣孢子了。另有研究表明，越冬后的厚垣孢子之萌发率仅有千分之几。因此，人工接种时，采用内层组织块法对头年保存在冰箱中的稻曲球进行分离，或用当代菌种产生的厚垣孢子作为接种体，将有助于提高发病率。

2.4 稻曲病菌与伴随菌稻黑孢霉的区分

稻黑孢霉 *Nigrospora oryzae*（Berk & Br.）Petch是稻曲病菌最主要的一种伴随菌。

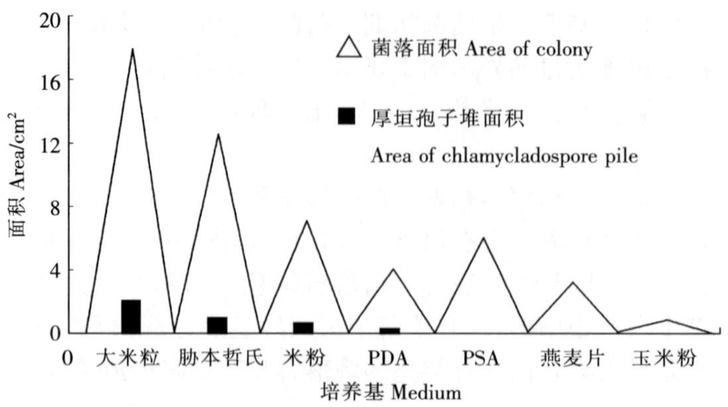

图 1　不同固体培养基上稻曲病菌的菌落大小和产孢量

Fig. 1　The size of colony and sporulation on different solid media.

魏景超（1975）报道江苏、浙江、福建和湖北曾有发生。我们对采自辽宁、吉林、云南、江西、广西、天津、北京和湖北的 115 份标样外层组织块分离的结果表明，稻黑孢霉在上述地区也有发生，出现频率达 90％以上。为了便于与稻曲病菌区分，将两者在胁本哲氏培养基上的培养性状和分生孢子特征列于表 1。

表 1　*Ustilagonoidea virens* 和 *Nigrospora oryzae* 在培养基上的生长性状比较

Table 1　Comparision of cultural traits between *Ustilagonoidea virens* & *Nigrospora oryzae* on the medium

病原菌 Pathogen	菌落 Colony		孢子 Spore		
	形态 Shape	色泽 Color	形成时间 Sporulation time	形状和大小 Shape and size	色泽 Color
稻曲病菌 *U. virens*	凸起、气生菌丝致密	初白色、1 个月中央黄色、黄绿色或橄榄绿	20～30d 形成橙黄色厚垣孢子堆	厚垣孢子球形、壁具有疣突（4～5）μm×（5～7）μm	初形成淡黄色、成熟时黄色
稻黑孢霉 *N. oryzae*	平铺、毡状，气生菌丝稀疏	初灰白，1 周时中央橄榄绿色，边缘灰色或灰黑色	10d 左右	分生孢子球形和扁形，壁光滑，<（12～18）μm×（14～19）μm	初无色、渐变为黄色、橄榄绿色，成熟时黑色

3　结语

本研究初步表明，稻曲病菌在培养性状和产孢能力以及在培养基中产生的孢子类型存在分化现象。王疏等（1997）在田间观察到有白色稻曲球发生，其菌株的形态、生理特性和培养特征与绿色稻曲球菌株明显不同。章琦等（1997）的研究表明，水稻品种对稻曲病的抗性存在着明显差异，种植抗病品种应视为经济有效的防治措施。鉴于稻曲病菌生活史比较复杂，在其整个侵染循环中的繁殖体，包括厚垣孢子、分生孢子和子囊孢子，迄今还不能确定有效的接种体。人工培养大量接种体和接种方法还不够成熟，这正是系统研究水稻对稻曲病抗性遗传基础和应用于育种的主要障碍，也是未来值得进一步研究的内容。

原载：中国水稻科学，1999，13（3）：186～188.

由农杆菌介导将白叶枯病抗性基因 *Xa21* 转入我国 5 个水稻品种

程文学[1]　李晓兵[1]　田文忠[1]　周永力[2]　潘学彪[3]
曹守云[1]　赵显峰[1]　赵　彬[1]　章　琦[2]　朱立煌[1]
（1. 中国科学院遗传研究所；2. 中国农业科学院；3. 扬州大学农学院）

The Bacterial Blight Resistance Gene *Xa21* was Mediated by *Agrobacterium* and Transferred into 5 Rice Cultivars

摘要　利用农杆菌介导的转化系统将已克隆的 *Xa21* 基因转入我国 5 个水稻主栽品种，获得 110 个独立的转基因系。对转基因植株进行的 PCR 和 Southern 分析，显示了 *Xa21* 基因已被整合到受体基因组。已整合的 *Xa21* 基因能稳定遗传，单拷贝整合的转化体在自交 T_1代呈 3∶1 抗感分离。转基因 T_0 植株和 *Xa21* - PCR 阳性 T_1植株高抗白叶枯病。经筛选的 *Xa21* 纯合的优质抗病转基因系可以作为品种直接种植，或用于杂交稻育种。

关键词　转化　水稻　*Xa21*　农杆菌　白叶枯病抗性

由革兰氏阴性水稻黄单胞杆菌（*Xanthomonas oryzae* pv. *oryzae*，*Xoo*）引起的白叶枯病是世界水稻生产中最严重的细菌性病害。将抗性基因导入水稻品种是控制此病的最经济有效的方法。迄今至少已经鉴定了 19 个白叶枯病抗性基因（Kinoshita，1995）。通过传统的回交育种方法已将少数抗性基因导入一些优良水稻品种。由于传统的育种方法艰巨、费时，并经常受不良性状连锁的累赘。近年，采用定位克隆战略，克隆了两个白叶枯病抗性基因 *Xa21*（Song et al.，1995）和 *Xa11*（Yoshimura et al.，1998），为通过转化改良植物抗性的基因工程提供了新工具。其中，广谱抗性基因 *Xa21* 在改良杂交水稻白叶枯病抗性方面具有重要应用前景。

Wang 等（1996）、Tu 等（1998）、Zhang 等（1998）利用 *Xa21* 通过基因枪方法转化水稻品种。期间，我们也曾用相同方法转化水稻品种（未出示结果），发现基因枪转化所获得的转基因植株经常出现不育，其他农艺性状也容易发生变异，实际上，这种现象也见诸于 Zhang 等所获得的转 *Xa21* 基因植株中（Zhang 和 Song，私人通信）。近年，农杆菌介导的转化系统已经成功地用于水稻（Hiei et al.，1994；Rashild et al.，1996）。我们在本研究中用农杆菌转化系统将 *Xa21* 基因转入 5 个水稻品种，包括两个籼型杂交稻恢复系和一个保持系。结果还表明，转化的 *Xa21* 基因已被整合到受体基因组并且稳定遗传。

1 材料与方法

1.1 质粒和菌株

含有 *Xa21* 基因的质粒 pC822 由 Song 等（1995）构建。pCAMBIA1301 质粒是澳大利亚 CAMBIA 研究中心构建的一个双元载体，含有潮霉素抗性基因、*GUS* 基因和细菌卡那霉素抗性基因。采用的农杆菌菌株为 EHA105。

1.2 水稻品种

本研究中转化的水稻品种列于表 1。

表 1 利用 *Xa21* 基因通过农杆菌系统转化的水稻品种

品　种	PCR/Southern 分析	抗性分析	所获转基因系总数[a]	转化效率[b]（%）	变异转基因植株数[c]
明恢 63（籼稻杂交稻恢复系）	＋	＋	26	4.3	0
珍汕 97B（籼稻杂交稻保持系）	＋	＋	24	4.0	2
盐恢 559（籼稻杂交稻恢复系）	＋	＋	18	3.6	1
太湖粳 6 号（粳稻品种）	＋	＋	15	7.5	0
中花 11（粳稻品种）	＋	＋	27	10.5	2

[a] 一个转基因系可能包括一个以上转基因植株；

[b] 转化效率＝（获得的转基因系总数/感染的愈伤总数）×100%；

[c] 包括所有的形态变异株和不育株。

1.3 培养基

基本培养基：NB（Li et al.，1993），或 MS（Murashige，1962），或 CC（Potrykus，1979）。

共培养培养基：基本培养基＋100μmol/L 乙酰丁香酮。

选择培养基：基本培养基＋2mg/L 2,4－D，300～500mg/L 噻孢霉素，潮霉素 30～50mg/L。

预分化培养基：基本培养基＋5mg/L ABA，2mg/L BAP，2mg/L NAA，50mg/L 潮霉素。

分化培养基：基本培养基＋3mg/L BAP，0.5mg/L NAA，1.0mg/L IAA，2.5mg/L KT，50mg/L 潮霉素。

1.4 转化

采用改进 Hiei 等（1994）、Rashild 等（1996）的转化方法，对 5 个品种的成熟胚愈伤组织进行转化。种子去壳，70%乙醇消毒 2min，0.1%$HgCl_2$溶液消毒 30min，无菌水漂洗 3 次，置于含 2.0mg/L 2,4－D 的基本培养基培养，在暗处于 26℃培养 2 周后，切下诱导产生的盾片愈伤组织，选择致密的愈伤组织颗粒（直径 3～5mm）用于转化。

带有 pCXK 1301 的农杆菌在含利福平 10mg/L，卡那霉素 50mg/L 和潮霉素 50mg/L 的 YEP 培养基中培养 24h 至 A_{595} 为 0.8. 离心收集农杆菌细胞，用 AAM 培养基洗涤 1 次，再悬浮在 AAM 中至 A_{595} 为 0.5（约 10^9 细胞/mL）。

将待转化的愈伤颗粒放入 AAM 农杆菌悬液中浸泡 15min 后，置于一叠干灭菌纸上，除去过多的菌液。浸染愈伤无需进一步吸干，转移置共培养基上，于 26℃暗培养 2～3d。在共培养后，用无菌水彻底洗涤愈伤几次后，在含噻孢霉素 500mg/L 无菌水中 120r/min 振荡洗涤 2h，将洗涤后的愈伤铺在选择培养基上培养 3 周进行第 1 次选择，然后转移到新选择培养基上继续培养 2～3 周进行第 2 次选择，经两次选择培养后，生长旺盛的愈伤在预分化培养基上，于 26℃暗培养 2～3 周后，转移到分化培养基上持续光照下 26℃继续培养 2～3 周至分化出苗，当再生苗长至 10cm 高时移植到土壤中，在温室中生长至成熟。

1.5 PCR 分析

根据 *Xa21* 序列设计了特异扩增引物 Z_2 和 Z_3。引物的序列如下：Z_2 5′-ATTGAATAATTCACTGGGTATTGG，Z_3 5′-GTCTTGCCTTGCACTTCTGCACGA。每一 25μL PCR 反应物中含 50ng 基因组 DNA，10mmol/L Tris-HCl pH 9.0，50mmol/L KCl，1.8mmol/L $MgCl_2$，0.1%Triton X-100，200μmol dNTPs，每一引物 50ng，1Utaq DNA 聚合酶，PCR 反应在 PE-480 热循环仪上，按以下程序进行：预变性 94℃ 5min；然后 94℃ 1min，56℃ 1min，72℃ 2min，共 35 个循环；最后保温 72℃ 10min。PCR 产物在 1%琼脂糖凝胶上于 1×TAE 缓冲液中电泳检测。

1.6 DNA 提取与 Southern 杂交

DNA 提取按 MeCoueh 等的方法进行，5μgDNA 用于 *Kpn* Ⅰ或 *Hind*Ⅲ酶切，然后在 0.8%琼脂糖凝胶上电泳。按厂商（Amersham）提供的方法进行 Southern 杂交，所用探针是 1.4kb *Xa21* PCR 扩增片段。

1.7 抗性分析

转基因 T_0 和 T_1 植株栽培在温室中，在分蘖盛期，用剪叶法对已伸展的叶片接种 P6 或其他白叶枯菌株。接种物在 PSA 培养基上于 28℃培养 72h，调节浓度至 10^9 细胞/mL，接种后两周当病斑面积明显而稳定时，调查每株抗病情况。

2 结果与讨论

2.1 转化载体的构建

质粒 pC822 带有一个包含 *Xa21* 完整基因的 9.6kb *Kpn*I 片段。将 9.6kb *Kpn*I 片段切下 回收后重新插入到双元载体 pCAMBIA1301 的 *Kpn* I 位点。新构建的质粒在 *Xa21* 基因两侧分别有 HYG 和 GUS 选择标记，该质粒命名为 pCXK1301（图 1）。pCXK1301 可以直接用于水稻 基因枪转化或导入农杆菌本研究中用电激法将 pCXK1301 导入农杆菌菌株 EHA105，然后用于水稻转化。

2.2 水稻转化

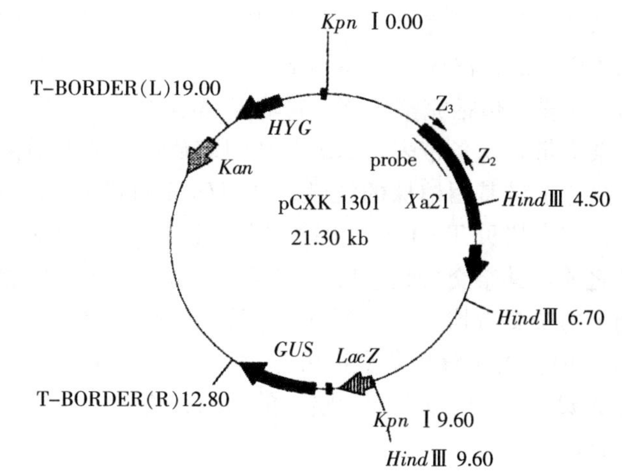

图1 转化载体 pCXK1301 的基本框架

T-BORDER（L）示 T-DNA 左边界，T-BORDER（R）示 T-DNA 右边界，*HYG* 为潮霉素磷酸转移酶基因，*GUS* 为 β-葡萄糖苷酸酶基因，*LacZ* 为 β-半乳糖苷酶 α 链基因，*Kan* 为细菌卡那霉素抗性基因

利用农杆菌介导的转化将 *Xa21* 转入明恢 63，珍汕 97B，盐恢 559，太湖粳 6 和中花 11 五个品种。经分子检测和抗性鉴定共获得 110 个独立转基因系。表 1 中列出已转化的我国主栽籼稻和粳稻品种及各品种的转化效率。虽然适当调整了不同品种的培养条件，但籼稻的转化效率仍明显低于粳稻。转化效率的差异也反应了受体品种遗传背景的不同。发现在农杆菌转化所获得的转基因植株中，很少出现形态变异和不育，而基因枪转化的转基因植株中却出现高比例的形态变异和不育（结果未示）。本研究转化的品种有明恢 63，籼稻杂交稻恢复系，所配制的杂交稻组合曾覆盖我国水稻种植面积的 20%；珍汕 97B 是籼稻杂交水稻中最常用的保持系。因此获得的所有转基因系具有重要实用价值。

2.3 分子鉴定

为了研究 *Xa21* 在受体基因组中的整合，对转基因植株进行 PCR 和 Southern 杂交分析。转基因植株的 PCR 分析见图 2。转基因植株 DNA 经引物 Z_2 和 Z_3 扩增后产生 1.4kb *Xa21* 特

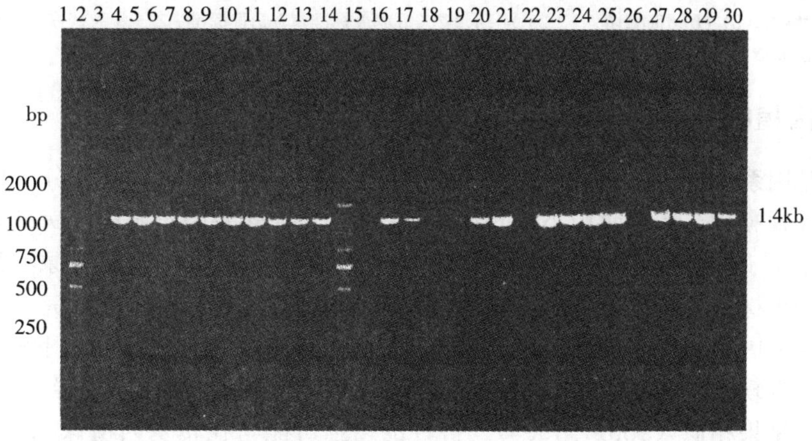

图2 转基因 To 植株的 PCR 分析

1、13 为 DL-2000DNA 标准；2 为明恢 63 非转基因对照；3～12 为明恢 63 转基因植株；14 为盐恢 559 非转基因对照；15、16 为盐恢 559 转基因植株；17 为珍汕 97B 非转基因对照；18～20 为珍汕 97B 转基因植株；21 为太湖粳 6 非转基因对照；22～25 为太湖粳 6 转基因植株；26 为中花 11 非转基因对照；27～30 为中花 11 转基因植株。

异扩增带，与设计完全一致，而非转基因对照植株中没有相应的扩增带。转基因植株 DNA 与 1.4kb *Xa21* 特异探针 Southern 杂交结果见图 3。所有测定的转基因植株都含有 9.6kb *Kpn* Ⅰ片段，包括完整的 *Xa21* 基因（图 3，a）。而非转基因对照植株只含 1 条约 4.0kb 的背景杂交带，这条带来自 *Xa21* 基因家族的感病同源拷贝。*Xa21* 基因在受体基因组的完整整合决定了转基因植株的抗性。当 *Hind* Ⅲ酶切的转基因植株 DNA 与 *Xa21* 探针杂交时，如果没有多拷贝 T-DNA 串联整合，根据 pCXK1301 的酶切图谱（图 1），除 4.0kb 背景杂交带之外，其杂交带数应反映 *Xa21* 在受体基因组中整合的拷贝数。除背景杂交带外，所有检测出的杂交带均大于 6.8kb，6.8kb 是根据 pCXK1301 酶切图谱预计的最小杂交片段长度（图 3，b）。由于背景杂交带可作为内对照，通过 Southern 分析，大体上可以估计在绝大多数转 *Xa21* 的水稻植株中所整合的 *Xa21* 基因，为单拷贝或双拷贝，本实验中没有发现多拷贝整合。

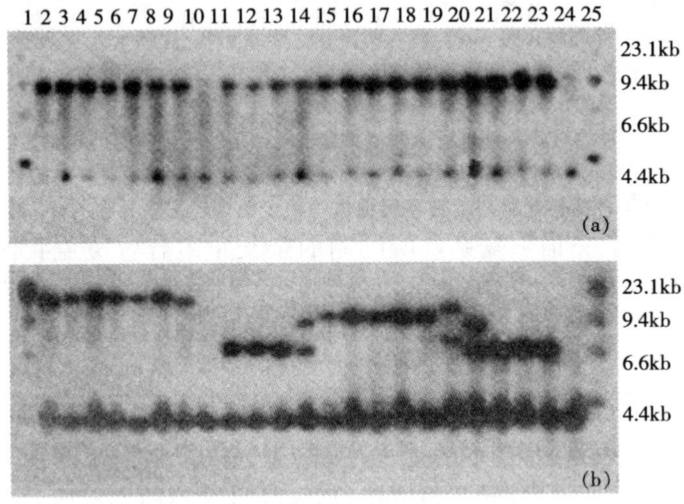

图 3　转基因 T_0 植株的 Southern 分析

(a) *Kpn* Ⅰ酶切；(b) *Hind*Ⅲ酶切。1、25 为 DL-2000DNA 标准/*Hind*Ⅲ标准；2～8 为盐恢 559 转基因植株；9 为盐恢 559 非转基因对照；10～23 为明恢 63 转基因植株；24 为明恢 63 非转基因对照

2.4　转基因植株的抗性分析

所有转基因系在分蘖盛期接种 *Xa21* 的白叶枯病专化小种 P6，所有分子检测阳性的转基因植株均表现高抗，病斑面积小于叶片面积的 5%。非转基因对照植株病斑面积为 30%～70%。另外用了 19 个包括 9 个菲律宾小种，3 个日本小种和 7 个中国病原型代表菌株，接种部分转基因植株分析其抗谱，结果显示，转基因植株对 19 个菌株表现完全抗性，没有发现被检测的转基因植株中有部分抗性（表 2），表明 *Xa21* 在不同遗传背景下仍然保持其广谱抗性。我们还发现转基因系的抗性略强于 *Xa21* 基因供体 IRBB21，与在台北 309 转基因植株中所观察的结果相似（Song，1995）。抗性的增强可能与在转基因植株 *Xa21* 基因的拷贝相对增加，而基因家族中其他同源拷贝相对减少有关。

表 2 抗性分析中采用的白叶枯病菌株

来源	小种/病原型	代表菌株	来源	小种/病原型	代表菌株
菲律宾	PI	PXO61	日本	TI	T7174
	P2	PXO86		T2	T7147
	P3	PXO79		T3	T7133
	P4	PXO71	中国	C1	HLJ72
	P5	PXO112		C2	HBI7
	P6	PXO99		C3	NX42
	P7	PXO145		C4	ZHE173
	P8	PXO280		C5	GDI358
	P9	PXO330		C6	LN57
				C7	JS49－6

2.5 *Xa21* 基因在转基因植株中的遗传

Xa21 基因在自交 T_1代中的遗传可在 DNA 和表型两个水平进行分析，表 3 列出其中 8 个独立转化体在自交 T_1代的分离情况。由 PCR 分析和接种分析揭示的抗感分离比均为 3∶1，与 Southern 杂交所揭示的 T_0转化体中 *Xa21* 单拷贝整合的结果是一致的。其他转基因系的分离方式，取决于 *Xa21* 基因的整合位点与拷贝。也进行了 T_1代转基因植株的抗谱测定，所有 PCR 阳性 T_1代植株对 19 个菌株都产生抗性，抗性表现与 T_0代无差异，没有观察到明显的基因沉默，表明 *Xa21* 基因的广谱抗性在新的遗传背景下是稳定遗传的。田间试验还表明转基因植株及其后代的其他性状均未见到可察觉的变异。

表 3 自交 T_1 代中转基因植株中的分离

转化体	转化体 *Xa21* 拷贝数	*Xa21* PCR 分析				抗性分析			
		植株总数	＋	－	＋/－比值	植株总数	R	S	R/S 比值
MH1	1	39	29	10	2.9	22	17	5	3.4
MH2	1	77	58	19	3.0	45	34	11	3.1
MH3	1	74	54	20	2.7				
MH4	1	34	25	9	2.8				
MH5	1	21	16	5	3.2				
MH6	1	38	29	9	3.2				
YHI	1	54	41	13	3.2	42	31	11	2.8
YH2	1	33	28	10	2.8	25	19	6	3.2

致谢 美国加利福尼亚州立大学 Davis 分校 P. Ronald、宋文源和王国梁博士的支持和帮助，澳大利亚堪培拉 CAMBIA 研究中心 R. A. Jefferson 教授提供农杆菌 EHA105 和质粒 pCAMBIA1301，中国水稻研究所钱前博士提供水稻品种；新加坡分子农业研究所尹中朝博士，中国科学院遗传研究所李良才教授和杨敏教授在水稻转化方面给予的建议和帮助。本实验室郑先武、刘国庆、李晶诏、郭官强和刘巧泉等同事在实验中给予的协助。

原载：中国科学（C 辑）SCIENCE IN CHINA（Series C)，30（2)：200～206.

Xa21 转基因杂交稻组合汕优 63 和 汕优 559 对白叶枯病的抗性

周永力[1]　章　琦[1*]　王春莲[1]　邢全党[1]
翟文学[2]　潘学彪[3]　朱立煌[2]
（1. 中国农业科学院作物育种栽培研究所；2. 中国科学院遗传研究所；3. 扬州大学）

Resistance of *Xa21* Transgenic Hybrid Rice Shanyou 63 and Shanyou 559 to Bacterial Blight

Abstract　Seventy-three F_1 population of *Xa21* transgenic hybrid rice Shanyou 63 and 13 F_1 population of Shanyou 559 derived from the *Xa21* transgenic homozygous restorer lines Minghui 63 and Yanhui 559 crossed with sterile line, Zhenshan 97 A common used were inoculated with 12 international differential races and seven phathotypes of bacterial blight pathogen in China to evaluate their resistance. The results indicated that the transgenic hybrid combinations maintained broad-spectrum resistance to bacterial blight. and the resistance level was as same as the donor *Xa21* showed.

Key words　gene transfer; hybrid rice; bacterial blight; resistance

摘要　用包括与 *Xa21* 相匹配的菌株在内的 12 个国际鉴别小种和中国 7 个病原型代表菌株，接种由珍汕 97A 分别与明恢 63、盐恢 559 的 *Xa21* 转基因单拷贝抗性纯合系杂交所获得的转基因杂交稻汕优 63（汕优 63 - *Xa21*）F_1群体 73 个，和转基因杂交稻汕优 559（汕优 559 - *Xa21*）F_1群体 13 个。测试转基因杂交稻对白叶枯病的抗性。结果表明这 2 个转基因杂交稻组合具有广谱抗性，其抗谱与 *Xa21* 基因供体 IRBB21 相同，对照非转基因杂交稻表现感病。

关键词　转基因；杂交稻；白叶枯病；抗性

水稻白叶枯病（*Xanthomonas oryzae* pv. *oryzae*，*Xoo*）是我国水稻生产中的重要病害之一。种植抗病品种是防治此病的有效措施。Song 等（1995）成功地克隆了 Khush 等（1990）从长药野生稻（*O. longistaminata*）中鉴定的广抗谱基因 *Xa21*。Tu 等（1998，2000）采用基因枪法将该基因导入籼稻 IR72 中，并进行了田间抗病性鉴定。1997 年，我们首次采用农杆菌介导，将 *Xa21* 导入应用最广的杂交水稻恢复系明恢 63 和新推广的恢复系盐恢 559。通过分子检测和接种鉴定选择单拷贝的高抗的 T_0代植株（翟文学等，2000）。对 T_1和 T_2代进行分子标记辅助选择和抗病表型鉴定，获得了一批高抗的明恢 63 和盐恢 559 转基因单拷贝抗性纯合系。于 1998 年配制了杂交稻。本研究采用包括与 *Xa21* 基因对应的鉴别小种 P6 和中国 7 个病原型代表菌株分析了转基因杂交稻的抗性，为利用基因工程加速改良杂交稻白叶枯病抗性提供资料。

1 材料与方法

1.1 转基因杂交稻的配置

将珍汕 97A 分别与明恢 63、盐恢 559 的 *Xa21* 转基因单拷贝抗性纯合系杂交，获得 73 个转基因杂交稻汕优 63（汕优 63 - *Xa21*）F_1群体和 13 个转基因杂交稻汕优 559（汕优 559 - *Xa21*）F_1群体。

1.2 菌株和接种

采用包括与 *Xa21* 相对应的鉴别小种在内的 12 个国际鉴别小种和中国 7 个病原型，接种和分析转基因杂交稻的抗性。菌株使用前，复壮保存于一20℃的原菌种，在鉴别品种上接种测试毒力稳定性，转数管斜面菌种，保存于 4℃冰箱供全年使用。接种体在 PSA 培养基上于 28℃培养 48h，配制成 1×10^9/mL 菌悬液接种。

采用人工剪叶法接种，在隔离网室/温室中进行。以对 P6 不携有白叶枯病抗性基因的品种 IR24 为感病对照。接种后 14～20d，当感病对照品种的病情发展趋于稳定时进行调查，每株调查 3～5 张叶片，以病斑占叶面积百分率为抗性反应参数，根据病斑占叶面积百分率划分抗感反应型（章琦等，1996）：≤5%为高抗（HR）；5%～10%为抗（R）；10%～15%为中抗（MR）；15%～30%为中感（MS）；≥30%为感（S）。

1.3 明恢 63、盐恢 559 与珍汕 97B 的抗性

用国际鉴别小种包括菲律宾小种 1～9、日本小种 1～3，及中国 7 个病原型，在植株分蘖初期和孕穗期对明恢 63、盐恢 559 与珍汕 97B 各接种 1 次，分析其抗性。

1.4 转基因杂交稻汕优 63 - *Xa21* 和汕优 559 - *Xa21* 对 P6 的抗性反应

全部转基因杂交稻汕优 63 - *Xa21* 和汕优 559 - *Xa21* F_1群体（各 15 株）于分蘖盛期接种菌株 P6（PXO99）。以 *Xa21* 的供体 IRBB21 和 IR24 分别为抗病和感病对照，同时接种基因受体汕优 63 和汕优 559。计算 2 个杂交稻组合所有植株的病斑面积百分率平均值，分析 *Xa21* 在杂交稻中的抗性表达效应。

1.5 转基因杂交稻的抗谱

随机选取汕优 63 - *Xa21* 和汕优 559 - *Xa21* 各 5 株，用上述全套菌株，按同株分蘖分组法在分蘖盛期接种，分析 *Xa21* 转基因杂交稻的抗谱及其抗性水平。

2 结果和分析

2.1 明恢 63、盐恢 559 与珍汕 97B 的抗性

结果显示，明恢 63 和珍汕 97B 从分蘖期到成株期对国际鉴别小种都呈感病反应，对中

国主体抗性基因 *Xa4* 的鉴别小种 P1 和中国 7 个病原型也均无抗性，表明明恢 63 和珍汕 97B 对上述菌株都无抗病功能。盐恢 559 对 P2、P3、P6 和 T3 感病，中感 C2、C3、C7、T2、P4、P5，中抗 C1、C4，C5、C6、T1、P1、P7 和 P8。

2.2 转基因杂交稻对 P6 的抗性反应

2 个转基因杂交稻组合的各植株对 P6 都表现抗病。汕优 63 - *Xa21* 和汕优 559 - *Xa21* 的病斑面积百分率的平均值分别为 8.7%±3.3%和 7.7%±3.0%，受体对照汕优 63 和汕优 559 分别为 49.9%±4.9%和 49.3%±5.6%，*Xa21* 基因供体 IRBB21 的病斑为 7.5%±2.6%。显然，转基因杂交稻抗白叶枯病，其抗性水平与基因供体 IRBB21 相近（图 1）。

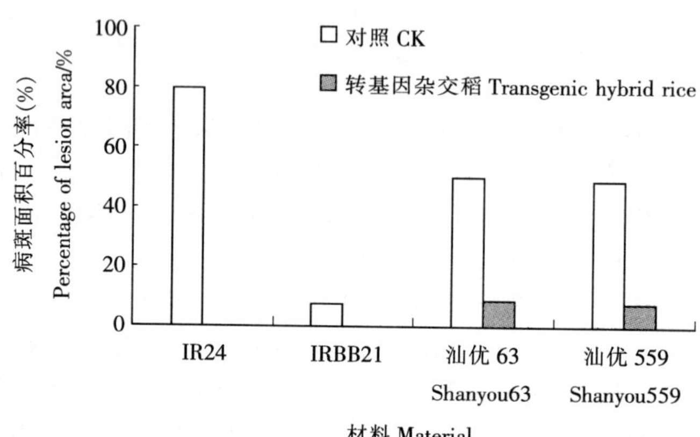

图 1　转基因杂交稻对菌株 PXO99（P6）的抗性水平

Fig. 1　Resistance level of transgenic hybrid rice to *Xoo* strain PXO99

2.3 转基因杂交稻抗谱

转基因杂交稻汕优 63 - *Xa21* 和汕优 559 - *Xa21* 和对照汕优 63 和汕优 559 对 12 个菌株的抗性反应列于表 1。汕优 63 - *Xa21* 和汕优 559 - *Xa21* 的抗谱与供体 IRBB21 相同，抗性强度相近。两个转基因杂交稻组合对国际鉴别小种和我国病原型都呈抗性反应，并且对大部分菌株都高度抗病，病斑面积小于 5%。

汕优 63 高感全部菌株，病斑面积范围达 30%～70%；汕优 559 的感病程度略低，病斑面积范围 20%～60%。

表 1　转基因杂交稻对 19 个白叶枯病菌株的抗性反应

Table 1　Reaction of transgenic hybrid rice to 19 strains of *Xoo*

菌株 (strain)	来源 (Origin)	小种/病原型 Race/pathotype	抗性反应（Reaction）				
			汕优 63 - *Xa21*	汕优 559 - *Xa21*	汕优 63	汕优 559	IRBB21
PX061（P1）	菲律宾（Philippine）	1	R	HR	S	S	HR
PX086（P2）	菲律宾	2	HR	R	S	S	HR
PX079（3）	菲律宾	3	HR	HR	S	S	HR
PX071（P4）	菲律宾	4	HR	HR	S	S	HR
PX0112（P5）	菲律宾	5	R	R	S	S	HR

（续）

菌株 (strain)	来源 (Origin)	小种/病原型 Race/pathotype	抗性反应（Reaction）				
			汕优 63 - *Xa21*	汕优 559 - *Xa21*	汕优 63	汕优 559	IRBB21
PX099 (P6)	菲律宾	6	MR - R	MR - R	S	S	R
PX0280 (P7)	菲律宾	7	HR	HR	S	S	HR
PX0145 (P8)	菲律宾	8	HR	HR	S	S	HR
PX087 (P1)	菲律宾	9	HR	HR	S	S	HR
T7174 (T1)	日本（Japan）	Ⅰ	R	HR	S	S	HR
T7147 (T2)	日本	Ⅱ	HR	HR	S	S	R
T7133 (T3)	日本	Ⅲ	MR - R	R	S	S	R
HLJ72 (C1)	中国（China）	Ⅰ	HR	HR	S	S	HR
HB17 (C2)	中国	Ⅱ	R	HR	S	S	R
NX42 (C3)	中国	Ⅲ	R	R	S	S	R
ZHE173 (C4)	中国	Ⅳ	R	R	S	S	HR
GD1358 (C5)	中国	Ⅴ	R	R	S	S	R
LN57 (C6)	中国	Ⅵ	R	R	S	S	HR
JS49 - 6 (C7)	中国	Ⅶ	R	R	S	S	HR

HR=高抗（Highly resistant）；R=抗（Resistant）；MR=中抗（Moderately resistant）；S=感（Susceptible）。

3 讨论

20 世纪 70 年代以来，我国杂交水稻的研究和利用取得了举世瞩目的成就。然而，我国杂交稻不育系基本不携带抗病基因，恢复系大多采用了一对不完全显性基因 *Xa4*（章琦，1995）。在品种选育过程中因未跟踪测定抗性而漏选抗性基因材料。李宣铿等（1993）曾测定全国水稻育种协作组的杂交稻新组合的白叶枯病抗性，44 个组合中仅有 2 个表现中抗。汕优 63 是栽培面积最广的杂交稻组合，1991 年我们在湖北等地调查时就发现汕优 63 对白叶枯病非常感病。改良杂交稻的白叶枯病抗性是迫在眉睫的工作。当前，*Xa21* 是国际命名的抗性基因中抗谱较广的一个显性基因，携有该基因的近等基因系 IRBB21 在育种中的抗性转育效应不够理想，限制了 *Xa21* 在常规育种中的应用。本研究表明，采用农杆菌介导法，将被克隆的 *Xa21* 基因导入主栽杂交水稻恢复系后，*Xa21* 转基因恢复系的纯合系的抗性特点，包括抗谱的广度、强度，都可达到基因供体，携有 *Xa21* 的品系 IRBB21 之水平。显示了通过基因工程，利用克隆的 *Xa21* 基因对于改良杂交稻的白叶枯病抗性的前景广阔。

原载：中国水稻科学，2002，16（1）：93～95.

Ⅱ 水稻白叶枯病抗性遗传研究

Part 2 Studies on Genetic Resistance to Bacterial Blight

水稻品种对菲律宾白叶枯病菌株 PXO61 的抗性遗传研究

章 琦[1] 林世成[1] G. S. Khush[2] T. W. Mew[2]

（1. 中国农业科学院作物育种栽培研究所；2. 国际水稻研究所）

Genetics of Resistance to Bacterial Blight (*Xanthomonas campestris* pv. *oryzae*) in Two Rice Cultivars

Abstract The inheritance and allelic relationships of the genes for resistance to the Philippine strain, PX061 of *Xanthomonas campestris* pv. *oryzae* in two rice cultivars, Shi-zu and-Yindunou, were studied. The study of F_1, F_2 and F_3 derived from crosses between resistant parents and the susceptible cultivar Kinmaze revealed that the resistance in Shi-zu and Yindunou is under monogenic control to strain PXO61 at all growth stages, but there are two genes controlling the strain at the booting stage in Yindunou. The resistance in the two cultivars is recessive. Allele tests with known genes for resistance indicated that the recessive gene expressing resistance is *xa5* which is independent to the two recessive resistant genes, *xa8* and *xa9*.

日本报道了水稻抗本国白叶枯病菌株的 4 个显性基因 *Xa1*、*Xa2*、*Xa3*（*Xa－w*）和 *Xa-kg*（Sakagushi，1967；Ezuku et al.，1975；Ogawa et al.，1978；Sidhu et al.，1978）国际水稻研究所（IRRI）的遗传研究表明，迄今籼稻品种对菲律宾菌株的抗性至少由 8 个位点上的基因控制（Librojo et al.，1976；Petpsist et al.，1977；Sidhu et al.，1976；1978；

Singh et al.，1983；Yoshimura et al.，1983；1984)。这些基因不能与日本报道的4个基因进行比较。因为它们都是用本地菌株鉴定的。这些基因的等位性测定也必须通过交换各自试验中曾采用的菌株和品种来进行。

国内也进行了大量抗性遗传研究（朱立宏等，1979；高山林等，1982；张端品等，1982；曾左葵等，1983)。其中都是用携有国外已知基因的抗源对我国菌株的抗性遗传分析。章琦从20世纪70年代进行水稻抗性研究工作中，发现两个品种多年来对北方白叶枯病和稻瘟病菌株表现稳定的广谱抗性，故在IRRI合作研究期间，用菲律宾菌株PXO61分析了这两个品种的抗性遗传及其与已知基因的关系。

1 材料与方法

1.1 供试材料

选用于本试验的8个品种（系）列于表1。在遗传研究中，以对菲律宾菌株PXO61没有抗性的日本品种金南风作为感病亲本。朝鲜粳稻品种矢租和籼稻品种印度诺为抗病亲本。孟加拉国品种DZ192及源出于DZ192的品系IR1545－339（携 *xa5*)，美国品种PI 231129（携 *xa8*）和两个老挝品种Sateng，Khao Lay Nhay（携 *xa9*)，分别用作抗性基因等位性测验。

1.2 抗性遗传分析的杂交配组及试材种植

研究矢租和印度诺的抗性遗传中，分别将这两个品种与金南风杂交。根据这两个组合的F_1对PXO61的抗性表现为隐性，便将矢租分别与携带隐性基因的IR1545－339、PI 231129、Sateng和印度诺杂交。将印度诺分别与PI 231129和Khao Lay Nhay杂交。分析这些组合的F_1、F_2以确定矢租和印度诺之间以及与这3个隐性抗病基因的等位关系。

将金南风/矢租和金南风/印度诺两个组合的P_1、P_2、F_1、F_2和F_3种植于网室。P_1、P_2和F_1各栽20株，F_2约400株，将F_2群体各株按序编号挂牌，从中随机选择64个F_3株系，每株系种植一行，每行20～22株。作等位性测试的各组合的P_1、P_2、F_1和F_2的种植同上。按IRRI常规田间管理，于接种前一周，每公顷加施30kg氮肥。在最高分蘖期（播后45d）和孕穗期接种两次。孕穗期接种剑叶和剑叶下数第1张叶片。因F_2群体的生育期有分离，孕穗期增加接种一次。用剪叶法接种。

1.3 采用的菌株及接种、调查

采用曾用于抗性基因鉴定的菲律宾白叶枯病小种1的代表菌株PXO61。经分离的单菌落，被回接到鉴别品种上，确认其毒性无异常变化。保存斜面菌种于4℃冰箱中，供整个试验应用。接种菌在胁本哲氏改良培养基上，在30℃温箱培养3d，菌液浓度10^9/mL。

接种两周后调查发病面积占叶面积百分率。亲本和F_1调查30张叶片，F_2和F_3每株调查3张叶片，F_3各株系调查抗、分离、感数，根据以下IRRI标准进行调查：

病　级	1（抗）	3（中抗）	5（中感）	7（感）	9（高感）
病斑面积/叶面积百分率（%）	≤5	<5～25	>25～40	>40～75	>75

全部试验在 IRRI 进行，1982 年 6 月至 1984 年 6 月，每年 1 个旱季、2 个湿季，共 6 季。

2 结果

2.1 亲本品种的抗性反应

用 PXO61 对 8 个品种（系）的接种鉴定结果如表 1 所示。在不同季节对所有品种最高分蘖期和孕穗期的多次鉴定，矢租表现稳定的全生育期抗性，对季节和生育期的感应范围很小。印度诺也表现全生育期抗性。感病品种金南风在不同生育期和旱、湿季的病情有起伏，但始终保持其感病性。携有 *xa5* 的 IR1545－339 和 DZ129 在全生育期都表现高抗。携有 *xa8* 的品种 PI 231129 在最高分蘖期为中感、孕穗期为中抗，到开花期才充分表现其抗性。携有 *xa9* 的品种 Sateng 和 Khao Lay Nhay 在最高分蘖期为中抗，孕穗期为高抗。

表 1　亲本品种对白叶枯病菌株 PXO61 的抗性反应[a]

品种（系）	来　源	抗性基因	病斑面积占叶面积百分率（%）	
			最高分蘖期	孕穗期
矢租	朝鲜	?	4.0±0.8（抗）	2.0±0.8（抗）
印度诺	南亚	?	4.5±1.5（抗）	2.4±0.5（抗）
金南风	日本	无	55.5±5.5（感）	45.5±4.8（感）
IR1545－339	IRRI	*xa5*	5.0±2.0（抗）	3.4±0.8（抗）
DZ192	孟加拉	*xa5*	3.0±0.7（抗）	2.0±0.5（抗）
PI 231129	美国	*xa8*	28.5±2.8（中感）	15.9±1.9（中抗）[b]
Sateng	老挝	*xa9*	22.0±4.5（中抗）	3.0±0.6（抗）
Khao Lay Nhay	老挝	*xa9*	13.0±3.6（中抗）	4.0±1.2（抗）

[a] 1982—1984 年测试的平均值。

[b] PI 231129 在开花期高抗。

2.2 抗病与感病品种杂交组合 F_1、F_2和 F_3的抗性表现

金南风/矢租和金南风/印度诺的 F_1 植株对 PXO61 在两个生育期均表现感病，矢租和印度诺对 PXO61 的抗性为隐性现象。

金南风/矢租的 F_2 群体对 PXO61 在最高分蘖期和孕穗期的病斑变异范围很广，分布曲线表现明显的双峰，峰槽位置出现在病斑面积 17%和 20%附近（图 1）。按 F_2分布曲线的峰槽位置作为划分抗、感标准，在两个生育期的抗病和感病分离比均符合 1 抗∶3 感，P 值均大于 0.05（表 2）。表明矢租对 PXO61 的全生育期抗性由一对隐性基因控制。

金南风/印度诺的 F_2 群体对 PXO61 在两个生育期的病斑分布曲线也表现明显的双峰，峰槽位置在病斑面积 20%～25%（图 1）。按峰槽位置划分，该组合的抗感分离比在最高分蘖期符合 1∶3，但在孕穗期则符合 7∶9，P 值均大于 0.05，表明印度诺对 PXO61 在最高分蘖期的抗性由一对隐性基因控制，而在孕穗期则由两对隐性基因控制（表 2）。

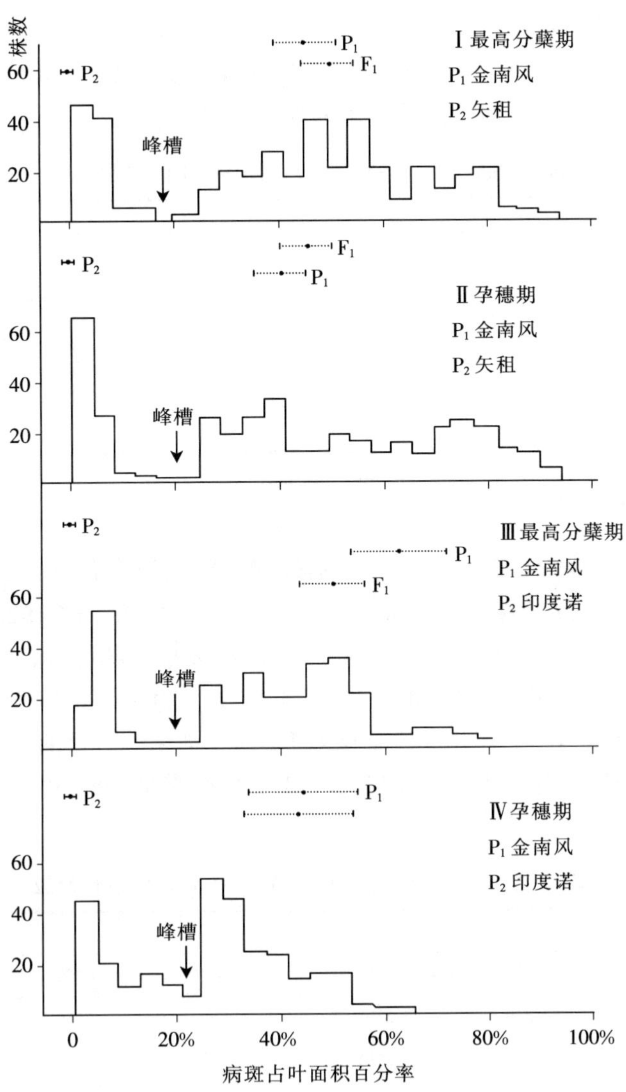

图 1　金南风/矢租和金南风/印度诺的亲本、F_1和F_2对菌株 PXO61 在两个生育期的病斑面积百分率的分布曲线

表 2　抗/感组合的 F_1、F_2 对白叶枯病菌株 PXO61 的抗性反应（%）

杂交组合	接种时期	F_1（%）	F_2 抗病株	感病株	总株数	χ^2（1∶3）	P	χ^2（7∶9）	P
金南风/矢租	最高分蘖期	49.0±5.0	96	278	374	0.089	0.75～0.95		
	孕穗期	45.0±4.6	101	266	367	1.24	0.25～0.50		
金南风/印度诺	最高分蘖期	50.0±6.4	80	192	272	2.82	0.05～0.10		
	孕穗期	43.0±9.6	111	179	290			3.152	0.05～0.10

从两个杂交组合 F_2群体各随机抽取 64 个 F_3株系的抗性反应结果见表 3。金南风/矢租 F_3株系在两个生育期抗性表现期望值比例为 1 抗∶2 分离∶1 感。金南风/印度诺 F_3在最高

分蘖期和孕穗期的抗性分别表现1∶2∶1和7∶8∶1。经χ^2测验，P值均大于0.05。根据对F_3株系的鉴定，证实了矢租由一隐性单基因控制全生育期抗性，印度诺由一对隐性基因控制全生育期抗性，其生殖生长期的抗性则由两对隐性基因控制。

表3　感/抗组合的F_3株系对PXO61的抗性反应

杂交组合	接种时期	抗	分离	感	总数	χ^2 (1∶2∶1)	P	χ^2 (7∶8∶1)	P
金南风/矢租	最高分蘖期	15	33	16	64	0.125	0.90～0.95		
	孕穗期	14	35	15	64	0.428	0.50～0.75		
金南风/印度诺	最高分蘖期	13	36	15	64	1.120	0.50～0.75		
	孕穗期	15	35	4	64			0.64	0.50～0.70

2.3　抗性基因等位性测验

矢租和印度诺与携有*xa5*、*xa8*和*xa9*基因的品种进行等位性测验的7个组合的F_1和F_2对菌株PXO61的抗性反应结果见表4。IR 1545－339/矢租、DZ192/印度诺和矢租/印度诺的F_1植株在两个生育期都抗病，这3个组合的F_2群体也全部抗病，没有分离出感病植株，表明矢租和印度诺的抗病基因与IR 1545－339和DZ192的*xa5*是等位的，

PI231129/矢租和Sateng/矢租的F_1单株对PXO61都感病。这两个组合的F_2群体分离的抗感植株为7∶9，表明矢租的隐性基因与PI231129携带的*xa8*和Sateng的*xa9*不等位，并独立遗传。

表4　等位性测交组合的F_1、F_2对菌株PXO61的抗性反应（%）

杂交组合	F_1		F_2								
	抗病株	感病株	抗病株	感病株	总数	χ^2 (7∶9)	P	χ^2 (4∶12)	P	χ^2 (37∶27)	P
IR1545/矢租	22	0	319	0	319						
印度诺/矢租	15	0	319	0	319						
Sateng/矢租	0	10	109	154	263	0.567	0.25～0.50				
PI 231129/矢租	0	11	143	174	317	0.238	0.50～0.75				
DZ192/印度诺[a]	10	0	233	0							
PI 231129/印度诺											
分蘖期	0	10	63	191	254			0.05	0.75～0.90		
PI 231129/印度诺											
孕穗期	0	10	144	104	248					0.006	0.90～0.95
Khao Lay Nhay/印度诺　分蘖期	0	10	89	144	233	2.912	0.05～0.10				
Khao Lay Nhay/印度诺　孕穗期	0	10	130	82	212					1.069	0.25～0.50

[a] 包括两个生育期。

因为在印度诺对PXO61的抗性遗传分析中，在不同生育期表现不同模式，上文已经表

明印度诺的抗病基因与 *xa5* 等位，此外，还有另一对隐性基因。因此等位性测验也观察了两个生育期的反应。PI231129/印度诺的 F_1 单株在全生育期都表现感病，但其 F_2 群体在最高分蘖期的分离比不符合预期的两对独立隐性基因分离比（7 抗∶9 感），感病株偏多，其原因是由于 PI231129 这个亲本在最高分蘖期对 PXO61 表现中感（表 1），而不是抗，因此这一组合 F_2 的分离比应为 4 抗∶12 感；在孕穗期的抗性分离比为 37 抗∶27 感（$\chi^2=0.006$）；Khao Lay Nhay/印度诺的 F_1 单株也全部感病，其 F_2 群体在最高分蘖期分离比符合两对独立隐性基因的 7 抗∶9 感（$\chi^2=2.91$）。在孕穗期符合 3 对独立基因分离比 37 抗∶27 感（$\chi^2=1.069$）。表明印度诺对 PXO61 菌株在最高分蘖期和孕穗期的抗性基因与 PI 231129 和 Khao Lay Nhay 的抗性基因不等位，并独立遗传。

3　讨论

3.1　水稻品种对白叶枯病的抗性，按生育期可分为全生育期抗性和成株期抗性。本研究所用的 8 个亲本，除 PI 231129 外，其他品种不论在哪一个生育期鉴定的结果都表现抗病。特别是矢租、印度诺、DZ192、IR1545－339，两期的抗性反应相当一致，显示了全生育期抗性的特征，此结果与 IRRI 已经确定的 *xa5* 基因是全生育期抗性相一致。PI 231129 是成株抗性品种，它携有隐性抗性基因 *xa8*，在最高分蘖期表现感病，至孕穗期才表现抗病。分析这类品种的抗病基因型要比较谨慎，很容易被不同生育期的表现型所迷惑，而得出错误结论。这可能涉及生育期与抗性关系的实质性问题。由此可见，对某些品种来说，所谓由显性基因或隐性基因控制抗性并不是绝对的，需要联系生育期来考虑。

3.2　本研究所用的亲本矢租和印度诺在最高分蘖期都含有一对隐性基因 *xa5*，当它们分别与对 PXO61 不具抗病基因的金南风杂交后，其 F_1、F_2、F_3的抗性都表现了典型的孟德尔单基因遗传规律。但当矢租与含有隐性抗病基因 *xa8* 或 *xa9* 的亲本杂交，以及金南风/印度诺 F_2群体在孕穗期鉴定抗性时，由于不同抗性基因之间存在互补作用（Complementary effect），便出现 7 抗∶9 感（2 对基因）和 37 抗∶27 感（3 对基因）的分离比。这说明只有当起互补作用的显性基因（2 个或 3 个）同时存在时表现感病。亦即当任何一对隐性基因为纯合时，不论显性基因存在与否，植株均表现抗病。但印度诺与具有成株抗性的品种 PI231129 杂交 F_2在最高分蘖期鉴定的抗感分离比不是 7∶9 而是 4∶12（表 4）。因为在最高分蘖期鉴定，纯合的 *xa8*、*xa8* 植株不是抗而是感，如表 1 所示。而两对隐性基因为纯合的个体（*xa5 xa8*，*xa5 xa8*）表现的是抗而不是感。用对白叶枯病菌具成株期抗性的亲本进行抗性遗传研究，在后代中出现如此异常的抗感分离比及其解释是否确切，还有待进一步验证。

3.3　矢租不但是全生育期抗性的品种，而且抗性稳定、广谱。1984—1985 年我们再次用国内很多菌株测定都表现高抗，但对少数菌株表现成株抗性，并高度感染江苏省农科院提供的菌株浙 173。矢租是粳稻育种抗白叶枯的良好抗源，不过，其株、穗型有较大的缺点。经进一步观察确定其专化抗性后，有希望成为我因白叶枯病的一个鉴别品种。

3.4　著名的美国春小麦品种 Thatcher 被应用以来，经 25 年以上还保持高抗秆锈（*Puccinia graminis* Pers. f. sp. *tritici*）某些小种的特性，因为该品种兼具苗期抗性基因 *Sr* 和成株抗性基因（Nazareno et al.，1981）。印度诺在不同生育期也可能有不同的抗性基因控制，其中一对是 *xa5*，控制其全生育期抗性，印度诺的另一对隐性基因可能只控制其成株期的抗性，是否为新基因，有必要进一步测试。这一假设在印度诺与 Khao Lay Nhay 的等位性测验中得到了证实。该组合分蘖期的 F_2 群体出现 7 抗∶9 感的两对独立基因分离比，在孕穗期则表现 37 抗∶27 感的 3 对基因分离比。该品种高抗白叶枯病，兼抗稻瘟病，也是一个较好的抗源。

3.5 水稻对白叶枯病的抗、感划分标准在国内外是一致的（方中达等，1978；Kauffman et al.，1973）。Kauffman 建议的 5 级与我国采用的 1～5 级的调查标准，都将抗感界限定在病斑长度占叶片长度的 1/4 左右。按此标准目测分级简便易行，但不足的是剪叶接种后的病斑多半从中脉延伸，形成上宽下窄的剑状斑，因此，将病斑长度为依据的误差较大。另外水稻的株型、叶型的差异很大，长叶型和短叶型品种的剑叶长度差异悬殊，尤其在抗性遗传分析中，要同时测量病斑的长度和面积，相当费时。IRRI 病理学家 T. W. Mew 在原来的调查基础上，提出调查病斑面积占叶面积百分率。本研究即采用的这一调查方法，结果显示，F_2 群体病斑面积占叶面积百分率的分布曲线的双峰之峰槽位置，一般都在发病面积的 20%左右。此法简单易行，比较准确，所反映的病情程度与国内普遍采用的方法也基本一致。

原载：中国农业科学，1986（5）：19～23.

水稻品种对三个菌株的抗性遗传分析

章　琦　施爱农　王春莲　阙更生

（中国农业科学院作物育种栽培研究所）

Genetics of Resistance to Bacterial Blight (*Xanthomonas oryzae* pv. *oryzae*) in Five Rice Cultivars

Abstract The inheritance and allelic relationship of genes resistant to three strains of *Xanthomonas oryzae* pv. *oryzae* in five rice cultivars Pei-Zao15, Bang-Zhu-Mang, Nan-Geng 15, Zhong-Bai 4, and Qing-Hua-Ai 6 were studied. The analysis of F_1, F_2 and B_1 F_1 populations from the crosses of the five cultivars with susceptible cuitivar, She-Nong 1033. It was concluded that resistance in Pei-Zao 15 and Bang-Zhu-Mang is under a single dominant gene controlling the 3 strains at booting stage. Both Nan-Geng 15 and Zhong-Bai 4 have two linked dominant genes with crossover value 30% operating in the 3 strains at booting stage. Qing-Hua-Ai 6 is under a single incomplete dominant gene controlling the 3 strains at all growth stages. Allellic tests with known genes for resistance indicate that the dominant gene expressing resistance in adult, Pei-Zao 15 and Bang-Zhu-Mang is *Xa3*. One of the two dominant genes in Nan-Geng 15 and Zhong-Bai 4 is *Xa3* too. Qing-Hua-Ai 6 has an allelic gene at the *Xa4* locus.

Key words bacterial blight of rice; resistance; allelic test; gene

摘要　研究了5个水稻品种邳早15、蚌珠芒、南粳15、中百4号、青华矮6号对3个水稻白叶枯病菌株的抗性遗传和抗性基因等位关系。这5个品种分别与感病品种沈农1033杂交。对各组合的F_1、F_2和B_1 F_1群体的抗性测试表明，邳早15和蚌珠芒由一对显性基因控制其抗性。南粳15和中百4号均由两对连锁的显性基因控制，交换值为30%。等位性测验表明，邳早15和蚌珠芒的一对显性基因与*Xa3*等位，南粳15和中百4号也均有一对显性基因与*Xa3*等位；青华矮6号的一对不完全显性基因与*Xa4*是等位的。

关键词　水稻白叶枯病；抗性；等位测定；基因

日本和国际水稻研究所（IRRI）曾各自用当地菌株研究和命名了12个水稻白叶枯病抗

性基因（Horino et al.，1980；Ogawa et al.，1988；Petpisit et al.，1977）。近年双方合作，用菲律宾和日本的多个菌株重新研究了早期命名的抗性基因并有所订正。最近整理的 14 个抗病基因是 *Xa1*、*Xa1*^{h}、*Xa2*、*Xa3*、*Xa4*、*xa5*、*Xa7*、*xa8*、*Xa10*、*Xa11*、*Xa12*、*Xa12*^{h}、*xa13* 和 *Xa14*。以前命名的 3 个成株抗性基因 *Xa6*、*xa9* 和 *Xa4*^{b}被订正为 *Xa3*。

国内水稻白叶枯病的抗性遗传模式和转育效应研究的报道较多。只有一篇关于基因等位性测定和基因分析的报道（章琦等，1986）。本研究分析了“成抗”和“全抗”水稻品种对 2～3 个白叶枯病菌株的抗性遗传模式，并测验这些品种与已知基因的等位关系，了解抗源的抗性基因类型，为抗病育种提供信息。

1　材料和方法

1.1　供试的品种和菌株

本试验采用的 10 个品种列于表 1。以高感品种沈农 1033 作为感病亲本。抗病品种邳早 15、蚌珠芒、南粳 15 和中百 4 号是广抗谱的成株抗性粳稻，青华矮 6 号是全生育期抗病籼稻品种。IR20（*Xa4*）、早生爱国 3 号（*Xa3*）、Zenith（*Xa6*）和 DV 85（*xa5*、*Xa7*）用作基因等位性测试品种。Cas209（*Xa10*）对我国所测试的菌株都没有抗性功能，故未采用。

采用 3 个菌株，菌株 P1 和 T 1 曾用来测定上述已知基因。HB17 为我国北方一个稳定的代表菌株。测定由 IRRI 命名的基因时，用菌株 P1 和 HB17，测定日本命名的基因时则用 T1 和 HB17。

1.2　杂交组合的配制

在遗传研究中，5 个抗病品种分别与沈农 1033 杂交，5 个组合的 F_1对 3 个菌株呈显性抗性现象，遂分别与沈农 1033 回交获 B_1F_1种子。5 个抗病品种又分别与 4 个显性基因代表品种杂交，分析这些等位性测试组合的 F_1和 F_2群体的抗性反应，以确定这些品种所携有的抗性基因与已知基因的等位关系。

1.3　田间种植设计

供试抗病品种的各套待测试材，包括感×抗组合的两个亲本 P1、P2、F_1、F_2、B_1F_1和与已知基因进行等位性测试的 P 1、P2、F_1和 F_2群体都安排在同期测定。保证外界条件一致。试材种植于隔离严密的网室。P1、P2 和 F_1各栽 10 株，$F_2$200～300 株，B_1F_1 50～100 株，植株生长正常。

1.4　接种与调查

采用（Ogawa，1986）多菌株接种和“单株分蘖分组”法，即按所采用的菌株数，用不同色泽的塑料带将各稻株的分蘖进行分组。不同颜色表示接种不同的菌株，以测定各个体对不同菌株的实际反应。

在接种前半个月，将低温保存的冻干菌种在 PSA 斜面培养基上复壮，然后转移数管斜面菌种保存于 4℃冰箱，供整个生长季节使用。接种菌液的菌龄 2d，浓度约 10^{9}/mL。

全生育期抗病的试材在最高分蘖期和孕穗期人工剪叶接种各世代植株的完全展开叶片。

成株抗性品种组合则在孕穗期接种。接种后 20d 左右调查单株对各菌株的抗性反应。亲本和 F_1 各调查 10 株（30 张叶片）。

F_2 和 B_1 F_1 群体单株的各个处理，均调查 3 张叶片。用病斑面积占叶片面积百分率作为抗性表现参数。以 F_2、B_1F_1 群体的病斑百分率分布所呈现的峰槽作为划分抗、感界限。抗、感分界值为病斑面积占叶面积的 20%左右。

全部试验在 1985—1988 年进行，包括 1986 年冬季在海南和 IRRI 共 6 季。

2 结果和分析

2.1 抗性遗传模式分析

2.1.1 亲本品种的抗性表现 10 个品种对 3 个菌株的抗性表现如表 1。感病亲本沈农 1033 在全生育期对 P1、T1 和 HB17 三个菌株都高度感病。邳早 15、蚌珠芒、南粳 15、中百 4 号、早生爱国 3 号和 Zenith 在成株期高抗 3 个菌株。青华矮 6 号、IR20 和 DV85 在全生育期对 3 个菌株都表现十分稳定的抗性。

2.1.2 抗病品种与感病品种杂交组合 F_1、F_2、B_1F_1 的抗性表现 沈农 1033 分别同邳早 15、蚌珠芒、南粳 15 和中百 4 号杂交的 F_1 植株对 3 个菌株在成株期均表现高抗，表明上述 4 个品种对 3 个菌株的成株抗性受完全显性基因控制。沈农 1033/青华矮 6 号 F_1 植株全生育期对 3 个菌株的抗病面积占叶面积百分率 12%～14%，为中等抗性，表明其全生育期抗性功能受不完全显性基因控制（表 2）。

5 个感×抗杂交组合的 F_1、F_2 和 B_1 F_1 群体对 3 个菌株的病情分布频率见图 1、2。沈农 1033/邳早 15 和沈农 1033/蚌珠芒 F_2 群体的病斑变异分布范因很广，都形成一个清晰的抗感双峰分布，峰槽出现在病斑面积 20%左右。沈农 1033/邳早 15 和沈农 1033/蚌珠芒 F_2 植株对 P1、HB17 和 T1 三个菌株的反应数值分别为：171 RRR∶64 SSS（$\chi^2=0.626$）和 156 RRR：48 SSS（$\chi^2=0.235$），符合 3 抗∶1 感的一对显性分离比例（表 3）。这两个组合的 B_1 F_1 植株对 3 个菌株的抗感分离比分别为 53 RRR∶44 SSS（$\chi^2=0.835$）和 46 RRR∶51 SSS（$\chi^2=0.258$），符合 1 抗∶1 感的期望比例（表 4）。据 χ^2 测验，验证了邳早 15 和蚌珠芒都由一对显性基因控制其成株抗性。

表 1 亲本对 3 个水稻白叶枯病菌株的抗性反应*

Table 1 Reaction of parents to 3 strains of *X. o.* pv. *oryzae*

品种 Cultivar	类型 Type	基因 Gene	病斑占叶片面积百分率（%）(Percentage of lesion area，%)					
			最高分蘖期（Max. tillering stage）			孕穗期（Booting stage）		
			P1	HB17	T1	P1	HB17	T1
沈农 1033	粳	无	55±5.1	60±5.3	50±6.3	48±6.4	70±7.0	37±5.6
邳早 15	粳		25±2.1	25±3.5	20±3.8	5±0	5±0	5±0
蚌珠芒	粳		25±2.1	25±3.7	25±1.5	5±0	5±0	5±0
青华矮 6 号	籼		5±0	5±1.1	1±1.2	5±0	5±0	5±0
南粳 15	粳		30±3.1	35±0	35±1.3	3±0	3±0	3±0

（续）

品种 Cultivar	类型 Type	基因 Gene	病斑占叶片面积百分率（%）(Percentage of lesion area, %) 最高分蘖期（Max. tillering stage）			孕穗期（Booting stage）		
			P1	HB17	T1	P1	HB17	T1
中百 4 号	粳		35±3.1	30±1.5	30±1.5	8±0	7±1.1	8±3.0
早生爱国 3 号	粳	*Xa3*	30±3.5	40±1.2	30±5.2	3±0	5±0	3±0
IR20	籼	*Xa4*	5±0	5±0	3±0	3±0	3±0	
Zenith	粳	*Xa3*	30±3.1	30±3.4	35±1.2	3±0	3±0	
DV85	爪哇	*xa5*，*Xa7*	3±0	3±0	3±0	3±0	3±0	

* 接种 20d 后调查 10 个单株的 30 张叶片病斑面积百分率平均值

Average percentage of lesion areas in 30 inoculated leaves from 10 plants at 20 days after inoculation

表 2 各组合 F_1 植株孕穗期对 2～3 个水稻白叶枯病菌株的抗性反应*

Table 2 Reaction of F_1 plants from each cross to 2～3 strains of *X. o.* pv. *oryzae* at booting stage

组合 Cross	菌株 Strain	抗性反应 Reaction	组合 Cross	菌株 Strain	抗性反应 Reaction
沈农 1033/邳早 15	P1	5±0	蚌珠芒/Zenith	P1	5±0
	HB17	5±0		HB17	6±1.3
	T1	5±0			
沈农 1033/蚌珠芒	P1	9±3.9	蚌珠芒/DV85	P1	5±0
	HB17	11±3.3		HB17	5±0
	T1	9±3.2			
沈农 1033/青华矮 6 号	P1	14±3.2	青华矮 6 号/IR20	P1	5±0
	HB17	12±6.7		HB17	5±0
	T1	12±3.0			
沈农 1033/南粳 15	P1	5±0	南粳 15/早生爱国 3 号	T1	3±0
	HB17	5±0		HB17	3±0
	T1	5±0			
沈农 1033/中百 4 号	P1	5±0	南粳 15/IR20	P1	5±0
	HB17	5±0		HB17	5±0
	T1	5±0			
邳早 15/早生爱国 3 号	T1	5±0	南粳 15/Zenith	P1	5±0
	HB17	5±0		HB17	5±0
邳早 15/IR20	P1	6±1.9	南粳 15/DV85	P1	5±0
	HB17	8±1.4		HB17	5±0
邳早 15/Zenith	P1	3±0	中百 4 号/早生爱国 3 号	T1	3±0
	HB17	3±0		HB17	3±0
邳早 15/DV85	P1	6±1.5	中百 4 号/IR20	P1	7±2
	HB17	5±1.5		HB17	7±2
蚌珠芒/早生爱国 3 号	P1	5±0	中百 4 号/Zenith	P1	5±0
	HB17	5±0		HB17	5±0
蚌珠芒/IR20	T1	5±0	中百 4 号/DV85	P1	7±2
	HB17	5±0		HB17	7±2

* 接种 20d 后调查各组合 F_1 植株（10～15 株），各处理 2～3 张叶片的病斑面积百分率平均值。

Average percentage of lesion areas in inoculated leaves from 10～15 plants at 20 days after inoculation.

表 3 感/抗组合 F_1、F_2 植株孕穗期对 3 个水稻白叶枯病菌株的抗性表现

Table 3 Reaction of F_1, F_2 plants from susceptible parent crossed with the 5 resistant cultivars to 3 strains of *X. o.* pv. *oryzae* at booting stage

组合 Cross	F_1 对 3 个菌株* 的抗性反应 F_1 reaction to 3 strains	F_2对菌株 P1、HB17、T1 的抗性反应 F_2 reaction to strains P1，HB17，T1					
		RRR	SSS	Total	Exp.	χ^2	*P*
沈农 1033/邳早 15	RRR	171	64	235	3∶1	0.626	0.25～0.50
沈农 1033/蚌珠芒	RRR	156	48	204	3∶1	0.235	0.50～0.75
沈农 1033/青华矮 6 号	RRR	167	53	220	3∶1	0.097	0.75～0.90
沈农 1033/南粳 15	RRR	218	36	254	14.04∶1.96	0.874	0.25～0.50
沈农 1033/中百 4 号	RRR	176	21	197	14.04∶1.96	0.463	0.25～0.50

* R=抗，S=感，第 1、2、3 个字母分别代表对菌株 P1、HB17、T1 的抗性反应。

R=Resistant，S=Susceptible，The first letter stands for reaction to strain P1，2nd to HB17，3th to T1.

沈农 1033/青华矮 6 号 F_2群体对 3 个菌株的病情变异分布范围也很广，分布曲线的峰槽明显（图 2）。F_2群体抗、感分离比数值为 167 RRR∶53 SSS，符合 3 抗∶1 感的分离比（χ^2=0.097），表明青华矮 6 号由一对显性基因控制全生育期抗性（表 3）。该组合的 B_1F_1 植株全生育期对 3 个菌株的抗感分离比为 50 RRR∶39 SSS（χ^2=1.360），符合 1 抗∶1 感的期望比例，证实了上述结果（表 4）。

沈农 1033/南粳 15 和沈农 1033/中百 4 号的 F_2群体在成株期对 P1、HB17 和 T1 三个菌株的病情变异呈明显偏态分布（图 2）。其抗、感分离比分别为 218 RRR∶36 SSS 和 176 RRR∶21 SSS，均符合交换值为 30%的两对连锁抗性基因分离比 14.04 抗∶1.96 感（χ^2=0.874 和 χ^2=0.463）。两个组合的 B_1F_1植株抗感比分别为 36 RRR∶12 SSS（χ^2=2.100）和 47 RRR∶15 SSS（χ^2=3.183），符合 2.6 抗∶1.4 感的期望比例，验证了南粳 15 和中百 4 号对 3 个菌株的抗性由交换值为 30%的两对连锁显性基因控制（表 4）。

表 4 感/抗组合 B_1F_1 植株孕穗期对 3 个菌株的抗性表现*

Table 4 Reaction of B_1F_1 plants from 5 back-crosses to 3 strains of *X. o.* pv. *oryzae* at booting stage

组合 Cross	B_1F_1 对菌株 P1、HB17、T1 的抗性反应 Reaction of B_1F_1 to strains P1，HB17，T1					
	RRR	SSS	Total	Exp.	χ^2	*P*
沈农 1033/邳早 15//沈农 1033	53	44	97	1∶1	0.835	0.25～0.50
沈农 1033/蚌珠芒//沈农 1033	46	51	97	1∶1	0.258	0.50～0.75
沈农 1033/青华矮 6 号//沈农 1033	50	39	89	1∶1	1.360	0.10～0.25
沈农 1033/南粳 15//沈农 1033	36	12	48	2.6∶1.4	2.110	0.10～0.25
沈农 1033/中百 4 号//沈农 1033	47	15	62	2.6∶1.4	3.183	0.00～0.10

* R=抗，S=感，第 1、2、3 个字母分别代表对 菌株 P1、HB17、T1 的抗性反应。

R=Resistant，S=Susceptible，The first letter stands for reaction to strain P1，2nd to HB17，3th to T1.

表 5　抗/抗组合 F_1，F_2 植株孕穗期对菌株 HB17、P1 或 T1 的抗性表现

Table 5　Reaction of F_1，F_2 plants from crosses between resistant cultivars to 3 strains of *X. o.* pv. *oryzae* at booting stage

组合 Cross	F_1 对 2 个菌株*的抗性反应 F_1 reaction to 2 strains	F_2对菌株 HB17 和 P1 或 T1 的抗性反应 F_2 reaction to 2 strains HB17，P1 or T1					
		RR	SS	Total	Exp.	χ^2	*P*
邳旱 15/早生爱国 3 号	RR	248	0	248			
邳旱 15/IR20	RR	177	9	186	15∶1	0.632	0.25～0.50
邳旱 15/Zenith	RR	298	0	298			
邳旱 15/DV85	RR	198	25	223	61∶3	21.240	<0.01
蚌珠芒/早生爱国 3 号	RR	238	0	190			
蚌珠芒/IR20	RR	164	26	190	15∶1	17.921	<0.01
蚌珠芒/Zenith	RR	175	0	175			
蚌珠芒/DV85	RR	127	8	135	61∶3	0.463	0.25～0.50
青华矮 6 号/IR20	RR	219	0	219			
南粳 15/早生爱国 3 号	RR	300					
南粳 15/IR20	RR	108	6	114	62.04∶1.96	1.860	0.10～0.25
南粳 15/Zenith	RR	175	0	175			
南粳 15/DV85	RR	196	2	198	250.12∶5.88	1.460	0.10～0.25
中百 4 号/早生爱国 3 号	RR	150	0	150			
中百 4 号/IR20	RR	229	9	238	62.04∶1.96	0.414	0.50～0.75
中百 4 号/Zenith	RR	264	0	264			
中百 4 号/DV85	RR	214	9	223	250.12∶5.88	3.005	0.05～0.10

*　R=抗，S=感，第 1、2 个字母分别代表对 HB17、P1 或 T1 两个菌株的抗性反应。

R=Resistant，S=Susceptible，The first，2nd letters stand for reaction to strains HB17 and P1，or T1.

2.2　抗性基因等位性测验

5 个品种同已知显性基因品种之间进行等位测验的 17 个组合的 F_1、F_2对 T1 和 HB17 两个菌株的抗性反应结果列入表 5。邳旱 15、蚌珠芒、南粳 15 和中百 4 号分别与早生爱国 3 号和 Zenith 的 8 个杂交组合 F_1植株在成株期对 2 个菌株均表现抗病，其 F_2植株也全部抗病，没有分离出感病株，表明邳旱 15、蚌珠芒、南粳 15 和中百 4 号对 T1 和 HB17 的成株抗性基因同早生爱国 3 号的 *Xa3*、Zenith 的 *Xa6*（已订正为 *Xa3*）是等位的。邳旱 15/IR20 的 F_1抗病，F_2抗感分离比为 177RR∶9SS（χ^2=0.632），符合 15R∶1S 的期望分离比，表明邳旱 15 的一对显性基因与 IR20 的 *Xa4* 不等位，且独立遗传。邳旱 15/DV85 的 F_1抗病，F_2的抗感分离比为 198RR∶25SS，不符合 61R∶3S 的两对显性、一对隐性基因的独立遗传分离比，需进一步研究再作分析。蚌珠芒/IR20 的 F_1抗 P1 和 HB17 两个菌株，F_2抗感分离比为 164RR∶26SS，也不符合 15R∶1S 的两对独立遗传基因分离比，需进一步测定。蚌珠芒/DV85 的 F_1抗 HB17 和 P1，F_2植株的抗感分离比为 127RR∶8SS，符合 61R∶3S（χ^2=0.463）两对显性一对隐性基因的独立遗传分离比，表明蚌珠芒的一对显性基因与 DV85 的

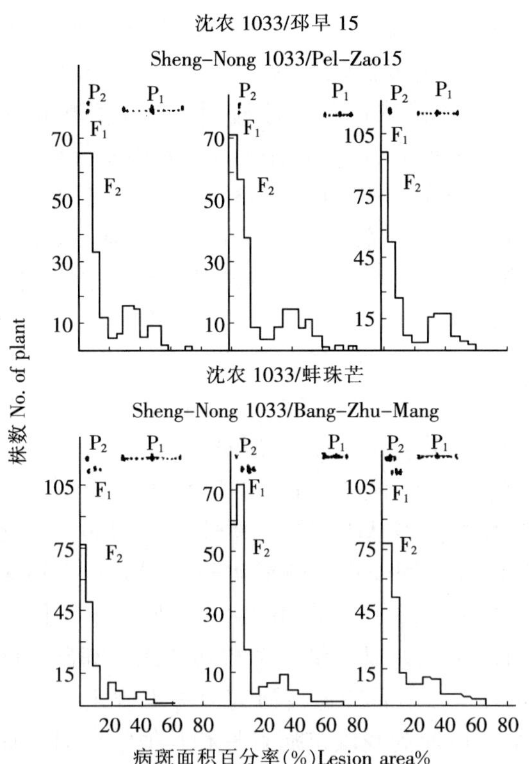

图 1 沈农 1033/邳早 15 和沈农 1033/蚌珠芒双亲、F_1 和 F_2 对 P1，HB17 和 T1 三个菌株的抗性反应分布

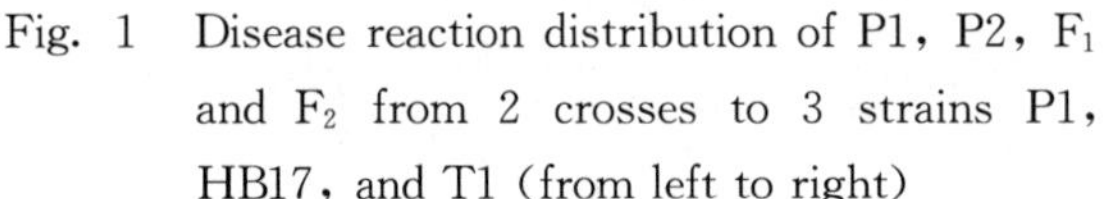

Fig. 1 Disease reaction distribution of P1，P2，F_1 and F_2 from 2 crosses to 3 strains P1，HB17，and T1 (from left to right)

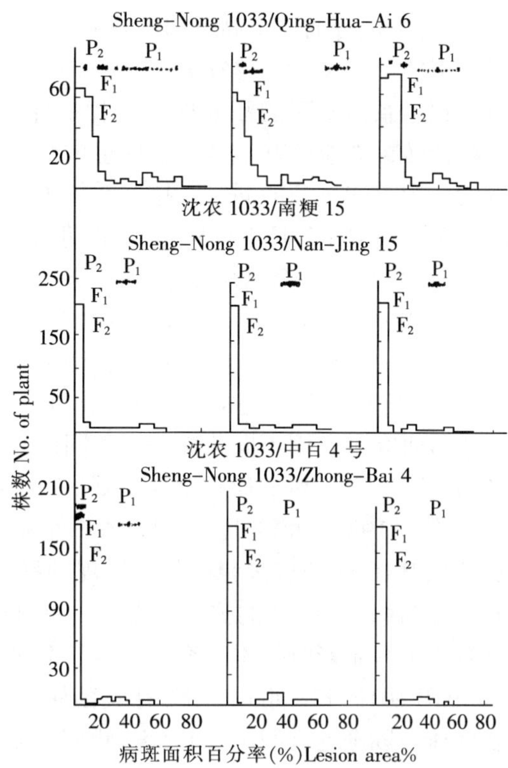

图 2 沈农 1033/青华矮、沈农 1033/南粳 15 和沈农 1033/中百 4 号 3 个组合对 P1，HB17 和 T1 三个菌株的抗性反应分布

Fig. 2 Disease reaction distribution of P1，P2，F_1 and F_2 from 3 crosses to 3 strains P1，HB17，and T1 (from left to right)

xa5、*Xa7* 不在同一位点，并独立遗传。南粳 15/IR20 和中百 4 号/IR20 两个组合的 F_1 均抗 HB17，F_2 群体抗感分离比分别为 108RR∶6SS（62.04R∶1.96S，$\chi^2=1.860$）和 229RR∶9SS（62.04R∶1.96S，$\chi^2=0.414$），均符合交换值为 30%的两对连锁显性基因和一对独立显性基因的分离比例。表明南粳 15 和中百 4 号的两对连锁显性基因与 IR20 的 *Xa4* 不等位。南粳 15/DV85 和中百 4 号/DV85 两个组合的 F_1 抗 HB17 和 P1，其 F_2 群体抗感分离比分别为 196RR∶2SS（250.12R∶5.88S，$\chi^2=1.460$）和 214RR∶9SS（250.12R∶5.88S，$\chi^2=3.005$），均符合交换值为 30%的两对连锁显性基因、一对独立隐性基因的分离比例。表明南粳 15 和中百 4 号的两对连锁显性基因同 DV85 的 *xa5*、*Xa7* 均非等位。青华矮 6 号/IR20 组合的 F_2 群体对 HB17 和 P1 全部抗病，没有分离出感病株，表明青华矮的全抗基因同 IR20 的 *Xa4* 是等位的。

3 讨论

3.1 在 4 个成株抗性品种中，邳早 15、蚌珠芒、南粳 15 和中百 4 号的抗性均受一对显性

基因 *Xa3* 的控制。追溯其系谱渊源，邳早 15 由南粳 15 辐射育成；中百 4 号是从喜峰×南粳 15 组合中选育的一个高抗白叶枯病、高产、优质的粳稻新品种。这些品种均有南粳 15 的亲缘。南粳 15（农垦 57/农垦 51）有旭的渊源。均有旭或爱国的血缘（章琦等，1988）。这些抗源包括系谱不明的江苏地方品种蚌珠芒，该品种在国内已被广泛应用，由于其抗谱很广、抗性转育效应较好，是一个理想抗源。本研究还表明，我国粳稻抗源的遗传变异较窄，在继续应用上述抗源的同时，挖掘新抗源，如野生稻和我国其他地方品种，实为当务之急。

3.2 日本和 IRRI 近年合作报道早期命名的 *Xa6* 和 *Xa3* 系同一个位点（Ogawa et al.，1986）。我们在另一个试验中验证了这一结果（章琦等，1991）。在本试验等位性测定中，邳早 15、蚌珠芒、南粳 15 和中百 4 号分别与早生爱国 3 号和 Zenith 杂交组合的 F_2 植株全部抗病，无感病植株，进一步间接地验证了 *Xa*6 与 *Xa*3 系同一位点。

3.3 植株抗病性反应是寄主—病原菌在一定环境条件下相互作用的产物。在水稻白叶枯病抗性遗传研究中，用不同菌株分析相同的品种，得出不同结果的报道屡见不鲜。Petpisit 等（1977）分析 IR26 对菲律宾菌株 PXO25 的抗性遗传受一对不完全显性单基因 *Xa4* 控制。而 Horino 等（1980）报道 IR26 对日本 5 个小种的抗性则受多基因控制。章琦等（1986）用菲律宾菌株 PXO61 研究矢租的抗性发现其携有一对隐性基因 *xa5*；周泰初等（1982）报道矢租对江苏菌株 KS1－3 的抗性，则由一对显性基因控制。除了所采用的菌株不同外，不同遗传背景产生一定影响。本研究采用当年命名基因时的菌株，以提供一个相同的比较基础，趁在与 IRRI 协作研究期间，用菲律宾菌株进行测定工作，今后的遗传研究应考虑确定我国的代表菌株，逐渐建立我国一套统一研究与命名方案，以便为我国抗病育种和其他学科研究提供科学基础。

3.4 本研究中有关南粳 15 和中百 4 号的遗传模式，还需要进一步作基因定位研究，方能验证文中初步的分析。

原载：中国水稻科学，1990（10）：1～8.

两个水稻白叶枯病成株抗性基因 *Xa6* 与 *Xa3* 的等位性分析*

章　琦[1]　施爱农[1]　王春莲[1]　阙更生[1]　T. W. Mew[2]

（1. 中国农业科学院作物育种栽培研究所；2. 国际水稻研究所）

Allelism Analysis of Two Adult Plant Resistance Genes *Xa6* and *Xa3*

Abstract　The inheritance and allelic relationships of the genes for resistance to 4 races of *Xanthomonas oryzae* pv. *oryzae* in the Philippines in two cultivars, Wase Aikoku 3 and Zenith with adult-plant resistance were studied. The study of F_1, F_2 and B_1F_1 populations of crosses between resistant parents and the susceptible cultivar, Shennong 1033 revealed that the resistance in Wase Aikoku 3 and Zenith is controlled by a dominant gene at the booting stage. Allelism tests between Zenith and Wase Aikoku 3 directly and indirectly by making crosses, Wase Aikoku 3/Zenith, Peizao15/Zenith, Bangzhu-Mang/Wase Aikoku 3, and Bangzhu-Mang/Zenith showed no susceptible plants segregated in those F_2 populations from the 5 crosses. It was indicated that the adult-plant resistance of the 4 cultivars were controlled by a same dominant gene. It is also confirmed that Zenith has allelic gene at the *Xa3* locus for resistance.

Key words　inheritance of resistance; adult-plant resistance; bacterial blight

摘要　研究两个品种早生爱国 3 号和 Zenith 对菲律宾水稻白叶枯病 4 个小种 1～4、日本小种 1 和我国菌株的成株抗性遗传和抗性基因等位关系。这两个抗病品种与感病品种沈农 1033 杂交所得的 F_1、F_2和 B_1F_1群体的成株抗性反应，表明它们均携有一对显性基因。将早生爱国 3 号同 Zenith 杂交进行直接等位性测验，并将邳早 15 和蚌珠芒分别与 Zenith 和早生爱国 3 号杂交进行间接等位测验。这 5 个组合的 F_2群体都没有分离出感病植株，表明这 4 个品种的成株抗性都由同一基因控制，也验证了 Zenith 在 *Xa3* 位点上具有一等位基因。

关键词　抗病性遗传，成株抗性，水稻白叶枯病

* 全部试验于 1986—1988 年在海南、北京和 IRRI 进行。

Ezuka 等（1975）报道抗日本水稻白叶枯病Ⅰ、Ⅱ和Ⅲ菌群的早生爱国 3 号抗性品种群是由一对显性成株抗性基因支配的，被命名为 *Xa3*。Sidhu 等（1978）研究了 Zenith 等品种对菲律宾小种 1 的抗性由一对显性成株抗性基因支配，当时将该基因命名为 *Xa6*。近年，Ogawa 等（1989）的研究，表明 *Xa6* 与 *Xa3* 是在同一个位点上。

Zhang 等（1985）曾较系统地研究了水稻白叶枯病成株抗性特性，注意到 *Xa6* 和 *Xa3* 由感病转为抗病的表达方式有所不同。国内有关这两个基因的代表品种 Zenith 和早生爱国 3 号的抗性遗传研究的结果也不一致（周毓珍等，1985；张端品等，1982；赵新平，1985）。鉴于我国粳稻资源中白叶枯病的抗源比较丰富，其中多数具有成株抗性。进一步研究和验证 *Xa6* 与 *Xa3* 两个抗病基因的遗传表现及其等位关系，是我国成株抗性品种抗性遗传研究工作的一部分，为水稻抗病育种提供确切的依据。为此，在与国际水稻所（IRRI）合作期间，用菲律宾多个菌株和我国菌株分别在两地研究了 *Xa6* 和 *Xa3* 的抗性遗传表现和两者的等位关系。

1　材料和方法

1.1　供试品种

本研究分析的 4 个抗病品种为 Zenith、早生爱国 3 号、蚌珠芒和邳早 15。后两个粳稻品种在另一个试验中表明它们都有一对显性成抗基因与 *Xa3* 在一个位点，用来作为间接等位测定品种。沈农 1033 对菲律宾 6 个白叶枯病小种和我国 72 个菌株高度感染，用作感病亲本。

B_1F_1种子种植于 IRRI 网室，用 4 个菲律宾小种测定；间接等位性测验在北京隔离网室进行，用中国菌株 HB17、P1（菲律宾小种 1）和 T1（日本小种 1）接种。

1.2　杂交组合的配制及种植

沈农 1033 分别与 Zenith 和早生爱国 3 号杂交。根据这两个组合 F_1植株对供试病菌小种呈显性抗性遗传，遂分别与沈农 1033 回交，获得两个组合的 B_1F_1种子。用这套试材进行抗性遗传模式分析。

配制 5 个组合用作基因等位性测验。早生爱国 3 号直接与 Zenith 杂交，将邳早 15 和蚌株芒分别同 Zenith 和早生爱国 3 号杂交，获得这些组合的 F_1和 F_2种子。

用以分析抗性遗传模式和直接等位性测验的材料，包括亲本、每个组合的 F_1和 B_1F_1种子，种在 IRRI 网室，用 4 个菲律宾小种测定。间接等位性测定的试材种植在北京隔离网室内，用中国菌株 HB17、P1（菲菲律宾小种 1）和 T1（日本小种 1）接种测定。

1.3　接种鉴定和调查记载

多菌株接种，采用同株分蘖分组法。按所用菌株数用异色带将稻株的分蘖分组，不同色泽表示接种不同菌株，以测定各个体对不同菌株的实际抗性反应。在植株孕穗期接种，鉴于 F_2群体的生育期分离的差异较大，再进行两次成株期接种。

接种前半个月将保存的冻干菌种转移于 PSA 斜面培养基上复壮，然后转移数管斜面菌种保存于冰箱中，供整个生长期使用。接种菌液的菌龄 48h，浓度约 10^9/mL。接种后 14～

20d 调查发病面积占叶面积百分率。各亲本和 F_1 调查 30 张叶片，F_2 和 B_1F_1 每株每个菌株接种 2～3 张叶片。参照以下标准逐株调查全部接种试材：

抗性反应	抗	中抗	中感	感
病斑面积占叶面积百分率（%）	<5	>5～20	>20～35	>35

2 试验结果

2.1 亲本对 6 个水稻白叶枯病菌株的抗性反应

在预备试验中经多次接种鉴定的 5 个亲本品种于 1987—1988 年再次同其他试材同期测定（表 1）。Zenith 和早生爱国 3 号在成株期高抗 6 个菌株，病斑面积都小于 5%。邳早 15 和蚌珠芒对所测的 3 个菌株也表现抗病，表明抗病亲本的抗性是稳定的。感病亲本沈农 1033 的发病很重，与抗病亲本的病情差距很大，便于分析鉴定后代材料。是个适宜的感病亲本。

表 1 亲本对 6 个水稻白叶枯病菌株的抗性反应

Table 1 Resistance reaction of parents to 6 strains of *X. o.* pv. *oryzae*

品种 Cultivar	基因 Gene	病斑占叶片面积百分率（%）(Percentage of lesion area, %) 菌株（Strain）					
		PXO61	PXO86	PXO79	PXO71	HB17	T1
沈农 1033	无	50±6.4	35±3.5	36±3.6	70±5.3	70±5.3	37±5.5
早生爱国 3 号	*Xa3*	5±1.2	3±0	3±0	3±0	3±0	3±0
Zenith	*Xa6*?	3±0	3±0	3±0	3±0	5±0	
邳早 15	*Xa3*	5±0				5±0	5±0
蚌珠芒	*Xa3*	5±0				5±0	5±0

2.2 早生爱国 3 号和 Zenith 的抗性遗传模式

沈农 1033/早生爱国 3 号和沈农 1033/Zenith 两个组合的 F_1 植株对菲律宾 4 个小种在成株期表现高抗。F_2 群体的病斑面积百分率的变异范围很广，分别为 3%～70%和 3%～60%。沈农 1033/早生爱国 3 号和沈农 1033/Zenith F_2 群体单株对菲律宾 4 个小种的抗感分离比分别为 178RRRR（依次抗菲律宾小种 1～4）∶68SSSS（依次感小种 1～4）和 244RRRR∶71SSSS，χ^2 测验符合 3 抗∶1 感的一对基因分离的比例（χ^2 值分别是 0.916 和 1.017，*P* 值都在 0.25～0.75 之间）。表明这两个品种对菲律宾 4 个小种的成株抗性均受一对显性基因支配。这两个组合的 B_1F_1 群体对 4 个小种的抗感比均为 1∶1，验证了上述假定（表 2）。

表 2 感/抗组合的双亲、F_1、F_2 及 B_1F_1 群体在孕穗期对菲律宾 4 个小种的抗性表现[a]

Table 2 Resistance reaction of parents, F_1, F_2, and B_1F_1 populations from cross of S/R to Philippine 4 races of *X. o.* pv. *oryzae* at booting stage

组合 Cross	对 4 个小种抗性反应 Reaction to 4 races			F_2对 4 个小种的反应型 F_2 reaction to 4 races					B_1F_1 对 4 个小种的反应型[c] B_1F_1 reaction to 4 races				
	P1	P2	F_1	RRRR	SSSS	Total	χ^2 3∶1	*P*	RRRR	SSSS	Total	χ^2 3∶1	*P*
沈农 1033/早生爱国 3 号	SSSS	RRRR	RRRR	178	68	246	0.916	0.25～0.50	40	47	87	0.563	0.25～0.50
沈农 1033/Zenith	SSSS	RRRR	RRRR	244	71	315	1.017	0.25～0.50	43	37	80	0.450	0.50～0.75

[a] R=抗（Resistant），S=感（Susceptible）；

依次对菲律宾 4 个小种的抗性反应（The combined four capitals stand for order of reaction to the Philippine four races, 1～4 from left to right, respectively）。

2.3 *Xa6* 与 *Xa3* 的等位性测验

早生爱国 3 号/Zenith F_1植株孕穗期抗 4 个菲律宾小种，F_2群体中没有分离出感病株（表 3）。表明 Zenith 携有的主效显性单基因与早生爱国 3 号的 *Xa3* 在同一位点。在间接等位性测定中，邳早 15/早生爱国 3 号、邳早 15/Zenith、蚌珠芒/早生爱国 3 号和蚌珠芒/Zenith 这 4 个组合的 F_1植株对 3 个菌株 HB17、P1 和日本小种 T1 表现高抗，各组合 F_2群体全都表现抗病，没有分离出感病植株，并与两个亲本的抗病反应类似。表明这 4 个抗病品种的抗性基因都是等位的，与 Ogawa 等（1993）的初报结果相同，即 *Xa6* 与 *Xa3* 是等位基因。

表 3 抗病品种间杂交组合的亲本、F_1、F_2群体对水稻白叶枯病几个菌株的抗性反应[a]

Table 3 Reaction of parents, F_1, and F_2 populations from cross between resistant cultivars to strains of *X. o.* pv. *oryzae*

组合 Cross	抗性反应 Reaction to 4 races			F_2的抗性反应型 F_2 reaction pattern					
	P1	P2	F_1	抗 RRRR	感 SSSS	总数 Total	抗 RR	感 SS	总数 Total
早生爱国 3 号/Zenith	[a]RRRR	RRRR	RRRR	282	0	282			
邳早 15/早生爱国 3 号	[b]RR	RR	RR				248	0	240
邳早 15/Zenith	[c]RR	RR	RR				298	0	298
蚌珠芒/早生爱国 3 号	[b]RR	RR	RR				238	0	298
蚌珠芒/Zenith	[c]RR	RR	RR				175	0	175

[a] 依次对菲律宾 4 个小种的抗性反应（The combined four capitals stand for order of reaction to the Philippine four races, 1～4 from left to right, respectively）. R 抗（Resistant），S 感（Susceptible）；

[b] 对 HB17 和日本小种 T1 的抗性反应（The capitals stand for order of reaction to 2 strains, HB17 and T1）；

[c] 对 HB17 和菲律宾小种 P1 的抗性反应（The capitals stand for order of reaction to 2 strains, HB17 and P1）。

3 讨论

Kaku 和 Kimura（1978）曾研究携有 *Xa3* 基因的抗性品种群对日本和菲律宾水稻白叶枯病小种的抗性反应相似，表现抗谱广、侵染型均呈褐色病变反应。Ogawa 等（1989）又用菲律宾 4 个小种测定早生爱国 3 号、中国 45、Java 14 和 Zenith 等品种，发现它们也表现类似的病斑侵染型。便研究其抗病基因的等位关系。采用了携有 *Xa3* 抗病基因的品种中国 45 和 Java14 来测试 Zenith。在两个测交组合的 F_2群体分别对日本小种 3 和菲律宾 4 个小种的抗感分离反应中，均没有出现感病株，提出了 Zenith 所携有的早期被命名为 *Xa6* 基因与 *Xa3* 系同一个位点。在我们早期的研究中，也注意到早生爱国 3 号和 Zenith 都具有广谱抗性，高抗我国多数菌株。但是它们之间对菲律宾 4 个小种的成株抗性由感病转为抗病的方式略微不同，Zenith“转抗”快速而清晰，早生爱国 3 号逐渐转抗。因此本试验用携有 *Xa3* 的代表品种早生爱国 3 号直接与 Zenith 比较，同时用两个已知携有 *Xa3* 的中国粳稻品种作间接测验，也证明了早生爱国 3 号与 Zenith 不仅对菲律宾 4 个小种、并对日本小种 1 和我国华北菌株 HB17 的抗性均由同一对成株抗性基因支配。它们的成株抗性由感转抗的方式与抗性基因的功能之间似乎没有联系。

由于水稻白叶枯病寄主与病原菌之间有明显的特异性互作，也有非特异性互作。因此同一个品种与毒力不同菌株的互作结果不同或者相同，其遗传分析结果也可能有不同或相同的结果。赵新平等（1985）报道早生爱国 3 号和 Zenith 对湖北菌株江陵 691 的成株抗性各由一对显性抗病基因支配，这两对抗病基因彼此是独立遗传的。张端品等（1982）报道 Zenith 对江陵 691 菌株的抗性由一显性抗性基因支配。而周毓珍等报道（1985）Zenith 对江苏 KS－8－4 等菌株的抗性则是由一隐性抗病基因支配。为了进行系统比较，本试验采用的几个菌株中包括了当年研究和命名 *Xa3* 和 *Xa6* 的菲律宾菌株（P1）和日本小种 1 的代表菌株（T1）进行遗传模式和基因等位性分析，在相同基础下的结果应该具有比较性。

原载：作物学报，1991，17（3）：233～237.

籼稻品种对三个白叶枯病菌株的抗性遗传分析

施爱农　章　琦　王春莲　白建法　杨文才

（中国农业科学院作物育种栽培研究所）

Genetic Analysis of Resistance to Bacterial Blight in Indica Rice

Abstract The inheritance of resistance to 3 strains of *Xanthomonas oryzae* pv. *oryzae*, HB17, PXO61 (P1) and T7174 (T1) in five indica rice cultivars. Erjiu-Feng, Sanhuang-Zhan 2, Chuan-Zhi 5, Yang-Dao 1 and 83077 was studied. The analysis of F_1, F_2 and B_1F_1 populations derived from crosses between the five resistant cultivars and susceptible cultivar, Shen-Nong 1033 or Jin-Gang 30, revealed that one dominant gene was responsible for the cultivars resistance to HB17 and PXO61, and the resistance to T7174 was controlled when an additional independently two dominant genes, among the two genes, one gene was responsible for all the three strains, and the other one just controlled the resistance to the strain T7174.

Key words bacterial blight of rice; genetics of resistance

摘要　分析和研究了 5 个全生育期抗水稻白叶枯病的籼稻品种二九丰、三黄占 2 号、川植 5 号、扬稻 1 号和 83007 对 3 个水稻白叶枯病菌株 HB17、PXO61 和 T7174 的抗性遗传。用感病品种沈农 1033 或金刚 30 分别与这 5 个抗病品种杂交。根据各组合 F_1、F_1和 B_1F_1群体植株对 3 个菌株的抗性反应和分离世代抗感分离比例的分析结果表明，这 5 个抗病品种对菌株 HB17 和 PXO61 的抗性均由一对显性基因控制，对菌株 T7174 的抗性均由二对独立遗传的显性基因控制，其中一对控制对全部 3 个菌株的抗性，另一对只控制对菌株 T7174 的抗性。

关键词　水稻白叶枯病，抗病性遗传

在国内，水稻白叶枯病抗性遗传研究以往大多采用当地的不同菌株接种，以致所得结果不尽相同，且互相之间也难以进行比较（余功新等 1990；曾左葵等 1983；章琦 1991）。为了逐渐形成我国白叶枯病抗性遗传研究的标准化程序，同时便于与国际系统鉴定基因研究相比较，本研究除采用我国北方代表菌株 HB17 外，还用国际鉴定抗性基因时所用的菲律宾小种 1 代表菌

株 PXO61（P1）和日本小种 1 代表菌株 T7174（T1），来研究 5 个籼稻抗病品种的抗性遗传行为。按章琦引用 Ogawa（1986）的同株分蘖分组法测定，探讨它们对各个菌株的抗性遗传表现和分析其对各个菌株的抗性是由同一基因还是由不同基因控制，以及基因之间的关系。

1 材料和方法

1.1 供试品种

以高感白叶枯病的品种沈农 1033 和金刚 30 为感病亲本。所研究的籼稻抗病品种为二九丰、三黄占 2 号、川植 5 号、扬稻 1 号和 83007（表 1）。以感病品种×抗病品种方式配制杂交组合。将各组合 F_1 自交获得 F_2 代种子，同时用各杂交组合 F_1 与感病亲本回交获得 B_1F_1 种子。在田间同时种植各杂交组合的 P1、P2、F_1、F_2、B_1F_1 植株。P1、P2 和 F_1 各种植 10 株，种植 200～300 株 F_2，种植 50～100 株 B_1F_1。常规管理，植株生长正常。

1.2 接种鉴定

采用我国北方白叶枯病代表菌株 HB17、国际抗性基因鉴定菌株 PXO61（P1）和 T7174（T1）。用多菌株同株分蘖分组法接种，即按所用菌株数用不同色泽塑料带将水稻单株的分蘖分组，不同色泽表示接种不同菌株。分别在分蘖期和孕穗期剪叶接种两次。菌液浓度为 10^9/mL。接种后 14～20d 调查植株对 3 个菌株的病斑面积占叶面积的百分率，以病斑面积占叶面积的 20%为抗感分界值。

2 结果和分析

2.1 亲本品种的抗性表现

7 个品种对 3 个供试菌株 HB17、P1 和 T1 的抗性反应结果列于表 1。沈农 1033、金刚 30 在全生育期对 3 个菌株表现高感，作为感病亲本是合适的。二九丰、三黄占 2 号、川植 5 号、扬稻 1 号和 83007 在全生育期对 3 个菌株都表现高抗。

表 1 亲本对 3 个白叶枯病菌株的抗性反应*

品 种	类型	最高分蘖期			孕 穗 期		
		HB17	P1	T1	HB17	P1	T1
沈农 1033	粳	82.8±3.63	76.0±16.58	59.0±9.62	75.5±16.58	75.8±3.63	41.2±6.42
金刚 30	籼	58.0±9.20	68.0±9.20	43.3±6.23	55.0±4.63	55.0±4.63	32.5±8.66
二九丰	籼	5±0	5±0	3±0	5±0	5±0	3±0
三黄占 2 号	籼	3±0	3±0	3±0	3±0	3±0	3±0
川植 5 号	籼	5±0	5±0	3±0	5±0	5±0	3±0
扬稻 1 号	籼	5±0	5±0	3±0	5±0	5±0	3±0
83007	籼	3.6±0.97	3.6±0.97	3±0	3±0	3±0	3±0

* 病斑占叶面积百分率。

2.2 抗性遗传

抗、感杂交各组合 F_1、F_2和 B_1F_1群体对 HB17、PXO61 和 T7174 菌株的抗性反应及分离世代植株的抗性分离结果列于表 2、表 3。

表 2 感/抗各组合 F_1 和 F_2 群体对 HB17、PXO61 和 T7174 的抗性表现

组 合	* F_1对 3 个菌株的抗性反应	F_2对 P1、HB17、T1 三个菌株的抗性反应									
		RRR	SSR	SSS	总数	HB17 或 P1		T1		HB17、P1、T1	
						χ_c^2 (3∶1)	P	χ_c^2 (15∶1)	P	χ_c^2 (12∶3∶1)	P
沈农 1033/二九丰	RRR	156	31	10	197	1.626	0.10～0.25	0.285	0.50～0.75	1.474	0.25～0.50
沈农 1033/三黄占 2 号	RRR	214	35	19	268	3.109	0.05～0.10	0.195	0.50～0.75	5.290	0.05～0.10
沈农 1033/川植 5 号	RRR	209	42	22	273	0.275	0.50～0.75	1.231	0.10～0.25	2.697	0.25～0.50
金刚 30/扬稻 1 号	RRR	165	28	15	208	1.853	0.10～0.25	0.185	0.50～0.75	3.463	0.10～0.25
沈农 1033/83007	RRR	165	47	17	229	0.910	0.25～0.50	0.357	0.50～0.75	0.857	0.50～0.75

R=抗，S=感，第 1、2、3 个字母分别代表对 P1、HB17、T1 三个菌株的抗性反应。

5 个组合的 F_1对 3 个菌株均高抗，抗性呈显性遗传。F_2和 B_1F_1群体植株对菌株 HB17 和 P1 的抗感株数分离比经 t 测验均符合一对显性基因支配抗性时的期望比例 3 抗∶1 感和 1 抗∶1 感；而对菌株 T1 的抗感株数分离均为 15 抗∶1 感和 3 抗∶1 感，符合由两对显性基因支配抗性，且这两对基因呈独立遗传时的期望比例。二九丰、三黄占 2 号、川植 5 号、83007 和扬稻 1 号 5 个品种对菌株 HB17、P1 的抗性均由一对显性基因控制，而对 T1 的抗性均由二对独立遗传的显性基因控制，这两对基因之间存在重叠作用。

表 3 感/抗各组合回交一代群体对 3 个菌株的抗性表现

组 合	B_1F_2对 P1、HB17、T1 三个菌株的抗性反应									
	RRR	SSR	SSS	总数	HB17 或 P1		T1		HB17、P1、T1	
					χ_c^2 (1∶1)	P	χ_c^2 (3∶1)	P	χ_c^2 (2∶1∶1)	P
沈农 1033/二九丰//沈农 1033	35	14	13	62	0.790	0.20～0.50	0.344	0.50～0.75	0.718	0.50～0.75
沈农 1033/三黄占 2 号//沈农 1033	23	14	17	54	0.907	0.25～0.50	0.889	0.25～0.50	1.120	0.50～0.75
沈农 1033/川植 5 号//沈农 1033	50	23	36	109	0.587	0.25～0.50	3.330	0.05～0.10	3.307	0.10～0.25
金刚 30/扬稻 1 号//金刚 30	21	11	6	38	0.237	0.50～0.75	1.263	0.25～0.50	1.171	0.25～0.50
沈农 1033/83007//沈农 1033	27	9	13	49	0.327	0.50～0.75	0.007	0.90～0.95	0.786	0.50～0.75

R=抗，S=感，第 1、2、3 个字母分别代表对 P1、HB17、T1 三个菌株的抗性反应。

为了探索品种对多个毒性不同菌株的抗性是由不同基因或同一基因控制的问题，根据用多个菌株接种 F_1和 B_1F_1群体植株的单株分蘖分组接种后的抗性反应型，来进行遗传剖析。

下文以二九丰为例进行分析（表 4）。假设二九丰对菌株 HB17 和 P1 的显性抗性基因分别为 A1 和 A2，对菌株 T1 的两个独立遗传显性抗性基因为 A3 和 B，则：

1）A1 与 A2 之间的关系有 3 种可能：A1 与 A2 独立遗传；A1 与 A2 连锁；A1 与 A2 等同（A1＝A2）。从理论上分析，F_1 和 B_1F_1 群体植株对 HB17 和 P1 两个菌株的抗感反应型 RR、RS、SR 和 SS 表现如下 3 种理论比例：

①当 A1 与 A2 独立遗传时，F_2 为 9RR∶3RS∶3SR∶1SS，B_1F_1 为 1RR∶1RS∶1SR∶1SS。

②当 A1 与 A2 连锁，交换值为 h 时，F_2 为（$3-2h+h^2$）RR∶（$2h-h^2$）RS∶（$2h-h^2$）SR∶（$1-2h+h^2$）SS，B_1F_1 为（$1-h$）RR∶hRS∶hSR∶（$1-h$）SS。

③当 A 1＝A2 时，F_2 为 3 RR∶1 SS，B_1F_1 为 1 RR∶1 SS。而沈农 1033/二九丰和 F_2 和 B_1F_1 群体植株对 HB17 和 P1 两个菌株的抗感反应型实际观察数值为 156 RR∶41 SS 和 35 RR∶27 SS，均符合 A1＝A2 时的抗感反应型理论比例。

2）A1 与 A3 或 B 之间的关系也有 3 种可能性：A1 与 A3 和 B 独立遗传；A1 与 A3 或 B 连锁；A1 与 A3 等同（A1＝A3）。从理论上分析，F_1 和 B_1F_1 群体植株对 HB17 和 T1 两菌株的抗感反应型 RR、RS、SR 和 SS 有以下 3 种理论比例：

①当 A1 与 A3 和 B 独立遗传时，F_2 为 45RR∶3RS∶15SR∶1SS，B_1F_1 为 3RR∶1RS∶3SR∶1SS。

②当 A1 与 A3 或 B 连锁，交换值为 h 时，F_2 为（$12-2h+h^2$）RR∶（$2h-h^2$）RS∶（$3+2h-h^2$）SR∶（$1-2h+h^2$）SS，B_1F_1 为（$2-h$）RR∶h RS∶（$1+h$）SR∶（$1-h$）SS。

③当 A 1＝A3 时，F_2 为 12 RR∶3 SR∶1 SS，B_1F_1 为 2 RR∶1 SR∶1 SS。而沈农 1033/二九丰 F_2 和 B_1F_1 群体植株对菌株 HB17 和 T1 的反应型实际观察数值为 156 RR∶31 SR∶10 SS 和 35 RR∶14 SR∶13 SS，均符合 A1＝A3 的理论比例。因此 A1＝A2＝A3，即由同一个抗性基因控制对 3 个菌株的抗性（表 4）。

表 4 沈农 1033/二九丰组合 F_2 和 B_1F_1 植株孕穗期对 3 个菌株之间的抗性分析

	HB17、P1					HB17(P1)、T1				
	RR	RS	SR	SS	Total	RR	RS	SR	SS	Total
F_2 Obs	156	0	0	41	197	156	0	31	10	197
1)EXP.	9∶	3∶	3∶	1	16	45∶	3∶	15∶	1	64
χ^2			154.440					28.40<0.001		
p			<0.001					<0.001		
2)EXP.	($3-2h+h^2$)∶	($2h-h^2$)∶	($2h-h^2$)∶	($1-2h+h^2$)	4	($12-2h+h^2$)∶	($2h-h^2$)∶	($3+2h-h^2$)∶	($1-2h+h^2$)	16
χ^2			1.626(h=0)					1.474(h=0)		
p			0.10～0.25					0.25～0.50		
3)EXP.	3∶	∶		1	4	12	∶	3∶	1	16
χ^2			1.626					1.474		
p			0.10～0.25					0.25～0.50		
B_1F_1 Obs	35	0	0	27	62	35	0	14	13	62
1)EXP.	1∶	1∶	1∶	1	4	3∶	1∶	3∶	1	8
χ^2			60.129					28.436		
p			<0.001					<0.001		

（续）

	HB17、P1					HB17(P1)、T1				
	RR	RS	SR	SS	Total	RR	RS	SR	SS	Total
B_1F_1OBs	35	0	0	27	62	35	0	14	13	62
2)EXP.	(1－h)：	h：	h：	(1－h)	2	(2－h)：	h：	(1＋h)：	(1－h)	4
χ^2			0.790(h=0)					0.718(h=0)		
p			0.25～0.50					0.50～0.75		
3)EXP.	1	：	1	1	2	：2	：	1 ：	1	4
χ^2			0.790					0.718		
p			0.25～0.50					0.50～0.75		

h =连锁交换值；1）独立关系；2）连锁关系；3）等同关系。

从沈农 1031/二九丰 F_2 和 B_1F_1 对 HB17、P1 和 T1 三个菌株的抗感反应型来分析，实际观察值 156RRR：31SSR：10SSS 和 35RRR：14SSR：13SSS，符合由一对基因控制对 3 个菌株的抗感反应型理论比例 12RRR：3SSR：1SSS 和 2RRR：1SSR：1SSS（表 2、表 3）。

同理，可得三黄占 2 号、川植 5 号、扬稻 1 号和 83007 的结果。试验表明在这 5 个籼稻抗病品种携有两对独立遗传的显性抗性基因中，一对能控制对 3 个菌株的抗性，另一对只控制对 T1 的抗性，而对 HB17 和 P1 是感病的。

3 讨论

分析水稻品种对白叶枯病多菌株的抗性遗传基础，我们用多菌株进行同株分蘖分组法是目前可以达到这一目的的方法之一。日本曾采用此法鉴定了抗性基因 *Xa1*、*Xa2*、*Xa3* 和 *Xa12*（*Xa-kg*）(Ezuka et al.，1975；Sakaguchi，1967)，并报道了黄玉群品种携有的 *Xa1* 和 *Xa12* 分别控制日本菌群Ⅰ和Ⅴ的抗性，并以 2％的交换值紧密连锁。Rantai-emas Ⅱ 抗性品种群携有的 *Xa1*、*Xa2* 分别控制日本菌群Ⅰ、Ⅱ的抗性，两基因紧密连锁，交换值为 3％。早生爱国 3 号品种群携有的 *Xa3* 控制日本菌群Ⅰ、Ⅱ、Ⅲ的抗性。水稻品种对若干个病菌小种的抗性，可能只由一对基因控制，或由几对基因分别控制不同小种。我国白叶枯病抗性遗传研究也有类似的报道（万建民等，1990；谢岳峰等，1989）。本研究也得出类似结果。同一个品种采用不同菌株测定，得出不同的结果，已有详细论述。目前对于建立全国标准化系统鉴定白叶枯病抗性基因的建议已趋一致，其中涉及采用统一的鉴定菌株、基因鉴别品种、感病亲本标准化等方面，以及与国际上相同领域的研究达到可比性等。

原载：华中农业大学学报，1993，12（1）：215～220.

感病品种背景对水稻白叶枯病抗性遗传的影响

施爱农　章　琦　王春莲　白建法

（中国农业科学院作物育种栽培研究所）

Effect of Genetic Background of Susceptible Parents on Expression of Resistance to Bacterial Blight of Rice

Abstract The inheritance behaviors for resistance to 3 strains of bacterial blight (*Xanthomonas oryzae* pv. *oryzae*, *Xoo*) HB17, P1 and T1 in two japonica cultivars with adult-plant resistance, Pei Zao 15, Bang Zhu Mang, and four indica cultivars with over-all resistance, Yang Dao 1, 83007, Chuan Zhi 5, and DV85, under condition of genetic background of various susceptible parents were studied. The analyses of F_1, F_2 populations from crosses of the six resistance cultivars with susceptible cultivars, Shen Nong 1033, Jin Gang 30, TN1 and Zao Xian 141, revealed that the inheritance behaviors of resistance to 3 strains of *Xoo* in six resistance cultivars under genetic background of the susceptible parents were basically similar, but existed the specific genetic behaviors of resistance, showing the change from the dominant to the recessive one and the difference of inheritance model, which were related to the plants growth stage and the *Xoo* strains used.

Key words bacterial blight of rice; genetics of resistance.

摘要　分析了邳早 15、蚌珠芒 2 个粳稻成株抗性品种和川植 5 号、扬稻 1 号、83007 和 DV85 4 个籼稻全生育期抗性品种在不同感病品种背景下，对 3 个白叶枯病菌株 HB17、P1 和 T1 的抗性遗传行为。分析这 6 个抗病品种与粳稻感病亲本沈农 1033、籼稻感病亲本金刚 30 和 TN1 或早籼 141 杂交后代 F_1 和 F_2 对 3 个菌株的抗性反应。结果表明，这些抗病品种在不同感病品种背景下的抗性遗传行为基本一致，但是存在抗性显隐性的变化和遗传方式不一致的异常现象，异常表达与植株的生育期和所用的菌株有关。

关键词　水稻白叶枯病，抗病性遗传

水稻白叶枯病抗性表现是水稻品种和病原菌相互作用的结果，同时还受遗传背景的影响。有关水稻抗病品种在不同感病品种背景下的抗性遗传行为的结果报道不一。章琦等

(1990)、周泰初等(1982),伍尚忠等(1990)报道蚌珠芒、南粳15和青华矮6号的抗性遗传结果都不相同。高山林等(1982),张端品等(1982)报道的Zenith抗性遗传行为也不一致。其原因除了所采用的菌株不同外,不同的感病品种背景也可能是一个因素。张林等(1988)报道成抗品种在不同感病品种背景下的抗性表达基本一致,但全生育期抗性品种籼粳交后代却分离出成抗植株个体。万建民等(1985;1990)的研究表明抗病品种在不同感病品种背景下有抗性特异性表达,特异性发生在亚种间杂交后代、也发生在亚种内杂交的后代中。还有报道认为抗病品种在不同感病品种背景(包括亚种籼籼交、粳粳交亚种内和籼粳交)下的抗性遗传行为有明显差异(朱立宏等,1990;张耀钢等,1990;周泰初等,1982)测定的基础上,选择了抗性稳定、不同类型的高抗材料和几个高感的籼、粳稻品种,用作亲本配制了成套组合。探讨不同感病品种遗传背景下的抗性表达和遗传行为,以便为抗病育种提供依据和探索我国水稻白叶枯病抗性遗传研究中确定感病亲本的标准化问题。

1 材料和方法

1.1 供试品种

粳稻沈农1033,籼稻金刚30、早籼141或TN 1为感病亲本。粳稻成株抗性品种邳早15、蚌珠芒和籼稻全生育期抗性品种川植5号、扬稻1号、83007和DV85(表1)。

1.2 配制组合和种植

以感病亲本为母本、抗病品种为父本配制杂交组合,获F_1、F_2种子。亲本、F_1各10株、$F_2$200～300株种植于网室或大田。常规管理。

1.3 采用菌株和接种

我国菌株HB17、国际鉴定基因菌株P1和T1共3个。分别在植株分蘖期和孕穗期剪叶接种。菌液浓度10^9/mL,接种后14～20d调查植株病斑面积占叶面积的百分率。F_2分离群体抗感划分以病斑占叶面积的20%为分界值。

2 结果和分析

2.1 亲本品种的抗性表现

10个品种对菌株HB17、P1和T1的抗性反应列于表1。沈农1033、金刚30和早籼141全生育期都高感3个菌株,是理想的感病亲本。TN1虽然在全生育期抗T1,但对P1和HB17都是高度感染的,本研究用其作为对这2个菌株的感病亲本。粳稻品种邳早15和蚌珠芒对T1在分蘖期中抗,至孕穗期高抗;对HB17和P1表现成株抗性,分蘖期中感、孕穗期高抗。川植5号、扬稻1号、83007和D85在全生育期都高抗3个菌株。

2.2 F_1的抗性表现

各组合F_1对3个菌株的抗性反应列于表2。邳早15、蚌珠芒和川植5号与不同感病亲本

杂交 F_1之间对 HB17、P1 和 T1 的抗性反应基本是一致的。扬稻 1 号和 83007 分别与不同感病亲本杂交的 F_1植株均全生育期高抗 T1；对 HB17 和 P1 的抗性反应与金刚 30 和 TN1 或早籼 141 的籼籼交 F_1之间也是一致的，基本上都在全生育期高抗，但是，上述品种与沈农 1033 杂交 F_1的抗性则到成株期才表达。DV85 与 3 个感病亲本杂交 F_1均在全生育期高抗 T1；而对 HB17 和 P1 的抗性则在成株期表达，如 TN1/DV85 的籼籼交 F_1在分蘖期感病，孕穗期抗病，符合 DV85 的 *xa5* 和 *Xa7* 基因的预期表达；而与早籼 141/DV85 的籼籼交 F_1及沈农 1033/DV85 的粳籼交 F_1均在全生育期感病，其携带的显性成抗基因 *Xa7* 未能得到表达，在 3 年中测试沈农 1033/DV85 F_1的重复结果是一致的。

2.3　F_2代的抗性表现

各组合的 F_2对 3 个菌株的抗性反应结果列于表 3。邳旱 15 与 3 个感病亲本杂交后代 F_2抗感株数比值均符合 3 抗∶1 感的 1 对显性基因期望值，并与 F_1相符。蚌珠芒与沈农 1033 和金刚 30 杂交后代 F_2抗性表现基本一致，为 1 对成抗显性基因的表达，而与 TN1 杂交 F_2则为 2 对显性基因互补，并呈近正态分布。川植 5 号、扬稻 1 号和 83007 的杂交后代 F_2在孕穗期对 HB17 和 P1 的抗性反应均表现为 1 对显性基因，对 T1 则为 2 对显性基因且独立遗传。川植 5 号与沈农 1033 的籼粳交和与金刚 30 的籼籼交 F_2之间的抗性表现基本一致，但是与 TN1 的籼籼交 F_2不一致。在 TN1/川植 5 号组合的 F_2群体中，分离出许多成株抗性植株，分蘖期与孕穗期的抗感株数分离比值也不相同。扬稻 1 号、83007 与金刚 30 和 TN1 的籼籼交 F_2之间的抗性反应基本上是一致的。而在与沈农 1033 的籼粳交 F_2中产生许多成抗植株，因此在分蘖期和孕穗期的抗感株数的分离比值便不相同了。籼籼交 F_2群体在两个生育期的抗性表现是一致的，与 F_1符合。DV85 的抗性遗传表达比较复杂，与沈农 1033、TN1 和早籼 141 杂交组合 F_2的抗感株数比均不一致，分别表现为 2 对显性互补、1 显 1 隐独立遗传及 2 对隐性遗传的差别。

表 1　亲本对 3 个水稻白叶枯病菌株的抗性反应

品种	类型	抗性类型	病斑占叶片面积百分率%					
			最高分蘖期			孕穗期		
			HB17	P1	T1	HB17	P1	T1
沈农 1033	粳	感	82.8±3.6	75±16.6	59±9.6	75±16.6	75±16.6	59±9.6
金刚 30	籼	感	68±9.2	68±9.2	42.3±6.2	55±4.6	55±4.6	32±8.7
TN1	籼	感	78.8±4.8	78.8±4.8	5±0	67.5±5.4	59±9.7	3±0
早籼 141	籼	感	87±5.4	87±5.4	76±5.2	81±7.8	76.5±7.5	76±8.1
邳旱 15	粳	成株抗性	25±2.1	25±3.5	8.8±2.0	5±0	5±0	5±0
蚌珠芒	粳	成株抗性	27±7.2	23±5.4	10.7±3.3	5±0	5±0	5±0
川植 5 号	籼	全生育期抗性	5±0	5±0	5±0	5±0	5±0	3±0
扬稻 1 号	籼	全生育期抗性	5±0	5±0	3±0	5±0	5±0	3±0
83007	籼	全生育期抗性	3.6±1.0	3.6±1.0	3±0	3±0	3±0	3±0
DV85	爪哇	全抗+成抗	3±0	3±0	3±0	3±0	3±0	3±0

表 2　各组合 F_1 对 3 个菌株的抗性表现

组　　合	F_1 对菌株 HB17、P1、T1 的抗性反应					
	HB17		P1		T1	
	MTS[a]	BS[b]	MTS	BS	MTS	BS
沈农 1033/邳早 15	27.5±5.3	5±0	26.2±7.4	5±0	5±0	3±0
金刚 30/邳早 15	23.6±8.0	6±1.8	28.6±8.0	6±1.8	5±0	3±0
TN1/邳早 15	28.1±6.5	5.9±1.5	25.9±9.9	5.9±1.8	5±0	3±0
沈农 1033/蚌珠芒	50±5.5	11±3.3	45.3±6.0	5±0	5±0	3±0
金刚 30/蚌珠芒	52.4±6.0	11.9±3.1	50.3±6.7	5±0	5±0	3±0
TN1/蚌珠芒	53.3±5.8	8.2±3.0	45±12.9	5±0	5±0	3±0
沈农 1033/川植 5 号	11.1±2.2	7.2±2.2	6.3±2.3	5±0	5±0	5±0
金刚 30/川植 5 号	7±2.7	5±0	7±2.7	5±0	5±0	5±0
TN1/川植 5 号	7±2.7	5±0	7±2.7	5±0	3±0	3±0
沈农 1033/扬稻 1 号	25.5±7.6	5.7±1.6	20±7.8	5.4±1.4	5±0	3±0
金刚 30/扬稻 1 号	10±5.8	5±0	6.3±2.3	5±0	5±0	5±0
TN1/扬稻 1 号	7.8±2.6	5±0	7.8±2.6	5±0	3±0	3±0
沈农 1033/83007	23±2.7	10.6±2.5	15±3.5	8.9±4.3	1±0	1±0
金刚 30/83007	8.3±2.9	5±0	5±0	5±0	5±0	5±0
TN1/83007	5.7±1.9	5±0	5.7±1.9	5±0	5±0	5±0
早籼 141/83007	6.6±2.0	5±0	6.6±2.0	5±0	5±0	5±0
沈农 1033/DV85	55.5±1.6	35.5±16.1	55.5±1.6	37.5±8.2	5.7±1.9	2.6±1.7
TN1/ DV85	50.9±6.9	3±0	50.9±6.9	3±0	5±0	3±0
早籼 141/ DV85	55.6±11.3	38.1±6.5	39.2±13.9	30.8±8.0	5±0	2.2±1.8

[a] MTS=最高分蘖期，[b] BS=成株期。

表 3　各组合 F_2 对 HB17、P1 和 T1 3 个菌株的抗性表现

组　　合	所用菌株	接种时期[a]	F_2对 3 个菌株的抗性反应				
			抗病株	感病株	总株数	χ^2	P
沈农 1033/邳早 15	HB17	BS	167	70	237	2.364 (3∶1)	0.10～0.25
金刚 30/邳早 15	HB17	BS	130	46	176	0.068 (3∶1)	0.75～0.90
TN1/邳早 15	HB17	BS	188	78	266	2.426 (3∶1)	0.10～0.25
沈农 1033/蚌珠芒	HB17	BS	153	61	214	1.221 (3∶1)	0.25～0.50
金刚 30/蚌珠芒	HB17	BS	142	58	200	1.500 (3∶1)	0.10～0.25
TN1/蚌珠芒	HB17	BS	161	111	272	0.840 (9∶7)	0.25～0.50
沈农 1033/川植 5 号	HB17	MTS	199	74	273	0.538 (3∶1)	0.25～0.50
		BS	209	64	273	0.275 (3∶1)	0.50～0.75
	P1	MTS	211	62	273	0.646 (3∶1)	0.25～0.50
		BS	213	60	273	1.173 (3∶1)	0.25～0.50
	T1	MTS	248	25	273	3.458 (15∶1)	0.05～0.10
		BS	251	22	273	1.231 (15∶1)	0.25～0.50

（续）

组　合	所用菌株	接种时期[a]	F_2对 3 个菌株的抗性反应				
			抗病株	感病株	总株数	χ^2	P
金刚 30/川植 5 号	P1	MTS	176	63	239	0.169（3∶1）	0.50～0.75
		BS	185	54	239	0.615（3∶1）	0.25～0.50
TN1/川植 5 号	T1	MTS	141	101	242	0.321（9∶7）	0.50～0.75
		BS	178	64	242	0.198（3∶1）	0.50～0.75
沈农 1033/扬稻 1 号	HB17	MTS	122	91	213	0.054（9∶7）	0.75～0.90
		BS	153	60	213	0.978（3∶1）	0.25～0.50
	HB17	BS	237	86	323	0.373（3∶1）	0.50～0.75
	P1	BS	248	81	329	0.069（3∶1）	0.90～0.95
	T1	BS	286	34	320	9.720（15∶1）	<0.005
金刚 30/扬稻 1 号	HB17	MTS	166	42	208	2.314（3∶1）	0.10～0.25
		BS	165	43	208	1.853（3∶1）	0.10～0.25
	P1	MTS	168	40	208	3.391（3∶1）	0.05～0.10
		BS	168	40	208	3.391（3∶1）	0.05～0.10
	T1	MTS	191	17	208	1.005（15∶1）	0.25～0.50
		BS	193	15	208	0.185（15∶1）	0.50～0.75
TN1/扬稻 1 号	HB17	MTS	194	66	260	0.005（3∶1）	0.90～0.95
		BS	180	65	245	2.230（3∶1）	0.10～0.25
沈农 1033/83007	HB17	MTS	89	74	163	0.119（9∶7）	0.50～0.75
		BS	107	43	150	0.889（3∶1）	0.25～0.50
	P1	BS	166	63	229	0.642（3∶1）	0.25～0.50
	T1	BS	212	17	229	0.357（15∶1）	0.50～0.75
TN1/83007	HB17	MTS	55	21	76	0.158（3∶1）	0.50～0.75
		BS	86	16	102	4.235（3∶1）	0.025～0.05
早籼 141/83007	HB17	MTS	169	63	232	0.466（3∶1）	0.25～0.50
		BS	159	64	223	1.436（3∶1）	0.10～0.25
沈农 1033/DV85	P1	MTS	59	191	250	0.192（1∶3）	0.50～0.75
		BS	149	101	250	1.008（9∶7）	0.25～0.50
	HB17	MTS	57	216	273	2.255（1∶3）	0.10～0.25
		BS	64	209	273	0.275（1∶3）	0.50～0.75
TN1/DV85	HB17	MTS	66	151	217	3.111（1∶3）	0.05～0.10
		BS	190	39	229	0.339（13∶3）	0.50～0.75
早籼 141/DV85	HB17	MTS	17	230	247	0.078（1∶15）	0.75～0.90
		BS	104	140	244	0.084（7∶9）	0.75～0.90

[a] MTS＝最高分蘖期，BS＝成株期。

3 讨论

水稻品种对白叶枯病的抗性在不同感病亲本背景下的遗传行为不一致的原因可能是多方面的。不同感病亲本之间的遗传基础差异是一个因素。在本研究中，全生育期抗性籼稻品种扬稻 1 号和 83007 与粳稻感病亲本杂交后代能分离出成株抗性植株，这可能与籼粳稻遗传基础存在差异有关。张林等（1988）也有类似现象。而 DV85 在 3 个感病品种背景下，F_1 有显隐性之差，F_2 抗感株数分离比也都不一致，不仅在亚种间、而且在亚种内的杂交后代中也出现异常现象，万建民等（1985，1990）曾报道类似结果。品种的抗病性是可以转育的，可以将抗性基因导入不同的感病品种。但抗性转育效应存在差异，受到感病品种背景的影响。因而在水稻白叶枯病抗性遗传研究中，选择感病亲本的标准化问题值得引起重视。

原载：北京农学院学报，1993（1）：90～94.

水稻花粉株系白叶枯病抗性遗传

张承妹[1]　陆家安[1]　章振华[1]　章　琦[2]

（1. 上海市农业科学院作物育种栽培研究所；
2. 中国农业科学院作物育种栽培研究所）

The Inheritance of Resistance to Bacterial Blight in Rice Pollen Lines

Abstract Inheritance of resistance to bacterial blight (BB) in rice pollen lines was analyzed. A susceptible japonica cultivar, Shennong 1033 was crossed with seven BB resistance cultivars including three indica, 75 - 34, Zenith, and IR20, three japonica, Wase Aikoku, Peizao 15, Bangzhu-mong, and one javanica, Java 14, respectively. The F_1 s of the 7 crosses were anther-cultured and advanced. The bacterial blight resistance transferring performance was observed in the pollen planes, H_2, H_3, and H_4 inoculated with BB strain HB17. There were 599 H_2 pollen strains. Among those, 572 (95.5%) were either resistant or susceptible uniformly and stably, 376 (62.8%) were resistant (including highly and moderately R), 196 (32.7%) were susceptible, and 27 (4.5%) showed to be segregated.

The resistance reaction range of variation of H_2 and F_2 groups varied from different crosses. There was significant difference between days of maturity and BB resistance in pollen strains in the same cross, and the resistance of longer growing period ones exhibited stronger.

The traits such as days of maturity, plant height, panicle length and grain weight in pollen progenies H_2 showed quite variety. Inheritance analysis of resistance indicated that the BB resistance was controlled by a dominant epistatic gene and the broad heritability was over 80%.

The BB resistance stability and other traits in H_2, H_3, and H_4 showed durably, uniformity and no variable with increase of the generations.

Key words rice anther culture, bacterial blight, resistance, rice pollen plant gene transfer

摘要 用高感白叶枯病的粳稻品种沈农 1033 分别与 7 个抗病品种，籼稻 75 - 54、

Zenith、IR20，粳稻早生爱国3号、邳早15、蚌珠芒和爪哇稻Java 14杂交，各组合的F_1进行花药培养。观察花粉株系第二代（H_2）的抗性转移及其遗传表现以及花粉株系第三代、第四代的抗性稳定性。572个株系的抗性都整齐一致，占95.5%。其中有376个株系抗病（高抗和中抗），占62.9%；196个株系感病，占32.7%。表现分离的杂合株系约占4.5%。在不同组合的H_2与F_2群体中的抗、感反应的变异幅度不同。对同一组合不同熟期花粉株系的生育期与抗病性之间存在极显著相关，生育期长的株系抗性较强。花粉再生植株后代H_2在生育期、株高、穗长、粒重等性状的多样性很大。对白叶枯病抗性遗传分析的结果表明，在4个组合中，抗性均由1个显性上位基因控制，抗病性广义遗传力在80%以上。杂种花粉植株第二、三、四代的农艺性状以及白叶枯病抗性表现持续稳定，其抗性的整齐度和强度不因世代的增加而减弱。

关键词 花药培养，水稻白叶枯病，成株抗性，抗性基因转移

近年来，水稻白叶枯病抗性育种和遗传研究取得了一定进展，并获得了许多抗源（高山林等，1982；章琦等，1986）。在目前推广的良种中普遍存在高产低抗的问题，迫切需要提高品种的抗病性。通过杂交育种途径转育抗性的年限较长，水稻花药培养作为快速有效的育种方法，已在我国广泛开展，并育成了若干品种（沈锦华等，1982）。证明了花药培养显、隐性基因均可早期纯合，缩短育种年限，并可创造出大量新类型。育种家用杂种花药培养来加速获得纯合系和提高选择效率。随着花培育种工作的深入发展，科学家们对来源于品种和杂种的花粉植株之性状遗传做了一些研究（刘进等，1983）。已报道了水稻花培后代农艺性状及抗瘟性遗传研究（朱德瑶等，1983；李梅芳等，1983），水稻花培后代的白叶枯病抗性遗传的行为还鲜有报道。因此，本试验对水稻白叶枯病抗性基因在花粉再生植株后代中的转移和抗病性的遗传特点方面进行初步研究，为抗病育种提供参考依据。

1 材料和方法

1.1 供试材料

以白叶枯病粳稻感病品种沈农1033为母本，分别与7个抗病品种杂交，其中3个籼稻：75-34、Zenith、IR20；3个粳稻：早生爱国3号、邳早15、蚌珠芒；1个爪哇稻Java 14，获7个组合的F_1。

1.2 培养基和接种方法

1.2.1 诱导培养基 以N_4为基本培养基，附加2,4-D 2mg/L、丝瓜伤流液20%（V/V）、蔗糖6%、琼脂粉0.6%、pH5.8。

1.2.2 分化培养基 以MS为基本培养基，附加KT 2mg/L、NAA 0.5mg/L、IAA 0.5mg/L、蔗糖3%、琼脂粉0.6%、pH5.8。花药培养按常规方法操作。

1.3 各世代种植和管理

在海南岛种植当代花粉再生植株（H_1）。成熟后按不同组合的单株、单穗收藏，翌年种

植在本院抗病鉴定圃隔离区，5 月 27 日同时种植 P_1、P_2、F_1、F_2、H_2，7 月 1 日移栽，用作人工接种白叶枯病鉴定。各组合 P_1、P_2各 20 株、F_2约 130 株，130 个 H_2小区（单穗区），各小区 24 株，株行距为 16.5cm×19.8cm，中等肥水条件，常规管理，植株生长正常，接种前未发生白叶枯病。

将 H_2种子等分为二，一半在海南加代、另一半贮存，翌年 5 月按上述种植设计，在上海本院病圃隔离区同时种植 H_3、H_4，进行接种鉴定，考察各系世代之间性状的整齐度和抗性稳定性。

1.4 接种鉴定

均在植株剑叶期进行接种。采用北方代表菌株 HB17。由中国农科院作物所分离。供试菌株接种前在 PSA 斜面培养基上于 30℃培养箱生长 72 h，接种菌液浓度为 10^9/mL。采用剪叶法接种，每株接种 3～5 片叶，接种后 15～20d 按国际水稻研究所“SES”分级标准评价病害反应。

2 结果与讨论

2.1 杂种花粉植株 H_2 白叶枯病抗性表现及其纯合率

在本试验中的花粉植株 H_2代的抗性稳定，整齐一致。在 7 个杂交组合的 599 个 H_2株系中，有 572 个株系的抗、感表现整齐一致，同其他农艺性状在花培后代中的纯合表现一样。其中抗病（高抗和中抗）株系有 376 个，占 62.7%；感病株系有 196 个，占 32.7%；抗感分离的杂合系约 4.5%（表 1）。通常来源于花粉的单倍体植株，经自然或人工加倍，遗传上应该是纯合的，但在本试验中出现了少量的变异或分离。这可能是再生植株来自非单倍体或继代培养无性系变异所致。试验显示，花粉植株后代的二倍体植株有 95%纯合，其余 2%～5%植株分离，此结果与章振华等（1982）的报道一致。在多数组合的花粉植株 H_2株系中抗病（包括中抗）系的百分率高于感病系，有些组合甚至没发现感病系，如沈农 1033/蚌珠芒和沈农 1033/早生爱国 3 号。

表 1 7 个杂交组合的花培 H_1 株系对白叶枯病的抗性反应及其纯合性

Table 1 Reaction and homozygosity of anther H_1 lines derived from 7 crosses to bacterial blight

组合 (Cross)	抗性反应 (Reaction)*					纯合系 (Homozygous line)		杂合系 (Heterozygous line)	
	R	MR	S	R/S	总数	株系数	百分率（%）	株系数	百分率（%）
沈农 1033/邳早 15	50	16	50	7	123	116	94.3	7	5.7
沈农 1033/Wase Aikoku 3	14	17	0	4	35	31	88.6	4	11.4
沈农 1033/蚌珠芒	25	6	0	0	31	31	100	0	0
沈农 1033/Java 14	49	22	35	4	110	106	96.4	4	3.6
沈农 1033/75-34	52	18	52	5	127	122	96.1	5	3.9
沈农 1033/Zenith	52	33	53	7	145	138	95.2	7	4.5

（续）

组合 (Cross)	抗性反应 (Reaction)*					纯合系 (Homozygous line)		杂合系 (Heterozygous line)	
	R	MR	S	R/S	总数	株系数	百分率（%）	株系数	百分率（%）
沈农 1033/IR20	20	3	5	0	28	28	100	0	0
合计	262	114	196	27	599	572		27	
百分率	43.7	19.0	32.7	4.5		95.5		4.5	

* R=抗（Resistant），MR=中抗（Moderately resistant），S=感（Susceptible），R/S=抗/感（Resistant/susceptible）。

感病品种沈农 1033 分别与 4 个具有 1 对白叶枯病显性成株抗性基因品种 75－34、Zenith、邳旱 15、Java 14 进行杂交。各组合 F_2 群体的抗病株率，均高于 H_2 株系（表 2）。表现出配子体（花粉株系）与孢子体各自分离的不一致性。这是杂种 F_2 抗性植株的遗传表现型，含有显性纯合体和杂合体，因此其抗性植株百分率必然大于 H_2，而 H_2 株系中的抗病系则是纯合的，因此小于 F_2 群体中的抗病率。抗性的遗传型和遗传表现型的不一致是造成品种改良中误选的因素。

表 2　花培与常规杂交后代的白叶枯病抗性遗传表现型比较

Table 2　Comparison of inheritance phenotype for bacterial blight resistance between progenies by anther culture and conventional hybrid in rice

组合 Cross	抗病株（或株系）率（%） Percentage of resistant plant or line			植株（或株系）数 No. of plant or line		
	F_1	F_2	H_2	F_1	F_2	H_2
沈农 1033/邳旱 15	100	80.9	53.2	10	105	123
沈农 1033/75－34	100	74.0	55.1	10	100	127
沈农 1033/Zenith	100	75.4	58.7	10	118	145
沈农 1033/Java 14	100	77.2	54.8	10	110	135

2.2　花粉植株 H_2 株系的抗性变幅

我们观察到花粉植株 H_2 的抗性反应之变幅（病斑占叶面积百分率）大于 F_2，如组合沈农 1033/75－34 的 H_2 抗性反应范围为 5%～100%，而 F_2 则为 5%～45%（图 1），其他组合也有类似趋势。这是因为 F_2 的遗传表现型受到抗、感的显隐性干扰所致。虽然 H_2 的抗感变幅大，但其遗传型与表现型是一致的，对抗性的选择具有较高的准确率。

2.3　花粉株系抗白叶枯病与生育期的关系

花药培养后获得了不同熟期的株系，其中早熟与晚熟株系之间的生育期相差 20 多天。经用白叶枯病菌接种鉴定的结果表明，生育期与抗病性之间的相关性达极显著水平，4 个组合都相一致。经过相关系数 Z 转换后，合并为 F=0.572，也达到极显著（表 3）。该结果表明，生育期的长短对白叶枯病的抗性有很大影响，生育期长、成熟较晚的株系抗性较强。

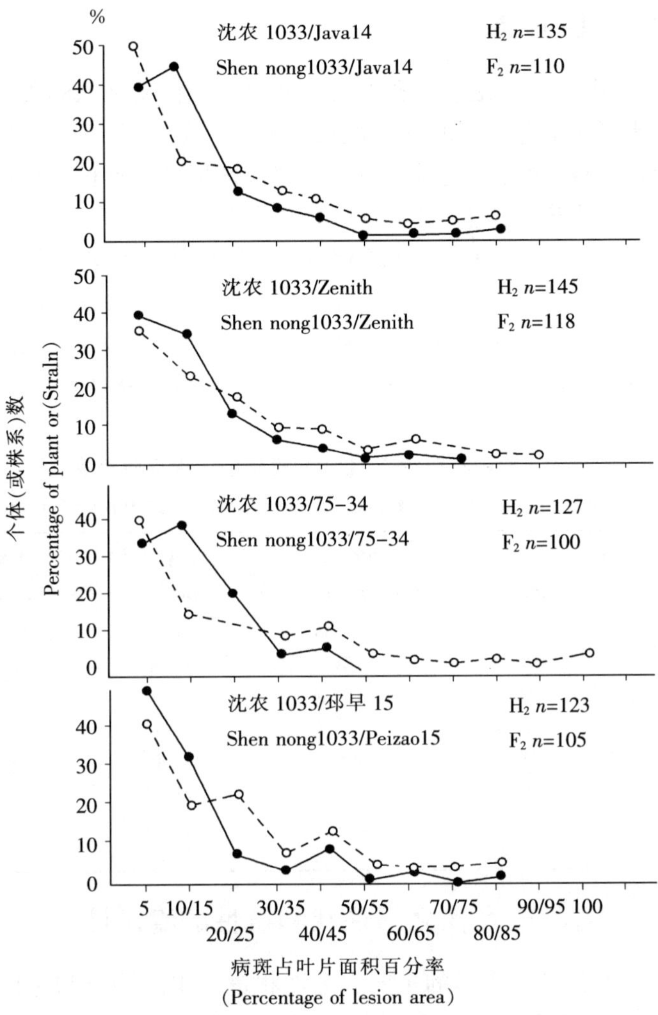

图 1 4 个抗/感杂交组合 H_2 和 F_2 群体对白叶枯病抗性反应范围的差异

Fig. 1 Difference of resistance reaction range to bacterial blight between H_2 and F_2 populations in 4 crosses

表 3 花粉株系的生育期与其对白叶枯病的抗性相关性

Table 3 The correlation coefficient of maturity days and bacterial blight resistance in rice anther lines

组合 Cross	相关系数 Correlation coefficient	合并相关系数 Correlation coefficient combined
沈农 1033/邳早 15	−0.489**	ſ=−0.572**
沈农 1033/75-34	−0.534 2**	ſ=−0.572**
沈农 1033/Zenith	−0.641 8**	ſ=−0.572**
沈农 1033/Java 14	−0.606**	ſ=−0.572**

2.4 花粉植株 H_2 抗性的转移

从4个组合中进一步证明了花培株系中具有综合双亲优良性状的类型，并有丰富的变异类型。有些 H_2 株系还出现了超双亲的优良性状（表4）。这4个组合白叶枯病抗性的转移并不受籼粳亚种的影响。粳稻的抗性基因能转移到感病粳稻中，籼稻的抗性基因也可以转移到粳稻中。在粳/籼交组合的花粉再生植株中，出现了完全类似沈农1033半矮生型的理想株型抗病株系。如沈农1033/75-34组合的花粉植株 H_2 株系内，在完全类似母本型中，抗（R）为3.5%，感（S）为18.26%；偏母本型中，R为15.65%，S为7.8%。在中间型中，R为24.34%，S为7.8%；偏父本型中，R为6.1%，S为6.1%；在完全父本型中，R为10.43%。说明花粉植株后代中的抗性反应是多样的。有可能从花培后代中选出经济性状好的抗病籼/粳中间型材料。

表4 水稻花粉株系 H_2 不同类型的抗性反应频率

Table 4 The frequency of resistance reaction in different types of anther lines

组合 Cross	抗性反应 Reaction	母本型（%） Female type	偏母本型（%） Deviation female type	中间型（%） Mid type	偏父本型（%） Deviation male type	父本型（%） Male type
沈农1033/邳早15	R	2.60	12.17	5.22	16.52	23.34
	S	6.96	2.60	4.35	20.87	4.30
沈农1033/75-34	R	3.50	15.64	23.34	6.09	10.43
	S	18.26	7.80	7.80	6.09	0
沈农1033/Zenith	R	0	0	58.40	6.19	0
	S	0.88	0.88	38.10	0	0
沈农1033/Java 14	R	0	4.59	39.50	8.26	8.26
	S	0.91	0.91	29.36	7.34	0.91

2.5 花粉植株 H_2、H_3、H_4 世代之间性状和抗性的稳定性

对花粉植株 H_2～H_4 各世代之间的主要农艺性状进行田间调查和考种的结果表明，无论是粳/粳或粳/籼交组合，其花粉植株后代性状的整齐度是一致的，并不因为植株世代增加或经过异地种植加代而使性状整齐度下降。

花粉植株 H_2～H_4 各世代之间的白叶枯病的抗性是一致的，不因世代增加而抗性减弱或改变（表5）。各株系世代之间的抗、中抗或感的反应相同，病斑面积占叶片的百分率保持不变。可见花粉植株后代的水稻白叶枯病抗性也是纯合的。选择抗白叶枯病亲本杂交组合进行花药培养可获得抗性持续稳定的后代。

表5 水稻花粉植株后代白叶枯病抗性的稳定性

Table 5 The bacterial blight resistance stability in generations of anther culture lines*

组合 Cross	世代 Generation	编号 No.	抗病 （病斑面积%） R (lesion area percent)	编号 No.	中抗 （病斑面积%） MR (lesion area percent)	编号 No.	感病 （病斑面积%） S (lesion area percent)
沈农1033/Zenith	H_2	89-3710	R (3)	89-3724	MR (15-20)	89-3744	S (80)
	H_3		R (2)		MR (15-20)		S (80-100)
	H_4		R (2)		MR (15-20)		S (80-100)

（续）

组合 Cross	世代 Generation	编号 No.	抗病（病斑面积%） R (lesion area percent)	编号 No.	中抗（病斑面积%） MR (lesion area percent)	编号 No.	感病（病斑面积%） S (lesion area percent)
沈农 1033/75－34	H_2	89－3841	R（5）	89－3873	MR（15）	89－3861	S（100）
	H_3		R（5）		MR（15）		S（100）
	H_4		R（5）		MR（15）		S（100）
沈农 1033/Java 14	H_2	89－3753	R（3）	89－3924	MR（10－15）	89－4115	S（80）
	H_3		R（5）		MR（10－15）		S（80）
	H_4		R（5）		MR（10－15）		S（80）
沈农 1033/邳早 15	H_2	89－3795	R（5）	89－3819	MR（15）	89－3824	S（55）
	H_3		R（5）		MR（10－15）		S（55）
	H_4		R（5）		MR（10－15）		S（60）

* R＝抗（Resistant），MR＝中抗（Moderately resistant），S＝感（Susceptible）。

3 结论

3.1 本研究证明水稻白叶枯病的抗性与其他性状一样，其花粉植株后代表现为多样性和一致性，即在相同组合的不同花粉植株之间表现为多样性，不同株系间表现不同程度的抗性。花粉植株 H_2代经人工接种鉴定，有高抗、中抗、高感、中感等表现。但在同一株系内的抗性表现则高度一致。花粉植株后代的性状表现持续稳定。对白叶枯病的抗性也不因世代递增而减弱。所以花粉植株早代接种鉴定是可靠的。尤其是白叶枯病抗性可以和生育期、株型、穗、粒、谷等其他性状组合成优良的株系。

3.2 水稻白叶枯病的抗性基因不仅可在不同粳稻品种之间互相转移，也能在籼、粳品种间相互转移，这在育种上极为有利，表明推广花培途径可以加速改良和利用抗源。用常规手段进行籼粳交育种时，常常因为其分离年限长，后代性状不易稳定，育种困难。而籼粳杂种花培后代的抗性则可能互相转移并且稳定，是转移水稻白叶枯病抗性基因的一个快速有效的途径。

3.3 H_2代中的晚熟株系抗性较强，有利于选育中、晚熟品种。但在同一组合的 H_2代中，生育期的长短与抗性强度的相关，是否具有普遍性值得进一步探讨。

原载：遗传学报，1993，20（1）：88～95.
洛克菲勒基金会资助

九个水稻品种对水稻白叶枯病（*Xanthomonas oryzae* pv. *oryzae*）的抗性遗传研究

章　琦　施爱农　王春莲　白建法　杨文才

（中国农业科学院作物育种栽培研究所）

Genetic Studies of Resistance to Bacterial Blight (*Xanthomonas oryzae* pv. *oryzae*) in Nine Rice Cultivars

Abstract　The inheritance and allelic relationship of genes for resistance to strains P1 and HB17 of *Xanthomonas oryzae* pv. *oryzae* in eight indica rice cultivars, Chuanzhi 4, Chuanzhi 5, Sanhuangzhan 2, Erjiufeng, 83007, 929, 75 - 34, NT02 and one japonica cultivar Xianghu 25 were studied. Analyses of F_1, F_2 and B_1F_1 populations from crosses of the nine cultivars with susceptible cultivar Shennong 1033 or Jingang 30, respectively, revealed that the resistance in the eight indica rice cultivars at the all growth stages was controlled by a single incomplete dominant gene, Allelism tests with known resistance genes indicated that the gene in the eight cultivars is *Xa4*, and the resistance of japonica cultivar, Xianghu 25 is controlled by a dominant gene which is one of the alleles at the *Xa3* locus.

Key words　bacterial blight of rice; inheritance of resistance

摘要　研究了 8 个籼稻抗病品种川植 4 号、川植 5 号、三黄占 2 号、二九丰、扬稻 1 号、75 - 34、83007、NT02 和一个粳稻品种湘虎 25 号对水稻白叶枯病菌株 P1、HB17 或 T1 的抗性遗传。这些抗病品种分别与感病品种沈农 1033 或金刚 30 杂交，所得的 F_1、F_2 和 B_1F_1 群体的抗性反应，表明 8 个籼稻品种对菌株 P1、HB17 的全生育期的抗性均由一对不完全显性基因控制，该抗性基因与 *Xa4* 等位；粳稻品种湘虎 25 的成株抗性则由一对显性基因控制，该基因与 *Xa3* 相等位。

关键词　水稻白叶枯病；抗病性遗传

Ogawa（1993）概述了水稻白叶枯病主效抗性基因研究，系统地介绍了日本与 IRRI 在 1982—1987 年的合作研究进展。根据双方的鉴别品种对日本和菲律宾水稻白叶枯病小种的抗性反应，重新将早期各自命名的抗性基因进行了系统的遗传分析。统一命名了 14 个基因。近年又命名了 *xa15*、*Xa16*、*Xa17*、*Xa18*、*xa19*、*xa20* 和 *Xa21*（Nakai et al.，1988；

Noda et al.，1989；Ogawa et al.，1989；Khush et al.，1990）。章琦等（1988；1990；1991）报道我国多数籼稻主栽品种的抗性不佳，在全国水稻育种协作组1987年和1989年提供统一抗病鉴定的211个籼稻新品种（系）中，仅有的30多个抗性较好的品种也都由*Xa4*基因所控制。因此近年我们用15～21个包括中国、日本和菲律宾的代表菌株评价了345份（次）品种和资源，鉴定出抗性稳定、抗谱较广的8个籼稻、1个粳稻品种，其中多数是育种抗源。为了解这些品种的抗性遗传基础，配制了感/抗和抗/抗组合，采用国际命名基因时所采用的和我国代表菌株分析这些品种的抗性遗传模式，并测验与基因*Xa4*、*Xa3*、*Xa7*、*xa5*的等位性关系。

1 材料和方法

1.1 供试品种和菌株

采用15个品种：遗传分析9个抗病品种，包括8个全生育期抗病的籼稻品种：川植4号、川植5号、三黄占2号、二九丰、扬稻1号、75-34、83007和NT02，和1个成株抗性粳稻品种湘虎25。两个感病亲本沈农1033（粳）和金刚30（籼）。4个作为基因等位性测交的已知基因品种：IR20，全生育期携有一对不完全显性基因*Xa4*；IRBB4是携有*Xa4*的近等基因系；早生爱国3号，携成株抗性基因*Xa3*，DV85携全生育期隐性抗性基因*xa5*和显性成株抗性基因*Xa7*。

用3个菌株：PXO61（P1），系IRRI鉴定*Xa4*、*xa5*和*Xa7*基因的菲律宾小种1的代表菌株；T1为日本鉴定成株抗性基因*Xa3*所用，HB17为我国北方一个稳定代表菌株。

1.2 杂交组合的配制及种植

沈农1033分别与川植4号、川植5号、三黄占2号、二九丰和湘虎25杂交；金刚30分别与扬稻1号、75-34、83007和NT02杂交，根据9个组合F_1植株对供试的菌株表现显性抗性，遂与各自的感病亲本回交获B_1F_1种子。分析这些组合的F_1、F_2和B_1F_1群体的抗性反应结果及抗病品种的抗性遗传模式。

9个抗病品种与IR20或IRBB4、DV85以及早生爱国3号杂交。分析这些组合F_1和F_2群体的抗性反应结果，以确定抗病品种与上述品种所携基因的等位关系。

为了在外界条件一致的情况下接种评价，各抗病品种的抗性遗传模式分析和基因等位性测定的整套材料，包括亲本、F_1、F_2、B_1F_1都在同期种植和抗性鉴定。材料都种植于隔离严密的网室。亲本和F_1各栽10株，F_2 200～300株，B_1F_1 50～100株。常规田间管理。

1.3 接种鉴定和调查

在接种前1个月，将保存于－20℃冰箱的冻干菌种在PSA斜面培养基上复壮，然后转移数管斜面菌种保存于4°C冰箱，供整个生长季节使用。接种液菌龄2d，浓度约10^9/mL。参考Ogawa（1986）所用的多菌株接种和单株分蘖分组法，以测定各个体对不同菌株的实际抗性反应。人工剪叶接种植株的完全展开叶片。除成株抗性品种湘虎25的杂交后代在孕穗期接种外，其他品种杂交组合植株均在最高分蘖期和孕穗期各接种一次。有些组合F_2群体的生育期分离悬殊，则在孕穗期再接种一次。

接种后 14～20d，当感病对照品种病情发展稳定时，调查单株对各菌株的抗性反应。亲本和 F_1 各调查 10 株（30 张叶片）。F_2 和 B_1F_1 群体的单株的各处理均调查 3 张叶片。用病斑面积占叶面积百分率作为抗性表现参数。以 F_2、B_1F_1 群体的病斑百分率分布所呈现的峰槽，作为划分抗、感分界。本研究抗、感分界值为病斑面积占叶面积的 20%左右。

全部试验于 1988—1991 年进行，包括北京、海南、IRRI 和唐山市唐海县共 8 季。

2 试验结果

2.1 亲本品种的抗性表现

表 1 所示 15 个包括 2 个感病、9 个抗病、4 个已知基因品种，对菌株 P1、HB17 或 T1 的抗性反应。11 个籼稻品种对 P1 和 HB17 表现全生育期高抗，病斑面积小于 5%。湘虎 25 和早生爱国 3 号在成株期分别抗菌株 P1、HB17 和 T1、HB17。感病亲本沈农 1033 和金刚 30 在最高分蘖期和孕穗期的病斑占叶面积 55%以上，与抗病亲本的差异很大，是研究抗性遗传的理想感病亲本。

表 1 亲本对两个白叶枯病菌株的抗性反应

Table 1 Resistance reaction of parents to 2 strains of *X. o.* pv. *oryzae*

品种 Cultivar	类型 Type	基因 Gene	病斑占叶片面积百分率% Percentage of lesion area%			
			最高分蘖期 Max. tillering stage		孕穗期 Booting stage	
			P1	HB17	P1	HB17
沈农 1033	粳	无	69±7.3	70±6.3	73±3.6	75±16.6
金刚 30	籼	无	68±9.2	68±9.2	55±4.6	55±4.6
川植 4 号	籼		3±0	3±0	3±0	3±0
川植 5 号	籼		5±0	5±0	3±0	3±0
三黄占 2 号	籼		3±0	3±0	3±0	3±0
二九丰	籼		5±0	5±0	3±0	3±0
扬稻 1 号	籼		5±0	5±0	3±0	3±0
75－34	籼		9±1.7	9±1.7	3±0	3±0
83007	籼		4±0.97	4±0.97	2±0	2±0
NT02	籼		5±0	5±0	3±0	3±0
湘虎 25	粳		25±7.97	[illegible]±4.18	5±0	5±0
早生爱国 3 号	粳	*Xa3*	31±3.1	35±4.23	3±0	3±0
IR20	籼	*Xa4*	5±0	5±0		
IRBB4	籼	*Xa4*	3±0	3±0		
DV85	籼	*xa5*，*Xa7*	3±0	3±0		

2.2 抗病品种的抗性遗传模式

9 个抗病品种与感病品种沈农 1033 或金刚 30 杂交组合的 F_1、F_2 和 B_1F_1 群体对供试菌株的抗性反应结果见表 2 和表 3。8 个籼稻抗病品种与感病亲本杂交组合的 F_1 植株对菌株 P1 和 HB17 在分蘖期和孕穗期表现中抗，表明这些品种的全生育期抗性由不完全显性基因所控制。它们的 F_2 群体在分蘖期的抗感分离比均符合 3 抗：1 感的一对显性抗性基因期望值；这些 F_2 群体到孕穗期的抗病株数有所增加，但其抗、感比仍然为 3 抗：1 感的一对显性基因期望值；其 B_1F_1 群体除了缺少 NT02 的数据以外，其余 7 个组合在两个生育期的抗、感分离比均符合 1 抗：1 感的一对显性基因期望值。经 χ^2 测验，P 值均大于 0.05，表明 8 个籼稻品种对 P1 和 HB17 的全生育期抗性由一对不完全显性抗性基因所制约。

沈农 1033/湘虎 25 的 F_1 植株对 2 个菌株在孕穗期表现高抗，表明其成株抗性由完全显性基因所控制。其 F_2 群体对两个菌株的抗感株分离比为 178RR ： 54SS，χ^2 测验符合 3 抗：1 感的一对显性抗性比例。该组合的 B_1F_1 群体在成株期的抗、感比为 1 抗：1 感，验证了湘虎 25 对菌株 P1 和 HB17 的成株抗性由一对显性基因控制（表 2、表 3）。

表 2 感/抗组合的 F_1、F_2 植株在最高分蘖期和孕穗期对两个菌株的抗性表现[a]

Table 2 Resistance reaction of F_1 and F_2 plants from the crosses of susceptible parents with the 9 resistant cultivars to 2 strains of *X. o.* pv. *oryzae* at the maximum tillering and booting stages

组合 Cross	生育期[b] Growth stage	F_1 的抗性反应 F_1 reaction		F_2 对 P1、HB17 两个菌株的抗性反应[c] F_2 reaction to strains P1 & HB17				
		P1	HB17	RR	SS	Total	χ^2 3：1	P
沈农 1033/川植 4 号	MTS	MR	MR	193	69	262	0.183	0.50～0.75
	BS	MR	MR	201	60	261	0.461	0.25～0.50
沈农 1033/川植 5 号	MTS	MR	MR	212	64	276	0.391	0.50～0.75
	BS	MR	MR	209	64	273	0.275	0.50～0.75
沈农 1033/三黄占 2 号	MTS	MR	MR	198	59	257	0.468	0.25～0.50
	BS	MR	MR	196	61	257	0.157	0.50～0.75
沈农 1033/二九丰	MTS	MR	MR	169	50	219	0.440	0.50～0.75
	BS	MR	MR	165	48	213	0.565	0.25～0.50
金刚 30/扬稻 1 号	MTS	MR	MR	155	42	197	1.234	0.25～0.50
	BS	R	R	154	41	195	1.438	0.10～0.25
金刚 30/75 - 34	MTS	MR	MR	175	63	238	0.202	0.50～0.75
	BS	R	R	176	62	238	0.090	0.75～0.90
金刚 30/83007	MTS	MR	MR	187	55	242	0.551	0.25～0.50
	BS	R	R	187	52	239	1.173	0.10～0.25
金刚 30/NT02	MTS	R	R	169	63	232	0.466	0.25～0.50
	BS	R	R	171	61	232	0.144	0.50～0.75
沈农 1033/湘虎 25	BS	R	R	178	54	232	0.282	0.50～0.75

[a] R=抗（Resistant），MR=中抗（Moderately resistant），S 感（Susceptible）。

[b] MTS=在最高分蘖期接种（Inoculation at Max. tillering stage）。

BS=在孕穗期接种（Inoculation at booting stage）。

[c] 依次对 P1 和 HB17 两个菌株的抗性反应（The combined two capitals stand for order of reaction to strains P1 and HB17 from left to right，respectively）。

表 3 8个回交组合 B_1F_1 植株在最高分蘖期和孕穗期对两个菌株的抗性表现[a]

Table 3 Resistance reaction of B_1F_1 plants from the 8 backcrosses to 2 strains of *X. o.* pv. *oryzae* at the maximum tillering and booting stages

组合 Cross	生育期[b] Growth stage	B_1F_1 对 P1、HB17 两个菌株的抗性反应[c] B_1F_1 reaction to strains P1 & HB17				
		RR	SS	Total	χ^2 3∶1	*P*
沈农 1033/川植 4 号//沈农 1033	MTS	31	35	66	0.136	0.50～0.75
	BS	35	32	67	0.060	0.50～0.75
沈农 1033/川植 5 号//沈农 1033	MTS	51	53	104	0.010	>0.90
	BS	53	50	103	0.039	0.75～0.90
沈农 1033/三黄占 2 号//沈农 1033	MTS	24	30	54	0.463	0.25～0.50
	BS	23	31	54	0.907	0.25～0.50
沈农 1033/二九丰//沈农 1033	MTS	35	27	62	0.790	0.25～0.50
	BS	35	27	62	0.790	0.25～0.50
金刚 30/扬稻 1 号//金刚 30	MTS	21	17	38	0.237	0.50～0.75
	BS	21	17	38	0.237	0.50～0.75
金刚 30/75-34//金刚 30	MTS	19	24	43	0.372	0.50～0.75
	BS	21	22	43	0.000	>0.95
金刚 30/83007//金刚 30	MTS	12	9	21	0.190	0.50～0.75
	BS	12	9	21	0.190	0.50～0.75
沈农 1033/湘虎 25//沈农 1033	BS	48	54	102	0.245	0.50～0.75

* 注释同表 2（The notes are the same as table 2）。

2.3 抗性基因间的等位关系

9 个品种同已知基因品种之间进行等位性测试的 27 个组合 F_1、F_2群体对所测已知基因相应菌株 P1 或 T1 的抗性反应见表 4。

表 4 抗/抗组合的 F_1、F_2 植株对菌株 P1 或 T1

Table 4 Resistance reaction of F_1、F_2 plants from crosses between resistant cultivars to strain of *X. o.* pv. *oryzae* at the maximum tillering and booting stages

组合 Cross	菌株 Strain	F_1 抗性反应 F_1 reaction	F_2 的抗性反应（F_2 reaction）				
			抗病 R	感病 S	总数 Total	χ^2	*P*
川植 4 号/IR20*	P1	R	292	0	292		
川植 4 号/IRBB4*	P1	R	270	0	270		
川植 5 号/IRBB4*	P1	R	83	0	83		
三黄占 2 号/IRBB4*	P1	R	330	0	330		
二九丰/IRBB4*	P1	R	490	0	490		

（续）

组合 Cross	菌株 Strain	F_1 抗性反应 F_1 reaction	F_2 的抗性反应（F_2 reaction）				
			抗病 R	感病 S	总数 Total	χ^2	P
扬稻 1 号/IR20*	P1	R	365	0	365		
75-34/IR20*	P1	R	283	0	283		
83007/IR20*	P1	R	220	0	220		
NT02/IRBB4*	P1	R	280	0	280		
川植 5 号/早生爱国 3 号	P1	R	283	17	300	0.089（15∶1）	0.75～0.90
三黄占 2 号/早生爱国 3 号	P1	R	244	14	258	0.175（15∶1）	0.50～0.75
75-34/早生爱国 3 号	P1	R	217	18	235	0.574（15∶1）	0.25～0.50
川植 4 号/早生爱国 3 号	P1	R	242	8	250	2.582（26%）	0.10～0.25
二九丰/早生爱国 3 号	P1	R	145	5	150	1.549（26%）	0.10～0.25
扬稻 1 号/早生爱国 3 号	P1	R	234	6	240	0.523（26%）	0.25～0.50
83007/早生爱国 3 号	P1	R	156	4	160	0.288（26%）	0.50～0.75
NT02/早生爱国 3 号	P1	R	253	7	260	1.027（26%）	0.25～0.50
川植 4 号/DV85	P1	R	233	17	250	2.047（61∶3）	0.10～0.25
川植 5 号/DV85	P1	R	271	12	283	0.046（61∶3）	0.75～0.90
三黄占 2 号/DV85	P1	R	144	9	153	0.258（61∶3）	0.50～0.75
二九丰/DV85	P1	R	119	9	128	1.093（61∶3）	0.25～0.50
75-34/DV85	P1	R	244	16	260	0.093（61∶3）	0.25～0.50
83007/DV85	P1	R	204	15	219	1.832（61∶3）	0.10～0.25
扬稻 1 号/DV85	P1	R	314	25	339	4.894（61∶3）	0.025～0.050
						0.552.（15∶1）	0.25～0.50
NT02/DV85	P1	R	170	38	208	8.863（61∶3）	<0.005
						0.008（13∶3）	0.75～0.90
湘虎 25/早生爱国 3 号	T1	R	250	0	250		
湘虎 25/DV85	P1	R	239	48	287	90.403（61∶3）	<0.005
						0.645（13∶3）	0.25～0.50

* 最高分蘖期接种（Inoculated at maximum tillering stage），其余为孕穗期接种（The others were inoculated at booting stage）。

川植 4 号、川植 5 号、三黄占 2 号、二九丰、扬稻 1 号、75-34、83007 和 NT02 八个籼稻品种同 IR20 或 IRBB4 杂交的组合 F_1植株在最高分蘖期都高抗 P1。其 F_2植株也全部抗病，没有分离出感病株，表明这些全生育期抗病的籼稻品种对 P1 的抗性基因都与 IR20 或 IRBB4 的基因 *Xa4* 是等位的。

用菲律宾小种 1（P1）测定这 8 个籼稻品种与早生爱国 3 号之间的关系，获得两种结果：川植 5 号、三黄占 2 号和 75-34 与早生爱国 3 号杂交组合 F_2群体的抗感分离符合 15 抗∶1 感的比值，其抗性基因与 *Xa3* 独立遗传。川植 4 号、二九丰、扬稻 1 号、83007 和 NT02 与早生爱国 3 号杂交各组合 F_2群体的抗感分离结果表明，它们的抗性基因与 *Xa3* 连锁，其实际交换值分别为 35.8%、36.5%、31.6%、31.6%和 32.8%。川植 4 号、川植 5 号、三黄占 2 号、二九丰、75-34 和 83007 分别与 DV85 进行等位测验的 F_2群体对 P1 的抗感分离比符合 61 抗∶3 感的 2 对显性、1 对隐性基因独立遗传。而扬稻 1 号和 NT02 与 DV85 测验的 F_2群体对 P1 的抗感分离比不符合 61 抗∶3 感的比值，却分别符合 15 抗∶1 感的 2 对显性和 13 抗∶3 感的各 1 对显、隐性基因独立遗传，该结果还需要进一步验证。

湘虎 25/早生爱国 3 号的 F_2群体在成株期对 T1 表现全部高抗，叶片病斑面积均小于 5%。没有分离出感病株，表明湘虎 25 的抗性与 *Xa3* 等位。湘虎 25 与 DV85 测交 F_2群体对菌株 P1 的分离比不符合 61 抗∶3 感的 2 对显性 1 对隐性的独立遗传，却符合 13 抗∶3 感的一显一隐性基因的独立遗传。

3 讨论

本研究分析的 8 个全生育期抗病籼稻品种的抗性均由 *Xa4* 所控制。这些品种与感病亲本杂交组合 F_1植株在最高分蘖期呈现中等抗性，为不完全显性抗性反应。追溯其系谱渊源，扬稻 1 号由 BG90－2 系选育而成。由 BG90－2 衍生的扬稻 2 号、扬稻 3 号和金陵 57 四个品种都抗白叶枯病，推测它们可能携有相同的抗性基因（章琦等，1988）。二九丰（IR29/原丰早）的抗性源出于 Tadukan。早籼品种 83007（矮南早 1 号/GE451F_3//GE451/广陆矮 4 号）的抗性可能来自 GE451。因为源自矮脚南特和广场矮的矮南早 1 号和广陆矮 4 号虽然都是优良育种亲本，但都不抗病。川植 4 号、川植 5 号、三黄占 2 号、NT02 和 75－34 的系谱有待查明。这些携有 *Xa4* 的品种的抗谱较广，除了我国Ⅴ群菌（GD1358）外，抗其他 6 群菌，仍可用于常规育种。

湘虎 25（湘粳 8 号/农虎 6 号）系浙江农大育成的优良中熟晚粳品种。其白叶枯病抗性源自农垦 58。农垦 58 综合性状优良、适应性广，抗白叶枯病，是 20 世纪 60 年代中期至 70 年代初期我国南方稻区推广面积最大的良种之一，也是晚粳育种中的主要亲本，已有 121 个衍生品种，其中多数抗白叶枯病（林世成等，1991；章琦等，1988），但对农垦 58（世界稻）的抗性遗传基础尚不清楚。

水稻白叶枯病抗性基因系统研究的进展虽然较快。但是在迄今鉴定的 21 个基因中，有些基因的抗谱很窄，例如：*Xa1*（黄玉）、*Xa10*（Cas209）、*Xa11*（IR8）、*Xa18*（IR24）等，只抗日本或菲律宾的个别小种，对我国绝大多数菌株是不抗的。*xa5*、*xa8*、*xa13*、*xa19* 和 *xa20* 等基因的抗谱虽然较广，但在常规育种中利用隐性基因有一定限制，更不宜直接用于杂交水稻。因此，在育种中一直集中利用 *Xa4* 和 *Xa3*。本研究进一步证实我国水稻抗病育种所利用的抗源也都局限于这两个基因，在生产中有丧失抗性的风险。进一步发掘和利用其他稻种中可能蕴藏的新基因必然成为国内外研究的热点。

因为水稻与白叶枯病原菌之间存在着特异性互作，而目前国际系统鉴定和命名新基因多数采用日本菌株测定。日本稻种资源较为贫乏，其白叶枯病菌小种主要采集于粳稻。我国有丰富的籼、粳品种、地方品种和野生稻种资源。因此，在我国统一采用几个不同致病型菌株，建立鉴定抗性基因的标准化程序，系统挖掘和利用我国丰富的稻种资源中所蕴藏着的新抗源，拓宽遗传变异基础，对促进水稻抗病育种具有重要意义。

原载：作物学报，1994，20（1）：84～92.

六个籼稻品种对水稻白叶枯病的抗性遗传研究

杨文才　章　琦　王春莲

（中国农业科学院作物育种栽培研究所）

Inheritance of Resistance to Bacterial Blight (*Xanthomonas oryzae* pv. *oryzae*) in six Indica Cultivars

Abstract The Resistance of six indica rice cultivars, Cuba2301, Cuba2196, Cuba4010, Shuang-kang 12, Shuang-kang 18 and San-jiang to 14 strains of *Xanthomonas oryzae* pv. *oryzae* has been analysed. 1t was shown that these six cultivars were resistant to 12 strains, but susceptible to Philippine strains P2 and P3 at the all growth stages, indicated that these cultivars have a relatively broad spectrum of resistance. The inheritance and allelism test of resistance to 3 *Xoo* strains P1, HB17 and JL691 in six cultivars were contacted. The analyses of F_1, F_2 and B_1F_1 data from crosses between the six cultivars and susceptible cultivar Jingang 30 or Zaoxian 141 revealed that the resistance in the six indica cultivars is controlled by a single dominant nuclear gene at all growth stages. Allelism tests with the reported resistant genes indicated that the dominant gene in these six cultivars is *Xa4*.

Key words indica rice; inheritance of resistance; bacterial blight

摘要　分析了 Cuba 2301、Cuba 2196、Cuba 4010、双抗 12、双抗 18、三江 6 个籼稻抗病品种对 14 个水稻白叶枯病不同毒性菌株的抗性表现，进行了对 P1、HB17 和江陵 691 三个菌株的抗性遗传分析并测试了其与已知抗性基因的等位关系。结果表明，这 6 个抗病品种均在全育期抗除了菲律宾小种 2（P2）和 3（P3）之外的其余 12 个菌株；它们对 3 个菌株的全生育期抗性均由一对显性基因控制，受细胞核影响。与已知基因的等位性测定表明，这 6 个品种携有的一对显性基因与 IR20 或 IRBB4 所携有的 *Xa4* 基因是等位的。

关键词　籼稻；抗病性遗传；白叶枯病

由 *Xanthomonas oryzae* pv. *oryzae*（*Xoo*）引起的水稻白叶枯病是我国主要水稻病害之一。利用抗病品种是防治水稻白叶枯病最经济有效的方法。筛选抗源、分析其抗谱和抗性，了解我国主要抗源的基因构成以及与一些已知基因的等位关系，为抗病育种提供参考资料。在单菌株鉴定的基础上，从国外引进种质和国内育种材料中选出6个抗病籼稻品种，进一步用中国、日本和菲律宾的14个代表菌株进行抗性评价，表明其抗谱较广，为了进一步了解其抗性遗传基础，用国际命名基因时所用的小种和中国致病型代表菌株，分析其抗性遗传模式；根据多菌株鉴定和抗谱比较的结果，确定这些品种需要测试与 *Xa4* 基因的等位关系。

1 材料和方法

1.1 供试品种和菌株

采用10个品种。感病亲本为籼稻金刚30和早籼141，6个抗病籼稻是古巴品种 Cuba 2301、Cuba 2196、Cuba 4010、日本品种三江和中国品种双抗12、双抗18。与携有全生育期抗性基因 *Xa4* 的品种 IR20 及近等基因系 IRBB4 进行等位性测验。

用14个菌株进行鉴定，菲律宾小种1（P1）、2（P2）、3（P3），日本小种1（T1）和10个中国菌株，江苏 KS6－6、KS1－21、浙江 ZHE173、辽宁 LN－57、LN－44、河北 HB17、宁夏 NX－42、湖北江陵691、湖南 JS49－6和黑龙江 HLJ72。用 P1、HB17和江陵691三个菌株进行抗性遗传分析。

1.2 杂交组合的配制

金刚30或早籼141分别与 Cuba 2301、Cuba 2196、Cuba 4010、双抗12、双抗18和三江进行正交和反交，获感/抗组合 F_1 和抗/感组合 F_1；上述抗病品种分别与 IR20或 IRBB4进行等位性测验，3个古巴品种之间进行间接等位性测交，获得 F_1。所有组合的 F_1 种子分为两份，一份在下季播种，一份于海南岛加代获 F_2。由于不宜在海南进行人工接种鉴定，因此将感/抗组合 F_1 植株都分别与抗病亲本和感病亲本回交，获得 B_1F_1 种子。

1.3 种植和田间管理

1991年夏季进行多菌株鉴定的材料，各品种2行，每行14株。1992年的试验分两次进行，第一次播亲本和 F_1 各10株，感/抗组合的 F_2 群体300株。根据接种鉴定所有 F_1 均抗供试3个菌株的显性抗性结果，确定种植感/抗//感组合的 B_1F_1 约80株，同时种植用以作为等位性测试的抗/抗组合 F_2 群体300株。各套材料均设有感病对照。单本栽插，常规水肥管理，接种后保持3cm浅水层，不喷施任何杀菌剂农药。

1.4 接种鉴定与调查

接种前将保存的冻干菌种转到胁本哲氏斜面培养基上复壮后，转移数管斜面菌种保存于4℃冰箱，供整个生长季节用。接种液菌龄2～3d，浓度约 10^9/mL。

在最高分蘖期和孕穗期人工剪叶接种各世代植株的完全展开叶片。多菌株鉴定的材料，各菌株接种2个单株，每株剪4～5张叶片。F_1、F_2 和 B_1F_1 群体的各个单株都分别用3个菌株接种。引用 Ogawa（1986）的单株分蘖分组法，分别接种 P1、HB17和江陵691菌株。各处理接种4～5张叶片。由于6个品种都是全生育期抗性，等位性测试的 F_2 只需在最高分蘖期接种。接种后14～20d，当感病对照品种病情发展趋于稳定时，调查各单株对各菌株的抗性反应。以病斑面积占叶面积的百分率作为抗性反应参数，每株每菌株调查3张叶片，取平均值。以F_2、B_1F_1群体的病斑面积百分率分布出现的峰槽，作为抗感划分界限参考。本研究抗、感分界值为病斑面积占叶面积的20%。

全部试验于1991—1992年在北京、海南和广西共进行4季。

2 结果与分析

2.1 6个籼稻对14个菌株的抗性分析

1991年用14个菌株对6个抗病品种、抗、感对照品种IRBB4和金刚30接种鉴定。8个品种对14个菌株的抗、感反应列于表1。金刚30全生育期对14个菌株都感病，6个抗病品种和IRBB4对14个菌株的抗感表现是一致的，均在全生育期抗其中的12个菌株，感菲律宾小种P2和P3，抗谱较广。推测这6个品种可能携有与IRBB4相同的抗性基因*Xa4*。

表1 8个品种对14个水稻白叶枯病菌株的抗感反应

Table 1 Reactions of 8 cultivars to 14 strains of *Xoo*

菌株 Strain	金刚30	Cuba230	Cuba2196	Cuba1010	双抗12	双抗18	三江	IRBB4
P1	S	R	R	R	R	R	R	R
P2	S	S	S	S	S	S	S	S
P3	S	S	S	S	S	S	S	S
T1	S	R	R	R	R	R	R	R
KS6-6	S	R	R	R	R	R	R	R
KS1-21	S	R	R	R	R	R	R	R
HB17	S	R	R	R	R	R	R	R
NX42	S	R	R	R	R	R	R	R
LN44	S	R	R	R	R	R	R	R
JS49-6	S	R	R	R	R	R	R	R
LN57	S	R	R	R	R	R	R	R
HLJ72	S	MR	MR	MR	MR	MR	MR	MR
ZHE173	S	R	R	R	R	R	R	R
江陵691	S	R	R	R	R	R	R	R

R=抗（Resistant），MR=中抗（Moderately resistant），S=感（Susceptible）。

2.2 6个籼稻品种对3个菌株P1、HB17和江陵691的抗性遗传

2.2.1 亲本品种的抗性表现 根据用多菌株鉴定的结果，选用了P1、HB17和江陵691 3个菌株对6个籼稻品种进行复测，并对感病亲本金刚30、早籼141和携有*Xa4*基因的品种IR20和IRBB4进行抗性鉴定（表2）。6个籼稻品种和IR20、IRBB4全生育期对3个菌株都表现高抗。感病亲本金刚30和早籼141在全生育期始终高度感病。

表 2 亲本对 3 个菌株 P1、HB17 和江陵 691 的抗性反应

Table 2 Resistance reaction of parents to three strains，P1、HB17 and JL 691 of *Xoo*

品种 Cultivar	病斑占叶片面积百分率％（Percentage of lesion area ％）					
	最高分蘖期（Max. tillering stage）			孕穗期（Booting stage）		
	P1	HB17	江陵 691	P1	HB17	江陵 691
金刚 30	86.7±6.1	85.0±10.0	50.0±4.5	73.8±4.8	77.5±6.5	45.0±5.8
早籼 141	85.0±7.7	73.3±8.8	47.5±5.2	93.3±4.1	93.6±6.1	51.7±8.2
Cuba2301	3.2±0.6	3.0±0	3.5±0.9	3.0±0	3.0±0	3.0±0
Cuba2196	3.4±0.8	3.1±0.4	3.5±0.9	3.0±0	3.0±0	3.0±0
Cuba1010	3.5±0.9	3.6±0.9	3.7±1.0	3.0±0	3.0±0	3.0±0
双抗 12	5.0±0	5.0±0	5.8±1.0	3.0±0	3.0±0.0	3.0±0
双抗 18	5.0±0	5.0±0	3.0±0	3.0±0	3.0±0	3.0±0
三江	5.0±0	5.0±0	3.0±0	3.0±0	3.0±0	3.0±0
IRBB4	8.7±1.2	16.7±2.7	5.0±0	3.0±0	3.0±0	3.0±0
IR20	4.1±1.0	4.2±1.0	4.0±1.1	3.2±0.6	3.6±1.0	3.4±0.8

2.2.2 抗病品种的抗性遗传模式 6 个抗病品种分别与感病品种金刚 30 或早籼 141 正反交组合的 F_1 植株在最高分蘖期和孕穗期对 3 个菌株均表现抗病，表明这 6 个品种对 3 个菌株的全生育期抗性由显性基因控制，受细胞核影响。各组合 F_1 植株对 3 个菌株的抗性反应列与表 3。

表 3 各组合 F_1 植株对 3 个菌株的抗性反应

Table 3 Reaction of F_1 plants from all crosses to 3 strains of *Xoo*

组合 Cross	病斑占叶片面积百分率％（Percentage of lesion area ％）					
	最高分蘖期（Max. tillering stage）			孕穗期（Booting stage）		
	P1	HB17	江陵 691	P1	HB17	江陵 691
金刚 30/Cuba 2301	4.7±0.8	4.4±1.0	4.4±1.0	3.0±0	3.0±0	3.0±0
Cuba 2301/金刚 30	5.0±0	3.0±0	5.0±0	3.0±0	3.0±0	3.0±0
金刚 30/Cuba 2196	5.0±0	5.0±0	5.0±0	3.0±0	3.0±0	3.0±0
Cuba 2196/金刚 30	5.0±0	5.0±0	5.0±0	3.0±0	3.0±0	3.0±0
金刚 30/Cuba 1010	3.0±0	4.0±1.0	4.3±1.0	3.0±0	3.0±0	3.0±0
Cuba 1010/金刚 30	3.0±0	5.0±0	3.0±0	3.0±0	3.0±0	3.0±0
金刚 30/双抗 12	5.0±0	5.5±1.6	5.0±0	3.0±0	3.0±0	3.0±0
双抗 12/金刚 30	5.0±0	5.0±0	5.0±0	3.0±0	3.0±0	3.0±0
金刚 30/双抗 18	5.0±0	5.0±0	5.0±0	3.0±0	3.0±0	3.0±0
双抗 18/金刚 30	5.0±0	5.0±0	5.0±0	3.0±0	3.0±0	3.0±0
金刚 30/三江	5.0±0	5.0±0	5.0±0	3.0±0	3.0±0	3.0±0
三江/金刚 30	5.0±0	5.0±0	5.0±0	3.0±0	3.0±0	3.0±0
Cuba 2301/IRBB4	3.0±0	5.4±2.1	5.4±2.1	3.0±0	3.6±1.0	3.6±1.0

（续）

组合 Cross	病斑占叶片面积百分率%（Percentage of lesion area %）					
	最高分蘖期（Max. tillering stage）			孕穗期（Booting stage）		
	P1	HB17	江陵 691	P1	HB17	江陵 691
Cuba 2301/IR20	3.0±0	3.0±0	3.0±0	3.0±0	3.0±0	3.0±0
Cuba 2196/IRBB4	3.0±0	3.0±0	3.0±0	3.0±0	3.0±0	3.0±0
Cuba 2196/IR20	3.0±0	3.0±0	3.0±0	3.0±0	3.0±0	3.0±0
Cuba 1010/IRBB4	3.0±0	3.0±0	3.0±0	3.0±0	3.0±0	3.0±0
Cuba 1010/IR20	3.0±0	3.0±0	3.0±0	3.0±0	3.0±0	3.0±0
Cuba 2301/Cuba 1010	3.0±0	3.0±0	3.0±0	3.0±0	3.0±0	3.0±0
Cuba 1010/Cuba 2301	3.0±0	3.0±0	3.0±0	3.0±0	3.0±0	3.0±0
Cuba 2196/Cuba 2301	3.0±0	3.0±0	3.0±0	3.0±0	3.0±0	3.0±0
Cuba 1010/Cuba 2196/	3.0±0	3.0±0	3.0±0	3.0±0	3.0±0	3.0±0
双抗 12/IRBB4	3.0±0	3.0±0	3.0±0	3.0±0	3.0±0	3.0±0
双抗 18/IRBB4	3.0±0	3.0±0	3.0±0	3.0±0	3.0±0	3.0±0
三江/IRBB4	3.0±0	3.0±0	3.0±0	3.0±0	3.0±0	3.0±0
三江/IR20	3.0±0	3.0±0	3.0±0	3.0±0	3.0±0	3.0±0

金刚 30 或早籼 141 分别与 6 个抗病品种杂交组合的 F_2 和 B_1F_1 植株群体对供试菌株的抗性反应见表 4 和表 5。各组合的 F_2 群体在最高分蘖期的抗感分离均符合 3 抗∶1 感的一对显性基因分离比例；这些 F_2 植株到孕穗期的抗病株数除个别组合有所增加外，大部分组合的 F_2 抗病株数几乎没变化，仍符合 3 抗∶1 感的一对显性基因分离比例；早籼 141/三江 F_2 在孕穗期抗病株数虽有所增加，但经 χ^2 测验表明，其仍符合 3 抗∶1 感的一对显性基因分离比。各组合的 B_1F_1 群体在两个生育期的抗感分离比均符合一对显性基因期望比值 1 抗∶1 感，表明这 6 个籼稻品种对 P1、HB17 和江陵 691 的全生育期抗性由一对显性基因控制。

表 4 感/抗组合 F_1、F_2 植株对 3 个菌株的抗性表现

Table 4 Reaction of F_1、F_2 plants from S/R to 3 strains of *Xoo*

组合 Cross	F_1 对 3 个菌株的抗性反应（F_1 reaction to 3 strains）	F_2 对 P1、HB17、江陵 691 三个菌株的抗性反应*（F_2 reaction to P1，HB17，JL691 strains）									
		最高分蘖期（Max. tillering stage）					孕穗期（Booting stage）				
		RRR	SSS	Exp.	χ^2	*P*	RRR	SSS	Exp.	χ^2	*P*
金刚 30/Cuba 2301	RRR	225	70	3∶1	0.254	0.50～0.75	235	70	3∶1	0.254	0.50～0.75
金刚 30/Cuba 2196	RRR	235	71	3∶1	0.527	0.25～0.50	235	72	3∶1	0.392	0.50～0.75
金刚 30/Cuba 1010	RRR	219	39	3∶1	0.167	0.50～0.75	220	68	3∶1	0.296	0.50～0.75
金刚 30/双抗 12	RRR	218	71	3∶1	0.029	0.75～0.90	218	71	3∶1	0.029	0.75～0.90
金刚 30/双抗 18	RRR	223	75	3∶1	0.004	>0.90	223	75	3∶1	0.004	>0.90
早籼 141/三江	RRR	221	70	3∶1	0.284	0.50～0.75	234	66	3∶1	1.44	0.10～0.25

* R=抗（Resistant），S=感（Susceptible），第 1、2、3 个字母分别代表对 P1、HB17、江陵 691 三个菌株的抗性反应（The first letter stands for reaction to strain P1，2nd to HB17，3rd to JL691）。

表 5　感/抗组合 B_1F_1 植株对 3 个菌株的抗性表现

Table 5　Reaction of B_1F_1 plants from S/R to 3 strains of *Xoo*

组合 Cross	最高分蘖期（Max. tillering stage）					孕穗期（Booting stage）				
	* RRR	SSS	Exp.	χ^2	*P*	RRR	SSS	Exp.	χ^2	*P*
金刚 30/Cuba 2301//金刚 30	38	42	1∶1	0.20	0.50～0.75	38	42	1∶1	0.20	0.50～0.75
金刚 30/Cuba 2196//金刚 30	24	26	1∶1	0.08	0.75～0.90	24	26	1∶1	0.08	0.75～0.90
金刚 30/Cuba 1010//金刚 30	28	30	1∶1	0.069	0.75～0.90	30	31	1∶1	0.016	>0.90
金刚 30/双抗 12//金刚 30	36	34	1∶1	0.057	0.75～0.90	36	34	1∶1	0.057	0.75～0.90
金刚 30/双抗 18//金刚 30	36	38	1∶1	0.054	0.75～0.90	37	37	1∶1	0	1
早籼 141/三江//早籼 141	29	30	1∶1	0.017	>0.90	30	30	1∶1	0	1

* 第 1、2、3 个字母分别代表对 P1、HB17、江陵 691 三个菌株的抗性反应（The first letter stands for reaction to strain P1，2nd to HB17，3rd to JL691）。

2.2.3　抗性基因的等位性测验　6 个抗病品种与已知基因品种之间进行等位性测验的 10 个组合的 F_1、F_2和 3 个古巴品种之间间接等位性测验的 4 个组合的 F_1、F_2群体对 3 个菌株 P1、HB17 和江陵 691 的抗性反应结果列入表 6。

这 6 个品种与 IR20 或 IRBB4 杂交组合的 F_1植株在最高分蘖期对 3 个菌株都表现高抗，其 F_2植株也全部抗病，没有分离出感病植株，表明这些全生育期抗病的籼稻品种对 3 个菌株的抗性基因均与 IR20 或 IRBB4 的基因 *Xa4* 等位。

3 个古巴品种间杂交组合的 F_1植株在最高分蘖期对 3 个菌株均表现抗病，其 F_2植株也全部抗病，没有分离出感病植株，表明 Cuba2301、Cuba2196 和 Cuba4010 所具有的抗性基因是等位的。

表 6　抗/抗组合 F_1、F_2 植株对 3 个菌株的抗性表现

Table 6　Reaction of F_1，F_2 plants from R/R to 3 strains of *Xoo*

组合 Cross	F_1 对 3 个菌株的抗性反应* F_1 reaction to 3 strains	F_2 对 P1、HB17、江陵 691 三个菌株的抗性反应 F_2 reaction to P1，HB17，JL691 strains	
		RRR	SSS
Cuba 2301/IRBB4	RRR	280	0
Cuba 2301/IR20	RRR	300	0
Cuba 2196/IRBB4	RRR	300	0
Cuba 2196/IR20	RRR	300	0
Cuba 1010/IRBB4	RRR	300	0
Cuba 1010/IR20	RRR	300	0
Cuba 2301/Cuba 1010	RRR	300	0
Cuba 2196/Cuba 2301	RRR	300	0
Cuba 1010/Cuba 2301	RRR	300	0
Cuba 1010/Cuba 2196	RRR	300	0
双抗 12/IRBB4	RRR	300	0
双抗 18/IRBB4	RRR	300	0
三江/IRBB4	RRR	300	0
三江/IR20	RRR	300	0

* 同表 4（as same as the table 4）。

3 讨论

抗病基因 *Xa4* 的抗谱较广，它抗日本所有的小种，也抗我国除Ⅴ群（代表菌株 GD1358）以外的其他 6 群菌。我们分析的 6 个全生育期抗病品种与感病亲本杂交组合 F_1 植株在最高分蘖期和孕穗期均呈现高度抗病，为完全显性抗性反应。但等位性测验表明，该抗性基因与 IR20 或 IRBB4 所携带的 *Xa4* 是等位的。说明这 6 个品种均携有基因 *Xa4*，但抗性强于 *Xa4*，其原因尚不明确。这些均携有 *Xa4* 的品种分别来自中国、日本、古巴，进一步证实了目前我国抗病育种集中利用了 *Xa4*。在生产上丧失抗性的风险很大。因此在抗白叶枯病育种中，应将这些品种与携有其他抗性基因的品种搭配说明。需要不断发掘利用新的抗性基因，培育出抗性寿命较长的品种。

原载：华中农业大学学报，1994，13（3）：234～240.

对水稻广亲和品种抗白叶枯病的评价和抗性基因分析

杨文才　章　琦　王春莲

（中国农业科学院作物育种栽培研究所）

Evaluation and Analysis of BB Resistance（gene）in Wide Compatibility Rice Cultivars

水稻广亲和品种（Wide compatible variety）是一种特殊类型的稻种资源。它在水稻超高产育种和亚种间杂种优势利用方面具有重要作用。袁隆平提出了利用广亲和基因实现籼粳亚种间杂种优势利用的设想。近年来，国内外学者围绕着广亲和品种资源的发掘和利用作了较多工作，筛选出一批广亲和品种，研究了广亲和性的遗传基础，选育了一些广亲和品系、广亲和恢复系和广亲和光敏核不育系（李新奇，1987；邹江石等，1989；罗孝和等，1989），但对于广亲和品种抗白叶枯病的研究至今鲜有报道。因此，我们从1991年开始，选用中国、日本和菲律宾的22个代表菌株，对29个广亲和品种进行了多次重复接种鉴定，以了解我国主要广亲和品种的抗性状况，并通过遗传分析明确抗病品种的抗病性遗传基础，为在抗病育种计划中利用广亲和品种提供参考。

1　材料和方法

1.1　供试品种和菌株

供试品种包括29个广亲和品种、感病和抗病对照品种各一个（表1）。用于遗传分析的感病品种为金刚30、南京11、02428、珍珠矮和沈农1033；抗病品种为Dular、Aus375、69232-2、鉴12和029。携有抗性基因品种为IRBB5、IRBB8、DV85和早生爱国3号。

采用的22个菌株包括菲律宾白叶枯病6个小种的代表菌株：P1、P2、P3、P4、P5和P6；日本4个主要小种代表菌株：T1、T2、T3和T4，以及中国的包括7个致病型的12个菌株：KS6-6（苏）、KS1-21（苏）、HB17（冀）、NX42（宁）、LN43（辽）、LN44（辽）、JS49-6（湘）、LN57（辽）、HLJ72（黑）、ZHE173（浙）、J1691（鄂）和GD1358（粤）。

表1 供试的29个品种及抗病对照（IR20）与感病对照品种（金刚30）

品种	类型	来源或系谱	品种	类型	来源或系谱
Aus 373	籼	印度	Bellement	爪哇	C19881/PI331581
CA537	籼	印度	CPSL 0-17	爪哇	CP231/SL017
CA544	籼	四川农业大学	E164	籼	CP231/SL017 的杂交后代
Calotoc	籼	四川农业大学	JW-9	粳	02428 的杂交后代
Dular	籼	印度	Ketan Nangka	爪哇	印度尼西亚
PC311	粳	培迪/C57	Lement	爪哇	Lebonnet//C19881/PI331581
69232-2	籼	籼粳复交后代	CP231	爪哇	Texas Patna//Rexoro/Tainan-iku 487
花广 518	籼	四川农业大学	Pecos	粳	C19545//Gulfrose/Tainan-iku 487
轮回 422	爪哇	培迪/C57	Pelde	籼	澳大利亚
培迪	爪哇	印度尼西亚	02428	粳	螃蟹谷/汲滨稻
鉴 12	籼	美国	029	粳	陆矮//5009/IR36//ASD7
Mcp231-4	粳	CP231 辐射	华师大 7009	粳	02428 的杂交后代
Mcp231-7	籼	CP231 辐射	早 AB78	籼	CPSL0-17 的杂交后代
金刚 30	籼	感病对照	迟 AB78	籼	CPSL0-17 的杂交后代
IR20	籼	抗病对照	培矮 64	籼	培迪/矮黄米//测 64
测 09	粳	02428 杂交后代			

1.2 杂交组合的配制与种植

以感病亲本作为母本与抗病品种杂交。根据F_1的抗性反应确定回交组合的配制方式，呈感病反应的F_1与抗病亲本回交，抗病反应的F_1与感病亲本回交，以抗病亲本作母本与已知抗性基因品种进行测验。所有F_1分成两组，一组种植获得F_2，留出一组与F_2同期种植。用多菌株鉴定的材料在每年4月中旬播种，5月中下旬移栽至严密隔离的网室水泥池中，每个品种播2行，每行22株。用于遗传分析的双亲、F_1、F_2和B_1F_1在同期播种，亲本和F_1各10株，$B_1F_1$100株，$F_2$200～300株。以上均设置感病对照，单株栽插，常规水肥管理，接种后，保持3cm浅水层，不喷施任何农药。

1.3 接种鉴定与调查

接种前将保存的冻干菌种转至胁本哲氏斜面培养基上进行复壮，经毒力稳定性测试后，转移数管斜面菌种，保存于4℃冰箱，供整个生长季节用。接种菌种在28℃温箱中培养2～3d，接种菌液浓度约10^9/mL。采用人工剪叶接种法，在植株最高分蘖期和孕穗期分别接种完全展开的叶片。每个菌株接种2株，每株接种4～5张叶片。接种后20 d，当感病对照品种充分发病、病情发展趋于稳定时，调查单株对各菌株的抗性反应。以病斑面积占叶面积的百分率和病斑侵染型作为抗性反应参数，每株每菌株调查3张叶片，取平均值。分离群体以病斑面积百分率分布出现的峰槽趋势作为抗感划分的参考，本研究抗感分界值为病斑面积占叶面积的20%。

试验于1991—1994年在本所试验网室进行。

2　结果与分析

2.1　供试品种对若干菌株的抗性分析

在 29 个供试品种中，8 个抗病品种对 22 个菌株的结果列于表 2。感病对照品种金刚 30 对 22 个菌株均表现感病，抗病对照 IR20 对各菌株的抗感反应符合 *Xa4* 的典型反应型，即抗除了菲律宾小种 P2、P3、P6 和中国Ⅴ型菌以外的 18 个菌株。

在 29 个广亲和品种中，有 8 个表现抗病，占总数的 27.6%（表 2）。其中 Dular 在全生育期抗除菲律宾菌株 P6 以外的 21 个菌株，Aus 373 全生育期抗除 P4 和 P6 以外的 20 个菌株。这两个品种的抗性好，抗谱也广。Pecos 在全生育期抗除了 P6 以外的 21 个菌株，也表现广谱抗性，但对部分菌株为中等抗性。品种华师大 7009 在全生育期抗除了 P3 和 P6 以外的 20 个菌株，对部分菌株表现中抗。69232－2 和鉴 12 在成株期抗除了 P6 和 T4 以外的所有菌株，抗谱也较广。029 和培矮 64 这两个品种对部分菌株的抗性稳定，而对另一部分菌株的抗性在不同年份、季节和生育期表现不同，分别为抗和中抗、甚至感病，其原因尚不清楚。其余 21 个品种对 22 个菌株均感病，占供试品种的 72.4%。

从各供试品种的系谱或来源看，广谱、全生育期抗病的 Dular 和 Aus373 都来自印度；Pecos 和鉴 12 来自美国。Pecos 的抗源可能是 Tainaniku 487；培矮 64 的亲本之一是测 64，选自 IRRI 的抗白叶枯病品系 IR9761－19－1，029 的亲本之一是抗白叶枯病的 IR36，为此，培矮 64 和 029 的抗性可能均由 *Xa4* 基因所控制。表明我国主要广亲和品种中的抗病品种，或直接来源于国外品种，或通过引进品种（系）杂交育成，这与我国主栽品种中的抗源来源情况是一致的。

表 2　8 个抗病广亲和品种和抗病对照品种（IR20）对 22 个水稻白叶枯病菌株的抗性反应

品　种 Cultivar	所用菌株											
	P1[c]	P2 C5	P3	P4	T4	P5 C2 C4 KS－1－21	P6	C1 JI691	C3	C6	C7	T1 LN44	T2 T3
Dular	R[b]	R	R	R	R	R	S	R	R	R	R	R	R
Pecos	MR	R	MR	R	R	R	S	R	R	R	R	R	R
Aus373	R	R	R	S	R	R	S	R	R	R	R	R	R
华师大 7009	R	R	S	R	R	R	S	R	R	R	R	R	R
69232－2[a]	R	R	R	R	S	R	S	R	R	R	R	R	R
鉴 12[a]	R	R	R	R	S	R	S	R	R	R	R	R	R
培矮 64[a]	R	R	S	S	R	R	S	S	R/S	R	R/S	R	R
029[a]	R	R	S	R/S	S	R	S	R	R/S	R/S	S/R	S/R	S
IR20	R	S	S	R	R	R	S	R	R	R	R	R	R

[a] 为成株期抗病品种，其他为全生育期抗性品种。

[b] R 抗；MR 中抗；S 感。

[c] P1～P6 菲律宾小种；T1～T4 日本小种；C1～C7 中国致病型。

2.2 5个抗病品种的抗性遗传分析

根据多菌株鉴定的结果，分析Dular、Aus 373、69232-2、鉴12和029五个抗病品种的抗性遗传。这些品种分别与感病亲本和抗病品种杂交后代F_1、F_2的抗性反应结果列于表3和表4。Dular和Aus373在苗期携有一对隐性基因，在孕穗期有两对隐性基因控制其对菌株P1和HB17的抗性；029在成株期携有一对显性基因控制其对P1和HB17的抗性；69232-2和鉴12在成株期由一对显性基因控制其对T3和HB17的抗性。等位性测验表明，Dular所携有的一对全生育期抗性隐性基因与IRBB5和DV85所携带的隐性基因*xa5*等位，IRBB5/Dular F_1是感病的，但是Dular在成株期所携带的一对隐性基因与成抗基因*xa8*不等位；69232-2和鉴12所携带的成株期抗性基因与早生爱国3号的*Xa3*基因等位；Aus373和029这两个品种的基因等位性研究尚在进行之中。

表3 各组合F_1、F_2植株的抗性表现

组 合	所用菌株	接种时期[a]	F_1抗性反应[b]	F_2抗性表现					
				抗病	感病	总数	Exp.	χ^2	P
金刚30/Dular	P1、HB17	MTS	SS	76	196	272	1∶3	1.00	0.50～0.25
		BS	SS	129	143	272	7∶9	1.35	0.25～1.0
IRBB5/Dular	P1、HB17	MTS及BS	RR	250	0				
IRBB8/Dular	P1、HB17	MTS及BS	SS						
Dular/DV85	P1、HB17	MTS	RR	280	0				
南京11/Aus373	P1、HB17	MTS	SS	39	146	285	1∶3	1.31	0.50～0.25
		BS	SS	84	101	185	7∶9	0.21	0.75～0.50
02428/69232-2	T3、HB17	MTS	SS	0	240	240			
		BS	RR	179	60	239	3∶1	0.001	＞0.90
69232-2/早生爱国3号	T3、HB17	BS	RR	250	0	250			
珍珠矮/鉴12	T3、HB17	MTS	SS	0	300	300			
		BS	RR	207	73	300	3∶1	0.12	0.75～0.50
鉴12/早生爱国3号	T3、HB17	MTS	SS	0	250	250			
		BS	RR	250	0	250			
沈农1033/029	P1、HB17	MTS及BS	RR	236	76	312	3∶1	0.04	0.90～0.75

a MTS=最高分蘖期，BS=孕穗期；R=抗，S=感；

b 第1、2个字母分别代表对P1、HB17或T3、HB17两个菌株的抗性反应。

表4 各组合B_1F_1植株的抗性表现*

组 合	所用菌株	接种时期	抗性表现					
			抗病	感病	总株数	Exp.	χ^2	P
金刚30/Dular//Dular	P1、HB17	MTS	40	38	78	1∶1	0.01	＞0.90
		BS	58	19	77	3∶1	0.004	＞0.90
南京11/Aus373//Aus373	P1、HB17	MTS	48	50	98	1∶1	0.09	0.90～0.75
		BS	73	25	98	3∶1	0	1

（续）

组　　合	所用菌株	接种时期	抗性表现					
			抗病	感病	总株数	Exp.	χ^2	P
02428//02428/69232-2	T3、HB17	MTS	0	100	100			
		BS	55	45	100	1∶1	0.81	0.50～0.25
珍珠矮/鉴12//珍珠矮	T3、HB17	MTS	0	100	100			
		BS	51	49	100	1∶1	0.01	＞0.90

* MTS=最高分蘖期，BS=孕穗期。

3　讨论

张林等（1988）报道全生育期抗病品种在籼粳交后代中能分离出成株抗性植株。万建民等（1990）报道抗病品种在不同的感病品种背景下有特异抗性表达，特异性可发生在亚种间杂交后代，也可发生在亚种内杂交后代。本研究中的抗病品种029的亲本之一是携有全生育期抗性基因 *Xa4* 的IR36，经过与其他品种杂交育成的029，其抗性在成株期表达，可能就是上述特异性表达的结果，也受到了遗传背景的影响，其抗性基因有待进一步研究。

广亲和品种在克服籼、粳亚种间杂种优势利用中的育性障碍、选育超高产新品种、培育不育系和恢复系等方面具有广阔的应用前景，有必要对其进行其他性状的了解，为育种家提供更多的信息。本研究结果表明，在目前的广亲和品种中，抗白叶枯病的品种不多，因此在选育广亲和品种时，改良抗病性需要引起重视。目前，我国所用的抗白叶枯病基因过于集中于 *Xa3* 和 *Xa4*，丧失抗性的风险是存在的。在本研究所鉴定的8份抗病材料中，两个全生育期抗病品种Dular和Aus373的抗性好、抗谱广。由于它们所携带的是隐性抗性基因，一般可用于常规育种，而在杂交水稻育种中，显性抗性基因更便于利用。

此外，感病的广亲和品种可用作抗病性遗传研究中的感病亲本，以获得足够的亚种间杂交后代的种子进行遗传分析，或作为轮回亲本进行抗性改良，培育抗病的育种中间试材。

原载：北京农业科学，1995，13（6）：13～16.

我国水稻抗白叶枯病（*Xanthomonas oryzae* pv. *oryzae*）性状遗传研究的标准化问题的商榷*

章　琦[1**]　杨文才[1]　施爱农[1]　王春莲[1]　朱立宏[2**]　沙学延[2]
诸培新[2]　陆志强[2]　谢岳峰[3**]　张端品[3**]　林兴华[3]　余功新[3]

（1. 中国农业科学院作物育种栽培研究所；2. 南京农业大学；3. 华中农业大学）

Discussion on Standardization of Genetic Study for Resistance to Bacterial Blight（*Xanthomonas oryzae* pv. *oryzae*）in China

Abstract The standardization of genetic study methods for resistance to bacterial blight in China have been explored cooperatively from 1991 to 1994. Contents：1) Materials and methods，including key points of selection of susceptible parents and deciding resistant materials，scientific basis for strains adoption，time of inoculation and identification，and investigation standard. 2) Methods of genetic analysis for resistance，including combination type and its capacity of progenies，contents of genetic analysis，methods and procedure for identification，location and designation of new genes. A primary suggestion has been proposed in the paper for further consummation. It will be useful for excavating new available resistant gene（s）and advancing rice breeding for resistance.

Key words bacterial blight of rice；genetics for resistance；standardization

摘要 1991—1994 年，通过协作探索了我国水稻对白叶枯病抗性遗传研究方法标准化的有关问题。内容是：（1）取材和方法，包括选择感病亲本和确定待测抗病材料的基本要求、采用菌株的依据、接种鉴定的时间、调查标准等。（2）抗性遗传分析方法，包括配组方式及其群体样本容量、遗传分析内容，以及新基因的鉴定、定位及命名等方法和步骤。我们在本文提出了一个初步的设想，拟与本领域的科学家们进行研讨，便于进一步完善，这将对我国今后科学地鉴定发掘可利用的抗性基因，促进水稻抗性育种和实现高产稳产具有积极意义。

关键词 水稻白叶枯病；抗性遗传；标准化

* 国家自然科学基金重点资助项目和农业部重点资助项目。 ** 主要撰稿和定稿人

水稻白叶枯病（*Xanthomonas oryzae* pv. *oryzae*，*Xoo*）是一种寄生性较强的病害。稻类对白叶枯病的抗性遗传所表现的寄主与病原菌之间的互作，多半是典型的基因对基因的关系，也就是水稻白叶枯病菌对于品种的毒性是不同的。因此，品种携有的主效抗病基因与所控制的病原菌是相匹配的。

日本和国际水稻研究所（IRRI）分别于20世纪60和70年代用各自的菌株进行水稻品种的抗性主效基因鉴定。由于采用不同菌株和方法所鉴定的抗性基因无法直接相比，为此双方自1982年开始合作，采用统一的方案，历经5年，整理和统一鉴定了21个白叶枯病抗性基因，并且建立了一套国际水稻白叶枯病单基因鉴别系统（Ogawa et al.，1989）。

我国自70年代以来已从云南、太湖流域地方资源和国内外野生稻种资源及引进品种中评价出大量抗病材料，对部分材料进行了抗性遗传研究与已知基因等位测定和基因定位（王昌明 1986；周毓珍等，1985；谢岳峰等，1990；林兴华等，1992；章琦，1986），为抗病育种提供了很有价值的科学依据。但是由于研究分散，采用的菌株、方法及调查标准不同，对相同的品种却得出不同的遗传解释（周泰初等，1987；章琦等，1990），结果难以比较。随着水稻生产和遗传育种的发展，迫切需要深入开发利用新基因源和进行国内外交流，以提高抗性基因鉴定水平。为此，中国农业科学院、华中农业大学和南京农业大学于1991年至1994年，共同协作，制定了统一的研究方案，鉴定和分析了地方资源和外引品种的抗性基因，探讨了建立抗性遗传研究标准化的有关问题，以实现立项目标，并与国际标准接轨。今将合作结果公之于众，共同研讨，以期进一步完善。本文仅就稻类对白叶枯病主基因抗性遗传研究中的取材与方法的标准化程序，提出几点商榷意见。

1 试验材料标准化

1.1 感病亲本的选择

抗病性遗传研究中的感病亲本的4个基本要求是：1）植株在全生育期对绝大多数菌株高度感病，尤其必须对所用的或者致病力不同的菌株无抗病功能；2）较好的农艺性状和适应性，生育期适中；3）群体内遗传上达到相对纯化；4）籼粳稻兼备，一般用于亚种内配置抗感交组合。

根据上述要求，自1991年预试以来，在北京、武汉和南京前后采用了10～20个与包括国际命名的抗性基因相对应的鉴别菌株和国内代表菌株，鉴定了预选的5个籼稻品种，即珍珠矮、早籼141、金刚30、南京11以及国际籼稻白叶枯病抗性基因近等基因系的轮回亲本IR24和3个粳稻品种，即南粳33、沈农1033和02428的侵染反应。结果表明，5个籼稻品种在全生育期都高度感病；就生育期来看，早籼141太短，与南京11配组的后代有些植株生育期太晚，IR24的生育期太长，珍珠矮和金刚30较为适中；在粳稻中较难于选到全生育期对多数菌株高度感病的品种；02428的感病程度较高，沈农1033的感病性近年不大稳定，而南粳33在成株期对多数的菌株表现中抗（表1）。

在用多个菌株预选的基础上，选中的感病亲本不少于20个单株套袋自交留种，在选择过程中继续用多菌株测定，套袋自交单收。最后，我们选择了感病性相对纯合的籼稻珍珠矮、南京11（原种）、粳稻02428及沈农1033。

表 1 感病亲本对白叶枯病菌株的测试反应及其他性状

Table 1 Reaction of susceptible parents to strains of *Xoo* and other characters

品种 Cultivar	类型 Type	生育期[a] Growth Duration (d)	测试反应（Test reaction）			
			毒性菌株数[b] No. of virulent strain	占总数百分率（%） % of the total	病斑占叶片面积百分率%（Percentage of lesion area）	
					MTS[c]	BS[d]
珍珠矮	籼	125～130	20	100	50	40
金刚 30	籼	130	20	100	85	60
早籼 141	籼	106	20	100	90	90
南京 11	籼	130～135	20	100	80	50
IR24	籼	142	20	100	85	45
02428	粳	140	20	100	80	50
南粳 33[b]	粳	145	6	60	75	15～40
沈农 1033	粳	155	18	90	80	10～35

[a] 所列各品种的生育期系根据《中国水稻品种及其系谱》专著［Growth duration (days) of each cultivar listed based on the book 〈Rice Variety and Their Genealogy in China〉］。

[b] 除南粳 33 用 10 个菌株测定外，其余均用 20 个菌株测定（Nangeng 33 was tested with 10 strains, the others were tested with 20 strains）。

[c] MTS＝最高分蘖期（Maximum tillering stage）。

[d] BS＝孕穗期（Booting stage）。

1.2 抗病亲本的确定

进行主基因抗性遗传研究的抗病亲本，应表现稳定的抗性反应，经过多菌株或某一特试菌株的多次和不同环境条件下的鉴定。抗、感亲本对供试菌株的反应差异显著。

1.3 基因等位性测定在已知抗病基因品种（系）的应用

鉴定新基因，须有一套国际命名的已知抗性基因的代表品种及其携有单基因的近等基因系。国际已推出一套包括 18 个材料的鉴别品种，其中有 11 个单基因已分别导入轮回亲本 IR24，育成近等基因系。其他近等基因系还在培育中（Ogawa，1993）（表 2）。

我们于 1994 年在北京用 48 个包括中国和国际鉴别菌株测定了 18 个国际鉴别基因品种（表 2），以选择适用于我国基因等位性测定的材料。初步认为：

1）在显性基因的鉴别品种中，IRBB2（*Xa2*）、IRBB4（*Xa4*）、IRBB7（*Xa7*）、IRBB21（*Xa21*）和 Asominori（*Xa17*）可以利用。*Xa3* 广泛分布于粳稻品种，可用于进行粳稻品种的基因等位性测试。IRBB3 携有的 *Xa3* 是由 Wase Aikoku 3（早生爱国 3 号）（粳）导入 IR24（籼）转育而成，但在多次测试中，其抗性不够稳定，为此可采用我国培育的粳稻近等基因系 CBBB3（章琦等，1996）或早生爱国 3 号。IRBB11（*Xa11*）和 IRBB14（*Xa14*）对我国华北和辽宁的菌株具有专化抗性，可用作专化抗性材料的等位性测试。此外，IRBB1（*Xa1*）、IRBB10（*Xa10*）和丰锦（*Xa18*）仅分别对日本小种 1、菲律宾小种 2 和缅甸的 2 个菌株具有专化抗性，*Xa1* 和 *Xa10* 对我国菌株基本上没有抗性功能，*Xa18* 对我国部分菌株的抗性中等或无抗性，为此不宜采用这些品种。

2）在携有隐性基因的鉴别品种中，IRBB5（*xa5*）和IRBB13（*xa13*）携带全生育期抗性基因，它们的抗谱不同，可作为籼稻中未知的全生育期隐性基因测试品种。IRBB8（*xa8*）携带一个来自籼稻的成株抗性基因，可测试籼稻成株抗性的未知隐性基因。籼稻感病品种IR24通过化学诱变，获得的两个突变体XM5（*xa19*）和XM6（*xa20*），以及由粳稻Harebare诱变而来的M41（*xa15*），其抗性的类型及抗谱尚需作进一步测试，才能确定。

表2 推荐用以监测水稻白叶枯病小种分布及新基因鉴定的国际鉴别品种

Table 2 Proposed international differentials for monitoring the race distribution and identifying new genes of the bacterial blight pathogen

鉴别品种 Differential	类型 Type	抗病基因 Resistance gene	分析的品种 Cultivar analyzed
Toyunishiki	粳（Japonica）	*Xa18*	Toyunishiki，Miyang 23，IR24
Tetep	籼（Indica）	*Xa1*，*Xa2*，*Xa16*	Tetep
IRBB1	籼	*Xa1*，*Xa12*	Kogyoku
IRBB2	籼	*Xa2*	Rantai-Emas 2
IRBB3	籼	*Xa3*	Wase Aikoku 3
IRBB4	籼	*Xa4*	IR20
IRBB5	籼	*xa5*	IR1545－339
IRBB 7	籼	*Xa7*	DZ278
IRBB8	籼	*xa8*	PI 231129
IRBB10	籼	*Xa10*	GAS209
IRBB11	籼	*Xa11*	IR8
IRBB21	籼	*Xa21*	*O. longistaminata*
BJ1	籼	*xa5*，*xa13*	BJ1
TN1	籼	*Xa14*	TN1
Asominori	粳	*Xa17*	Asominori
选择性鉴别品种 Optional differentials			
M41	粳	*xa15*	Harebare
XM5	籼	*xa19*	IR24诱变体（Mutant）
XM6	籼	*Xa20*	同上

1.4 所用菌株的确定

在白叶枯病原菌分化研究基本清楚的情况下，水稻与菌株（小种）之间存在的基因对基因关系比较明显。如果鉴定新的抗病基因，应采用能克服全部已知抗性基因的小种。因此，广致病谱或有特异毒性的小种或菌株是可选的目标。在缺乏这种菌株（小种）时，应进行杂交鉴定，根据在遗传分离中未知与已知基因是否独立分离进行分析和推断。进行未知基因的基因等位测试时，应采用与已知抗性基因相匹配的病菌小种。迄今国际已命名的水稻白叶枯病抗性基因，主要与日本或菲律宾的鉴别小种对应。因此，确定所用的菌株，一方面考虑与国际鉴别系统接轨，同时根据我国的需要与基础，选择若干与常用已知基因相对应的菌株。

我国目前菌株分化的研究还没有如同菲律宾或日本那样存在清晰的寄主—病原菌的互作关系。由此，确定我国的菌株要考虑：1）不同生态区的优势菌株；2）引起生产上推广的抗病品种丧失抗性的菌株；3）具有特殊毒性的菌株。初步推荐：

（1）中国代表菌株 JL691（特殊毒性菌）、HB17（华北优势强毒菌）、ZHE173（长江流域强毒菌）、LN44（特殊毒性菌）等。

（2）国际抗性基因相对应的菌株：菲律宾小种 1（PXO61），测定 *Xa4*、*xa5*、*xa8* 和 *Xa7*；小种 5（PXO112），测定 *Xa14*；小种 6（PXO99），测定 *Xa21* 和 *xa13*。日本：小种Ⅰ（T7174），测定 *Xa1*；小种Ⅱ（T7147），测定 *Xa2*；小种Ⅲ（T7133），测定 *Xa3*。

2 接种和鉴定

抗性遗传研究的重要工作是抗病鉴定及调查分析。接种物（病菌生活力、接种数量、纯化度）和被测试材（生育期、群体与单株），以及外界环境条件之间的相互作用和调查标准，直接左右结果的可信程度。为此，提出以下几点：

2.1 接种物

采用稳定的鉴别致病菌（使用前应进行复壮，并在鉴别品种上测试验证）。毒性变化者需经过单孢分离，回接到植株上，再分离、测定等程序。菌种在 28～30℃培养 48～72h，接种液的浓度 10^8～10^9/mL。

2.2 接种时期和方式

水稻在不同生育期对白叶枯病的抗性表现不同，全生育期抗病品种的遗传分析群体在分蘖和孕穗期各接种一次，以确定不同生育期的抗性基因数；成株抗性品种，一般在孕穗期接种。测试品种的全套遗传分析材料包括两个亲本、F_1、F_2、B_1F_1 或 F_3 应安排在同期接种，避免在不同年份的不同外界环境条件下所得的结果无法比较。

多菌株接种同株分蘖分组法：水稻抗白叶枯病对不同菌株的抗性可以由一个或几个不同的基因所控制。因此，日本遗传学家鉴定国际鉴别品种抗性基因时，均采用几个菌株分析品种的抗性遗传基础。即将参试材料的各个体，同时用几个菌株接种。根据所采用的菌株数，用不同色泽的塑料带将各稻株的分蘖分组，每种颜色标示接种某一菌株，以便同时测定群体中各个体对不同菌株的实际反应，我们称之为多菌株同株分蘖分组法。以上均用剪叶接种法。

2.3 调查标准

2.3.1 一般在接种后 14～21d 进行调查，接种与调查的间隔时间因时、因地制宜，以参试的感病对照品种已经充分发病、病情发展相对稳定时进行调查，必要时可适当延后。

2.3.2 调查数量和记载各植株对各菌株的抗性反应，调查 3～5 张叶片。一般以病斑占叶片面积的百分率或病斑长度（cm）为抗性反应的鉴定指标。在抗性遗传分析中应根据分离世代群体的抗感分布来确定抗、感群界限。可供参考的标准是病斑面积占叶片面积的 20%～25%，长度以 5cm 左右为界。

3 抗性遗传分析

3.1 配组方式及其群体样本容量

3.1.1 配组方式 一般以感病亲本为母本，抗病供体为父本。抗性基因主要由雄配子传递配组效果较好，具有良好农艺性状的感病亲本作母本配组，易于获得性状较好的后代。B_1F_1的配置：抗病亲本的抗性如为显性，其F_1与感病亲本回交；如为隐性，其F_1与抗病亲本回交。

3.1.2 各世代群体的容量 两个亲本及基因型一致的F_1，10～20株；分离世代群体，在不清楚基因对数的情况下，至少要考虑两对基因的F_2群体，300～500株，亲本生态型差别较大时，则需要扩大群体。用来验证F_2分离结果的B_1F_1一般为80～100株；F_3家系数约占F_2群体容量的20%为妥，各个家系需种植30株左右。

3.2 遗传分析

根据F_1和亲代的抗感反应，可以作出抗性为显性或部分显性或隐性的遗传判断。将F_2和B_1F_1群体植株的抗感反应制出抗感频数分布图，并可参照F_3家系的抗感分布，对抗性有关的基因数、基因互作方式，进行综合的遗传分析。在主基因效应明显的情况下，不论基因是独立分离或连锁遗传，都须进行分离比的统计测验和有关参数估计。

4 新基因的鉴定、定位与命名

4.1 抗性反应型比较

采用本文提出的与已知基因相对应的国际鉴别菌株和我国的代表性菌株，鉴定和比较参试的抗源与国际单基因鉴别品种的抗性反应型。假如该材料分别与一系列鉴别菌株互作后的反应型清晰地表现出与国际抗性基因代表品种完全不同，初步推测它可能携有新的抗性基因。

4.2 遗传鉴定

4.2.1 按3.1和3.2所述，对新的抗病材料的抗病性进行显性还是隐性、遗传模式和基因数的分析判断。

4.2.2 等位性测定 新的抗病材料的抗病性为显性时，须与已知携带显性基因的品种进行等位性测交，F_1应表现抗病，F_2群体若全部表现抗病，证明两者的抗病基因呈等位关系；若出现抗感株分离，则表明新的抗病材料带新的抗病基因。新的抗病材料的抗病性为隐性时，须与已知携带隐性基因的品种测交。若测交F_1表现抗病，表明两者的抗病基因互为等位。若F_1感病，F_2出现抗病株分离，表明彼此非等位，所测试的抗病材料携带新的抗病基因。

抗病材料携带两对以上抗病基因时，须经过基因分离和选育，然后逐一进行基因等位性测试，才可能准确判定各个基因的等位关系和基因性质。

4.3　基因定位

目前水稻基因定位有以下几种方法：形态标记法、初级三体法、相互易位系法和DNA分子标记法等，根据实际条件选用。

4.4　基因命名

我国尚未成立基因命名委员会，确定新基因后，我们建议按国际水稻遗传基因命名委员会有关章程，申请命名登记。

5　讨论

5.1　感病亲本在抗性遗传研究中的影响

水稻抗病品种的抗性遗传行为除受到所采用的菌株影响之外，很大程度上受稻种生态型的左右。以全生育期抗性的籼稻与感病的粳稻配组杂交的抗性遗传行为尤为明显（章琦，1991）。为此，水稻白叶枯病的抗性遗传采用亚种内杂交为好。如探索亚种间、种间、野栽间配组抗性遗传行为则另当别论。此外有些品种可能存在光（温）敏的问题，杂交后代群体的生育期变态分离，干扰遗传分析，为此须慎重选择感病亲本。

此外，由于自交作物群体内常混有遗传异质性个体，天然杂交的存在，突变和微小变异的发生，以及早代谱系的混合与扩大繁殖，都足以造成品种群体内遗传上的不纯。作为遗传研究的感病亲本，需要一个相对纯合的群体。为此，采用自交，以降低群体内杂合子百分率，结合定向选择，育成标准系，是必须的经常工作。

5.2　病原菌的纯化意义

自然界的菌株在水稻上生长的群体，在遗传上是一个杂合体，存在着原有的和潜在的毒性基因，遗传分析采用的菌株应强调是一个具有相对纯合的群体。为此，所用的菌株，必须进行严格的分离、保存、复壮和活体毒性的检测，这是实现标准化的必须手段。

5.3　建立鉴别菌株和品种基础

一套包括病原菌生理小种与遗传基础清晰的鉴别品种是实现抗性遗传研究标准化及基因命名规则建立的科学基础。我们尚需从病理生理、分子生物学方面作进一步努力，广泛筛选有明晰鉴别作用的品种，分析我国病原群体的遗传结构，检测新的菌群和特殊广致病菌株、菌群的检测，以及广致病菌株的发现，均是未来很有探索意义的课题。这一基础的建立，使我国抗性遗传研究从目前基因的细胞遗传学水平向基因的分子生物学发展。

5.4　我们的合作研究，尚不足以代表我国全部水稻生态地区

各地在进行此类研究中，尚须根据当地具体情况掌握，其共同目的是相同的，就是鉴定发掘可利用的抗性基因，为育种所用。

原载：中国农业科学，1996，29（4）：85～92.

水稻抗白叶枯病基因 *Xa14* 在分子标记连锁图上的定位

谭震波[1,2]　章　琦[1*]　朱立煌[2]　王春莲[1]

（1. 中国农业科学院作物育种栽培研究所；2. 中国科学院遗传研究所）

RFLP Mapping of a Rice Bacterial Blight Resistance Gene *Xa14*

Abstract In order to map a rice bacterial blight resistance gene *Xa14*, a F_2 population containing 99 plants was generated from a cross between Zhenzhuai (susceptible to all six Philippine races) and CBB14 with gene *Xa14* (resistant to race 5). It was observed that resistant and susceptible plants were segregated in a ratio of 3∶1 (χ^2=0.570). Based on RFLP analysis in the population, a molecular linkage map of chromosome 4 was constructed, and the gene *Xa14* was located between RFLP markers RG 620 and G 282.

Key words rice; bacterial blight; gene mapping; RFLP analysis

摘要 选用对菲律宾水稻白叶枯病菌 6 个小种均感病的籼稻品种珍珠矮为母本，携带抗病基因 *Xa14* 的粳稻近等基因系 CBB14（抗菲律宾小种 5，P5）为父本配制杂交组合。F_1植株对相应菌株 P5 表现全生育期抗病，F_2群体在分蘖前期的抗、感分离符合 3∶1 显性单基因比率。根据对 F_2群体中选择的 99 个单株进行的 RFLP 分析的结果，构建了水稻第 4 连锁群的分子图谱，并把抗病基因 *Xa14* 定位于 RG620 和 G282 之间。

关键词 水稻，白叶枯病，基因定位，RFLP 分析

由 *Xanthomonas oryzae* pv. *oryzae* 引起的水稻白叶枯病，是危害我国乃至亚洲水稻生产最为严重的病害之一。利用寄主的抗性是控制水稻白叶枯病危害的最经济有效的手段。为此，国内外在水稻白叶枯病抗性的遗传研究方面做了大量工作（章琦等，1996；Ogawa et al.，1993）。通过经典遗传分析，至今鉴定了至少 19 个抗病基因（Ogawa et al.，1993）。近年来，由于分子标记应用的不断深入，水稻白叶枯病的抗性基因 *Xa1*、*Xa2*、*Xa3*、*Xa4*、*xa5*、*Xa7*、*Xa10*、*xa13* 和 *Xa21* 都已定位在水稻的连锁图谱上（Kinoshita，1993；McCouch et al.，1988，1991；Yoshimura et al.，1992；Ronald et al.，1991）。抗病基因的

定位提高了抗病基因在品种选育上的利用效率，从而可能聚集多个抗病基因，避免因利用单一基因给水稻生产带来的潜在危害。

水稻抗白叶枯病基因 *Xa14* 是 Taura 等（1992）从水稻品种 TN1 中鉴定出来的，其对菲律宾小种 5 具有专化抗性，前人已利用水稻三体技术，将其定位于第 4 染色体（Kinoshita et al.，1993；Taura et al.，1992）。本文报道 *Xa14* 在水稻分子图谱上的定位结果。

1 材料和方法

1.1 试验材料及抗性鉴定

1995 年，用本课题组培育的抗白叶枯病近等基因系 CBB14（携有一对抗菲律宾小种 5，P5）的显性抗性基因（章琦等，1998）为父本，用对菲律宾小种 1～6 都高度感病的籼稻品种珍珠矮为母本，配制杂交组合。杂种 F_1 套袋，获 F_2 种子。1996 年将两个亲本、F_1 及 F_2 群体种植于网室，在分蘖初期用剪叶法同时接种。接种前将保存的冻干菌种 P5 在 PSA 培养基上复壮，在鉴别品种上检测其具有标准毒性后，转移数管斜面菌种，保存于 4℃冰箱内备用。接种前转管培养接种用菌，在 28℃温箱培养 2d，接种物浓度为 10^9/mL。接种后两周，待感病品种珍珠矮的病情趋于稳定时，调查单株的抗性反应，包括亲本和 F_1 各 10 株及 99 个 F_2 单株，每株 3 张叶片，以病斑面积占叶片面积的 20%为抗、感分界值。

1.2 RFLP 分析

水稻总 DNA 的提取、酶切、Southern 吸印及分子杂交采用 McCouch（1988）的方法用 8 种常用的限制性内切酶酶切亲本、F_1 及选出的 99 株 F_2 的各单株，总 DNA（*Bam*HⅠ、*Bgl*Ⅱ、*Eco*RⅤ、*Dra*Ⅰ、*Hind*Ⅲ、*Dra*Ⅰ、*Sca*Ⅰ和 *Xba*Ⅰ）。Southern 吸印后，选择水稻第 4 染色体上的探针检测亲本间的多态性，用能够揭示亲本多态性的探针/酶组合检测 F_2 群体各单株的 RFLP 标记基因型的分离。所用的探针由美国 Cornell 大学 Tanksley 和 McCouch 及日本水稻基因组计划（RGP）提供。

1.3 分子标记和抗病基因的连锁分析

根据 X 光片显影结果，显示具有母本、父本相同带型的个体分别赋值 1 和 3，具有双亲带型（杂合体）赋值为 2。采用 Mapmaker 软件（Lander et al.，1987），在 Macintosh Ⅱ型计算机上进行第 4 连锁群分子图谱的构建。用“Compare”命令，选择 LOD>2.5 的多个标记，推测其最佳排列顺序，建立基本框架；用“Try”命令，将其他标记放入框架中，找出这些标记的最佳位置。

根据抗性鉴定结果，表现抗、感病的单株分别赋值 4 和 1，用“Try”命令，找出 *Xa14* 在第 4 连锁群上的最佳位置。采用 Kosambi 函数将重组值转换成图距单位（cM）。

全部试验于 1995—1998 年进行。

2 结果与分析

2.1 亲本的多态性分析

利用 8 种限制性内切酶对亲本珍珠矮和 CBB14 的 DNA 进行酶切，选择在第 4 连锁上已定位的 9 个 RFLP 探针对亲本的多态性进行检测，其中 RZ569、G177 和 C975 在亲本间没有多态性，于是将能揭示亲本间多态性的 6 个探针 RG620、G282、C513、G271、RG329 和 RG684 进一步用于 F_2分离群体的连锁分析。

2.2 *Xa14* 与分子标记在 F_2群体中的分离

用菌株 P5 对亲本及珍珠矮/CBB14 的 F_2群体中各单株接种。接种 18d 后，感病品种珍珠矮的发病趋于稳定时，调查单株对菌株 P5 的抗性反应。抗和感的单株数分别为 78 和 21，经 χ^2测验符合 3∶1 显性单基因分离比（$\chi^2=0.570<\chi^2_{0.005,1}=3.84$）。第 4 连锁群的各分子标记的分离，经 χ^2测验，也没有发现严重的偏分离，符合 1∶2∶1 的分离比（表 1）。

表 1　珍珠矮/CBB14 F_2 群体中第 4 染色体各分子标记分离的检测

探针	样本数	基因型分类			χ^2 1∶2∶1	基因频率
		1	2	3		
RG620	94	20	50	24	0.723	0.473
G282	93	22	46	25	0.204	0.484
C513	90	20	47	23	0.378	0.483
G271	89	20	45	24	0.424	0.478
RG329	97	23	49	25	0.292	0.490
RG684	91	20	47	24	0.450	0.478

2.3 水稻第 4 连锁群分子图谱的构建及 *Xa14* 的分子定位

用具有多态性的 6 个探针对 F_2群体各单株进行 RFLP 分析（表 1）。利用 Mapmaker 软件，构建了水稻第 4 连锁群的分子图谱。这些探针在第 4 连锁群上的排列顺序与 Causse 等（1994）、Kulrata 等（1994）和 Tan 等（1997）发表的图谱相一致。根据 F_2各植株的抗、感分离情况，将抗病基因 *Xa14* 定位于 RG620 和 G282 之间，且分别相距 20.1cM 和 19.1cM（图 1）。

3 讨论

Taura 等（1992）曾用三体技术将 *Xa14* 定位于水稻第 4 连锁群上。本研究进一步用 RFLP 标记将 *Xa14* 定位于第 4 连锁群的分子图谱上，与两侧的探针 RG620 和 G282 分别相距 20.1 cM 和 19.1cM，此结果为进一步精细定位打下了基础。根据在这一染色体区域内已经定位的水稻抗白叶枯病基因 *Xa1*、*Xa12* 和 *Xa2*（Causse，1994），我们以图 2 来显示这 4 个抗性基因在第 4 连锁群上的定位情况，说明同一类型的抗病基因有聚集在同一染色体相近区域的倾向，可能与抗病基因的起源有关。虽然抗病基因 *Xa1*、*Xa12* 和 *Xa2* 在该区域内的相对位置已被确定，但在本实验中由于 *Xa14* 近侧的分子标记 RZ569 和 G177 在亲本间没有多态性，因此，*Xa14* 与其他 3 个抗病基因的位置还很难确定，需作进一步研究。

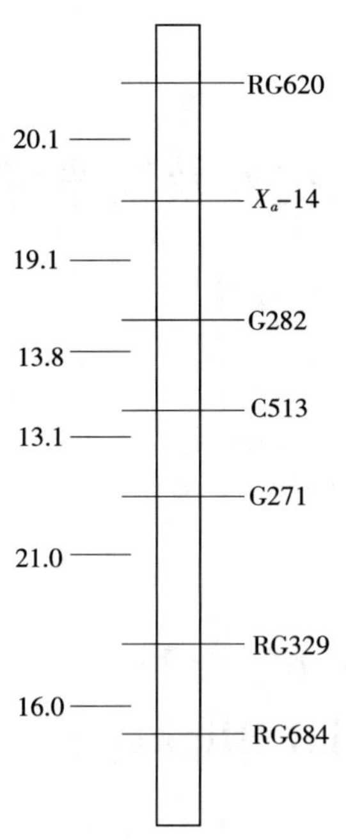

图 1　*Xa14* 在水稻第 4 连锁群上的定位

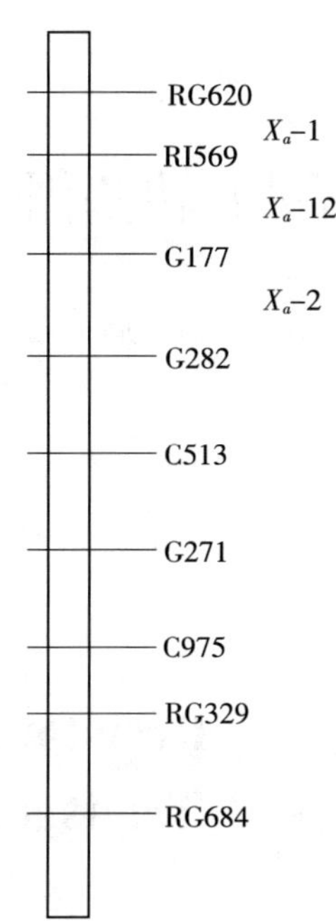

图 2　*Xa1*、*Xa12*、*Xa2* 和 *Xa14* 在水稻第 4 连锁群上的位置

原载：遗传，1998，6（4）：155～158.

Xa21 转基因水稻对白叶枯病的抗性及其遗传

周永力[1]　翟文学[2]　章　琦[1]　王春莲[1]
田文忠[2]　潘学彪[3]　朱立煌[2]

（1. 中国农业科学院作物育种栽培研究所；
2. 中国科学院遗传研究所；3. 扬州大学农学院）

Resistance and Inheritance of *Xa21* Transgenic Rice to Bacterial Blight

Abstract *Xa21* transgenic rice lines derived from 8 cultivars with different genetic background through the Agrobacterium-mediated transformation system and their generations T_0 and T_1 were inoculated with 12 international differential races and 7 Chinese pathotypes of *Xanthomonas oryzae* pv. *oryzae*. Resistance analysis showed that *Xa21* gene expressed dominantly, and inherited stably in transgenic progenies with a broad resistance spectrum. The results also revealed that different genetic background showed certain effect on the resistance expression level of *Xa21*.

Key words *Xa21*; transgenic rice; bacterial blight; resistance

摘要 采用包括与 *Xa21* 基因相对应的白叶枯病鉴别菌株 P6 在内的 12 个国际鉴别小种和中国 7 个病原型的代表菌株，对通过农杆菌介导的 8 个不同遗传背景的 *Xa21* 转基因品系及其不同世代的植株，进行接种和抗性分析。结果表明，*Xa21* 在不同的遗传背景下呈现显性，在转基因后代中可以稳定遗传，并保留其广谱抗性。不同遗传背景对 *Xa21* 的抗性表达有所影响。

关键词 *Xa21*；转基因水稻；白叶枯枯病；抗性

Xa21 是国际上迄今命名的抗谱较广的白叶枯病显性抗性基因之一，源于非洲马里长药野生稻（*Oryza longistaminata*）(Khush et al.，1990)，并由 Song 等（1995）克隆了该基因。从而开辟了基因工程培育抗白叶枯病水稻品种的新领域。Zhang 等（1998）、Tu 等（1998）先后采用基因枪将 *Xa21* 导入水稻。我们用基因枪法也获得了转基因植株，然而这些植株存在不育率高、后代的农艺性状严重分离以及抗性不稳定等问题。近年来，农杆菌介导转化系统的应用已扩展到单子叶植物，水稻已建立了农杆菌转化系统（Hiei et al.，1994）。我们在前期的研究中，采用农杆菌介导将 *Xa21* 导入水稻品种，获得了抗病转基因

植株，从分子水平上初步证实了 *Xa21* 基因的整合和遗传。本文分析 *Xa21* 在不同转基因品种及其不同世代中的抗性表现，揭示 *Xa21* 在不同转基因品系中对白叶枯病的抗性的稳定遗传信息，为水稻 *Xa21* 转基因品种的推广与产业化提供依据。

1　材料与方法

1.1　菌株与接种

采用与 *Xa21* 相对应的鉴别菌株 P6 在分蘖盛期接种，分析 *Xa21* 在不同的转基因水稻品种 T_0～T_2代植株的抗性。用包括 P6 在内的 12 个国际鉴别小种和我国 7 个病原型的代表菌株接种（表 1），分析转基因植株的抗谱和抗性水平。

表 1　采用的白叶枯病菌小种及病原型

Table 1　Races and pathotypes of bacterial blight used for resistance analysis

来源 Origin	小种 Race	菌株 Strain	来源 Origin	小种/病原型 Race/pathotype	菌株 Strain
菲律宾 Philippine	1（P1）	PXO61	日本 Japan	Ⅰ（T1）	T7174
	2（P2）	PXO86		Ⅱ（T2）	T7147
	3（P3）	PXO79		Ⅲ（T3）	T7133
	4（P4）	PXO71	中国 China	C1	HLJ72
	5（P5）	PXO112		C2	HB17
	6（P6）	PXO99		C3	NX42
	7（P7）	PXO280		C4	ZHE173
	8（P8）	PXO145		C5	GD1358
	9（P9）	PXO87		C6	LN57
				C7	JS49－6

菌种保存于－20℃的甘油中，每年在 PSA 培养基上进行复壮，于鉴别品种上测试毒力，将毒力稳定的菌种保存于 4℃冰箱，供全年使用，保证毒力一致。接种前将病菌转管到 PSA 上，于 28℃培养 48 h，配制成 10^9/mL 菌液备用。在植株分蘖盛期用人工剪叶法接种。所有试验均以 IRBB21（携有 *Xa21* 基因）和 IR24 分别作为抗病和感病对照，同时接种转基因受体品种，每株接种 5～6 张叶片。在隔离温室或网室中进行试验。接种后 14～20d，当感病对照品种 IR24 的病情发展趋于稳定时，以病斑占叶面积的百分率为抗性反应参数，调查各植株的抗病反应。

1.2　不同受体品种 T_0代植株的抗性导入效应和抗谱分析

在前期的研究中，用农杆菌介导将 *Xa21* 导入到明恢 63、珍汕 97B、盐恢 559、太湖粳 6、中花 11、C418、8706 和培矮 64S 8 个不同遗传背景的水稻品种（翟文学等，2000），获得 156 个独立转化系，共 580 株转基因株。本文将来源于同一块愈伤组织的转基因植株称为一个系，不同受体品种的分化能力不同，多数品种的 1 个独立转化系仅包括 1 株转基因株，少数品种的 1 个系可能包括几个到几十个植株。这些植株均经过潮霉素严格筛选，并经

PCR 和 Southern 检测。将 *Xa21* 基因单拷贝整合的各转基因植株种植于温室或网室中，当植株进入分蘖盛期接种菌株 P6，分析不同遗传背景转基因植株的抗性表现。

选择各转基因品种的抗性植株，采用同株分蘖分组法，即在同一植株的不同分蘖上接种包括 P6 在内的国际鉴别小种和我国的病原型代表菌株，分析转基因植株的抗谱。

1.3 不同品种转基因植株后代的抗性分析

将高度抗病的 T_0 植株在开花前套袋自交，单株收种，获得各品种多个 T_1 代群体。随机种植明恢 63、盐恢 559、中花 11 各 3 个 F_1 群体，培矮 64S、太湖粳 6、C418、珍汕 97B 和 8706 各 1 个 F_1 群体。各群体 120～250 株，在分蘖盛期接种 P6，当感病品种上的病情趋于稳定时，调查各株的病斑面积占叶片面积的百分率。

将各转基因品种的 T_1 群体中的高抗植株在开花前套袋自交，单株收获抗病植株 T_2 系。各 T_2 系随机接种 15 株，筛选到 73 个明恢 63、8 个珍汕 97B、13 个盐恢 559、5 个太湖粳 6、4 个 C418、6 个 8706 和 30 个中花 11 的抗病纯合系。在分蘖盛期用 P6 接种，每个纯合系接种 10 株，分析携带 *Xa21* 基因的抗性纯合系的反应，并与抗性杂合系进行比较。

2 结果与分析

2.1 受体品种对白叶枯病的抗性背景

在植株的分蘖初期和孕穗期，分别用 19 个菌株接种 8 个受体品种。结果表明，除盐恢 559 外，各受体品种对各菌株都表现感病，病斑面积 30%～90%（表 1），与不携带 *Xa21* 基因的这些品种的系谱分析的结果一致；盐恢 559 在分蘖初期对测试的菌株均感病，对 C1、C4、C5、C6、T1、P1、P7 和 P8 在孕穗期为中抗，病斑面积 10%～15%，对其他菌株则表现感病。

2.2 不同转基因品种 To 代植株的抗性导入效应及抗谱分析

在相同条件下，对 8 个转基因材料的单拷贝整合的转基因系及每个系的部分转基因植株接种。当感病对照 IR24 的病情发展稳定时（病斑占整个叶片面积的 80. 7%±2. 45%），进行调查。结果表明，转基因植株的病斑面积占叶片面积百分率为 3%～10%，非转基因对照为 30%～80%，基因供体品种 IRBB21 为 6. 5±1. 2%（表 2）。

表 2 不同受体水稻品种 *Xa21* 转基因 To 代植株对 P6 的抗性反应

Table 2 Resistance reaction of *Xa21* transgenic To plants in different acceptors to P6 of *Xoo*

品种 Cultivar	类型 Type	转基因系/株数 No. trans. line/plant	抗性系/株数 No. resistant line/plant	病斑面积（%） % of lesion area	
				转基因植株 Trans. plant	受体植株 Accept. plant
明恢 63 - *Xa21*	籼稻	26/35	26/30	5. 8±1. 1	57. 86±2. 44
珍汕 97B - *Xa21*	籼稻	24/28	24/25	3. 90±0. 92	35. 83±2. 19
盐恢 559 - *Xa21*	籼稻	18/18	18/18	5. 06±0. 54	65. 71±2. 49
培矮 64S - *Xa21*	籼稻	28/40	28/36	4. 60±1. 22	47. 86±3. 06

（续）

品种 Cultivar	类型 Type	转基因系/株数 No. trans. line/plant	抗性系/株数 No. resistant line/plant	病斑面积（%） % of lesion area	
				转基因植株 Trans. plant	受体植株 Accept. plant
C418 - *Xa21*	粳稻	6/6	5/5	9.29±2.32	43.33±2.96
太湖粳 6 - *Xa21*	粳稻	17/20	15/17	3.33±0.58	37.86±2.45
8706 - *Xa21*	粳稻	10/16	8/12	7.60±2.63	48.75±2.37
中花 11 - *Xa21*	粳稻	26/417	26/239	3.57±0.82	33.40±2.99

供试的 8 个转基因受体品种包括我国水稻生产中重要的籼稻和粳稻。不同遗传背景的转基因植株抗性水平略有不同。如珍汕 97B、太湖粳 6 和中花 11 转基因植株的病斑长度为 0.5～1cm，病斑面积百分率小于 5%，抗性水平高于基因供体 IRBB21；而明恢 63 等品种的抗性水平与 IRBB21 相近。但 *Xa21* 在籼稻和粳稻背景下，均呈现显性，表现高度抗病（表 2）。

采用多菌株接种，进行抗谱测定，8 个品种的转基因植株对 19 个白叶枯病代表菌株均表现抗病，病斑面积为 5%～10%，抗谱与供体 IRBB21 相同。结果表明 *Xa21* 在不同遗传背景下仍保持广谱抗性（表 1）。

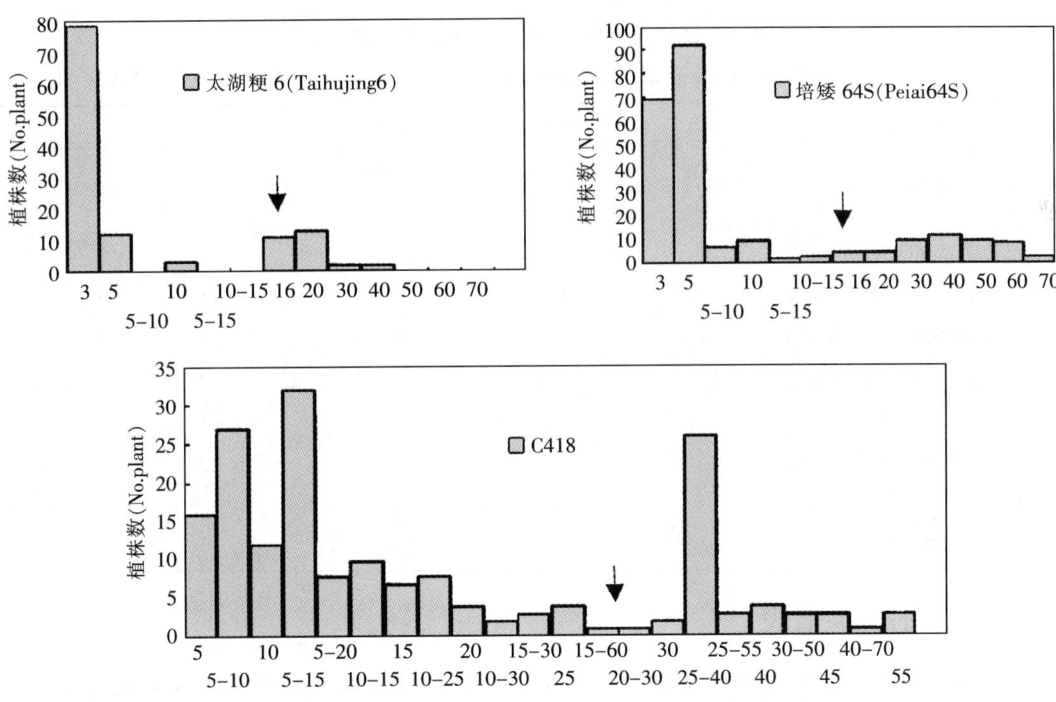

同一植株不同叶片病斑面积（%）（% of lesion area of different leaves from the same plant）

图 1 不同遗传背景的 *Xa21* 转基因 T_1 群体的抗性反应

Fig. 1 Resistance reaction of T1 populations with different genetic background

↓：抗感临界值（Critical value of resistance and susceptible）

2.3 不同转基因品种的抗性反应

在植株分蘖盛期用P6接种，抗病株的一般表现过敏性反应型，病、健交界明显，感病反应株的病、健交界不明显，呈典型的水渍状感病型。在调查之前，根据病斑反应型并结合分子检测，不同品种的病斑面积百分率：太湖粳6、中花11、培矮64S和珍汕97B为15%～20%，明恢63、盐恢559和8706为20%～25%，C418为15%～40%。P6对3个明恢63、3个盐恢559和3个中花11的T_1群体各植株的结果表明，相同遗传背景的T_1群体的抗性反应是一致的。不同遗传背景的T_1代群体，抗感分离都很明显。根据抗病植株的反应，被分为3种类型（图1），1）同一植株接种的各张叶片病斑面积百分率大致相同，抗病株表现高抗，如太湖粳6、中花11和珍汕97B；2）多数植株的病斑面积相同，少数植株不同叶片的病斑面积相差5%，多数抗病株的抗性水平较高，如培矮64S、明恢63、盐恢559、8706；3）抗病株的病斑面积百分率呈连续分布，少数植株不同叶片的病斑面积百分率相差5%～20%，差异很大，如C418，这类品种的抗感评价，还需结合病斑反应型来确定。

2.4 *Xa21* 在不同遗传背景下的抗性遗传

用P6接种8个不同遗传背景的T_1群体，按2.3的方法和指标调查各植株的抗病反应，分析*Xa21*在不同背景下的抗性遗传行为。结果表明，各T_1群体的抗感分离比均为3R∶1S，符合一对显性基因的孟德尔遗传规律（表3）。

表3 不同品种转基因T_1代植株的抗感分离

Table 3 Segregation of transgenic T_1 plants derived from different cultivars to P6 of *Xoo*

转基因系T_1 T_1 transgenic lines	植株数（No. of plants）				
	总数 Total	抗病株 R plants	感病株 S plants	分离比 Segregation ratio	χ^2 (3∶1)
明恢63－*Xa21*（MH63－*Xa21*）	212	155	57	2.72	0.402 5
珍汕97B－*Xa21*（ZS97B－*Xa21*）	203	154	49	3.14	0.080 4
盐恢559－*Xa21*（YH559－*Xa21*）	178	132	46	2.86	0.067 5
培矮64S－*Xa21*（PA64S－*Xa21*）	246	184	62	2.97	0.005 5
C418－*Xa21*	184	133	51	2.61	0.724 7
太湖粳6－*Xa21*（THJ 6－*Xa21*）	121	93	28	3.32	0.223 2
8706－*Xa21*	138	101	37	273	0.241 6
中花11－*Xa21*（ZH 11－*Xa21*）	205	159	46	340	0.717 1

2.5 纯合系的抗病表现

源自相同遗传背景的纯合系之抗性水平相近。来自单拷贝高度抗病的（病斑面积≤5%）T_0植株的T_2纯合系之平均抗性水平也较高。如图2所示，中花11、珍汕97B、太湖粳6的平均病斑面积百分率均小于5%，抗性高于IRBB21；明恢63和盐恢559的抗性与IRBB21相近；C418的抗性略低于IRBB21。多菌株接种结果表明，各抗病纯合系对12个国际鉴别小种和7个中国的病原型都高度抗病，病斑面积百分率为5%～10%。此外，没有发现在各

转基因品系的 *Xa21* 纯合型与杂合型的抗性存在差异，表明 *Xa21* 在转基因植株中为完全显性。

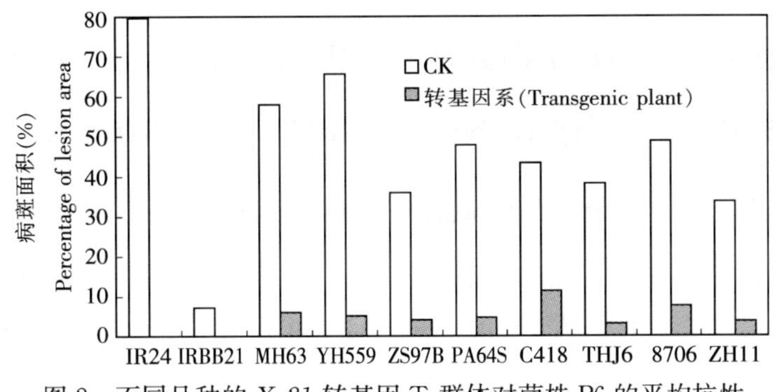

图 2 不同品种的 *Xa21* 转基因 T_2群体对菌株 P6 的平均抗性

Fig. 2 Average resistance to strain P6 in T_2 population of different transgenic cultivars

3 讨论

种植抗病品种是防治水稻白叶枯病的最经济有效方法。我国水稻抗白叶枯病育种存在的问题是抗性基因使用集中，粳稻和籼稻分别使用 *Xa3* 和 *Xa4*；杂交稻不育系基本不携有抗白叶枯病基因（章琦等，1988）。在品种选育过程中多数未采用鉴别小种进行抗性跟踪测定选择，造成抗性基因丢失；杂交稻的抗性水平较低或缺乏抗性。长期大规模地种植单一抗源的品种，导致我国 80 年代初在广东出现了能侵袭 *Xa4* 的 V 型菌，并从华南扩展至华中地区，潜在极大风险（章琦，1995）。应尽快利用新的抗病基因改良杂交稻的白叶枯病抗性。

Xa21 是国际上迄今已命名的抗性基因中的广抗谱显性基因。IRRI 以 IR24 为轮回亲本转育成携有该基因的近等基因系 IRBB21。我国育种家发现其抗性转育效应较差，在籼稻背景下的抗性表达呈不完全显性遗传（徐建龙等，1996），因而限制了其在常规育种中的应用。我们采用农杆菌介导 *Xa21* 导入到 8 个不同遗传背景的籼稻和粳稻品种中，从分子水平上分析的 *Xa21* 在受体基因组中，可以在不同位点以及不同的拷贝整合，接种结果也表明了整合位点对 *Xa21* 抗性表达水平的影响很小（翟文学，1999），因此在本研究中忽略整合位点对 *Xa21* 抗性表现的影响，进一步对单拷贝整合的不同遗传背景的 *Xa21* 转基因植株的抗性进行了分析。本试验表明，克隆的 *Xa21* 在不同的遗传背景下抗性水平虽略有差异，但均表现完全显性，高度抗病，抗谱与基因供体 IRBB21 相同；不同遗传背景的自交 T_1代的抗、感分离比均为 3∶1，与 Southern 杂交揭示的 T_0 转化体中 *Xa21* 单拷贝整合的结果是一致的；T_2纯合系的抗性水平和抗谱与 T_0植株相同，表明 *Xa21* 在后代中能稳定遗传。因此，利用克隆的 *Xa21* 基因，通过基因工程改善我国水稻白叶枯病抗性具有广阔的应用前景。

原载：植物病理学报：2001，31 (2)：123～129.

Evaluation of Near-Isogenic Rice Lines with 8 Genes for Bacterial Blight Resistance to Strains in China

Zhang Qi[1] Wang Chunlian[1] Shi Ainong[1] T. W. Mew[2]

(1. Institute of Crop Breeding and Cultivation CAAS, Beijing 100081, China; 2. IRRI, P. O. Box 933, Manila, Philippines)

A set of near-isogenic rice lines with monogenic resistance to bacterial blight were developed by IRRI. The cultivar IR24 was used as the recurrent parent. They were evaluated with 6 races of *Xanthomonas campestris* pv. *oryzae* (*Xco*) in the Philippines at the maximum tillering and the booting stages by ZHANG and MEW at IRRI in 1989. The results showed clear-cut differentiation of the lines and *Xco* races as well as specificity of reaction of the lines and races. The results were similar to what IRRI reported earlier (Table 1) .

Table 1 Interaction between near-isogenic rice lines (NILs) for bacterial blight and races of *X. c.* pv. *oryzae* in the Philippines at two growth stages, IRRI, April, 1989

NILs	Growth stage[a]	Disease reaction[b]					
		PXO61	PXO86	PXO79	PXO71	PXO112	PXO99
IRBB1	MTS	S	S	S	S	S	S
	BS	S	S	S	S	MR	S
IRBB3	MTS	S	S	S	S	S	S
	BS	R	MR	R	MR	MR	S
IRBB4	MTS	MR	S	S	S	MS	S
	BS	R	S	S	MR	MR	S
IRBB5	MTS	R	R	R	S	R	S
	BS	R	R	R	S	R	S
IRBB7	MTS	S	R	R	S	R	S
	BS	MR	R	R	MR	R	S
IRBB8	MTS	S	S	S	S	S	S
	BS	MR	MR	MR	MR	MR	—
IRBB10	MTS	S	R	S	S	R	S
	BS	S	R	S	S	R	S
IRBB11	MTS	S	S	S	S	S	S
	BS	S	S	S	S	MR	S

[a] Average lesion area percent from 10 leaves inoculated at 14th day, MTS=Maximum tillering stage, BS=Booting stage.

[b] R=Resistant <5% lesion area percent, S=Susceptible (>20% lesion area percent), MR=Moderately resistant (>5%~<20% lesion area percent).

The near-isogenic rice lines were evaluated by strains in China in Summer season 1989 in Beijing also.

Eight near-isogenic rice lines with the *Xa1*, *Xa3*, *Xa4*, *xa5*, *Xa7*, *xa8*, *Xa10* and *Xa11*, respectively, were used namely: IRBB1, IRBB3, IRBB4, IRBB5, IRBB7, IRBB8, IRBB10 and IRBB11. IR24 was used as susceptible check. Seeds of the 9 materials were sown in the rice nursery. Forty seedlings of each line were transplanted with one plant per hill at a spacing of 15cm×20cm.

Sixteen strains of *Xco* representing the 7 pathotypes except the 7th in China were used in the study. The inoculum of each strain was prepared from 48-hr culture on PSA slants. The inoculum was adjusted to a concentration of 10^9 cells/mL .

Fully expanded leaves were inoculated at the maximum tillering (MTS) and the booting stages (BS) by clipping method. Twenty leaves on two plants at MTS were inoculated with each strain and trimmed after scoring. About 15 flag leaves and the first leaves below the flag leaf were inoculated at the booting stage.

Disease reactions were assessed at 20 days after inoculation based on percent lesion area.

Table 2 Interaction between near-isogenic rice lines for bacterial blight and strains of *X. o.* pv. *oryzae* in China at two growth stages, Beijing, 1989

Strain	Disease reaction																	
	IRBB1		IRBB3		IRBB4		IRBB5		IRBB7		IRBB8		IRBB10		IRBB11		IR24	
	MTS	BS	MTS	BS	MTS	BS	MTS	BS	MTS	BS	MTS	BS	MTS	BS	MTS	BS	MTS	BS
BJ4	S	S	S	MR	MR	R	R	R	R	R	S	MR	S	S	S	S	S	S
HB17	S	S	S	MR	MR	R	R	R	R	R	MR	MR	S	S	S	S	S	S
NX42	S	S	S	MR	MR	R	R	R	R	R	S	MR	S	S	S	S	S	S
JL67	S	S	S	MR	R	R	R	R	R	R	S	MR	S	S	S	S	S	S
LN57	S	S	S	MR	MR	R	R	R	R	R	S	MR	S	S	S	S	S	S
HLJ72	S	S	S	MR	MR	R	R	R	R	R	S	MR	S	S	S	S	S	S
P1	S	S	S	MR	R	R	R	R	R	R	S	MR	S	S	S	S	S	S
ZHE173	S	S	S	MR	R	R	R	R	R	R	S	S	S	S	S	S	S	S
LN59	S	S	S	MR	R	R	R	R	R	R	MR	MR	S	S	S	S	S	S
JS92-2	S	S	MR	MR	MR	MR	R	R	S	S	MR	MR	S	S	S	S	S	S
HN31	S	S	S	MR	R	R	R	R	R	R	MR	MR	S	S	S	S	S	S
ZJ13	S	MR	S	S	MR	R	R	R	R	R	MR	MR	S	S	S	S	S	S
JL76	S	S	S	MR	MR	MR	R	R	R	R	MR	MR	S	MR	S	S	S	S
GZ9	S	MR	R	R	R	R	R	R	R	R	R	R	S	S	S	S	S	S
GD37	S	MR	R	R	S	MR	R	R	R	R	MR	MR	S	MR	S	S	S	S
T1	R	R	S	MR	R	R	R	R	R	R	MR	MR	S	S	S	S	S	S

The notes are the same as Table 1, but the average lesion area percent from 10 leaves inoculated at 20th day.

The interactions between the lines and Chinese strains did not show specificity of reactions of the lines and pathotypes. Some lines showed broad spectrum resistance to Chinese strains. For example, IRBB5 was of resistance to all strains tested at two growth stages. IRBB4 was of resistance to all strains at MTS and BS except GD37 at the MTS. IRBB7 was of resistance to 15 strains and susceptible to strain JS92－2 at both MTS and BS. The resistant function of IRBB3 and IRBB8 to all Chinese strains appeared at the booting stage except strains ZJ13 and ZHE173. On the other hand, we could not find out the resistant function in lines IRBB1, IRBB10 and IRBB11 to the most of the Chinese strains tested (Table 2).

Among these near-isogenic rice lines and 16 strains in China, two combinations of lines and strains, IRBB3 with ZHE173 and IRBB4 with GD37 indicating strong interaction were shown at the MTS (Table 2).

Published in: Chinese Rice Research Newsletter, 1990, 1: 3－5.

携有不同主效白叶枯病抗性单基因 6个粳稻近等基因系的选育*

章　琦　施爱农　杨文才　王春莲　阙更生　赵炳宇　邢全党

（中国农业科学院作物育种栽培研究所）

Breeding of Six Near-isogenic Japonica Rice Lines with Different Major Genes for Resistance to Bacterial Blight

Abstract A set of near isogenic japonica rice lines for resistance to bacterial blight were developed and designated as CBB2、CBB3、CBB4、CBB7、CBB12 和 CBB14 from 1986 to 1996. The cultivars carrying different major genes, *Xa2*, *Xa3*, *Xa4*, *Xa7*, *Xa12* and *Xa14*, respectively, were used as resistance donor; and a japonica cultivar, Shennong 1033 with improved plant type but highly susceptible to bacterial blight was used as recurrent parent in the study. The progenies possessing each one of the genes, were developed by back-crossing, selfing until BC_5F_5 or BC_6F_4 generations. Evaluation of resistance to bacterial blight and selection of plant type being successively till obtaining the lines with stable resistance and common agronomic characters. Comparison of reaction for resistance to bacterial blight between the near-isogenic lines and their donors was made. Genetic analysis of resistance to bacterial blight and investigation of agronomic characters were made. The six lines are valuable in rice breeding and biotechnology studies.

Key words near-isogenic line; resistance; rice bacterial blight

摘要　自1986至1996年，选用6个分别携有抗水稻白叶枯病基因 *Xa2*、*Xa3*、*Xa4*、*Xa7*、*Xa12* 和 *Xa14* 的品种为供体亲本，以一个不携有抗病基因的改良型感病粳稻品种沈农1033为轮回亲本。通过回交、自交直至 BC_5F_5 或 BC_6F_4 代。各世代经抗性测定和株型选择，直至各系的抗性稳定、农艺性状趋于一致。育成6个粳型抗水稻白叶枯病近等基因系，CBB2、CBB3、CBB4、CBB7、CBB12 和 CBB14。进行了6个近等基因系与其供体基因的抗性反应型比较、农艺性状考查以及抗性遗

* 国家“八五”水稻育种攻关应用基础项目。

传分析。通过育种和电泳分析研究应用，证实了这批近等基因系具有良好的利用价值。

关键词 近等基因系；抗性；水稻白叶枯病

作物的不同品种因为遗传背景不同彼此之间无法精确比较某个性状或基因在不同基因型间的差异或表现。近等基因系（Near-isogenic lines，NILs）是指仅某个性状或基因的基因型不同而其遗传背景则近于相同的一系列品系，用它们来研究或检测基因效应的异同就可以克服上述干扰；近等基因系也适于用作鉴定病菌小种异同的鉴别品种。近 30 年来国内外学者都在利用对重要病、虫害具单基因抗性的近等基因系来鉴定新的抗性基因和检测新的病虫小种（生物型）。由于近等基因系的遗传学特点，采用分子标记检测时，只要在目标基因所在区域显示不同，就可以通过与目标基因紧密连锁的 DNA 分子标记对目标基因进行快速定位。因此，近等基因系又是开展基因的分子定位、基因克隆和抗病性、致病性机理研究的重要工具。而且近等基因系的株型都已经过遗传改良，也是抗病育种的理想抗性供体。

Kuhara 等（1958）首先报道水稻白叶枯病菌（*Xanthomonas oryzae* pv. *oryzae*，*Xoo*）存在致病特异性，此后若干研究报道了 *Xoo* 的致病性分化和相对应的抗性品种群。日本和 IRRI 各有一套水稻白叶枯病鉴别品种。建立一套国际鉴别体系，则需要具备统一的鉴别品种。1982 至 1987 年间，日本热带农业研究中心（TARO）和国际水稻研究所（IRRI）曾合作培育了 3 套不同遗传背景的携有水稻白叶枯病不同抗性基因的 NILs（Ogawa，1993）。这三套 NILs 的遗传背景分别是 IR24（籼稻）、密阳 23（Milyang 23）（籼/粳交）和丰锦（Toyonishiki）（粳稻），分别携带 10 个主效抗性基因。章琦等在上世纪 80 年代在 IRRI 合作研究时，曾多次用菲律宾 *Xoo* 鉴别小种的代表菌株测定这三套 NILs 的抗性反应。其中以 IR24 为遗传背景的 NILs 各基因的专化抗性清晰、稳定，农艺性状一致。但是其他两套 NILs 的专化抗性不够清晰，多数品系的抗性反应呈中抗偏感，不太稳定，甚至仍保留了抗性供体的某些表型性状，未达到纯合一致，后来又在 IRRI 和北京测试多次，都未能获得预期的抗性结果（Zhang et al.，1990）。因此，在 1986 至 1996 年间，章琦参考了 Ogawa 的基本培育方法，并在轮回亲本的选择和目标基因转入后的验证方面作了改进，育成了一套具粳稻遗传背景的近等基因系。

本文重新编辑前后发表的选育两批携有不同单基因的 6 个近等基因系，以及对其抗性稳定性的测试、抗性遗传分析、农艺性状的考查和在遗传育种、抗性机制研究利用等方面的结果。

1 材料和方法

1.1 确定目标抗性基因及其供体

在当时国际命名的 12 个抗病基因中，排除了对中国白叶枯病致病型没有抗性功能的显性基因 *Xa1*、*Xa10* 和 *Xa11* 以及传统育种中不常用的隐性基因 *xa5* 和 *xa8*，选择了对我国水稻生产上有利用价值的 6 个抗性基因及其代表品种：*Xa2*（Tetep，兼抗稻瘟病）、我国主体抗源 *Xa3*（早生爱国 3 号）及 *Xa4*（IR20）、抗谱较广的 *Xa7*（DV85）及 *Xa12*（Java14）

和对中国菌株有鉴别性能的 *Xa14*（TN1）。

1.2 选择轮回亲本

作为选育抗病 NILs 的轮回亲本，必须具有典型的感病性。由于粳稻品种对白叶枯病菌一般都具有由主效或主效加微效基因控制的成株期抗性，较难选出对国内外主要鉴别菌株都高度感病的品种。本研究采用杨守仁教授馈赠的株型优良的粳型品种沈农 1033 作为轮回亲本。该品种全生育期对当时菲律宾全部 4 个小种、日本主要 3 个小种和属于中国 7 个致病型的 65 个菌株高度感病。避免了由于感病亲本自身具有一定程度的“抗性”而干扰目标抗病基因的导入。

1.3 菌株与接种

根据导入不同的目标抗病基因，采用鉴定上述 6 个抗病基因时所用的专化菌株。例如，在转育携有抗日本小种的基因 *Xa3* 和 *Xa12* 的近等基因系过程，采用日本小种Ⅰ、Ⅲ（T1、T3）测试这些品系各世代的材料；转育 *Xa4* 的品系时，用菲律宾小种 1 的代表菌株 PXO61。同时各增加 1 个中国的强致病力菌株 HB17。

使用菌株之前，提前将保存的冻干原菌种在 PSA 培养基上复壮。每年在国际鉴别品种上接种测试其鉴别毒力的稳定性。培养数支各菌株斜面菌种，保存于 4℃冰箱，供全年整个生长期间使用，以保证毒力一致。接种物在 PSA 培养基和 28℃下培养 2d，菌液浓度 10^9/mL。

用剪叶法接种，转育全生育期抗性的基因 *Xa4*、*Xa14* 和成株抗性的基因 *Xa2*、*Xa3*、*Xa7* 和 *Xa12* 的各世代植株，分别在苗期和成株期接种。接种后 14d 或 20d 左右，当同时被接种的轮回亲本的病情发展趋于稳定时，调查抗性反应。选择各世代抗病（病斑占叶面积的 5%以内）、株型趋于轮回亲本的单株。

1.4 近等基因系的选育

采用常规回交法。以轮回亲本沈农 1033 为母本，分别与 6 个抗病基因的代表品种杂交、第一次回交，获 BC_1F_1 植株。采用与目标基因相对应的鉴别菌株接种鉴定。根据二项式分布原理（莫惠栋，1979）按通常认为的置信概率估算每次回交时的最小群体，即从第二次回交开始，每个组合的各世代至少选择 7～14 株抗病单株分别与轮回亲本回交，则可能获得携有各抗病基因的概率大于 99%的后代。连续进行这一程序，直至 $BC_{5\text{-}6}F_1$，然后自交多次，以缩短位点互换染色体异质部分的进程。继续对各代进行抗性鉴定、农艺性状筛选，直至 $BC_{5\text{-}6}F_{5\text{-}4}$。当入选后代的农艺性状与轮回亲本基本一致，而所携带的抗病基因各异，才育成了各个目标基因对相应小种的专化抗性清晰、稳定的近等基因系。

1.5 近等基因系的命名

近等基因系以 CBB 和编号表示，冠以“CBB”为中国水稻白叶枯病抗性近等基因系，编号指导入的抗性基因。如携有基因 *Xa2*、*Xa3*、*Xa4*、*Xa7*（来自 DV85，未定）、*Xa12* 和 *Xa14* 的近等基因系，分别被命名为 CBB2、CBB3、CBB4、CBB7（暂定）CBB12

和 CBB14。

1.6 近等基因系的抗性反应型检测

从 1991 年至 1993 年，采用与 *Xa4*、*Xa3* 和 *Xa12* 三个基因相匹配的鉴别菌株日本小种Ⅰ、Ⅲ的代表菌株 T1、T3 和菲律宾小种 P1 测试，将首批育成的 NIIs，CBB3、CBB4 和 CBB12 分别与其供体基因品种间的抗性反应型进行比较。第二批从 1994 年至 1996 年，用与 *Xa2*、*Xa7* 和 *Xa14* 三个基因相匹配的鉴别菌株日本小种Ⅱ（T2）和菲律宾小种 P1 测试随后育成的 NIIs，CBB2、CBB7 和 CBB14 的抗性反应型，分别与各自的供体相比较。携有全生育期抗性基因 *Xa4* 和 *Xa14* 的植株在苗期和最高分蘖期接种，成株抗性基因 *Xa2*、*Xa3*、*Xa7* 和 *Xa12* 的植株在最高分蘖期和成株期各接种一次。每次都有轮回亲本作为感病对照。病斑占叶面积小于或等于 5%为抗病（R），大于 30%为感病（S）。

1.7 近等基因系的抗性遗传分析

1991—1993 年、1995 年和 1996 年先后测试了 6 个近等基因系与其导入的抗性基因的等位性。以早生爱国 3 号、IRBB4、Java 14、Tetep、DV85 和 TN1 为已知抗性基因亲本，分别与 CBB3、CBB4、CBB12、CBB2、CBB7 和 CBB14 杂交获 F_1，经自交加代、回交，获 F_2 和 B_1 F_1全套遗传模式测定材料。用相对应的菌株接种上述群体，分析其遗传模式，以确认目标基因的导入效应。其中 CBB7 与沈农 1033 杂交的 F_1、F_2群体同时提供中国科学院遗传所朱立煌研究员实验室进行分子标记定位。

1.8 近等基因系与轮回亲本农艺性状考查

1991—1993 年和 1994—1996 年，先后考查第一批育成的 NILs，CBR3、CBB4 和 CHB12，和第二批育成的 CBR2、CBB7 和 CHB14 的农艺性状。每年从各近等基因系高代材料中选择 5～6 个株系与其轮回亲本沈农 1033 种植于本所昌平试验田，观察和比较各系之间以及与沈农 1033 的农艺性状。单本栽插。小区面积为 20m^2，3 次重复，各小区随机取样 15 株，测量株高和有效分蘖，室内考种穗部性状及千粒重等组分，取 3 年考查平均值。

2 结果与分析

2.1 携有不同抗水稻白叶枯病单基因的 6 个粳稻近等基因系的选育

从 1986 至 1996 年，以粳型品种沈农 1033 为轮回亲本、以已知抗病基因代表品种为供体，通过采用与抗白叶枯病目标基因相应的鉴别菌株逐代接种鉴定、汰选，同步进行农艺选择的选育、评价等工作（图 1），育成了携有不同抗性单基因的 6 个近等基因系，CBB3、CBB4、CBB12、CBB2、CBB7（暂名）和 CBB12。这些品系又经抗性反应型病理学比较鉴定、抗性遗传分析验证、农艺性状考查以及遗传育种和抗性机制研究应用，获得了科学数据（图 2）。

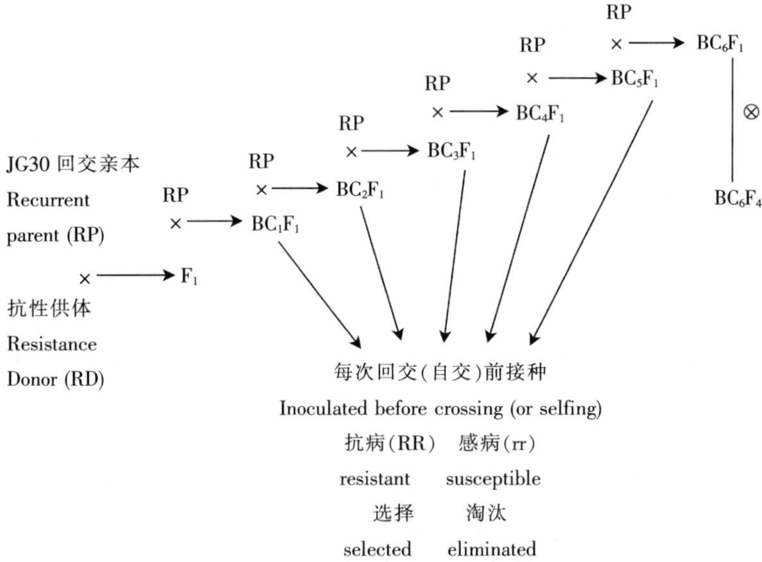

图 1 水稻白叶枯病抗性近等基因系选育程序［根据 Ogawa（1991）增补］

Fig. 1 Breeding procedure of rice near-isogenic lines for resistance to bacterial blight（supplemented based on Ogawa，1991）

图 2 中国粳稻抗白叶枯病近等基因系

Fig. 2 The Chinese near-isogenic japonica rice lines for bacterial blight resistance

上图，6 个 CBB 系统的粳稻白叶枯近等基因系的田间表现（Performance of six near-isogenic japonice rice lines for BB resistance in the field）

下图，该套近等基因系的轮回亲本（沈农 1033）及其中两个品系 CBB2 和 CBB3［The recurrent parent（SN1033）of the NILs，and 6 out of the two lines，CBB2 and CBB3］（另见彩图Ⅱ-14-2）

2.2 近等基因系与其抗病供体品种的白叶枯病抗性表达比较

鉴于这套粳稻近等基因系中的多数基因由籼稻导入粳稻背景，是否会随着多次回交、继代选择而丢失目标基因或削弱抗性效应，抗性基因的专化性是否稳定、清晰，则需要经过病理接种检测。我们连续 3 年采用与目标基因相对应的鉴别菌株，测定和比较育成的 6 个 NILs 之间以及与各自供体植株在不同生育期的抗性反应。结果表明，5 个系的专化抗性是稳定、清晰的。携有成株抗性基因的 CBB3、CBB12、CBB2 和 CBB7 的 4 个近等基因系在分蘖期表现感病，成株期转抗，各自与其供体早生爱国 3 号、Java 14、Tetep 和 DV85 都呈现典型的成株抗性；CBB14 与其抗性供体 TN1 对其匹配的菌株 P5 也都表现全生育期抗性。但是，CBB4 虽然具有清晰的专化抗性，却在分蘖期才开始表达，比其全生育期抗性基因 *Xa4* 的供体 IR20 滞后一段时间。事实上，在整个选育过程中，很难在沈农 1033/IR20 这个亚种间杂交组合的世代中发现全生育期抗性的单株，在各代精选过程中，也只发现到分蘖盛期表现中抗的株系。这种籼粳交后代中，籼稻的全生育期抗性发生特异性变异的现象曾在张林（1988）、万建民等（1985）的研究中有所报道，但张耀纲等（1991）、章琦等（1990）报道的籼稻全生育期抗性在籼、粳两种遗传背景下无明显差异。这可能与所用试材的抗性的遗传稳定性有关，其遗传本质与变化的原因还不清楚（表 1）。

表 1 中国粳稻近等基因系与其抗性供体对白叶枯病抗性反应比较[a]

Table 1 Comparison of resistance reaction to bacterial blight between Chinese japonica rice near-isogenic lines and their donors

测试材料 Material 轮回亲本 RP	测试年份 Year tested							
	1994		1995			1996		
	所用菌株							
	T2、P1 或（or）P5		T2、P1 或（or）P5			T2、P1 或（or）P5		
	TS[b]	BS		TS	BS		TS	BS
沈农 1033	S	S		S	S		S	S
供体 Donor								
Tetep（*Xa2*）	S	R		S	R		S	R
DV85（*Xa7*，*xa5*）	R	R		R	R		R	R
TN1（*Xa14*）	R	R	R	R		R	R	
近等基因系								
CBB2（*Xa2*）	S	R		S	R		S	R
CBB7（*Xa*? 未知）	S	R		S	R		S	R
CBB14（*Xa14*）	R	R	R	R		R	R	

[a] 携有 *Xa2*、*Xa7* 和 *Xa14* 的试材分别用菌株 T2（日本小种Ⅱ）、P1 或 P5（菲律宾小种 1 或小种 5）接种；RP=轮回亲本。[NILs carrying *Xa2*, *Xa7* and *Xa14* were inoculated with strains T2 (Japanese raceⅡ), P1, P 5=Philippine races 1 & 5, respectively; RP=Recurrent parent]。

[b] 接种时的生育期（Inoculation at growth stage），TS=分蘖期（Tillering stage），BS=孕穗期（Booting stage）。

R=抗（Resistant），S=感（Susceptible）。

2.3 近等基因系及其轮回亲本的农艺性状考查

连续6年在田间和室内考查了这套近等基因系与轮回亲本沈农1033包括生育期、株高、有效分蘖以及穗部性状等主要农艺性状。用LSR法比较了彼此间的株高、有效分蘖、穗长、结实率及千粒重参数。差异在统计学上均不显著（表2），表明这套近等基因系与其轮回亲本的遗传背景已经基本相同，彼此间的株型、穗形等田间农艺性状极为一致（图2）。这套NILs已在国内育种及相关研究中得到有效利用。如，浙江省农科院利用CBB7育成了3个全生育期抗性、广抗谱、高产的晚粳品系D601、D602和D603，它们成为周边地区育种单位的优良抗源（徐建龙等，1996）；中国农科院作物所李梅芳用CBB4育成了抗病、高产粳稻品种中花14和中花15。中国科学院遗传所朱立煌、陆朝福进行的分子检测表明，这套NILs基因组只在抗病基因区域存在差异，是水稻白叶枯病抗性基因定位、克隆和抗性机理研究的理想材料，他们已在克隆CBB7的抗性基因的进程中获得了两个候选基因（Lu et al.，1998），值得进一步进行基因鉴定并了解其功能属性。

表2 中国抗水稻白叶枯病粳稻近等基因系的农艺性状表现（1994—1996年）

Table 2 Performance of agronomic characters in Chinese japonica rice near-isogenic lines resistant to bacterial blight

轮回亲本 近等基因系 Recurrent parent NIL	播种-抽穗期 Seeding to heading (day)	株高 Plant height (cm)	有效分蘖数 Effective tillers/plant	穗长（cm） Panicle length	结实率 Grain setting (%)	千粒重 1000－grain weight (g)
沈农1033	122	83.9	9.8	15.8	89.1	22.5
CBB2	124	81.6	10.8	16.2	89.5	23.1
CBB7	124	83.0	10.9	15.6	87.6	22.7
CBB14	127	85.4	11.5	15.6	87.0	21.4
CBB3	126	84.3	9.8	16.4	89.2	23.6
CBB4	119	85.7	10.1	16.4	87.6	21.9
CBB12	128	85.4	9.9	15.0	86.4	22.4
平均值$\bar{x}$	124	84.2±2.6	10.5±2.2	15.9±0.7	87.9±1.2	22.2±1.6

2.4 近等基因系的抗性遗传分析

沈农1033/CBB3的F_1植株对与*Xa3*相对应的专化菌株T3在成株期抗病，其F_2群体在成株期的抗、感分离比符合3抗∶1感，*P*值大于0.05，表明CBB3对T3的成株期抗性由一对完全显性基因控制。沈农1033/CBB4的F_1植株对菌株P1在最高分蘖期为中抗，成株期高抗，该组合的F_2群体在分蘖期的抗、感分离比也符合3抗∶1感，*P*值大于0.05，表明CBB4在分蘖期对P1的抗性由一对不完全显性基因控制。沈农1033/CBB2和沈农1033/CBB7两个组合的F_1植株分别对相对应的菌株T2和P1在成株期高度抗病，其F_2群体在成株期抗、感分离比均符合3抗∶1感，*P*值均大于0.05，表明CBB2对T2和CBB7对P1的成株抗性各由一对完全显性基因控制。沈农1033/CBB14的F_1植株对相对应的菌株P5表现全生育期抗病，其F_2群体在分蘖前期的抗、感分离比符合3抗∶1感，*P*值大于0.05，表明 CBB4 对 P5 的全生育期抗性由一对显性基因控制（表 3）。

CBB3、CBB4 和 CBB12 分别同其相应的基因供体，早生爱国 3 号、IR20 和 Java 14 之间进行等位性测试，结果见表 4。CBB3/早生爱国 3 号和 CBB12/Java 14 的 F_1 植株对 T3 的抗性在成株期均为高抗。CBB4/IR20 的 F_1 对 P1 在分蘖期也抗病，在这 3 个组合的 F_2 群体中，均没有分离出感病植株，表明 CBB3、CBB4 和 CBB12 所携有的抗性基因同其供体是等位的。CBB2 和 CBB14 的目标性状在选育中纯合较快，抗病性稳定，最后其与目的基因的等位性测试也表明同其供体是等位的。

这套以 CBB 命名的 NILs 经抗性遗传分析并与供体基因进行等位性测试后表明，CBB2、CBB3、CBB4、CBB12 和 CBB14 各携有一对抗性基因，除 CBB7 以外，分别与其供体所携有的基因是等位的。但是当采用与 *Xa7* 基因相对应的菌株 P1 测试 CBB7 与 IRBB7（携有 *Xa7*）的等位性时，两者却不等位，表明 CBB7 携有的来自 DV85 的成株抗性基因与 *Xa7* 基因是不同的。证实这个结论的进一步依据有，（1）Lu（陆朝福，1998）将 CBB7 所携有的抗性基因定位于水稻第 11 染色体，而 Kaji 等（1995）将 *Xa7* 定位于第 6 染色体；（2）Kansas 州立大学 Leach 植物病理实验室的测试结果是：含有对 *Xa7* 基因无毒的 *Xoo* 菌无毒基因 *avrXa7* 的两个菌株能侵染 CBB7（感）（个人通讯）。这些结果表明 CBB7 所携有的这个显性成株抗性基因是来自 DV85 的另外一对未知基因，该基因的抗谱很广，至 2003 年仍没有发现可用以进行基因等位性测验的特异性菌株，值得继续寻找适合的鉴别菌株来揭示这个未知的广谱抗性基因。

表 3　粳稻近等基因系对相应白叶枯病菌株的抗性遗传分析

Table 3　Inheritance analysis of near-isogenic lines for resistance to the specific strains of *Xoo*

组合 Cross	接种时间 Growth stage inoculated	F_1 抗性反应[a] F_1 Reaction	F_2 抗性反应				
			抗病株 R plants	感病株 S plants	总株数 Total plants	χ^2 (3∶1)	P
沈农 1033/CBB2[b]	孕穗期	抗	228	72	300	0.111 1	0.75～0.50
沈农 1033/CBB3[c]	孕穗期	抗	197	66	263	0.001	0.90
沈农 1033/CBB4[d]	分蘖期	中抗	188	73	261	1.227	0.50～0.25
沈农 1033/CBB7[d]	孕穗期	抗	182	63	245	0.034	0.90～0.75
沈农 1033/CBB14[e]	分蘖期	抗	195	66	261	0.001 3	>0.90
CBB3/早生爱国 3	孕穗期	抗	251	0	251		
CBB4/IR20	分蘖期	抗	257	0	257		
CBB12/Java14	孕穗期	抗	261	0	261		

[a] 10～15 个 F_1 植株的叶片病斑面积百分率平均值（Average percent age of lesion areas inoculated leaves from 10～15 F_1 plants）。

[b] 用菌株 T2 接种（Inoculated with strain T2），[c]用菌株 T3 接种（Inoculated with strain T3），[d]用菌株 P1 接种（Inoculated with strain P1），[e]用菌株 P5 接种（Inoculated with strain P5）。

3　讨论

3.1　近等基因系的利用

我国这套粳稻抗白叶枯病近等基因系 CBB2、CBB3、CBB4、CBB7、CBB12 和 CBB14

是采用回交方法育成的。通过多年选育，改良了供体的某些不良性状，具有轮回亲本优良的株型和农艺性状。在选育过程中，同步严格采用与各抗性基因相对应的病菌小种进行特异抗性鉴定；在各回交世代结合抗性选择，直至育成的抗性品系除了带有目标基因的染色体片段外，近似于轮回亲本，其遗传背景是一致的，抗性也很稳定，是一套抗性连锁分子标记研究可利用的理想遗传工具，也是抗性机制和抗病育种的良好材料。截止 1996 年，已向全国 31 所大学、科研单位提供了种子、植株。中科院朱立煌实验室采用沈农 1033/CBB7 的 F_2 植株为作图群体，已将 CBB7 携有的基因定位于第 11 染色体，鉴定出在该基因两侧定位的分子标记 G1465 和 RG1109，与 G1465 的图距为 2.0 cM，通过染色体步移（Chromosome working）起点，已调出候选基因（陆朝福、朱立煌，1998）。南京农业大学进行了蛋白质双向电泳分析，在可清晰观察的近 300 个蛋白质斑点范围内，轮回亲本与近等基因系之间基本没有差异，表明它们的遗传背景高度一致，并且，它们对白叶枯病不同毒性菌株的抗性反应是明显不同的（朱友林，1993）；他们并以此探究水稻对白叶枯病的抗性分子机制，从稻叶中发现与抗病性有关的质膜蛋白 PR 蛋白，该结果表明它们具有明显的抑菌作用，并且与有关防卫基因的表达有密切关系。湖南农业大学选用该套 NIL 研究对白叶枯病的抗性生理指标。鉴于试材遗传一致，所测定与抗性密切相关的几种酶活性的表达与生理指标的数据准确，取得了可信的结果（李爱贞，1999）。

育种利用的反馈信息认为有些基因系的抗性导入效应较好，兹摘要将部分结果列于表 4。由于轮回亲本沈农 1033 存在后期早衰和不抗稻瘟病的不足，提请注意。这套近等基因系将在建立我国水稻白叶枯病寄主和病原菌之间的鉴别系统、检测新基因、在基因水平上研究抗性遗传和机制等方面，具有很好的利用价值。

表 4 我国白叶枯病抗性基因粳稻近等基因系在育种中的应用

Table 4 Utilization of near-isogenic japonica rice lines developed in China for resistance to bacterial blight

组合[a] Cross	所用抗源 R source used	抗性反应 Resistance reaction		利用的研究所 Institute for using
		MTS[c]	BS[d]	
CBB3/r8004	CBB3	S	R	安徽省农科院
CBB4/r8004	CBB4[b]	MR	R	同上
CBB4/r8004	CBB4	S	R	同上
CBB12/r8004	CBB12	S	R	同上
CBB3/8411	CBB3	S	R	浙江省农科院
CBB3/丙 89－2，CBB3/丙 89－17	CBB3	S	R	同上
CBB12/丙 89－2，CBB12/丙 89－17	CBB12	S	R	同上
86－106/CBB3	CBB3	S	R	湖南省农科院
87－249/CBB3	CBB3	S	R	同上
湘晚籼 1 号/CBB4	CBB4	MS	R	同上
辽粳 287/CBB3	CBB3	S	R	辽宁省农科院
79－3/CBB4	CBB4	MS	R	同上

（续）

组合[a] Cross	所用抗源 R source used	抗性反应 Resistance reaction		利用的研究所 Institute for using
		MTS[c]	BS[d]	
87-249/CBB12	CBB12	S	R	辽宁省农科院
CBB4//秋风/长白7号	CBB4[b]	MS	R	吉林省农科院
CBB4//秋风/双152	CBB4[b]	MR	MR	同上

[a] 所有与抗源配组的亲本均为高感品种（All of the parents which were crossed with the resistance sources are highly susceptible cultivars）。

[b] CBB4的抗性的表达在杂种后代中趋于孕穗期（The resistance exhibited at the booting stage in hybrid progenies of CBB4）。

[c] MTS=最高分蘖期（Max. tillering stage），[d]BS=孕穗期（Booting stage）。

3.2 暂名CBB12的说明

Java 14与早生爱国品种群的代表品种早生爱国3号对水稻白叶枯病有类似的广抗谱性。两者都抗日本小种Ⅰ、Ⅱ和Ⅲ以及菲律宾小种1、2、3和4。故而Java 14最初归属早生爱国品种群（Ogawa et al.，1991）。Yamamoto等（1977）鉴定Java 14抗一个印尼菌株Xo-7306（后来将其归为日本小种Ⅴ），并从早生爱国品种群中划分出来，成为Java群品种。Ogawa等（1978，1991）前后用日本和菲律宾小种研究了Java 14、早生爱国3号等品种的抗性遗传，认为两个品种都由*Xa3*控制日本小种3和菲律宾4个小种的抗性。除此，Java 14对日本小种Ⅰ和Ⅴ的抗性分别由与*Xa3*独立遗传的*Xa1*和*Xa12*（原名为*Xa-kg*）所控制。因此用Java 14作为供体的品系，必须用与其相应的日本小种Ⅴ的代表菌株鉴定，由于尚没有获得日本小种Ⅴ。故暂时命名为CBB12。

原载：作物学报，1996，22（2）：135～141；1998，24（6）：799～804.

携有抗白叶枯病新基因 *Xa23* 水稻近等基因系的构建及应用*

章　琦[1]　王春莲[1]　赵炳宇[1]　杨文才[1]
周永力[1]　赵开军[1]　乔　枫[1]　江祺祥[2]

（1. 中国农业科学院作物育种栽培研究所；2. 江苏省武进市稻麦育种场）

Development of Near-isogenic Line CBB*23* with a New Resistance Gene to Bacterial Blight in Rice and Its Application

Abstract　The near-isogenic line (NIL) with a new gene *Xa23* for resistance to bacterial blight (BB) was developed and designated as CBB23 in 1993—1998. A susceptible indica cultivar JG30 with improved plant type was used as a recurrent parent. The progenies derived from the cross JG30/H_4 harboring the gene *Xa23* were developed by back-cross until BC_5, and selfing for 4 times. Evaluation of the resistance gene *Xa23* with specific strain PXO99 (P6) and selection of plant type successively were carried out till the line showing stable resistance and agronomic characters similar with the recurrent parent. Comparison of reaction for resistance spectrum to 20 strains including 10 races in the Philippines, 3 Japanese races, and 7 Chinese pathotypes between CBB23 (carrying *Xa23*) and IRBB21 (carrying *Xa21*) was made. The *Xa23* is resistant to all 20 strains, and *Xa21* is resistant to 19 and highly susceptible to the Philippine race 10. The NIL CBB23 was used to establish segregation populations of JG30/CBB23 for mapping the gene *Xa23*. The gene was located on rice chromosome 11 in the preliminary tagging by screening SSR. Three AFLP markers screened are closely linked with *Xa23*. The genetic distance between APKj23 and *Xa23* is about 1.0 cM. The NIL is also valuable in rice breeding.

Key words　bacterial blight; new resistance gene; near-isogenic line; development and application

* Grant Items: Supported by the Rockefeller Foundation (No. 94001 # 253, 97001 # 258); the National Natural Science Foundation of China (No. 39670508) and a grant from the Ministry of Science and Technology of China (J00 – A – 002).

摘要 1993—1998年，构建了携有抗水稻白叶枯病新基因 *Xa23* 的近等基因系，被命名为CBB23。以全生育期高度感病并具有改良株型的籼稻品种JG30为轮回亲本，与携有 *Xa23* 的抗性供体 H_4 杂交，JG30/H_4 的 F_1 通过回交、自交直至 B_5F_4，各世代均用与目标基因 *Xa23* 相应的专化菌株P6进行人工接种鉴定，同步进行农艺性状选择，直至抗性稳定和农艺性状类似轮回亲本。比较了CBB23（携 *Xa23*）和IRBB21（携 *Xa21*）对20个包括10个菲律宾小种，3个日本小种和中国7个病原型代表菌株的抗谱。携带 *Xa23* 的品系抗所有20个菌株；携带 *Xa21* 的品系抗19个菌株，高感菲律宾小种10。用近等基因系CBB23构建了JG30/CBB23组合的 F_2 分离群体，通过SSR标记筛选，初步将 *Xa23* 定位于水稻第11染色体。进一步筛选出3个与 *Xa23* 更紧密连锁的AFLP标记。其中的APKj23与 *Xa* 23之间的图距约为1.0 cM. 并报道了在抗病育种中已有效的应用了携有 *Xa23* 的近等基因系CBB23。

关键词 稻白叶枯病；新抗性基因；近等基因系；构建和应用

Bacterial blight, caused by *Xanthomonas oryzae* pv. *oryzae* (*Xoo*) is a world wide disease and particularly destructive in Asia and China. Host resistance is considered a sound approach to control the disease. The resistance gene *Xa4* introduced from TKM6 into improved cultivars provided to most of the rice growing countries in Asia. Over the 20 - year period following the release of cultivars carried the *Xa4* in Southeast Asia, pathogen races were able to overcome the gene and increasingly frequent in the pathogen population (Mew et al., 1992). In China, the gene *Xa4* also has been widely adopted in development of indica hybrid and conventional rice for two decades. The compatible pathotype V to *Xa4* was originally emerged from Guangdong and then extended to Fujian, Zhejiang and Jiangsu provinces (Zhang, 1991).

Since 1987, we made efforts to identify new resistance resources from 871 accessions of wild rice species (Zhang et al., 1994). One accession, RBB16 from *Oryza rufipogon* showed high resistance to a broad spectrum pathogenic strain P6 which was virulent to all identified resistance genes to bacterial blight from cultivated rice (*O. sativa*). It might carry a new resistance gene. RBB16 was crossed with JG30, a susceptible indica cultivar, and the F_1 plants segregated in different resistance level. The F_1 plants with high resistance to P6 were selected and anther-cultured to obtain doubled haploid, which would be homozygous for resistance. The resistant H_2 plants were advanced to H_4 (Zhang et al., 2000).

A new resistance gene was identified in the H_4 and designated as *Xa23* based on comparison of resistance pattern and spectrum, inheritance analysis, and allele tests between *Xa23* and all the dominance genes designated (Zhang et al., 2000). The *Xa23* is a single complete dominance gene, with overall resistance and broad spectrum to BB (Zhang et al., 1998). The present paper deals with the breeding procedure, characteristics, and application of CBB23 in mapping of the gene and rice breeding.

1 Materials and Methods

1.1 Resistant donor and recurrent parent

The resistant donor, H_4, the anther-cultured line with the gene *Xa23*. The recurrent parent, JG30, a indica cultivar with improved plant type is susceptible to all the strains of bacterial blight tested during 1984 - 1989.

1.2 Bacterial strains

Twenty strains of *Xoo* were used for testing resistance spectrum in the study. Including ten Philippine strains, P1, P2, P3, P4, P5, P6, P7, P8, P9 and P10 representative the International differential races 1 - 10, respectively; three Japanese strains, T1, T2 and T3, representative races 1, 2 and 3, respectively; seven strains, HLJ72, HB17, NX42, ZHE173, GD1358, LN57, and JS49 - 6, representative to 7 Chinese pathotypes, C1 - C7, respectively. The Philippine strain P6 with specific avirulence to the gene *Xa23* was used to screen the materials with gene *Xa23* among each progeny during the development of the NIL. The two strains, P6 and P10 were used specially for comparing the different reactions of genes *Xa23* and *Xa21*.

1.3 Inoculation

The inoculum of each strain was prepared from 48 - hr culture grown on PSA slants. Inoculum was adjusted to a concentration of about 10^9 cells/mL.

For developing the NIL with the overall resistance gene *Xa23*, all of each progeny of plants were inoculated at the seedling stage. The leaf-clipping method was used for inoculation. Disease reactions were measured 14 - 20 days after inoculation. The disease was measured based on lesion percentage of the leaf area, the lesion percentage $\leqslant 5\%$ was scored as resistant (R), $>30\%$- 50% as susceptible.

1.4 Breeding of the NIL

The NIL was developed by making single cross of JG30/H_4, and back-crossing, selfing until BC_5F_4 generation. Evaluation of resistance to the specific strain P6 and selection of plant type have been making successively till the line with stable resistance and agronomic characters similar to that of JG30.

1.5 Comparison of resistance reaction between CBB23 and IRBB21

The tester, IRBB21 (carrying *Xa21*), with broad resistance spectrum was used for comparing the spectrum resistance with CBB23 (carrying *Xa23*). Since IRBB21 (*Xa21*) is susceptible to the strain P6 at seedling stage, then becomes resistant at tillering stage, the inoculations were done at maximum tillering and adult stages for comparing the spectrum resistance between *Xa23* and the tester *Xa21* by using two Philippine races , P6 and P10.

1.6 Tagging *Xa23* gene by using AFLP DNA markers

In 1999 – 2000, the NIL CBB23 was crossed with the recurrent parent JG30. A set of F_1, F_2 and BC_1F_1 populations derived from the cross JG30/CBB23 was used for obtaining the data of disease reactions as parameter and resistance inheritance analysis. The F_2 population was used for fine-mapping the gene. Genomic DNA of the plants were extracted as we did previously (Zhang et al. , 1998) . Amplified fragment length polymorphism (AFLP) was carried out following Pieter et al. (1995) except the gels were analyzed by silver staining method. Restriction enzymes *Eco*RI , *Mse*I and T4 – ligase used are New England Biolabs products. The adapters and primers were synthesized by Sangon, Shanghai. In the pre-amplification, the *Eco*RI-primer with no selective nucleotide, while the *Mse*I-primer with selective nucleotide C. In the second amplification, the *Eco*RI-primers with two selective nucleotides (E+2), while the *Mse*I-primers with three selective nucleotides (M+3) . The linkage analysis was performed with Mapmarker 3. 0.

1.7 Application of the NIL CBB23 in rice breeding

Since 1996, a japonica cultivar, 9520, with resistance to blast and high yielding but susceptible to bacterial blight was used as a receptor and recurrent parent. A cross between 9520 and CBB23 was made. The BC_3F_5 generations of 9520/CBB23 with double resistance to blast and bacterial blight were obtained by back-crossing and selfing. Resistance to blast and bacterial blight were evaluated as well as plant type successively was selected in each generation. The inoculum of blast was mixed with 5 Chinese races, Zhong A, C, D, E, and G with spraying method at rice seedling stage. The plants with resistance to blast were selected, then, inoculated with the strain of *Xoo*, P6 to obtain double resistance to the two diseases and with *Xa23* gene. The resistance to bacterial blight of six japonica lines were tested with 18 important strains selected including the 10 Philippine races, P1 – P10, 7 Chinese pathotypes , C1 – C7, and the Japanese race 3, T3 in 2001.

2 Results

2.1 Breeding procedure of the near-isogenic line (NIL)

Since 1993, initially for the development of near-isogenic line with the new gene for resistance to bacterial blight, the BC_5F_4 line was bred with the target gene for resistance in 1998. The basic procedure is shown in Fig. 1.

The original cross was made between the recurrent parent and the resistance donor. The first back-cross with the F_1 hybrid was completed. The selection of resistant plants was started by inoculation using the specific strain P6. From the second back-cross, at least 7

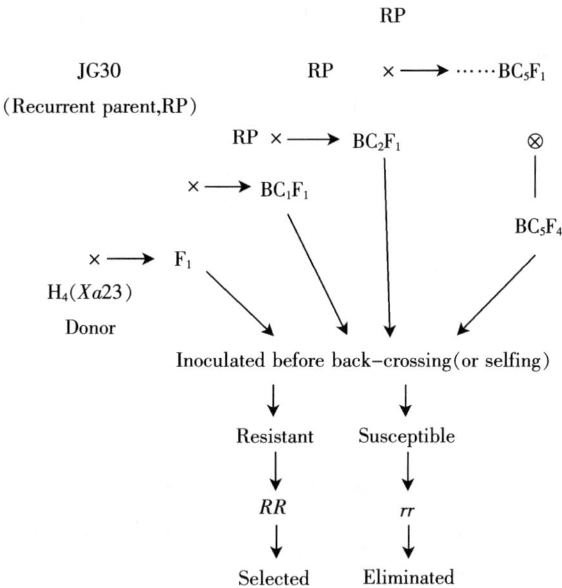

Fig. 1 Procedure steps in breeding for near-isogenic line to bacterial blight resistance.

resistant plants of the line were crossed with JG30 to obtain progeny carrying the resistance gene at a probability of more than 99%. This process was repeated until BC_5. The plants were essentially similar to JG30 and resistant to the specific strain P6, then selfing for 4 times. Evaluation of resistance to bacterial blight with the strain P6 and selection of plant type successively were made in the duration of development. The near-isogenic line was developed with stable resistance and uniform agronomic characters. The near-isogenic line is designated as CBB23 based on it carries the new gene *Xa23*.

2.2 Comparison of resistance spectrum to bacterial blight

During 1998 - 2000, the NIL CBB23, its recurrent parent JG30 and a tester IRBB21 were inoculated with 19 strains including 9 Philippine races, 7 Chinese strains and 3 races in Japan at the seedling, tillering and booting stages. The CBB23 was resistant to the 19 strains from seedling to booting stages, the results were similar to the donor, H_4 tested in 1992 - 1995 (Zhang et al., 1998). The IRBB21 was susceptible to all the strains at seedling stage, but showed resistance at tillering stage; the JG30 was susceptible to the strains at all growth stages. In 2001, the new Philippine race 10 (P10) obtained from IRRI was added. The CBB23 showed high resistance to all the 20 strains including the P10. The recurrent parent JG30 was highly susceptible to all the strains. The tester IRBB21 was resistant to all the strains except the P10 (Table 1). It is obviously that the resistance spectrum of CBB23 is broader than IRBB21. The broadest spectrum of resistance to BB at all growth stages in CBB23 inherited completely from its donor H_4.

Table 1 Resistance reactions of CBB23, H_4, JG30 and IRBB21 at tillering stage

Material		CBB23 (*Xa23*)		H4 (donor)	IRBB21 (*Xa21*)		JG30 (RP)	
Year tested		1998 - 2000	2001	1992 - 1995	1998 - 2000	2001	1998 - 2000	2001
Philippine race	P1	R	R	R	R	R	S	S
	P2	R	R	R	R	R	S	S
	P3	R	R	R	R	R	S	S
	P4	R	R	R	R	R	S	S
	P5	R	R	R	R	R	S	S
	P6	R	R	R	R	R	S	S
	P7	R	R	R	R	R	S	S
	P8	R	R	R	R	R	S	S
	P9	R	R	R	R	R	S	S
	P10		R			S		S
Chinese pathotype	C1	R	R	R	R	R	S	S
	C2	R	R	R	R	R	S	S
	C3	R	R	R	R	R	S	S
	C4	R	R	R	R	R	S	S
	C5	R	R	R	R	R	S	S
	C6	R	R	R	R	R	S	S
	C7	R	R	R	R	R	S	S
Japanese	T1	R	R	R	R	R	S	S
	T2	R	R	R	R	R	S	S
	T3	R	R	R	R	R	S	S

2.3 Agronomic characters of the NIL

The analysis of variance of the data indicated that the agronomic characters between CBB23 and its recurrent parent, JG30 have no significant differences (Table 2).

Table 2 Comparison of agronomic characters between CBB23 and JG30

Material	Gene	Seedling-heading period (days)	Plant height (cm)	Effective tillers/plant	Length of panicle (cm)	Percentage of seed setting	1 000 - grain weight/g
JG30	non	132	87.5	10.5	21.2	75.8	22.8
			88.5	12.0	21.5	77.4	24.2
			88.0	11.5	21.0	76.1	23.7

（续）

Material	Gene	Seedling-heading period (days)	Plant height (cm)	Effective tillers/plant	Length of panicle (cm)	Percentage of seed setting	1 000 - grain weight/g
CBB23	*Xa23*	133	87.5	10.0	20.9	76.4	23.5
			88.5	11.5	20.6	75.1	23.7
			87.0	11.5	20.9	78.8	22.5
F value			0.40	0.062	6.0	0.08	2.75

$n=3$. $F_{0.01}=7.71$.

2.4 Utilization of CBB23 for fine-mapping of *Xa23*

In the previous study, inheritance of the resistance gene *Xa23* was analyzed based on several sets of F_1, F_2, and B_1F_1 populations derived from crosses between WBB1 (CBB23, carrying *Xa23*) and different susceptible cultivars. The results showed that the resistance in WBB1 is controlled by a single, complete dominant gene (Zhang et al., 1998). One set of F_1 and F_2 populations derived from cross JG30/WBB1 was used for preliminary tagging the new gene and found it located on rice chromosome 11 by SSR analysis in 1998 (Zhang et al., 1998). In 1999 - 2000, a big F_2 population with 2 562 plants derived from the cross between NIL CBB23 and JG30 was established for fine-mapping the gene *Xa23*. Inoculation test showed that 1928 plants of the F_2 population were resistant (R) and 634 plants were susceptible (S) to the strain P6, and the R∶S ratio fits to 3∶1 very well ($\chi^2=0.075$, $P=0.75-0.90$) (Table 3). The disease reactions of the F_2 populations to P6 were provided an important parameter to mapping the gene.

Since 2001, one hundred and twenty eight combinations of AFLP E-primer/M-primer were surveyed. Three AFLP markers closely linked with *Xa23* were found. Among them, APKj23 (E-AC/M-CAG) is only 1.0 cM from *Xa23*.

Table 3 Resistance reactions of F_1, F_2 and BC_1F_1 populations from cross JG30/CBB23 to P6 at seedling stage (Beijing, 2000)

Cross	F_1 plant	F_2 or BC_1F_1 population			χ^2 (expected) ratio	*P* value
		R	S	Total		
JG30/CBB23	R	1 928	634	2 562	0.075 (3∶1)	0.75 - 0.90
JG30//CBB23/JG30		35	32	67	0.060 (1∶1)	0.25 0.5

R=resistant; S=susceptible.

2.5 Application of CBB23 in rice breeding for bacterial blight resistance improvement

Since 1996, a japonica cultivar, 9520, with resistance to blast and high yielding but susceptible to bacterial blight was used as a receptor and recurrent parent. We made a cross between 9520 and an advanced line of CBB23 (BC_3F_3). The BC_3F_5 generations of 9520/CBB23 with double resistance to blast and bacterial blight were obtained by back-crossing and selfing. Resistance to blast and bacterial blight was evaluated as well as plant type selected successively in each generation. Six japonica lines showed resistance to the two diseases with improved plant type. On the other hand, the lines also were resistant to all 18 strains including 10 Philippine races, 7 Chinese pathotypes and Japanese race 3 tested in 2001 (Table 4).

Table 4 Resistance reactions to blast and BB of six rice lines derived from 9520/CBB23

strains	Line	Zhong 5158	Zhong 5171	Zhong 5173	Zhong 5174	Zhong 5175	Zhong 5176	CBB23 (donor)
Philippine race	P1	R	R	R	R	R	R	R
	P2	R	R	R	R	R	R	R
	P3	R	R	R	R	R	R	R
	P4	R	R	R	R	R	R	R
	P5	R	R	R	R	R	R	R
	P6	R	R	R	R	R	R	R
	P7	R	R	R	R	R	R	R
	P8	R	R	R	R	R	R	R
	P9	R	R	R	R	R	R	R
	P10	R	R	R	R	R	R	R
Chinese pathotype	C1	R	R	R	R	R	R	R
	C2	R	R	R	R	R	R	R
	C3	R	R	R	R	R	R	R
	C4	R	R	R	R	R	R	R
	C5	R	R	R	R	R	R	R
	C6	R	R	R	R	R	R	R
	C7	R	R	R	R	R	R	R
Japanese	T3	R	R	R	R	R	R	R
Mixture with 5 blast races*		R	R	R	R	R	R	S

P1～P10=Philippine races; C1～C7= Chinese pathotypes; T3=Japanese race 3.

* The mixture with 5 blast races: Chinese Zhong A, C, D, E, and G

R=Resistant; S=Susceptible.

3 Discussions

The near-isogenic line CBB23 with the new gene *Xa23* is highly resistant to the 20 strains including the main International differential races and all 7 Chinese pathotypes. It is therefore clearly that the *Xa23* has the broadest spectrum among the bacterial blight resistance genes identified so far. The gene *Xa23* not only has the broadest spectrum resistance to bacterial blight , but also shows high level of resistance at all growth stages. The overall resistance with high level and broad spectrum to bacterial blight derived from gene *Xa23* has been inherited strongly into the advanced rice lines. The NIL CBB23 and the lines (carrying *Xa23*) with double resistance to blast and bacterial blight have been provided to scientists and breeders and used in rice breeding or resistance mechanism programs in China. Since the resistance of *Xa4* has been broken down in Southeast and South countries (Mew et al., 1992; Adhikari et al., 1999), and some *Xoo* strains can overcome the resistance conveyed by *Xa21* in the Philippines, Korea, Indonesia (Bustamam, pers comm.) and China, the CBB23 and lines with *Xa23* will make contribute to widening the genetic background for bacterial blight resistance of rice in China and in the world.

Near-isogenic lines (NIL) with diverse genes for resistance to major diseases and insects can be used as powerful tools for identifying races of diseases and biotypes of insects as well as new genes for resistance. Moreover, NILs could serve as excellent materials for gene mapping and cloning. With the F_2 population of CBB23/JG30, the fine-scale mapping and cloning of *Xa23* is under going. Until recently, 250 SSR markers, 35 RFLP markers, 900 RAPD markers and 128 AFLP primer combinations have been surveyed and 11 molecular markers closely linked with *Xa23* have been screened out (Zhao et al., unpublished). Some of the linked markers have been used in breeding program for marker-assisted selection and 2～3 most closely linked markers will be used to screen genomic library for molecular cloning of *Xa23*. These results demonstrated that CBB23 is a pure and *elite* genetic material for molecular mapping and cloning of *Xa23*.

Published in: Chinese J. Rice Sci. 2002, 16 (3): 206－210.

Ⅲ 水稻白叶枯病抗病新基因的鉴定与应用

Part 3 Identification and Application of New BB Resistance Gene

广西普通野生稻 RBB16 抗白叶枯病育种初报

林世成[1] 阙更生[1] 邢祖颐[1] 章 琦[1] 王春莲[1] 李道远[2] 陈成斌[2]

（1. 中国农业科学院作物育种栽培研究所；2. 广西农业科学院品种资源研究所）

Preliminary Report on the Utilization of the Common Wild Rice RBB16 from Guangxi in Breeding for Resistance to Bacterial Blight

摘要 以广西普通野生稻（*Oryzae rufipogon*）编号 RBB16 为水稻白叶枯病抗源亲本，与粳稻品种杂交，在 F_6 代育成 2 个籼型、3 个粳型稳定新品系。研究表明，RBB16 的抗性传递力很强，在育种上有广阔的利用前景。新育成的 5 个品系中有 3 个对水稻白叶枯病具有广谱抗性，可作为改良型抗源，提供育种利用。其中丰产性好的可在生产中应用。以粳稻品种为母本，RBB16 为父本，通过回交、复交和花培技术，能有效地缩短野栽杂交育种年限，为利用野生稻资源开辟了好途径。

关键词 普通野生稻；抗稻白叶枯病；抗源筛选；野栽杂交育种

我国当前水稻主要栽培品种的白叶枯病抗源，籼稻品种的抗性基因主要为 *Xa4*，粳稻品种为 *Xa3*（章琦等，1988；章琦，1991），抗性资源的单一性，是抗性育种的限制因素。广西普通野生稻中存在高抗水稻白叶枯病的种质资源，其抗性及抗谱均较国际水稻研究所（IRRI）的 IR 系统更为理想，充分开发利用将为水稻育种开拓广阔前景。

本研究旨在通过野栽杂交、回交、复交及花培等途径，导入广西普通野生稻的抗白叶枯病基因，选育出农艺性状稳定、丰产性较好的抗病品系及其衍生抗源。

1 材料和方法

1.1 供试材料

抗源亲本，由广西农科院品种资源所提供的广西普通野生稻 17 个编号，RBB12、16、37、56、82、97、139、151、443、473、479、502、503、530、544、553、554（李道远，1986）。

栽培稻亲本，粳稻品种有中丹 2 号、中花 8 号、中作 180、辽粳 5 号、垦系 3 号、沈农 1033、中系 237、Zenith；籼稻品种有桂朝 2 号、特青 2 号、中铁 31、桂 D1 号。

白叶枯病菌株有 GX54、GX56、GX57、HB 17、ZHE173 五个。此外，还采用了菲律宾小种 1（P1）和日本小种 1（T1）。

1.2 试验方法

抗源筛选：1987—1989 年，对上述广西农科院提供的 17 个抗源野生稻编号进行接种筛选。用 GX54、GX56、GX57 及 HB17、ZHE173 等菌株分别在苗期及成株期接种，从中筛选出抗谱广、抗性强的 RBB16（1989 年在 IRRI 接种 P1 和 T1 菌株也表现抗病）。

野栽杂交配组：以籼型和粳型栽培稻品种为母本，以野生稻 RBB16 为父本，配制单交、回交和复交组合。

杂交后代的选择：用系谱法进行后代选择，每个世代选择优良单株、建立株系，并进行主要农艺性状的观察记载和白叶枯病抗性鉴定，对优良的中间材料进行花培。

2 试验结果

2.1 普通野生稻 RBB16 对水稻白叶枯病的抗性

RBB16 具有广谱抗性，其抗性遗传力很强，抗、感杂交 F_1 代高度抗病，表现为完全显性。经遗传分析，RBB16 所携带的基因与 *Xa4* 非等位。1987—1989 年以 RBB16 为抗源，与籼、粳品种杂交，共配制了 26 个组合，其中单交 11 个、回交 6 个、复交 9 个。单交组合 F_1 表现为白叶枯病显性抗性。F_2 表现为 15∶1 的抗感分离比，部分组合在 F_4～F_6 都分离出抗病株系。说明其抗性遗传力持久而稳定（表 1）。

表 1 野生稻 RBB16 在野栽杂交后代中对白叶枯病的抗性传递情况

杂交组合	各代抗病株出现状况			
	F_1	F_2	F_4	F_6
中作 180/RBB16	R	R	R	
中作 180/RBB16//中作 180		R	R	R
中花 8 号//中作 180/RBB16		R	R	R
垦系 3 号/RBB16	R	R	R	
垦系 3 号//垦系 3 号/RBB16		R		
中系 237//垦系 3 号/RBB16			R	

（续）

杂交组合	各代抗病株出现状况			
	F_1	F_2	F_4	F_6
中丹 2 号/RBB16	R		R	
中丹 2 号//中丹 2 号/RBB16				
垦系 3 号//中丹 2 号/RBB16		R	R	R

2.2 野生稻 RBB16 与栽培稻杂交育成的稳定品系和中间材料

RBB16 与栽培稻杂交后，通过回交、复交及花培育成 5 个稳定品系 T157、T167、T169、T200 和 T209，其中，T209 和 T169 为杂交后通过花培育成。这 5 个品系都是粳稻与普通野生稻配组，均在 F_6代稳定，育种年限为 3.5 年。用籼稻与普通野生稻配组，单交、回交、复交或花培均未获得稳定品系，以粳稻为亲本的单交组合未经花培也没有获得稳定品系。稳定品系的主要农艺性状表现见表 2。

表 2　野栽杂交育成的稳定育种品系主要农艺性状（广西，南宁）

品系	杂交组合	生育期（d）	株高（cm）	穗长（cm）	穗粒数	粒长（mm）	粒宽（mm）	[c]抗性
T167	[a]中作 180/RBB16//中作 180	128	106.4	24.8	173.0	7.6	3.5	R
T157	[a]中花 8 号//中作 180/ RBB16	130	103.0	25.4	136.5	8.3	3.1	R
T209	[b]垦系 3 号//垦系 3 号/RBB16	120	99.0	18.4	176.0	7.4	3.3	S
T200	[b]中丹 2 号//中丹 2 号/RBB16	120	92.0	20.7	155.6	9.1	3.5	S
T169	[b]垦系 3 号//中丹 2 号/RBB16	115	94.0	20.1	125.0	7.3	3.4	R
中作 180	亲本	120	95.0	24.3	154.0	7.6	3.7	S
中花 8 号	亲本	115	91.5	20.3	123.0	7.2	3.4	S
垦系 3 号	亲本	110	94.5	15.3	139.0	6.8	3.7	S
中丹 2 号	亲本	120	95.0	23.1	157.0	—	—	S
RBB16	亲本	感光型	122.0	20.1	52.0	8.4	2.6	R

[a] 籼型；[b] 粳型；[c] R=抗；S=感。

野栽杂交后代的生育期和株高一般表现为双亲中间型，穗长、每穗粒数、粒长都有较明显的超亲倾向，粒型有偏籼的趋势，在 5 个品系中有两个比原亲本明显增产。

品系 T167：籼型，茎秆粗壮、穗大粒多、米质中等，1991 年早造折亩产 452kg，比亲本中作 180 增产 21%，比对照籼稻高产品种双朝 25 增产 13.1%，抗白叶枯病。

品系 T157：籼型，茎秆粗壮、基部节间短、抗倒性极强。1991 年早造经受两次台风不倒，米质较好，抗白叶枯病。

品系 T209：粳型，茎秆粗壮、抗倒性强、着粒密、穗大粒多、结实率高、米质优。

1991 年晚造折亩产 448kg，比亲本垦系 3 号增产 59.3%，比中作 180 增产 15.2%。

品系 T169：粳型、早熟、穗较小、抗白叶枯病性好，可作抗病亲本。

品系 T200：粳型、株高 90cm，茎秆特别粗壮、抗倒性极强、粒型大，可作亲本利用。

在上述 5 个品系中，有 3 个为稳定的抗白叶枯病新品系。它们都是以不抗白叶枯病和农艺性状优良的粳稻品种为亲本，经回交及复交在 F_6代育成的。粳型品系 T169（垦系 3 号//中丹 2 号/RBB16）于 1990 年曾进行广谱抗性测定，用我国白叶枯病强毒菌株 HB17 和菲律宾小种 1（P1）和日本小种 1（T1）接种，其抗性反应与 RBB16 完全一致，病斑面积小于 5%。

1991 年晚造，以 HB17 菌株接种品系 T167 的 76 个 F_6株系和品系 T157 的 34 个 F_6株系，均表现为抗，抗性稳定。RBB16 与栽培稻的杂交后代中，变异类型非常丰富，从中选出一批具有特异性状的优良中间材料。其最突出的性状是穗大粒多，最长的穗可达 44cm，主穗粒数最多可达 391 粒。此外，还有株型好、矮秆抗倒、抗病、优质等优良性状（表 3）。

表 3　野栽杂交的大穗型中间材料

杂交组合	世代	秆型	芒性	主穗长（cm）	主穗粒数
垦系 3 号//沈农 1033 /RBB16	F_5	矮秆	无芒	29.2	391
桂 D1 号//垦系 3 号/RBB16	F_5	矮秆	无芒	28.2	309
垦系 3 号//垦系 3 号/RBB16	F_4	矮秆	无芒	28.0	250
桂朝 2 号//辽粳 5 号/RBB16	F_5	矮秆	无芒	25.7	268
双朝 25//垦系 3 号/RBB16	F_5	矮秆	无芒	32.0	270
特青 2 号//特青 2 号/RBB16	F_6	矮秆	无芒	27.2	259
中作 180/RBB16	F_3	高秆	有芒	44.0	302
Zenith/ RBB16	F_3	高秆	有芒	35.1	264

2.3　缩短野栽杂交育种年限的技术

野栽杂交在育种年限上遇到的最大的问题是杂种后代农艺性状稳定慢，育种年限长。本试验从 1987 年以来，观察了 26 个野栽杂交组合后代的性状变异和遗传情况，发现无论籼稻或粳稻亲本与 RBB16 的单交组合，F_3代以前，综合性状好的单株出现频率极低。籼稻的回交及复交组合其优良类型的单株比以粳稻为亲本的组合少，出现的世代也较迟。但是，以粳稻品种为亲本配制的回交或复交组合从 B_1F_1便出现矮秆、无芒、不落粒、结实率较高、综合性状较好的类型，而且优良的变异类型比较丰富，可以出现籼型、粳型，成熟期的变异大，有早、中、晚熟类型；理想的株叶型、穗型较多，粒型变异也较丰富，米质较优，杂交优势明显，粗秆大穗类型较多，耐寒性和叶片耐衰老性较强。这些在 B_1F_1代就出现的特殊变异株，往往就是育成优良品系的中间材料，特别是粳型和矮秆变异类型。本试验选出的 5 个稳定品系均为矮秆及粳型变异株的后代中育成的（表 4）。可见，以粳稻品种为亲本，并进行回交又复交，是缩短野栽杂交育种年限的一个有效方法。

表 4　在 B_1F_1 中出现的变异株

杂交组合	B_1F_1 出现的变异株		F_3～F_6 性状分离情况	品系
	株号	性状		
中作 180/RBB16//中作 180	21-1	籼型、矮秆、茎粗、穗大、稀短芒、不落粒、腹白小	苗期耐寒、腹白大小分离、有极少数全芒株	T167
中花 8 号//中作 180/RBB16	17-1	籼型、矮秆、茎粗、抗倒、无芒、颖尖紫色、落粒性弱	苗期耐寒、色素及腹白大小均有分离，少数有芒株和中秆株	T157
垦系 3 号//垦系 3 号/RBB16	13-31	粳型、植株矮小、叶片大、无芒、难脱粒、米质优	穗型和腹白大小均有分离，随世代增多，植株逐渐粗壮，穗大粒多	T209
中丹 2 号/RBB16//RBB16	7-1	粳型、茎粗、叶大而厚、无芒、难脱粒、粒长大、米质优	腹白大小和穗型均出现分离	T200
垦系 3 号//中丹 2 号/RBB16	14-5	粳型、矮小、无芒、米质优、易落粒	落粒性出现分离，植株随世代增高	T169

此外，在 B_1F_1 或 B_1F_2 代的群体中，选择矮秆优良单株进行花培，也是缩短育种年限的途径。但是，具有野生稻性状的籼型株系，其愈伤组织诱导率和绿苗率较低，而粳型优良中间材料，其愈伤组织诱导率和绿苗率则较高（表 5）。

表 5　野栽杂交后代优良中间材料花培情况

杂交组合	花培株系			获得绿苗数	育成的稳定品系
	株号	类型	世代		
垦系 3 号//垦系 3 号/RBB16	13-31	粳型	F_4	25	T209
垦系 3 号//中丹 2 号 /RBB16	14-5	粳型	F_3	14	T169
中花 8 号//中作 180/RBB16	17-1	籼型	F_4	5	
中作 180/RBB16//中作 180	21-1	籼型	F_4	0	

3　讨论

3.1　普通野生稻 RBB16 具有广谱抗性，是目前我国抗水稻白叶枯病的珍贵抗源。本试验在广西南宁及菲律宾分别采用我国白叶枯优势菌株 HB17、ZHE173、GX54、GX 56、GX 57 及 P1、T1 等菌株接种 RBB16 后，都表现为 R 型抗性反应。遗传分析表明，RBB16 的抗性基因与 *Xa4* 非等位，初步认为其抗性为 1 或 2 个显性基因所控制，其抗性传递力很强，持久而稳定。这一新抗源的发掘与利用，对我国水稻白叶枯病育种将产生重大影响。

3.2　野栽杂交育种是水稻选育新品种的重要新途径，容易获得优异类型。尤其是利用粳稻为亲本，更易于成功地获得类型丰富的矮秆、大穗的优良株系及其他优异变异类型，值得提倡。

3.3　野栽杂交育种成功的关键在于有效的促进性状稳定，以粳稻为亲本比籼稻更有利于远缘杂交的亲和性和性状稳定。回交、复交及花培技术都能缩短育种年限，选择 B_1F_1 代中出现的矮秆大穗特异变异类型也是获得稳定优良株系（品系）的重要途径。

原载：广西农业科学，1993，(1)：1～5.

野生稻抗白叶枯病性（*Xanthomonas oryzae* pv. *oryzae*）的评价*

章　琦[1]　王春莲[1]　施爱农[1]　白建法[1]
林世成[1]　李道远[2]　陈成斌[2]　庞汉华[3]

（1. 中国农业科学院作物育种栽培研究所；2. 广西农业科学院品种资源研究所；
3. 中国农业科学院品种资源研究所）

Evaluation of Resistance to Bacterial Blight（*Xanthomonas oryzae* pv. *oryzae*）in Wild Rice Species

Abstract　Eight hundred and seventy-one materials including thirteen wild rice species collected from six provinces，Jiangxi，Hunan，Yunnan，Fujian，Guangdong，Guangxi and IRRI，Thailand，Bangladesh，Upper Volta，were evaluated with three strains of *Xanthomonas oryzae* pv. *oryzae*，HB17，T1 and P1，using artificial clipping method in 1987—1992. All of the twelve Acc. No. of *O. meyeriana* were highly resistant to the three strains. Many Acc. No. of *O. punctata*，*O. latifolia*，*O. minuta*，*O. officinalis*，*O. longistaminata*，*O. alta*，*O. rufipogon* and *O. glaberrima* showed very resistant also. A few materials of *O. rufipogon* were highly resistant only. Percentages of highly resistant ones in both *O. officinalis* and *O. longistaminata* were 49. 8% and 38. 9%，respectively. Twenty-two of sixty-one materials tested repeatedly were highly resistant to a broad spectrum virulence strain，P6. Some of them might have new resistance genes，which were absent in cultivated rice. The performance of resistance was various in *O. rufipogon*. Approch of utilization of wild rice species with different genomes was discussed also.

Key words　wild rice；bacterial blight in rice；resistance

摘要　1987—1992 年，用水稻白叶枯病 3 个菌株 HB 84－17、T1 和 P1，以人工剪叶法鉴定了来自我国江西、湖南、云南、福建、广东、广西 6 个省（自治区）及 IRRI、泰国、孟加拉、上沃尔特的 13 个野生稻种共 871 份材料的抗性。在参试的 12 份疣粒野生稻（*O. meyeriana*）全部高抗，斑点（*O. punctata*）、阔叶

* 国家攻关和国家自然科学基金资助项目．彭绍裘先生提供小粒野生稻，杨文才同志参加部分工作，谨致谢忱。

（*O. latifolia*）、小粒（*O. minuta*）、药用（*O. officinalis*）、长药（*O. longistaminata*）、高秆（*O. alta*）、普通野生稻（*O. rufipogon*）和非洲栽培稻（*O. glaberrima*）等8个种内，也有许多编号的抗性颇佳。如药用和长药野生稻中的高抗材料分别占其参试总数的49.8%和38.9%，而普通野生稻中的高抗材料则不多。在重复测定的61份包括药用、长药和普通野生稻3个种中，发现22份材料高抗一个广谱致病菌株，该菌株能侵染当时已知栽培稻中的所有抗病基因，推测其中可能具有栽培稻中所缺乏的新基因。普通野生稻的抗性表现多种多样，极少数表现抗性一致和较稳定的全生育期抗性，多数材料表现同一编号不同株的抗感反应各异以及同株同生育期抗性异型。本文还讨论了不同染色体组野生稻种的利用途径。

关键词 野生稻；水稻白叶枯病；抗性

Ling，Khush等（1970，1977）从尼瓦拉野生稻（*O. nivara*）中发现迄今唯一抗草状矮缩病（grassy stunt）的显性单基因*GS*之后，IRRI育种家用该基因育成了IR28、IR29等许多抗该病的品种（Khush et al.，1977）。Khush等（1990）在来自印度的长药野生稻（*O. longistaminata*）中发现了一个高抗当时印度和菲律宾所有白叶枯病小种的新基因*Xa21*，进一步显示从野生稻资源中发掘抗病基因的潜在价值。

早在1933年，我国丁颖教授曾从广东普通野生稻种和栽培稻的天然杂交后代中，选育出对白叶枯病、矮缩病、褐飞虱具有中度抗性、优质和适应性强等特性的品种中山1号。该品种此后成为华南晚籼当家品种的骨干亲本（Chen et al.，1980）。彭绍裘等（1981）考察和测定了云南普通野生稻、药用野生稻和疣粒野生稻对湖南白叶枯病强毒菌株的抗性，其中疣粒野生稻的抗性接近免疫。孙恢鸿等（1985）和广东省农科院（1989）也鉴定了当地普通野生稻对本地菌株的抗性，评价出一些抗病编号。我们于1987—1992年采用国内外4个毒力各异的稻白叶枯病鉴别菌株，对来自我国6个省和国外包括13个野生稻种共871个编号进行了抗性评价、抗性类型分析、野生稻同栽培稻主要抗病基因的抗谱和抗扩展性比较等方面的研究。

1 材料与方法

1.1 供试材料

中国农业科学院作物品种资源研究所提供了来自江西、湖南、云南、福建、广西、广东6个省（自治区）及IRRI、泰国、孟加拉、上沃尔特等国家和研究所的13个野生稻种共580份。1991—1992年于北京本院温室育苗，移植于隔离网室。广西农科院品资所野生稻组于1987至1990年供试291份材料。主要是当地的普通、药用野生稻以及云南疣粒野生稻。全部试材种植于该所的野生稻保存圃。接种前两个月移植分蘖于隔离网室待测。

本试验测试的13个野生稻种的名称及其编号份数如下：普通野生稻（*O. rufipogon*）585份，药用野生稻（*O. officinalis*）237份，疣粒野生稻（*O. meyeriana*）12份，长药野生稻（*O. longistaminata*）18份，尼瓦拉野生稻（*O. nivara*）3份，斑点野生稻

（*O. punctata*）3 份，阔叶野生稻（*O. latifolia*）2 份，颗粒野生稻（*O. granulata*）2 份，高秆野生稻（*O. alta*）3 份，小粒野生稻（*O. minuta*）1 份，重颖野生稻（*O. grandiglumis*）2 份，非洲栽培稻（光稃稻）（*O. glaberrima*）2 份，澳洲野生稻（*O. australiensis*）1 份。

3 个全生育期广抗谱栽培稻品种 IR20（*Xa4*）、IR1545（*xa5*）和 DV85（*xa5*、*Xa7*）为抗病对照；金刚 30（籼）和沈农 1033（粳）为感病对照品种。

供试菌株 HB17 是我国白叶枯病致病型Ⅱ的代表菌株；国际鉴定和命名白叶枯病抗性基因时所用的代表菌株，菲律宾小种 1（P1）和日本小种 1（T1）。笔者曾将它们接种于中国鉴别品种，P1 和 T1 的毒性分别类似于我国的致病型Ⅱ和Ⅰ，P6（PXO99）为菲律宾小种 6，迄今只有来自长药野生稻的新基因（*Xa21*）具有抵御 P6 的功能。在国内尚未发现可以与之相比拟的广致病谱菌株，用作本试验鉴定新抗源的专化抗性菌株。

1.2　试验内容和方法

1.2.1　抗性鉴定　用 3 个菌株 HB17、P1（PXO61）和 T1（T7174）测定所有参试材料的抗性。用 3 种不同色泽塑料带将各植株的分蘖或叶片分成 3 组来标示接种的不同菌株。全部植株在分蘖盛期接种。分蘖盛期表现感病的材料到孕穗期再接种 1 次，测定其成株期的抗性反应。

接种前将保存的冻干菌种在 PSA 斜面培养基上复壮，在鉴别品种上测试毒力，转移数管斜面菌种，置于 4℃冰箱，供整个生长期使用毒力一致的菌种。接种前，将菌种移至 PSA 斜面培养基上，在 28℃温箱培养，菌龄 2～3d，浓度约 10^9/mL。人工剪叶接种，接种后 14～20d，当感病对照品种的病情发展趋于稳定时，进行调查。每个处理调查 3 张叶片。以病斑占叶面积百分率为抗性反应参数。根据病斑占叶面积百分率划分抗感反应型：≥5%抗，＞5%～＜20% 中抗，＞20%～30%中感，＞30%感。

1.2.2　野生稻与栽培稻的抗性比较　1990—1991 年分别在南宁和北京选择了初测表现抗性较好的 6 份普通野生稻、48 份药用野生稻和 7 份长药野生稻，与抗我国大多数菌株的 3 个来自栽培稻的抗性基因品种 IR20（*Xa4*）、IR 1545（*xa5*）和 DV85（*xa5*、*Xa7*）比较它们对菌株 HB17 和 P6 的抗性，以金刚 30 为感病对照品种。接种前用 2 种色带将植株的分蘖或叶片分成 2 组，于分蘖期接种。接种后 20d 进行第一次调查、60d 第二次调查，调查病斑占叶片面积百分率及反应型的变化。

1.2.3　普通野生稻的白叶枯病抗性类型分析　1987 年在南宁测定 48 份广西普通野生稻的抗性时，发现在 6 份抗病材料中，有 5 份材料的同一植株表现抗、感不同的反应。在 1991 年夏季，于北京评价 509 份来自我国 6 个省的普通野生稻对 HB17、P1 和 T1 三个菌株的抗性，重点观察其抗性类型。以金刚 30（籼）和沈农 1033（粳）为感病对照。植株于分蘖和孕穗期各接种 1 次，接种后 14～20d 调查。也采用植株分蘖分组法，接种菌龄、菌液浓度、方法以及调查标准等均相同于抗性鉴定部分。

2　结果与分析

2.1　不同野生稻种抗白叶枯病性状的评价

在不同野生稻种中，抗病材料的比重差异很大（表 1）。在 585 份普通野生稻中，感病者居多，仅有 24 份高抗 3 个菌株，占总数的 4.1%。此结果类似广东省农科院对该省普通野生稻抗白叶枯病鉴定的报道。在其测定的 1 597 份材料中，只有 14 份抗病（广东省农业科学院，1989）。参试的 12 份疣粒野生稻全部高抗；在药用和长药野生稻中的抗病材料比重也较大，高抗 3 个菌株的材料分别占总数的 49.8%和 38.9%。中抗者也分别达 40.5%和 44.4%。其他野生稻种参试量较少，不足以进行比较。

参试的 13 个野生稻种共 871 份材料对 3 个菌株 HB17、P1 和 T1 的抗性测定结果表明，多数野生稻种中具有抗性优良的资源（表 2）。就抗性水平而言，以疣粒野生稻最为突出，全生育期都表现高抗，甚至免疫，病斑呈褐色侵染型。斑点、阔叶、小粒、药用、长药、高秆、普通野生稻和非洲栽培稻（光稃稻）等 8 个野生稻种中，也不乏抗性上乘的材料。它们在全生育期高抗 3 个菌株，病斑占叶面积均不超过 5%。颗粒野生稻表现中等抗性。尼瓦拉野生稻只抗 3 个菌株中的 T1，呈专化抗性反应。重颖野生稻对菌株 T1 和 HB 17 具有中等抗性。只有参试的澳洲野生稻是感病的，但比感病对照金刚 30 和沈农 1033 的感病程度较轻。

表 1　野生稻对 3 个白叶枯病菌株的抗性鉴定

Table 1　Identification of resistance to three strains of *Xoo* in wild rice species

年份 (Year)	普通野生稻 *O. rufipogon*			药用野生稻 *O. officinalis*			长药野生稻 *O. longistaminata*			疣粒野生稻 *O. meyeriana*			其他种 Other species			总计 Total
	总数	抗	中抗	总数	抗	中抗	总数	抗	中抗	总数	抗	中抗	总数	抗	中抗	
1987[a]	48	2	7													48
1988[a]	17	1	2													17
1989[a]				200	89	91	15	6	7	1	1					216
1991[b]	509	20	22	37	29	5	3	1	1	11	11		16	10	5	576
1992[b]	11	1	1										3	1	1	14
总计 Total	585	24	32	237	118	96	18	7	8	12	12		19	11	6	871
抗病百分率（%）		4.1	5.5		49.8	40.5		38.9	44.4		100			57.9	31.6	

[a] 在南宁接种鉴定（Inoculation in Nanning）。

[b] 在北京接种鉴定（Inoculation in Beijing）。

表 2　13 个野生稻种对 3 个白叶枯病小种（致病型）代表菌株的抗性

Table 2　Resistance to 3 representative strains of *Xoo* races（pathotypes）in 13 wild rice species

种名 Species	染色体组 Genome	对 3 个 *Xoo* 菌株的抗性反应[a b] Reaction of resistance to 3 *Xoo* strains					
		HB17（中国Ⅱ型） Chinese pathotype Ⅱ		PXO61（菲小种 1） Phi. race 1		T7174（日本小种 1） Jap. race 1	
		MTS	BS	MTS	BS	MTS	BS
疣粒野生稻 *O. meyeriana*	GG	HR	—	HR	—	HR	—

（续）

种名 Species	染色体组 Genome	对 3 个 *Xoo* 菌株的抗性反应[a b] Reaction of resistance to 3 *Xoo* strains					
		HB17（中国Ⅱ型） Chinese pathotype Ⅱ		PXO61（菲小种 1） Phi. race 1		T7174（日本小种 1） Jap. race 1	
		MTS	BS	MTS	BS	MTS	BS
小粒野生稻 *O. minuta*	BBCC	HR	HR	R	R	R	R
斑点野生稻 *O. punctata*	BBCC，BB	R	HR	R	R	R	R
阔叶野生稻 *O. latifolia*	CCDD	R	R	R	R	HR	R
药用野生稻 *O. officinalis*	CC	R	R	R	R	R	R
长药野生稻 *O. longistaminata*	A^bA^b	R	R	R	R	R	R
普通野生稻 *O. rufipogon*	AA	R	R	R	R	R	R
高秆野生稻 *O. alta*	CCDD	R	R	R	R	R	R
非洲野生稻 *O. glaberrima*	A^gA^g	R	R	R	R	R	R
颗粒野生稻 *O. granulata*	GG	MR	—	MR	—	MR	—
重颖野生稻 *O. grandiglumis*	CCDD	MR	MR	MS	MS	MR	MR
尼瓦拉野生稻 *O. nivara*	AA	R	—	S	—	S	—
澳洲野生稻 *O. australiensis*	EE	S	—	S	—	S	—
金刚 30 Jingang 30（CK）	AA	S	S	S	S	S	S

[a] MTS 最高分蘖期（Maximum tillering stage），BS 孕穗期（Booting stage）.

[b] R 抗（Resistant）；MR 中抗（Moderately resistant）；S 感（Susceptible）；MS 中感（Moderately susceptible）.

2.2 野生稻与栽培稻对白叶枯病抗性的比较

从 61 份包括 3 个野生稻种的重复测定中，有 22 份材料的抗扩展性大大强于栽培稻中主要抗病基因代表品种（表 3）。感病对照金刚 30 对两个菌株高度感病，接种后 60d，其病斑面积分别高达 65%和 85%，病叶快速枯萎。栽培稻抗性基因代表品种 IR20（*Xa4*）、

IR1545（*xa5*）和 DV85（*Xa7*，*xa5*）高抗菌株 HB17，发病面积仅 3%～5%，但不抗 P6，病斑面积达 40%以上，病叶也迅速枯萎。表明抗性基因 *Xa4*、*xa5* 和 *Xa7* 对菌株 P6 均不具有抗性功能。在 48 份药用野生稻中，有 17 份不仅高抗 HB17，并且能抗毒性强的 P6，病斑面积仅 2%～5%，其中 9 份材料对 P6 的抗性尤其突出，在接种后 60d，先期形成的褐色斑纹将病斑面积限制于 15%左右，不再扩展；其余 8 份材料接种 P6 后 2 个月的病斑约为 20%。在复测的 6 份普通野生稻和 7 份长药野生稻中，分别有 3 份和 2 份是高抗 2 个菌株的。发病面积仅 2%～3%。接种后 60d，也只扩展到 15%左右，表现很强的抗扩展性。这部分参试的大多数抗病的野生稻植株始终保持青枝绿叶，野生稻特有的抗衰老性，也显示其对白叶枯病的抗扩展性。同栽培稻相比，野生稻中某些材料对白叶枯病的抗性水平、抗扩展等方面都有明显的优越性（图 1）。

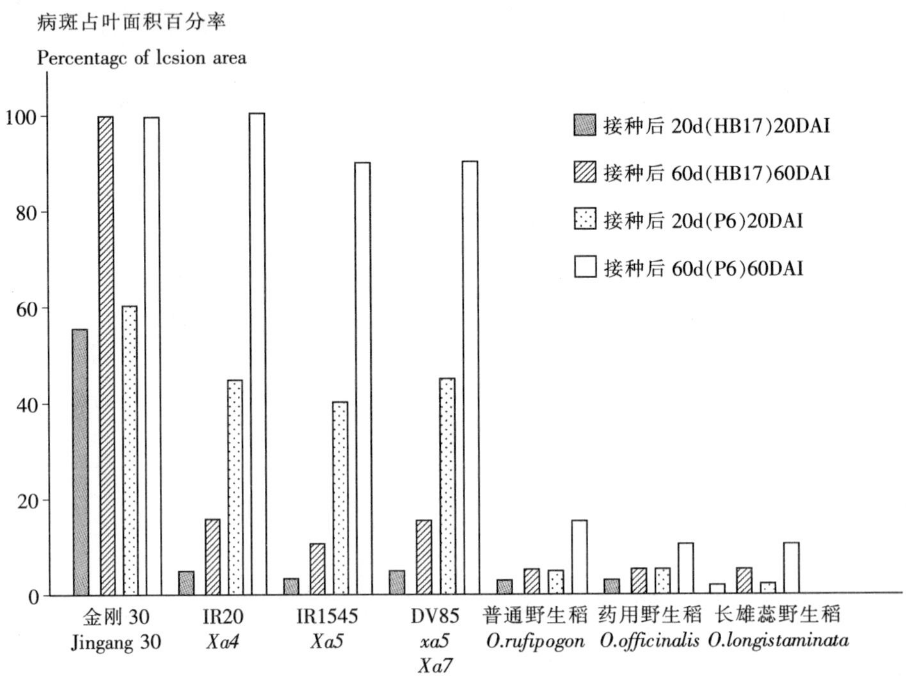

图 1　野生稻与栽培稻对白叶枯病的抗扩展比较

Fig. 1　Comparison of resistance to colonization to bacterial blight between wild rice and cultivated rice

表 3　野生稻与栽培稻对白叶枯病抗性的比较[a]

Table 3　Comparison of resistance to bacterial blight between wild and cultivated rice

材料 Material	基因 Gene	病斑占叶片面积百分率（%）[c] Percentage of lesion area（%） 接种后天数（DAI） HB 17 20d	HB 17 60d	P6 20d	P6 60d	注（Remark）
JG30（CK）	Non	55（S）		60（S）		叶片枯萎（Leaves blighted）
IR 20	*Xa4*	5（R）	15（M）	45（S）		叶片枯萎（Leaves blighted）

（续）

材料 Material	基因 Gene	病斑占叶片面积百分率（%）[c] Percentage of lesion area（%）				注（Remark）
		接种后天数（DAI）				
		HB 17		P6		
		20d	60d	20d	60d	
IR 1545	*xa*5	3（R）	10（M）	40（S）		叶片枯萎（Leaves blighted）
DV85	*xa*5，*Xa*7	5（R）	15（M）	45（S）		叶片枯萎（Leaves blighted）
药用野生稻	未知（Unknown）					保持青枝绿叶 Plants and leaves maintained verdant and greenery
O. officinalis No. 35		3（R）	5（R）	5（R）	10（M）	
O. officinalis No. 14		2（R）	5（R）	2（R）	15（M）	
O. officinalis No. 44		3（R）	5（R）	5（R）	15（M）	
O. officinalis No. 42		3（R）	10（M）	3（R）	15（M）	No. 8、18、20、25、27、31、32、33、41 具有相似反应[b]
O. officinalis No. 34		3（R）	10（M）	5（R）	15（M）	
O. officinalis No. 21		3（R）	12（M）	2（R）	15（M）	
O. officinalis No. 29		3（R）	12（M）	2（R）	15（M）	
O. officinalis No. 15		3（R）	10（M）	3（R）	18（M）	
普通野生稻	未知（Unknown）					
O. rufipogon RBB16		2（R）	5（R）	3（R）	5（M）	
O. rufipogon No. 52		2（R）	10（M）	2（R）	13（M）	
O. rufipogon No. 49		3（R）	13（M）	3（R）	20（M）	
长药野生稻	未知（Unknown）					
O. longistaminata No. 3		2（R）	5（R）	2（R）	10（M）	
O. longistaminata No. 55		2（R）	10（M）	2（R）	15（M）	

[a] 从 61 份中复测出 22 份野生稻与栽培稻对 P6 和 HB17 的抗性稳定性比较（Comparison of stability of resistance between 22 out of 61 wild rice and cultivated rice to *Xoo* strains P6 and HB 17）。

[b] Reaction of No. 8、18、20、25、27、31、32、33、41 were similar。

[c] R= 抗（Resistant）；M= 中抗（Moderately resistant）。

2.3 我国不同来源普通野生稻的白叶枯病抗性

来自我国 6 个省的普通野生稻对白叶枯病的抗性表现较为特殊。表 4 列出几种不同类型，1）同一编号不同植株的抗感反应不同。2）同株同生育期抗性异型反应，植株在相同生育期对同一菌株同时出现抗病和感病叶片，这两种类型的材料广泛存在于参试的普通野生稻中，可能是由于普通野生稻与栽培稻同属 AA 型染色体组，且普通野生稻自身花粉多不完全，柱头大而外露，异交结实率较高（丁颖，1983），造成所采集的种子并不都是 1 个抗性纯合体。3）全生育期抗性，这类为数不多的材料自苗期到抽穗期都表现高抗，抗性一致而稳定，这些可能是抗性相对纯合的材料，它们是否采集自远离栽培稻田、未开垦的低丘荒塘、沼泽地，值得进一步调查。4）成株期抗性，这类材料较多，同栽培稻中的多数粳稻类似，在苗期和分蘖期高度感病，到成株期转抗。有些材料还呈现小种专化抗性，对一些菌株在成株期抗病，对另一菌株则为全生育期抗病。某些寄主与病原菌之间还存在交叉反应。

表 4 我国不同来源的普通野生稻对白叶枯的同型/异型抗性表现

Table 4 Performance of resistance to bacterial blight in *O. rufipogon* from different origins in China

抗性类型 (Resistance type)	编号 Acc. No.	对 3 个 *Xoo* 菌株的病斑面积%/反应型[a] Reaction type of resistance to 3 strains (lesion area%)						产地 Origin
		T1		HB17		P1		
		MTS	BS	MTS	BS	MTS	BS	
同编号不同株 Different plants within same Acc. No.	0119 (1)	5 (R)	5 (R)	35 (S)	30 (S)	40 (S)	30 (S)	广西贵县
	0119 (2)	55 (S)	40 (S)	40 (S)	40 (S)	40 (S)	10/25 (M/MS)	
	0141 (1)	3 (R)	3 (R)	40 (S)	30 (S)	35 (S)	10 (M)	广西武宜
	0141 (2)	40 (S)	40 (S)	40 (S)	30 (S)	40 (S)	25 (MS)	
同株相同生育期 At the same growth stage on one plant	YD 6 - 0052	5/25 (R/S)		5/25 (R/S)		5/25 (R/S)		湖南茶陵
	029 (1)	5/35 (R/S)		5/35 (R/S)		5/35 (R/S)		广西桂平
	YD 5 - 0003	5/45 (R/S)		45 (S)		55 (S)		福建漳浦
全生育期抗性 Overall resistance	S - 2159	5 (R)	5 (R)	5 (R)	5 (R)	5 (R)	5 (R)	广东
	RBB16	3 (R)	3 (R)	3 (R)	3 (R)	3 (R)	3 (R)	广西
成株期抗性 Adult plant resistance	47 - 43	45 (S)	5 (R)	70 (S)	10 (M)	5 (R)	5 (R)	江西东乡
寄主-病菌交叉反应 Reversal reaction between host and pathogen	G103	5 (R)		35 (S)		30 (S)		湖南江永
		35 (S)		5 (R)		55 (S)		

[a] R= 抗 (Resistant), M= 中抗 (Moderately resistant), MS= 中感 (Moderately susceptible), S= 感 (Susceptible)。

3 讨论

3.1 野生稻具有十分宝贵的抗病资源

本研究进一步表明，野生稻中存在丰富的优良白叶枯病抗源。在参试的 13 个野生稻种中，9 个拥有高抗编号，其余的因参试量太少或未能测定，尚不能定论。在药用、长药、普通野生稻中，有 22 个高抗栽培稻都不能抵御的菲律宾小种 6 的材料。其中一个普通野生稻 RBB16，我们在另一个试验中，初步分析了其对 P1 的抗性遗传，表明该材料对 P1 的全生育期抗性，由完全显性基因所控制，并与 *Xa4* 非等位。在未曾与更多的已知基因进行等位性测定之前，虽不能确定为新基因，但其特有的全生育期的完全显性抗性，就优于不完全显性抗性基因 *Xa4*，适用于改良杂交水稻的抗性。我们评价的这些优异抗源，为白叶枯病抗性遗传改良、拓宽抗性遗传背景提供了丰富的材料和进一步研究的依据。

3.2 评价野生稻白叶枯病抗性的几个问题

同栽培稻相似，野生稻的白叶枯病抗性也存在寄主与病原菌之间的特异性互作。着眼于新基因的发掘，应采用致病谱广以及不同小种（病原型）的代表菌株进行鉴定。那些抗性表现异型的种，如普通野生稻，宜进行单株分蘖（或叶片）分组，用多菌株鉴定，选择全生育期表现抗性一致和稳定的个体。

野生稻的株型特殊，多数种的叶片很长，野栽杂交后代“疯狂”分离，叶片的长、短差异很大，如以病斑长度作为评价抗性的参数，可能会将抗性较好的材料汰除。病斑占叶面积百分率则能反映试材的实际抗性水平。另外，必需设置抗、感对照品种。

3.3　野生稻抗病基因的利用

野栽杂交的不亲和性和野生稻不良性状的难于排除，是水稻育种中利用野生稻抗性基因的主要问题。那些相同于栽培稻 AA 型染色体组的野生稻种，虽然也存在亲和性的差异，一般与栽培稻杂交是容易成功的。上文提及的抗白叶枯病新基因 *Xa21* 和现今唯一的抗草状矮缩病基因*GS*，都分别由具有 AA 型染色体组的长药和尼瓦拉野生稻与栽培稻杂交育成改良型抗病品系后，才进行遗传分析和用于育种。但野生稻和栽培稻杂交毕竟属于种间远缘杂交，存在着不良性状难以汰除及后代定型时期长的问题。成功的经验是通过多次回交、复交，评价各代目标性状、进行筛选，或结合花培方法，可有效地缩短选育纯合改良型中间材料的进程（吴妙燊，1990）。利用自然杂交率较高的普通野生稻，进行连续套袋自交或花培手段，以取得纯合性状的材料是十分必要的。

非 AA 型染色体组的野生稻种具有许多抗性优异的材料。但用常规有性杂交的方法不容易成功。因此，不同染色体组的稻属种间杂交的探索，逐渐成为国内外这一领域的热门课题。Khush（1985）曾采用栽培稻和药用野生稻有性杂交和幼胚离体培养方法，获得有药用野生稻染色体附加系的杂种后代，国内也进行了这方面的研究，并提出运用组织培养和染色体工程技术、理化因素诱导染色体结构、或基因位点突变后的花药与栽培稻杂交，以及外源 DNA 导入等生物技术，提供利用野生稻有利基因的可能性。

由于野生稻具有包括抗病虫性在内的许多特异有利基因，必须组成涉及的多学科协作，才能在促进水稻育种、遗传研究和生物技术等方面发挥重要作用。

原载：中国农业科学，1994，27（5）：1～9.

Identifying of a New Gene for Resistance to Bacterial Blight from *O. rufipogon*

Q Zhang[1]　S C Lin[1]　B Y Zhao[1]　C L Wang[1]　W C Yang[1]
Y L Zhou[1]　D Y Li[2]　C B Chen[2]　L H Zhu[3]

(1. Institute of Crop Breeding and Cultivation, CAAS Beijing 100081, China ;
2. Institute of Gerrnplasm Resources, GAAS, Naining 530007, China;
3. Institute of Genetics, Academia Sinica, Beijing 100101, China)

The gene *Xa4* for resistance to bacterial blight (BB) has been widely used in the development of all commercial hybrids and conventional cultivars of indica rice in the main rice growing regions of China which encompass more than 90% of the national rice area (Min, 1992) . Such narrow genetic basis has increased the genetic vulnerability of hybrid rice and conventional indica rice to BB, since a strong selection on the pathogen population has increase the frequency of pathotype V (virulent to rice carrying *Xa4*) . This pathotype is the most virulent one among the 7 Chinese BB pathotypes (Zhang et al. , 1996) . To overcome this problem, diversification of rice cultivars with different resistance gene (s) is necessary.

Eight hundred seventy-one accessions of 13 wild species were evaluated with 3 strains of BB from China, the Philippines and Japan. Twenty-one resistant accessions were re-evaluated with strains of 7 Chinese, 6 Philippine and 3 Japanese races at both seedling and adult stages. One accession (RBB16) of *O. rufipogon* showed high level of resistance to all the strains used (Zhang et al. , 1994) .

An inter-specific cross between RBB16 and Jingang 30 (JG30), a susceptible cultivar was made, the F_1 plants showed different levels of resistance to several strains of BB including PXO99. Since *O. rufipogon* was heterozygous for resistance, the F_1 plants showing high resistance to BB were anther-cultured to obtain doubled haploids that would be homozygous for resistance. The anther-cultured progenies were inoculated with 3 strains: GD1358 (Chinese pathotype V), PXO99 (Philippine race 6), and T7133 (Japanese race 3) .

Lines showing high resistance and improved plant-type in H_2 populations were selected and advanced to H_4 which were collectively designated as WBB1 (Zhang et al. , 1994) .

Among all the known genes for resistance to BB, *Xa21* is the only dominant gene from a wild species (*O. longistaminata*) resistant to all 6 races of BB in the Philippines (Khush et al. , 1989, 1990) . Therefore, *Xa21* was used as tester for identifying the resistance gene of *O. rufipogon*.

Comparison of resistance reactions of WBB1 and 6 testers (with different major dominant resistance genes) using 9 Philippine races of BB at both seedling and adult stages.

WBB1 showed highly resistance to the 9 races at all growth stages; IRBB21 (*Xa21*) showed resistance to the 9 races at the tillering stage; and *Xa3*, *Xa4*, *Xa7*, *Xa10* and *Xa14* showed no resistance to race 6 (PXO99), but showed resistance to some races. This suggests that the resistance of WBB1 to 9 Philippine races is different from *Xa21* and the other testers (Table 1).

Table 1 Comparison of reaction to 9 Philippine races of BB between WBB1 and 6 tester genes

Material	Resistance gene	Race									* Growth stage inoculated
		1	2	3	4	5	6	7	8	9	
WBB1	*Xa* -?	R	R	R	R	R	R	R	R	R	SS
IRBB21	*Xa21*	R	R	R	R	R	R	R	R	R	TS
IRBB3	*Xa3*	R	R	R	R	R	S	S	S	R	AS
IRBB4	*Xa4*	R	S	S	M	R	S	R	R	S	SS
IRBB7	*Xa7*	R	R	R	S	R	S	R	R	R	AS
IRBB10	*Xa10*	S	R	S	S	R	S	R	S	S	SS
IRBB14	*Xa14*	S	S	S	S	R	S	R	S	R	SS
JG30	None	S	S	S	S	S	S	S	S	S	

* SS, TS, AS=Resistant at seedling, tillering and adult stage, respectively.

Inheritance of BB resistance in WBB1 to the Philippine race 6 (PXO99)

WBB1 was crossed with 2 susceptible Chinese cultivars, JG30 (indica) and 02428 (japonica), and IRBB21 (*Xa21*). The F_1, F_2 and B_1F_1 populations were inoculated with PXO99 at the seedling stage. As shown in Table 2, all the F_1s were resistant. The F_2 populations from the crosses of WBB1 with susceptible parents segregated in a ratio of 3R∶1S. The backcross progenies segregated in a ratio of 1R∶1S. The χ^2 values for the 15R∶1S segregation expected for two independently segregating genes in reciprocal crosses of IRBB21 and WBB1 were significant indicating that the dominant gene of WBB1 may be linked to *Xa21*. The results suggest that a single dominant gene form *O. rufipogon* confers resistance to PXO99 in WBB1, and the gene is nonallelic but probably linked to *Xa21*.

Table 2 Reaction to PXO99 of F_1, F_2 and B_1F_1 populations from the crosses of WBB1 with susceptible cultivars and IRBB21 at seedling stage

Cross	Reaction of F_1	F_2 or BC populations			χ^2 (Expected ratio)	*P* value
		R	S	Total		
JB30/WBB1	R	183	52	235	0.887 (3∶1)	0.25～0.50
02428/WBB1	R	199	58	257	0.686 (3∶1)	0.25～0.50

(continued)

Cross	Reaction of F_1	F_2 or BC populations			χ^2 (Expected ratio)	P value
		R	S	Total		
JG30/WBB1//JG30		35	32	67	0.06 (1 : 1)	0.25 - 0.50
02428/WBB1/WBB1		24	30	54	0.463 (1 : 1)	0.25 - 0.50
IRBB21/WBB1	R	136	31	167	46.73 (15 : 1)	0.001
WBB1/IRBB21	R	164	28	192	25.08 (15 : 1)	0.001

R=Resistant; S=Susceptible.

PCR analysis of WBB1 with *Xa21* specific primers

Since only *Xa21* confers resistance to 9 races including race 6among all known dominant genes for resistance to BB, we checked the allelism of the gene from O. *rufipogon* with *Xa21* at the molecular level. Two *Xa21* primers U1 (U15′- CGA TCG GTA TA CAG CAA ACC TA CAG CAA AAC - 3′) and I1 (3′- ATA GCA ACT 5′) were used to test the allelic relationship between *Xa21* and WBB1, JG30 was used as check. Only IRBB21 (*Xa21*) showed the 1.4 Kb fragment which is specific to *Xa21* locus, whereas WBB1 and JG30 showed 1.3 Kb fragments (Fig. 1) . This again indicates that the resistance gene of WBB 1 is different from *Xa21*.

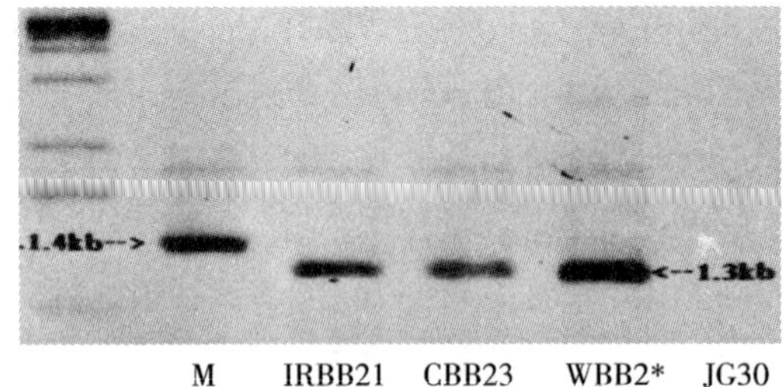

Fig. 1 Polymorphism was detected between IRBB21 and CBB23, JG30 using STS primers U1/I1specific for *Xa21* locus.

Lane 1, M DNA marker; lane 2, IRBB21;

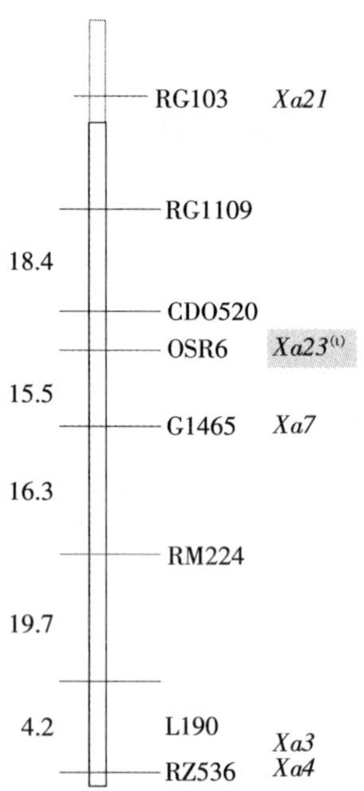

Fig. 2 Preliminary linkage map constructed from a population consisting of 40 F_2 plants using Mapmaker 3.0. Stringency is set at LOD=3.0. Map distance is in cM Based on Kosambi function.

Tagging the new gene by using DNA markers

After screening 94 rice Microsatellite (SSR) primers, two markers linking with the resistance gene were identified. The linkage analysis was performed by Mapmaker 3. 0. Via the SSR markers mapped, the resistance gene was located on rice chromosome 11 (5. 4 cM to OSR6 and 27. 7cM to RM224) (Fig. 2) . Fine mapping is on going.

Thus, the gene for resistance to BB from *O. rufipogon* appears to be distinct from all the known genes for resistance. We have designated the new gene as *Xa23* (*t*) .

Published in: Rice Genetic Newsletter, 1998 (15): 138 - 142.

普通野生稻抗水稻白叶枯病（*Xanthomonas oryzae* pv. *oryzae*）新基因 *Xa23*（*t*）的鉴定和分子标记定位*

章　琦[1]　赵炳宇[1]　王春莲[1]　杨文才[1]　林世成[1]
阙更生[1]　周永力[1]　李道远[2]　陈成斌[2]　朱立煌[3]

（1. 中国农业科学院作物育种栽培研究所；2. 广西农业科学院品种资源研究所；
3. 中国科学院遗传研究所）

Identifying and Mapping a New Gene *Xa23t* for Resistance to Bacterial Blight（*Xanthomonas oryzae* pv. *oryzae*）from *O. rufipogon*

Abstract　A new resistance gene to bacterial blight *Xa23t* from *O. rufipogon* was identified based on the bacterial blight resistance spectrum, comparison of resistance type, inheritance analysis and testing the allelism relationship between *Xa23t* and *Xa21*. The results indicated that it is conditioned by a new complete dominance overall resistance gene with the widest spectrum resistance, and highest efficiency of resistance transfer among all the known genes identified so far. One hundred and sixty plants were selected from the F_2 population derived from a pair of Near-isogenic line（JG30^{6*} // WBB1 ）for mapping the gene by using 160 SSR markers. The resistance gene was located on rice chromosome 11（5. 3cM to OSR06，27. 7 cM to RM244）.

Key words　*O. rufipogon*；bacterial blight；new resistance gene；molecular mapping

摘要　通过抗谱评价、抗性类型比较、遗传分析鉴定出一个来自普通野生稻的抗白叶枯病新基因 *Xa23t*。它是迄今已知基因中抗谱最广、抗性导入效应很强的一个完全显性的全生育期抗性基因。从携有新基因 *Xa23t* 的一对近等基因系（JG30^{6*} // WBB1）构建的 F_2 群体中，选出 160 个单株用 SSR（微卫星）标记进行基因定位，初步将该基因定位于水稻第 11 染色体 SSR 标记 OSR06 和 RM224 附近，图距分别为 5. 3cM 和 27. 7cM。

* 本研究由国家自然科学基金、洛克菲勒基金和中国农业科学院重点项目资助。

Supported by the National Nature Sciences Foundation; the Rockefeller Foundation and the Chinese Academy of Agricultural Science Key Program.

关键词　普通野生稻；稻白叶枯病；新抗性基因；分子标记定位

水稻白叶枯病（*Xanthomonas oryzae* pv. *oryzae*）是一个全球性病害。从野生稻中发掘可利用的抗病新基因和进行分子标记筛选和定位，是近年国内外这一领域的研究热点。Khush 等（1990）从长药野生稻（*O. longistaminata*）中鉴定出一个广抗谱基因 *Xa21*，Ronald 等（1996）将 *Xa21* 定位于水稻第 11 染色体，Song 等（1995）成功地分离出该基因。国内自 80 年代就开始进行了野生稻抗源筛选，彭绍裘等（1991）用湖南省白叶枯病强菌株测定了云南普通野生稻（*O. rufipogon*）、药用野生稻（*O. officinalis*）和疣粒野生稻（*O. meyetiana*）的抗性，结果表明疣粒野生稻的抗性接近免疫。孙恢鸿等（1992）报道了广西普通野生稻和药用野生稻中有一批抗中国病原型Ⅰ～Ⅴ的资源。我们自 1987 年以来，用国内外 4 个鉴别菌株评价了包括 13 个野生稻种的 871 个编号的抗性，于重复测定的 61 份抗病材料中，发现 22 份包括普通、药用和长药野生稻中高抗栽培稻抗病基因都不能抵御的一个国际广致病菌株 P6（PXO99），推测其中可能存在新抗源（章琦等，1994）。经初步分析了其中一个普通野生稻编号为 RBB16 的抗性遗传，其抗性与 *Xa4* 非等位（林世成等，1992）。为了加速该普通野生稻抗性的纯合性，将 RBB16 与感病品种金刚（JG30）杂交，其 F_1 植株通过花培、自交至 H_4，获抗性纯合系。再以 JG30 为轮回亲本，与 H_4 回交，经农艺性状选择、各世代抗病性测定直至 BC_5F_3，育成携有该抗病基因的近等基因系 WBB1。

本研究自 1994 年以来，采用国际全套鉴别菌株进行了抗谱比较、抗性遗传分析，SSR、RFLP 标记筛选和染色体定位，确认了 WBB1 携有新的抗病基因，被命名为 *Xa23t*。本文报道上述研究结果，为拓宽白叶枯病抗性基因源和开展分子标记辅助选择育种提供信息。

1　材料与方法

1.1　抗谱和抗性类型分析

比较携有该基因的近等基因系 WBB1 与由日本和 IRRI 育成的携有 6 个不同显性抗性基因的近等基因系 IRBB3、IRBB4、IRBB7、IRBB10、IRBB14 和 IRBB21（Ogawa et al.，1991）的抗谱和抗性类型，感病对照品种为 JG30。

1.2　抗性遗传和导入效应分析及配组

1.2.1　JG30（籼）和 02428（粳）对全部供试菌株高度感病，在抗性遗传分析中用作感病亲本。将其分别与 WBB1 杂交，根据 2 个组合的 F_1 植株对供试菌株 P6 表现抗病，遂与各自的感病亲本回交，获 B_1F_1 种子。根据两个组合的 F_1、F_2 和 B_1F_1 群体对 P6 的抗性反应参数，进行遗传分析。由于 IRBB21（*Xa21*）与 WBB1 都抗菲律宾小种 6，因此将其与 WBB1 杂交，依据其 F_1、F_2 群体对 P6 的抗性，测试两者间的等位关系。

1.2.2　用 14 个包括感病品种 JG30、珍珠矮、南京 11、02428、桂潮 2 号、辽粳 5 号、中丹 2 号、R917、中系 237、中花 9 号、中作 180、TN1、IR24 和沈农 1033 分别与 WBB1 杂交。选择其中前 4 个感病品种，再分别与 IRBB21 杂交，观察和比较 WBB1 与 *Xa21* 的抗性导入效应。

1.3 采用的菌株与接种

用包括 9 个小种在内的全套国际鉴别菌株 P1～P9，对参试的抗病基因进行抗谱和反应型比较。用所有正式命名的显性基因均不能抵御的鉴别菌株 P6 进行 WBB1 的抗性遗传分析。

菌株保存于－20℃的甘油营养液中，使用前在 PSA 培养基上复壮。于鉴别品种上测试其专化毒性，然后培养数管斜面菌种，保存于 4℃冰箱，供整个生长期使用毒力一致的菌种。接种物菌龄 2d、菌液浓度为 10^9/mL，用人工剪叶接种法。根据不同研究内容，供试植株分别在苗期、分蘖期和成株期接种 1 至 2 次。接种后 2 周或 20d 左右，当感病对照品种的病情稳定时进行调查。以病斑面积占叶面积的百分率为抗感反应参数。试材各调查 10 株，每株 2～3 张叶片。遗传分析的 F_2 和 B_1F_1 群体每株调查 2～3 张叶片，以分离群体的抗性反应参数曲线分布所形成的峰槽为抗感分群界限。

1.4 DNA 提取、酶切和 Southern 分析

用携有抗性基因的近等基因系 WBB1 与感病亲本 JG30 杂交的 F_1 和 F_2 分离群体，在苗期接种菌株 P6，接种后 16d 调查，获 245 株 F_2、各 10 株 F_1 和双亲的抗感反应数据。

采用改进的 Dellaporta（1983）方法提取抗感亲本 WBB1、JG30、BC_5F_1 和中选的 160 株 F_2 叶片总 DNA，提取液中用 0.38%（w/v）的亚硫酸钠代替 10mmol/L 的 β-巯基乙醇。用 7 个内切酶 *Bam*HⅠ、*Bgl*Ⅱ、*Eco*RⅠ、*Eco*RⅤ、*Hind*Ⅲ、*Sac*Ⅰ、*Xba*Ⅰ酶切双亲和 F_1 DNA。最后选用 *Bgl*Ⅱ对 160 株 F_2 酶切。按 McCouch 等（1988）和 Feinberg 等（1993）的标准程序和标记法进行电泳及 Southern 分析。

1.5 SSR（Simple Sequence Repeats，微卫星）标记选择

按常规 PCR 反应进行 SSR 标记筛选，反应体积为 25μL，反应液组成为 1×PCR 缓冲液（10mmol/L Tris-HCl pH 8.3，50mmol/L KCl，0.01% Triton X－100），1.5mmol/L $MgCl_2$，2.0mmol/L dNTP，两个方向的引物各 50ng，100ng 基因组 DNA 和 1 单位的 *Taq* 酶。反应程序为 95℃预变性 5 分钟，然后每个循环：94℃变性 60 秒，55℃复性 60 秒，72℃适温延伸 2 分钟，共循环 35 次，最后 72℃保温 10 分钟。扩增产物在 3.0%的琼脂糖凝胶中电泳分离，EB 染色，紫外灯下照相。

1.6 SSR 分子标记和 WBB1 抗病基因的连锁分析

根据 PCR 扩增结果，结合分离群体抗性表型参数，应用 Mapmaker 软件（Lander et al.，1987），在 MacintoshⅡ型计算机上进行连锁群分子图谱的构建。采用 Kosambi 函数将重组值转换成图距单位（cM）。

1.7 RFLP 标记与 WBB1 抗病基因的连锁分析

为确认 *Xa23t* 染色体位点，选择了 15 个邻近 SSR 标记区域内的 RFLP 探针，探测双亲、F_1 及 F_2 中的抗病和感病池，选出呈多态性的 RFLP 标记，分析 148 株 F_2 群体。

2 结果与分析

2.1 WBB1 与参试显性抗病基因间的抗谱及抗性类型比较

WBB1 与 6 个抗性基因对 9 个国际鉴别小种的抗性反应显示，WBB1 自苗期到成株期高抗包括广致病菌 P6 在内的 9 个国际鉴别小种。在 6 个抗病基因中，仅 *Xa21* 与 WBB1 的抗谱相同，然而 *Xa21* 在苗期是感病的，分蘖盛期之后才表达抗性。其余 5 个基因均不抗 P6，且抗谱各异。在这些显性抗性基因中，只有 WBB1 的抗性基因与 *Xa21* 具有广谱抗生，但两者的生育期抗性类型各不相同（表 1）。

表 1 WBB1 与 6 个显性基因对 9 个国际水稻白叶枯病鉴别小种的抗性反应

Table 1 Resistance reaction of WBB1 and 6 dominant genes to 9 differential races of *Xoo*

材料 Material	抗病基因 Resistance Gene	对 9 个小种的抗性反应 Resistance reaction to 9 races									抗性表达时期 Phase of resistance expression
		P1	P2	P3	P4	P5	P6	P7	P8	P9	
WBB1	*Xa23t*	R	R	R	R	R	R	R	R	R	全生育期 All growth stage
IRBB21	*Xa21*	R	R	R	R	R	R	R	R	R	分蘖盛期 Maximum tillering stage
IRBB3	*Xa3*	R	R	R	R	R	S	S	S	R	成株期 Adult stage
IRBB4	*Xa4*	R	S	M	M	R	S	R	R	S	全生育期 All growth stage
IRBB7	*Xa7*	R	R	R	S	R	S	R	R	R	成株期 Adult stage
IRBB10	*Xa10*	S	R	S	S	R	S	R	S	S	全生育期 All growth stage
IRBB14	*Xa14*	S	S	S	S	R	S	R	S	R	全生育期 All growth stage
JG30 (CK)	non	S	S	S	S	S	S	S	S	S	全生育期 All growth stage

R=抗（Resistant），M=中抗（Moderately resistant），S=感（Susceptible）.

2.2 WBB1 的抗性遗传分析

JG30/WBB1 和 02428 /WBB1 两个组合的 F_1 植株对专化小种 P6 在苗期高度抗病，其 F_2 群体在苗期的抗感分离比例均符合 3 抗∶1 感，P 值大于 0.05。两个组合与各自的感病亲本回交的 B_1F_1 也均以 1 抗∶1 感分离，表明 WBB1 对 P6 的抗性由一对完全显性抗性基因控制（表 2）。

WBB1 与 *Xa21* 等位性测定的正反交组合 F_1 植株苗期对 P6 均高度抗病，其 F_2 于分蘖期表现抗感分离。表明 WBB1 的抗性基因与 *Xa21* 不等位（表 2）。暂命名为 *Xa23t*。

表 2 感/抗、抗/抗和感/抗//感杂交组合 F_1 和 F_2、B_1F_1 的抗性反应

Table 2 Resistance reactions of F_1, F_2 and B_1F_1 plants from crosses of S/R, R/R and S/R//S

组合 Cross	F_1 抗性反应 Reaction of F_1	F_2 或 BC_1F_1 群体反应 (Reaction of F_2 or BC_1F_1 populations)			χ^2 Expected ratio	*P* 值 *P*. value
		R	S	总株数（Total）		
JG30/ WBB1	R	183	52	235	0.887（3∶1）	0.25～0.50
02428/WBB1	R	199	58	257	0.683（3∶1）	0.25～0.50

（续）

组合 Cross	F_1 抗性反应 Reaction of F_1	F_2 或 BC_1F_1 群体反应 (Reaction of F_2 or BC_1F_1 populations)			χ^2 Expected ratio	P 值 P. value
		R	S	总株数（Total）		
JG30/WBB1//JG30		35	32	67	0.06（1∶1）	0.25～0.50
02428/WBB1 //02428		24	30	54	0.463（1∶1）	0.25～0.50
IRBB21/WBB1	R	136	31	167		
WBB1/IRBB21	R	164	28	192		

R=抗（Resistant），S=感（Susceptible）.

2.3 WBB1（*Xa23t*）和 IRBB21（*Xa21*）的抗性导入效应比较

WBB1 分别与 14 个感病推广品种杂交，其 F_1 在苗期对 P6 都高度抗病，病斑面积均小于 5%；繁殖了其中 5 个组合的 F_2 群体，其抗感分离明显，抗病株中的多数为高抗，病斑长度仅 0.5～1cm，约占叶面积的 3%～5%，而感病株多数高感，病斑占叶面积 50%～100%（表 3）。在 F_2 中呈现分离丰富的高抗植株，极有利于进行种选择。

IRBB21（*Xa21*）与 14 个感病品种的 4 个亲本杂交 F_1 植株在苗期是感病的，至分蘖后期才开始转抗，各组合 F_1 的抗性水平不一，多数为中等抗性。各选 4 个组合的结果列于表 3。两者相比，*Xa23t* 不仅全生育期高度抗病，其抗性导入效应也较强。

表 3 *Xa23t* 和 *Xa21* 对 P6 抗性导入效应比较

Table 3 Comparison of efficiency of transfer for resistance to P6 between *Xa23*（*t*）and *Xa21*

组合 Cross	F_1 病斑面积百分率（%） % of lesion area on F_1	* 抗性表达生育期 Phase of resistance expression	F_2 群体抗性分离百分率（%） Percentage of segregation of resistance in F_2				
			<5%	10%～15%	20%～50%	55%～70%	>70%
JG30/WBB1	3%	SS	74.3	0.2	1.1	4.5	19.9
Zhenzhuai/WBB1	5%	SS	71.3	1.8	7.2	10.5	9.2
NJ 11/WBB1	3%	SS	72.7	2.8	7.5	7.7	9.3
02428/WBB1	3%	SS	66.2	6.3	3.5	4.7	19.3
JG30/IRBB21	10%	MTS	15.4	54.8	14.3	5.7	9.8
02428/IRBB21	5%	MTS	11.5	65.7	10.8	7.5	4.5
Zhengzuai/IRBB21	10%	MTS					
NJ 11/IRBB21	10%	MTS					

* SS=苗期（Seedling stage），MTS=最高分蘖期（Maximum tillering stage）.

2.4 与 *Xa23t* 位点相连锁的 SSR 分子标记选择

共筛选了 160 对 SSR 引物（Susan et al.，1997），其中 OSR06 和 RM224 两对引物能在亲本和抗感 DNA 池之间检测到多态性。用 OSR06 和 RM224 分别对选择的 160 个 F_2 单株进行共分离分析（图 1），采用作图软件 Mapmaker（3.0）进行遗传图距分析，发现 OSR06 与 *Xa23t* 基因紧密连锁，图距为 5.3cM，RM224 与 *Xa23t* 的遗传图距为 27.7cM。由于

OSR06 和 RM224 均被定位于水稻第 11 染色体的长臂上，*Xa23t* 也被定位于该区域（图 4）。

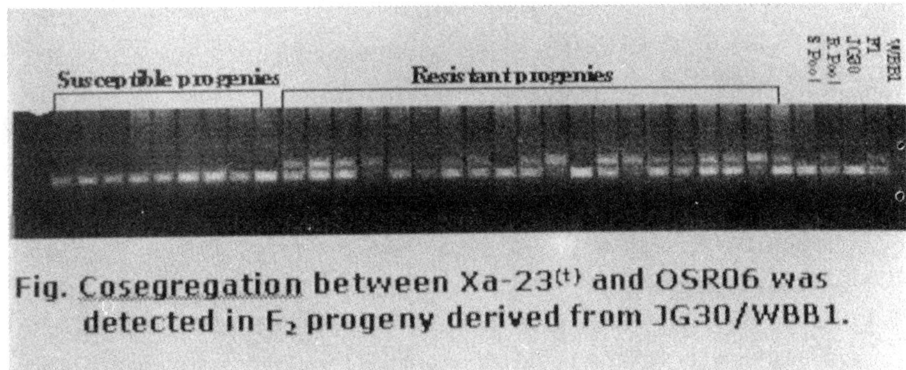

图 1　自 JG30/WBB1（*Xa23t*）F_2群体中测定基因 *Xa23*（*t*）与 SSR 标记 OSR06 的共分离图

Fig. 1　Co-segregation between *Xa23t* and OSR06 was detected in F_2 progeny derived from JG30/WBB1

2.5　RFLP 分子标记与目的基因的连锁分析

在测试的 15 个 RFLP 探针中，仅有 G1465 在抗感亲本及 F_1之间呈现多态性（图 2）。用 G1465 测定 148 个 F_2单株进行共分离分析（图 3），经作图软件 Mapmarker（3.0）分析了 G1465 与 *Xa23t* 之间的遗传图距，两者相距 16.7cM（图 4）。

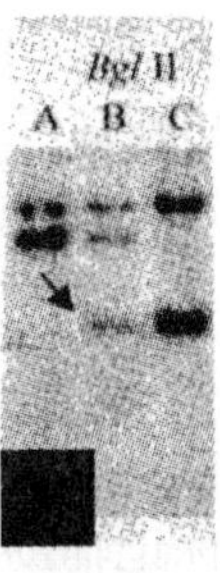

图 2　G1465 在 *Xa23t* 近等基因系 WBB1、JG30 /WBB1 F_1和 JG30 之间显示多态性（用 *Bgl*Ⅱ酶切 DNA）

Fig. 2　Atuora-diograph showing polymorphism of the RFLP marker G1465，genomic DNA of WBB1，F_1 of JG30/WBB1 and JG30 were digested with *Bgl* Ⅱ.

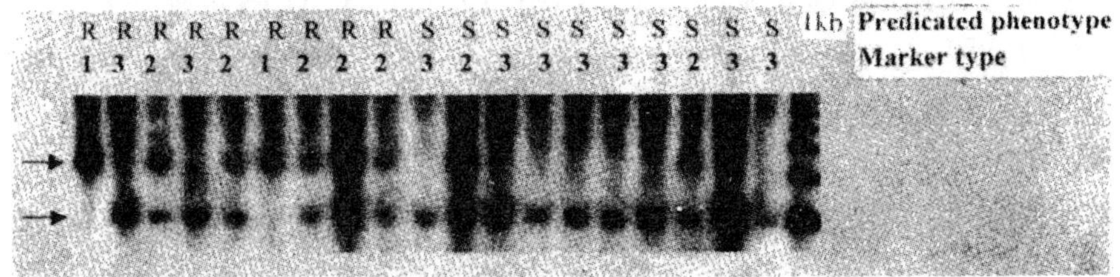

图 3　G1465 标记揭示的 JG30/WBB1（*Xa23t*）F_2群体植株共分离

Fig. 3　Co-segregation between *Xa23*（*t*）and G1465 was detected in F_2 progeny derived from JG30/WBB1

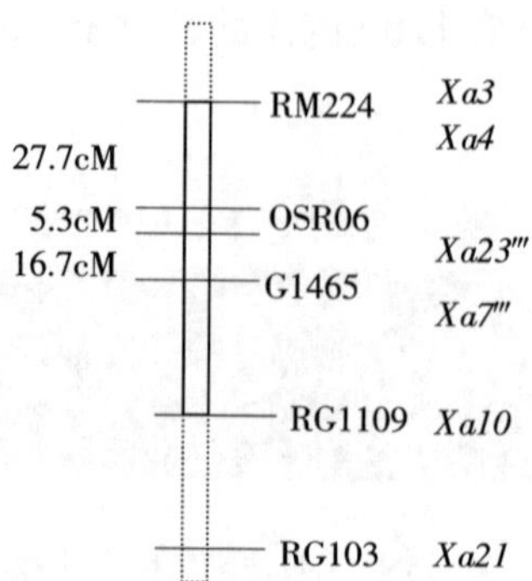

图 4 *Xa23t* 及另外 5 个白叶枯病抗性基因在水稻第 11 染色体上的 DNA 分子标记位点

Fig. 4 The molecular location of *Xa23t* on rice chromosome 11 was constructed using the JG30/WBB1 F_2 population. The distances between markers are presented in Kosambi contimorgans (cM); and the distal part of chromosome 11 mapped by Causse et al. (1994) showing the positions of other 5 bacterial blight resistance genes.

3 讨论

3.1 *Xa23t* 的利用价值

在一系列预备试验中，*Xa23t* 及其抗性供体 RBB16 与正式命名的显性基因对其相应病菌小种的抗性等位测定表明，其抗性特点同所有的显性基因明显不同。在迄今正式命名的抗病基因中，*Xa23t* 抗谱最广、抗性导入效应强、并且是全生育期高度抗病的完全显性基因，这是个十分罕见的优异基因。这一新基因的问世，将在改良国内外杂交水稻和常规稻品种的抗病性上发挥重要作用。该基因的分子定位也将促进我国水稻分子标记辅助选择育种的开展。

3.2 SSR 分子标记的利用

植物重要的有利基因通过 DNA 分子标记辅助选择（MAS）在育种中加以应用，有助于育种效率的提高和精确性。近年，许多报告多半限于理论上的探讨，实际应用较少。目前影响 MAS 普及利用的主要制约因子是：1）基因分子作图的速度不够快。由分子标记定位的控制水稻重要性状的主效基因仅有 30 多个，可用于 MAS 的基因有限，许多作图的基因与其连锁的分子标记的图距太远而无法用于 MAS；2）可用于 MAS 的便于操作的 PCR 标记不多（Zheng，1995）。

SSR（微卫星 DNA），意指动植物基因组中存在大量的串联简单重复序列（Li，1989），具有高度的变异性，它们较均匀地分布于动植物基因组中，为此，SSR 成为一种较为理想的基因分子标记定位手段（Bachman，1991）。它不仅具有 PCR 技术的简单易行、快速、廉价、所需 DNA 量很少等优点，而且有 RFLP 技术准确、共显性等优点。与 RFLP 相比，SSR 具有更多的多态性，为此，SSR 广泛用于人类、老鼠、大麦、大豆、水稻等动植物基因组研究中（Chen et al.，1997）。

本研究通过筛选 160 对水稻微卫星标记，成功地将新基因 *Xa23t* 初步定位于水稻第 11

染色体上，与其中的 1 个标记之图距为 5.3cM。随着更多的 SSR 标记序列的发表，将有可能缩短标记与目标基因的图距，为进一步依图谱分离基因和开展标记辅助育种打下基础。由于目前筛选的 2 个 SSR 标记不在同一张遗传图谱上，因此无法确定两个标记与目标基因的方向性。为了进一步将 *Xa23t* 定位于更小的区域中，以提高标记辅助选择育种的效率和为染色体步进分离克隆基因提供起始点，我们正在本研究的基础上进行精细定位。

3.3 普通野生稻抗病性遗传分析的特点

普通野生稻中高抗白叶枯病的编号虽然不很多，但是它具有与栽培稻相同的 AA 型染色体组，两者杂交易于成功。为此我们曾较系统地研究了普通野生稻的抗性特点。它对白叶枯病的抗性表现是多种多样的，大多数材料在同 1 个编号内兼有抗、感植株，或在同一植株上于相同生育期具有抗、感不同的叶片，仅有极少数编号的抗性表现相对一致（章琦等，1994）。*Xa23t* 的抗性供体 RBB16 经多次重复接种鉴定，它是测试的编号中唯一表现抗性相对一致的材料。我们于 80 年代末直接分析了 RBB16 对 P1（菲律宾小种 1）、HB17（我国Ⅱ型菌）和 T1（日本小种 1）的全生育期抗性的遗传模式，初步表明为显性基因，但是还需要进一步分析（林世成等，1992）。鉴于普通野生稻的有利性状一般都处于杂合或不稳定状态，必须采用抗性纯合系进行遗传分析，才能得到合理的结果。为此，我们通过逐代鉴定 JG30/RBB16 组合 F_1 花培后代直至 H_4 时，才获得了抗性纯合系。随后又转育成近等基因系 WBB1。本试验分析了 WBB1 对国际广致病小种 P6 的苗期抗性，系由 1 对完全显性基因制约。根据基因对基因学说，我们还将继续进行 WBB1（*Xa23t*）对 P1、HB17 和 T1 菌株的抗性遗传分析。

原载：作物学报，2000，26（5）：536～542.

Identification and Tagging a New Rice Bacterial Blight Resistance Gene from *Oryzae rufipogon* *

Zhao Bingyu [1] **Leung H**[2] **Zhang Qi**[1]

(1. Institute of Crop Breeding and Cultivation, CAAS, Beijing, China;
2. International Rice Research Institute, Manila, Philippines)

Breeding of resistance rice cultivars has been the most affective approach for the management of rice bacterial blight (BB) caused by *Xanthomonas oryzae* pv. *oryzae* (*Xoo*). The gene *Xa4* for BB resistance has been widely adopted in the development of hybrid and conventional indica rice in China, which encompass more than 90% of national rice area. Such narrow based genotype is prone to the genetic vulnerability of cultivars to bacterial blight. Thus, to explore new resources for resistance must be identified continuously. An accession, No. 95 – 238 (WBB2) from 269 *O. rufipogon* accessions showed high resistance to 2 strains with broad virulence spectrum in China, and 7 out of the 9 races including race 6 (except race 1 and race 4) of *Xoo* in the Philippines. An interspecific cross was made between WBB2 and JG30, a susceptible indica cultivar. Genetic analysis showed that the resistance was conditioned by a dominant gene in No. 95 – 238. Of the 21 designated BB resistance genes, *Xa21* is the only dominant gene showing resistance to the race 6 (PXO99) in the philippines, it is important to know whether the gene is allelic with *Xa21*. Sequence-tagged site (STS) analysis was used to test the allelism between *Xa21* and the resistance gene derived from WBB2. Distinct polymorphism was observed between the alleles amplified in IRBB21 (*Xa21*) and WBB2, suggesting that the resistance gene from *O. rufipogon* (No. 95 – 238) is different from *Xa21*. In order to fix the the resistance and gradually eliminate the undesired traits, selfing, backcross and triple cross were done. In triple cross population, one plant, BC 1005 – 27, showed high resistance to strain PXO99 with improved plant type was selected. To tag the new resistance gene, RAPD (Random Amplification polymorphism DNA), AFLP (Amplified Fragment Length Polymorphism) and RGA (Resistance Gene Analogue) primers were used to analyze F_2 progeny bulk segregation analysis. A high level of polymorphism (18% – 40%) was detected between the parents. At least 8 polymorphism RGA motifs (LRR and kinases) were found to be

* 由洛克菲勒基金项目资助，赵炳宇在 IRRI Dr. Leung H 指导下，进行的硕士研究论文摘要，1997 Sixth International Symposium on Rice Molecular Biology Oct. 31st– Nov. 1998 , Shanghai, Nov. 2nd 1998 Hangzhou, RF。

associated with the resistance gene. These markers will be used in marker-aided selection to accumulate BB resistance genes in cultivars for use in northern China.

摘要 本研究旨在挖掘普通野生稻中的抗白叶枯病新抗源，拓宽抗白叶枯病的遗传基础。用 1 个国际广致病菌株，菲律宾小种 6 的代表菌株 PXO99 接种评价 269 份普通野生稻编号，筛选出 5 份高抗材料，推测其中可能携有新的抗病基因。聚焦于其中 1 个农艺性状较好的编号 95－238。将其分别与金刚 30（JG30）和中作 8957 进行杂交获 F_1，加代收 F_2种子，进一步选择株型优良的 F_2套袋自交，获 7 个 F_3家系各 200 株，经用 PXO99 鉴定，获 2 个纯合抗病系 3002 和 3004。将 IR24 与 3002 杂交、自交，经专化抗性鉴定、农艺性状考察，选出一个高抗 PXO99、株型紧凑、株高适中、结实率高、不落粒的品系 BC1005－27，被命名为 WBB2。通过与已知白叶枯病显性抗性基因抗谱比较、抗性遗传分析、分子标记检测，确认了 WBB2 携有一个新的显性抗性基因。

1 抗谱比较

根据水稻与白叶枯病菌之间存在基因对基因互作的原理，首先，用 9 个菲律宾国际鉴别小种对 WBB2 和迄今已命名的 10 个抗白叶枯病显性基因进行接种鉴定，比较其间的抗谱。结果表明，*Xa21* 和 *Xa23* 抗所有 9 个小种，WBB2 感小种 1 和 4 ，抗包括小种 6 在内的小种 2、3、5、7、8 和 9 共 7 个小种，而其他显性抗性基因都不抗小种 6（表 1），表明 WBB2 的抗谱与所有已知基因都不相同，推测 WBB2 可能携有一个新的基因。

表 1 WBB2 与所有抗白叶枯病显性基因对国际 9 个鉴别小种的抗谱比较

Table 1 Comparison of resistance between WBB2 and the 11 dominant genes to 9 differential races of *Xoo*

材料	抗性基因	小种									接种生育期
		1	2	3	4	5	6	7	8	9	
WBB2	?	S*	R*	R	S	R	R	R	R	R	TS*
CBB23	*Xa23*	R	R	R	R	R	R	R	R	R	SS*
IEBB21	*Xa21*	R	R	R	R	R	R	R	R	R	TS
IRBB1	*Xa1*	S	S	S	S	S	S	S	S	S	AS*
IRBB2	*Xa2*	S	S	S	S	S	S	S	S	S	AS
IRBB3	*Xa3*	R	R	R	R	R	S	S	R	R	AS
IRBB4	*Xa4*	R	S	S	M	R	S	R	S	S	SS
IRBB7	*Xa7*	R	R	R	S	R	S	R	R	R	AS
IRBB10	*Xa10*	S	R	S	S	R	S	S	S	S	SS
IRBB11	*Xa11*	S	S	S	S	S	S	S	S	S	AS
IRBB14	*Xa14*	S	S	S	S	R	S	R	R	R	SS
IR24	non	S	S	S	S	S	S	S	S	S	SS

* R=抗（Resistant），S=感（Susceptible），SS=苗期（Seedling stage），TS=分蘖期（Tillering stage），AS=成株期（Adult stage）。

2 抗性遗传分析

分析 WBB2 分别与 2 个感病品种，JG30 和 NJ11 杂交组合 JG30/WBB2 和 NJ11/WBB2 的 F_1、F_2和 B_1F_1群体对菌株 PXO99 的抗性。两个 F_1 和 F_2分离群体分别均符合 1 抗：1 感和 3 抗：1 感比例，两个组合的回交群体分离比为 1 抗：1 感，确认 WBB2 对 PXO99 的抗性由一对显性基因控制。

表 2 WBB2 野栽杂交组合 F_1、F_2和 BC_1 群体对菌株 PXO99 的抗性反应*

组合 Cross	F_1 抗性反应 F_1 reaction		χ^2	P	F_2 抗性反应 F_2 reaction		χ^2	P	BC_1F_1 抗性反应 BC_1F_1 reaction		χ^2	P
	抗病 R	感病 S	(1：1)		抗病 R	感病 S	(1：1)		抗病 R	感病 S	(1：1)	
JG30/WBB2	15	21	0.69	0.25～0.50	127	49	0.61	0.25～0.50	16	24	1.23	0.25～0.50
NJ11/WBB2	11	16	0.59	0.25～0.50	169	48	0.81	0.25～0.50	18	25	0.12	0.50～0.75

R=抗（Resistant），S=感（Susceptible）。

3 分子标记检测 WBB2 携有的基因与 *Xa21*

在抗谱比较测试中，已经明确了在显性抗性基因中，只有 *Xa23*、*Xa21* 和 WBB2 能控制菲律宾小种 6 的毒性，WBB2 与 *Xa23* 的抗谱及抗性类型均不相同。因此应检测 WBB2 携带的抗性基因与 *Xa21* 的等位关系，鉴于 *Xa21* 已经克隆，可用 *Xa21* 位点的特异引物 U1/I1 扩增该位点的特异带型，见图，WBB2 及感病亲本 JG30 均没有显示 *Xa21* 的特异带型，表明两个基因是不相同的。

4 WBB2 携带的抗性基因位点相连锁的 SSR 分子标记选择

采用 RAPD、AFLP、RGA 和 SSR 结合 DNA R/S 分池法。第一阶段在筛选的 160 个 RAPD、64 对 AFLP 和 19 对 RGA 引物中，未发现与 WBB2 抗性基因共分离的 DNA 标记。2000 年进一步用 JG30 和 JG30/WBB2 的 F_3家系中的一个抗病系为抗感亲本筛选了 318 对 SSR 引物，并检测了 57 株 F_2分离群体 DNA，测定出位于第 4 染色体上的一个分子标记 FR372 与 WBB2 携有的抗性基因有连锁关系，初步将 WBB2 携有的新基因定位于水稻第 4 染色体上。

1998 年第 6 届国际水稻分子生物学会议论文，P. 80

水稻白叶枯病新抗源 Y238 的鉴定及其近等基因系培育

王春莲[1]　赵炳宇[1]　章　琦[1]　赵开军[1,2]　邢全党[1]

（1. 中国农业科学院作物育种栽培研究所，北京　100081；
2. 国际水稻研究所驻中国办事处，北京　100081）

Identification of a New Rice Resistant Resource to Bacterial Blight and Breeding a Pair of Near-isogenic Line

Abstract　A new resistant resource, Y238, to rice bacterial blight（BB）was identified from 269 *Oryza rufipogon* accessions. It harbors a new BB resistance gene, WBB2, based on the resistance spectrum analysis, and comparison with all the known BB resistance genes. According to resistance reaction and genetic analysis at adult stage of progeny from JG30/Y238, the gene of WBB2 is a completely dominant resistance gene. The new gene has been transferred into cultivated rice variety JG30 by cross and backcross for breeding a pair of near-isogenic line.

Key words　*O. rufipogon*；bacterial blight；new gene；near-isogenic line

摘要　从 269 份普通野生稻中鉴定出一个高抗白叶枯病的新抗源，编号为 Y238。通过多菌株鉴定与目前国际已知基因的抗谱比较，证明该新抗源携有一个新基因，暂命名为 WBB2。对 JG30/Y238 杂交后代成株期接种鉴定、遗传分析表明，WBB2 为完全显性基因。通过杂交和回交，已将 WBB2 导入栽培稻中，构建了一对近等基因系。

关键词　野生稻；白叶枯病；新基因；近等基因系

水稻白叶枯病（*Xanthomonas oryzae* pv. *oryzae*）是一个世界性的水稻细菌病害，也是威胁我国水稻生产的主要病害。随着大量水稻改良品种的育成和推广，水稻产量得到提高，但是由于多数品种具有相同或相近的抗性遗传来源，导致抗性遗传基础狭窄，原来处于弱势的致病菌上升为优势菌，甚至出现新的毒性菌株，从而使抗病品种出现感病化（章琦，1995）。因此，发掘新的抗病基因源是抗病育种的长期需要。野生稻中蕴藏着丰富的优异种质，在抗病虫方面尤为突出。IRRI 在长药野生稻中，发现具有广谱的抗白叶枯病基因 *Xa21* 已被克隆（Song et al.，1995），该基因正在通过转基因及常规手段在育种中应用。章琦等从普通野生稻（*Oryza rufipogon*）中发掘的抗白叶枯病基因 *Xa23*，具有高抗、全生育期抗性、完全显性以及抗性转移效应强等特点，已经培育成近等基因系 CBB23，并已育成兼抗稻瘟病的中野 5112、中野 5114 品系，正被育种单位应用（章琦等，2000；Zhang et al.，2001）。赵炳宇等（1998）从 1995—1997 年评价了 269 份普通野生稻对白叶枯病的抗性，鉴定出一个抗谱较广的抗源（编号 Y238），抗参试的两个中国病原型和菲律宾 9 个小种（1～9）中的 7 个（包括抗强毒小种 6），初步抗性遗传分析结果表明，与 *Xa21* 不等位。本研究在此基础上，进一步对新抗源 Y238 做了较系统的研究，并进一步培育了一对近等基因系。

1 材料与方法

1.1 水稻材料

269 份普通野生稻由中国农科院作物品种资源所庞汉华副研究员提供；4 个感病品种为金刚 30（籼稻）、南京 11（籼稻）、沈农 1033（粳稻）、中作 8957（粳稻）；已知含抗白叶枯病基因品种：IRBB1（*Xa1*）、IRBB2（*Xa2*）、IRBB3（*Xa3*）、IRBB4（*Xa4*）、IRBB5（*xa5*）、IRBB7（*Xa7*）、IRBB8（*xa8*）、IRBB10（*Xa10*）、IRBB11（*Xa11*）、IRBB13（*xa13*）、IRBB14（*Xa14*）、M41（*xa15*）、Tetep（*Xa16*）、Asominori（*Xa17*）、Milyang（*Xa18*）、XM5（*xa19*）、XM6（*xa20*）、IRBB21（*Xa21*）、扎昌龙（*Xa22*）和 CBB23（*Xa23*）。

1.2 供试菌株

菲律宾国际白叶枯病 10 个鉴别小种的代表菌株，P1、P2、P3、P4、P5、P6、P7、P8、P9 和 P10。原菌种均保存在甘油中（-20℃），在接种前用肋本哲氏培养基复壮，置于 28℃ 培养 48h，以无菌水配制接种菌液，浓度调至 1×10^9/mL。

1.3 接种和调查

分别在植株分蘖和成株期进行人工剪叶接种。接种后 14～20d，当感病品种病情稳定时调查，用病斑面积占叶片面积百分率作抗感反应参数，以 20%为抗感界限，≥20%为感病，15%～20%为中感，10%～15%为中抗，<10%为抗。

1.4 杂交组合的配制及抗性遗传分析

以栽培稻为母本，普通野生稻为父本配制杂交组合。选择 F_1 的抗病株与感病亲本回交（B_1F_1），随后，各世代包括 F_2、B_1F_1、B_1F_2、B_1F_3、B_2F_2、B_3F_1、B_4F_1 均同步接种菌株 P6，设置感病对照，对杂交后代进行常规遗传分析。

2 结果与分析

2.1 WBB2 的鉴定及与已知抗白叶枯病基因的抗谱比较

在前期研究的基础上，1995 年再次用白叶枯病强毒性的国际鉴别小种 P6，在最高分蘖期对 269 份普通野生稻接种。将抗 P6 的 Y238 与其他抗白叶枯病基因进行抗谱比较测定（表 1），结果进一步表明 Y238 与这些已知抗性基因的抗谱均不相同，推断 Y238 可能为一新的抗源，暂被命名为 WBB2（Zhao et al.，1998）。

表 1 Y238 与已知水稻白叶枯病基因的抗谱比较

Table 1 Comparison of resistance spectrum to bacterial blight between Y238 and the known genes

试材 Material	R 基因 Resistance gene	鉴别小种 Differential races										鉴定年份 Identifying year		
		P1	P2	P3	P4	P5	P6	P7	P8	P9	P10			
Y238	WBB2	S	R	R	S	R	R	R	R	R	/	1997		
IRBB1	*Xa1*	S	S	S	S	S	S	S	S	S	/	1997	1994	
IRBB2	*Xa2*	S	S	S	S	S	S	S	S	S	/	1997	1994	
IRBB3	*Xa3*	R	R	R	R	R	S	R	R	R	R	1997	2000	2001
IRBB4	*Xa4*	R	S	S	R	R	S	R	R	S	R	1997	2000	2003
IRBB5	*xa5*	R	R	R	S	R	S	R	R	R	R	1997	2000	2001
IRBB7	*Xa7*	R	R	R	S	R	S	R	R	R	R	1997	2000	2003
IRBB8	*xa8*	S	S	S	MS	R	S	R	R	/	/	1997	1994	
IRBB10	*Xa10*	S	R	S	S	R	S	S	S	S	/	1997	1994	
IRBB11	*Xa11*	S	S	S	S	S	S	S	S	/	/	1997	1994	
CBB12	*Xa12*	MR	R	R	R	R	S	R	R	R	/	1997	1994	
IRBB13	*xa13*	R	R	R	R	MS	R	R	S	/	/	1997	1994	
IRBB14	*Xa14*	S	S	S	S	R	S	R	S	R	/	1997	1994	
M41	*xa15*	R	MS	MS	R	R	S	1	/	/	/	1994		
Tetep	*Xa16*	S	S	S	S	R	R	MR	MR	S	/	2000		
Asominori	*Xa17*	R	R	R	R	R	R	/	/	/	/	1994		
Milyang	*Xa18*	S	S	R	R	R	S	/	/	/	/	1994		
XM5	*xa19*	S	MS	R	R	R	MS	/	/	/	/	1994		
XM6	*xa20*	S	R	R	R	R	R	/	/	/	/	1994		
IRBB21	*Xa21*	R	R	R	R	R	R	R	R	R	S	1999	2000	2003
扎昌龙	*Xa22*（*t*）	S	MR	S	MR	R	S	R	S	MR	MS	2003		
CBB23	*Xa23*	R	R	R	R	R	R	R	R	R	R	2000	2001	2003
JG30（CK）		S	S	S	S	S	S	S	S	S	S	1994	1997	2001
													2000	2003

R=抗（Resistant）；S=感（Susceptible）；MR=中抗（Moderately resistant）；MS=中感（Moderately susceptible）。

2.2 Y238 的白叶枯病抗性遗传分析

用 2 个高感水稻白叶枯病籼稻品种 JG30 和南京 11，2 个粳稻品种沈农 1033 和中作 8957 为母本分别与 Y238 杂交，对 F_1进行主要农艺性状考查，并接种菌株 P6。结果显示 F_1抗病，表明 Y238 对 P6 的抗性由显性基因控制，F_1植株呈散状株型、落粒性强、长芒、种皮褐色等野生稻性状（表 2）。

在回交后代群体中，A、B、C 和 D 四个 B_2F_2群体，分别来自不同的 B_2F_1植株，在其成株期进行接种鉴定，高抗植株（病斑面积 3%～5%）数和感病株（病斑面积＞25%）明显多于中抗（10%～15%）和中感（15%～20%）的植株数，抗、感界限分明（图 1）。虽然这 4 个群体的株数较少，但抗感比接近于 3 抗：1 感，表明 Y238（WBB2）成株期的抗性由一对显性基因控制。

表 2 野栽杂交亲本及其 F_1 植株的农艺性状及对菲律宾小种 6（P6）的抗性反应

Table 2 Reaction to Philippine race P6 and agronomic characters of donor and F_1 from crosss between Y238 and cultivated rice varieties

试材 Material	株型 Plant type	落粒性 shattering	芒 Arista	柱头色泽 Stigma color	外稃色泽 Lemma color	种皮色泽 Seed skin color	抗性反应 Reaction
Y238（WBB2）	散生	强	长芒	紫	黑色	褐	R
金刚 30	紧凑	弱	无芒	紫	秆黄色	白	S
南京 11	紧凑	弱	无芒	白	秆黄色	白	S
沈农 1033	紧凑	弱	无芒	白	秆黄色	白	S
中作 8957	紧凑	弱	无芒	白	秆黄色	浅红	S
金刚 30/Y238（F_1）	散生	强	长芒	紫	黑	褐	R
南京 11/Y238（F_1）	散生	强	长芒	紫	黑	黑	R
沈农 1033/Y238（F_1）	散生	强	长芒	紫	黑	黑	R
中作 8957/Y238（F_1）	散生	强	长芒	紫	紫	褐	褐

R=抗（Resistant）；S=感（Susceptible）。

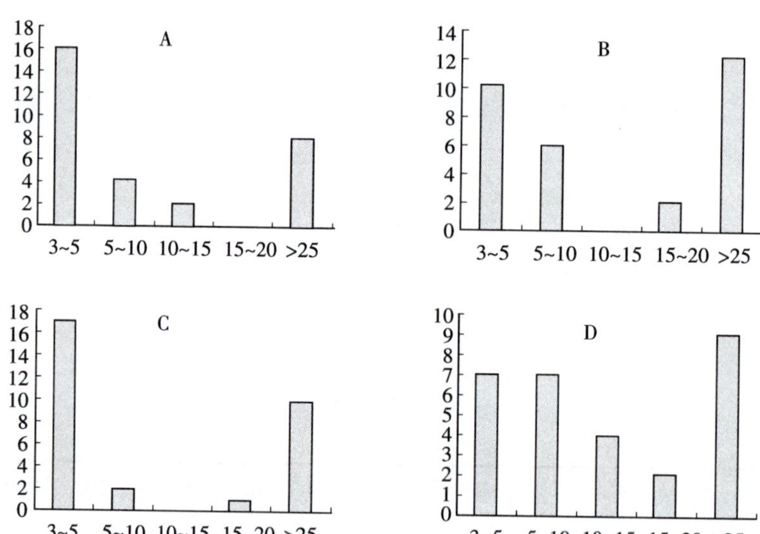

图 1 金刚 30/Y238 的 4 个 B_2F_2 群体（A、B、C、D）对国际鉴别小种 P6 的抗性反应

Fig. 1 Resistance reaction of 4 B_2F_2 population（A、B、C、D）from cross JG30/WBB2 to P6

横坐标为病斑面积占叶片面积的百分数；纵坐标为植株个数

Abscissa as percentage of lesion area；Ordinate as no. of plant

2.3 近等基因系的培育

从 1996 年开始，以感病品种金刚 30 为轮回亲本，以 Y238（WBB2）为供体，通过杂交、回交、自交，每代用专化小种 P6 进行接种鉴定，选择抗病且株型类似金刚 30 的植株，

淘汰不良性状，目前已获得 B_4 F_1种子，选育过程如图 2。

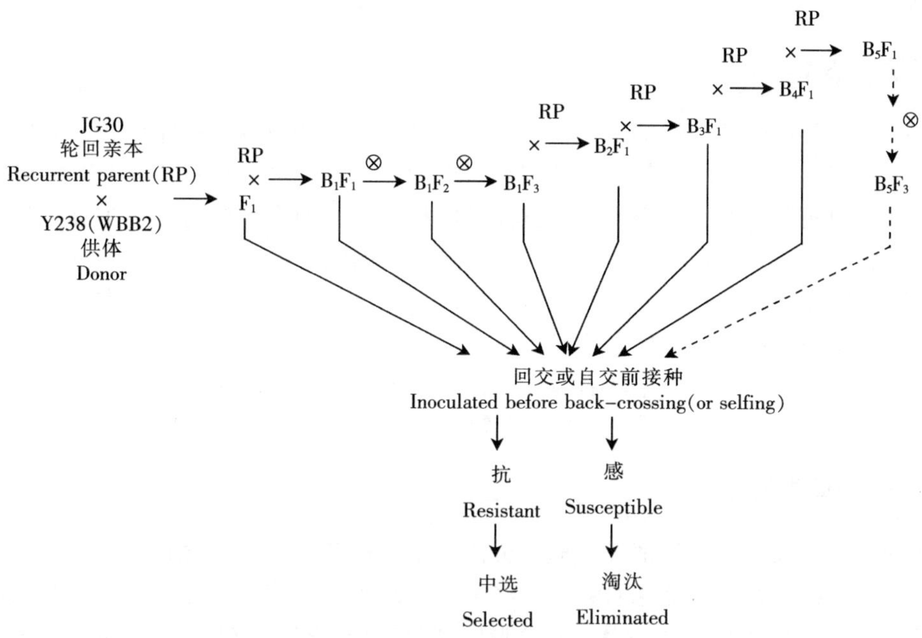

图 2　近等基因系的培育程序

Fig. 2　Procedure in breeding for near-isogenic line to BB resistance

2.4　WBB2 与 *Xa23*、*Xa21* 基因的等位性测定

CBB23（*Xa23*）和 IRBB21（*Xa21*）是目前水稻抗白叶枯病育种的极有价值的基因源，对国际鉴别小种强毒菌株 P6 都高度抗病，均为显性抗型遗传，因此将 JG30/Y238 后代（B_1 F_4）的有芒紫稃尖、无芒紫稃尖、无芒无色三种类型与 CBB23 和 IRBB21（为母本）进行杂交，在其 F_1、F_2的成株期用 P6 接种，以测定这些基因间的等位关系。所有 F_1都抗，F_2 抗性均出现分离（表 3），表明 WBB2 携有的抗性基因与 *Xa23*、*Xa21* 为非等位。

表 3　WBB2 三种株型与 IRBB21 和 CBB23 分别杂交组合的 F_1和 F_2对 P6 的抗性反应

Table 3　Reaction of F_1 and F_2 drived from 3 plant types of WBB2 crossed with IRBB21and CBB23，respectively

组合（Cross）	接种时期（Growth stage inoculated）	F_1 抗性反应（Reaction of F_1）	F_2 群体抗性反应（Reaction of F_2）		
			抗病株（R）	感病株（S）	总株数（Total）
CBB23/有芒紫稃尖	成株期	R	134	15	149
CBB23/无芒紫稃尖	成株期	R	129	19	148
CBB23/无芒无色	成株期	R	45	5	50
IRBB21 有芒紫稃尖	成株期	R	114	36	150
IRBB211 无芒紫稃尖	成株期	R	135	15	150
IRBB21/无芒无色	成株期	R	130	20	150

R=抗病（Resistant）；S=感病（Susceptible）。

3 讨论

3.1 新抗源鉴定

鉴定一个抗源是否携有新基因，首先与所有已知抗性基因进行抗谱比较。应选用有鉴别力的菌株，一套国际上通用的与抗性基因相对应的鉴别小种最为理想。对于寄主与病菌存在基因互作关系的病害来说，完整的一套已知基因材料和生理小种作为分析工具尤为重要，根据两者互作后抗谱的表现的差异，可初步了解抗性基因的异同。二是与已知抗性基因进行等位性测试，以确定所鉴定基因与已知基因的异同。在抗谱比较的基础上，重点选择对专化抗性表现相似的基因进行测试。分子检测是鉴定新抗源的一个快速方法。采用已知基因特异标记，对未知基因进行检测，根据两者有无同源性或差异，来确定基因是否相同。本研究通过抗谱比较、等位性测定可初步推测 Y238 为新抗源，正在进行分子检测。

3.2 培育近等基因系

野栽种间远缘杂交，后代都发生疯狂分离，很难稳定。由于普通野生稻与栽培稻有相同的 AA 型染色体组，种间转移相对容易，在减数分裂中染色体可以正常配对，利用常规育种技术就能实现目的基因的转移，而且普通野生稻与栽培稻杂交成功已早有报道（Khush 等，1990；章琦等，2000；林世成等，1992）。本研究也利用普通野生稻新抗源 Y238 通过常规育种方法，包括回交、自交，结合抗性鉴定，将抗性基因转移到栽培稻金刚 30 中。将其培育成近等基因系的目的是为以后的基因定位、基因克隆打基础，提供育种家利用。

在选育过程中，从 B_1F_1之后，没有及时回交，进行了两次自交至 B_1F_3后，才选择抗病且株型像轮回亲本的植株又进行回交，致使回交代数到目前仅获得 B_4F_1种子，要想获得稳定的近等基因系，还需要增加代数，目前这方面工作正在进行中。

原载：植物遗传资源学报，2004，5（1）：26～30.

The Effectiveness of Advanced Rice Lines With New Resistance Gene *Xa23* to Rice Bacterial Blight

Q. ZHANG[1]　CL WANG[1]　YL ZHOU[1]　Vera Cruz C. M[2]
XD ZHU[3]　DY LI[4]　QX JIANG[5]

(1. Institute of Crop Breeding and Cultivation, *CAAS*, Key Laboratory of Crop Genetics & Breeding, ministry of Agriculture, the People's Republic of China . Beijing 100081, China;
2. International Rice Research Institute, DAPO 7777 Metro P. O. Box 933, Manila, Philippines ;
3. China National Rice Research Institute, Hangzhou 310006, China ;
4. Institute of Germplam Resources, *GAAS* Naining 530007, China ;
5. Institute of Rice and Wheat Breeding, Wujin 213149, China)

Bacterial blight (BB) of rice is found world wide and particularly destructive in southeast Asia and China. The gene *Xa4* has been incorporated in many rice cultivars and provided resistance to the disease in most of the rice growing countries in Asia. The gene also has been widely adopted in the development of indica hybrid and conventional rice for two decades in China. All of the restoring lines of indicahybrid rice carry the incomplete dominance gene *Xa4* resulting to almost all of the combinations of indica hybrid being infected due to high disease pressure (Zhang , 1991) . Strains of BB that overcome the resistance of *Xa4* has been reported in countries like the Philippines (Mew et al. , 1992), Indonesia (Bustarnam, pers. comm.); Eastern and Northern India (3/7 & 6/11 races virulent) Vietnam (Quang, pers. comm.); Nepal (Adhikari et al. , 1999) in these countries except Eastern and Northern India such strains represent the dominant population. Although *Xa4* has residual resistance, breeders prefer to use the gene *Xa21* with broader spectrum resistance in rice when it was identified by Khush et al. (1990), and cloned by Song et al. (1995) . Unfortunately, for *Xa21*, certain groups of strains can overcome the resistance conveyed by *Xa21*, such as race 10 of the Philippines and some strains in Korea, Indonesia, Japan, and China.

We identified and tagged a new gene for resistance to BB and designated as *Xa23* from *O rufipogon* (Zhang et al. , 1998) . This new gene demonstrated the broadest spectrum of resistance to BB. To further compare the spectrum of resistance to BB between the two genes, *Xa21* and *Xa23*, 20 strains including 10 Philippine races, 7 Chinese pathotypes, and 3 Japanese races were used to inoculate plants at maximum tillering stage (MTS) in 2001 in Beijing. The CBB23 (carrying *Xa23*) is highly resistant to all the 20 strains including Philippine race 10, while IRBB21 (carrying *Xa21*) is resistant to 19 strains, but highly susceptible to the Philippine race 10. The results on *Xa21* confirmed the findings at IRRI (Madamba et al., 2000) and are similar to IRRI tested by Vera Cruz C. M in 2000. It is therefore clearly that *Xa23* has the broadest spectrum resistance to BB designated so far (Table 1).

Table 1 Comparison of reactions between IRBB21 and CBB23 to 10 Philippine races of *Xanthomonas oryzae* pv. *oryzae* (*Xoo*)

Line	Resistance Gene	Philippine *Xoo* races										Types of resistance
		1	2	3	4	5	6	7	8	9	10	
CBB23	*Xa23*	R	R	R	R	R	R	R	R	R	R	AGS*
IRBB21	*Xa21*	R	R	R	R	R	R	R	R	R	S	MTS*
JG30	non	S	S	S	S	S	S	S	S	S	S	AGS

* AGS= resistance at all growth stages; MTS=Resistance at maximum tillering stage.

The gene *Xa23* not only has the broadest resistance spectrum to races in the Philippines, Japan and China, but also shows highly level of resistance at all growth stages and has strong efficiency of inheritance for resistance. Since 1989, the resistance donor, RBB16, an accession from *O. rufipogon*, and H_4, the anther-cultured line derived from cross of JG30/RBB16 were introduced into some susceptible cultivars: Japonica Zhongzuo 180 with high yielding ability, 9520 with high yielding, good quality and resistance to blast; and indica cultivars, IR24, NJ11, and Zhengzhuai with improved plant type, and a restoring line of hybrid rice, Longtep. The resistance evaluation was made with the specific strain PXO99 for all progenies of the crosses until advanced line. The lines were highly resistant to the strain at all growth stages (Table 2).

Table 2 Reaction of resistance to bacterial blight of advanced rice lines with gene *Xa23*

Cross developed	Generation (3 homozygous lines)	Reaction[a]	Organization	Year
Longtep/JG30/RBB 16 (lL)	F_3	HR	CNRRI[c]	1999
Zhongzuo 180/RBB 16	BC_6F_4	HR[b]	CAAS[d] & GAAS[e]	2001
NJ 11 //JG30/RBB 16 (H_4)	BC, F_3	HR	CAAS	2001
IR24//JG30/RBB 16 (H_4)	BC_4F_2	HR	CAAS	2001
Zhengzhuai//JG30/RBB 16 (H_4)	BC_4F,	HR	CAAS,	2000
ZY5112 (Wuxiangnuo//9520/CBB23/^{3}/ 99-1)	F_8	HR	CAAS, and WWRI[f]	2000

[a] Resistant to 20 strains including 10 Philippine races, 7 Chinese pathotypes and 3 Japanese races tested;

[b] HR=Highly resistant;

[c] China National Rice Research Institute;

[d] Chinese Academy of Agricultural Sciences;

[e] Guangxi Academy of Agricultural Sciences;

[f] Wujin Wheat Rice Institute.

The resistance with high level and broad spectrum to BB derived from *Xa23* has been inherited strongly into all of the progenies and advanced lines. Obviously, the new gene *Xa23* will contribute to widening the genetic background for resistance to BB and serves as additional resistance genetic resource by turns in rice breeding program in the world.

Acknowledgements

This project is supported by grants from the Rockefeller Foundation (No 97001 ＃258) and the National Natural Sciences Foundation of China (No. 39670508)

Published in: Rice Genetics Newaletter, 2001 (18): 71－72.

水稻抗白叶枯病基因 *Xa23* 的 PCR 分子标记定位及辅助选择

潘海军[1] 王春莲[1] 赵开军[1] 章 琦[1]
樊颖伦[1] 周少川[2] 朱立煌[3]

（1. 中国农业科学院作物育种栽培研究所，北京；2. 广东省农业科学院，广州；3. 中国科学院遗传研究所，北京）

Molecular Mapping by PCR-based Markers and Marker-assisted Selection of Bacterial Blight Resistance Gene *Xa23* in Rice

Abstract A F_2 population, with 2562 individuals derived from the cross JG30/CBB23 was constructed for molecular mapping of the bacterial blight resistance gene *Xa23* in rice. After screening 1200 random amplified polymorphic DNA (RAPD) primers, we found RAPD markers, RpdH5 and RpdS1184, were linked to the *Xa23* gene, with genetic distance of 7.0 cM and 7.6 cM, respectively. Two simple sequence repeats (SSR) markers, RM187 and RM206, were also found to be linked to the *Xa23* gene, the genetic distance were 7.1 cM and 1.9 cM, respectively. The marker-assisted selection (MAS) efficiency of RpdS1184, RM206 and RpdH5 was evaluated using two breeding F_2 populations. The results showed that the MAS efficiency of RpdH5 and RpdS1184, in population of Fengfu 23, were 91.0% and 87.3%, respectively, but the MAS efficiency reached as high as 99% when these two markers were co-used . For population Luyou 23, the MAS efficiency of RpdH5 , RpdS1184 and RM206 were 77.1%, 81.1% and 80.8% , respectively. When RpdH5 and RpdS1184 were co-used, the MAS efficiency was 90.3% ; when RpdH5 and RM 206 were co-used, the MAS efficiency was 91.3%; when RpdH5, RM206 and RpdS1184 were co used, the MAS efficiency was 90.8%.

Key words *Xa23*; rice bacterial blight; SSR; RAPD; MAS

摘要 用携带 *Xa23* 基因的近等基因系 CBB23 及其感病轮回亲本 JG30 构建的包含 2562 个单株的 F_2作图群体。通过分析 571 个感病单株，找到 2 个新的与 *Xa23* 基因连锁的 SSR 标记 RM187 和 RM206，它们与 *Xa23* 之间的遗传图距分别为 7.1 cM 和 1.9 cM。通过筛选 1 200 个 RAPD 引物，获得 2 个与 *Xa23* 基因连锁的

RAPD 标记 RpdH5 和 RpdS1184，与 *Xa23* 之间的遗传图距分别为 7.0 cM 和 7.6 cM。利用丰富占/CBB23（丰富 23）、绿油占/CBB23（绿油 23）两个育种组合的 F_2 群体，结合人工接种抗性鉴定，测算了 RpdS1184、RM206 和 RpdH5 用于分子标记辅助选择（MAS）的可能性。结果表明，对于丰富 23 群体，RpdH5 和 RpdS1184 的 MAS 准确率分别为 91.0%和 87.3%，同时使用这两个标记，可使 MAS 的准确率达 99%。对于绿油 23 群体，RpdH5、RpdS1184 和 RM206 的 MAS 准确率分别为 77.1%、81.1%和 80.8%。同时使用 RpdH5 和 RpdS1184 标记的 MAS 准确率为 90.3%；同时使用 RpdH5 和 RM206 的准确率为 91.3%；同时使用 RpdH5、RM206 和 RpdS1184 标记的准确率为 90.8%。

关键词 *Xa23*；水稻白叶枯病；SSR；RAPD；MAS

水稻白叶枯病是水稻生产中最严重的病害之一，世界各产稻区均有发生，一般减产 10%～30%，严重时甚至绝收（Mew，1987）。自 20 世纪 60 年代以来，国际上统一鉴定命名了 19 个水稻白叶枯病抗性基因（Ogawa，1993），其中能在育种上有效利用的并不多，仅 *Xa3*、*Xa4* 已广泛用于育种（章琦，1991），另外，来自长药野生稻的广谱抗性基因 *Xa21* 已被分离克隆（Song et al.，1995）也正在普遍应用。人们还通过基因工程和杂交转育途径开发利用新基因（翟文学等，2000；陈升等，2000）。近 10 年来的白叶枯病抗性新基因的序号已排至 *xa24*（Khush et al.，1999），章琦等（1998）从我国普通野生稻中鉴定发掘的全生育期显性抗性基因 *Xa23*，迄今对国内外所有鉴别小种都表现高抗，对于杂交稻的改良具有广阔的应用前景。Xα23 基因的优异特性已引起育种家的重视，正被应用于水稻育种计划。章琦等（2000）已将 *Xa23* 定位在水稻第 11 染色体的长臂上，与 SSR 标记 OSR 6、RM224 和 RFLP 标记 G1465 之间的遗传图距分别为 5.3 cM、27.7 cM 和 16.7 cM。但是这 3 个标记均位于 *Xa23* 基因近端粒一侧，并且离 *Xa23* 基因较远，无法用来分离克隆 *Xa23*。并且，能鉴定 *Xa23* 基因的专化菌株 P6 和 P10 都引自 IRRI，在国内必须在严格隔离的条件下进行人工接种，因此，在育种运作中，*Xa23* 基因的分子标记辅助选择尤为重要。两年来我们采用 SSR、RAPD、AFLP 及 RFLP 等多种途径寻找与 *Xa23* 基因更为紧密连锁的分子标记，以期用于 *Xa23* 基因的图位克隆和分子标记辅助选择（MAS）。

1 材料与方法

1.1 材料

1.1.1 植物材料 携有 *Xa23* 基因的近等基因系 CBB23、培育 CBB23 的轮回亲本金刚 30（JG30）及其 F_2 分离群体由本实验室制备。绿油 23（绿油占/CBB23）、丰富 23（丰富占/CBB23）两个育种群体由广东省农科院水稻所提供。所有提取 DNA 的植株种植在本所温室或网室，常规管理。

1.1.2 水稻白叶枯病菌株 水稻白叶枯病专化菌株 P6 引自 IRRI，由本实验室保存。

1.1.3 RAPD 引物 购自 OPERON 公司的 10 碱基的 RAPD 引物 20 套共 400 个（编号：OPA、OPB、OPC、OPD、OPE、OPF、OPG、OPH、OPI，OPJ、OPK、OPL、OPM、OPN、OPO、OPU、OPV、OPW、OPX、OPY）；购自上海生工生物工程技术服务有限公

司（上海生工）的 10 碱基 RAPD 引物 800 个（编号：S181－S220、S321－S340、S461－S520、S1001－S1520、S2001－S2160）。

1.1.4　SSR 引物　根据 Temnykh et al.（2000）选用水稻第 11 染色体长臂上包括 RM206 和 RM187 的 10 个 SSR 标记。引物由上海生工合成。其中 RM206 和 RM187 的引物序列为：

RM206F：5′-CCCATGCGTITAACTATICT-3′，

RM206R：5′-CGTTCCATCGATCCGTATGG-3′；

RM187F：5′-CCAAGGGAAAGATGCGACAATTG-3′，

RM187R：5′-GTGGACGCTITATATTATGGG-3′

1.2　方法

1.2.1　植株抗病接种鉴定　采用人工剪叶接种法，用与 *Xa23* 的专化菌株 P6 在水稻植株苗期进行抗性鉴定。

1.2.2　基因组 DNA 提取　对 JG30、CBB23 及它们的 F_2 分离单株，参考 McCouch 等（1988）的方法提取叶片总 DNA。对绿油 23、丰富 23 两个 F_2 群体的分离单株，采用基因组 DNA 的快速提取法。具体步骤：取一小片嫩叶放入盛有 100μL NaOH（0.5 mol/L）的 1.5 mL Eppendorf 管中，用磨砂玻璃棒研磨成匀浆，取 5 μL 上清液与 245μL Tris-HCl（0.1 mol/L，pH 8.0）混合，4℃下保存，尽快用于 PCR。

1.2.3　PCR 反应及电泳　SSR 标记筛选及其用于 MAS 时，反应体积 25μL，反应液组成为 1 x PCR 缓冲液（10mmol/L Tris—HCl，pH 8.3，50mmol/L KCl，0.1% Triton－100），2.5mmol/L $MgCl_2$，0.25mmol/L dNTP，每个引物 50ng，100ng 基因组 DNA，1U *Tag* 酶。反应程序为：94℃预变性 5min，然后每个循环：94℃变性 1min，55℃复性 1min，72℃延伸 2min，共循环 35 次，最后 72℃保温 10min。

RAPD 标记筛选及其用于 MAS 时，反应体积 20μL，反应液组成为 1 x PCR 缓冲液（10mmol/L Tris—HCl，pH 8.3，50mmol/L KCl，0.1% Triton－100），2.0mmol/L $MgCl_2$，0.125mmol/L dNTP，引物 30ng，1 U *Taq* 酶，基因组 DNA 50ng（标记筛选和定位）或 3μL（MAS）。反应程序为：95℃预变性 5min，然后每个循环：95℃变性 30 s，37℃复性 45 s，72℃：延伸 1min，共循环 45 次，最后 72℃保温 7min。

PCR 扩增产物在 3.0%（SSR）或 1.0%（RAPD）琼脂糖凝胶中电泳分离，经溴化乙锭（EB）染色后，紫外灯下扫描照相。

1.2.4　分子标记与 *Xa23* 基因的连锁分析　根据分子标记多态性，结合分离群体抗性表型参数，用 Mapmaker/EXP3.0 作图软件进行基因和分子标记位点的遗传连锁分析。LOD ≥ 3.0。重组值经 Kosambi 函数转换为遗传图距（cM）。

2　结果与分析

2.1　构建 F_1 作图群体

1999 年将 JG30 与 CBB23 杂交，其 F_1 植株在海南岛繁殖加代，单株收种。2000 年在北京种植 F_2 分离群体，共 2 562 个单株。用 P6 接种抗性鉴定，结果严格符合 3 ∶1 的抗感分离（表 1），可以用于 *Xa23* 基因的分子标记定位。

表 1 JG30/CBB23 的 F_1 和 F_2 群体对 P6 的抗性反应

Table 1 Resistance reactions of F_1 and F_2 populations from cross JG30/CBB23 to strain P6

组合 (Cross)	F_1 植株 (F_1 Plant)	F_2 群体 (F_2 population)			χ^2 (Expected ratio) (3∶1)	P value
		抗病株 (R)	感病株 (S)	总株数 (Total)		
JG30/CBB23＃1	R	1 046	341	1 387	0.106	0.50～0.75
JG30/CBB23＃2	R	882	293	1 175	0.000	>0.90
总计 (Total)		1 928	634	2 562	0.075	0.75～0.90

R＝抗 Resistant，S＝感 Susceptible。

2.2 *Xa*23 基因的 RAPD 标记定位

2.2.1 多态性 RAPD 引物的筛选 从 JG30/CBB23 的 F_2 分离群体中，选择表现典型的抗和感的单株各 10 株，每株取等量 DNA 分别混合成抗病 DNA 池（R-pool）和感病 DNA 池（S-pool）。以抗、感池 DNA 为模板，从 1 200 个 RAPD 引物中，筛选到 2 个在抗、感池之间出现多态性带的引物 H5 和 S1184，它们在抗病 DNA 池中分别扩增出 0.65 kb 和 0.45 kb 特异片段，记为 RpdH5 和 RpdS1184（图 1）。

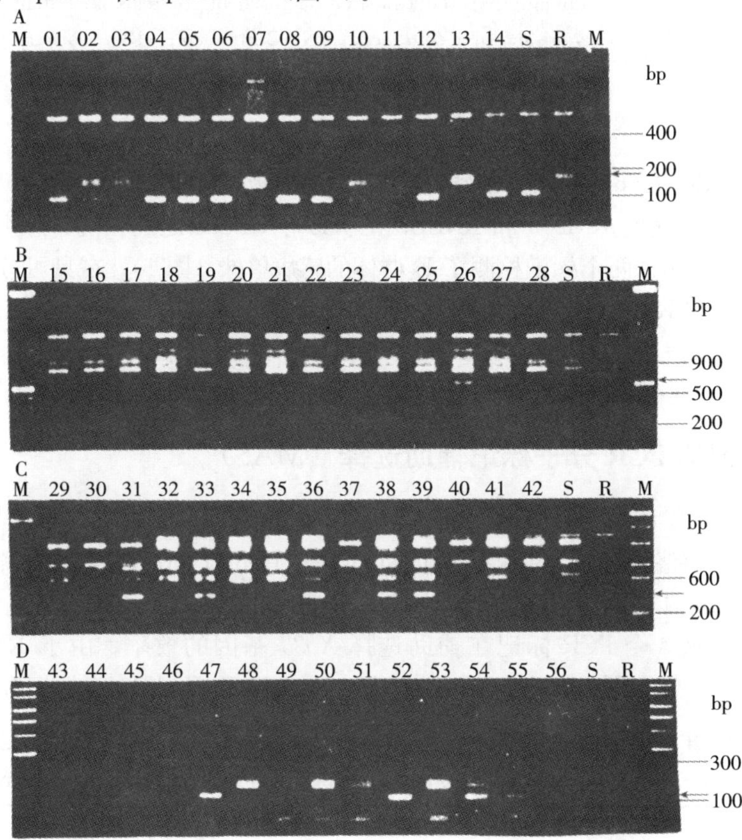

图 1 分子标记在 F_2 群体中与 *Xa23* 基因的连锁分析

M：分子量 R：抗病池 S：感病池 01－56：F_2 感病植株

Fig. 1 Co-segregation between molecular markers and *Xa23* in F_2 population

M：Molecular weight；R：R-pool；S：s-pool；01－56：F_2 plants susceptible to P6 A：RM187；B：RpdH5；C：Rpds1184；D：RM206

2.2.2 F_2感病单株检测及遗传图距 鉴于 *Xa23* 为显性基因，JG30/CBB23 的 F_2群体中表型抗病的植株包含纯合和杂合两种基因型，抗病植株是杂合的，感病植株为隐性纯合体，其在理论上完全真实。因此，在分子标记检测中，感病植株的扩增带型可真实地反映分子标记与 *Xa23* 基因间的遗传交换情况。我们用 RAPD 引物 H5 和 S1184，对 F_2群体的感病单株逐一检测（图 1）。由于部分单株的 DNA 量不足，最后获得 571 个感病单株的可分析数据。其中，H5 和 S1184 分别检测到 78 和 81 个单株在 *Xa23* 基因与分子标记之间发生了遗传交换。用 Mapmaker 3 .0 软件对上述结果进行连锁分析表明，RpdH5 和 RpdS1184 标记均与 *Xa23* 基因连锁，其中 RpdH5 位于 *Xa23* 基因的靠着丝粒一侧，遗传距离为 7.0 cM；RpdS1184 位于 *Xa23* 基因的靠端粒一侧，遗传距离为 7.6 cM。由此便将 *Xa23* 基因框在特定的染色体区间内（图 2）。

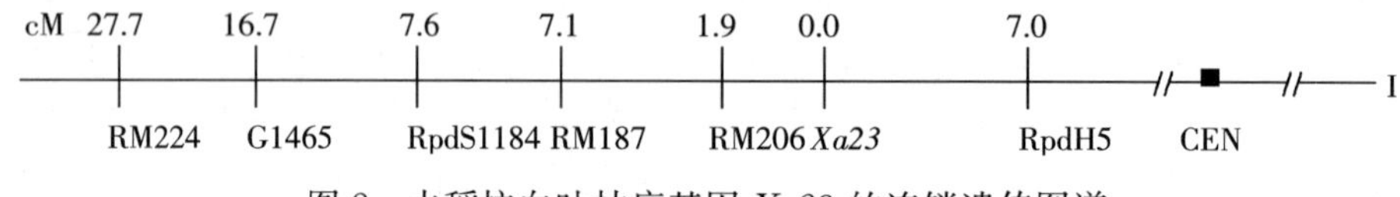

图 2 水稻抗白叶枯病基因 *Xa23* 的连锁遗传图谱

Fig. 2 The genetic linkage map flanking rice bacterial blight resistance gene *Xa23*

New markers identified in this study are shown in *boldface*. CEN: Centromere; 11S: Short arm of chromosome 11

2.3 *Xa23* 基因的 SSR 标记定位

利用 JG30、CBB23、F_2抗病池、感病池 DNA 为模板，筛选了水稻第 11 染色体长臂上的 10 个 SSR 标记。结果表明 RM206 和 RM187 在 JG30、CBB23 之间和抗、感 DNA 池之间扩增出多态性条带。所以用这两个标记检测了 F_2群体的感病单株（图 1），在具可分析数据的 571 个感病株中，RM206 和 RM187 分别检测到 17 和 70 个单株在 *Xa23* 基因与分子标记之间发生了单交换，2 和 3 个单株发生了双交换。Mapmaker 3 .0 软件分析表明，RM206 和 RM187 标记均位于 *Xa23* 基因的靠端粒一侧，与 *Xa23* 的遗传距离分别为 1.9 cM 和 7.1 cM（图 2）。

2.4 *Xa23* 基因的 PCR 分子标记辅助选择（MAS）

由广东省农业科学院水稻研究所提供了丰富 23、绿油 23 两个育种材料的 F_2分离群体，在用 P6 进行人工接种鉴定每个单株抗性的同时，选用位于 *Xa23* 基因两侧的 RAPD 标记 RpdHS5 和 RpdS1184 以及目前找到的离 *Xa23* 基因最近的 SSR 标记 RM206，对各单株进行分子标记检测，评估这 3 个 PCR 标记在辅助选择 *Xa23* 基因的效率。由于 RM206 在丰富占、CBB23 之间没有多态性，所以对丰富 23 群体只有 RpdH5 和 RpdS1184 标记选择的资料。

表 2 丰富 23 和绿油 23 两个 F_2群体分子标记检测与田间抗感表型的符合率

Table 2 Marker-assisted selection efficiency in two F_2 breeding populations

组合 Cross	分子标记 Marker	检测株数 Plant tested	抗、感表型株 (Reaction)		分子扩增带型株 (Amplified pattern)				分子检测符合率 MAS efficiency 平均符合率 (Mean %)
					R	S	R%	S%	
丰富占/CBB23 (丰富 23)	RpdH5	328	抗病株	245	227	18	92.7		91.0
			感病株	83	9	74		89.2	

（续）

组合 Cross	分子标记 Marker	检测株数 Plant tested	抗、感表型株 (Reaction)		分子扩增带型株 (Amplified pattern)				分子检测符合率 MAS efficiency 平均符合率 (Mean %)
					R	S	R%	S%	
丰富占/CBB23 (丰富 23)	RpdS1184	235	抗病株	175	151	24	86.3		87.3
			感病株	60	7	53		88.3	
绿油占/CBB23 (绿油 23)	RpdH5	273	抗病株	193	172	21	89.1		77.1
			感病株	80	28	52		65.0	
	RpdS1184	276	抗病株	196	178	18	90.8		81.1
			感病株	80	23	57		71.3	
	RM206	278	抗病株	197	187	10	94.9		80.8
			感病株	81	27	54		66.7	

从表 2 可以看出，对于丰富 23 群体，在 RpdH5 标记检测的 32 个单株中，245 株在接种鉴定中表现抗，在分子标记检测中，有 227 株呈抗病带型，符合率为 92.7% ；在接种鉴定表现感的 83 个单株中，有 74 株在分子标记检测中呈感病带型，符合率为 89.29%，平均符合率为 91.0%。同样，可算出 RpdS1184 标记对丰富 23 群体的 MAS 平均符合率为 87.3%。由于 RpdH5 标记位于 *Xa23* 基因靠着丝粒一侧 7.0 cM 处，RpdS1184 标记位于 *Xa23* 基因另一侧 7.6 cM 处（图 2），说明对于 *Xa23* 基因，利用相距 7.0cM 的单个分子标记进行辅助选择，其准确率约为 90.0%。对于丰富 23 群体，在 RpdH5 和 RpdS1184 标记检测中同时表现为抗病带型的 148 个单株中，仅 1 株在接种鉴定中表现感病，表明同时使用位于 *Xa23* 基因两侧的 RpdH5、RpdS1184 标记，可使 MAS 的准确率达到 99%（表 3）。

对于绿油 23 群体，RpdH5、RpdS1184 标记的平均 MAS 符合率均有所降低，分别为 77.1%和 81.1%。距 *Xa23* 基因仅 1.9cM 的 RM206 标记，其平均 MAS 符合率也只有 80.8%（表 2）。但是抗病植株的 MAS 符合率并不低，3 个标记分别为 89.1%、90.8%和 94.9%（表 2），之所以造成平均 MAS 符合率偏低的原因是感病植株的符合率低。对照原始记录发现，绿油 23 群体的部分单株在接种鉴定中的抗感反应居中，当时主要按育种要求取舍，其中有 12 个单株的病斑面积在 15%～20%之间，被归入感病植株 ，因此绿油 23 群体的感病株比例远多于 1/4，例如用 RpdH5 检测的 273 株中，被划分为 R 的有 193 株，划为 S 的为 80 株，感病株比例为 1.7/4（表 2），以致使该群体中 MAS 符合率的估计值偏低。对于绿油 23 群体，同时使用 RpdH5、RpdS1184 标记的 MAS 准确率为 90.3%；同时使用 RpdH5、RM206 标记的 MAS 准确率为 91.3%；同时使用 RpdH5、RM206、RpdS1184 标记的 MAS 准确率为 90.8%（表 3）。可见，同时使用位于 *Xa23* 基因两侧的 2 个或 3 个分子标记可提高 MAS 的准确率，但使用 3 个分子标记的 MAS 符合率并不比使用 2 个分子标记的 MAS 符合率提高，所以在实际应用中，选择 2 个位于 *Xa23* 基因两侧的分子标记开展 MAS 就已足够。

表 3 同时使用 *Xa23* 基因两侧的分子标记的 MAS 符合度（R%）

Table 3 MAS efficiency when 2 or 3 marker flanking *Xa23* were co-used

F_2群体 (population)	RpdH5 - RpdS1184				RpdH5 - RM206				RpdH5 - RM206 - RpdS1184			
	植株数	R	S	R%	植株数	R	S	R%	植株数	R	S	R%
丰富 23	148	147	1	99.3								
绿油 23	175	159	17	90.3	184	168	16	91.3	174	158	16	90.8

3 讨论

3.1 分子标记定位

为使 *Xa23* 基因的分子标记定位足以精细到可以进行图位克隆，课题组构建了一个有 2 562个单株的 F_2大群体。参照 Song 等（1995）的方法，用感病植株进行连锁分析。其理论基础是显性基因控制的抗病表型包含纯合和杂合两种基因型，一般 F_2植株的抗病表型都是杂合的，其中的感病植株为隐性纯合体，无论出现纯合的抗性亲本带型还是杂合带型，都能清晰地揭示分子标记与抗病基因间的遗传交换情况。

通过分析 571 个感病单株，找到 2 个新的与 *Xa23* 基因连锁的 SSR 标记 RM187 和 RM206，并测得它们与 *Xa23* 之间的遗传图距分别为 7.1 cM 和 1.9 cM，确证了 *Xa23* 在水稻第 11 染色体长臂上的相对位置（图 2）。我们并用前期找到的 SSR 标记 OSR6 检测该群体，其扩增带型都与 RM187 的一致，通过比较引物碱基序列发现，RM187F 包含了 OSR6F 的全部 20 个碱基，只是在 3′端多了 TTG 碱基；RM187R 包含了 OSR6R 的全部 20 个碱基，仅在第 8 位多了 T 碱基（Temnykh et al.，2000；Akagi et al.，1996），因此，RM187 与 OSR6 在本群体中的扩增位点是相同的，从而测得 OSR6 与 *Xa23* 之间的遗传图距也为 7.1 cM。章琦等（2000）曾用包含 160 个单株的 JG30/WBB1（前期培育 CBB23 中的 BC_5F_3代）F_2群体测得 OSR06 与 *Xa23* 的遗传图距为 5.3 cM，这可能由于所用不同群体所致。

至今发现的与 *Xa23* 基因连锁的 SSR 标记 OSR06、RM224、RM187、RM206 及 RFLP 标记 G1465 均位于 *Xa23* 基因靠染色体端粒的同一侧（图 2），为了精细定位 *Xa23* 基因，必须在 *Xa23* 的另一侧找到标记，将其框在特定的染色体区段内。本研究通过筛选 1 200 个 RAPD 引物，获得 2 个与 *Xa23* 基因连锁的 RAPD 标记 RpdH5、RpdS1184，其中 RpdH5 位于 *Xa23* 基因靠着丝粒的一侧，这样 *Xa23* 就被框在了 RpdH5 与 RM206 之间 8.9 cM 的染色体区段内（图 2）。我们正集中力量寻找处在该染色体区段内的各种分子标记，包括 RFLP、AFLP、EST、YAC 亚克隆标记等。

3.2 分子标记辅助选择

从实验成本和技术可行性考虑，开展 MAS 首选基于 PCR 的分子标记。郑康乐等（1997）认为，用于 MAS 的分子标记与目标基因的遗传距离最好小于 5.0 cM。我们目前找到的位于 *Xa23* 基因两侧的最近的基于 PCR 的分子标记为 RM206 和 RpdH5，距 *Xa23* 分别为 1.9 cM 和 7.0 cM，因此我们利用丰富 23、绿油 23 两个育种 F_2群体，结合人工接种抗性鉴定，测算 RM206 和 RpdH5 用于 MAS 的可靠性。由于 RM206 在丰富占、CBB23 之间没有多态性，所以又纳入 RpdS1184 进行测算。RpdS1184 与 RM206 位于 *Xa23* 基因的同一侧，但遗传图距为 7.6 cM（图 2）。

对于丰富 23 群体，RpdH5 和 RpdS1184 的 MAS 平均符合率分别为 91.0%和 87.3%，如果同时使用这两个标记，可使 MAS 的准确率达 99%。我们曾利用 JG30/WBB1 的 F_2 植株，JG30/WBB1、9520/WBB1、南京 11/WBB1、IR24/WBB1 的 14 个 F_3 家系，估算 OSR6 的 MAS 准确率为 92.2%（Zhao et al.，2000），从这些结果来看，上述测算数据是可靠的。

对于绿油 23 群体，RpdH5、RpdS1184、RM206 的平均 MAS 符合率分别为 77.1%、81.1% 和 80.8%。同时使用 RpdH5、RpdS1184 标记的 MAS 准确率为 90.3%；同时使用 RpdH5、RM206 标记的准确率为 91.3%；同时使用 RpdH5、RM206 和 RpdS1184 标记的为 90.8%（表 3）。这些测算数据较实际效率偏低，由于当时该群体的抗性鉴定中，是按育种的要求，将 12 个抗、感反应介于中抗/中感的单株归为感病群，而使群体中感病植株数远多于 1/4，感病植株的分子检测符合率仅为 65%～71.3%，从而拉低了 MAS 的平均符合率。3 个标记对抗病植株的分子检测符合率较高（分别为 89.1%、90.8% 和 94.9% ）也反证了这一点（表 2）。

鉴于用 3 个分子标记的 MAS 符合率并不比用 2 个分子标记的 MAS 符合率有明显提高，因此在应用中，选择 2 个位于 *Xa23* 基因两侧的分子标记就已足够。就目前情况看，选择 RpdH5 标记与 *Xa23* 基因另一侧的 RM206、OSR6、RM187 和 RpdS1184 标记之一联合使用，可使 *Xa23* 基因的 MAS 准确率达 90%～99%。在实际应用中，应根据亲本间的多态性，选择位于 *Xa23* 基因两侧最近的 2 个标记开展 MAS，不仅可提高准确率，而且可增加中选个体与受体亲本的遗传背景同质性。

由于目前利用 MAS 对 *Xa23* 基因进行选择的准确率为 90%～99%，所以建议在育种利用的早期分离世代（F_2～F_4）进行 MAS，到后期的基本稳定世代阶段，在中选的株系中可取少量种子（15～30 粒）在控制的条件下用专化菌株进行人工接种鉴定抗性，既不影响育种进程，又确保 *Xa23* 基因不致丢失。

原载：作物学报，2003，29（2）：501～507.

水稻抗白叶枯病基因 *Xa23* 的 EST 标记及其在分子育种上的利用

王春莲[1,2]　戚华雄[3]　潘海军[1]　李进波[3]
樊颖伦[1,2]　章　琦[1]　赵开军[1,2,4]

（1. 中国农业科学院作物科学研究所/农业部作物遗传育种重点实验室；
2. 农作物基因资源与基因改良国家重大科学工程，北京；
3. 湖北省农业科学院粮食作物研究所，武汉；4. IRRI 中国办事处，北京）

EST-Markers Flanking the Rice Bacterial Blight Resistance Gene *Xa23* and Their Application in Marker-Assisted Selection

Abstract　A F_2 population, with 2 562 individuals of JG30/CBB23 was constructed for molecular mapping of the bacterial blight resistance gene *Xa23* in rice. Inoculation test of the F_2 plants showed that the segregation ratio of resistant and susceptible fits to 3∶1. Based on the authors previous mapping of *Xa23* gene, 12 EST markers from Rice Genome Program（RGP）database were surveyed in the susceptible F_2 plants and two markers, C189 and CP02662 were found flanking the *Xa23* gene with genetic distance of 0.8 cM and 1.3 cM, respectively. These two EST markers were successfully used in rice breeding program. The results showed that efficiency of C189 marker-assisted selection（MAS）for *Xa23* gene was almost 100%. Three elite hybrid rice restore lines with *Xa23* gene and improved bacterial blight resistance have been developed.

Key words　rice；*Xa23*；bacterial blight；EST；MAS

摘要　用携有水稻抗白叶枯病基因 *Xa23* 的近等基因系 CBB23 与其感病轮回亲本 JG30 杂交，构建了包含 2 562 个单株的 F_2 作图群体。抗性鉴定表明，F_2 植株抗感分离比严格符合 3∶1。根据日本水稻基因组计划 RGP 数据，筛选并合成 12 个 EST 标记的引物，进行亲本间多态性检测，找到 2 个在 CBB23 与 JG30 间有多态性的 EST 标记，C189 和 CP02662。用该标记对 F_2 群体中的 571 个感病单株进行分子检测和连锁分析，结果表明这两个 EST 标记位于 *Xa23* 基因的两侧，C189 靠着丝粒一侧，与 *Xa23* 的遗传距离为 0.8cM；CP02662 靠端粒一侧，与 *Xa23* 的遗传距离为 1.3cM。将 C189 成功用于水稻分子育种实践，标记辅助选择的正确率接近 100%，已培育出 3 个

将 *Xa23* 基因与高产、优质、抗褐飞虱等性状聚合的水稻恢复系。

关键词　水稻；*Xa23*；水稻白叶枯病；EST 标记；分子育种

水稻白叶枯病是世界水稻主要病害。一般引起减产 10％～30％（Mew，1987）。利用寄主抗性是控制该病的最有效途径。20 世纪 60 年代以来，国际上鉴定出水稻抗白叶枯病基因近 30 个（Ogawa，1993；Lee et al.，2003；谭光轩等，2004），其中 *Xa1*、*xa5*、*Xa21* 和 *Xa26* 已被分离克隆（Song et al.，1995；Yoshimura et al.，1998；Sun et al.，2004；Lyer et al.，2004）。在众多的水稻抗白叶枯病基因中，由于有些隐性基因不便利用或因抗谱狭窄容易丧失抗性，在育种中利用的并不多。国内外大面积种植的水稻品种大多含有 *Xa3* 或 *Xa4* 基因，这些品种因多年、长期被广泛地利用，在许多稻区已经丧失抗性（章琦等，1988；1991）。来自长药野生稻的广谱抗性基因 *Xa21*，近年正通过基因工程和分子标记辅助育种途径被开发利用（Zhai et al.，2000）。从中国普通野生稻中鉴定发掘的 *Xa23* 基因，对国内外鉴别菌株都表现高抗，且完全显性、全生育期抗病，因此对杂交稻的改良具有广阔的应用前景（章琦等，1998；Zhang et al.，2001）。章琦等（2002）已将 *Xa23* 转育到栽培稻中，育成了近等基因系 CBB23。潘海军等（2003）曾将 *Xa23* 定位在水稻第 11 染色体长臂上的 SSR 标记 RM206 和 RAPD 标记 RpdH5 之间，遗传图距分别为 1.9 cM 和 7.0 cM。为了克隆 *Xa23* 基因和促进该基因在分子标记辅助育种上利用，我们又采用了 EST、AFLP 及 RFLP 等多种途径寻找与 *Xa23* 基因更为紧密连锁的分子标记。本文报道 2 个与 *Xa23* 基因紧密连锁的 EST 标记筛选及其在分子辅助育种上的利用。

1　材料与方法

1.1　材料

1.1.1　植物材料　用于分子标记定位的携有 *Xa23* 基因的近等基因系 CBB23、培育 CBB23 的轮回亲本金刚 30（JG30）及其 F_2 分离群体，均由本实验室制备。定位群体和分子标记辅助育种的抗病接种鉴定材料，均种植在中国农科院作物科学研究所网室，常规水肥管理。用于分子标记辅助育种的水稻品种（系）9311、M087、1826、B5 和 Q1347 等及其杂交后代均由湖北省农科院粮作所提供，种植在该所试验农场，常规水肥管理。

1.1.2　水稻白叶枯病菌株　采用由 IRRI 提供的广致病菌株 P6，原菌种保存于－70℃，用前在胁本哲氏培养基上复壮。

1.1.3　EST 标记及引物　根据潘海军等（2003）将 *Xa23* 所定位的区域，参照日本 RGP（Rice Genome Program）EST 数据（Wu et al.，2002），选择了 12 个 EST 标记（表 1），其 24 条引物由上海生物工程技术服务有限公司合成。

表 1　12 个 EST 标记的引物序列

Table 1　Primers of the 12 EST markers screened

标记名称 Marker	5′-引物序列 5′- Primer	3′-引物序列 3′- Primer
CP10653	5′- CAG TAT CGG TCC GGG AAA AC - 3′	5′- TGC AGG GTT TAG TTC GAT TG - 3′

（续）

标记名称 Marker	5′-引物序列 5′- Primer	3′-引物序列 3′- Primer
CP03890	5′- GGA AGG GTC TGC TCA GGG TG - 3′	5′- ATT CCA CGG TTC CAC AGA AG - 3′
CP05426	5′- AAC AGG TTC CCC AAG CAG AG - 3′	5′- GGG GTC AAA ACA TAT GCA AC - 3′
C189	5′- TAA GTT CTA CAT CGA CCC CA - 3′	5′- CAC ATG AAG AGC TGG AAA CG - 3′
CP10914	5′- ATG TCG CCA GCT GAG ATC AC - 3′	5′- AAA CAC AAC CTG CCA GCT CC - 3′
CP05682	5′- GGC AGG AAA CAG GCA TCA TC - 3′	5′- ATA AAA GCG GTT TAC CCT CG - 3′
CP05693	5′- CCT CCT GCT GAA CCT TCT GG - 3′	5′- AAT CCA CGG TAC TTT GAG CG - 3′
CP05602	5′- ACG TAC GAC AGG GAA CAT GG - 3′	5′- CCG GAA GAC TCT CGT GAA TG - 3′
CP02662	5′- AAG GGA GAA GAG TAG AGC GG - 3′	5′- CAG TTT AGC GGC CAA TAT GC - 3′
CP08533	5′- TGA AGA CAG CAT CAA AGC TC - 3′	5′- CAC CAC TTT TGT TCC TAT GC - 3′
CP03371	5′- TGG GTT GGA TCT GGA CTA CG - 3′	5′- ACA GAG TGC CAA CCT CTT CC - 3′
CP06337	5′- TGC AGA AAG CAT TGG CAT TG - 3′	5′- ACA TAC GGC GAG TGC ACG AC - 3′

1.1.4　试剂　提取 DNA、PCR 反应、电泳等所用试剂为进口或国产分析纯产品，其中 Taq 聚合酶购自华美公司，dNTPs 为 Pharmacia 公司产品，华美公司分装。

1.2　方法

1.2.1　植株抗病接种鉴定　亲本 JG30、CBB23 及 F_2群体植株至 5～6 片叶时，进行苗期接种鉴定（章琦等，1998），采用人工剪叶接种法，到分蘖期再接种 1 次，以保证各植株抗性反应的真实性。接种后 2 周左右，感病对照品种（JG30）病情趋于稳定时进行调查，用病斑面积占叶片面积的百分率作为抗感反应参数，以分离群体的抗性反应参数分布的峰槽作为抗感分群界限。本研究所用的分离群体单株的抗、感分明，病斑面积占叶片面积的百分率≤10%为抗，≥20%为感，没有中间类型。对分子标记辅助育种抗病接种鉴定材料只在分蘖期接种 1 次，接种方法和抗感标准同上。

1.2.2　基因组 DNA 提取　对 JG30、CBB23 及它们的 F_2单株，参考 McCouch 等（1988）的方法提取叶片总 DNA。DNA 纯化后，用分光光度计测定浓度，用于 PCR 反应的 DNA 浓度调至 100ng・μl^{-1}左右。对于分子标记辅助育种的材料，采用基因组 DNA 快速提取法提取叶片总 DNA（潘海军等，2003）。

1.2.3　PCR 反应及电泳　PCR 反应体积 20μl，反应液组成为：1×PCR 缓冲液（10mmol・L^{-1} Tris-HCl，pH8.3，50mmol・L^{-1} KCl，0.1%Triton－100），2.5mmol・$L^{-1}$$MgCl_2$，dNTP 0.25mmol・$L^{-1}$，每个引物 50ng，100ng 基因组 DNA，1 U Taq 酶。PCR 反应程序为：94℃预变性 5min，然后每个循环：94℃变性 1min，55℃（或 58℃）复性 1min，72℃延伸 2min，共循环 35 次，最后 72℃保温 10min。

PCR 扩增产物在 1.0%或 2.5%（2.8%）琼脂糖凝胶中电泳分离，经溴化乙锭（EB）染色后，紫外灯下扫描照相。

1.2.4　分子标记与 *Xa23* 基因的连锁分析　根据分子标记多态性，结合分离群体抗性表型

参数，用 Mapmaker/EXP3.0 作图软件进行基因和分子标记位点的遗传连锁分析。LOD≥3.0。重组值经 Kosambi 函数转换为遗传图距（cM）。

2　结果与分析

2.1　构建 F_2作图群体及亲本间多态性 EST 标记的筛选

将 JG30 与 CBB23 杂交，收获 F_1单株。采用其中 1 个 F_1单株繁殖的 F_2共 2 562 个单株的大群体。用 P6 进行抗性鉴定，F_2代单株抗、感界限分明，抗病株 1 928 株，感病株 634 株，明显符合 3∶1 抗感分离，可用于 *Xa23* 基因的分子标记定位。

以 JG30、CBB23 的基因组 DNA 为模板，用 12 对引物分别进行 PCR 扩增，筛选在亲本间具有多态性的 EST 标记。结果表明，在 55℃退火温度、1%的凝胶电泳下，标记 C189 在双亲间出现多态性条带，但带型不够清晰，将退火温度提高到 58℃时，带型清晰。在 55℃退火温度、2.8%的凝胶电泳下，标记 CP02660 在双亲间产生明显的差异带型。而其他 10 对引物即使调整了退火温度和凝胶浓度，两个亲本间均无多态性条带。标记 C189 和 CP02660 引物扩增双亲的多态性片段分别在 0.8 kb 左右和 0.35 kb 左右（图 1）。

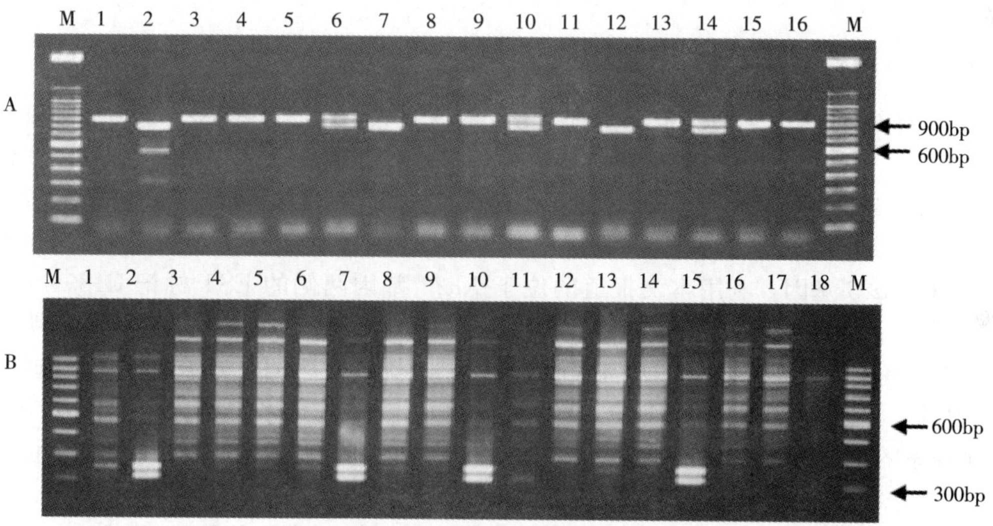

图 1　分子标记在 F_2群体中与 *Xa23* 基因的连锁分析（A：C189，B：CP02662）

Fig. 1　Co-segregation between molecular markers and *Xa23* in F_2 population（A：C189，B：CP02662）

M. 分子量；1. JG30（S）；2. CBB23（R）；3～18. F_2感病植株

M. Molecular weight ladder；1. JG30（S）；2. CBB23（R）；3～18，Susceptible F_2 plants

2.2　F_2群体单株的分子检测及 *Xa23* 基因定位

用中选的 2 个 EST 标记 C189 和 CP02662 对 571 个 F_2代中的感病单株进行分子检测（田间鉴定感病株共 634 株，因部分严重感病植株死亡，实获 571 个单株 DNA）。另外随机选择 66 个抗病单株用作分子检测，表现杂合带型的单株留种作为分子检测的备用群体。用 C189 标记引物鉴定 F_2代感病单株，出现 3 种带型，相同于 JG30 带型的有 564 株，相同于 CBB23 带型的有

2株，具双亲带型的有5株。表明有5株发生了单交换，2株发生了双交换（图1-A）。

CP02662在CBB23和JG30之间是一个显性标记，用其检测F_2代感病单株时出现了2种带型，与JG30带型相同的有551株，与CBB23带型相同的有20株（图1B）。用CP02662的引物扩增时，出现多条非特异带，但并不影响特异条带的观察。

用Mapmaker3.0软件进行连锁分析表明，标记C189和CP02662位于*Xa23*基因的两侧，C189标记靠着丝粒一侧，与*Xa23*的遗传距离为0.8 cM，CP02662标记靠端粒一侧，与*Xa23*的遗传距离1.3 cM，从而将*Xa23*框在2.1 cM的范围内（图2）。

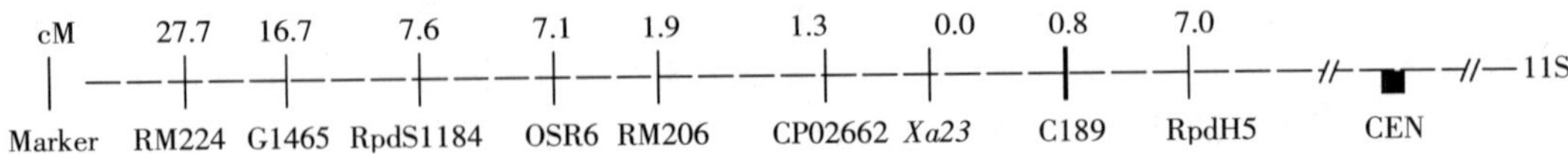

图2 水稻抗白叶枯病基因*Xa23*的连锁遗传图谱

Fig. 2 The genetic linkage map flanking rice bacterial blight resistance gene *Xa23*

黑体为本研究新筛选的标记；CN. 着丝粒；11S. 11染色体短臂

New markers identified in this study are shown in boldface; CEN. Centromere; 11S. Short arm of chromosome 11

2.3 EST标记在分子育种上的应用

常规品种9311为著名超级稻组合两优培九的恢复系，M087和1826是湖北省农科院近年培育的高产、优质籼稻恢复系，但有待改良病虫抗性。因此，从2001年开始，我们用抗白叶枯病的CBB23、抗褐飞虱的B5及抗稻瘟病的Q1347抗源材料，与9311、M087和1826进行杂交和回交，结合分子标记辅助选择，以期培育出将高产、优质与多抗性聚合的水稻恢复系。针对*Xa23*基因，利用本文报道的位于*Xa23*基因两侧的EST标记C189、Cp02662和翟文学等（2000）报道的SSR标记RM206进行分子标记辅助选择。实践表明，由于C189离*Xa23*基因很近（0.8 cM），仅利用该标记，就足以达到用分子标记辅助选择*Xa23*基因的目的。我们在2003年用C189分子标记检测了180份株系材料，2004年从中随机选择16个株系的自交种子进行抗病性接种鉴定的结果见表2，利用C189单个分子标记辅助选择的11个*Xa23*基因纯合株系的*Xa23*基因的正确率接近100%；5个*Xa23*基因杂合株系后代抗性接种鉴定，也符合预期的1对基因分离（$\chi^2 \leqslant 0.1$，$P \geqslant 0.75$），表明C189检测的*Xa23*基因型与实际基因型相吻合。利用该技术已选育出C601、C627和C664三个双抗白叶枯病和褐飞虱、农艺性状优良的恢复系（经863课题鉴定），正被用于组配杂交稻。

表2 分子标记C189辅助选择*Xa23*基因的效果

Table 2 Marker-assisted selection efficiency of EDT marker C189

株系编号 Lined	2003年田间编号（组合） Cross	C189检测*Xa23*基因型 *Xa23* genotype tested by Marker C189	自交后代植株接种鉴定抗性反应 Resistance reaction of self-cross progeny			分子检测与接种鉴定符合程度 Coincidence of molecular test with inoculation identification（%）
			R	S	Total	
1	HC1841（9311/CBB23//1826/B5）	纯合 Homozygous	17	1	18	94.1

（续）

株系编号 Lined	2003年田间编号（组合） Cross	C189检测 *Xa23* 基因型 *Xa23* genotype tested by Marker C189	自交后代植株接种鉴定抗性反应 Resistance reaction of self-cross progeny			分子检测与接种鉴定符合程度 Coincidence of molecular test with inoculation identification（%）
			R	S	Total	
3	HC1881（9311/CBB23//1826/B5）	纯合 Homozygous	20	0	20	100
4	HC1882（9311/CBB23//1826/B5）	纯合 Homozygous	20	0	20	100
5	HC2211（Q1347//9311/CBB23）	纯合 Homozygous	20	0	20	100
6	HC2212（Q1347//9311/CBB23）	纯合 Homozygous	20	0	20	100
9	HC2241（Q1347//9311/CBB23）	纯合 Homozygous	20	0	20	100
10	HC2242（Q1347//9311/CBB23）	纯合 Homozygous	20	0	20	100
11	HC2261（Q1347//9311/CBB23）	纯合 Homozygous	20	0	20	100
12	HC2262（Q1347//9311/CBB23）	纯合 Homozygous	20	0	20	100
13	HC2271（Q1347//9311/CBB23）	纯合 Homozygous	20	0	20	100
14	HC2272（Q1347//9311/CBB23）	纯合 Homozygous	20	0	20	100
15	HC2291（Q1347//9311/CBB23）	杂合 Heterozygous	17	5	22	3∶1（χ^2=0.00）
16	HC2292（Q1347//9311/CBB23）	杂合 Heterozygous	17	5	22	3∶1（χ^2=0.00）
2	HC1842（9311/CBB23//1826/B5）	杂合 Heterozygous	16	4	20	3∶1（χ^2=0.07）
7	HC2221（Q1347//9311/CBB23）	杂合 Heterozygous	16	4	20	3∶1（χ^2=0.07）
8	HC2222（Q1347//9311/CBB23）	杂合 Heterozygous	22	9	31	3∶1（χ^2=0.10）

3　讨论

3.1　水稻 EST 图谱对基因定位的重要作用

水稻抗白叶枯病优异基因 *Xa23* 来自野生稻，构建精细的分子定位图谱是分离克隆基因的关键。我们曾根据国际上发表的 RFLP、SSR 等图谱，将 *Xa23* 定位在水稻第 11 染色体的长臂上，找到的与 *Xa23* 基因连锁的 SSR 标记有 RM206、OSR6、RM224 和 RFLP 标记 G1465 等，均位于 *Xa23* 基因的近端粒一侧，随后，从 1 200 个 RAPD 引物中，仅筛选到 1 个位于 *Xa23* 另一侧的 RAPD 标记 RpdH5，但位于 *Xa23* 基因两侧的标记都比较远，其中 RM206 和 RpdH5 的遗传图距分别为 1.9 和 7.0cM（潘海军等，2003）。本研究根据 Wu 等（2002）构建的水稻 EST 图谱，找到 2 个位于 *Xa23* 基因两侧的 EST 标记 C189 和 CP02662，它们与 *Xa23* 之间的遗传图距分别为 0.8 cM 和 1.3 cM，从而将 *Xa23* 框在2.1cM 的染色体区段内（图 2），将原定位区间缩短了 6.8 cM，为精细定位 *Xa23* 基因提供了重要基础。

3.2　分子标记辅助选择 *Xa23* 基因

水稻抗白叶枯病基因 *Xa23* 的优异特性已引起育种家的重视。目前已被 30 多个育种单位引用。由于 *Xa23* 基因的专化菌株 P6 和 P10 均引自 IRRI，用于国内人工接种必须在严格隔离的条件下进行，因此在育种中，应用分子标记辅助选择（MAS）*Xa23* 基因非常重要。郑康乐等（1997）认为，用于 MAS 的分子标记与目标基因的遗传距离应小于 5.0 cM。潘海军等（2003）曾利用丰富 23、绿油 23 两个育种 F_2群体，测试了位于 *Xa23* 基因两侧的分子标记 RM206 和 RpdH5 的 MAS 准确率。结果表明，MAS 准确率与群体遗传背景有关，距 *Xa23* 基因 1.9cM 的 RM206 的 MAS 准确率在 81%～95%之间；距 *Xa23* 基因 7.0 cM 的 RpdH5 的 MAS 准确率在 77%～91%之间。本研究将距 *Xa23* 基因 0.8 cM 的 EST 标记 C189 用于不同育种群体的 MAS，结果表明，C189 辅助选择 *Xa23* 基因的准确率接近 100%，完全可以用于分子育种实践。可见，利用单一分子标记进行 MAS 时，分子标记与目的基因的遗传距离在 1.0 cM 以内更为可靠。

从试验成本和技术的可行性来看，应首选基于 PCR 的分子标记开展 MAS。在分子标记辅助育种实践中，通常遇到与育种群体遗传背景有关的另一个问题是，目标基因附近的 PCR 分子标记在育种亲本之间没有多态性。本文报道的与 *Xa23* 基因紧密连锁的 EST 标记 C189 和 CP02662，则提供了更多的基于 PCR 的分子标记选择机会，可促进利用 *Xa23* 基因的水稻分子育种工作。

原载：中国农业科学，2005，38（10）：1996～2001.

水稻抗白叶枯病基因 *Xa23* 的 RFLP 标记定位及其 STS 标记的转化

樊颖伦[1,2]　陈学伟[3]　王春莲[1,2]　朱立煌[3]　章　琦[1]　赵开军[1,2,4]

（1. 中国农业科学院作物科学研究所，农业部作物遗传育种重点实验室，北京；
2. 农作物基因资源与基因改良国家重大科学工程，北京；
3. 中国科学院遗传与发育生物学研究所，北京；4. IRRI 中国办事处，北京）

Mapping the Rice Bacterial Blight Resistance Gene *Xa23* with RFLP Markers and Converting RFLP to STS Marker

Abstract　A F_2 population derived from cross of the *Xa23* near-isogenic lines，JG30 and CBB23，was used for fine mapping of the rice bacterial blight resistance gene *Xa23*. Inoculation of the F_2 plants with a specific resistance *Xanthomonas oryzae* pv. *oryzae* strain P6 showed 3 ∶ 1 ratio for the segregation of resistant and susceptible as expected. One hundred and forty-five susceptible plants were selected from the F_2 population for mapping gene *Xa23* and fifteen restriction fragment length polymorphism (RFLP) markers anchored on the rice high-density map. A RFLP marker C1003A was found to be linked to the gene *Xa23* by 0. 4cM. This result constitutes an important basis for map-based cloning of the gene *Xa23*. The RFLP marker C1003A was transformed into sequence-tagged-site (STS) marker，which can be applied for molecular marker-assisted selection (MAS) of *Xa23* in rice breeding programs.

Key words　*Xa23*；rice bacterial blight；RFLP markers；STS marker

摘要　用水稻抗白叶枯病基因 *Xa23* 的近等基因系 CBB23 与其感病轮回亲本金刚 30（JG30）杂交，构建了包含 2 562 个单株的 F_2 作图群体。用水稻白叶枯病专化菌株 P6 进行抗性鉴定表明，F_2 植株抗感分离比严格符合 3∶1。根据日本水稻基因组计划 RGP 水稻高密度图谱上的 RFLP 探针对 F_2 群体中的 145 个感病单株进行 RFLP 检测和连锁分析，获得 6 个与 *Xa23* 紧密连锁的 RFLP 分子标记。其中 RFLP 标记 C1003A 靠向丝粒一侧，与 *Xa23* 的遗传图距为 0. 4 cM，为 *Xa23* 的图位克隆提供了重要基础。并将标记 C1003A 成功地转化为 STS 标记，为分子标记辅助选择育种（MAS）提供了有效的分子标记。

关键词　*Xa23*；水稻白叶枯病；RFLP 标记；STS 标记

水稻白叶枯病是由 *Xanthomonas oryzae* pv. *oryzae*，（*Xoo*）引起的一种细菌性维管束病害。是世界水稻生产中严重的病害之一，可造成水稻减产 10%～30%，严重时甚至绝收（Mew，1987）。利用品种的抗性是控制其最经济有效的方法。发掘新的抗水白叶枯病基因一直是水稻抗病育种的热点之一。20 世纪 60 年代以来，国际上约鉴定出水稻抗白叶枯病基因 30 个（Ogawa，1993；Lee et al.，2003；Chen et al.，2003；章琦，2005）。长期以来广泛种植的多数抗白叶枯病的水稻品种携有 *Xa4* 或 *Xa3* 基因，有些品种在国内外许多稻区已经丧失抗性（章琦，1995；Zhang et al.，2002）。章琦等从中国普通野生稻（*Oryzae rufipogon*）中鉴定发掘的 *Xa23* 基因具有对国内外鉴别菌系都表现高度抗病、完全显性和全生育期抗病等优异特点，对水稻的抗性遗传改良具有广阔的应用前景（章琦等，2000；Zhang et al.，2001）；*Xa23* 也已被转育到栽培稻金刚 30（JG30）中，育成了近等基因系 CBB23（Zhang et al.，2002）。为了克隆 *Xa23* 基因和促进该基因在分子标记辅助育种上的利用，前期已用 CBB23 和 JG30 构建了含 2 562 个单株的 F_2作图群体，并将 *Xa23* 定位在水稻第 11 染色体长臂上的 SSR 标记 RM206 和 RAPD 标记 RpdH5 之间。遗传图距分别为 1.9 cM 和 7.0 cM（Pan et al.，2001）。本研究采用 RFLP 分子探针寻找与 *Xa23* 基因更紧密连锁的分子标记，并将其转化为 STS 标记，以便用于 *Xa23* 基因的图位克隆和分子标记辅助育种。

1 材料和方法

1.1 水稻材料

水稻近等基因系 CBB23 和轮回亲本 JG30 及其 F_2代的分离群体

1.2 抗白叶枯病鉴定

用与 *Xa23* 对应的国际白叶枯病鉴别菌株 PXO99（P6），对 F_2群体及其亲本于苗期和分蘖期分别进行抗性鉴定。采用人工剪叶接种法，菌株 P6 引自国际水稻研究所（IRRI），由本实验室保存。

1.3 杂交探针制备与同位素标记

根据 Harushima 等（1998）建立的水稻遗传图谱，选取位于第 11 染色体长臂 G257 至 G1465 区间的 15 个 RFLP 分子标记进行 JG30 与 CBB23 之间的 Southern 多态性分析，探针分别为 C6、C1003A、C10775SA、C11521S、C1172、E61360S、G257、G1465、G4001、R2458、R3341、S723、S10928、S11988 和 S16073，由日本 MAFF DNA Bank 提供质粒 DNA。将质粒 DNA 进行热激转化大肠杆菌 DH5α，保存菌株备用。应用的探针载体为 pBluescriptⅡSK+，碱裂解法提取质粒 DNA，根据 MAFF DNA Bank 提供的数据，通过酶切来检测这些重组质粒是否含有预期外源片段。鉴于该载体 DNA 片段较小、载体序列与水稻基因组无同源性，本研究便直接使用这些重组质粒进行探针标记。所有探针都采用^{32}P-dCTP（Amersham，UK）随机标记（Harushima et al.，1989）。

1.4 Southern 杂交

根据分蘖期接种后调查的抗性结果，对 JG30/CBB23 组合的 F_2全部感病植株、部分抗病植株及两个亲本的各个单株，参考 McCouch 等的酚氯仿法提取叶片总基因组 DNA (McCouch et al.，1988)。按 Sambrook 等（1989）的 Southern 杂交操作方法，进行了部分改进。分别用 *Bam*HⅠ、*Bgl*Ⅱ、*Dra*Ⅰ、*Eco*RⅠ、*Eco*RⅤ、*Hind*Ⅲ、*Sca*Ⅰ和 *Xba*Ⅰ等限制性内切酶，对上述植株样品基因组 DNA 各 5μg 进行完全酶切。琼脂糖凝胶电泳（0.8% 凝胶，电压 1～2 V/cm）后，用 0.25 mol/ L HCl 溶液进行脱嘌呤处理凝胶 10min，用 0.4 mol/L NaOH 碱法吸印，将 DNA 转移到尼龙膜（Hybond - N+，Amersham，UK)。吸印约 16 h，用 2×SSC（0.3 mol/L NaCl，0.03 mol/L$C_6H_5Na_3O_7$）洗去碱液并固定 DNA 分子，分别制备亲本 JG30 和 CBB23 的亲本杂交膜和 F_2感病株及部分抗病株杂交膜。将杂交膜置于盛有磷酸杂交液［0.36 mol/L Na_2HPO_4、0.1mol/L NaH_2PO_4（$pH_{7.2}$)、1% BSA、1mmol/L EDTA、7% SDS 和 100 mg/mL 变性鱼精 DNA］的杂交管中，5℃预杂交 3～4 h，将标记好的探针碱变性后加入预杂交液中 65℃杂交过夜，弃杂交液，于 65℃下分别用含有 0.1% SDS 的 1×SSC、0.5×SSC 和 0.25×SSC 3 个梯度洗液，各洗 20min（根据杂交信号的强弱，来确定洗膜时间，故在洗膜中，测杂交信号的强弱）。用滤纸吸干尼龙膜水分，用保鲜膜包好，置于 X 光暗盒中，在其上放置 X 光片，在－80℃下，曝光 5～10 d（据杂交信号的强弱来确定曝光时间）。在暗室依次用显影液、定影液冲洗 X 光片直至出现清晰的杂交信号带。

1.5 STS - PCR 分析

探针 C1003A 是从日本晴的 cDNA 片段克隆到 pBluescriptⅡSK＋载体上的重组质粒，使用 pBluescriptⅡSK＋载体上的通用引物 T7 和 T3 引物进行 C1003A 测序，根据 C1003A 序列用 Primer3（http：//frodo. Wi. Mit. edu/cgi-bin/primer3/primer3—www. cgi）在线引物设计软件设计 2 对特异 PCR 引物。PCR 反应体积 25μL，含 1×PCR 缓冲液（10mmol/L Tris - HCl，pH8.3，50mmol/L KCl，0.1% Triton - 100)、2.5mmol/L $MgCl_2$、0.25mmol/L dNTP，每个引物 50ng、200ng 基因组 DNA、1 U *Taq* 酶。在 1.5%琼脂糖凝胶（含溴化乙锭 EtBr）中电泳分离 PCR 反应产物，紫外灯下扫描照相。

1.6 遗传作图

用 MAPMAKER3.0 软件对分离群体单株的白叶枯病抗性和分子标记的分离数据进行连锁分析。并用 Kosambi 函数将重组率转化为遗传图距（cM)。

2 结果与分析

2.1 亲本间 RFLP 多态性分析

碱裂解法提取质粒 DNA 后限制性内切酶消化重组质粒，发现 15 个 RFLP 探针中只有 S723 没有外源插入片段。探针 DNA 分子进行同位素标记后进行亲本间多态性分析，Southern 杂交结果表明 C1003A、C10775SA、C1172、C11521S、R2458 和 S10928 等 6 个标记在亲本之间存在多态性（图 1)，多态性的检出率为 42.8%。

2.2 F_2代群体 RFLP 分析及 *Xa23* 遗传定位

利用 JG30 和携有 *Xa23* 的近等基因系 CBB23 进行杂交构建了共 2 562 个单株的 F_2作图

群体，通过水稻白叶枯病菌株 P6 进行白叶枯病抗性鉴定，后代植株的抗感反应非常清晰，抗病植株为 1 928 株，感病植株为 634 株，抗、感植株的分离比例严格符合 3∶1 的抗感分离比（$\chi^2=0.075$，$0.75<P<0.90$），表明此群体非常适合 *Xa23* 基因的分子标记定位。由于感病植株为隐性纯合体，在分子标记检测中，感病植株的带型可真实地反映分子标记与 *Xa23* 基因间的遗传交换情况而选取感病植株进行分子定位。根据前期定位结果（Pan et al.，2003），本试验选取了 RAPD 标记 RpdH5 和 SSR 标记 OSR6 之间的 145 个 F_2感病交换单株用于 Southern 分析。将以上 6 个在亲本之间存在多态性的 RFLP 标记进行 F_2感病植株 Southern 杂交，统计结果表明，RFLP 标记 R2458 和 S10928 与 OSR6、RM206 标记均位于 *Xa23* 的近端粒一侧；RFLP 标记 C1003A、C1172、C10775SA 和 C11521S 在 *Xa23* 的另一侧。其中探针 C1003A 标记在所测 F_2代的感病植株中，共有 5 个交换单株。C1003A 在 F_2代部分感病植株中的分离情况见图 2。用 Mapmaker 3.0 软件进行分析表明，将 *Xa23* 定位在 RM206 和 C1003A 之间，其中 *Xa23* 距 C1003A 的遗传图距为 0.4 cM（图 3）。

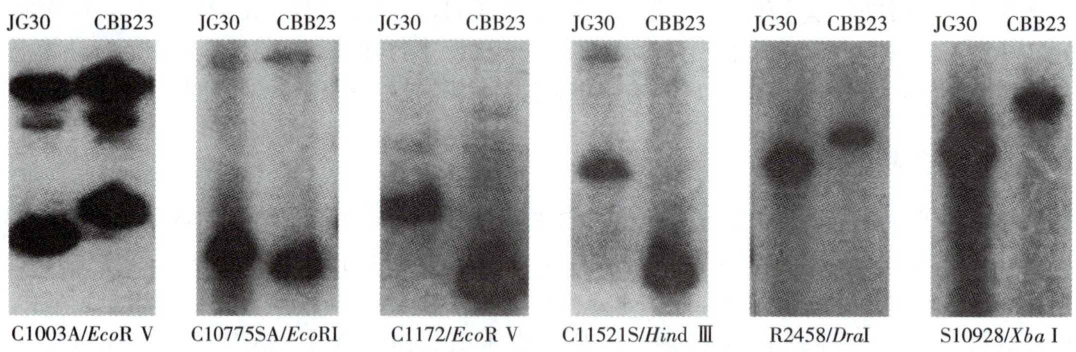

图 1　6 个 RFLP 探针/限制性内切酶组合揭示亲本（JG30，CBB23）间多态性

Fig. 1　RFLPs between the near iso-genic lines JG30 and CBB23, revealed by six probe/enzyme combinations

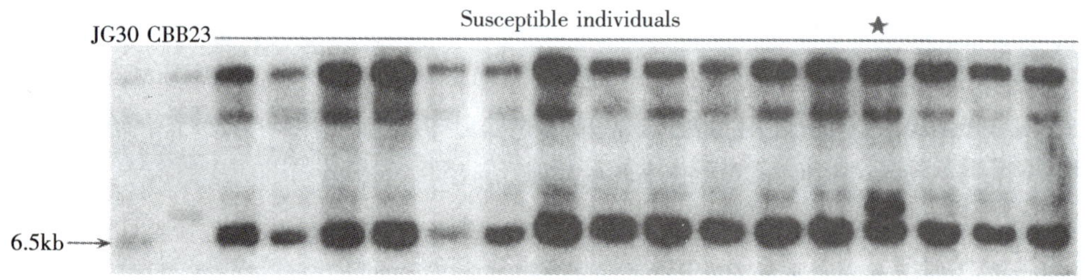

图 2　RFLP 探针 C1003 A 在 F_2部分感病植株中的分离情况

Fig. 2　RFLP patterns of the near isogenic lines and F_2 susceptible individuals revealed by RFLP marker C1003 A

★：Individual with an exchange event between RFLP marker C1003 A and *Xa23* gene

2.3　RFLP 标记 C1003 A 转化为 STS 标记

在 F_2代群体中共有 634 株感病植株，如果全部进行 RFLP 分析，不但费用高、操作繁琐，还要用放射性同位素标记技术，因此将 RFLP 标记转化为 STS 标记，以便提供稳定、可靠、简便的分子标记用于 *Xa23* 基因转育过程中的辅助选择。

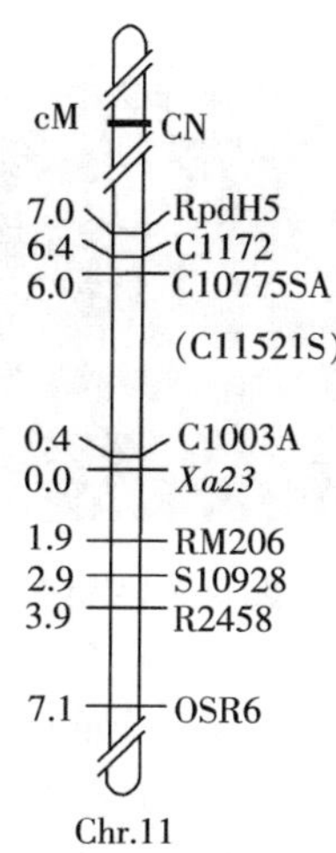

图 3 *Xa23* 基因在水稻第 11 染色体上的连锁遗传图谱

Fig. 3 Genetic linkage map of *Xa23* gene on rice chromosome 11

2.3.1 STS－PCR 反应体系的建立 根据 C1003 A 序列设计了 2 对特异 PCR 引物，其序列为：03STS1F：5′－CTCGGTTT－CCGTCTTCTCAG－3′；03STS1R：5′－GACTTTGCGTGCTTTCCAGC－3′；03STS2F：5′－GCGTCCGCCTTGATACAT－3′；03STS2R：5′CTACCCAAACCCAGGCATACC－3′。以 JG30 和 CBB23 基因组 DNA 为模板进行了亲本间多态性分析，引物 03STS1F/R 在亲本间表现出多态性（图 4）。引物 03STS2F/R 在亲本间未表现出多态性（结果未显示）。由于探针 C1003 A 来自日本晴（Nipponbare）的部分 cDNA 序列，在 PCR 扩增中以日本晴 DNA 作为对照（图 4）。引物 03STS1F/R 之间的 C1003 A 序列片段长度为 473 bp，因扩增片段中存在内含子，因此在 Nipponbare、JG30、CBB23 中实际扩增出的多态性片段长度均比预期片段大。为了减少引物 03STS1F/R 的非特异性扩增，将退火温度提高为 62℃。PCR 反应程序为 94℃预变性 3min，然后 94℃变性 50 s，62℃退火复性 50 s，72℃延伸 1min 20 s，共循环 35 次，最后 72℃保温 6min。

2.3.2 引物 03STS1F/R 在 F_2 群体中 STS－PCR 分析与 RFLP 结果验证 为了进一步验证该 STS 标记是否转化成功，用引物 03STS1F/R 对上述 145 个 F_2 感病植株进行 PCR 扩增。结果表明，STS－PCR 标记的交换单株与 RFLP 标记 C1003A 的交换单株完全相符。图 4 中 F_2 部分感病植株与图 2 中 F_2 部分感病植株编号和顺序相同。证实了本试验中由 RFLP 标记 C1003 A 转化为 STS 标记是可靠的。

3 讨论

RFLP 标记的多态性来源于 DNA 序列的缺失、插入与限制酶切位点的改变。其优点是标记数量大，等位基因呈共显性，可以区分纯合与杂合基因型。但工作量大、费用高，而 PCR 标记技术操作简便、快速、费用低。对 RFLP 标记使用的探针进行测序，然后根据序列信息设计 PCR 引物，对基因组 DNA 进行特异扩增，有可能将 RFLP 标记转化为 STS 标

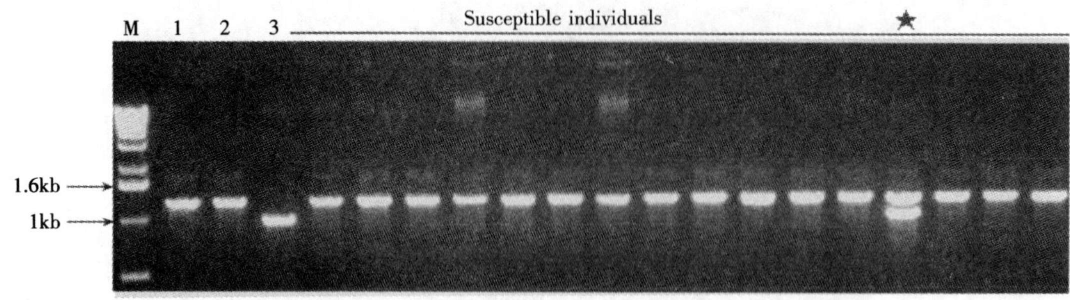

图 4　引物 03STS1F/R 在亲本及 F_2 部分感病植株中的扩增结果

Fig. 4　STS - PCR patterns of the near isogenic lines and F_2 susceptible individuals revealed by 03STS1F/R
M：1 kb DNA Ladder（Invitrogen）；1：Nipponbare；2：JG30；3：CBB23；
★：individual with an exchange event between the PCR primers 03STS1F/R and *Xa23* gene.

记。然而并非所有 RFLP 标记都能转化为相应的 STS 标记，在转化时需注意以下几点，1）STS - PCR 引物的扩增片段长度应尽量包括 RFLP 探针的全长，可增加发现多态性的机会；2）如果扩增片段大小在抗、感材料之间没有差异，则可分析两亲本材料扩增片段的 DNA 序列在个别核苷酸序列上的差异，根据双亲材料之间酶切位点的差异进行酶切，找到多态性或者在此序列基础上再次设计引物进行第二轮 PCR 来找多态性，比如单核苷酸多态性（Single Nuclear Polymorphism）；3）如有必要，可用分辨率较高的聚丙烯酰胺变性胶来区分扩增产物间的细微差异。携 *Xa23* 基因的材料已提供了 30 多个育种单位进行培育广谱抗性水稻品种。STS 标记的开发，为育种家提供了方便快捷的分子标记辅助选择手段，可满足田间育种大量筛选的需求。

本研究旨在寻找与 *Xa23* 基因更加紧密连锁的分子标记，用于 *Xa23* 基因的图位克隆。图位克隆是基于遗传作图和物理作图的一种克隆分离基因的有效方法。在水稻中已经克隆的 4 个抗白叶枯病基因 *Xa1*、*xa5*、*Xa21* 和 *Xa26* 均通过图位克隆战略完成（Song et al.，1995；Yoshimura et al.，1998；Sun et al.，2004；Lyer et al.，2004）。进行图位克隆首先要有一个根据目的基因构建的遗传分离群体，然后找到与目的基因紧密连锁的分子标记，建立精密的遗传图谱。这样，就可以用紧密连锁的分子标记去筛选基因组文库，构建物理图谱和遴选目的基因。遗传图谱的精密程度和准确性与作图群体的大小和表型鉴定的准确性有关，并取决于揭示双亲间多态性分子标记与目标基因的连锁程度。在 *Xa23* 基因前期定位上，近着丝粒一端只有 RAPD 标记 RpdH5，并且遗传距离有 7.0 cM（图 3），我们在 RpdH5 和 *Xa23* 位点之间尝试用 SSR、AFLP 和 RAPD 标记都没有找到距 *Xa23* 位点更近的标记。本研究利用高密度 RFLP 图谱上的分子标记进行 RFLP 分析，获得了一个离着丝粒端更为紧密连锁的标记 C1003A，与 *Xa23* 基因的遗传图距只有 0.4cM，这样就将 *Xa23* 定位在标记 RM206 和 C1003A 之间。按水稻基因每厘摩的物理距离 300kb 计算，C1003A 距 *Xa23* 仅有 120kb 左右，这无疑为 *Xa23* 基因的克隆奠定了重要基础。

本研究中使用的分子探针由日本 MAFFDNA Bank 提供，特此致谢。

原载：作物学报，2006，32（6）：931～935.

Ⅳ 水稻白叶枯病病理、群体遗传结构研究

Part 4 Studies on Phytopathology and Population Gonetic Stracture of *Xoo* Pathogen

应用噬菌体法估测水稻叶片组织中白叶枯菌（*Xanthomonas campestris* pv. *oryzae*）数量研究

章 琦[1] T. W. Mew[2]

（1. 中国农业科学院作物育种栽培研究所；2. 国际水稻研究所）

The Bacteriophage Technique for Estimating *Xanthomonas campestris* pv. *oryzae* in Leaf Tissues

Zhang Qi[1] T. W. Mew[2]

(1. CAAS 2. IRRI)

Abstract The bacteriophage technique was tested to enumerate the bacterial multiplication in leaf tissues of rice. In the absence of a selective medium, the method proved to be efficient and reliable to establish the bacterial number in leaves of different positions of Zenith after infection. A good correlation between phage count and viable count (direct isolation) by the dilution plating was also demonstrated.

Key words bacteriophage technique; *Xanthomonas campestris* pv. *oryzae*

摘要 本试验应用噬菌体法估测水稻叶片中白叶枯病菌的增殖。试验表明溶菌斑数同直接皿测稀释菌液存活细菌数之间表现高度相关。成株抗性品种 Zenith 接种后不同时间检测其稻株各个叶位的叶片组织中的菌量与其病斑的发展趋势是一致的。在目前分离生长较为缓慢的白叶枯病菌缺乏选择性培养基的情况下，这一方法是有

效和可信的。

关键词 噬菌体技术；稻白叶枯病菌

Bacterial blight, caused by *Xanthomonas campestris* pv. *oryzae* (Ishiyama, 1922) Dye, is one of the serious rice diseases in Asia. Current commercial cultivars of rice now in use in China are susceptible to the disease. Severe infection leads to yield losses and poor quality of grain in many of the principal rice growing areas. Until now as no efficacious chemicals are available, planting resistant cultivars remain the most effective and economical control measure of the disease. Research effort has therefore been directed to evaluate rice germplasm and to identify better and diverse resistance sources useful in the varietal improvement program (Mew et al., 1981).

Early studies demonstrated the adult-plant resistance of rice to bacterial blight, partial characterization of adult-plant resistance of cultivars such as Zenith and Malagkit Sungsong was also reported (Zhang et al., 1984; 1985). It was also noted that susceptibility of plant tissue to bacterial diseases was correlated to bacterial multiplication in host tissues (Chand et al., 1963). For instance, Chamberlain (1962) found that *X. phaseoli* var. *sojensis* multiplied in the leaves of resistant var. CNS of soybeans but at a lower rate than in leaves of susceptible var. Lincoln. Whether *X. c.* pv. *oryzae* multiplies in host tissues following a similar manner is not clear. Furthermore, leaves at different positions of cultivar Zenith showing adult-plant resistance respond differently to infection, but whether the bacteria also increase in different rate in the different leaf positions has not been investigated.

Direct plating of *X. c.* pv. *oryzae* from plant tissues on general laboratory media, such as peptone sucrose agar, was always hampered with contamination. This was due to the slow-growing character of the genus. Alternative method was needed. The bacteriophage technique was therefore studied for its efficiency and reliability in estimation and enumeration of the bacteria in plant tissues.

This paper describes the method for estimating multiplication of *X. c.* pv. *oryzae* in infected leaf tissue of 4 leaf positions of rice cultivar Zenith which possesses adult-plant resistance by the bacteriophage of the bacterial pathogen.

Materials and Methods

Phage strains and host bacteria The experiments were conducted in the greenhouse and the laboratory at IRRI from June 1983 to May 1984. Initially, host range of 125 phage isolates from the Philippines was tested against the bacterial strains. As a result, based on specific phage-bacterium interactions, phage strain P80 and bacterial strain PXO61, race 1 of *X. c.* pv. *oryzae* were chosen. A one-step growth, experiment was done to further characterize the phage-bacterium relationship. The latent period (about 56 min.) was obtained from the experiment (Di et al., 1982). For reference, in subsequent studies titration of PXO61 in suspension of various densities with a standard phage concentration of

P80 was established by serial dilution (Fig. 1).

Test cultivar and plant management Zenith carrying *Xa6* gene for resistance to bacterial blight was used in the experiments. Earlier study indicated that 6th, 9th and 10th leaf positions were susceptible to infection while the 12th was resistant. Seeds were sown at different time intervals in the seedboxes. Transplanting of the 12 - day - old seedlings to 6 - inch diameter clay pots at 2 seedlings/pot was staggered to synchronize inoculation. The soil was fertilized with N - P - K at 9 - 6 - 6 g/m^3. The leaf positions of the main culm were marked in order from the 2nd to the flag leaf.

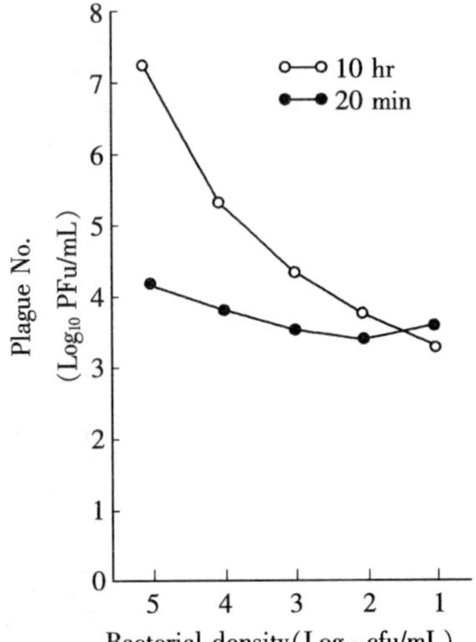

Fig. 1 The titration of *X. c.* pv. *oryzae* strain PX061 with a standard phage concentration (ca. 10^6) of P80.

Inoculation and leaf sampling Inoculum of indicator strain was prepared from 72 - hr - old culture of PXO61 of *X. c.* pv. *oryzae* grown on peptone sucrose agar slants, scraped and suspended into sterile distilled water and adjusted to a concentration of ca. 10^9 cells/ml. Inoculation was done by leaf clipping method. Sixty leaves of each leaf position were inoculated. Six 5 - cm long specimens from each of the 4 leaf positions were sampled for tissue maceration and bacterial estimation at 1, 24, 48, 72, 120, and 168 h after inoculation when lesion development was measured (Fig. 2). The final disease measurement was done 14 days after inoculation.

Leaf extraction The leaf samples were disinfected with 70% ethanol for few seconds and rinsed several times in sterile distilled water. The leaf specimens were macerated in a tissue homogenizer and brought to a final volume of 5 ml by filter sterilized PSB (Peptone Sucrose Broth).

Bacterial enumeration from infected leaves Bacterial number was estimated by both direct plating of leaf extracts on PSA plates and by the bacteriophage method.

Direct plating For viability count, the extract was serially diluted in PSB medium and 0.1 ml of the suspension was surface seeded on PSA. Counts were made 3 to 4 days after incubation at 30℃.

Phage method Fresh phage suspension of P80 was prepared by culturing the phage in PXO61 bacteria in Cavfch medium at 27℃ by aeration for 20 - 24 h. After centrifugation at 6000 rpm for 10 minutes, the supernatant was serially diluted to determine the suitable phage concentration for bacterial titration. The leaf extracts were mixed with standard phage concentration at ca. 10^6/ mL. Ten minutes after incubation by aeration at 27℃, the mixtures were centrifuged (6000 rpm) for 10 minutes to remove the excess phage (free phage). The procedure was repeated once. The precipitates were then resuspended with the same volume of Cavfch medium and incubated for 20 min. to determine the initial phages being adsorbed to

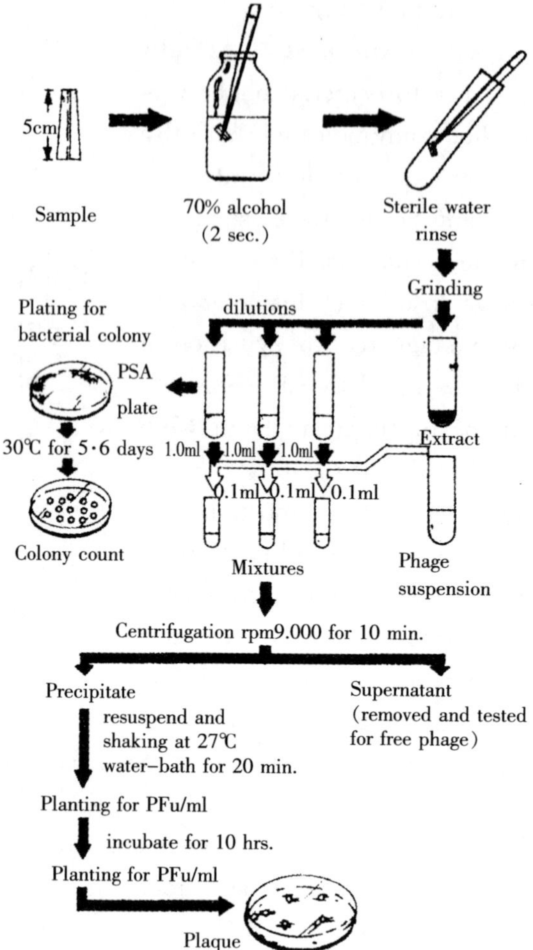

Fig. 2 Scheme for estimating bacterial multiplication In rice leaves by the bacteriophage method

the bacterial cells. Final phage count to indicate the relative increase in number of the bacteria was done 10 h later by which time, from a preliminary test, all the bacterial cells in the leaf extract were lysed and phage released.

Results

Detection of the bacteria In pure culture, the titration curve of strain PXO61 with a standard phage concentration of P80, indicated that the method was efficient when the bacterial suspension was at ca. 10^3 cfu/mL (Fig. 1). To apply it as a base-line reference, the bacterial cells in the leaf extracts, from 24 h to 168 h after inoculation, appeared to be greater than 10^3 cfu/mL by phage count (Table 1). When the phage count was compared to direct isolation of the bacteria by dilution plating from leaf extracts, a positive correlation was obtained (Fig. 3). When the lesion progressed, the phage count also increased. The phage plating where plaques were observed fell within 12 h while the viable count for bacterial count by dilution plating took 48 h to 72 h.

Table 1 Multiplication of *X. c.* pv. *oryzae* strain PXO61 in different leaf Positions of Zenith and enumerated by the bacteriophage technique

Time after inoculation (hr)	log pfu/cm² leaves[a]			
	6th	9th	10th	12th
1	5.6±0.01	5.64 ±0.02	5.7±0.01	5.7±0
24	5.73 ±0.02	5.7±0.01	5.92±0.03	5.7±0.01
48	7.53 ± 0	7.5±0	7.5± 0	6.3±0.01
72	8.2±0.05	8.0±0	7.6±0.01	7.3±0.02
120	7.4±0.02	7.3±0.01	7.4±0	6.4±0.08
168	7.1±0.02	6.9±0.02	7.0±0	6.2±0.02

[a] Based on plaque formation by the phage strain P80 with PXO61.

The results also showed that lesion progressed faster in leaves at the 6th, 9th and 10th position. At the 12th leaf position, the lesion did not develop more than 3 cm 14 days after inoculation (Fig. 4) .

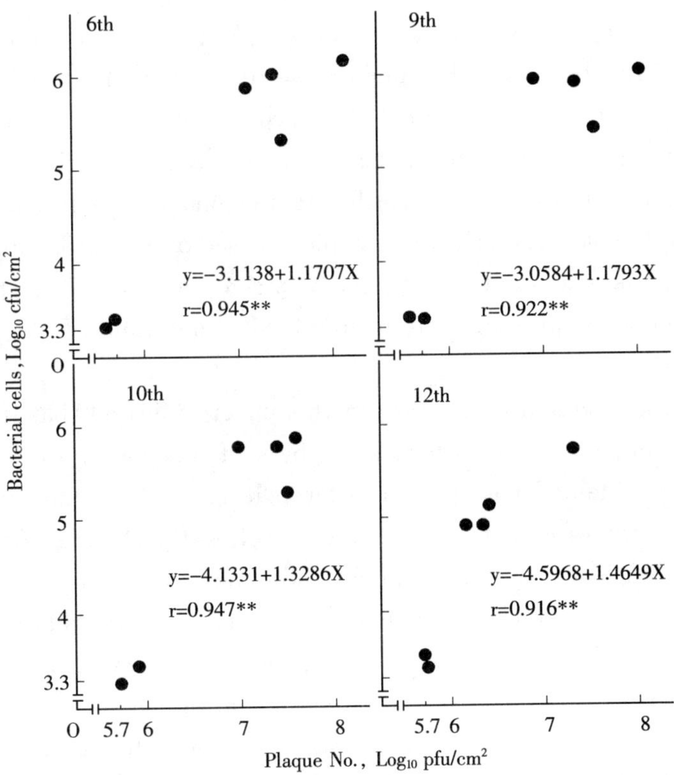

Fig. 3 Relationship between viable count (cfu/cm², and of *X. C.* pv. *oryzae*, PXO61 and its plaque P80 (pfu/ cm²) of leaf specimens with bacterial blight lesion at different leaf positions of Zenith.

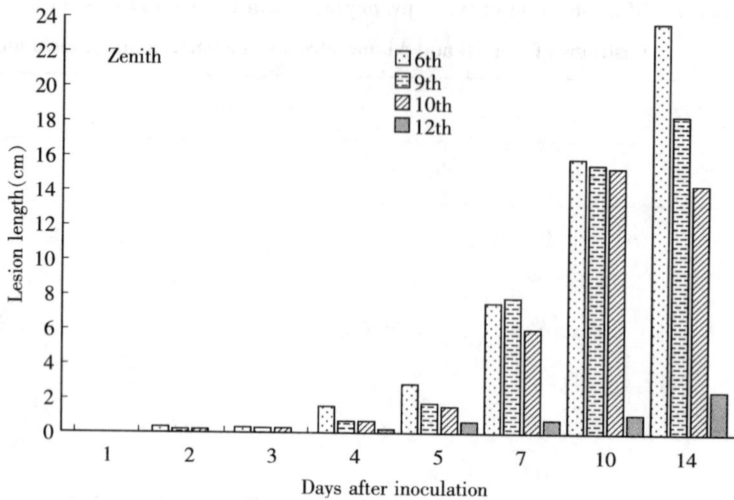

Fig. 4 The lesion development of bacterial blight in different leaf positions of Zenith. in the greenhouse

Discussion

Bacteriophage of *X. c.* pv. *oryzae* is known to be highly specific to its host bacterium (Goto, 1965). When suspension of a compatible phage strain is added to the bacterial suspension, phages are adsorbed and in due course the infected bacterial cells are lysed and new phage progenies released (Adam). Unlysed bacteria will be infected again. The infectious cycles thus continue and the phage number increases in each cycle by a factor equal to the burst size until all bacteria susceptible to the phages are lysed. The bacteriophage method of detecting the bacterial cells in a sample is based on the initial and final phage count of an experiment. It takes accounts of the initial phage particles added to the sample in a system where bacteria are present and the burst size and time of the phage in its host bacteria.

The bacteriophage technique described in this paper indicated that it is a sensitive and reliable method to enumerate the bacterial numbers of *X. c.* pv. *oryzae* in rice leaves. The information could be obtained within 12 h after plating while it took more than 72 h to complete the direct plating for bacteria. The rate as well as the chances of contamination were thus reduced. The efficiency of the method, however was around 10^3 cfu/mL of leaf extract. The method may be improved further for its reliability by addition of ferric or ferrous sulfate solution to neutralize the free phage being added to the system (Di et al., 1982).

Our results also indicated that *X. c.* pv. *oryzae* multiplied differently in different leaf positions of Zenith which has adult-plant resistance. The rate of bacterial multiplication apparently relates to the resistance of the cultivar expressed in different leaf positions conditioning Zenith's adult-plant resistance as shown earlier (Zhang et al., 1984).

原载：中国水稻科学，1987，1 (3)：177～183.

中国水稻白叶枯病菌致病型的研究

方中达[1]　许志刚[1]　过崇俭[2]　殷尚智[2]
伍尚忠[3]　徐羡明[3]　章　琦[4]　王春莲[4]

（1. 南京农业大学；2. 江苏省农业科学院；
3. 广东省农业科学院；4. 中国农业科学院）

Studies on Pathotypes of *Xanthomonas campestris* pv. *oryzae* in China

Abstract Eight hundred thirty five isolates of *Xanthomonas campestris* pv. *oryzae* Dye have been collected from different rice-cropping regions in China and tested during the last 4 years. The clipping method was adopted for inoculating at the booting stage. Seven pathotypes are divided in according to their disease reaction on five basic differentials，IR26，Java 14，Nangeng 15，Tetep and Jingang 30. Their reaction models in the order are：Ⅰ，RRRRR；Ⅱ，RSSSS；Ⅲ，RRSSS；Ⅳ，RSSSS；Ⅴ，SRRSS；Ⅵ，RRSRS；Ⅶ，RSSRS. The "R" means resistant with lesion length（ or area）less than 25%（or 20%）of inoculated leaf. The "S" means susceptible reaction and the lesion Length（or area）is longer（or larger）than 25%（or 20%）of inoculated leaf. In China，pathotype Ⅱ and Ⅰ of the pathogen are epidemical forms in Northern japonica rice growing regions，pathotype Ⅳ is dominant one in Southern indica rice growing area，while pathotype Ⅱ to Ⅳ are prevailing in rice fields along the Yangtze River Valley；pathotype Ⅴ occurs sporadically in southern provinces，such as Guangdong and Fujian.

摘要　在1975—1980年初步测定的基础上，近4年来先后从国内各病区又采集了835个菌株，分别在北京、南京、扬州和广州等地，在30个鉴别品种上接种测试其致病性差异，根据在5个最基本的鉴别品种上的反应特征，可将供试的菌株区分为7个致病型。它们在IR26，Java14，南粳15，Tetep和金刚30上成株期的反应模式分别为（Ⅰ）：RRRRS；（Ⅱ）：RRRSS；（Ⅲ）：RRSSS；（Ⅳ）：RSSSS；（Ⅴ）：SRRSS；（Ⅵ）：RRSRS；（Ⅶ）RSSRS。在北方粳稻区的菌株，多属Ⅱ型和Ⅰ型，在南方籼稻区的菌株以Ⅳ型为最多，还有少量Ⅴ型菌存在，在长江流域籼粳混栽稻区的菌株，则以Ⅰ、Ⅳ型为多。

有关水稻白叶枯病菌致病力差异的研究最早是由日本的 Kuhara（久原重松）在 1958 年开始的。根据 Kozaka（1969）有关病菌致病力分化的研究，日本的白叶枯病菌可以区分为三群，Ezuka 和 Horino（1974）在增加了鉴别品种后进一步分为五群。Reddy 和 Ou（1976）Mew（1977）从 1975 年开始测定菲律宾白叶枯病菌的致病性分化，最初分为 4 个毒性群，Mew 后来发现一个具有很好鉴别性能的品种 CAS209，能将第一群菌再区分为两群，并称为“生理小种（race）”，近年报道共有 6 个生理小种（Mew，1987；Vera Cruz et al.，1986）。

在 Mew、伍尚忠（1985）和 Horino（1982）进行联合试验以后，认为在不同国家或地区之间病菌的致病类型不同，不同国家所用的鉴别品种只适用于本地区，如日本的鉴别体系不能区分菲律宾的小种群，反之亦然。

为了探明中国稻区病菌的致病力分化，南京农业大学、江苏省农科院和广东省农科院从 1975 年开始进行研究。南京农业大学和江苏省农科院先后在 30 个鉴别品种上测试了 600 多个菌株。广东省农科院在 43 个鉴别品种上测试了 410 个菌株。在 1980 年前初步鉴定了国内病菌可区分为 4～5 个菌系群（方中达等，1981；许志刚等，1987；徐羡明等，1981）。云南、湖南、浙江等院校也相继研究了本省菌株的致病力分化问题。但是由于各单位采用的鉴别品种不一，接种方法、调查标准也不同，难以进行相互比较和了解我国白叶枯病菌株分化的全貌。

从 1985 年开始，南京农业大学、江苏省农科院、广东省农科院和中国农科院共同组成全国白叶枯病菌致病型研究组，开展进一步的研究。广东省农科院负责南方稻区病菌的致病型，南京农业大学和江苏省农科院测试长江流域稻区病菌的致病型，中国农科院负责北方稻区新菌株的采集和测试，共同制订了统一的试验方案，经过 3 年分片预选试验，于 1988 年在南京进行了联合鉴定。

1　材料和方法

1.1　供试菌株及其来源

在 1986—1987 年间，协作组分别在南京、扬州、广州和北京对长江流域稻区、南方稻区以及北方稻区的菌株的致病力分化进行了测试，在南京和扬州点测试的菌株共有 409 个，鉴别品种有 26 个；在北京点测试的菌株共有 146 个，鉴别品种为 23 个，在广州点测试的菌株有 280 个，鉴别品种为 20 个。

在上述测试的基础上，1988 年集中在南京进行联合试验，供试的菌株有 91 个，分别由广东省农业科学院、中国农业科学院、江苏省农业科学院和南京农业大学提供，详见表 1。

表 1　1988 年供试菌株的来源*

稻区	菌株来源
北方稻区：	黑龙江（2），吉林（3），辽宁（4），北京（2），天津（1），河北（4），河南（2），山东（1），陕西（2），宁夏（3）；共 24 个
长江流域：	安徽（2），江苏（8），上海（1），浙江（6），湖南（4），湖北（1），江西（3），四川（1）；共 26 个
南方稻区：	福建（5），贵州（4），广西（11），广东（4），云南（8）；共 32 个
其他稻区：	共 9 个

* 括号内为菌株数。

1.2　鉴别品种及田间设计

供试的鉴别品种为 6 个，即 IR26、Java14、南粳 15、Tetep、窄叶青和金刚 30。每个品种设两个试验小区，重复 2 次，每个品种每重复种植 100 行，每行栽插 10 穴，单本栽插，株行距为 30cm×30cm。5 月 7 日浸种育秧，6 月 11 日移栽，8 月 6 日接种（孕穗期），8 月 27 日记载。肥水管理水平中等，拔节期喷药治虫（稻飞虱和螟虫）2 次。

1.3　接种体的准备及接种方法

供试菌种大多数冻干保存，用前取出移植 2 代后备用，用无菌水稀释至 3×10^{8} 细胞/ml。采用剪叶法接种。每菌株在每一重复的品种上各接 10 穴稻株，约 30 片叶，配制的菌液在 2h 内接完。

1.4　记载及分级标准

①记载：接种后 21d，每处理测量 20 张叶片的病斑长度或目测病斑占接种叶面积的百分率，同时测量接种叶（20 片）的平均长度。

②分级标准：参照 1976 年试订的标准（方中达等，1981，1978；许志刚等，1987；徐羡明等，1981），略作适当修改，病情反应及代表值如下（表 2）：

表 2　病情分级标准及抗性类型的划分

病情数值	抗性类型		病斑大小或反应
0	抗（R）	HR	感剪口处无明显病斑
1		R	病斑纵向扩展的长度 2～3cm，或者仅有褐斑面积小于 10%
3		MR	病斑长度小于接种叶长的 1/4，或病斑面积少于接种面积的 20%
5	感（S）	MS	病斑长度达到 1/4，但小于 1/2，或病斑面积在 20%～49%之间
7		S	病斑长度达叶长 1/2，但小于 3/4，或病斑面积在 50%～74%之间
9		HS	病斑长度达 3/4，或面积大于 75%

2　结果及分析

2.1　鉴别品种

经过多年测试与选择，已确定了一套具有较好鉴别性能的品种，三个籼稻（IR26、Tetep、金刚 30）和两个粳稻（Java14、南粳 15）。它们的生育期在长江流域为 130d 左右，8 月上旬孕穗，9～10 月成熟，它们对中国的菌株有稳定的抗性反应。

（1）金刚 30（JG30）是一个高感类型的品种，除少数菌株外，接种的病斑长度都达叶长的 $\frac{1}{2}$ 左右。类似的品种还有沈农 1033、TN1、南京 11、珍珠矮 11 等。

（2）南粳 15（NG15）是我国粳稻型的较好抗源，也是多年来一直沿用的鉴别品种，它是能抵抗第Ⅰ、Ⅱ群菌的粳稻代表品种。

(3) IR26 除对个别南方菌株外，它都具有很强的抗性，生产上种植多年，抗性仍很稳定。

(4) Java 14 是原产印尼的高秆型品种，对绝大多数菌株都表现为抗性反应。

(5) Tetep 对大多数菌株均缺乏抗性，但对某些能侵染南粳 15 的菌株表现为抗性反应，因此具有较特殊的鉴别能力。

2.2 南方稻区病菌的致病型

广东省农科院从 1985 年以来在 IR26、Java14、南粳 15、金南风、Tetep、窄叶青和金刚 30 这 7 个鉴别品种上，测定了南方 5 省 280 个菌株的反应类型，可分为Ⅰ～Ⅶ个致病型，结果如表 3。

表 3　南方稻区白叶枯病菌的致病型（1985—1988 年）

致病型	金刚 30	窄叶青	Tetep	金南风	南粳 15	Java14	IR26	菌株数	%
0	R	R	R	R	R	R	R	14	5.0
ⅠA	S	R	R	R	R	R	R	26	9.3
ⅠB	S	S	R	R	R	R	R	18	6.4
ⅡA	S	S	S	R	R	R	R	29	10.3
ⅡB	S	S	S	S	R	R	R	60	21.4
Ⅲ	S	S	S	S	S	R	R	25	9.0
Ⅳ	S	S	S	S	S	S	R	102	36.4
Ⅴ	S	S	S	S	R	R	S	2	0.7
Ⅵ	S	S	R	S	S	R	R	1	0.4
Ⅶ	S	S	R	S	S	S	R	3	1.1
合计								280	

2.3 长江流域稻区病菌的致病型

南京农业大学和江苏省农科院在 1985—1987 年间，先后收集并测定了国内 24 省市稻区 409 个菌株，在 IR26、Java14、南粳 15、Tetep 和金刚 30 五个鉴别品种上接种测定，结果可分为Ⅰ～Ⅶ 7 个致病型，测试结果如表 4 所示。

表 4　长江流域病菌的致病型

（1985—1988 年，南京、扬州）

致病型	金刚	Tetep	南粳 15	Java14	IR26	菌株数	
						1986 年	1987 年
O	R	R	R	R	R	16	12
Ⅰ	S	R	R	R	R	29	29
Ⅱ	S	S	R	R	R	25	25
Ⅲ	S	S	S	R	R	9	9

（续）

致病型	金刚	Tetep	南粳 15	Java14	IR26	菌株数	
						1986 年	1987 年
Ⅳ	S	S	S	S	R	0	11
Ⅴ	S	S	R	R	S	2	10
Ⅵ	S	R	S	R	R	14	2
Ⅶ	S	R	S	S	R	45	2

2.4 北方稻区病菌的致病型

在 1985—1987 年间，中国农科院作物所收集和测定了北方 10 个省市稻区的 146 个菌株。根据在 6 个选择鉴别品种上测定筛选出来的 82 个菌株的反应型，可以区分为 7 型。其中优势群为Ⅱ和Ⅳ型，相当于全国统一鉴别寄主上的Ⅰ、Ⅱ致病型（表 5）。

表 5 1985—1987 年北方稻区白叶枯病菌在鉴别品种上的致病型

致病型	金南风	金刚 30	Tetep	南粳 15	Java14	IR26	菌株数	占总菌数（%）	相当于全国的致病型
O	R	R	R	R	R	R	6	7.3	0
Ⅰ	S	R	R	R	R	R	8	9.8	
Ⅱ	R	S	R	R	R	R	3	3.6	
Ⅲ	S	S	R	R	R	R	24	29.3	Ⅰ
Ⅳ	S	S	S	R	R	R	30	36.6	Ⅱ
Ⅴ	S	S	R	S	R	R	4	4.9	Ⅵ
Ⅵ	S	S	S	S	R	R	6	7.3	Ⅲ
Ⅶ	S	S	S	S	S	R	1	1.3	Ⅳ

2.5 中国各稻区病菌的致病型

根据前几年在各基点的分别测试和 1988 年在南京联合试验的结果，参试的菌株可分为七个致病型，详见表 6。

表 6 中国白叶枯病菌的致病型

致病型	在鉴别品种上的反应					供试菌株数（%）		
	金刚 30	Tetep	南粳 15	Java14	IR26	北京（87）	南京（88）	广州（87）
O	R	R	R	R	R	17	16	5.0
Ⅰ	S	R	R	R	R	32.9	3.5	15.7

（续）

致病型	在鉴别品种上的反应					供试菌株数（%）		
	金刚 30	Tetep	南粳 15	Java14	IR26	北京 (87)	南京 (88)	广州 (87)
Ⅱ	S	S	R	R	R	36	28	31.7
Ⅲ	S	S	S	R	R	7.8	(9)	9.0
Ⅳ	S	S	S	S	R	1.3	10	36.4
Ⅴ	S	S	R	R	S	0	1	0.7
Ⅵ	S	R	S	R	R	4.9	4.4	0.4
Ⅶ	S	R	S	S	R	0	3	1.1

2.6 病菌致病型的地理分布

从上述北方、南方和长江流域分别测试的情况结合以前历年测定的结果，可以看出致病型的分布有一定的地理特点，总体看来是长江流域以北以Ⅱ型和Ⅰ型为主，长江流域以Ⅱ型和Ⅳ型为多，而南方稻区则以Ⅳ型菌为最多，在广东和福建省还有少量的Ⅴ型菌（表 7）。

表 7　1986—1988 年间测试菌株致病型的地理分布

地点	O	Ⅰ	Ⅱ	Ⅲ	Ⅳ	Ⅴ	Ⅵ	Ⅶ	M_1
广东	6	15	50	29	58	I			150
广西	8	14	17	6	14		1	1	61
福建	2	4	13	4	3	1			27
贵州	3	9	15	3	8				38
云南	9	7	12	3	9				40
西藏		1							1
四川		2							2
湖南		1	2	2					5
湖北			1						1
江西		1	2						3
上海			1						1
浙江			1		1		1	1	4
江苏		5	5		2		1	1	14
安徽					2				2
河南		1	2	2					5
河北	1		6	1			1		9
陕西			1	1					2
宁夏	2	3	1	1					7
山东		1					1		2

（续）

地点	O	Ⅰ	Ⅱ	Ⅲ	Ⅳ	Ⅴ	Ⅵ	Ⅶ	M_1
天津		1		1					2
北京	2	1	4	1			1		9
辽宁	2	4	3				2		11
吉林		1	3	1	1				6
黑龙江		2	1	1			1		5
M_2	35	73	140	57	98	2	9	3	417

2.7 各致病型的代表菌株

经各点历年测试的结果，许多菌株的反应都较稳定，下列各菌株可作为各致病型的代表菌株（表8）。

表8 不同致病型的代表菌株

类型	代表菌株及来源		
	南方稻区	长江流域	北方稻区
Ⅰ	GD1329 粤	OS－200（云），JS97－2（泸）	HLJ85－72（黑龙江）
Ⅱ	GZ1098 黔	OS－40，KS－6－6（江苏）	HB84－17（河北）
Ⅲ	FJ856 闽	GX－274 JS158－2（浙）	NX85－42（宁夏）
Ⅳ	GX878 桂	OS－86（桂），浙173（浙江）	JL86－76（吉林）
Ⅴ	GD 粤 1358		
Ⅵ		OS－198（桂）	
Ⅶ		OS－225（浙），JS49－6（湘）	LN85－57（辽宁）

3 讨论

3.1 中国白叶枯病菌的致病型必须用适合的鉴别品种来鉴别

我国的白叶枯病菌致病型分化有着自身特点，它既不完全相同于菲律宾的6个小种，又与日本的类型有一定差别。根据在中国这套基本鉴别品种上的反应特征，可以分为7个型。它们在日本的鉴别品种和IRRI的品种上的反应如表9所示。

表9的反应也可看出，适于在IRRI鉴别生理小种的一些鉴别品种，对中国病菌的致病性有鉴别能力，如DV85、IR1545－339、CAS209等，而日本的一些品种，如金南风、黄玉，对我国病菌致病型的反应又不够稳定。这里选出的五个品种是基本的鉴别品种，各地（省）在进一步的测试时，还可根据各自的要求，适当增补一些当地认为合适的辅助品种。

表 9 供试菌株在三套鉴别品种上的反应

（1986，广州、南京、北京）

中国菌株致病型	中国					日本					IRRI				
	JG30	Tetep	NG15	Java14	IR26	金南风	黄玉	Tetep	早爱 3	Java14	IR8	IR20	IR1545	CAS209	DV85
0	R	R	R	R	R	R	R	R	R	R	R	R	R	R	R
Ⅰ	S	R	R	R	R	S/R	S/R	R	R	R	S/R	R	R	S	R
Ⅱ	S	S	R	R	R	S	S	S	R	R	S	R	R	S	R
Ⅲ	S	S	S	R	R	S	S	S	S	R	S	R	R	S	R
Ⅳ	S	S	S	S	R	S	S	S	S	S	S	R	R	S	R
Ⅴ	S	S	R	R	S	S	S	S	R	R	S	S	R	S	R
Ⅵ	S	R	S	R	R	S	S	R	S	R	S	R	R	S	R
Ⅶ	S	S	S	R	R	S	S	S	S	S	S	R	R	S	R

3.2 接种及记载标准

考虑到一些鉴别品种在不同生育阶段抗性不一致的特点，因此采用孕穗期剪叶的接种方法。从接种到病情记载的间隔以 21d 为宜。这是由于长江流域及其以北地区的气候条件，在 8 月中旬以后气温下降快，病斑扩展的程度常受影响。病情记载时，测量病斑长度与目测病斑面积（%）结果相近，试验单位可根据试验目的选用。品种抗性反应是以病斑长度（或面积）占接种叶长度（面积）的比值来判断的。

在大多数场合，抗、感的界线很明确，但在某些品种上，只好以病斑长度（面积）占接种叶的长度（面积）的比例来区分，即病级值在 3 以下（当病斑长度小于叶长四分之一或病斑面积小于 20%时）列为抗病反应，大于 5（5～9）时划为感病反应。

3.3 致病型的地理分布

从表 3，表 4，表 5 的结果看出，北方稻区的菌株，大多属致病型Ⅱ和Ⅰ型，这可能与北方粳稻区的生态特点有关，一是气温低，发病期短，二是粳稻上有粳稻专化型，这与日本的菌型较为相近；长江流域稻区是籼粳混栽区，品种类型多，栽培条件复杂，菌株以Ⅱ型和Ⅳ型为多，而南方籼稻区则以Ⅳ型为最多，还有少量的Ⅴ型存在，这与 IRRI 的Ⅰ群和Ⅱ群小种相仿。各地在鉴定品种抗性和抗病育种计划中，应该广泛采用当地优势致病型的代表菌株作为供试菌株。

3.4 病菌的致病型和品种的抗病性

从历年测定的结果可以看出，在各稻区流行中占优势的白叶枯病菌的致病型是相对稳定的。许多水稻品种在田间的抗性表现也是比较稳定的，无论是在 50 年代开始引种的农垦 57 和农垦 58，还是从 70 年代引种的 IR26，BG90－2，或是由它们衍生出来的许多品种，其抗性多比较稳定，如华南地区的青华矮 6 号、晚华矮 1 号，长江流域的扬稻 3 号、扬稻 2 号、金陵 57、南粳 15、农虎，北方稻区的秋光、中百 4 号等。它们在白叶枯病的综合防治中，都起着极其重要的作用，在生产上已取得了极大的经济效益和社会效益（许志刚等，1987；伍尚忠等，1980；章琦，1988；过崇俭等，1986）。但在近几年来，在各稻区也发现了一些新的致病型，如广东和福建的Ⅴ型，应引起足够重视。随着生产的发展，各地主栽品种的结构会有相应的变化，因此，应加强对各地病菌致病型消长的监测，以掌握其变化动态，为抗病育种与抗病品种的合理布局提供可靠的依据。

3.5　病菌与水稻品种之间的互作关系

从白叶枯病菌与水稻品种之间的致病反应来看，可分为抗病与感病两种，如果从两个、或多个菌株与两个以上品种的互作模式来看，则可以表现为强互作与弱互作两种形式（Zadoks，1979）。在 70 年代的试验中，供试菌株与品种间的反应主要为弱互作形式，虽已注意到在南粳 15 等品种上菌株间致病力的差异比较明显，但在金刚 30、江宁糯和农垦 57 上，供试的许多菌株在一个品种上的病斑长度是连续的，是数量程度的差异，因此当时认为菌株间致病力的差异主要是强弱的差异（方中达等，1981）。过崇俭等（1981 年）发现菌株与品种间存在强互作反应，并且指出白叶枯病菌的致病力特异性，随着测试菌株的增加，尤其是随着一些有鉴别能力的品种参加试验，如矢租、Tetep 等，这种反应更为明显。在最近 3 年的试验中，南粳 15 对菌株 GD1358，表现高抗并有褐斑出现，对 OS－225 和 JS49－6 则表现为高度感病，相反，IR26 对 OS－225 和 JS49－6 表现高度抵抗，而对 GD1358 表现为感病反应。这种反应模式表明病菌致病力的分化不能单用致病力强弱来衡量，而是以强互作和弱互作两种形式表现出来。这种关系在近几年来也为各地的抗病育种实践所证实（Mew，1987；章琦等，1988；过崇俭等，1986；Ezuka et al.，1978）。

3.6　有关病菌致病性分化的研究方法很多

有的从噬菌型、血清型或生理生化反应方面进行试验（Choi et.al.，1981；Waskimoto，1960），最近还有从分子生物学方面进行致病型研究的（Leach et al.，1988），但最基本的似乎还是与接种反应相结合。

原载：植物病理学报，1990，20（2）：81～88.

Genetic Diversity of *Xanthomonas oryzae* pv. *oryzae* in Asia

T. B. Adhikari[1] C. M. Vera Cruz[1] Qi. Zhang[2] R. J. Nelson[3] D. Z. Skinner[4] T. W. Mew[1] J. E. LEACH[1]*

(1. Department of Plant Pathology and *USDA-ARS*; 2. Institute of Crop Breeding and Cultivation, Chinese Academy of Agricultural Sciences, Beijing, ' China; 3. Division of Entomology and Plant Pathology, the International Rice Research Institute, Manila, the Philippines; 4. Department of Agronomy, Throckmorton Hall, Kansas State University, Manhattan, Kansas 66506 - 5502)

Restriction fragment length polymorphism and virulence analyses were used to evaluate the population structure of *Xanthomonas oryzae* pv. *oryzae* (*Xoo*), the rice bacterial blight pathogen, from several rice-growing countries in Asia. Two DNA sequences from *X. oryzae* pv. *oryzae*, IS1112, an insertion sequence, and *avrXa10*, a member of a family of avirulence genes, were used as probes to analyze the genomes of 308 strains of *Xoo* collected from China, India, Indonesia, Korea, Malaysia, Nepal, and the Philippines. On the basis of the consensus of three clustering statistics, the collections formed five clusters. Genetic distances within the five clusters ranged from 0.16 to 0.51, and distances between clusters ranged from 0.48 to 0.64. Three of the five clusters consisted of strains from a single country. Strains within two clusters, however, were found in more than one country, suggesting patterns of movement of the pathogen. The pathotype of *Xoo* was determined for 226 strains by inoculating five rice differential cultivars. More than one pathotype was associated with each cluster; however, some pathotypes were associated with only one cluster. Most strains from South Asia (Nepal and India) were virulent to cultivars containing the bacterial blight resistance gene *xa5*, while most strains from other countries were avirulent to *xa5*. The regional differentiation of clusters of *Xoo* in Asia and the association of some pathotypes of *Xoo* with single clusters suggested that strategies that target regional resistance breeding and gene deployment are feasible.

The use of resistant crop cultivars is an inexpensive and environmentally friendly approach to crop protection. Because some resistance genes are effective only against particular pest or pathogen subpopulations, it is important to understand the structure of pest or pathogen populations to determine the best strategy for deployment of resistance. Information on pathogen population structure would include knowledge of pathogen diversity, phylogeny, and the partitioning of variation in time and space (Leung et al., 1993). Knowledge of the spatial distribution of pathogen subpopulations, for example, can aid in the selection of disease resistance sources for a regional crop breeding program. Unfortunately, detailed information on pathogen populations is rarely available. The present study was undertaken to provide information about the population structure of *X. oryzae* pv. *oryzae* (Swings et al., 1990), the bacterial blight pathogen of rice, in Asia.

Bacterial blight is a widespread and destructive disease of rice in irrigated and rainfed environments in Asia (Mew, 1987). The disease can cause 30% to 50% yield loss (Exconde et al., 1971; Reddy, 1989). Previous studies have reported assessment of the pathotypic structure of *Xoo* on the basis of differential interactions with rice cultivars containing different resistance genes (Adhikari et al., 1994; Eamchit et al., 1982; Gupta et al., 1986; Hartini et al., 1976; Horino et al., 1983; Mew et al., 1982; 1987). Because relatively few genetic loci are assessed by such virulence analyses and these loci may be subject to strong selection, the genetic diversity and evolutionary history of *Xoo* cannot be inferred from these studies (Leung et al., 1993).

Molecular techniques have provided abundant genetic markers that can be used to assess the genetic structure of field populations of *Xoo* (Leach et al., 1991; Leung et al., 1993). Molecular probes suitable for DNA fingerprinting and phylogenetic analysis of *Xoo* have been identified and characterized (Hopkins et al., 1992; Leach et al., 1990; Nelson et al, 1994)., Leach et al. (1992) used a probe (pJEL101) carrying the insertion sequence IS*1112* (Yun, 1991), isolated from *Xoo*, for restriction fragment length polymorphism (RFLP) analysis of a collection of strains of *Xoo* from the Philippines (Leach et al., 1990). They evaluated the diversity of sets of strains collected over defined time periods and from different regions and determined the relationship between pathogenic races and phylogeny of the pathogen. Nelson et al. (1994) analyzed a similar set of strains with four mobile, repetitive elements and an avirulence gene, *avrXa10* (Hopkins et al., 1992), isolated from *Xoo*. Of the five probes tested, analysis with probe IS*1112* resulted in the most robust phylogeny. Probe *avrXa10*, which is of interest because of its role in host pathogen interactions, gave the least robust phylogeny (Nelson et al., 1994). Although details of the population structure derived from RFLP data were probe dependent, the basic phylogenetic structure that emerged from their study was consistent with that inferred from the study of Leach et al. (1992).

The broad-scale spatial partitioning of pathogen variation has implications for crop breeding programs that serve wide areas and for the utilization of resistant germ plasms in local breeding and deployment efforts. Furthermore, *Xoo* is thought to be seed borne (Mew et al., 1988), so international tracking of pathogen populations would be useful for quarantine programs. Although *Xoo* populations have been or are being studied in individual countries (Leach, et al., 1992; Nelson, et al., 1994), comparative studies of molecular variation at an international level have not been carried out previously. In this study, the population structure of *Xoo* collected from major rice-growing countries of Asia is assessed by using RFLP and virulence analyses.

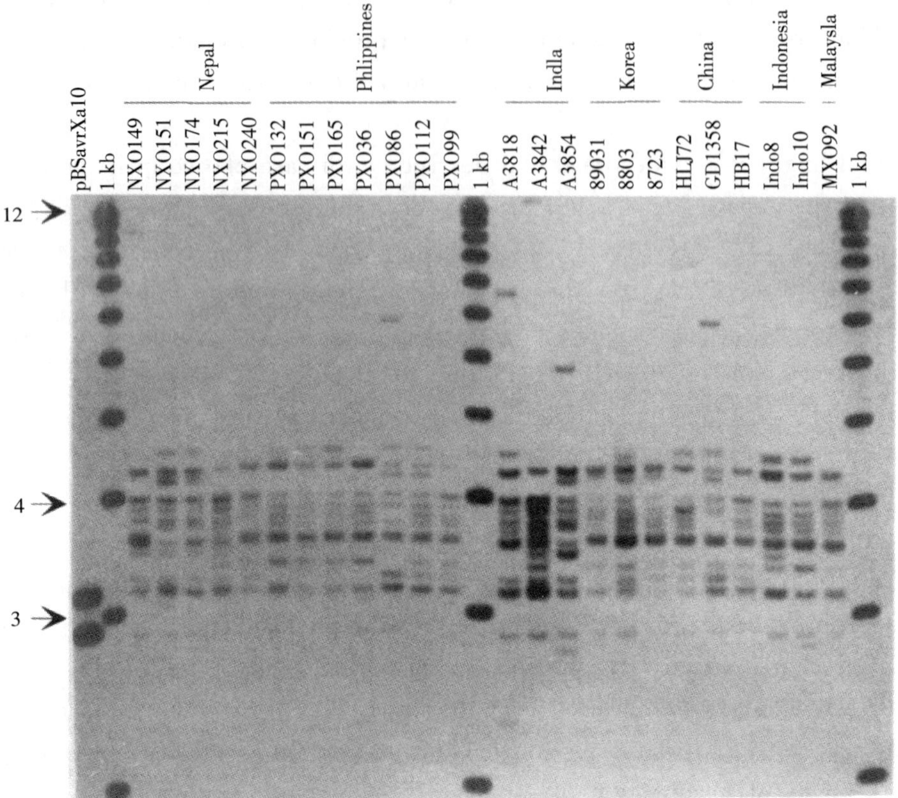

Fig. 1　DNA fingerprint patterns revealed by probe pBSavrXa10 for selected strains of *Xoo* from major rice-growing countries in Asia. Total DNA from each strain was *Bam*H1 digested, and the probe was biotin labeled. Numbers at left are the sizes (kilobases) of fragments indicated by arrows.

Materials and Methods

Bacterial strains　A total of 308 bacterial strains collected from various countries in Asia were used in this study. Strains from China (76 strains from 17 provinces, collected from 1981 to 1986 and provided as DNA), India (17 strains from six states, collected from 1982 to 1991), Malaysia (17 strains from 5 provinces, collected from 1982 to 1989), Korea (24 strains from 5 provinces, collected from 1987 to 1989), Indonesia (34 strains from six islands, collected from 1976 to 1993, 91% from 1992 to 1993), Nepal (45 strains from 18 districts, collected from 1987 to 1991), and the Philippines (95 strains from 23 provinces, collected from 1972 to 1988) were provided by Q. Zhang (Chinese Academy of Agricultural Sciences, Beijing, China), S. Gnanamanickam (University of Madras, Madras, India), K. S. Lum (Malaysian Agriculture Research and Development Institute, Serdang,

Malaysia), K. S. Jin (Office of Rural Development, Suweon, Korea), R. H. Hartini (Bogor Research Institute of Food Crops, Bogor, Indonesia), T. Adhikari (Institute of Agriculture and Animal Science, Kathmandu, Nepal), and T. W. Mew (International Rice Research Institute, Manila, the Philippines), respectively. Information for each strain, which includes the site of collection within a country, the rice cultivar from which the strain was isolated, the date of collection, the pathotype, and the haplotype, is available upon request from the corresponding author. The strains were cultured on peptone sucrose agar and preserved in 15% glycerol at −80℃.

DNA probes Two DNA clones, pJEL101 and pBSavrXa10, were used as probes in this study. Probe pJEL101 is plasmid pUC18 with a 2. 4 - kb *Eco*RI-*Hind*Ⅲ fragment that carries all but three nucleotides of insertion element IS*1112* (Yun, 1991) derived from *Xoo* (Leach et al. , 1990) . More than 80 copies of IS*1112* have been estimated to occur in some strains of *Xoo* (Leach et al. , 1990) . Probe pBSavrXa10 contains a 3. 1 - kb *Bam*Hl fragment that is internal to an avirulence gene from *Xoo* in vector pBluescript Ⅱ (Hopkins et al. , 1992). *Xoo* contains multiple copies of genes related to *avrXa10*; at least two of these (*avrXa10* and *avrXa7*) function as avirulence genes (Hopkins et al. , 1992) . Plasmid DNA was extracted by the cleared lysate method and purified by cesium chloride-ethidium bromide density gradient centrifugation (Maniatis et al. , 1982) .

RFLP analysis Genomic DNA from each bacterial strain was extracted by the lysozyme-sodium dodecyl sulfate lysis method of Owen and Borman (1987) as modified by Leach et al. (1992) . Bacterial DNA was digested to completion with *Bam*HI or *Eco*RI (2 U//μg of DNA) at 37℃ for 2 to 3 h in buffer supplied by the enzyme manufacturer (Promega Biotech, Madison, Wis. or Bethesda Research Laboratories, Gaithersburg, Md.) . Digested DNA (5 to 10μg per well) was separated by electrophoresis, and fragments were transferred to Magna NT nylon membrane as described by the manufacturer (MSI, Westboro, Mass.). A size standard (1 - kb ladder; Bethesda Research Laboratories) was included in each gel. The DNA probes were labeled with [^{32}p] dCTP or biotin-labeled dUTP with nick translation kits (Bethesda Research Laboratories and Tropix, Inc. , Bedford, Mass. , respectively) . Hybridization and washing conditions were of high stringency, as described by Vera Cruz et al. (1994) for biotin-labeled probes or Leach et al. (1990) for ^{32}P-labeled probes. Hybridized fragments were detected by autoradiography on Cronex film (Du Pont, Wilmington, Del.) at −80℃.

Virulence analysis Virulence data were obtained for a total of 226 Asian strains of *Xoo*, with the following regional representation: India, $n = 12$; Indonesia, $n = 34$; Korea, $n = 24$; Malaysia, $n = 17$; Nepal, $n = 44$; the Philippines, $n = 95$. Virulence data for the 95 Philippine strains were reported previously (Leach et al. , 1992) . Virulence data were not obtained for the 76 *Xoo* strains from China.

Virulence of *Xoo* was assessed by inoculating five differential rice cultivars, IR8 (with the bacterial blight resistance gene *Xa11*), lR20 (*Xa4*), CAS209 (*Xa10*), IR1545 - 339 (*xa5*), and DV85 (*xa5* and *Xa7*), as established by Mew (1987). The experimental design was a split plot, with the rice cultivar as the main plot and the bacterial strain as the subplot (Gomez et al., 1984). The experiment was conducted two times. Three to four rice seeds of each cultivar were planted in plastic pots (8.9cm×8.9cm) containing Bacto Potting Soil (Michigan Peat Co., Houston, Tex.). Plants were kept at 26 to 30℃ in a greenhouse and fertilized weekly with 20:20:20 NPK liquid fertilizer (Peters professional plant food; W. R. Grace, Cambridge, Mass.). Bacteria for inoculum were grown on a peptone sucrose agar slant for 48 h at 28℃, suspended in sterile distilled water, and adjusted to approximately 10^8 CFU/ml (100 Klett units) by using a photoelectric colorimeter (Klett Mfg. Co. Inc., New York, N. Y.). Six fully expanded leaves of each cultivar per pot were clip inoculated 50 days after sowing (Kauffman et al., 1973). Lesions were measured 14 days after inoculation. For the purposes of assigning pathotypic designations, plants having lesions of 0 to 3 cm were classified as resistant, and those longer than 3 cm were rated as susceptible.

Data analysis Each distinct RFLP banding pattern produced by the combined data from both probes was regarded as a haplotype. Binary data were derived from the restriction fragment profiles of the different strains by scoring the presence or absence of each band for each haplotype. By use of NTSYS-PC version 1.8 (Rohlf, 1992), a matrix was generated from the binary data by calculating the fraction of shared bands for each pair of haplotypes by using Jaccard's coefficient of similarity: $a/(n-d)$, where a is the number of bands in common (++) between two haplotypes, n is the total number of possible matches (++ and −−) and mismatches (+− and −+) for the two haplotypes, and d is the number of (−−) matches between the two haplotypes (reference 10, as cited in reference, Sneath et al., 1973). The data were converted to coefficients of dissimilarity (1 - Jaccard coefficient) before statistical analysis as recommended previously (SAS Institute, Inc. SAS/STAT user's guide 1989). The positions of the strains on a three-dimensional graph were determined from the first three dimensions of a multiple correspondence analysis [PROC CORRESP (SAS Institute, Inc. 1989. SAS/STAT user's guide)], which accounted for about 78.3% of the variation. The number of clusters was determined by consensus among three clustering statistics, the cubic clustering criterion, pseudo F, and pseudo t^2 as described previously; these criteria are appropriate for compact or slightly elongated clusters. Cluster assignments were then generated by use of the two-stage option of PROC CLUSTER ($k=25$) (SAS Institute, Inc. 1989. SAS/IML software: usage and reference). Average genetic distances between and within clusters were calculated with routines written in PROC IML of SAS.

The diversity of strains from each country, and of the entire collection, was analyzed by the approach of Nei and Tajima (1981). The genetic diversity (H) for each population considered was calculated by the equation $H=[n/(n-1)][1-\sum Xi^2]$, where Xi is the proportion of the ith haplotype within a group and n is the number of strains tested in each

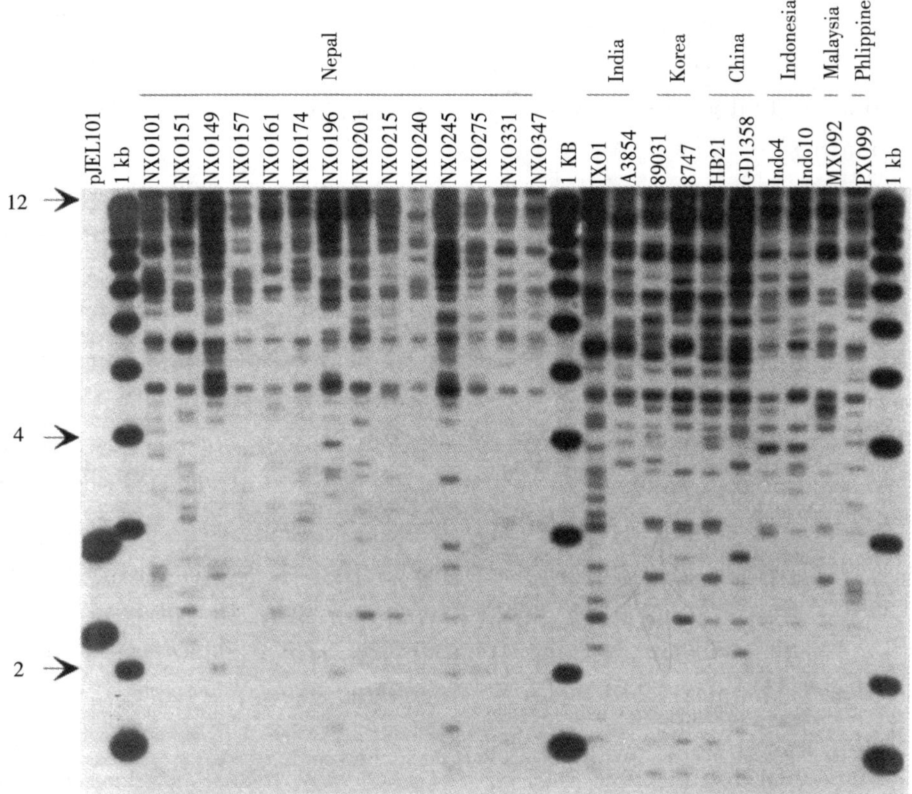

Fig. 2 DNA fingerprint patterns revealed by hybridization of *Eco*RI-digested DNA of representative strains of *Xoo* with probe pJEL101. The probe was 32 p labeled. Numbers at left are the sizes (kilobases) of fragments indicated by arrows.

group (Nei, 1987) .

Results

RFLP analysis Two probes, an avirulence gene (*avrXa10*) and an insertion sequence (IS*1112*), were used to analyze 308 strains of *Xoo* collected from diverse locations in Asia. Probe pBSavrXa10 hybridized with 10 to 14 differentsized DNA fragments per strain, with an average of 11 bands; 26 band positions were scored for the entire collections (Fig. 1) . Probe pJEL101 hybridized with 9 to 23 different-sized DNA fragments per strain, with an average of 17 bands; 38 band positions were scored (Fig. 2) . In general, probe pJEL101 resolved more haplotypes (81 haplotypes) than did pBSavrXa10 (70 haplotypes). For example, in strains from India, pJEL101 resolved nine haplotypes compared with five haplotypes resolved by pBSavrXa10, and in the Philippines, 21 pJEL101 haplotypes were resolved compared with 27 pBSavrXa10 haplotypes.

When the data from both pJEL101 and pBSavrXa10 were combined, a total of 146 haplotypes were observed. Twenty five haplotypes were detected in the collection of strains from China, 13 were detected from India, 16 were detected from Indonesia, 17 were detected from Korea, 7 were detected from Malaysia, 30 were detected from Nepal, and 38 were detected from the Philippines. No combined-probe haplotype was common to strains from different countries.

On the basis of consensus of three clustering statistics, the 308 strains constituted five clusters (Fig. 3). There was a general correspondence between the country of origin and five groups formed by analysis of RFLP data (Fig. 3). Cluster 1 consisted of strains from China (74 strains including 23 haplotypes) and two strains (two haplotypes) from Indonesia (A^1 in Fig. 3). Some strains, such as the two Indonesian strains indicated as A^1 are almost equidistant between two clusters (e. g., the A^1 strains are 0.444 and 0.443 from clusters 1 and 3, respectively) and thus are shown outside the cluster they were assigned to by the clustering program (cluster 1). Clusters 2 and 4 were composed of strains from the Philippines (17 and 16 haplotypes, respectively). Cluster 3 consisted of strains from Indonesia and Malaysia (15 haplotypes). Cluster 5 contained the most diverse group of strains, with strains from Nepal, India, Korea, and the Philippines represented (65 haplotypes). A few strains from China (two strains and two haplotypes), Indonesia (two strains and two haplotypes), and Malaysia (four strains and four haplotypes) also grouped in cluster 5.

The average genetic distance within clusters ranged from 0.16 (cluster 3) to 0.51 (cluster 5), and that between clusters ranged from 0.48 and 0.64 (Table 1). Particularly intriguing was the high average distance between clusters 2, 4, and 5, which all contained strains from the Philippines.

Haplotypic diversities (H) calculated by the method of Nei and Tajima (1981) for each country were as follows: Malaysia, 0.23; Korea, 0.83; Indonesia, 0.87; China, 0.91; India, 0.92; the Philippines, 0.96; and Nepal, 0.98.

Table 1 Genetic distance within and between clusters of *X. oryzae* pv. *oryzae* in Asia

Cluster	Genetic distance from cluster				
	1	2	3	4	5
1	0.27				
2	0.64	0.25			
3	0.48	0.64	0.16		
4	0.64	0.53	0.63	0.21	
5	0.56	0.64	0.60	0.61	0.51

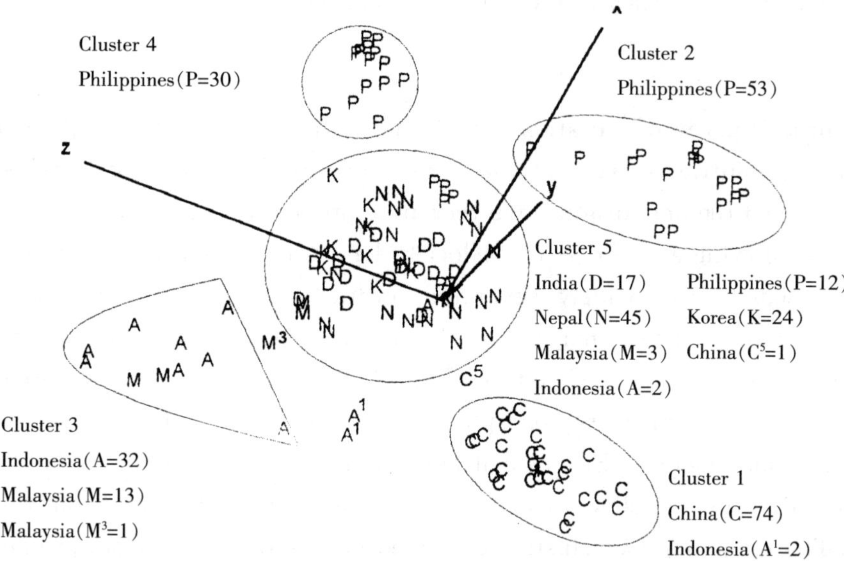

Fig. 3 Cluster analysis showing the relationships of *Xoo* strains as derived from DNA band data obtained with probes pBSavrXa10 and pJEL101.

Numbers of strains from the China (C), India (D), Indonesia (A), Korea (K), Malaysia (M), Nepal (N), and the Philippines (P) are shown. The number of clusters that best fit the data was determined by the consensus of three clustering statistics; positions of strains on a three-dimensional graph were determined by multiple correspondence analysis. The x axis accounts for the most variation (50.5%), while the y axis and z axis account for 17.5 and 10.3% variation, respectively. Strains shown outside the boundaries are similar in distance from the two bordering clusters. For example, the A^1 strains from Indonesia were almost equidistant from clusters 1 and 3 (distances of 0.444 and 0.443, respectively). The superscript numbers by these strains indicate the clusters to which the strains were assigned.

Virulence analysis When the virulence data on five differential rice cultivars were considered as qualitative information (reactions classified as resistant or susceptible), seven pathotypes of *X. oryzae* pv. *oryzae* were detected (Fig. 4). Of these, pathotype 6 was virulent on all host differentials, while the other pathotypes were incompatible with at least one of the hosts. pathotype 1 dominated in the collection from Malaysia (10 of 17 strains, all in cluster 3) but was also detected in the Philippines (cluster 4) and Korea (cluster 5). Pathotypes 2 and 5 were detected only in the Philippines and consisted only of strains found in cluster 2. All strains from Indonesia were of pathotype 3. Pathotypes 4 and 6 were common to strains in cluster 5 and included strains from Nepal, India, and the Philippines. Pathotype 0, which consisted of a few weakly virulent strains from Korea, Malaysia, and Nepal, was likely an artificial group of strains that had lost virulence in culture. Of the five cultivars tested, this group of strains was weakly virulent only on cultivars carrying the *Xa10* resistance gene. It is important to note that strains used in the collection from the Philippines were selected to represent the range of pathotypes defined by the differential rice cultivars used; therefore, the frequency of pathotypes within the Philippine collection is not representative of the frequency in the native population.

Discussion

In this broad survey of variation among strains of the bacterial blight pathogen from several Asian countries， we assessed the relationships among strains， virulence characteristics， and regional origin. Two repetitive probes were used for DNA fingerprinting of each strain， allowing an average of 14 loci to be scored for each *Xoo* strains used in this study were isolated from widely separated locations within each country. Thus， it is intriguing that after cluster analysis of the RFLP data， only five clusters （based on clustering statistics） were formed and that four of the five groups consisted predominantly of strains from a single country or geographic region （Malaysia and Indonesia are considered the same geographic region）. Strains collected from 16 different provinces in China formed a single cluster. Two distinct clusters were formed from strains collected from throughout the Philippines. The remaining two clusters consisted largely of strains from proximal but large geographic regions. For example， one clusters consisted of strains from two Southeast Asian countries， Indonesia （represented by strains from six different islands） and Malaysia， while the other clusters contained strains from South Asia （representing six states in India and 18 districts in Nepal）. With the exception of cluster 5， there is a tight correspondence between clusters and national or regional origin. The clonal populations within a country or region are likely a consequence of at least two factors， physical geographic barriers and the similarity of rice varieties grown on a national basis.

Two of the five clusters contained subclusters of strains from more than one geographic region. Cluster 5， the least defined of the clusters （average within-cluster genetic distance， 0.51）， consisted of strains from South Asia （India and Nepal）， Northeast Asia （Korea）， and Southeast Asia （the Philippines）. A few strains from China， Indonesia， and Malaysia also grouped in this cluster. The strains which grouped in cluster 5 from Indonesia， Malaysia， and the Philippines were clearly distinct from other strains from those countries based on RFLP pattern and， in some cases， pathotype. The presence of strains in clusters occupied predominantly by strains from different regions suggests patterns of pathogen migration， perhaps due to international exchanges of contaminated germ plasm. However， in some cases， differences in genomes are such that it is not clear where strains originated. For example， although two strains from Indonesia grouped with cluster 1 （two A^1 strains in Fig. 3）， the strains were almost equidistant from clusters 1 and 3 （distances of 0.444 and 0.443， respectively）.

RFLP data were used to estimate the haplotypic diversities of the pathogen collections. The diversity estimates obtained for the different countries varied substantially. Diversity was relatively low for the strains from Malaysia. In contrast， haplotypic diversity was relatively high for the collections from India and Nepal. The large differences in the levels of diversity are not simply a result of differences in sample size;

fewer strains from India (17 strains from 6 states) than from Nepal (45 strains from 18 districts) or China (76 strains from 16 provinces) were tested, and yet the diversity estimated for India is similar to that of its neighbor Nepal. The high diversity in these two South Asian countries may reflect the relatively broad range of environmental conditions under which rice is cultivated and the wide use of diverse traditional rice cultivars. Systematic sampling of *Xoo* populations within the various countries is needed to determine whether the differences in pathogen diversity suggested by these results reflect true differences in diversity between regions.

The pathotype of the *Xoo* strains was determined by inoculation of five differential cultivars carrying the *Xa4*, *xa5*, *Xa7*, *Xa10*, and *Xa11* genes for resistance (Mew, 1987). Only virulence to cultivars with the *xa5* gene was regionally differentiated. Many South Asian rice cultivars possess *xa5* (Sidhu et al., 1978), and most of the strains from Nepal and India were virulent to cultivars carrying *xa5*. A smallminority of the Philippine strains are virulent to *xa5* (race 6) (Mew et al., 1992), and these strains are part of the same cluster as that of the South Asian strains (cluster 5). Strains from Malaysia, Indonesia, and Korea were not virulent *to xa5*. Thus, the deployment of cultivars with *xa5* could provide resistance in parts of Southeast and Northeast Asia but not in South Asia.

The collection of strains from six different islands in Indonesia was relatively homogeneous in the reactions of the strains to the cultivars tested. These strains are pathotype 3. In contrast, the strains from the Philippines were relatively diverse; this was because the Philippine strains for this study were selected to represent the six pathotypes distinguished by the rice differential set utilized (Mew et al., 1979). Furthermore, the rice differential set was developed in the Philippines. A different set of host genotypes including other resistance genes likely would differentiate pathotypes among the Indonesian strains as well as strains from other parts of Asia. To facilitate the testing of this hypothesis, strains representing different clusters and haplotypes detected by RFLP analysis could be selected for inoculation of rice germ plasm collections indigenous to these regions. By this strategy, new pathotypes were identified previously among Philippine populations of *Xoo* (Nelson et al., 1994; Vera Cruz et al., 1992).

The presence of distinct clusters of *Xoo* in the Philippines is interesting from historical perspective. Cluster 4, which included strains of race 1, was the prevalent type of *Xoo* detected in the lowland areas prior to the deployment of the *Xa4* gene for resistance to bacterial blight in the early 1970s (Mew et al., 1992). After the deployment of *Xa4*, new, virulent pathotypes, predominantly race 2, were detected in the lowland areas (Mew et al., 1992). It is interesting that these strains are included in a different cluster (cluster 2) than the population they displaced (cluster 4), which suggests that *Xa4* was not, at least originally, colonized by variants of cluster 4. The collections used for this particular study are not sufficient to allow us to deduce if cluster 2 was present at low levels in the *Xoo* population prior to the deployment of *Xa4* or if it was introduced and established after *Xa4*

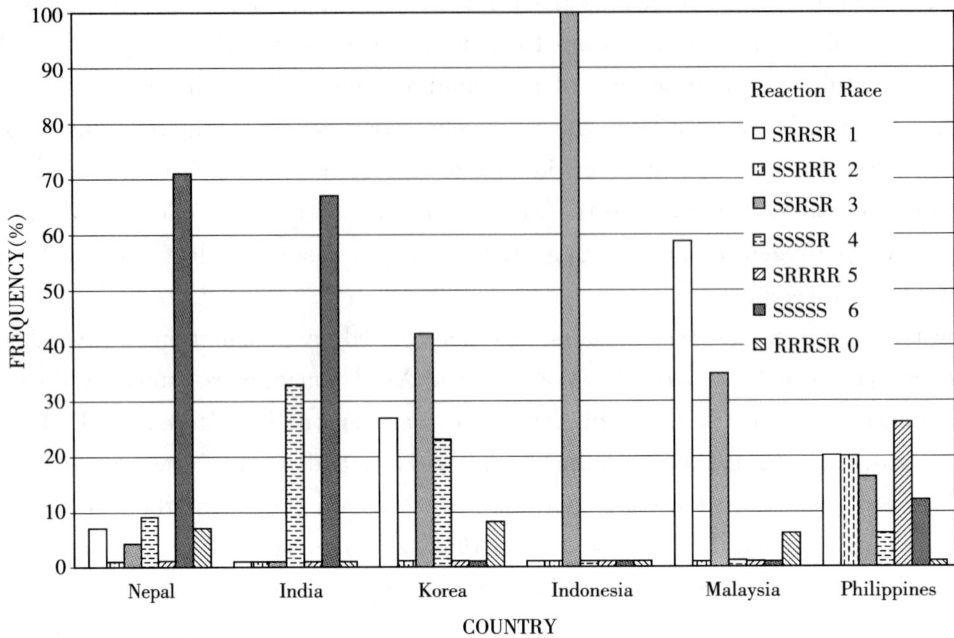

Fig. 4 Frequency (percentage of strains per country) of distribution and pathotypic (race) structure of *Xoo* in major rice growing countries in Asia. Pathotype was determined by inoculation to the rice cultivars IR8 (*Xa11*), IR20 (*Xa4*), IR1545 (*xa5*), CAS209 (*Xa10*), and DV85 (*xa5 and Xa7*). The host reactions of the five cultivars (R, resistant; S, susceptible) are shown from left to right, respectively, under "Reaction." *Xoo* populations from the Philippines are a selected subset of the strains surveyed by Mew et al. (1992).

deployment. However, we provided data previously that suggested that pathotype 2 strains arose from strains common to the highland areas of the Philippines (Nelson et al., 1994). The lack of similarity of cluster 2 strains to *Xoo* from other rice growing areas of Asia and the similarity to strains from the highlands (Nelson et al., 1994) support the notion that the site of origin for this group is within the Philippines.

Our results show that although populations of the bacterial blight pathogen of rice are very diverse, they exhibit regional differentiation. This information may provide a preliminary basis to design strategies to use different sources of resistance to the pathogen. For example, the resistance gene *xa5* might be useful in many countries but not in countries of South Asia. In fact, none of the resistance genes used in our inoculation tests were effective against the strains prevalent in South Asia. The results of this and further DNA fingerprinting analyses would be useful in the selection of strains for additional resistance screening.

Published in: Applied and Environmental Microbiology, Mar. 1995, 966-971.

中国水稻白叶枯病菌群体结构的初步分析*

章 琦[1] Jan E. Leach[2] R. J. Nelson[3] 王春莲[1] T. W. Mew[3]

(1. 中国农业科学院作物育种栽培研究所；2. 堪萨斯州立大学植物病理系，美国；
3. 国际水稻研究所，菲律宾)

Preliminary Analysis of the Population Structure of *Xanthomonas oryzae* pv. *oryzae* in China

Zhang Qi[1] Jan E. Leach[2] R. J. Nelson[3] Wang ChunLian[1] T. W. Mew[3]

(1. Institute of Crop Breeding and Cultivation, CAAS;
2. Plant Pathology Department, Kansas State University, USA;
3. IRRI, Philippines)

Abstract Seventy eight strains of *Xanthomonas oryzae* pv. *oryzae* (*Xoo*) collected mostly from North, Northeast and a few from Center and South China have been assessed the RFLP types for analyzing the population structure and genetic diversity by using two probes, pJEL101 and pBSavrXa10. Genomic DNA from each strain of *Xoo* was digested to complete by enzymes *Eco*R Ⅰ or *Bam*H Ⅰ for blots probed with pJEL101 or pBSavrXa10. Sixteen RFLP types were identified, respectively. Fourteen RFLP types were found in strains of Chinese pathotypes 1, 2 and 6, which were from North and Northeast China (using pJEL101). Among them, differed only at one or 3 - 4 banding positions. In comparison between those with RFLP types of strains from Guangdong and Yunnan provinces were distinct from one another at 7 or 8 positions varied. Cluster analysis based on RFLP banding showed six groupings at 85% similarity. The majority of strains from pathotypes 1, 2 and 6 were grouped in clusters 1 only. A few strains, LN43, GD 1358 and YN 5 were grouped in clusters 4, 5 and 6, respectively. The genetic diversities of the population of strains tested was 0.77 (for pJEL101) and 0.83 (for pBSavrXa10) respectively. Genetic diversities for each pathotypes tested as a subpopulation has been also

* Note: The project was cooperated with JEN Leach, Kansas State University. USA: Rebecca, and Mew at IRRI in 1999 and 1993.

calculated. The preliminary analysis showed that the genetic diversity of *Xoo* in China was various. It will be necessary to assess more strains collected from South, Center and Southwest China for analyzing the population constructure and genetic relationship of *Xoo* systematically.

Key words bacterial blight; genetic diversity; population structure; RFLP type; strains

摘要 用两个DNA探针pJEL101和pBSavrXa10，测定78个来自18个地区的水稻白叶枯病菌株的RFLP型。这些菌株主要来自华北、东北和少数来自华中、华南。用以分析病菌群体结构和遗传多样性。各菌株的总基因组DNA分别用限制性内切酶*Eco*RI和*Bam*HI酶切探针pJEL101或pBSavrXa10。通过分子杂交，分别鉴定出16种RFLP型。来自华北和东北的多数属病原型1、2和6的菌株，表现14种RFLP型。相互间大多只有一条带之差，少数有3～4个带位不同，但是与广东和云南的菌株相比，则差别较大，有7～8条带位不同。根据参试菌株的RFLP带型分簇（Cluster），以彼此之间的带位相似率85%为界，可分为6簇。多数属病原型1、2和6的菌株归为簇1，少数几个菌株诸如LN43、GD1358和YN5分别归为簇4、5和16。参试菌株群体遗传变异为0.77（用探针pJEL101）和0.83（用探针pBSavrXa10）。同时分析了亚群体各病原型内的遗传变异。初步分析表明，中国水稻白叶枯病菌株的遗传变异是不同的，但还需要测试更多华南、华中和西南地区的菌株，方能较系统地了解全国菌株的群体结构及其遗传关系。

关键词 水稻白叶枯病；遗传多样性；群体结构；RFLP型；菌株

In previous studies, *Xanthomonas oryzae* pv. *oryzae* (*Xoo*) population in China have been classified seven pathotypes on the pathogenicity patterns on a set of Chinese differential cultivars (Fang et al., 1990). The host-pathogen response rice-*Xoo* system was, however, still not clear-cut. The pathotype groupings provided little insight into the genetic structure of the bacterial population. Recently, restriction fragment length polymorphism (RFLP) analysis, which exploits the abundant variation in the DNA sequence, has been used to generate a large number of markers for the measurement of genetic diversity in population in order to better understand the population structure of *Xoo*, two probes were used to measure the genetic diversity of *Xoo* populations and relate the diversity to previous characterization of the pathogen in China.

1 Materials and Methods

1.1 Bacterial strains and DNA probes

Seventy eight strains were used in this study in RFLP analysis. Most of the strains were collected from North and Northeast China, a few from South and Southwest China. The data of collections, source cultivars, and pathotype grouping for the *Xoo* strains used are shown in Table 1. All stock cultures were kept as lyophilized cultures. The strains were revived in fresh slants of PBS medium for DNA isolation.

Two DNA probes were used in this study. One is pJEL101, a pUC18 plasmid containing a 2. 4 kb *Eco*R Ⅰ - *Hind*Ⅲ fragment derived from Philippine race 2 strain PXO86 of *Xoo*. The insert of pJEL101 contains a high-copy-number repetitive DNA element present in DNA from over 100 *Xoo* strains tested. Other one is plasmid pBSavrXa10, containing a 3. 1 kb *Bam*HI fragment internal to the *avrXa10* gene of *Xoo* (Hopking et al. , 1992) .

1. 2 DNA analysis

Genomic DNA from strains of *Xoo* was isolated as described by Leach et al. (1990), and digested with *Eco*R I or *Bam*HI for blots probed with pJEL101 or pBSavrXa10. DNA fragments were electrophoresed in 0. 7% agarose gel (20cm×21. 5 cm) in TBE buffer. A 1-kb ladder was included in gel as a size standard. Southern transfer onto a nylon membrane was done according to instructions of the manufacturer. The entire pJEL101 and pBSavrXa10 plasmids were labeled with ^{32}P by using a nick translation kit. Hybridization and washing conditions were high stringency. Autoradiography was done at -80℃ with Cronex film. Lightning-Plus intensifying screens. DNA from strains of same geographic origin analyzed in the same gel primarily. Distinct RFLP types were analyzed in a single blot to confirm that each RFLP type was unique.

1. 3 Data analysis

To determine the genetic relationships among RFLP types, the presence or absence of bands at 20 (for using pJEL101, based on appearing a difference in banding) or 19 (for using pBSavrXa10. appeared 19 total markers) different marker positions was converted into binary data; i. e. , the presence or absence of a band was coded as 1 or 0 respectively.

Cluster method, UPGMA was used to infer phylogenetic relationships among strains of *Xoo* based on RFLP data. Clusters that separated at the 85% similarity level were grouped and given designations, with cluster 1 being the cluster whose strains were most similar, cluster 2 being the cluster whose units were the second most similar, and so on. Genetic diversity (*H*) was calculated according to the following equation (Leach et al. , 1990) .

$H=[n/(n-1)]\times(1-Xi^2)$, i=1, where *Xi* is the frequence of the ith haplotype in the population, and n is the number of strains examined.

2 Results

2. 1 RFLP analysis

Seventy eight strains of *Xoo* for DNA polymorphisms were examined with two DNA probes pJEL101 and pBSavrXa10, sixteen RFLP types were identified respectively (Fig. 1, Fig. 2) . Most of the strains tested within Chinese pathotypes 1, 2 , and 6 were grouped in RFLP types 1 to 4 (Table 2) . Among them differed only at one banding position. The banding patterns among some RFLP types were distinct from one another at several banding positions. i. e, 7th or 8th positions vary between RFLP 11 and 10, 15, 6, respectively (Fig. 3). Similarly, using pBSavrXa10, 7th, 8th and 11th positions vary between RFLP type 11 and 10, 13 and 14, respectively (Fig. 2).

To determine the genetic relationships among strains of pathotypes, the banding patterns were analyzed by using cluster method, UPGMA. All strains were similar to each other at a level of 85% similarity, six clusters of strains were observed in sixteen RFLP types using pJEL101 probe (Fig. 4) .

Pathotypes represented in the six clusters were as follows: cluster 1, including pathotypes 1, 2 and 6; cluster 2, including pathotypes 3 and 4; cluster 3, including pathotype 1; cluster 4, 5 and 6, including pathotypes 6, 5 and unknown (strain YN5), respectively. Most of strains used were contained within one cluster: pathotype 1 (26 of 28 strains tested) and pathotype 2 (29 of 31 strains), and pathotype 6 (7 of 8 strains) in cluster 1 (Table 2) .

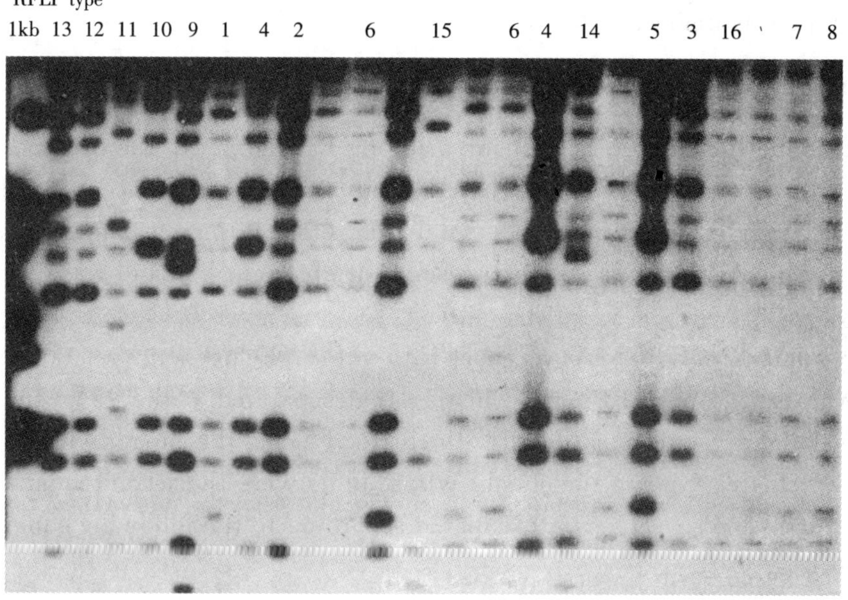

Fig. 1 Southern blot analysis of genomic DNA of strains of *X. o.* pv. *oryzae* in China showed 16 RFLP types (the probe used was ^{32}p-labeled pJEL 101)

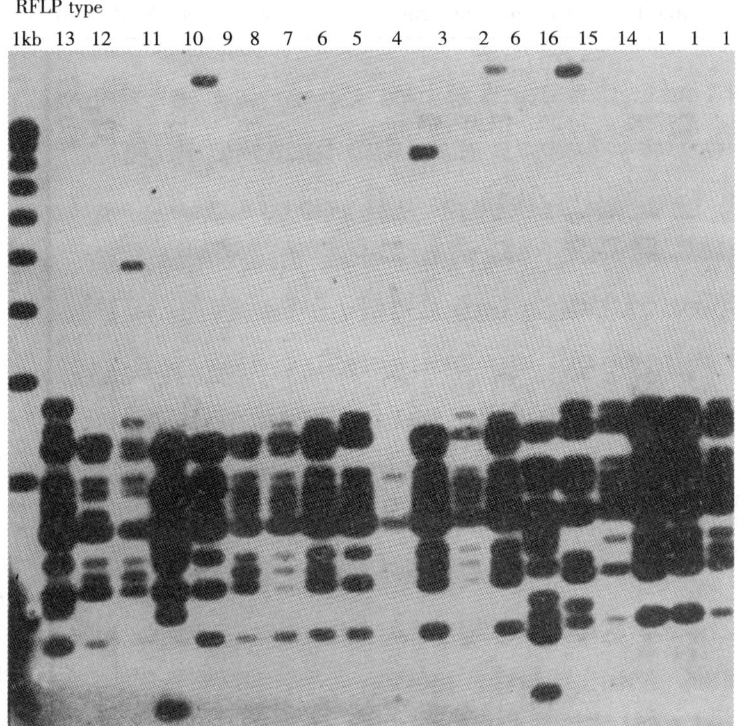

Fig. 2 Southern blot analysis of genomic DNA of strains of *X. o.* pv. *oryzae* in China showed 16 RFLP types (the probe used was ^{32}p-labeled pBSavrXa10)

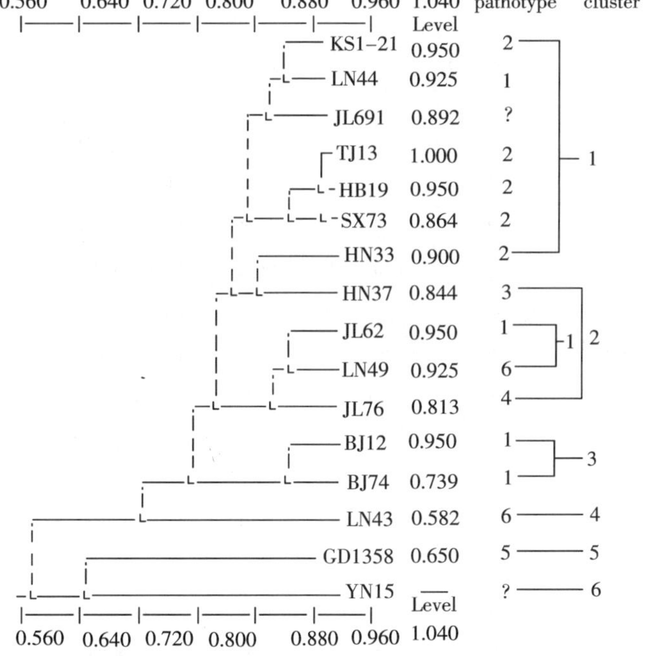

Fig. 3 Dendrogram displaying the relationships among the 16 RFLP types for strains of *X. o.* pv. *oryzae* after southern blot analysis using ^{32}P- labeled pJEL101 as a probe. Similarity coefficients were grouped by using the unweighted pair group method with arithmatic means

Table 1 X. o. pv. *oryzae* strains from China used in RFLP analysis

Strain	Pathotype[1]	RFLP type[2]			Year of collection	Geographic origin	Cultivars
		pBSavrXa10	pJEL101	Combined			
HU72	1	2.	2	1	1985	Ecbeng. Heilongjiang	Chengjian 6
HB20	1	2	2	1	1984	Chengde, Hebei	Jigcng 60
HB21	1	2	2	1	1984	Fengning. Hebei	Muzhun 2
NX39	1	2	2	1	1984	Yongning. Ningxia	Jingyin 35
NX40	1	2	2	1	1985	Wuzhong Ningxia	Xinghuozhan
LN51	1	2	2	1	1985	Donggou Liaoning	7530 - 4 - 3
LN52	1	2	2	1	1985	Donggou. Liaoning	Zhongdan 1
LN55	1	2	2	1	1985	Donggou. Liaoning	Jingyue 1
LN56	1	2	2	1	1985	Liaozhong. Liaoning	Liaogeng 5
BJ7	1	3	1	2	1985	Daxing, Beijing	78 - 10
HB30	1	3	1	2	1985	Zhuoxian. Hebei	Zhonghua 8
LN45	1	3	1	2	1985	Liaozhong. Liaoning	Liaogeng 5
HU69	1	3	1	2	1985	Wuchang. Heilongjiang	Jigeng 60
HU68	1	3	1	2	1984	Wuchang. Heilongjiang	Jigeug 60
LN44	1	3	2	3	1984	Kaiyuan. Liaoning	Liaogeng 5
BJ3	1	4	4	4	1984	Shunyi, Beijing	Nongken 5
JL62	1	4	4	4	1984	Ji'an. Jilin	Nuodao
HB25	1	14	2	5	1984	Baigezhuang. Hebei	Zhonghua 8
HB24	1	6	6	6	1984	Tangxian, Hebei	Qianjunbang
HB23	1	6	6	6	1984	Tangxiun. Hebei	no
HB32	1	6	1	7	1984	Henan	Zhengeng 107
BJ2	1	1	2	8	1984	Fangshan, Beijing	Yuefu
BJ1	1	1	1	9	1984	Fangshan. Beijing	Zhonghua 8
SC38	1	1	1	9	1977	Sichuan	—
BJ11	1	9	9	10	1984	Daxing. Beijing	Qiuyou 20
BJ12	1	9	9	10	1985	Huidiun. Beijing	Jingdao 2
BJ74	1	9	14	11	1986	Shuungqiuo. Beijing	Zhongzuo 180
KS6 - 6	2	1	1	9	—	Jıangsu	—
BJ4	2	1	1	9	1984	Haidian, Beijing	Yueshi
BJ9	2	1	1	9	1985	Daxiug. Beijing	Zhonghua 9
BJ10	2	1	1	9	1985	Daxing. Beijing	78 - 10
HB14	2	1	1	9	1984	Tunghai. Hebci	Jingyue / C57
HB17	2	1	1	9	1984	Fengrun. Hebei	7608
HB22	2	1	1	9	1984	Tangxian. Heber	720
HB26	2	1	1	9	1984	Baigezhuang. Hebei	Zhoughua 8

（续）

Strain	Pathotype[1]	RFLP type[2]			Year of collection	Geographic origin	Cultivars
		pBSavrXa10	pJEL101	Combined			
HB27	2	1	1	9	1984	Tanghai. Hebei	Kenfong：2
HB29	2	1	1	9	1984	Furuug. Hebei	Xingsdao
HLJ71	2	1	1	9	1985	Baoshan. Heilongjiang	Jingyin 127
HUN1	2	1	1	9	—	Hunan	—
SD38	2	1	5	12	1984	Linyi. Shandong	Jingyi 119
HN31	2	1	5	12	1984	Kaifeng，Henan	Zhoughuu 8
SX73	2	1	6	13	1985	Narizhen，Shaanxi	Zhenshan 97
BJ5	2	1	16	14	1984	Haidin n. Beijing	Zhongxi 8419
HB18	2	1	8	15	1984	Fengrun. Hebei	Zhonghua 9
HB19	2	1	8	15	1984	Fengrun，Hebei	Kengfeng 5
HN33	2	1	7	16	1984	Kaifeng. Henan	Xindao 68－11
JL64	2	1	2	8	1985	Zhangjiadian. Jilin	126
JL65	2	1	2	8	1985	Gudianzi. Jilin	126
JL67	2	1	2	8	1985	Yongji，Jilin	Jingyin 127
LN48	2	2	2	1	1985	Donggou. Liaoning	Zhonghua 9
NX42	2	6	1	7	1985	Wuzhong. Ningxia	Ninggeng
HN34	2	6	1	7	1984	Kaifeng. Henan	Zhengeng 107
JL63	2	7	1	17	1984	Ji'an，Jilin	Jigeng 60
HN35	2	7	1	17	1984	Kaifeng. Henan	Zhengeng 107
JL63	2	7	2	18	1984	Ji'an，Jilin	Zhongzuo 75
BJ8	2	8	1	19	1984	Daxing. Beijing	8－10
TJ13	2	8	1	19	1984	Jixian. Tianjin	Kenxi 2
KS1－21	2	13	13	20	—	Jiangsu	—
HN36	3	6	1	7	1985	Xinxiang. Henan	Zhengeng 107
HN37	3	7	5	21	1985	Xinxiang. Henan	68－11
LN57	4	I	5	12	1985	Liaozhong. Liaoning	Liaogeng 5
ZHE173	4	5	2	22	—	Zhejiang	—
FJ1	4	5	2	22	—	Fujian	—
ZHE18	4	5	2	22	—	Zhejiang	—
ZHE35	4	5	2	22	—	Zhejiang	—
JL76	4	16	16	23	1985	Tonghua. Jilin	Tongxi 22
GD1358	5	11	11	24	1982	Xinyi. Guangdong	Zhi 20－5
HB15	6	2	2	1	1984	Zunhua，Hebei	Jingyue 1/C 57
HB16	6	2	1	25	1984	Zunhua，Hebei	Jingyue 1
LN47	6	2	2	1	1985	Xjnrniu，Liaoning	292

（续）

Strain	Pathotype[1]	RFLP type[2]			Year of collection	Geographic origin	Cultivars
		pBSavrXa10	pJEL101	Combined			
LN53	6	2	2	1	1985	Donggou. Liaoning	Zhongdan 2
LN49	6	4	3	26	1985	Kaiyuan, Liaoning	Zhongzuo 59
LN50	6	4	3	26	1985	Kaiyuan, Liaoning	Tegeng 1
LN54	6	4	3	26	1985	Donggou. Liaoning	Jingyue 1
LN43	6	10	10	27	1984	Liaoning	Liaogeng 5
JS49－6	7	1	1	9	—	Changsha, Hunan	—
Jiangling	?	12	12	28	1981	Hubei	IR8
YN5	?	15	15	29	—	Yunnan	—

[1] Pathotype was determined by inoculation on the rice differential cultivars, IR26, Java14, Tetep, Nangeng15 and Jingang 30.

[2] RFLP type determined from Southern blot analysis of *Eco*RI digested total DNA and *Bam*HI digested total DNA from strains of *Xoo* with pJEL101 and pBSavrXa10 as probes respectively.

2.2 Genetic diversity

Genetic diversity values were calculated for each pathotype based on RFLP data derived with probes pJEL101 and pBSavrXa10 (Table 2) . Approximative values, 0.77 and 0.83 were estimated respectively. The genetic diversity calculated for each pathotype treated as a subpopulation (H_R) was the highest for pathotype 6 (0.79) and (0.84) and the lowest for pathotype 5 and 7 (both 0) using probes pJEL 101 and pBSavrXa10, respectively (Table 2) .

On the basis of the occurrence of RFLP types, most of the pathotypes were heterogeneous groups, For instance, by using pJEL101 probe, pathotype 1 (including 28 strains) and pathotype 2 (including 31 strains), presenting 7 and 8 RFLP types, respectively; by using pBSavrXa10 probe, pathotype 1 and 2 all presenting 7 RFLP types. The other pathotypes also expressed heterogeneous groups, even though in a few strains treated, i. e. , pathotypes 3, 4 and 6, with two, three, and four types in 2, 3, and 6 strains tested. The another two pathotypes, 5 and 7, with one each different type in one strain (Table 2) .

Table 2 Frequencies of RFLP types of *X. o.* pv. *oryzae* according to pathotype grouping and the estimated genetic diversity of each group (pJEL101 and PBSavrXa10)

RFLP type	Type strain	Total population		Frequency of RFLP type in each pathotype grouping													
				1		2		3		4		5		6		7	
		NS	NP	NS	%[1)]	NS	%[1)]	NS	%[1)]	NS	%[1)]	NS	%[1)]	NS	%[1)]	NS	%[1)]
pJEL101																	
1	JL13	29	37.2	8	28.6	18	58.0	1	33.3					1	12.5	1	100

（续）

RFLP type	Type strain	Total population		Frequency of RFLP type in each pathotype grouping													
				1		2		3		4		5		6		7	
		NS	NP	NS	%1)	NS	%1)	NS	%1)	NS	%1)	NS	%1)	NS	%1)	NS	%1)
2	LN44	24	30.1	12	42.9	5	16.1			4	66.6			3	37.5		
3	LN49	3	3.8											3	37.5		
4	JL62	2	2.6	2	7.1												
5	HN37	4	5.1			2	6.5	1	33.3	1	16.7						
6	HB23	3	3.8	2	7.1	1	3.2										
7	HN33	1	1.3		1	3.2											
8	HB19	2	2.6			2	6.5										
9	BJ12	2	1.3	2	7.1												
10	LN43	1	1.3											1	12.5		
11	GD1358	1	1.3									1	100				
12	JL691	1	1.3					1	33.3								
13	KS1-21	1	1.3			1	3.2										
14	BJ74	1	1.3	1	3.6												
15	YN15	1	1.3			1	3.2										
16	JL76	2	2.6	1	3.6					1	16.7						
Total no. of strains		78		28		31		3		6		1		8		1	
Genetic diversity		0.77		0.74		0.65		1.0		0.67		0		0.79		0	
pBSavrXa10																	
1	HB17	27	34.6	3	11.1	22	68.8							1	11.1	1	100
2	HLJ72	14	17.9	9	33.3	1	3.1							4	44.4		
3	LN44	6	7.7	6	22.2												
4	JL62	5	6.4	2	7.4									3	3.33		
5	ZHE173	4	5.1							4	80.0						
6	HB23	6	7.7	3	11.1	2	6.3	1	33.3								
7	HN37	4	5.1			3	9.4	1	33.3								
8	TJ13	2	2.6			2	6.3										
9	BJ12	3	3.8	3	11.1												
10	LN43	1	1.3											1	11.1		
11	GD1358	1	1.3									1	100				
12	JL691	1	1.3					1	33.3								
13	KS1-21	1	1.3			1	3.1										
14	HB25	1	1.3	1	3.7												
15	YN15	1	1.3			1	3.1										
16	JL76	1	1.3							1	20.0						
Total no. of strains		78		27		32		3		5		1		9		1	
Genetic diversity		0.83		0.83		0.52		1.0		0.4		0		0.75		0	

Note：NS=No. of strains，NP=No. of pathotypes，JL691=Jiangling 691. 1) Representing the percentage of pathotype.

3 Discussion

In *Xoo* in China, we observe much greater variations in molecular phenotypes than in pathogenic patterns as Leach et al. (*1990*) reported. Although avirulence loci and the genomic distribution of the repetitive element are not functionally related, our observation would suggest that the molecular phenotypes as revealed by pJEL101 or pBSavrXa10 hybridization patterns, change at a higher rate than avirulence loci. In general, the avirulence loci must take account that measurement of diversity at avirulence loci is limited by the number of rice differential cultivars available for difining race. Alternatively, the repetitive element may play an important role in race differentiation. Rabecca et al. (1994) based on DNA and pathotypic analysis, together with information on the spatial and temporal distribution of the pathogen types from their studies, a general picture of *Xoo* evolution in the Philippines is presented.

The relationship between pathotype and putative phylogeny was assessed with two probes that yielded relatively robust phylogenies. Similar groups of strains were inferred from the RFLP data obtained. Strains of pathotypes 1, 2 and 6 formed a large heterogeneous cluster based on the two probes. A single strain, LN43 of pathotype 6, strain, GD1358 of pathotype 5, and a strain YN5, of unknown pathotype appeared to be only remotely related to the other strains based on the both probes.

Based on the analysis above, it appeared that some pathotypes consisted of multiple lineages, and some lineages consisted of multiple pathotypes. Many strains with very different RFLP patterns proved to be good candidates for a new race group and were useful in screening new sources of resistance (Vera Cruz et al., 1992). The molecular technique will be a very available tool for explanation the host-pathogen response in rice-*Xoo* system in China and reconnoitring new races.

To determine whether the inferred population structure would depend on the particular probe utilized, we compared the results of RFLP analysis using two different element probes. Similar estimates of diversity were obtained for some of the pathotypes which were assessed a few strains between the two probes used. Somewhat variation of diversity however, was appeared for estimating pathotypes 1 and 2 between the two probes. The genetic diversities estimated based on pJEL101 and pBSavrXa10 were 0.74, 0.83 for pathotype 1, and 0.65, 0.52 for pathotype 2, respectively. The preliminary results suggested that the both DNA probes can be used for characterizing the population structure of *Xoo* in China.

Although the dendrogram constructed on the basis of RFLP banding patterns provides new insight into the evolutionary relationship among those strains predominantly from North and Northeast China and a few ones from other regions of China only. Therefore, an intensive analysis of the pathogen population structure and genetic diversities including the background regarding the time, location of collection and host origins of strains will be very necessary.

原载：中国水稻科学，1995，9 (1)：7～14.

中国水稻白叶枯病菌染色体DNA的RFLP谱型初析

章　琦[1]　Jan E. Leach[2]　王春莲[1]　R. J. Nelson[3]　T. W. Mew[3]

（1. 中国农业科学院，北京；
2. 堪萨斯州立大学植物病理系，美国；3. 国际水稻研究所，菲律宾）

Preliminary Analysis of Genome DNA RFLP of *Xanthomonas oryzae* pv. *oryzae* in China

Abstract　The RFLP types of 78 strains of *Xanthomonas oryzae* pv. *oryzae* have been assessed by using two probes, pJEL101and pBSavrXa10 to analyze the population structure and genetic diversity. Sixteen RFLP types were identified, respectively. Cluster analysis based on RFLP banding showed six groupings at 85% similarity. The genetic diversities of the population of strains tested were 0.77 (by using pJEL101) and 0.83 (by using pBSavrXa10) respectively. The analysis of RFLP types suggested that most of the Chinese pathotypes formed large heterogeneous clusters. It was observed that the variation of molecular phenotypes was much greater than that of the pathogenic patterns in *Xoo* in China. Both of DNA probes can be used for characterizing the population structure of *Xoo* in China. The molecular technique will be an useful tool for explicating the host-pathogen response in *rice-X. o.* pv. *oryzae* system in China and detecting new races.

Key words　bacterial blight; strains; RFLP types

摘要　用2个DNA探针pJEL101和pBSavrXa10对78个水稻白叶枯病菌株进行RFLP分型，以分析其群体结构和遗传多样性。分别鉴定出16种RFLP标记带的谱型。以彼此带位相似率达85%为界，被分为6簇。参试菌株的群体遗传多样性为0.77（用pJEL101）和0.83（用pBSavrXa10）。RFLP谱型分型表明，我国多数病原型是杂合组群。白叶枯病菌的分子表现型变异远远大于其致病型，采用的两个探针都能有效地分析我国菌株的群体结构，分子技术可以作为侦察我国水稻与白叶枯病菌之间的鉴别反应系统和新小种的一种手段。

关键词　稻白叶枯病；菌株；RFLP谱型

水稻白叶枯病是一种利用寄主的抗性能够有效控制为害的细菌病害。目前生产上能否成功地利用品种的抗病性，取决于两个因素：1）对病原菌的遗传多样性是否充分了解；2）有相应的抗性基因进行轮换。建立一个寄主与病原菌互作反应清晰的鉴别系统，是进行上述研究的重要基础。

菲律宾和日本都建立了一套能清晰地反映水稻—白叶枯病互作的鉴别系统，在此基础上，成功地鉴定了若干遗传背景清楚的抗白叶枯病基因。近年，IRRI 和美国合作，又进行了能揭示 DNA 序列中大量变异的 RFLP 分析，已经产生许多标记，用以测定白叶枯病菌的群体遗传多样性，可以在分子水平上探索病菌生态，流行病学以及寄主抗性机制等方面的问题，为遗传育种利用和基础研究提供了很有价值的资料（Gabriel et al.，1988；Leach et al.，1990）。

我国植物病理学家曾根据在 5 个鉴别品种上的反应型，将全国采集的白叶枯病菌株分为 7 个致病型。但是寄主与病原菌之间的互作反应不很清晰。因此在评价我国的丰富资源以及分析抗性基因时，还必须采用国际鉴定系统。为了解我国菌株遗传多样性及病原型之间的系统发育关系，本研究采用 2 个 DNA 探针，对我国白叶枯病菌株基因组 DNA 进行 RFLP 谱型分析，进行初步探索。

1 材料和方法

1.1 菌株和 DNA 探针

本试验在 RFLP 分析中，采用 78 个包括中国白叶枯病 7 个病原型的菌株。其中多数来自华北和东北、少量来自华中、华南和西南，共 18 个地区。其所属病原型、采集地区、时间和寄主品种等均列于表 1（见前一篇表 1 本书 264～266 页）。全部菌种保存于冻干状态。用新鲜 PBS 斜面培养基复壮，以分离菌株全基因组 DNA。

采用的 2 个 DNA 探针：pJEL101 为一个含有源出菲律宾白叶枯病菌生理小种 2（代表菌株 PXO86）的 2.4 Kb *Eco*RⅠ-*Hind*Ⅲ片段的 pUC18 质粒。pJEL101 的插入片断含有高拷贝的重复 DNA 单位，它存在于经测定过的 100 多个白叶枯病菌株的 DNA 中。另一个探针是 pBSavrXa10 质粒，含有一个在白叶枯病菌无毒基因 *avrXa10* 内的长为 3.1 Kb 的 *Bam*HⅠ片段（Hopkins，et al.，1992）。

1.2 DNA 分析

白叶枯病菌株全基因组 DNA 的分离方法如 Leach 等（1990）所述。用 pJEL101 为探针时，以 *Eco*RⅠ酶切，用 pBSavrXa10 为探针时，以 *Bam*HⅠ酶切。菌株 DNA 片断在 TBE 缓冲液中，通过浓度为 0.7%的琼脂糖凝胶（大小为 20cm×21cm）电泳。以 1 Kb 梯级 DNA 为大小标准，经 Southern 吸印，将凝胶上的所有 DNA 片段，转移到尼龙滤膜上。完整的 pEL101 或 pBSavrXa10 质粒，用缺刻转移试剂盒，以 32p 标记。均在严格条件下进行分子杂交和洗膜。用 Cronex 胶片和 Cronex Lightning-Plus 以感屏完成放射自显影，在－80℃进行。试验第一步，将来自同一地区的菌株，在同一块凝胶中测试，作为初步分析和归纳，随后选出不同 RFLP 谱型的代表菌株，在相同的条件下进一步测定和比较。

1.3 数据分析

为测定不同RFLP谱型间的遗传关系，在用探针pEJL101杂交时，以开始出现差异的带纹为起点，确定20条标记的带位；在用pBSavrXa10杂交时共有19条，将每个带位上有或无带，分别转换为1或0两个数据。根据RFLP分析数据，采用UPGMM（Vera Cruze et al.，1992）分簇法推断白叶枯病菌株之间的系统发育关系。本研究以不同的RFLP谱型标记带位的相似率达85%为界分簇，并根据下列公式计算遗传多样性：

$H=[n/(n-1)]\times(1-\sum Xi^2)$，$i=1$；$Xi$为群体中第$i$个样品的谱型出现的频率；$n$是所测菌株数。

2 结果

2.1 RFLP分析

用2个探针pJEL101和pBSavrX10测定的78个稻白叶枯病菌株及全基因组DNA的多态性表现极为相似。都有16种RFLP谱型（图1和图2）。病原型1、2和6的多数菌株为RFLP谱型1～4（表1和表2），这些主要来自华北、东北的多数菌株的RFLP谱型间的标记带位差异不大，仅有1条或3～4条带位之差。但有些谱型则明显不同，用pJEL101测试的RFLP型11，分别与1、15和6型有7或8条标记带位不同（图1）。同样，与pBSavrXa10探针杂交的RFLP型11与10、13和14型之间，分别有7、8和11条标记带位不同（图2）。结果表明来自广东（11型）、云南（15型）、湖北（12型）的菌株与北方菌株的差异很大，并发现了一个辽宁菌株LN43（10型）的RFLP谱型比较特殊（图1、图2）。

2.2 菌株病原型之间的遗传关系

根据UPGMA分簇，分析各菌株全基因组DNA的标记带谱，以探索病原型之间的遗传关系。因为用2个探针分别杂交的RFLP谱型相似，因此只分析了pJEL101的结果。本试验以全部参试菌株之间的带位相似率达85%为界，可将16个RFLP谱型分为6簇（图3）。来自华北、东北的绝大多数菌株，呈现8个RFLP谱型，属于中国致病型1、2和6，归为簇1；病原型3和4（河南、东北菌株）和属病原型1的北京菌株为簇2、簇3。与北方菌株有极少相似带位的广东和一个云南的未知病原型菌株（YN5）分别为簇5和簇6。初步结果表明，参试的绝大多数北方稻区的菌株的系统发育关系较近，而华南、西南、华中和一个辽宁菌株（LN43）与其他菌株的系谱则相距较远。

2.3 菌株群体的遗传多样性

根据参试的中国7个病原型菌株，分别用探针pJEL101和pBSavrXa10测定的RFLP数值计算的遗传多样值较为接近，它们分别为0.77和0.83（表2和表3）。用2个探针测定各病原型亚群内的遗传多样性值最高的是0.79（病原型6，用pJEL101），和0.83（病原型1，用pBSavrXa10）。

RFLP谱型分析表明，多数病原型是杂合组群，在pJEL101测定的病原型1的28个和

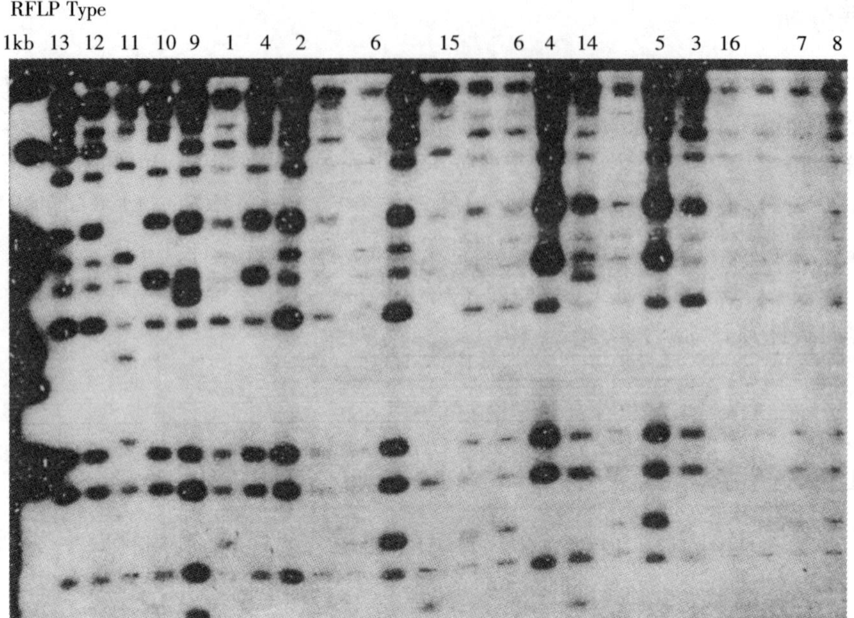

图 1　Southern 印迹分析来自我国 18 个地区的水稻白叶枯病菌系染色体 DNA 的 16 个 RFLP 型（^{32}p 标记探针为 pJEL 101）

Fig. 1　Southern blot analysis of genomic DNA of strains of *Xoo* from 18 rigions in China showed 16 RFLP types (the probe used was ^{32}p-labeled pJEL 101)

病原型 2 的 31 个菌株中，分别呈现 7 和 8 种不同的 RFLP 谱型。在用 pBSavrXa10 测定的病原型 1 的 27 个和病原型 2 的 32 个菌株中，均呈现 7 种 RFLP 谱型。其他病原型也表现杂合组群状态，例如，仅参试 2 个菌株的病原型 3 就有 2 种 RFLP 谱型。在分别参试的 3 个和 6 个菌株的病原型 4 和 6 中，分别测出 3 个和 4 个谱型。病原型 5 和病原型 7 各自参试的 1 个菌株，也都表现出不同的谱型。

3　讨论

3.1　水稻白叶枯病菌株全基因组 DNA 的分子表现型的变异远远大于其致病型

DNA 重复单位在研究病菌生理小种分化上，可能发挥了重要作用，虽然无毒基因的位点同基因组重复单位的分布，在功能上没有相关。本研究表明，由菌株全基因组 DNA 与 pJEL101、或 pBSavrXa10 两个探针杂交图形所揭示的分子表现型，比无毒基因位点的变异率更高。通常在衡量无毒基因时，应该考虑到对无毒基因位点多样性的测定，必然受到鉴定病原生理小种时所采用的水稻鉴别品种数量的限制。因此，DNA 重复单位在更为科学地划分生理小种方面可能发挥重要作用，因为它所揭示的结果更接近于生物的实质。Rebecca 等根据菌株 DNA 和小种的分析，结合小种在时空分布方面的研究资料，形成了一个较完整的菲律宾白叶枯病原菌演变的概貌，从而调整和划分了某些小种（个人通讯）。为抗病遗传育种提供了重要的依据。

表 2 水稻白叶枯病原菌群中参试菌株的 RFLP 谱型的频率及其遗传多样型估值（用 pJEL101 为探针）

Table 2 Frequencies of RFLP types of *Xoo* according to pathotype grouping and the estimated genetic diversity of each group（pJEL101used）

RFLP 谱型 (Type)	代表菌株 Strain	菌株数及所占% Stain No. & %		7 个病原型菌株 RFLP 谱型频率（Frequency of RFLP type in seven pathotypes） 1		2		3		4		5		6		7	
				菌株数 No.	% %	菌株数 No.	% %	菌株数 No.	% %	菌株数 No.	% %	菌株数 No.	% %	菌株数 No.	% %	菌株数 No.	% %
1	TJ13	29	37.2	8	28.6	18	58.0	1	33.3					1	12.5	1	100
2	LN44	24	30.8	12	42.9	5	16.1			4	66.6			3	37.5		
3	LN49	3	3.8											3	37.5		
4	TJ62	2	2.6	2	7.1												
5	HN37	4	5.1			2	6.5	1	33.3	1	16.7						
6	HB23	3	3.8	2	7.1	1	3.2										
7	HN33	1	1.3			1	3.2										
8	HB19	2	2.6			2	6.5										
9	BJ12		2	2.2	2	7.1											
10	LN43	1	1.3											1	12.5		
11	GD1358	1	1.3									1	100				
12	JANL691	1	1.3					1	33.3								
13	KS1－21	1	1.3			1	3.2										
14	BJ 74	1	1.3	1	3.6												
15	YN 15	1	1.3			1	3.2										
16	JL76	2	2.6	1	3.6					1	16.7						
合计 Total		78		28		31		3		6		1		8		1	
遗传多样性值（Genetic diversity）		0.77		0.74		0.65		1.0		0.67		0		0.70		0	

表 3 水稻白叶枯病原菌群中参试菌株的 RFLP 谱型的频率及其遗传多样型估值（用 pBSavrXa10 为探针）

Table 3 Frequencies of RFLP types of *Xoo* according to pathotype grouping and the estimated genetic diversity of each group（pBSavrXa10）

RFLP 谱型 (Type)	代表菌株 Strain	菌株数及所占% Stain No. & %		7 个病原型菌株 RFLP 谱型频率（Frequency of RFLP type in seven pathotype） 1		2		3		4		5		6		7	
				菌株数 No.	% %	菌株数 No.	% %	菌株数 No.	% %	菌株数 No.	% %	菌株数 No.	% %	菌株数 No.	% %	菌株数 No.	% %
1	HB17	27	34.6	3	11.1	22	68.6							1	11.1	1	100
2	HLJ72	14	17.9	9	33.3	1	3.1							4	44.4		
3	LN44	6	7.7	6	22.2									3	33.3		

（续）

RFLP 谱型 (Type)	代表菌株 Strain	菌株数及所占% Stain No. & %		7个病原型菌株RFLP谱型频率（Frequency of RFLP type in seven pathotype）													
				1		2		3		4		5		6		7	
				菌株数 No.	%	菌株数 No.	%	菌株数 No.	%	菌株数 No.	%	菌株数 No.	%	菌株数 No.	%	菌株数 No.	%
4	JL62	5	6.4	2	7.4												
5	ZHE173	4	5.1							4	80.0						
6	HB23	6	7.7	3	11.1	2	6.3	1	33.3								
7	HN37	4	5.1			3	9.4	1	33.3								
8	TJ13	2	2.6			2	6.3										
9	BJ12	3	3.8	3	11.1												
10	LN43	1	1.3													1	11.1
11	GD1358	1	1.3									1	100				
12	JANL691	1	1.3					1	33.3								
13	KS1－21	1	1.3			1	3.1										
14	HB25	1	1.3	1	3.7												
15	YN15	1	1.3			1	3.1										
16	JL76	1	1.3	1	3.6					1	20.0						
合计 Total		78		27		32		3		5		1		9		1	
遗传多样性值（Genetic diversity）		0.83		0.83		0.52		1.0		0.40		0		0.75		0	

3.2 进一步了解了我国水稻白叶枯病菌病原型之间的系统发育关系

根据参试菌株的RFLP数据，有些谱系包括了几个病原型。例如病原型1、2和6的多数北方稻区菌株，都在一个簇中，其DNA标记带位的相似率较高，表明其遗传系统发育关系较近。与北方菌株的RFLP谱型差距较大的广东、云南、湖北菌株，分别归属于不同的簇，它们与传统病理学的测定结果有较高相关性。但是因为这些地区的标样较少，需要增加样品作进一步研究。

有些病原型所包括的菌株，分别归属于不同簇，例如病原型1的菌株归属到簇1和簇3；病原型6的菌株，归属到簇1和4，显然，这些群体处于杂合状态。可以联想到，在新的鉴别品种上可能鉴别出新的病原型，这些都是有待筛选和发现的。

辽宁、北京的少数菌株如LN43、BJ12等呈现特异的RFLP谱型，这些菌株与那些特殊谱型的菌株，也许是新病原型的候选者。未来通过扩大筛选，可以用来鉴定新的抗源。因此，分子技术对于进一步完善我国在水稻与白叶枯病的寄主与病原物之间的鉴别反应系统、发掘新抗源和监测新小种，应该是一个有效的手段。

3.3 采用不同探针与测定中国菌株群体结构的关系

分别用2个探针杂交所形成的RFLP谱型结果表明，大部分菌株的遗传多样性估计值是类似的。有些病原型略为不同，如1型与2型，病原型1与pJEL101和pBSavrXa10的遗传多样性值分别为0.74和0.83。病原型2的遗传多样性值分别为0.65和0.52。表明这2个DNA探针均可用以描述我国水稻白叶枯病菌群体结构和遗传多样性。相对而言，用pBSavrXa10与中国的菌株基因组DNA杂交的RFLP谱型更为清晰可辨，便于归类。但要求较高的操作条件，需要在设备齐全的室温为4℃冷室（cold room）中进行病菌DNA片段的常规琼脂糖凝胶电泳，历时约70h。而用探针pEL101，则可在室温下电泳20～22℃小时，但其标记带位较多，不甚清晰，因此有条件的实验室，同时用2个探针进行比较分析更好。初步探索了我国菌株基因组DNA的RFLP谱型构成的谱系，和参试菌株之间的演变关系。但样品主要来自华北、东北，其他稻区的菌株较少。有必要进一步测定更多地区的菌株来研究我国白叶枯病菌的完整的群体结构和遗传多样性。

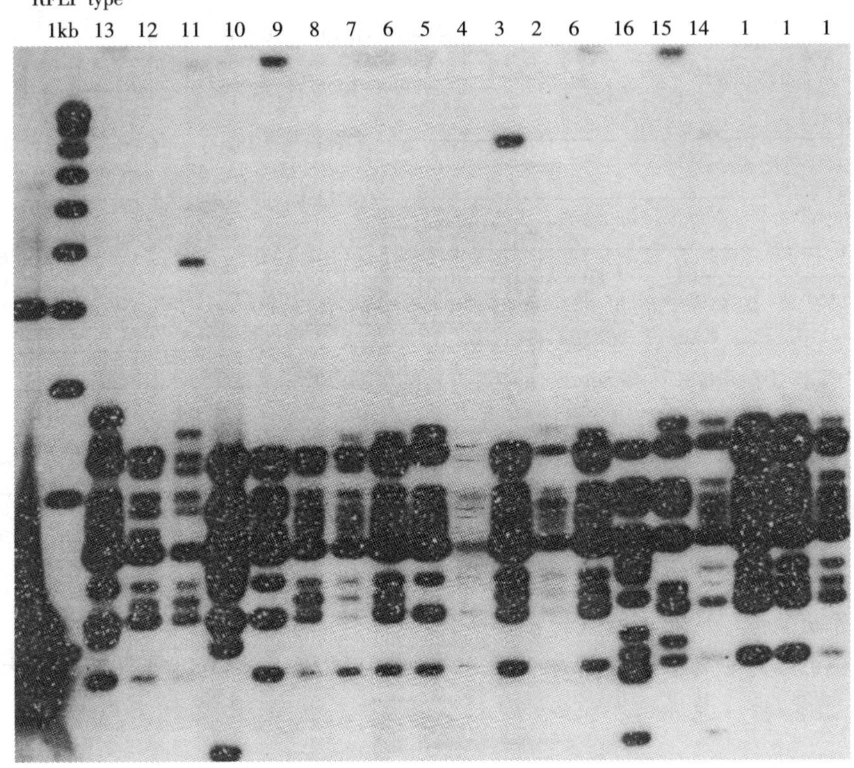

图2 Southern印迹分析来自我国18个地区的水稻白叶枯病菌株全基因组DNA的16个RFLP型（^{32}p标记探针为pBSavrXa10）

Fig. 2 Southern blot analysis of total genomic DNA of strains of *Xoo* from 18rigions in China showed 16 RFLP types (the probe used was ^{32}p-labeled pBSavrXa10)

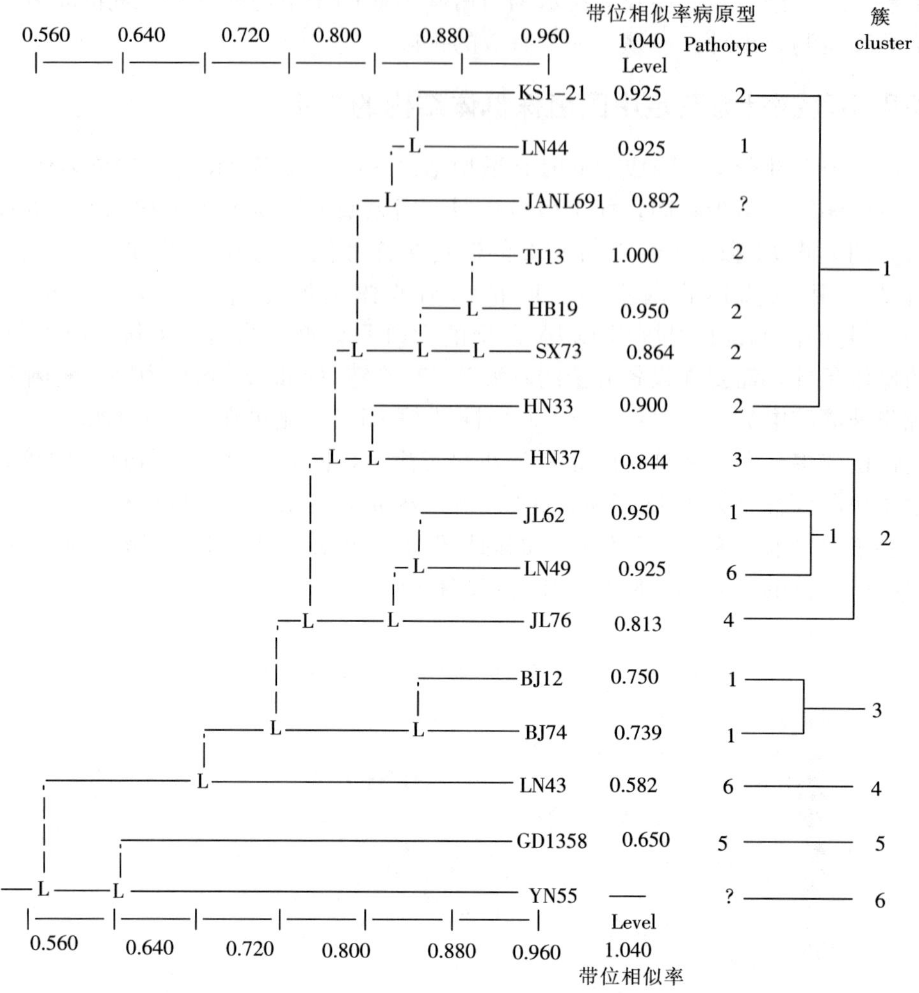

图 3 用 UPGM 计算 Southern 印迹分析的水稻白叶枯病的 16 种 RFLP 型所呈现的菌系间的树状系谱图

Fig. 3 Dendrogram displaying the relationships among the 16 types for strains of *Xoo* pv. *oryzae* after Southern blot analysis using [32]p-labeled pJEL 101 as a probe. Similarity coefficients were grouped by using the unweighted pair group method with arithmetic means

原载：作物学报，1996，22 (1)：13～19.

＊本试验由洛克菲勒基金资助，于 1989 年和 1993 年，在美国堪萨斯州立大学和 IRRI 合作进行

水稻白叶枯病菌的RAPD分型

朱　衡[1]　章　琦[2]　王春莲[2]　杨文才[2]　朱立煌[1]

（1. 中国科学院遗传研究所；2. 中国农科院作物研究所）

RAPD Typing of Rice Bacterial Blight Pathogen *Xanthomonas oryzae* pv. *oryzae*

Abstract　By using RAPD method to amplify 28 Chinese rice bacterial blight isolates-DNAs，primer A-14（Opron product）was used to assign them into 11 clusters. This result was comparable with their phenotypes characterization and RELP typing. Besides，random primer A-13 was also used to analyze some doubtful isolates.

Key words　RAPD；rice bacterial blight；typing

对植物病原菌的分型是生物技术应用的新领域。水稻白叶枯病（*Xanthomonas oryzae* pv. *oryzae*，*Xoo*）是亚洲地区水稻生产上的主要病害。方中达等（1990）曾根据*Xoo*菌株在5个水稻鉴别品种上的反应型将中国水稻白叶枯病菌分为7个致病型。近来，Leach等（1990）已开始利用基因组特异的重复顺序对白叶枯病菌进行基因组DNA指纹分析，Zhang等（1995）也采用相同的探针对我国部分菌株进行了RELP分型，其结果与传统病理学的分类有相符之处，但又有相当的不同，说明传统分类结果尚有许多值得探索之处。

与RFLP相比，RAPD（Random Amplified Polymorphic DNA）具有简便和快速的优点，而且相当灵敏，在动植物基因组分型（genome typing）上已得到较为广泛的应用（Welsh et al.，1991；Devos et al.，1992）。我们利用RAPD标记对我国28个菌株进行分型，发现随机引物A－14可用以区分所分析的28个菌株。进而对扩增带型进行了相似百分率分析，并利用另一随机引物A－13作了进一步分析。本文是用RAPD方法尝试对部分水稻白叶枯病菌分型的首次报道。

1　材料和方法

1.1　菌株和培养

应用中国农科院作物所提供的28个来自我国14个省市的属于7个致病型的水稻白叶枯

病菌株。细菌在5ml马铃薯液体培养基上，于28℃振荡培养15h。

1.2 DNA 提取

采用由朱衡等（1993）提出的氯化苄法。

1.3 RAPD 分析

RAPD反应参照Williams（1990）提供的条件并稍作调整。扩增反应体积为50μl，含10 mmol/L Tris-HCl，pH 8.3；50 mmol/L KCl，2.5 mmol/L $MgCl_2$；100μmol/L四种dNTP（Boehringer）；0.25μmol/L引物；5ng基因组DNA和1单位Taq酶（中国科学院遗传研究所），扩增在PE公司的DNA Thermel Cycler 480上进行：94℃ 2min、94℃ 1min、36℃ 1min、72℃ 2min共45个循环，最后72℃ 10min。所用引物序列如下（Opron公司）：A14：TCTGTGCTGG；A-13：CAGCACCCAC。

1.4 统计分析

RAPD电泳结果均进行了3次重复，大部分带型可以重复，少量不能重复者在统计时忽略不计。根据扩增条带的有无，共统计出37个出现扩增带的位置。对多态性片段的统计采用Nei和Li（1979）的公式：

$$S_{xy}=2N_{xy}/(N_x+N_y)$$

计算各菌株间的相似百分率。其中N_{xy}代表菌株x和y共有的条带数，N_x和N_y分别代表x和y菌株的总条带数。

2 结果和讨论

对我国28个水稻白叶枯病菌株的RAPD扩增结果见图1，反映出其基因组的复杂性和多样性。扩增带分布较广，数目从5条到20条不等。在所有28个菌株中无法找到一条所有菌株的共有带。28个菌株彼此间的Szy值列于表1。结果显示Szy值分布相当广泛，从96%～0。大部分组合间的Szy值分布于50%～80%，平均值为58%。

表1　28个白叶枯病菌株的相似百分率

	1	2	3	4	5	6	7	8	9	10	11	12	13	14	15	16	17	18	19	20	21	22	23	24	25	26	27
2	74																										
3	75	70																									
4	55	63	56																								
5	63	62	83	38																							
6	50	48	40	82	33																						
7	63	63	53	53	46	48																					
8	56	63	71	44	50	40	67																				
9	58	48	60	58	44	54	82	60																			
10	65	58	52	59	29	55	50	52	69																		

（续）

	1	2	3	4	5	6	7	8	9	10	11	12	13	14	15	16	17	18	19	20	21	22	23	24	25	26	27
11	76	78	82	67	67	61	56	59	70	69																	
12	70	70	74	70	69	64	70	74	80	71	91																
13	80	67	67	56	63	59	64	57	74	67	67	70															
14	73	63	78	64	63	42	53	56	58	80	76	70	80														
15	67	67	86	44	83	30	40	57	50	43	71	78	67	78													
16	83	74	78	64	63	58	63	56	67	57	86	80	80	82	67												
17	78	70	74	70	59	48	60	53	56	57	82	75	77	96	63	87											
18	76	78	82	57	67	43	67	59	52	54	80	73	57	76	71	67	77										
19	72	73	67	64	53	67	55	48	67	73	83	77	64	80	57	80	85	57									
20	59	71	77	59	73	42	57	62	53	55	75	67	60	71	62	71	67	63	60								
21	70	71	75	60	57	45	47	63	73	72	74	74	70	70	63	80	67	63	61	80							
22	75	67	70	58	67	54	57	60	69	69	87	88	74	58	60	80	72	70	74	63	64						
23	53	50	73	27	80	24	33	36	35	30	57	50	44	53	73	53	38	57	44	60	46	47					
24	60	59	63	40	57	36	35	38	45	48	63	57	43	50	63	60	57	63	61	40	56	55	46				
25	42	57	40	58	22	69	48	50	62	55	61	64	52	50	30	58	56	43	67	42	55	62	12	27			
26	15	9	9	15	0	29	35	27	29	39	8	15	28	23	9	15	22	16	28	0	17	14	0	17	21		
27	80	71	75	50	71	45	59	50	55	67	74	57	61	70	75	80	76	80	70	53	67	64	62	78	36	25	
28	63	62	83	38	80	22	31	46	44	38	67	71	53	63	83	63	59	67	53	55	57	56	67	71	22	10	71

根据结果，以 60％的相似率为界，将彼此间 S*zy* 值高于 60％的菌株归为一簇(Cluster)，结果列于表 1。菌株 ZHE35、LN43、BJ2、BJ4、BJ7 与其他菌株间的 S*zy* 值较低，平均只有 54％、51％、47％和 16％，因此单独组成一簇。在簇 1 的 11 个菌株中，均有 A、B、C 和 D 4 条带；簇 2 的 5 个菌株都有 A、B、C、D、E 和 F 6 条带。簇 1 中除了 JS49—6 属于致病型 7 和 JL691 的致病型尚未确定外，其余均属致病型 1。而 JS49—6 的 RFLP 分型结果也与其致病型不同（表 2），表明对这个菌株的传统分类需进一步探究。簇 2 中的 5 个菌株全属致病型 2，与传统结果是相符的。属致病型 1 的有些菌株还被划分入簇 3、簇 8 和簇 10，这与 RFLP 对致病型 1 分类的结果相类似（Zhang et al.，1995），说明传统分类确定的致病型 1 可能还存在若干其他组。在致病型 2 中也存在同样的情况。

表 2　28 个白叶枯病菌株的 RAPEE 分型

簇	菌　株
1	JS49－6、SC38、LN44、HW32、HB24、BJ11、BJ12、BJ74、JL691、HB25
2	HB17、NX42、BJ8、BJ9、KS1－21
3	HLJ72、BJ3
4	YUN5、BJ10
5	ZHE173、FJ1
6	ZHE35
7	LN43
8	BJ2
9	BJ4
10	BJ7
11	GD1358

北京地区菌株的致病型非 1 即 2，表现单一，但 RFLP 和 RAPD 分型结果均显示出其组成复杂，如表 2 和表 3 所示，其中 BJ7 的表现更为特异（图 1）。为此，我们又利用引物 A－13 作了进一步分析（图 2），结果表明其特异性仍很高。这一结果说明将 BJ7 单独划成一簇是合理的。同样，BJ10、BJ2 和 BJ4 也被证明可被单独划分出来。可见，北京地区白叶枯病菌的组成较为复杂。进一步的研究将扩大分析菌株的数量以证明引物 A－14 在白叶枯病菌分型上的广泛适用性。

表 3　28 个水稻白叶枯病菌的 RFLP 分析

菌株	序号	RLP 分型		
		致病型	pBSavrXa10	pJEL101
HB17	1	2	1	1
JS49－6	2	7	1	1
SC38	3	1	1	1
HLJ72	4	1	2	2
LN44	5	1	3	2
BJ3	6	1	4	4
ZHE173	7	4	5	2
ZHE35	8	4	5	2
FJ1	9	4	5	2
NX42	10	2	6	1
HB32	11	1	6	6
HW32	12	1	6	1
HB24	13	1	6	6
J8	14	2	8	1
BJ11	15	1	9	9
BJ12	16	1	9	9
BI72	17	1	9	9
LN43	18	6	10	10
GD1358	19	5	11	11
JL691	20	7	12	12
KS1－21	21	2	13	13
HB25	22	1	14	[illegible]
YUN5	23	3	15	15
BJ2	24	1	1	2
BJ4	25	2	1	1
BJ7	26	1	3	1
BJ9	27	2	1	1
BJ10	28	2	1	1

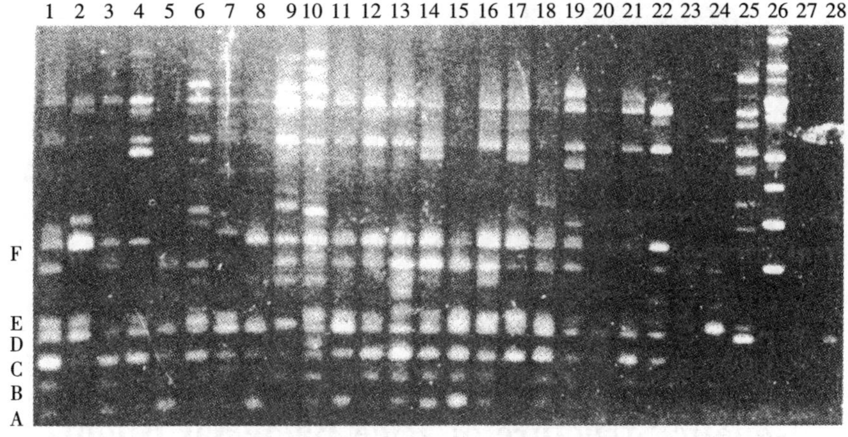

图 1　引物 A-14 对 28 个白叶枯病菌株进行 PAPD 扩增的结果图中各序号代表的菌株与表 3 相同，扩增的 DNA 在 1.8%的琼脂糖凝胶上电泳分离

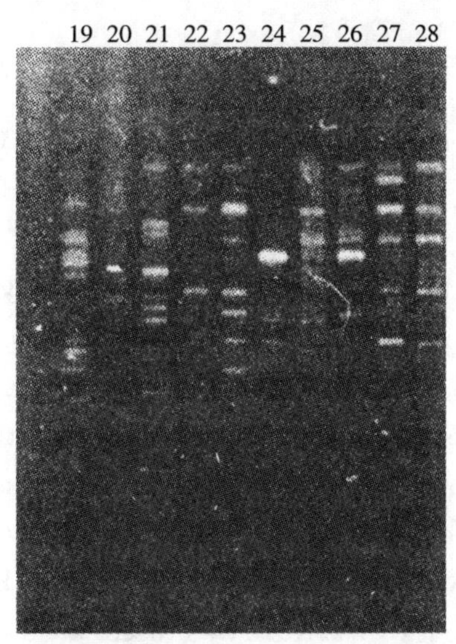

图 2　引物 A-13 对 11 个白叶枯病菌株进行 RAPD 扩增的结果图中各序号代表的菌株与表 3 相同，扩增的 DNA 在 1.8%的琼脂糖凝胶上电泳分离

原载：生物工程学报，1995，11（3）：281～284.

中国水稻白叶枯病原菌群体的遗传结构

章　琦[1]　J. E. Leach[2]　R. J. Nelson[3]　T. W. Mew[3]

（1. 中国农业科学院作物育种栽培研究所，北京；
2. 堪萨斯州立大学病理系，美国，堪萨斯；3. IRRI，菲律宾）

Genetic Structure of Rice Bacterial Blight Pathogen Population in China*

Zhang Qi[1]　J. E. Leach[2]　R. J. Nelson[3]　T. W. Mew[3]

（1. Institute of Crop Breeding and Cultivation，CAAS，China；2. Department of Plant Pathology，Kansas State University，Kansas；3. IRRI，Philippines）

Abstract　Rep-PCR and restriction fragment length polymorphism（RFLP）were used to analyze the genetic structure in a population of 143 *Xanthomonas oryzae* pv. *oryzae*（*Xoo*）strains. The strains were isolated from 96 sites in 20 different provinces of the Yangtze River Region，South，North and Northeast China. Rep-PCR was performed by using DNA amplification with primers based on short bacterial repetitive elements（ERIC，BOX and REP primers）and mobile，repetitive elements from *Xoo*（IS1112，IS1113）. For RFLP analysis，a member of a family of avirulence genes，*avrXa10* was used as a probe. Rep-PCR revealed higher haplotypic diversity（H ＝ 0.95）. Of the four rice growing regions，the highest genetic diversity（based on rep-PCR）was observed in South China（H＝0.96），followed by the Yangtze River Region（H＝0.93），Northeast China（H＝0.77），and North China（H＝0.69）. On the basis of the consensus of three clustering statistics，the 143 strains constituted three clusters. Strains from each different rice growing regions grouped into all clusters indicating no geographic differentiation.

Key words　rice bacterial blight；genetic structure；Rep-PCR；RFLP

摘要　用 Rep-PCR 和 RFLP 两种方法分析了 143 个水稻白叶枯病（*Xanthomonas*

---

* The project supported by the Rokefeller Foundation was cooperated with JEN Leach，Kansas State University，USA；Rebecca，and Mew at IRRI in 1989 and 1993.

oryzae pv. *oryzae*）菌株。这些菌株采集于长江流域、华南、华北和东北的 20 个省市的 96 个点。Rep-PCR 是利用一些基于细菌的短的重复序列引物（ERIC、BOX 和 REP）及来自稻白叶枯病原菌的移动重复单位（IS1112 和 IS1113）引物的 DNA 扩增特性。在 RFLP 分析中，以无毒基因家族中的一个成员 *avrXa10* 为探针。Rep-PCR 分析呈较高的单元型多样性（H＝0.75）。在 4 个主要水稻生产区中，用 Rep-PCR 分析的遗传多样性以华南稻区最高（H＝0.96），依次为长江流域（H＝0.93），东北（H＝0.77），华北（H＝0.69）。根据相似指纹型数据进行三维生物统计聚类分析，将 143 个菌株形成 3 个簇群。每个簇群中都有来自不同水稻产区的菌株，表明菌株的分簇并不受地理区域的影响。

关键词 水稻白叶枯病；遗传结构；Rep-PCR；RFLP

Bacterial blight, caused by *Xanthomons oryzae* pv. *oryzae* (*Xoo*), is particularly destructive in South China and the Yangtze River Region, which together encompass 90% of the rice growing area of China. Host resistance is considered a sound approach to manage the disease. Most of the resistance cultivars released during the last 15 years contain *Xa4* or *Xa3* genes for resistance to bacterial blight. However, due to the ability of the pathogen population to adapt to and overcome the resistance, the effectiveness of these genes to control the disease in the future is threatened. Furthermore some combinations of hybrid rice, incorporated with the incomplete dominant gene *Xa4*, are being cultivated in increasingly larger areas of China, and are susceptible to the pathogen, especially during years when the disease is epidemic (Zhang Qi. et al., 1994).

It has been postulated that information on the structure of *Xoo* can be used to refine the management strategies for bacterial blight (Leach et al., 1995). For example, the definition of different genetic lineages of *Xoo* and the discrimination of subpopulations of the pathogen that have different pathotypic properties by molecular typing enabled. Nelson and coworkers to detect three new races of the pathogen (Nelson et al., 1994). In addition, an understanding of *Xoo* population structure might allow predictions on what resistance genes might be deployed in different agroecosystems within a country, or at a regional level between countries (Adhiltari et al., 1995; Nelson et al., 1994). The structure of *Xoo* populations in a few countries, e.g., the Philippines (Leach et al., 1992) and in Asia as a whole have been published recently. Chinese strains were included with a limited sample set (78 strains) and predominantly from North and Northeast China. In this study, we investigated the genetic diversity of *Xoo* strains collected from different rice cultivars throughout China.

1 Materials and Methods

1.1 Bacterial strains

One hundred and forty three strains were collected from 96 locations in 20 provinces and sites located in the Yangtze River Region, South, North and Northeast China from 1984 to

1994. Single colonies were isolated from all cultures. The stocks were kept as lyphilized cultures and the strains were revived on PSA and preserved in 15% glycerol at−80℃.

1.2 DNA probe

pBSavrXa10 was used as the probe in RFIP analysis. pBSavrXa10 contains a 3.1-kb *Bam*H Ⅰ fragment that is internal to an avirulence gene from *Xoo* in vector pBluescript Ⅱ. *Xoo* contains multiple copies of genes related to *avrXa10*; at least two of these (*avrXa10* and *Xa7*) function as avirulence genes (Hopkins et al., 1992).

1.3 RFLP analysis

Genomic DNA from each bacterial strain was extracted by the lysozyme-sodium dodecyl sulfate lysis method of Owen and Broman (1987) as modified by Leach et al. (1992). Bacterial genomic DNA was digested to completion with *Eco*RI (2 U/μg of DNA) at 37℃ for 2 to 3h in buffer supplied by the enzyme manufacturer. Digested DNA (3 to 5μg per well) was separated by electrophoresis, and fragments were transfered to Magna NT nylon membrane as described by the manufacture (MSI, Westboro, Mass). A size standard (1 kb ladder) was included in each gel. The DNA probe was labelled using an ECL Kit (company name). Hybridized fragments were detected by autoradiography on Cronix film.

1.4 Primers

Five primers were used in this study. The primer sequences are as follows:

IS1112 JELIR1 (5′-CGAGTCCAGTCCAGCGGAGG-3′)
JELIR2 (5′-GCGGGGAGCTTCAGAAGAGG-3′)
ERIC ERIC1R (5′-ATGTAAGCTCCTGGGGATTCAC-3′)
ERIC2 (5′-AAGTAAGTGACTGGGGTGAGCG-3′)
IS1113 TX1 (5′-TGTAGTGGACCTTCGAA-3′)
TX2 (5′-ACGAGCGATTGATCATG-3′)
BOX BOXA1R (5′-CTACGGCAAGGCGACGCTGACG-3′)
REP REP1R-I (5′-IIIICGICGICATCTGGG-3′)
REP2-I (5′-ICGICTTATCIGGCCTAC-3′)

The sequence and use of primers ERIC, BOX and REP is described in Louws et al. (1994). Primers based on IS1112 and IS1113 were synthesized in the Kansas State University.

1.5 PCR conditions

Each 25μl PCR reaction contained 50 pmol each of two opposing, oligonucleotide primers, 50 - 100 ng of template DNA as described (Louws et al., 1994). The mixture containing water, reaction buffer, DMSO, primers, dNTP, and template DNA was overlaid with 25μl of mineral oil. PCR amplification were performed in an automated thermal cycle with an initial denaturation (95℃, 7 min) followed by 30 - 35 cycles that include denaturation (94℃ 1 min), annealing at 62℃, 52℃, 50℃, and 44℃ for 1 min with IS1112, ERIC, IS1113, BOX, and REP primers, respectively, and extension at 65℃ for 15 min followed by incubation at 4℃.

PCR amplification products were size-fractionated by agarose gel electrophoresis. An 8 - 10μl portion of each amplified PCR product was separated at 4℃ on 1.5% agarose gels in 0.5×TBE buffer for 10 hrs at 5V/cm. DNA was stained with 0.5μg/mL ethidium bromide and photographed on a UV transiluminator with Polaroid type 55 film.

1.6 Data analysis

For both RFLP and PCR, fingerprints generated from different strains were compared visually and scored for presence-absence of bands. Each distinct banding pattern produced by combined data from primer sets (IS1112, IS1113, ERIC, and BOX) was regarded as a haplotype. A combination of RFLP and rep-PCR generated

was analyzed also. The resulting data were analyzed using NTSYS-PC and the robustness of each lineage was determined by Bootstrap analysis. The genetic distance generated from RFLP and rep-PCR were compared.

The diversity of the strains from each rice-growing area, and of the entire collection, was analyzed by the approach of Nei and Tajima (1991). The genetic diversity (H) for each population considered was calculated by the equation $H= [n/ (n-1)] (1-Xi^2)$, where Xi is the proportion of the ith haplotype within a group and n is the number of strains tested in each group (Nei, 1987).

2 RESULTS

2.1 Rep-PCR analysis

Five primers, ERIC, BOX, IS1113, IS1112, and REP were used to analyze 143 strains of *Xoo* collected from diverse locations in China. Amplification with the four primers, ERIC, BOX, IS1113, and IS1112, revealed 17, 12, 9, and 14 haplotypes, respectively (Fig. 1: A, B, C, D). The REP primer gave the most conservative pattern.

When the data from four primers, ERIC, BOX, IS1113, and IS1112 were combined, a total of 54 haplotypes were observed (Table 1). Some combined probe haplotypes are found in particular regions and are not common between different rice-growing areas. For example, 9 haplotypes were detected only in South China, 8 haplotypes only in the Yangtze River Region, 12 in the North and 10 in Northeast China. More haplotypes (23) were detected in the collections from the Yangtze River Region than in South (13 haplotypes), North (14) and Northeast (10) China. In the Yangtze River Region, there a large areas are planted with hybrid rice, and other improved cultivars (indica and japonica) are also grown. In South China, both indica hybrid rice and improved cultivars are produced. In

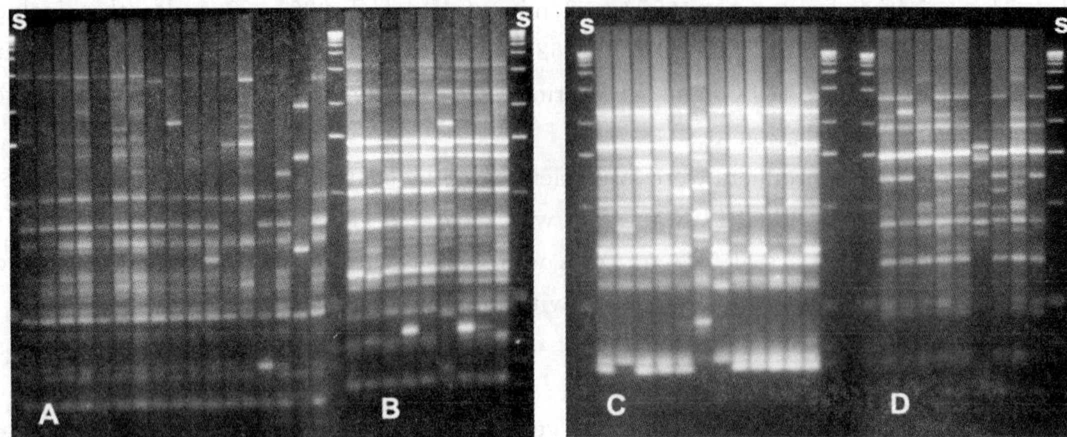

Fig. 1 Rep-PCR fingerpriting patterns from genomic DNA of *Xoo* strains. The representative patterns of ERIC (17), BOX (12), IS1113-PCR (9), IS1112-PCR (14) are shown in panels A, B, C, and D. The left and right lanes labeled "S" show the DNA molecular also marker (1 kb ladder); the sizes are indicated in base pairs.

North and Northeast China, only japonica cultivars are grown.

Table 1 Frequency and genetic diversity of rep-PCR fingerprinting patterns from genomic DNA of *Xoo* distributed in Yangtze Rever Rigion, South, and Northeast China (4 primers used)

Haplotype	Representative strain	Total population		Frequency of haplotype in different rice-growing regions							
				South China		Yangtze River Region		North China		Northeast China	
		No. of strain	% of strain	No. of strain	% of strain	No. of strain	% of strain	No. of strain	% of strain	No. of strain	% of strain
1	GD2270	2	1.4	2	10.5	0		0		0	
2	GD2340	1	0.7	1	5.3	0		0		0	
3	GD2400	2	1.4	2	10.5	0		0		0	
4	GD2090	1	0.7	1	5.3	0		0		0	
5	GD2370	3	2.1	3	15.8	0		0		0	
6	ZHE16	1	0.7	0		1	1.7	0		0	
7	GD2283	6	4.2	2	10.5	4	6.9	0		0	
8	ZHE48	5	3.5	0		5	8.6	0		0	
9	ZHE46	1	0.7	0		1	1.7	0		0	
10	JLN691	5	3.5	0		5	8.6	0		0	
11	GD2409	2	1.4	1	5.3	1	1.7	0		0	
12	ZHE34	8	5.6	0		8	13.8	0		0	
13	ZHE18	1	0.7	0		1	1.7	0		0	
14	FJ1	12	8.4	2	10.5	10	17.2	0		0	

(continued)

Haplotype	Representative strain	Total population		Frequency of haplotype in different rice-growing regions							
				South China		Yangtze River Region		North China		Northeast China	
		No. of strain	% of strain	No. of strain	% of strain	No. of strain	% of strain	No. of strain	% of strain	No. of strain	% of strain
15	AH6	3	2.1	0		3	5.2	0		0	
16	KS1 - 21	2	1.4	0		2	3.4	0		0	
17	HUN3	23	16.1	0		1	1.7	22	55	0	
18	HUN7	1	0.7	0		1	1.7	0		0	
19	HUN8	1	0.7	0		1	1.7	0		0	
20	HUN10	7	4.9	0		6	10.3	1	2.5	0	
21	HB30	1	0.7	0		0		1	2.5	0	
22	BJ1	2	1.4	0		0		2	5.0	0	
23	NX40	1	0.7	0		0		1	2.5	0	
24	NX42	1	0.7	0		0		1	2.5	0	
25	SX73	1	0.7	0		0		1	2.5	0	
26	HLJ72	5	3.5	1	5.3	0		0		4	15.0
27	YUN5	1	0.7	1	5.3	0		0		0	
28	GX325	1	0.7	1	5.3	0		0		0	
29	LN47	1	0.7	0		0		0		1	3.8
30	LN48	1	0.7	0		0		0		1	3.8
31	JL76	12	8.4	0		0		0		12	46.1
32	LN52	1	0.7	0		0		0		1	3.8
33	JL60	1	0.7	0		0		0		1	3.8
34	HLJ68	1	0.7	0		0		0	1	3.8	
35	HLJ71	1	0.7	0		0		0	1	3.8	
36	BJ74	3	2.1	0		0		3	7.5	0	
37	NX1	1	0.7	0		0		1	2.5	0	
38	NX39	1	0.7	0		0		1	2.5	0	
39	HB25	3	2.1	0		0		3	7.5	0	
40	HB16	1	0.7	0		0		1	2.5	0	
41	JL62	3	2.1	0		0		0		3	11.5
42	HUN1	1	0.7	0		1	1.7	0		0	
43	HUN9	1	0.7	0		1	1.7	0		0	
44	HUN2	1	0.7	0		1	1.7	0		0	
45	HUN5	1	0.7	0		1	1.7	0		0	
46	HUN4	1	0.7	0		1	1.7	0		0	
47	HB15	1	0.7	0		0		1	2.5	0	
48	GX1	1	0.7	1	5.3	0		0		0	

(continued)

Haplotype	Representative strain	Total population		Frequency of haplotype in different rice-growing regions							
				South China		Yangtze River Region		North China		Northeast China	
		No. of strain	% of strain	No. of strain	% of strain	No. of strain	% of strain	No. of strain	% of strain	No. of strain	% of strain
49	LN43	1	0.7	0		0		0		1	3.8
50	GD1358	1	0.7	1	5.3	0		0		0	
51	AH1	1	0.7	0		1	1.7	0		0	
52	KS3-1	1	0.7	0		1	1.7	0		0	
53	HUN6	1	0.7	0		1	1.7	0		0	
54	BJ3	1	0.7	0		0		1	2.5	0	
Total No. of strains		143		19		53		40		26	
Genetic diversity G= (HT−HX) / HT (HX=0.83)			0.95		0.93		0.96		0.69		0.77

Table 2 Frequency and genetic diversity of RFLP (avr Xa10 probe) fingerprinting patterns from genomic DNA of *Xoo* distributed in South, Yangtze River region, North, and Northeast China

Haplotype	Representative strain	Total population		Frequency of RFLP pattern in different rice-growing regions							
				South China		Yangtze River Region		North China		Northeast China	
		No. of strain	% of strain	No. of strain	% of strain	No. of strain	% of strain	No. of strain	% of strain	No. of strain	% of strain
1	GD2090	5	3.5	5	25	0		0		0	16
2	GD2340	5	3.5	5	25	0		0		0	32
3	GD2506	5	3.5	3	15	2	3.8	0		0	20
4	ZHE18	12	8.4	6		12	22.6	0		0	16
5	ZHE20	5	3.5	0		5	9.4	0		0	
6	HUB1	1	0.7	0		1	1.9	0		0	
7	HUB1	2	1.4	0		2	3.8	0		0	8
8	AH1	3	2.1	0		3	5.6	0		0	
9	KS3-1	1	0.7	0		1	1.9	0		0	
10	HUN3	3	2.1	0		1	1.9	2	4.4	0	4
11	HUN4	1	0.7	0		1	1.9	0		0	
12	HUN6	1	0.7	0		1	1.9	0		0	
13	HUN8	1	0.7	0		1	1.9	0		0	
14	GX1	1	0.7	1	5	0		0		0	
15	GX325	1	0.7	1	5	0		0		0	4

(continued)

Haplotype	Representative strain	Total population		Frequency of RFLP pattern in different rice-growing regions							
				South China		Yangtze River Region		North China		Northeast China	
		No. of strain	% of strain	No. of strain	% of strain	No. of strain	% of strain	No. of strain	% of strain	No. of strain	% of strain
16	KS6-6	38	26.6	1	5	14	26.4	19	42.2	4	
17	HLJ72	15	10.5	0		0		7	15.6	8	
18	LN44	7	4.9	0		0		2	4.4	5	
19	BJ3	5	3.5	0		0		1	2.2	4	
20	ZHE34	9	6.3	2	10	7	13.2	0		0	
21	NX42	6	4.2	0		0		6	13.3	0	
22	HN37	4	2.8	0		0		2	4.4	2	
23	TJ13	3	2.1	0		0		3	6.7	0	
24	BJ12	3	2.1	0		0		3	6.7	0	
25	LN43	1	0.7	0		0		0		1	
26	GD1358	1	0.7	1	5	0		0		0	
27	JL691	1	0.7	0		1	1.9	0		0	
28	KS1-21	1	0.7	0		1	1.9	0		0	
29	YUN5	1	0.7	1	5	0		0		0	
30	JL76	1	0.7	0		0		0		1	
Total No. of strains		143		20		53		45		25	
Genetic diversity G= (HT−HX) / HT (HX=0.85)			0.71		0.87 0.022		0.91 0.065		0.79 −0.075	0.83 −0.02	

In some provinces, a high degree of diversity was observed. Eight haplotypes were detected from Guangdong (13 strains) and 8 from Hunan province (10 strains).

On the basis of consensus of three clustering statistics, the 143 strains constituted three clusters (Fig. 2). There was not a general correspondence between the rice growing area and the three groups formed by analysis of rep-PCR data and RFLP. Two Yunnan strains were very different from other strains, and were used as the "outgroup" for calculation (Fig. 2). Haplotypic diversities (H) calculated by the method of Nei and Tajima (1981) for each rice growing region based on rep-PCR were as follows: South China, 0.96; Yangtze River Region, 0.93; North China, 0.69; and Northeast China, 0.77 (Table 1).

2.2 RFLP analysis

pBSavrXa10 was used to analyze 143 strains of *Xoo*. The probe was hybridized with 5 to 11 size-different DNA fragments per strain. Eighteen band positions were scored for the

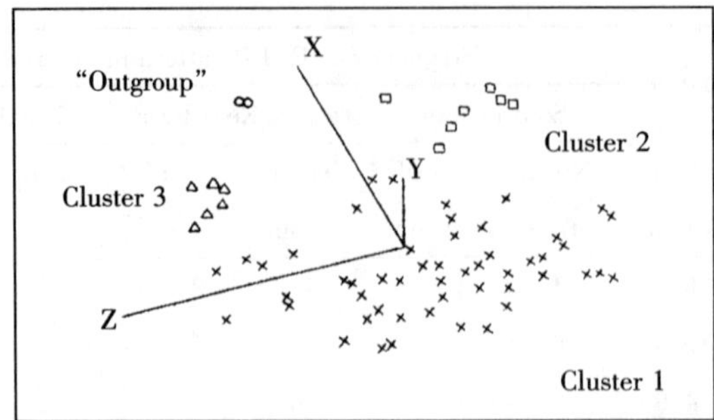

Fig. 2 Clustering analysis of Chinese 143 *Xoo* strains as derived from DNA band data obtained with rep-PCR (ERIC, BOX, IS1112 and IS1113 primers) and RFLP (pBSavrXa10 probe). The number of clusters that best fit the data was determined by the consensus of three clustering statistics. Positions of the strains on the three dimensional graph were determined by multiple correspondence analysis.

entire collections. Thirty haplotypes were observed. Fifteen were detected from the Yangtze River Region, seven from the Northeast China, nine from both South and North China.

Haplotypic diversities (H) calculated by the method of Nei and Tajima (1981) for each rice-growing region based on RFLP data were as follows: South China, 0.87; Yangtze River Region, 0.91; North China, 0.79; Northeast China, 0.83 (Table 2).

3 CONCLUSIONS

3.1 The application of rep-PCR for analysis of *Xoo* population is reasonably new. The advantages of rep-PCR over RFLP analysis are the reduction in cost and the increase in speed of sample processing. Rep-PCR also has several advantages over RAPD analysis. The primers for rep-PCR are 18 - 22 bp and have high homology to the conserved repetitive sequences whereas RAPD analysis relies on short primers (usually 10 bp) with arbitrary sequences.

3.2 In this preliminary survey of variation among strains of *Xoo* from the Yangtze River Region, South, North, and Northeast China, there is no relationship between haplotype of strains and region of origin. Based on clustering statistics, only three clusters were formed after analysis with rep-PCR and RFLP. Compared with strains collected from the Philippines, the genetic structure of the Chinese *Xoo* population seems not to be as complex as strains from the Philippines (Adhiltari et al., 1995). One factor influencing the structure of the Chinese population rice cultivars grown in China may be relatively narrow.

3.3 Based on rep-PCR data, the overall diversity estimate for the entire collection from China was H = 0.95. Haplotypic diversity was relatively low for the strains from North (40 strains, H = 0.69), and Northeast China (26 strains, H = 0.77), and high for the collections from South China (19 strains, H = 0.96) and the Yangtze River Region (53 strains, H = 0.93). The high diversity in South China and the Yangtze River Region may reflect the relative environmental conditions where is the mostly rice growing areas and the indica hybrid rice are widely planted and more susceptible than cultivars planted in North and Northeast China.

3.4 Among these primers used for analysis of the genetic structure of the Chinese *Xoo* strains, ERIC and IS1112 provided the clearest patterns and detected the most variation among strains. The use of sequences from IS elements for these studies has not yet been reported. Results from RFLP and rep-PCR analysis grouped strains showed that they are somewhat different, RFIP patterns of genomic DNA of the entire strains tested are more than rep-PCR analysis. It is necessary to be estimated by both of the methods in the future.

3.5 This preliminary work has revealed interesting insight into the Chinese *Xoo* population structure and has highlighted several questions to be answered with further studies:

a) Is the population of *Xoo* collected from indica hybrid rice which are more susceptible to the pathogen, more or less diverse than the population of strains collected from the improved rice cultivars?

b) How does the pathotypic structure of the Chinese *Xoo* to the genetic structure, as estimated by rep-PCR ?

c) Will selection of strains based on diversity of haplotypes aid in screening and identification of new resistance gene in rice ?

原载：作物学报，1997，23 (2)：150～158.

我国长江以南地区水稻白叶枯病原菌遗传多样性分析

王春莲　章　琦　周永力　赵炳宇

（中国农业科学院作物育种栽培研究所）

Genetic Diversity of Pathogen *Xanthomonas oryzae* pv. *oryzae* from Southern Regions of Yangtze River in China

Abstract　IS-PCR and rep-PCR were used to analyze the genetic diversity in a population of 128 strains of *Xanthomonas oryzae* pv. *oryzae* collected mostly from Center, South and Southwest China. Genomic DNA from 114 strains was amplified with four specific primers, J3, ERIC, REP and BOX by PCR and revealed 20, 12, 6 and 5 haplotypes, respectively. The genetic diversities of the population of strains tested were 0.82 (for J3), 0.49 (for ERIC), 0.35 (for REP) and 0.30 (for BOX), respectively. Dendrograms were generated from the data sets obtained by using UPGMA analysis. All strains tested were similar to each other at a level of 70% similarity, six and five clusters were grouped using primers J3 and ERIC, respectively. The strains classified into Chinese pathotypes 0 – Ⅴ and Ⅶ were grouped in Cluster 1 for primer J3; Strains from pathotypes 0 – Ⅶ, and a new pathotype were grouped in Cluster 1 for primer ERIC. One hundred and twenty-eight strains were classified into different pathotypes by assessing on the five differential cultivars. Most of the strains were classified into pathotypes Ⅳ, Ⅱ and 0. A new pathotype was identified based on its reaction pattern different from all the seven Chinese pathotypes.

Key words　rice; bacterial blight; pathotype; genetic diversity

摘要　用IS-PCR和rep-PCR法结合传统病理学分析了128个主要采自我国华中、华南和西南地区的水稻白叶枯病（*Xanthomonas oryzae* pv. *oryzae*）的群体遗传多样性。用4个专化引物J3、ERIC、REP和BOX对114个菌株基因组DNA进行PCR扩增反应，4个引物分别呈现20、12、6和5种谱型。遗传多样性值分别为0.82、0.49、0.35和0.30。以彼此间带位相似率达70%为界，用UPGMA

聚类分析法，可将114个菌株分为6簇（J3）和5簇（ERIC）。在用引物J3分析的簇1中，包括0～Ⅴ和Ⅶ6个病原型，ERIC引物的簇1中包括0～Ⅶ和一个新型共8个病原型。将128个菌株在我国5个鉴别品种上测定以进行病原型分群，参试菌株多数属于Ⅳ型、Ⅱ型和0型，并出现了一群系列反应型不同于所有病原型的新类型。

关键词 水稻；白叶枯病；病原型；遗传多样性

水稻白叶枯病（*Xanthomonas oryzae* pv. *oryzae*）是水稻生产上的重要病害之一，利用寄主的抗性是控制其危害的主要措施。鉴于水稻抗病基因对病菌的控制存在着明显的专化性，因此，多年广泛种植携有单一抗病基因的品种，必然促使新毒性菌的发展，致使抗病品种感病化。了解病原菌的群体结构可以为育种提供最佳组合选择和基因部署策略（章琦等，1997）。

寄主与病菌之间互作机理的研究长期以来多采用传统病理学手段，主要根据病菌菌株在一套鉴别品种上的一系列反应型进行生理小种（病原型）的划分，具有一定的生物学意义，但是反映不出病菌群体遗传结构中的系谱发育及遗传多样性的内涵。分子生物学的进展，将分子技术分析病原菌的遗传多样性推向国际研究前沿。美国和IRRI合作，用RFLP分析菲律宾不同生态区的病原菌群体多样性，在原属于1个小种的菌群中，鉴定出3个小种（Leach，1995；Nelson et al.，1994）。Rep-PCR（repetitive sequence based polymerase chain reaction）是基于基因组重复序列的一种DNA的多态性，操作简便，结果可靠。近年来在分析病菌群体结构中，逐渐取代RFLP法。章琦等于20世纪90年代初分别在IRRI和美国对我国主要来自华北和长江流域的白叶枯病菌株进行了RFLP和rep-PCR分析，筛选了适于我国菌株的DNA探针和rep-PCR专化引物，比较了参试菌株群体的分子遗传多样性与致病型的差异、采自不同水稻（亚种、杂交稻等）菌株的时空多样性的变化（章琦等1996；Zhang et al. 1995；1997）。但是，在上述研究中，来自主要水稻产区的华南、华中和西南地区的菌株较少，尚不能代表全面状况。因此，本文增加了上述地区的菌株，采用rep-PCR，并增加了IS-PCR（insertion sequence based polymerase chain reaction），结合病原型分群，进行我国白叶枯病菌群体结构分析，为抗病育种提供参考资料。

1 材料和方法

1.1 供试菌株

本试验参试菌株128个，其中69个在1996—1999年采自安徽（20）、江苏（14）、浙江（15）、湖北（34）、湖南（8）、江西（4）、广东（14）、广西（5）、福建（2）、四川（1）、云南（11）共11个省，其余59个菌株采集于1984—1988年。所有菌株经单孢分离，冻干或甘油保存于－20℃冰柜中，接种前用胁本哲氏培养基复壮菌株，于28℃下培养2d，用无菌水洗下菌苔，接种菌液浓度为1×10^9/mL。

1.2 鉴别品种及接种调查

全部供试菌株用人工剪叶接种法于苗期和成株期分别接种于我国5个水稻鉴别品种，金刚30、Tetep、南粳15、Java14和IR26上。接种后14～20 d，待感病品种病情趋于稳定时进行调查，用病斑面积占叶片面积百分率作为抗、感反应参数，根据各菌株在鉴别品种上的系列病情反应型进行病原型分群。

1.3 引物

采用IS－PCR和rep－PCR 4个引物，其序列如下：

J3 （5’-GCTCAGGTCAGGTGGCCTGG-3’）

ERIC ERIC1R（5’-ATGTAAGCTCCTGGGGATTCAC-3’）

ERIC2（5’AAGTAAGTGACTGGGGTGAGCG-3’）

REP REPIR-1（5’-IIIICGICGICATCTGGG-3’）（I ＝ inosine）

REP2－1（5’-ICGICTTATCIGGCCTAC-3’）

BOX BOXIR（5’CTACGGCAAGGCGACGCTGACG-3’）

1.4 DNA提取及IS-PCR、rep-PCR反应

各菌株经复壮后，取在28℃培养箱中培养了48h的新鲜菌，按George（1997）所述的方法提取病菌基因组DNA。

IS-PCR反应：引物J3为一个高拷贝重复单位。反应液体积为25μl（含标准PCR buffer，BSA、DMSO、dNTP、引物、Taq酶、DNA，用ddH_2O补足25μL，覆盖1～2滴矿物油），反应条件：开始变性94℃7 min，循环30次（94℃1 min、68℃1 min、68℃2 min），65℃15 min，保持4℃。

rep-PCR反应：采用3个特异的专化寡核苷酸引物ERIC、REP和BOX。PCR反应液体积为25μl（含5×Gitschier buffer，其他成分同IS-PCR）。反应条件：开始变性94℃7min，循环35次（变性94℃1 min、3个引物ERIC、REP、BOX的退火温度分别为52℃、44℃、53℃，延伸65℃8 min），65℃15 min，保持4℃。4个引物扩增出的产物DNA片段在1×TAE缓冲液中，通过浓度为1.5%琼脂糖凝胶电泳。

1.5 数据分析

根据PCR产物指纹带的有无，分别转换为1或0两个数码。用UPGMA聚类分析法，推断白叶枯病菌株之间的系统发育关系。本研究以不同的PCR谱型标记指纹带位的相似率达70%为界，进行分簇。根据下列公式计算遗传多样性：$H = [n/(n-1)](1-\sum Xi^2)$，$i=1$，$Xi$为群体中第$i$个样品的谱型出现频率，$n$是测定的菌株数。

2 结果与分析

2.1 PCR分析

2.1.1 IS-PCR分析 以引物J3扩增的片段大小范围为154 bp～ 2kb，指纹主带4～20个，有20种谱型，谱型间的带位差异较大（图1－A1，A2）。来自安徽、江苏、湖北、湖南、江西和浙江省的70个菌株呈现13种谱型；来自广东、广西和福建省的15个菌株呈现6种谱型；11个云南省的菌株有5种谱型。一些谱型只属于某个省，如：谱型9、11和13都是

广东的菌株，2 和 3 都来自安徽的菌株，谱型 20 仅来自湖北的菌株；但某些谱型却分布在几个省的菌株中，如谱型 1 和 5，分别出现在安徽、湖北及湖北、浙江和云南菌株中。83 个来自北方的菌株只有 4 种谱型，其中 2 种与南方的不同。总体来看，长江以南地区菌株的分子谱型多于北方菌株。

2.1.2 rep-PCR 分析　以引物 ERIC 扩增的片段大小范围为 200 bp～4 kb，有 6～20 条指纹主带，共 12 种谱型（图 1－B）。某些谱型之间仅有一条带之差，如谱型 1 与 6、10 与 11。来自安徽、江苏、湖北、湖南、江西和浙江的 70 个菌株呈现 7 种谱型，广东、广西和福建的 15 个菌株有 3 种谱型，云南的 11 个菌株呈现 4 种谱型。其中多数菌株的谱型相同，如图 1－B 中的谱型 1 包括 65 个菌株。本试验中某些菌株与 Zhang 等（1997）在美国测试的结果基本相同，如菌株 AH 21、GD1358、JL691、KS－3－1 等。在 83 个北方菌株表现的 2 种谱型中的一个，与长江以南地区的分子谱型相同。REP 引物扩增的片段大小范围为 300 bp 至 4kb，指纹主带 6～21 条，仅 6 种谱型（图 1－C）。BOX 引物扩增的片段大小范围为 344 bp 至 3 kb，指纹主带 6～19 条，有 5 种谱型（图 1－D）。这两个引物的分辨率较低，与章琦等（1995）的结果相同，不适用于分析我国病原菌的群体结构。

2.2　遗传多样性评价

用引物 J3、ERIC、REP 和 BOX 分别对 112 个菌株估测的遗传多样性值从 0.30 到 0.82，差异很大。最高值为 0.82（J3），依次为 0.49（ERIC）、0.35（REP）和 0.30（BOX）（表 1）。不同引物所扩增各菌株基因组 DNA 产物后估算的遗传多样性值也不同，用 J3、ERIC、REP 和 BOX 分别测试病原型Ⅰ、Ⅱ、Ⅲ、Ⅳ、Ⅴ型的遗传多样性值分别为 1.00～0.62、0.80～0.00、0.95～0.00 和 0.50～0.00。其中 J3 的多样性值大大高于 ERIC、REP 和 BOX。属于病原型Ⅵ和Ⅶ中的菌株的频率较小，分别为 1 个、2 个，没有进行多样性评价。

表 1　用 4 个引物扩增中国病原型的遗传多样性

Table 1　The genetic diversity of Chinese pathotypes by using four primers

病原型 Pathotype	菌株数 No. of strains	遗传多样性值（4 个引物）Genetic diversity for four primers			
		J3	ERIC	REP	BOX
0	17	0.91	0.59	0.95	0.50
Ⅰ	8	0.73	0.80	0.00	0.50
Ⅱ	21	0.80	0.40	0.00	0.00
Ⅲ	14	0.62	0.00	0.67	0.00
Ⅳ	40	0.71	0.37	0.46	0.22
Ⅴ	6	1.00	0.70	0.00	0.00
Ⅵ	1	0.00	0.00		
Ⅶ	2	1.00	1.00		
新型 New pathotype	5	1.00	0.67	0.00	0.67
总群体 Total population	114	0.82	0.49	0.35	0.30

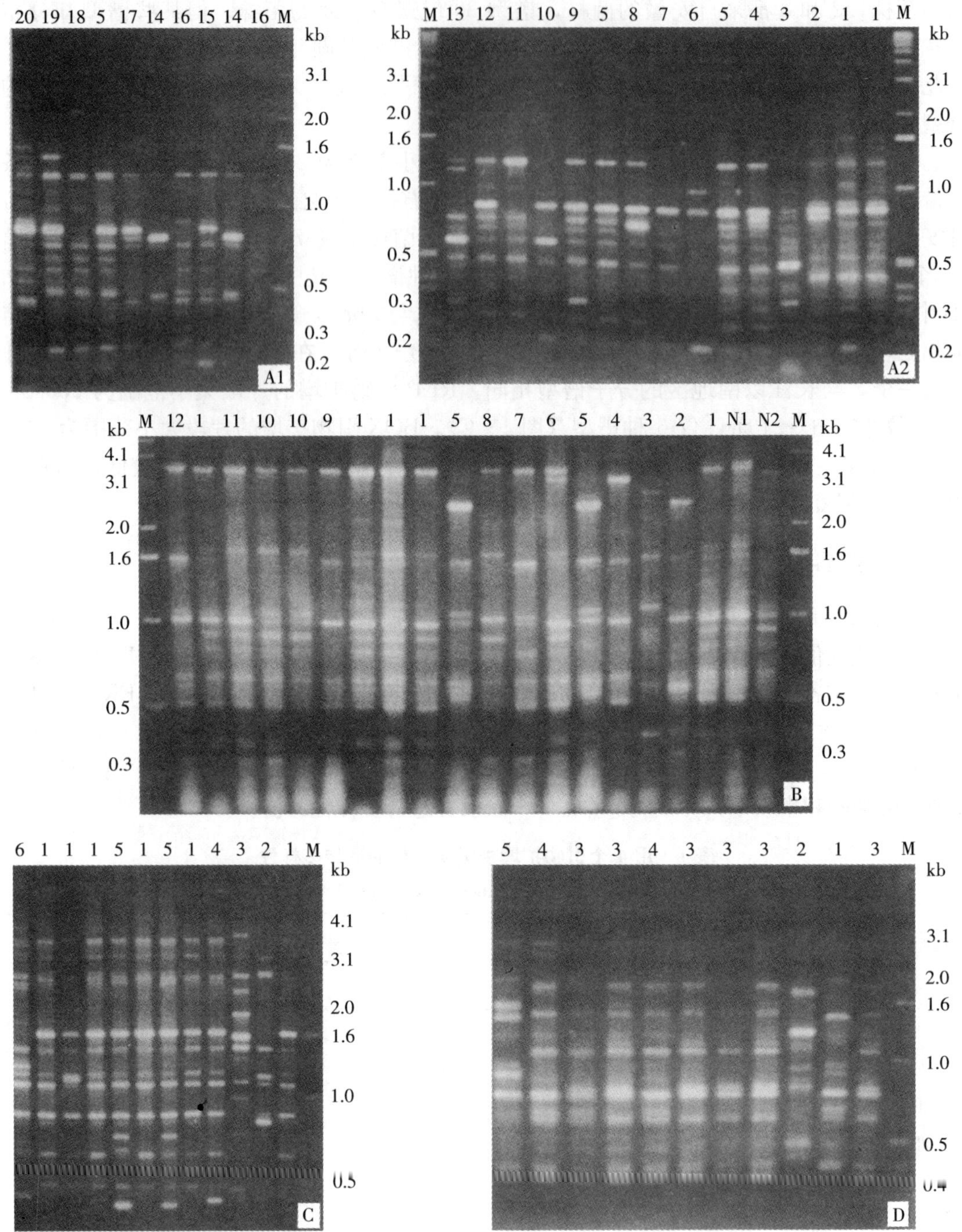

图 1　水稻白叶枯病原菌基因组 DNA 的 IS-PCR 与 rep-PCR 的指纹带型

A 为 IS-PCR（引物 J3）；B、C 和 D 分别为 rep -PCR 中的 ERIC、REP 和 BOX 引物，B 中的 N1 和 N2 为北方代表菌株的指纹；M 为 1 kb 分子标记。

Fig. 1　IS-PCR and rep -PCR fingerprinting patterns from genomic DNA of *Xoo*

The representative patterns of IS-PCR (primer J3) are shown in panels A. rep-PCR (ERIC, REP and BOX) are shown in panels B, C and D, respectively. N1, N2 in panels B are fingerprinting patterns of the typical strains of North China. The left and right lanes labeled "M" show the DNA molecular marker (1 kb ladder)

2.3 分子谱型与病原型的关系

根据 PCR 扩增 DNA 产物指纹带的有无，转换为数码 1 或 0，进行 UPGMA 聚类分析。4 个引物均以带位相似率 70%为界，用 J3 将 20 个 PCR 分子谱型分为 6 簇（图 2）。来自江苏、湖北、浙江、广东和云南的多数菌株和少数江西的菌株呈现 11 个分子谱型，包括 0～Ⅴ、Ⅶ 共 7 个病原型，归为簇 1；病原型为 0～Ⅳ、Ⅴ和新 1（多数来自安徽和湖南、少数为浙江菌株）归为簇 5；簇 2、簇 3 和簇 4 包括的病原型和菌株较少；一个云南的菌株较为特殊，是一个新的病原型，与其他菌株的谱型带位相似率仅有 20%，被归为簇 6。

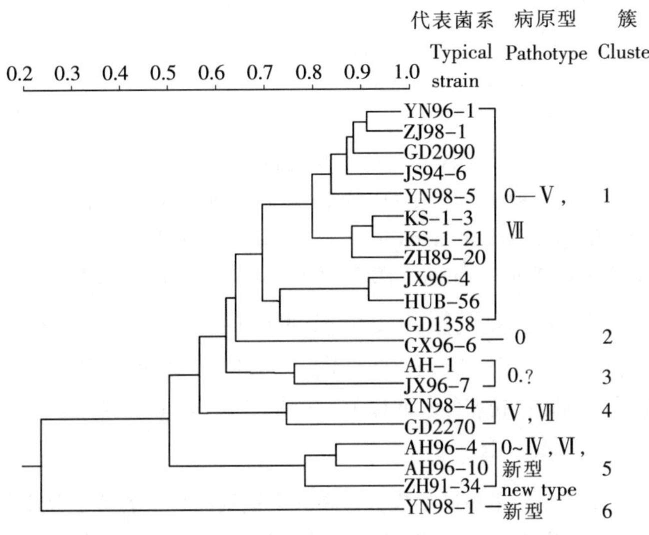

图 2 用 UPGMA 聚类分析法分析 114 个白叶枯菌株的 20 种 IS-PCR（引物 J3）谱型的树状结构图

Fig. 2 Dendrogram constructed with UPGMA on the basis of 20 haplotypes with IS-PCR (primer J3) for 114 strains of *Xanthomonas oryzae* pv. *oryzae*

用 ERIC 将 12 个 PCR 分子谱型分为 5 个簇（图 3），其结果与 J3 颇为类似。大多数来自安徽、江苏、湖北、江西、浙江、广东，部分来自湖南和云南的菌株呈现 7 个分子谱型，但均被归为簇 1，该簇包括了我国所有的 7 个病原型和新型；簇 2 均为来自云南的菌株；被归于簇 3 的菌株是病原型 0，这些多数来自广西及少数来自安徽和江西的菌株；有一个与其他菌株极不相似的安徽菌株被归为簇 4；少量属病原型Ⅳ的湖北菌株被归为簇 5。

从分簇结果来看，用了两个引物的簇 1 包括了多种谱型，表明大多数菌株的系统发育关系是密切的。病原型与谱型之间表现多种关系，属于一个病原型的菌株呈现不同的谱型、同一谱型中又包含了几个病原型，以及在一个谱型中，只包括相同的病原型等。

由于用 REP 和 BOX 引物的测试，分别仅呈现 6 和 5 种分子谱型，故不进行 UPGMA 聚类分析。

2.4 病原型分析

参试的 128 个来自长江流域和南方稻区的菌株在 5 个鉴别品种上测定，按其系列反应型进行分群。以Ⅳ型菌发生频率最高，为 32.8%；其次为Ⅱ和 0 型；Ⅴ型菌发生频率为

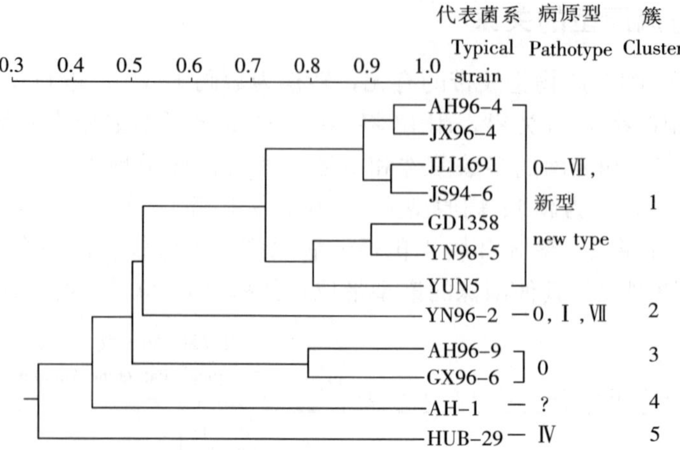

图 3 用 UPGMA 聚类分析法分析 114 个白叶枯病菌的 12 种 rep-PCR（引物 ERIC）谱型的树状结构图

Fig. 3 Dendrogram constructed with UPGMA on the basis of 12 haplotypes with rep-PCR (primer ERIC) for 114 strains of *Xanthomonas oryzae* pv. *oryzae*

5.5%；出现 1 群用现有鉴别品种无法归类的新的反应型（表 2）。这些新的反应型在金刚 30、Tetep、南粳 15、Java 14、IR 26 五个鉴别品种上的反应的趋势，分别为 R（抗）、R、R、S（感）、R，但个别菌株的抗感反应模糊不清，如菌株 YN 9823 的反应型分别为 R、MR（中抗）、MS（中感）、MS、R。

表 2 128 个水稻白叶枯病菌株在 5 个鉴别品种上的病原型归类

Table 2 Pathotypes of 128 strains classified according to the reaction patterns tested on the five differential cultivars

病原型 Pathotype	鉴别品种（Differential cultivar）					菌株数 No. of strains	占总菌数百分比 Percent of Strains（%）
	金刚 30	Tetep	南粳 15	Java14	IR26		
0	R	R	R	R	R	23	18.0
Ⅰ	S	R	R	R	R	9	7.0
Ⅱ	S	S	R	R	R	23	18.0
Ⅲ	S	S	S	R	R	16	12.5
Ⅳ	S	S	S	S	R	42	32.8
Ⅴ	S	S	R	R	S	7	5.5
Ⅵ	S	R	S	R	R	1	0.8
Ⅶ	S	R	S	S	R	1	1.6
新型 New pathotype	R	R	R	S	R	5	3.9

参考方中达等（1990）的研究，按各稻区菌株分布情况，我国南方稻区的白叶枯病菌致病型包括 0～Ⅴ、Ⅶ 及 1 个新的反应型（图 4）。Ⅳ 型菌为这一地区的优势群，发生频率为 25.0%；Ⅴ 型菌为 9.4%；归属于一个新型的菌株主要来自云南，发生频率为 15.6%，而在其他地区并未发现。长江流域菌株包括了 0～Ⅶ 8 个病原型。以Ⅵ型菌的发

生频率最高，达 35.4%；Ⅴ型菌发生频率为 4.2%；仅一个参试菌为Ⅵ型菌。在参试菌株中，江苏省未发现Ⅳ型，可能与种植了可以抵抗Ⅵ型菌的武育粳（*Xa3*）和扬稻 2 号（*Xa4*）品种有关（许志刚等，1998）；大面积种植的籼型杂交水稻汕优 63（感Ⅳ型菌）的湖北省有 60% 的菌株为Ⅳ型；浙江和江苏省均发现了原出于广东省的Ⅴ型菌，Ⅴ型菌频率已达到 5.1%，进一步确认了它已从广东向北扩展；江西省虽然仅参试 4 个菌株，但却有 3 种病原型。

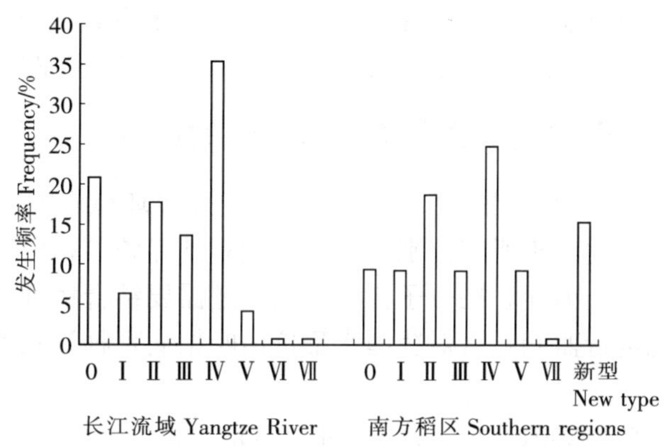

图 4 我国长江以南地区水稻白叶枯病原菌分布图

Fig. 4 Distribution of *Xoo* in Southern regions of Yangtze River of China

3 讨论

3.1 选择适宜我国菌株的引物

本研究用 4 个引物对 112 个 *Xoo* 菌株基因组 DNA 进行 PCR 扩增反应。选出 2 个适宜分析我国菌株的引物 J3 和 ERIC。章琦等（1995）早期研究认为适宜我国病原菌分析的引物为 ERIC，但与近年发展的新引物 J3 相比，J3 优于 ERIC。J3 的优点是：1）对有些特殊菌株的分辨率较高；2）试验流程时间短（J3 只需 3h，ERIC 需 7h）。其余 2 个引物 REP 和 BOX 虽然扩增带比较清晰，只适宜用于菲律宾菌株，对我国菌株的分辨率很低。

3.2 比较分子检测不同地区的病菌菌株

根据对分布在长江流域和华南稻区的菌株的 PCR 分析，虽然长江流域跨越的省份较多，菌株标样数多，但是其多态性并没有如预期的那样多，多数菌株间仅有 1～2 个条带之差，可能与大面积种植相同基因型的品种有关。但是，虽然南方稻区参加试验的标样较少，然而彼此间的差异却较大，尤其是云南菌株非常特殊，可能与该地区稻种资源丰富、低纬度、高海拔以及特殊的气候环境有关。但就我国白叶枯病菌总的群体而言，大多数菌株的相似率很高，归属于簇 1，差异并不明显，与我国长期大面积种植单一抗病基因的品种有密切关系。

UPGMA 聚类分析还表明多数病原型为杂合群体，在归属于 1 个病原型内的分子谱型变异率又较高，表明我国目前的一套白叶枯病鉴别品种需要进一步改善，这与章琦等（1996）的研究是一致的。

3.3 稻白叶枯病菌株病原型的变化

病理学检测表明，我国白叶枯病原菌已经发生了变化，云南省已出现了在20世纪80年代方中达等（1990）的研究中未发现的新类型，该新型竟占云南省菌株的41.7%，表明其优势菌群正处于明显变化之中。长江流域20世纪80年代以Ⅱ型菌居多，近年则以Ⅳ型菌占优势，这可能与大面积种植感病籼型杂交水稻组合汕优63有关。因此在育种计划中应注意基因轮换。新的病原型出现，表明现有的鉴别品种已不足以将我国所有病原菌加以分群，随着病原菌的变化，它们已不适应病原菌的动态变迁，需要筛选新的鉴别品种。

3.4 分子检测与病理学测定病原菌遗传多样性的比较

分子检测从病原菌本身的基因组DNA入手，不受外界环境影响，比较能反映病原菌的内涵。而病原型的归类，制约于所用鉴别品种的数量和外界条件，但这又是可以直接反应生产中抗性变化的直观手段。因此，在分子检测的同时还需要结合病理学测定，对于不同生态区的病菌时空消长状况进行监测，可对我国水稻抗白叶枯病育种计划的更新提供有效依据。

原载：中国水稻科学，2001，15（2）：131～136.

水稻白叶枯病抗性基因 *Xa23* 鉴别菌株突变体库的构建及无毒基因突变体的筛选

周永力[1]　潘雅姣[1,2]　翟文学[3]
徐建龙[1]　章　琦[1]　黎志康[1]
（1. 中国农科院作物科学研究所，农业部作物遗传育种重点实验室；
2. 沈阳农业大学生物技术学院；3. 中国科学院遗传与发育生物学研究所）

Construction of Mutant Population of Differential Race of *Xa23* Resistant to Rice Bacterial Blight and Avirulence Activity Identification of Mutants

Abstract The mutant population of *Xanthomonas oryzae* pv. *oryzae* strain differential to rice bacterial blight resistance gene *Xa23* has been constructed mediated by transposon in vivo. The results of PCR amplification with specific primers and analysis of flanking sequence of mutants indicated that the foreign DNA has been integrated into *X. oryzae* pv. *oryzae* genome. Four mutants with changed avirulent activity to *Xa23* gene have been identified by artificial inoculation. It is possible to clone genes that are required for *AvrXa23* avirulence activity using this new strategy.

Key words *Xanthomonas oryzae* pv. *oryzae*；*Xa*23；mutant

摘要 利用 in vivo 转座技术构建了白叶枯病抗性基因 *Xa23* 鉴别菌株的突变体库。特异性引物 PCR 扩增和转座子插入位点旁侧序列分析结果表明，转座子已插入白叶枯病菌基因组中。采用人工接种鉴定，筛选到 4 个毒力变化的突变体。为进一步克隆 *Xa23* 无毒基因提供了条件。

关键词 *Xanthomonas oryzae* pv. *oryzae*，*Xa*23，突变体

水稻白叶枯病（*Xanthomonas oryzae* pv. *oryzae*，*Xoo*）是世界水稻生产中的主要病害之一，也是我国水稻高产、稳产的重要限制因素。利用抗性基因选育抗病品种是防治白叶枯病的有效手段，目前国际上命名的水稻白叶枯病基因已达 28 个之多。Zhang 等（1999；2002）从普通野生稻（*Oryzae rufipogon*）中发掘鉴定的新基因 *Xa23* 是迄今发现的抗谱最广、抗性转移效应最强的全生育期完全显性白叶枯病抗性基因，不同遗传背景的 F_1 植株从

苗期到成株期均表现高度抗病，是改良水稻、特别是杂交水稻白叶枯病抗性的优异抗源。水稻与白叶枯病菌之间存在典型的特异性互作关系，克隆 *Xa23* 无毒基因将有助于阐明白叶枯病抗性基因的广谱抗性机理。本文采用一种新的 in vivo 转座技术构建了 *Xa23* 鉴别菌株的突变体库，并对突变体的毒力变化进行了初步鉴定。

1 材料与方法

1.1 材料

采用 *Xa23* 基因的水稻白叶枯病专化性的菲律宾小种 6 代表菌株 PXO99。携有 *Xa23* 的水稻品系 CBB23（轮回亲本 JG30）用于突变体鉴定，以对 PXO99 高感的品种 IR24 作感病对照。

1.2 菌株转化

将菌株 PXO99 振荡培养至 OD_{600}＝0.5～0.6，制备悬浮细胞。参考 Igor 等（1988）的方法，采用 Cell Porator Electroporation System（Life Technologies TM）将 EZ::TN＜KAN-2＞ Tnp Transposon（Epicentre®）电激转化至悬浮细胞中。悬浮细胞在 SOC 培养基中于 28℃振荡培养 1h，涂布于含有 30μg/mL 卡那霉素的 PSA 培养基表面，28℃培养 24～36 h后，将阳性菌落转移至装有 PS 培养基（含 30μg/mL 卡那霉素）的 384 孔培养板中。

1.3 突变体的 PCR 鉴定及转座子插入位点旁侧序列分析

随机取突变体进行 PCR 检测，引物序列为：5′gagccatattcaacgggaaa 3′（TS2-FW），5′cgagcatcaaatgaaactgc 3′（TS2-RV）。PCR 反应体系 25μL，内含 10×PCR 反应缓冲液 2.5μL，dNTPs（10mmol/L）2μL，引物（8μmol/L）2μL，Taq DNA 聚合酶（3u/μL）0.2μL，Mg2＋（25mmol/L）2μL，菌液 1μL。以未转化的 PXO99 的 DNA 作阴性对照。扩增程序为 95℃5min，（95℃30s，58℃45s，72℃1.5min）30 个循环，72℃7min。

采用 Inverse PCR 方法（Ochman et al.，1988）分析转座子在突变体基因组中插入位点旁侧序列。采用转座子内没有切点的限制酶 *Eco*RⅠ或 *Bgl*Ⅱ对突变体 DNA 进行酶切、连接，连接产物进行 Nested PCR 扩增，第一轮引物为 FP-2（5′-gtaacactggcagagcattacgc-3′）和 RP-2（5′-cacctgattgcccgacattatc-3′），第二轮引物为 FP-1（5′-acctacaacaaagctctcatcaacc-3′）和 RP-1（5′-gcaatgtaacatcagagattttgag-3′）。扩增程序为 95℃ 5min，（95℃30s，52℃ 45s，72℃1.5min）30 个循环，72℃7min。第二轮产物回收测序。

1.4 人工接种鉴定突变体

将保存于－80℃的突变体在含有 30μg/mL 卡那霉素的 PSA 培养基活化，在水稻分蘖期采用剪叶法接种携有 *Xa23* 的品系 CBB23，鉴定各突变体的毒力变化。不携带该抗性基因的品种 IR24 作为感病对照，接种后 12d 左右，当接种野生型菌株 PXO99 的感病品种病情趋于稳定时，调查 CBB23 各植株的抗性反应。

2 结果与讨论

2.1 突变体库构建

Xoo 感受态细胞经电激转化，经 SOC 培养基振荡培养后涂板，24～36h 开始在含有 30μg/mL 卡那霉素的 PSA 筛选培养基表面出现菌落，未经转化的受体菌株感受态细胞在筛选培养基表面无菌落产生。4 次试验白叶枯病菌感受态细胞的转化效率不同，其转化效率为 20ng EZ:: TN < KAN-2 > TnpTransposon 产生 10^2个以上的 Kan^R菌落。表 1 为 4 次转化的试验结果。阳性菌落转至 384 孔板培养后，保存于－80℃。采用 *in vitro* 转座技术获得了 4 700 个 *Xoo* 突变体，约覆盖白叶枯病菌基因组的 1.17 倍。

表 1 EZ:: TN<KAN-2>Tnp Transposon（20ng）电激转化 *Xoo* 获得的 Kan^R 菌落数

Table 1 Kan^R colonies of EZ:: TN <KAN-2> Tnp Transposon (20ng) produced by Electroporation

Experiment	1	2	3	4
No. of Kan^R colonies	430	213	600	284

2.2 转座子插入到 *Xoo* 基因组中的证据

采用编码卡那霉素抗性基因的序列设计特异性引物 TS2-FW 和 TS2-RV 随机扩增保存于－80℃的突变体。不同批次随机挑取的转化的各突变体扩增后，均产生 800bp 的目标片段（图 1），野生型菌株没有相应的扩增带产生；突变体继代 3 次后，PCR 检测仍有 100% 表现阳性。结果初步显示转座子已插入到突变体的基因组中。

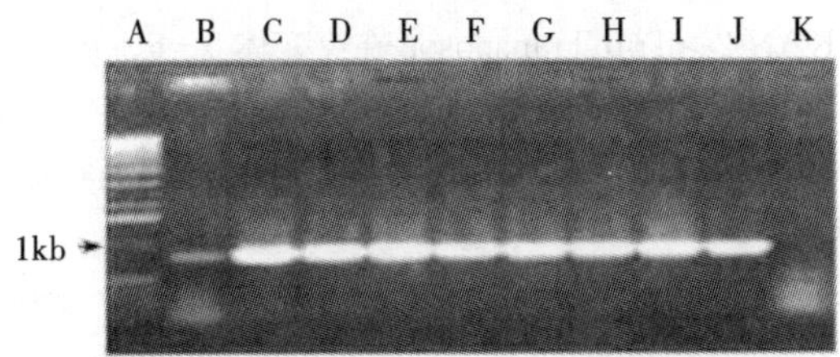

图 1 *Xanthomonas oryzae* pv. *oryzae* 的 PCR 检测

Fig. 1 PCR analysis of mutants of *X. oryzae* pv. *oryzae*. Primer pairs were TS 2-FW and TS2-RV. A: marker, 1kb ladder; B～J: mutants; K: *Xoo* wild strain

采用 Inverse PCR 方法分析了几个突变体转座子插入位点的旁侧序列（图 2）。所测得的序列包括转座子两端的 Mosaic end，进一步表明转座子已插入 *Xoo* 基因组中，并且不同突变体转座子的插入位点不同（图 3）。转座子在受体菌株基因组中随机插入为构建和筛选不同目标性状的突变体提供了可能。

2.3 无毒基因突变体的筛选

携带抗病基因 *Xa23* 的品系 CBB23 对 PXO99 菌株表现高度抗病反应，病斑长度为

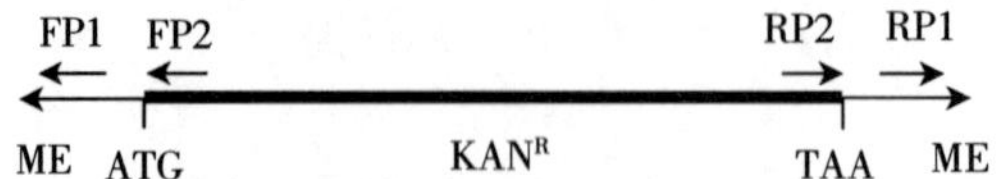

图 2 EZ:：TN< KAN-2> Tnp Transposon 结构图和 inverse PCR 引物位置

Fig. 2 Schematic map of EZ:：TN <KAN-2> Tnp Transposon and the locations of primers used in inverse PCR

ME = Mosaic End.

0.5～1.0cm。在所测定的突变体中，4 个突变体的致病力比野生型菌株致病力有明显提高（图 4）。病斑长度达到 5.0～7.0cm。*Xa23* 是一个高度抗病的广谱抗性基因，与病原菌存在清晰的特异性识别反应，对 28 个包括我国 7 个病原型、10 个菲律宾和 3 个日本的主要生理小种以及 8 个韩国代表菌株均表现高度抗病，病斑长度均小于 1.0cm（章琦，2002，个人通讯）。上述几个突变体经 PCR 扩增后均有转座子特异性片段产生。从 CBB23 的抗性反应来看，可以初步推断上述突变体的致病力发生变化可能与转座子插入到无毒相关基因位点有关。

← FP1

5-5 cggcgacggccgcgacgagtacgcgacgtg[gacagagaatatgtgtaga……*ccaactactctgaaacaacatcca*…

10-3 cgtaggtgttctacagcgagacgatgcccaac[gacagagaatatgtgtaga……*ccaactactctgaaacaacatcca*…

RP1 →

5-5 ……………*cgttacattgtagtctctaaaact*……tctacacatattctctgtc]cgcgtcgcttggtaagagttggtgg

10-3 ……………*cgttacattgtagtctctaaaact*……tctacacatattctctgtc]atgcccaacgcctgcgcacgttcg

5-5 gccgctccgctagtccg

10-3 cgccgcataccggaatt

图 3 EZ:：TN<KAN-2>Tnp Transposon 在突变体 5-5、10-3 和 2-16（9）*Xoo* 基因组中插入位点的旁侧序列

Fig. 3 Junction sequences of EZ:：TN <KAN-2> Tnp Transposon to flanking DNA of mutants 5-5，10-3 and 2-16（9）.

The sequence in [] is the sequence of transposon，the shaded letters are the sequence of mosaic end. The sequence outside [] is DNA of *Xoo*.

EZ:：TN<KAN-2>Tnp Transposon 转座系统是近年发展起来的一种转化细菌的新技术（Douglas et al.，2000），不需要进行细胞融合，也不需要相容的转座酶表达系统和自杀载体，可通过电激转化直接将转座子导入活细胞中，并且转座子在大多数突变体基因组中均为单拷贝插入（Igor et al.，2000）。

我们在探索 EZ:：TN<KAN-2>TnpTransposon 体系转化水稻白叶枯病菌有效性的结果表明，该系统可以用于黄单胞细菌的转化，与常规的转座子转化技术相比（Shen et al.，2002），具有操作简便、转化效率高的特点；本研究同时构建了广谱基因 *Xa23* 鉴别菌株的突变体库，为克隆 *AvrXa23* 基因提供了条件。目前我们正同时通过序列测定和 Southern 杂交分析 EZ:：TN <KAN-2> Tnp Transposon 在毒力发生变化的突变体基因组中的插入和分布情况，以期克隆 *Xa23* 的无毒相关基因。

图 4 毒力变化的突变体在叶片上的症状

Fig. 4 Leaves with increased symptoms produced by mutants from infection assays A：wild-type PXO99；B～E：the mutants.

原载：生物工程学报，2005，21（3）：486～488.

Characterizing Adult Plant Resistance to Bacterial Blight (BB)

Zhang Qi and T. W. Mew

(IRRI)

1 Adult plant resistance of selected Chinese cultivars

We evaluated 30 Chinese rice cultivars for BB resistance at IRRI. Nine cultivars appeared to have adult-plant resistance. This study seeks to understand how such adult-plant resistance is expressed.

Effect of plant growth stages The 9 cultivars were evaluated for BB resistance. 21 d and 45 d after sowing and at the booting stage. Their response to the four Philippine races of *Xanthomonas campestris* pv. *oryzae* (*Xoo*) differed with growth stage. All cultivars were very susceptible as seedlings but with age developed resistance to one or all of the races (see table). Guai Zhou Magu scores for race 1 decreased from 8. 5 as seedlings to 5. 5 at maximum tillering and 1. 0 at booting. Kwang-er-ai 5 disease scores decreased from 7. 5 as seedlings to 5. 7 at maximum tillering and at booting. Growth stage did not affect the response of Shi-zu and of susceptible check TN1to the four races.

Reaction patterns of adult plant resistance Among the cultivars with adult plant resistance, two disease development patterns were observed from the 6th leaf to the flag leaf. Resistance of most cultivars, such as Taichung Sen 5 and Kwang-er-ai 5, increased gradually as the leaf number increased from the 6th to flag leaves. Guai Zhou Magu, however, had a clear-cut susceptible and resistant response (Fig. 1).

Leaf response varied. Some cultivars showed resistance earlier than others. Guai Zhou Magu, Nangen 15, and Peng Chiu Moung were resistant from the 9th leaf and Taichung Sen 5 was resistant from the 11th to 12th leaf (Fig. 1). All flag leaves were BB resistant.

Six cultivars were resistant to the four *Xoo* races at booting stage. Wang Mi Hsiang, 3330 and Chen Guang Dao 3 were resistant only to some races at maturity (see Table1).

Table 1 Adult plant BB resistance of Chinese cultivars at seedling, maximum tillering, and adult stages. IRRI, 1984

Variety	[a]Resistances to given race											
	PXO 61			PXO 86			PXO 79			PXO 71		
	SS	MTS	BS	SS	MTS	BS	SS	MTS	BS	SS	MTS	BS
Guai Zhou Magu	8.5	5.5	1.0	8.5	3.1	1.1	8.0	5.2	1.0	7.0	5.0	1.4
Nangen 11	6.5	4.0	1.0	6.3	3.0	1.0	6.5	3.3	1.0	6.0	3.5	1.0
Peng Chiu Moung	6.5	3.1	1.2	6.1	1.3	1.0	6.0	3.0	1.0	6.2	3.2	1.0
Nangen 15	6.7	5.5	2.0	6.5	3.0	1.7	6.5	5.6	2.8	7.0	5.1	1.0
Kwang-er-ai 5	8.5	6.2	2.7	6.5	5.0	1.5	9.0	7.0	2.7	9.0	7.0	2.8
Taichung Sen 5	7.0	4.7	3.0	7.5	4.7	2.0	9.0	5.4	3.0	8.0	4.3	3.0
3330	7.5	5.7	3.5	6.7	3.3	1.6	7.2	5.4	2.7	7.2	5.5	4.0
Chin Kong Tao'	7.0	4.3	3.5	6.7	3.7	1.7	6.7	4.8	1.2	7.3	6.6	6.8
Wan Mi Hsiang	9.0	5.4	4.7	7.0	3.0	1.3	8.5	5.0	5.0	9.0	6.0	5.0
Shi-zu	1.0	1.3	1.0	3.0	1.0	1.0	2.5	1.3	1.0	1.0	1.2	1.0
IR1545 - 339	1.8	3.0	1.0	3.5	3.0	1.0	3.2	2.0	1.0	6.8	5.0	6.0
TN1	9.0	8.0	6.8	9.0	6.7	4.3	9.0	7.3	6.3	9.0	7.3	7.0

[a] By the 1980 Standard Evaluation System for Rice scale. SS=seedling stage, MTS=maximum tillering stage, BS=booting stage.

2 Greenhouse testing of *Xa6* and *Xa3* genes for BB resistance

Rices with adult plant disease resistance are disease susceptible as seedlings and develop resistance as they mature.

Zenith, Malagkit Sungsong, and lines IR1695 and IR944 derived respectively from them have the *Xa6* gene for BB resistance. Wase Aikoku 3 has the *Xa3* gene. We evaluated the rice for resistance expression in the IRRI greenhouse.

Four *Xoo* races were inoculated 4 - 5 times at the 6th to flag leaf of the same plants. TN1, the susceptible check was susceptible to the four races at all leaf positions (Fig. 2).

IRI545-339, a line with differential BB resistance, was resistant to races 1, 2, and 3 and susceptible to race 4 at all leaf positions. Zenith was resistant from the 11th to the flag leaf, and IR 1695 developed BB similarly. M. Sungsong was resistant from the 9th leaf and

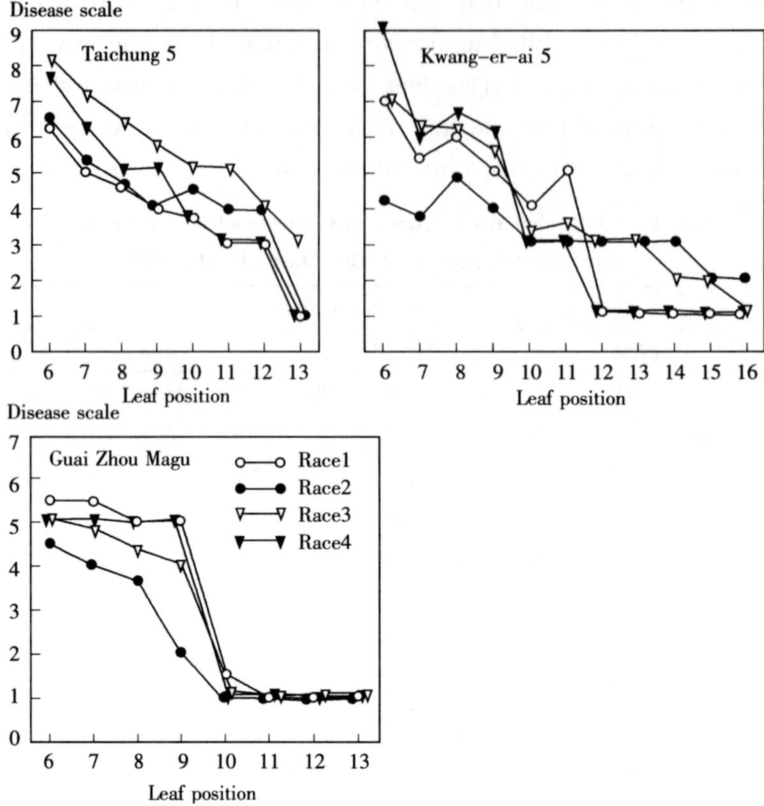

Fig. 1 Reaction of three cultivars to the 4 races from 6th to flag leaf

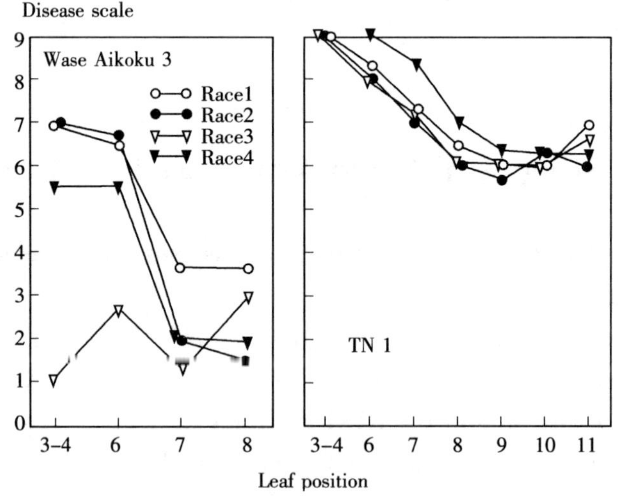

Fig. 2 Reactions of Wase Aikoku 3 and TN1 inoculated with 4 BB races at the same time. IRRI, 1984

IR944, from the 12th leaf.

When planting order was staggered to synchronize inoculation of different leaf positions,

Zenith and M. Sungsong developed resistance to the 4 races at similar leaf positions (Fig. 3).

Wase Aikoku 3 had adult plant resistance only to races 1, 2, 4, and was resistant to race 3 at all leaf positions, the susceptible check, was susceptible to the four races at all leaf positions.

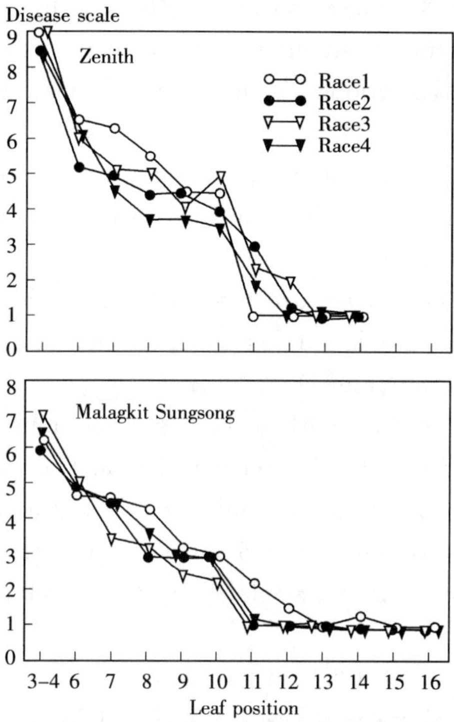

Fig. 3 Reactions of Zenith and Malagkit Sungsong inoculated with 4 BB races at the same time. IRRI, 1984

3 Influence of temperature on the *Xa6* gene for BB adult plant resistance

High temperature favors BB development, but it is not known whether adult plant resistance controlled by *Xa6* gene is stable in high temperature.

We evaluated BB resistance of Zenith, M. Sungsong, IR1695, and IR944 at 33 - 25℃, 29 - 21℃, and 25 - 20℃ ranges in growth chambers with 70% relative humidity. The four cultivars have the *Xa6* gene.

Zenith and M. Sungsong resistance was unaffected by temperature changes. However, resistance by leaf position was not consistent with greenhouse observations. Zenith and M. Sungsong expressed resistance best at booting stage (Fig. 4, Fig. 5). Lesions developed slowly, gradually yellowed, and stopped growing. The cultivars showed resistance earlier at 25 - 20℃ than at 29 - 21℃. Temperature affected BB development more in IR1695 and IR944. At higher temperatures, lesions were longer than those of their parents, but the difference was insignificant at booting. Temperature did not affect TN 1 lesion development.

4 Bacteriophage method for estimating *X. campestris* pv. *oryzae* multiplication in infected rice leaves

Taking direct counts of *X. campestris* pv. *oryzae* cells by plating on ordinary media is impossible because the bacterium is slowly growing and subject to contamination, especially in the tropics. We tested the bacteriophage method for estimating bacterial multiplication on rice leaf tissue.

A suitable bacterial strain and a phage strain were selected and a one-step growth experiment was done to characterize the phage-bacterium relationship. A titration curve of the bacteria densities with the phage also was established for reference. The estimate was based on plaque-forming units (pfu) caused by infection of the bacterial cells by the phage particles. We used a titration curve based on a serial dilution of isolate PXO61 of BB pathogen in response to phage strain P-80 (Fig. 6).

To estimate multiplication in Zenith leaves, six 5cm-long leaf samples infected with PXO61 were observed at different times after inoculation. The samples were disinfected with 70% ethanol and rinsed several times in sterile distilled water. The leaf specimens were homogenized in 5-ml filter-sterilized PSB medium. The leaf extract was diluted serially. For a viability count, 0.1 ml of the extract was plated on PSA medium. The rest was mixed with a standard phage concentration.

Ten minutes after incubation at 28℃ in aeration, the mixture was centrifuged to remove the excess phage. The precipitate was resuspended and incubated for 20 min to estimate phage absorption by the bacterial cells. Final phage plating was 10 h after incubation. The correlation between viability count for bacteria and phage count as an indication of bacterial multiplication is shown in Figure 7. The bacteriophage method seems reliable and efficient for detecting bacterial multiplication in pfu larger than 10^3/ml.

5 Relation of lesion development and bacterial multiplication in different leaf positions of Zenith inoculated with *X. campestris* pv. *oryzae*

Zenith, with *Xa6* gene for BB resistance, shows resistance from the 11th leaf. It is unclear if bacterial multiplication in susceptible and resistant leaf positions varies. We investigated bacterial multiplication in the 6th, 9th, 10th, and 12th leaf positions on Zenith using the bacteriophage method. Six samples from each leaf position were examined at 1, 24, 48, 72, 120 and 168 h after inoculation (HAI). Lesion development was measured at the same time.

On the susceptible 6th, 9th, and 10th leaf positions, lesions began to develop 48 HAI and grew very fast. On the 12th leaf, the lesions were visible 96 HAI, and developed slowly. Two weeks after inoculation, lesion length was 23.6, 18.2, 16.4, and 2.4 cm on leaves 6, 9, 10, and 12, respectively (Fig. 8).

The phage count, expressed as plaqueforming units/cm^2 leaf area, on leaves 6, 9, and 10 increased 24 HA1, indirectly indicating bacterial multiplication. Phage count at 48 HAl increased very fast. It peaked at 72 h, than slightly declined. Leaf tissue was totally infected at 120 and 168 h. The phage count on leaf 12 also increased 48 HAl, but was less than on other leaves and declined faster (Fig. 9).

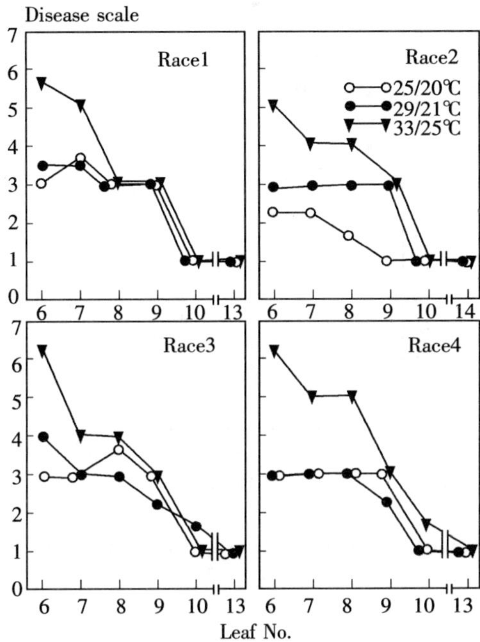

Fig. 4 Disease reactions of Zenith to BB races at 3 temperature ranges. IRRI, 1984.

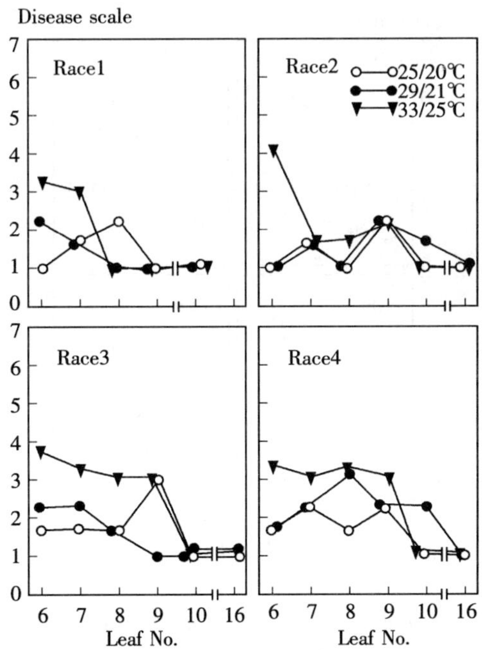

Fig. 5 Disease reactions of Malagkit Sungsong to the 4 BB races at 3 temperature ranges. IRRI, 1984.

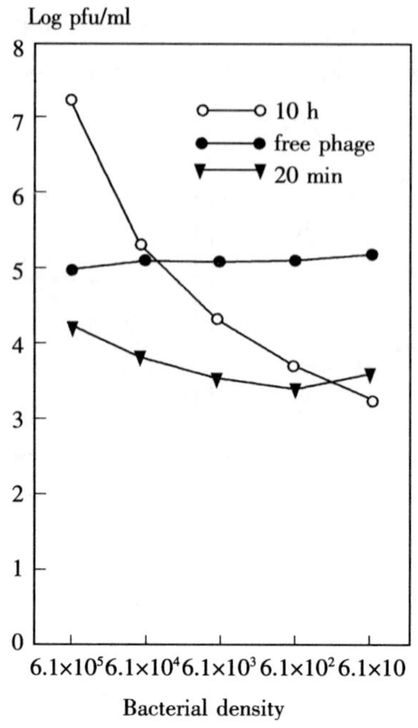

Fig. 6 The titration curve of bacterial density with a standard phage concentration

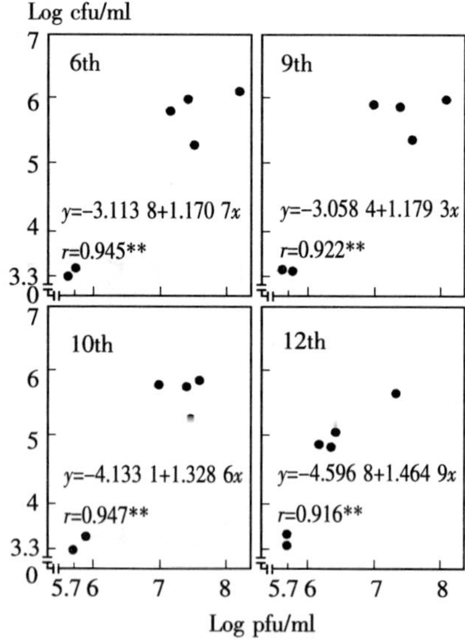

Fig. 7 Relationship between phage count (pfu/cm^2) and colony (cfu/cm^2) at different leaf positions of Zenith with BB lesions. IRRI, 1984

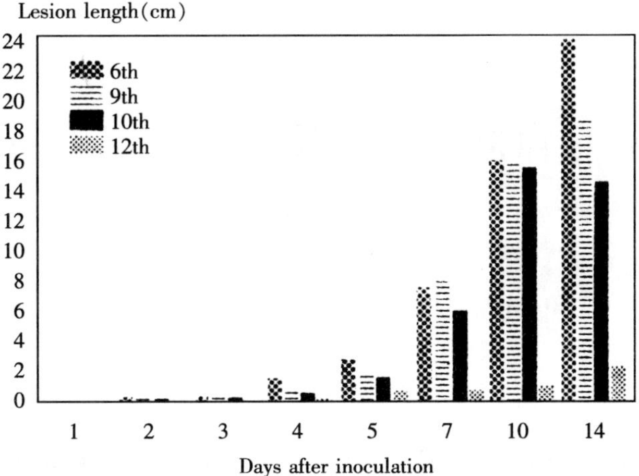

Fig. 8 BB lesion development on different leaf positions of Zenith in the greenhouse. IRRI, 1984.

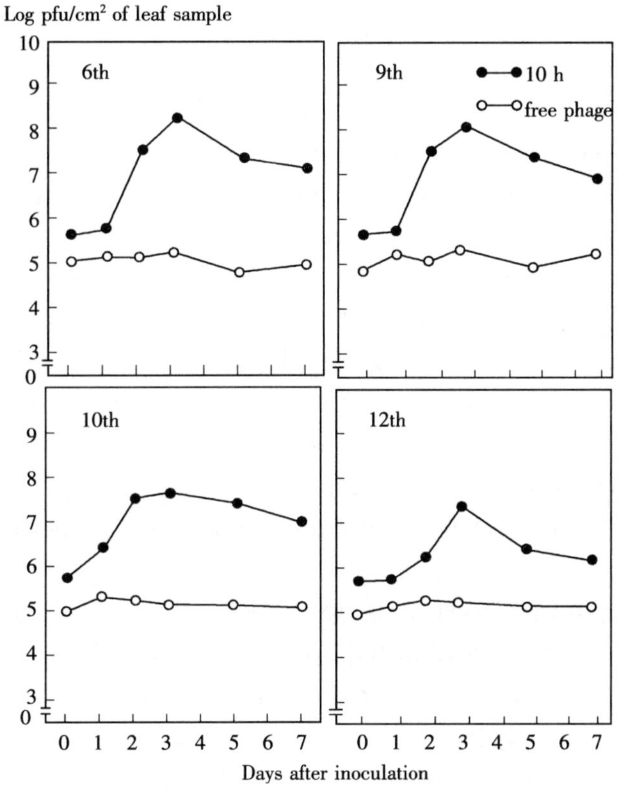

Fig. 9 Multiplication of *X. campestris* pv. *oryzae* on leaves 6, 9, 10, and 12 of Zenith estimated by the phage method. IRRI, 1984.

Guidelines and Style for IRRN Contributors IRRI, 1984, 9: 6

我国水稻白叶枯病抗性遗传的评价和利用

章　琦

（中国农业科学院作物育种栽培研究所）

Genetic Evaluation and Utilization of Resistance to Rice Bacterial Blight in China

Abstract Since 70's, genetic evaluation and utilization of resistance to rice bacterial blight has made advances in China. Numerous germplasms including local cultivars, wild rice and exotic rice sources have been evaluated for resistance to bacterial blight. Many materials have been identified and confirmed as resistance sources. Some of them with genes *Xa3* and *Xa4* are used as resistance donors to bacterial blight in hybridization programs widely. Some elite breeding lines or cultivars are released. Cultivars have been evaluated for adult plant resistance and overall resistance. Genetic analysis of bacterial blight resistance in rice with different types of resistance, progenies between subspecies hybridization and hybrid rice have been studied widely. Studies on systematic identification of resistance gene(s) and developing sources of resistance involving more than one or two genes of desired traits are proposed.

Key words Rice bacterial blight; Genetics of resistance; Types of resistance

摘要 70年代以来我国水稻对白叶枯病抗性遗传的评价和利用取得相当进展，评价了大量包括地方品种、野生稻和国外引进种质对水稻白叶枯病的抗性，鉴定出大批抗源。其中一些携有 *Xa3* 和 *Xa4* 基因的抗源广泛地用于杂交计划，已选育出一批优良的品种（系）。在应用基础研究方面，评价了品种成株抗性和全生育期抗性类型的特点，分析了白叶枯病抗性遗传，包括不同抗性类型的品种、亚种间杂交后代和杂交稻育种后代的抗性遗传行为。本文就系统研究抗性基因以及培育包括1～2个有利基因的改良抗源等方面提出建议。

关键词 水稻白叶枯病；抗性遗传；抗性类型

1884年首先在日本福冈发现水稻白叶枯病，日本从1923年开始采用杂交方法培育抗病品种（Khush et al.，1989）。20世纪50年代以来，此病危害严重，化学防治又难以奏效，引起了国内外学者研究利用品种抗性来防病的兴趣，在理论和应用方面的研究都取得相当进展。诸如病原菌毒性分化研究、单基因鉴别寄主和病菌生理小种的鉴别系统的建立、抗病性类型和抗性机制的探讨以及抗性遗传表现和基因分析与利用等。从细胞遗传学概念发展到分子遗传学水平来探索病原菌毒性的变异，用于改良寄主抗性。在作物细菌病害中，水稻白叶枯病的研究报道可能是最多的。伍尚忠等曾论述此病的有关研究。本文则侧重介绍我国水稻白叶枯病抗性遗传评价和利用进展。

1 水稻对白叶枯病抗性的评价和利用

日本最早利用寄主的抗病性防治稻白叶枯病。自第二次世界大战后，一方面利用自然发病，一方面采用人工针刺法接种评价品种抗性，逐渐形成了一套抗病性评价和利用方式（谢式坪钰，1971）。日本学者根据品种对菌群的病情反应型被分为不同群，至今在 5 个菌群上被分成 7 个品种抗性群（Mew，1987）。育种家便根据菌群的消长动态，来变换育种计划中所用的抗性品种群。1957 年，当黄玉群品种“朝风”因为推广所在地出现毒性菌株而丧失了抗性后，育种家便转而利用广抗谱品种早生爱国 3 号和 Lead rice，育成了中国 45 和 Chukei 314 两个品种，被广泛用于育种计划。20 世纪 70 年代以来，日本尤为注意结合“水平抗性”，用 Akamochi、Gomashirazu、Pelita-1、IR28 和 IR26 作为供体。60 年代末，国际水稻研究所（IRRI）在抗病虫育种上取得突破性成就。该所发挥了多学科组成的抗性遗传评价及利用的作用。首先广泛征集世界各地水稻种质进行抗病鉴定，经初筛和复测，研究其抗性类型和遗传表现，选出 100 多个抗白叶枯病的品种包括 7 个抗病基因的抗源，被用于杂交育种（Khush et al.，1989）。

IRRI 的育种计划也随菲律宾白叶枯病菌小种群体的消长而变动。1965 年首次用 TKM6、Tadukan、Sigadis 等抗源，于 1969 年育成第一批抗病品种 IR20 和 IR22。之后，育成的许多 IR 系统品种至 IR66，除 IR56 外，都携有 *Xa4* 基因。携带 *Xa4* 基因的品种，只抗菲律宾小种 1、不抗小种 2 和 3。在育种计划中，便开始利用抗谱广的 *xa5*、*xa7*，育成了 IR44962 等 9 个高代品系。近年，病理学家已鉴定出 6 个菲律宾生理小种，该所并鉴定出一个来自长药野生稻（*O. longistaminata*）的新抗源，该抗源在印度 17 个州都表现高抗白叶枯病，并且抗菲律宾全部 6 个小种，被育成了一个改良型抗源，即将用于育种计划（Khush et al.，1989）。

我国水稻对白叶枯病的抗性用于育种可追溯到 20 世纪 30 年代。丁颖于 1926 年从抗病虫和适应性强的普通野生稻与当地栽培稻的自然杂交种后代中筛选育成品种中山 1 号，之后，从中选出的衍生品种包选 2 号和包胎矮等，都遗传了中山 1 号的抗白叶枯病（兼抗矮缩病、褐稻虱）、优质及适应性强等特性，迄今一直是华南晚籼主要当家品种（中国农科院，1986）。随着近代育种目标由高产转向丰、抗、优，遗传改良品种抗病性已成为 70 年代以来水稻育种的主要目标之一。自 1968 年至 1989 年，我国鉴定了大量种质的抗性。据不完全统计，评价了包括地方品种、野生稻和国外引进种质在内的 78 500 多份（次）材料（表 1）。过崇俭、谢岳峰、黄继武和谢式坪钰等分别筛选了太湖流域粳稻，云南、贵州和台湾地方水稻资源对白叶枯病的抗性。筛选出千余份抗病材料（过崇俭等，1986；谢岳峰等，1989；黄继武，1989；广东省农科院，1989）。广东和广西农科院评价了 3 800 多份华南野生稻，筛选出 S1008 等 10 份优质、抗病和广谱高抗的 RBB16 等 50 份编号（表 1）。中国农科院品种资源所主持鉴定了 3500 多份引自亚、非、拉、欧、北美和大洋洲 82 个国家和 IRRI 的种质，其中 450 份材料的白叶枯病抗性颇佳。此外，粤、桂、苏、浙、蜀、滇、湘、鄂、皖等省（地区）农科院和大专院校也评价了大量引进品种，所鉴定出来的抗病品种多数来源于 IRRI（表 1）。

表 1　中国水稻品种和种质资源对水稻白叶枯病的抗性评价

Table 1　Evaluation of bacterial blight (BB) resistance in rice cultivars and germplasms in China

(1957—1993)

评价单位 Institute	总数 Total	抗病数 No. of R Acc.	主要抗病品种和种质资源 Main resistant cultivar & germplasm	参考文献 Reference
南京农业大学	1 209	279	瘦田赤、矮脚齐子、湖南早、一线红、红节川、太和粘、红米子、小冬稻、叶里培、陈家种、老大稻等	方中达等，植物病理学报，1963
湖北省农科院	2 992	59	鄂晚 5 号、嘉农选优 6 号、BW196，IRRI 系统品种	全国水稻白叶枯病防治研究资料第二辑，1976
福建省农科院	8 000	220	引自 IRRI 品种	同上
安徽省农科院	3 452	44	紫色粳、富光 16、大野中稻和 IRRI 系统品种（系）	全国稻病抗性鉴定会议资料 1976—1987
浙江省农科院	1 350	39	农试 4 号、厦引 11、辐包矮 22、福锦、神乐糯及 IRRI 系统品种	全国稻病抗性鉴定会议资料（1977）
江苏省农科院	2 382	738	奎稻、75－47 鉴选、7432-15、贵州麻谷、蚌珠芒、南京罗汉黄、金银粳、徐辐 4 号、贵农育等	江苏省农科院植保、粮食所，江苏农业科学，1980
福建农学院	5 000	59	梅非早 2 号、植科 1 号、十石、龙城 12、D32、C104、矢租等	夏怡厚等，福建农学院学报，1985
江苏省农科院	4 819	20	南粳 15、南粳 11、邳早 15、盐粳 2、农垦 58、紫金粳、筑紫晴、扬稻 1 等	江苏省农科院植保所、粮食所，江苏农业科学，1980
江苏省农科院	1 389	11	小爱 2 号、风景稻、大绿种、矮绿种、铁粳青、红芒糯、大稻头糯、红壳稻、长秸糯、早糯稻等	过崇俭等，江苏农业学报，1986
中国水稻所	970	32	惠谷占、玻美占、齐双占、二华早、青叶青丰、浙湖 2 号等	黄世文等，1986
浙江省农科院	366	11	丰阳矮 217、单 32、青封占 35、玻齐占、晚六早、齐双占等	“七五”全国水稻育种组新品种抗性鉴定，1987
中国农科院作物所	3 234	54	秋光、秀峰、早生爱国 3 号、中新 120、筑紫晴、京系酒、关东 60、银河、喜峰、黎明、银河、矢租、印度诺等	章琦等，1976—1987“75”全国水稻育种组新品种抗性鉴定。全国稻病抗性鉴定会议资料，1987；1986

（续）

评价单位 Institute	总数 Total	抗病数 No. of R Acc.	主要抗病品种和种质资源 Main resistant cultivar & germplasm	参考文献 Reference
中国农科院植保所	62	5	*O. laftifolia*（W14）、*O. minuta*（W17、W44） *O. officinalis*（W46）、*O. punctata*（W50）	韩玉梅，赵荣杰，植物保护学报，1988
华中农业大学	6 184	50	扎昌龙、中白米、黄牛尾香糯、安宁早、鲁士谷、毫梅、雾露谷、云南大粒、云香糯1号、八月糯、马罗糯谷、矮脚糯（1），长毛糯、中云香糯和毫双7号等	陈勇，谢岳峰等，1990
广东省农科院	5 361 2 064	162 10	早仔种、宫牌种、马尾齐、红梅糯、牛团糯、红壳糯、早水玻璃占、麻谷等 野生稻：*O. rufipogon*（S1008，S1115，S2165，S2178，S3116，S3375等）	徐羡明等，广东农业科学，1990；广东省农科院水稻所，植保所．(1989)
广西壮族自治区农科院	7 964 1 752	61（MR） 57（2份HR）	小颗油占、黑毛粳、大糯、香糯、白旱谷、白旱谷、白皮山糯谷等 野生稻：YD2-0363、YD2-0601、YD2-0795（RBB16）、YD2-0684等	孙恢鸿等，广西农业科学，1992 孙恢鸿等，植物保护学报，1991
贵州省农科院	3 759	103	西南175，大秆早、贵州麻谷、伊势景、孟关大麻谷、麻油占、香糯、红占谷、寸谷糯、三百棒等	张再兴等，中国稻种资源，应存山主编，1993
中国农科院品资所	3 500	458	IRRI品种为主	袁世传等，1993
湖南省农科院	4 615	457	假蚂蚱谷、背子谷、紫谷、二百号、双抗红、江水香糯、冬香糯、香稻、浏阳大糯、千禾糯、白麻谷、旱禾糯、麻糯谷、六十早、红其苏等	侯小华等，湖南农业科学，2001
合计 Total	70 424	2 929		

有些抗源已经用于育种计划并取得明显效果。用的最广的籼稻抗源引自IRRI，如IR20、IR26、IR30、IR29、IR2061等携有*Xa4*基因的品种已成为籼稻育种的主体抗源。1987—1989年，在全国水稻育种协作网“七五”期间统一鉴定的564个品种（系）中，对白叶枯病抗性上乘的籼稻多数带有IR系统的亲缘（表2）。斯里兰卡品种BG-90-2也已成为籼稻中的骨干抗源。李志正等由BG-90-2系选出高产、抗病品种扬稻1号和扬稻3号，仅在1983—1984年，就在长江中下游种植了434万hm^2以上（李志正等，1985）。南京农业大学最近选育的抗病品系882484和884003也有BG90-2的亲缘。章琦等曾用南北方6个菌株测定和分析了全国主栽品种的抗病状况，参试籼稻品种中有近80%的抗性较差（章琦等，1988）。时隔数年，抗病良种在新品种（系）中已占有一定比重，将为90年代提供一批优良的抗病品种。

表 2 我国近年培育的抗稻白叶枯病新品种*

Table 2 New rice cultivars (lines) with resistance to bacterial blight developed in China

品种 Cultivar	供体亲本 Donor	培育单位 Bred by Inst.	品种 Cultivar	供体亲本 Donor	培育单位 Bred by Inst.
青华矮 6 号	IR20	广东省农科院	桂陆早	IR26	广东省农科院
华竹 40	IR20	广东省农科院	早 36	IR30	广东省农科院
晚华矮 4 号	IR20	广东省农科院	乌珍 5 号	IR30、IR29	广东省农科院
珠包矮 385	IR20	广东省农科院	双二占	IR30、IR29	广东省农科院
秋青 2 号	IR20	广东省农科院	晚华 11 选	青华矮	广东省农科院
青封占 35	青华矮	广东省农科院	丰阳矮 217	?	广东省农科院
玻齐占	玻璃矮	广东省农科院	晚六早	晚华矮 1803	广东省农科院
齐双占	玻齐占	广东省农科院	三卢占 7 号	三黄占	广东省农科院
竹玻矮 2 号	玻惠占	广东省农科院	二华早 1 号	省糯保/HB580	广东省农科院
三五糯	青华矮 12	广东省农科院	晚青早 2 号	晚华 1 号	广东省农科院
特华早	青华矮 6 号	广东省农科院	晚桂早 1 号	晚华 1421	广东省农科院
二九丰	IR29	嘉兴地区农科所	5411	IR2061	江西省农科院
扬稻 1 号	BG90-2	里下河农科所	闽糯 580	—	福建稻麦所
扬稻 2 号	BG90-2	里下河农科所	籼 128	—	福建稻麦所
882484	BG90-2	南京农业大学	中 86-51	HA19317	中国水稻所
883013	843001	南京农业大学	中育 88-6	IR8850-229-3	中国水稻所
884003	BG90-2	南京农业大学	单 32	—	江西萍乡农科所
82-9	—	广西农科院	选 79-3	—	辽宁省农科院
82-1-660	—	广西农科院	中作 8538	城堡 1 号	中国农科院
中系 8714	喜峰	中国农科院	中作 8593	黎明	中国农科院
中百 4 号	南粳 15	中国农科院			

* 浙江省农科院整理，1987。1989 年全国水稻育种协作组新品种（系）抗性鉴定。

我国粳稻主栽品种多数对白叶枯病具有一定抗性。主要抗源为南粳 15、南粳 11、关东 60、日本晴、邳早 15、中新 120、筑紫晴、农垦 58 等。追溯系谱渊源，这些抗源大部分来自日本，与爱国、旭和农垦 58 有亲缘关系。这些抗源多数携有成株抗性基因 *Xa3*（章琦等，1988）。此外，台湾省品种改良场培育、推广的品种（系）中，具有抗病性的是以当地品种为亲本的蓬莱稻。主要有台东 25、台中 186、台南 1、2 号、嘉南 8 号和新竹 56 等（谢式坪钰，1971）。

籼粳交育种多数利用籼稻抗病基因。如 IR20、DV85、Tetep、Tadukan 和 Zenith 等。有些粳稻如喜峰、黎明、城堡 1 号等中抗稻瘟和白叶枯病的抗源也常用于育种。

我国水稻对白叶枯病抗性的评价和利用进展较快。存在一个明显的问题是抗源利用的单一化，籼稻中的主要抗病基因 *Xa4* 抗我国多数菌株，但不抗我国 V 型菌，事实上 *Xa4* 只抗菲律宾 6 个小种中的 1 和 5 两个小种，对南亚的小种不具有抗性。我国粳稻的抗源也较集中利用了一个小种专化抗性基因 *Xa3*，它只抗日本小种 1、2 和 3，不抗小种 4 和 5。抗源过于集中，遗传变异狭窄则容易出现抗性丧失。因此，日本和IRRI的育种计划中不断更新抗病基因。我国育种家也提出不同阶段的育种观点，提出实现品种更换达到高产稳产。发掘和研究地方资源野生稻以及引进新的抗性基因，是一项有远见和有价值的工作。

2 水稻对白叶枯病抗性类型的研究

品种的抗病反应是寄主与寄生物互作的结果。因此抗病性的研究应该是生物间遗传学的研究。这种自然属性决定了水稻对白叶枯病的抗性反应必然是多样的。可见，抗病性的评价应包括研究和分析，将丰富的抗源进行整理归类，方便于育种利用。

国外关于稻白叶枯病抗性类型的报道或多或少追随Vanderplank的垂直抗性与水平抗性的概念，质和量的抗性常用来代替上述名词。对于Vanderplank有关水平抗性的观点争论颇多。焦点是他将水平抗性的术语既用于遗传学又用于病害流行学，似乎引起了混乱，迄今在遗传学上还没有被证实，但是他提出的水平抗性理念，启迪了人们应该寻求稳定和持久抗性。

水稻对白叶枯病的抗性同其他作物的真菌病害类似。Mew等从植物生育期的角度，将水稻白叶枯病抗性分为苗期和成株抗性。Wakimoto等（1954）最早注意到有些品种在不同生育期的抗性反应不同。日本曾一度认为水稻在苗期的反应不稳，而提出唯以成株期鉴定为准。实际上不同生育期表现抗性的现象在水稻中是普遍的。湖南省农科院水稻所（1976）、章琦等（1978）鉴定的材料中，观察到水稻对白叶枯病有全生育期和成株期两种抗性类型。

全生育期抗性的特点是在苗期就表现抗性直至成株阶段，有明显的小种专化现象。如携有基因*xa5*的品系IR1545－339，在播种后23d时接种鉴定，对菲律宾的6个小种的抗性表现为，抗小种1、2、3和5，对小种4和6是感的（Mew，1987）。不过，抗性类型还取决于寄主基因型与病菌的互作。如印度诺和矢租全生育期高抗菲律宾4个小种（章琦等，1986），但矢租对菌株KS6－6却表现成株抗性。一般全生育期抗性易于传递到后代，这类抗源用作供体的育种后代，则可以在苗期就进行接种来选择抗病的植株。

成株期抗性品种在苗期感病，随着植株成长转为抗病。小麦锈病和燕麦冠锈病等许多作物病害都有这种现象。Ezuka等（1974）较早报道早生爱国3号抗性品种群的品种有成株抗性。Mew等（1981）发现Zenith和M. Sungsong具有典型的成株抗性。章琦等（1988）进一步研究并证实了水稻白叶枯病的成株抗性同其他作物病害的成株抗性是可以比拟的。其特点是成株抗性同样存在明显的小种专化现象。如陈光稻3号和BG90-2对菲律宾4个小种的成株抗性呈现交叉反应；不同品种对相同的菌株或同一个品种对不同菌株的抗性转变期不同；品种的抗性并非都到生殖生长阶段才表现，“转抗”期的迟、早同生育期并没有关系。M. Sungsong和Zenith的生育期分别为135d和100d，对菲律宾小种1的转抗期却分别为第八和第十一张叶位；由感病转为抗病的方式，有渐变、或清晰地在某叶位转抗的不同表现。成株抗性品种在同一植株上的不同抗、感反应叶片组织中的细菌增殖状况是不同的，抗病叶片内的细菌增殖量大大低于感病叶片，增殖速度也缓慢得多。电镜观察成株抗性品种在感病阶段的叶片中菌量聚集得很多，寄主组织出现变态，转抗叶片的导管分泌物将病菌包埋而使之变形（赵新平等，1987）。这些观察为病理生理、生化机制等方面的研究提出了值得探索的信息。如抗病组织中的分泌物是什么成分，抑制病菌生存的原因及如何利用于育种等。

3 水稻对白叶枯病的抗性遗传研究

Ou（1985）指出 Kariya 等在 1956 年关于稻白叶枯病抗性是显性或隐性、单基因或多基因的报道可能是较早的遗传研究。之后，日本和 IRRI 先后用本地菌株系统鉴定和命名了 16 个抗性基因。1982—1987 年双方合作，首先采用双方菌株进行重复比较研究，修正和命名为 14 个抗稻白叶枯病基因（表 3）（Qgawa et al.，1989）。

表 3 水稻白叶枯病系统鉴定的抗病基因

Table 3 Summary of bacterial blight resistance genes systematically identified

鉴定基因 Gene identified	原命名 Original designation	所用菌株 Strains used	代表品种 Representative cultivar	文 献 Reference
		日本菌株		
Xa1	*Xa1*	X－17	Kogyoku，Java14	Sakaguchi，1967
$Xa1^{h}t$	$Xa1^{h}$	T7141（1）	IR28，IR29，IR30	Yamada，1984
Xa2t	*Xa2t*	X－17，X－14	Rantai Emas 2，Tetep	Sakaguchi，1967
Xa3	*Xaw*	日本小种 T7174（1）， T7147（Ⅱ）， T7133（Ⅲ）	Wase Aikoku 3， Chugoku 45，Java 14	Ezuka et al.，1975
	$Xa4^{b}$ *Xa6* *xa9*		Semora Mangga Zenith Sateng	Librojo et al.，1976 Zidhu & Khush，1978 Singh et al.，1983
Xa11		（Ⅰ，Ⅱ，ⅢA，Ⅳ）	Elwee，IR8，RP9－3	Ogawa et al.，1987
$Xa12^{h}t$	$Xa-kg^{h}$	印尼菌株 H75304（Ⅴ）	Kogyoku，Java14	Ogawa et al.，1987
Xa4	$Xa4^{a}$	菲律宾菌小种 PXO25（1）	IR20，IR22， IR1529－680－3，TKM6	Petpisit et al.，1977
xa5		PXO25（1）	DZ192，IR1545 － 339， RP291－7	Petpisit et al.，1977
Xa7		PXO61（1）	DV85，DV86，DZ78	Singh et al.，1978
xa8		PXO61（1）	PI231129	Sidhu et al.，1978
Xa10		PXO61（1） PXO86（2） PXO79（3） PXO71（4）	CAS 209	Yoshimura et al.，1983
xa13		PXO99（6）	BJ1，Chinsurah BoroⅡ	Ogawa et al.，1987
Xa14		PXO112（5）	TN1	Taura et al.，1987

（t）＝临时基因符号（Tentative symbol pending further confirmation）。

近10年来，我国有关稻白叶枯病抗性的遗传研究内容比较广泛。包括不同抗性类型品种的基因分析、亚种间杂交后代抗性遗传行为以及常规稻和杂交稻育种后代的抗性表现等。反映了我国对水稻白叶枯病抗性的遗传研究紧密结合育种应用，也揭示了抗性遗传受抗性表达时期、寄主-病菌互作和不同遗传背景等诸多因子的影响而呈现错综复杂的情况。

综合现有报道，不同抗性类型品种一般由独立主效基因控制其抗性。主效基因间存在互补、重叠、上位性和抑制等互作关系。有些品种由主效与微效基因共同制约其抗性（祁祖白等，1987；周泰初等，1982；章琦等，1986；章琦等，1990）。

谢岳峰等（1989）研究了15个抗性稳定的云南地方品种对湖北菌株江陵691的抗性基因。结果表明大多数抗病的品种由1～2个主效基因控制，发现若干抗性基因与已知基因是非等位的，有可能从中鉴定出新基因。章琦等（1986，1990）在IRRI合作研究了7个不同抗性类型品种，对当年基因命名所采用的病菌小种的遗传模式和等位性测试，表明矢租和印度诺对菲律宾小种1的全生育期抗性由隐性基因 *xa5* 控制；印度诺成株期还携有另一对隐性基因，青华矮6号携有一对不完全显性基因 *Xa4* 控制其全生育期抗性；广泛用于育种的粳稻抗源南粳15、邳早15和地方品种蚌珠芒等都携有成株抗性基因 *Xa3*，能控制日本菌群1和我国北方菌株HB17的抗性（表4）。伍尚忠等（1989）报道了品种华竹矮、青华矮6号、玻璃矮和IR26选对广东优势的Ⅳ群菌的抗性受1～2个主基因控制，兼有微效基因作用，均有 *Xa4* 的亲缘。

表4 若干水稻品种水稻白叶枯病抗性基因分析

Table 4 Summary of bacterial blight resistance genes in some rice cultivars

品种 Cultivar	所用菌株 Strain used	遗传表现 Genetic behavior	基因 Gene identified	文献 Reference
矢租	菲律宾小种1～4	1对隐性基因	*xa5*	章琦等，1986
印度诺	菲律宾小种1～4	1对隐性基因控制全抗，2对隐性基因控制成抗	*xa5* *xa5*，*xa*?	章琦等，1986
青华矮6号	PXO61（1）	1对不完全显性基因（全抗）	*Xa4*	章琦等，1990
邳早15	T7174，HB17	1对显性基因（成抗）	*Xa3*	章琦等，1990
蚌珠芒	T7174，HB17	1对显性基因（成抗）	*Xa3*	章琦等，1990
南粳15	T7174，HB17	两对交换值30%连锁显性基因（成抗）	*Xa3*	章琦等，1990
中百4号	T7174，HB17	两对交换值30%连锁显性基因（成抗）	*Xa3*	章琦等，1990

赵新平等（1986）研究6个爪哇稻和3个粳稻的成株抗性遗传时，观察到成株抗性的显性反转现象，抗病基因在个体发育的不同生育阶段呈现显隐性不相一致，一个抗、感杂交组合 F_2 群体的分离比由前期的1（抗）∶3（感），在成株期转为3（抗）∶1（感），认为由于“成抗”基因处于纯合状态时由感病转为抗病的时期先于杂合状态所致。如果此见得到验证，且有普遍性，则提示人们在有“成抗”供体的育种世代材料中，可在早期选到纯合的抗病株，这是一个十分有意义的课题。

因为水稻与白叶枯病之间具有特异性或非特异性互作。因而同一个品种与不同毒力菌株的抗性遗传分析，有可能得出不同结果。周泰初等（1982）、章琦（1990）分别用不同菌株研究矢租和蚌珠芒的遗传的表现是不同的。IR26 对菲律宾小种 1 有一对显性基因控制，而对日本的 5 个菌群的抗性表现为多基因控制（表 5）（王昌明，1986；Ogawa et al.，1989；Yamada et al.，1981）。

表 5 水稻品种对白叶枯病菌不同小种（致病型）代表菌株的抗性遗传行为[a]

Table 5 Genetic behaviors of rice cultivars resistance to some *Xoo* representative strains of different races (pathotypes)

品　种 Cultivar	小种或菌株 Race or strain	遗传表现 Genetic behavior	参考文献 Reference
矢租	菲律宾小种 1～4[a]	一对隐性基因 *xa5*	章琦等，1986
	KS1－3	一对显性基因	周泰初等，1982
马罗糯谷和八月糯	江陵 691	一对显性基因	谢岳峰等，1990
	菲律宾小种 1	一对隐性基因	
蚌珠芒	P1、T3、C2	一对显性基因 *Xa3* [a]	章琦等，1990
	KS 1－3	两对显性互补基因	周泰初等，1982
IR26	菲律宾小种 1	一对显性基因 *Xa4*[a]	Petpisit et al.，1977
	日本 5 个小种	多基因	Horino et al.，1980
小爱 2 号	KS6－6、KS1－21、JS49－6	两对重复显性基因	万建民、朱立宏，1990
	Zhe 173	多基因	
风景稻	Zhe 173、KS1－21	一对隐性基因	万建民、朱立宏，1990
	KS6－6、JS49－6	多基因	

[a] 经等位性测定。

4 亚种间杂交后代白叶枯病抗性遗传行为

籼粳品种间在形态、生理、时间、生态演化以及遗传行为等方面具有明显差异。亚种间杂交后代的稻白叶枯病抗性遗传表现大致有以下三种：1）有特异性。万建民和朱立宏（1985）报道 DV85 在南京 11 和 TN1 的籼稻背景下，受一对显性和一对隐性基因控制，存在显性上位作用；而在粳稻农垦 57、桂花黄背景下，表现为两对基因的互补作用；早生爱国 3 号在与不同亚种杂交的遗传背景下，也出现 F_2 群体的抗、感分离比例不同。2）无特异性。张跃刚和朱立宏研究 7 个籼稻白叶枯病中间抗源与 7 个不同亚种的感病品种杂交后代的抗性遗传结果表明，籼稻抗病基因在籼粳两种遗传背景下的抗性表达和遗传行为没有明显差别，主效基因抗性的表现都是明显的，并遵照孟德尔遗传法则。章琦等分析籼稻品种青华矮 6 号的抗性遗传，以粳稻感病品种沈农 1033 为母本杂交，其后代呈明显单基因抗性表现。3）张林等（1983）分析亚种间和亚种内的杂交后代的成株抗性遗传表现基本是一致的，但是全生育期抗病亲本的亚种间杂交后代的成株抗性遗传表现与亚种内杂交基本一致，但在全生育期抗病亲本的亚种间杂交后代中，苗期的一部分感病植株到后期转为抗病。以上不同结果是否与试材的抗性遗传稳定性或其他原因有关，还值得进一步探讨。

5　杂交育种后代抗性遗传表现

若干报道分析数百个杂优组合子一代对当地菌株的抗性表现有显性、不完全显性和隐性三种。子一代的抗性表现取决于抗源亲本基因型。多数抗、感亲本正反交杂种一代的抗性反应趋于一致，表明抗性主要是由核基因控制（中山大学，1978）。万邦惠等（1980）报道正反交 F_1 的抗性表现有一定差异。以感病品种为母本配组的子一代抗病水平多数下降，认为抗病性除受主基因控制外，还可能受微效基因影响。

某些抗源因为伴随不良性状而难以直接利用，可培育改良中间抗源提供育种利用。周毓珍等（1985）观察高抗品种 DV85 等的衍生中间抗源在不同遗传背景下的抗性转育效果不同。认为可能是在连续世代单株选择过程中，发生了抗病基因的分离和重组以及基因体系的重建，加上抗病基因供体和受体的遗传背景不同的影响结果。

6　结语

大量研究提供了许多对抗病育种有价值的信息，同时又发现还有些问题有待进一步研究。大致有以下几个方面：

6.1　建立我国水稻对白叶枯病抗性遗传研究和基因系统分析鉴定的标准化程序。暂时采用方中达等（1990）最近提出的我国水稻白叶枯病各群代表菌株。

6.2　建立我国水稻白叶枯病的鉴别系统。包括具有不同亚种并携有单基因抗性的鉴别品种和相对应的专化菌株，为系统鉴定我国抗源基因组成提供基本条件。

6.3　创造抗性遗传基因多样性。系统分析挖掘地方资源和野生稻中的新抗源。

6.4　培育不同生态型的具 1～2 个有利基因的抗病中间材料。在研究遗传规律的同时为育种提供理想性状材料，将加速育种的进展。

原载：中国农业科学，1991，24：26～36.

我国水稻抗白叶枯病基因的应用及策略

章　琦

（中国农业科学院作物育种栽培研究所）

Utilization and Strategy of Gene for Resistance to Bacterial Blight in China

Abstract In recent years, the bacterial blight resistance of rice conventional cultivars has increased in China. Resistance genes, however, are used very singular. The cultivars planted in large areas and some new ones bred are all depended on the gene *Xa4* which is virulent to the Chinese pathotype V distributed in Guangdong Province. The frequency and distribution range of pathotype V are increasing and extending from Guangdong Province to other areas recently. It is necessary to be paid attention in the future.

The bacterial blight resistance of hybrid rice combinations planted in large areas is not good enough in China. The characters of resistance improvement in hybrid rice are analyzed in the paper. Strategy of rational arrangement and replacement of cultivars, widening utilization of resistance genes and exploring new resistance resources are suggested.

Key words rice bacterial blight; resistance gene; strategy

摘要 近年广泛开展了遗传育种与植物病理等多学科的协作，我国常规稻品种的白叶枯病抗性水平有较大提高。但是，所利用的抗性基因比较单一，大面积种植和新育成的抗病水稻品种（系）所携带的抗性基因几乎都是 *Xa4*。*Xa4* 是一个小种专化抗性基因，不抗我国广东、福建等省的Ⅴ型菌。近年Ⅴ型菌在广东的出现频率不断上升，分布范围扩大。我国大面积种植的杂交稻组合的抗性都不太理想。本文分析改良杂交稻抗性的特点、提出合理搭配和更新品种、拓宽抗性基因的利用范围，以及进一步发掘新抗源等对策的建议。

关键词 水稻白叶枯病；抗性基因；策略

水稻白叶枯病（*Xanthomonas oryzae* pv. *oryzae*）是威胁我国水稻生产的一个重要细菌性维管束病害。因为病菌主要通过水孔或伤口入侵，所以在台风登陆频繁的华南、长江流域主要稻区，受害尤为严重，有的年份能造成水稻大面积歉收。广东省 1978—1986 年因台风肆虐，晚稻平均每年发生白叶枯病、细菌性条斑病和稻瘟病的面积达 396 万 hm^2，损失稻谷 1.32 亿 kg（黄斌民，1988）。安徽省 1949 年以来，白叶枯病 7 次大流行。1974 和 1982 年两次发生面积 27 万 hm^2，损失产量分别达 1.65 亿 hm^2 和 2.45 亿 hm^2。种植抗病品种是经济有效的防治方法。本文重点分析我国近年新育成的和生产上主要种植的常规稻品种（系）、杂交稻组合的抗源及抗病基因的利用状况，商榷防治对策。

1 我国常规稻品种水稻白叶枯病抗性基因的应用

1.1 籼稻品种

作者等（1988）曾报道在我国主栽籼稻品种中，有 80%是高度感病的。因为它们多数源于几个不携带水稻白叶枯病抗性基因的矮源及其衍生品种，矮脚南特、广场矮和珍珠矮等。这些品种的年推广面积曾达 667 万 hm^2 左右（林世成等，1991）。白叶枯病一度曾十分猖獗，随着育种转向丰、抗、优目标，全国加强了改良抗病性研究。在国家“七五”和“八五”计划水稻育种协作攻关项目中，有相当一批新育成的籼稻品种（系）的白叶枯病抗性颇佳。抗病品种（系）占当年鉴定总数的比率，由 1986—1987 年的 7.1%上升到 1992 年的 19%。在 1991—1993 年国家水稻育种攻关上报成果或通过品种鉴定的品种中，抗病的占总数的 43.7%（表 1）。例如广东省农科院育成的高产、抗病的青华矮 6 号、华竹矮、晚华矮 1 号等晚稻品种，于 80 年代仅在两广地区就扩种了 37 万多 hm^2（林世成等，1991）。江苏省里下河地区农科所从 BG90 - 2 中，系选的扬稻 1 号和 2 号，两年内就在长江中下游种植了 41 万 hm^2 以上（李志正等，1985），都有效地降低了这些地区的危害程度。预计“八五”期间推出的优良品种，也将在今后几年内大面积种植。可以认为，80 年代以来，我国籼稻常规品种的抗病性有明显的改良，将在减少病害损失与稳定产量方面发挥重要作用。

但是未来应引起注意的问题是，包括新育成的抗病品种（系）所利用的绝大多数抗源，仍然来自 IRRI 的品系或有 IR 系统的亲缘。例如广东省农科院育成的抗病品种主要以青华矮 6 号为供体，该品种的抗性来自有 IR20 亲缘的华竹矮（伍尚忠等，1990）。章琦等（1994）和杨文才等（1994）的研究表明我国籼稻主要的抗病基因为 *Xa4*（表 2）。众所周知，IR 系统品种（系）的命名已从 IR20 至 IR66。其中除 IR56 外，都携有基因 *Xa4*。该基因仅能控制菲律宾小种 1 和 5，虽然对我国大多数病原型是抗的，但不抗华南地区的 V 型菌（方中达等，1990）。最初发现该毒性菌的徐羡明等报道，广东省 V 型菌的出现频率已由 1991 年的 6%上升为 1992 年的 10.2%，分布范围由原来的中山市扩展到佛冈、番禺和电白等县。福建省农科院于 1980—1984 年测定了 56 个县的 298 个菌株中，也发现了少数菌株对携有 *Xa4* 的品种具有毒力（林声和等，1986）。时隔 10 年，大面积连续种植单一抗性基因的品种，导致毒性菌株频率增高的可能性是很大的。寄主基因型的布局无疑是影响病原菌群体遗传结构变异的主导因子。品种抗性基因单一必然加速新的或原来处于劣势的病原菌的变迁而丧失抗性。例如，80 年代以来，菲律宾由于能侵染 *Xa4* 的新菌株（菲律宾小种 2）开始上升为优势小种，导致携有 *Xa4* 基因的品种在该国北部的 Isabela 省丧失了抗性。同样的情况也出现在南亚（Mew，1987；Vera et al.，1989）。因此，我国目前抗源遗传基础狭窄的问题必须引起注意。

表 1　我国水稻 1986—1993 育成的新品种（系）抗白叶枯病评价*

年　度	总数（份次）	籼稻（份次）	粳稻（份次）	抗病品种			
				籼稻	占总数（%）	粳稻	占总数（%）
1986—1987	366	198	168	14	7.1	22	13.1
1988	179	88	91	17	19.3	26	28.6
1989	180	97	83	13	13.4	17	20.5
1990	211	129	65	10	7.8	17	26.2
1992	160	79	52	16	19.0	7	13.5
1991—1993	31	16	15	7	43.7	9	60.0

* 根据浙江省农科院 1986—1993 年汇总的全国水稻育种协作组水稻新品种（系）抗病性鉴定整理。

表 2　水稻抗白叶枯病若干抗源的基因鉴定

品种	所用菌株	基因	文献	品种	所用菌株	基因	文　献
青华矮 6 号	P1、HB17	*Xa4*	章琦等，1990	双抗 18	P1、HB17	*Xa4*	杨文才等，1994
华竹矮	CⅣ	*Xa4*	伍尚忠等，1990	Cuba 2196	P1、HB17	*Xa4*	同上
玻璃矮	CⅣ	*Xa4*	同上	Cuba 2310	P1、HB17	*Xa4*	同上
三黄占 2 号	P1、HB17	*Xa4*	章琦等，1994	Cuba 4010	P1、HB17	*Xa4*	同上
二九丰	P1、HB17	*Xa4*	同上	邳早 15	T1、HB17	*Xa3*	章琦等，1990
扬稻 1 号	P1、HB17	*Xa4*	同上	蚌珠芒	T1、HB17	*Xa3*	同上
83007	P1、HB17	*Xa4*	同上	南粳 15	T1、HB17	*Xa3*	同上
75 - 34	P1、HB17	*Xa4*	同上	CN04	T3、HB17	*Xa3*	同上
NT02	P1、HB17	*Xa4*	同上	盐粳 902	T3、HB17	*Xa3*	同上
川植 4 号	P1、HB17	*Xa4*	同上	宁 86 - 14	T3、HB17	*Xa3*	同上
川植 5 号	P1、HB17	*Xa4*	同上	中百 4 号	T3、HB17	*Xa3*	同上
三江	P1、HB17	*Xa4*	杨文才等，1994	湘虎 25	T1、HB17	*Xa3*	同上
双抗 12	P1、HB17	*Xa4*	同上				同上

1.2　粳稻品种

我国粳稻品种的白叶枯病抗性较好。根据粳稻发展史来看，从 20 世纪 50 至 70 年代中期，我国数次引进农垦 58、农垦 57、丰锦、黎明、中新 120 等日本品种，直接用于生产或作亲本。培育了大批高产抗病品种。如农虎 6 号、鄂晚 5 号、鄂宜 105、秀水 48、矮粳 23 以及南粳 15、盐粳、中作、中花、辽粳系统等包括近年育成的新品种，几乎覆盖了全部粳稻种植地区。这些品种的亲缘关系归属于爱国、坊主和旭等（章琦等，1988；林世成等，1991）。作者等近年的研究表明，粳稻的主要抗源都携带 1 对显性成株抗性基因 *Xa3*，多年来在粳稻产区基本上控制了白叶枯病的流行。携带 *Xa3* 基因品种的抗谱虽然较广，但是其对分布在江浙一带的Ⅳ型菌是无效的。因此，粳稻育种中的抗性基因利用单一的问题，也不容忽视。

2　杂交水稻抗白叶枯病基因利用特点

杂交水稻一般具有比常规稻增产强的杂种优势。1992 年全国种植面积已达 1 730 多万 hm^2，占水稻种植面积的 50%以上。纵观大面积种植的籼稻杂交组合，其多数不育系不携有抗白叶枯病基因。如珍汕 97A、V20A、V41A 等。所采用的恢复系和配制的组合也少有抗性上乘者。李宣铿等（1993）于 1986—1990 年鉴定 3 000 多份湖南杂交水稻中心的组合及三系资源，在 175 份不育系和 185 份保持系中，只有 1 份不育系和 3 份保持系有中等抗性。在 2 346 份恢复系中，有一部分来自 IRRI 的品系的抗性较好。从全国水稻育种协作组所进行的杂交稻抗性鉴定结果来看，在 1986—1990 年的 27 个组合中，只有江优 594 表现中抗；在 1991—1993 年成果审定的 17 个籼型组合中，除博优桂为中抗外，其余组合的抗性均差（表 3）。20 世纪 70 年代以来，育种家在选育恢复系时，已经注意到抗病性的改良，采用了具有恢复基因又较抗病的 IR26、IR30、IR36 等品系。其中由珍汕 97A 与具有 IR30 亲缘的明恢 63 配制的组合汕优 63 是我国近年种植最广的组合，到 1988 年的种植面积达 470.7 万 hm^2；四川省 1989 年种植了 160 万 hm^2，占该省杂交稻面积的 85%以上；安徽省 1990 年 80%的中稻面积种植了汕优 63（林世成等，1991）。但是在白叶枯病流行的年份，尤其遇到风暴和涝灾，以汕优 63 为主的这些组合发病都很严重，安徽、湖北、浙江等省在 1991 年遭受洪涝灾害之后，又受到白叶枯病的肆虐。

表 3　1991—1993 年杂交籼稻新组合白叶枯病抗性评价*

组　合	抗性反应	鉴定时间	组合	抗性反应	鉴定时间
汕优桂 48	S	1986—1987	D 优 870424	S	1990
汕优桂 45	S	同上	江优 1126	S	同上
77ABA/明恢 63	S	同上	江优 K11	S	同上
汕优桂 8	S	同上	D 优 K25	S	同上
协优 2374	S	同上	D 优 48-2	S	同上
汕优桂 34	S	同上	针育 29	MS	同上
协优 2422	S	同上	江优 594	MR	同上
二九矮 4 号/明恢 63	S	同上	协优 432	S	1991—1993
V20A/萍恢 2 号	S	同上	D 优 1607	S	同上
珍 A/102	S	1988	K 优 3 号	S	同上
协优 46	S	1988	汕优 016	S	同上
959/明恢	S	1988	博优 49	S	同上
献改 A/明恢 63	MS	1989	博优 210	S	同上
D 优 123	S	1990	广优 4 号	S	同上
江优 04	S	同上	汕优 102	S	同上
协优 102	S	同上	I 优 323	S	同上
Ⅱ优 45-1	S	同上	K 优 48-2	S	同上
W6B45/测 49	S	同上	博优桂	MR	同上
D 优 46-5	S	同上	威优 77	S	同上
常优 87-88	S	同上	威优 72	S	同上

*　S=感，MS=中感，MR=中抗。

对于常规稻来讲，所采用的抗病基因，无论显性、不完全显性或隐性，只需通过各种育种手段，即能达到同质和改良抗性的目的。但对于主要利用杂种 F_1 优势的杂交稻而言，所配制组合的抗性水平，在遗传上要取决于抗病基因在 F_1 植株上的显性表现程度。我国所用的恢复系的抗病性，均由 *Xa4* 所控制。*Xa4* 为不完全显性基因，携带该基因的恢复系与感病不育系配制的杂种 F_1 的抗性，至多达到中抗甚至中感水平。实际上，抗病基因与不同遗传背景的感病亲本配组的 F_1 的抗性水平也是各不相同的。大面积种植的汕优 63 对白叶枯病的抵御性就较差。人为造成的品种基因型遗传上的脆弱性，加之其本身的低水平，引起病害大流行便在所难免。

3 抗性基因利用的策略

3.1 合理搭配、更新品种

长江中下游及东南沿海台风登陆频繁的地区，尽可能压缩感病或携带抗性基因单一的组合。种植高产和抗病性较稳定的常规稻品种，如近年推出的高产、优质并具有中度抗性的武育粳 2 号和武育粳 3 号等。同时还需要考虑合理搭配种植兼抗其他稻病的品种。

3.2 扩大利用其他抗病基因

在现有的 21 个白叶枯病抗性基因中，除 *Xa3*、*Xa4* 具有广抗谱外，*Xa7* 和 *Xa21* 也都是广抗谱显性基因。目前均已导入具有育性恢复基因的强恢复系 IR24 中，可能成为籼稻常规稻和杂交稻育种的中间材料。对于杂交稻和Ⅴ型菌的分布地区，都是值得采用的基因。

3.3 发掘和利用新的抗性基因

由中国农科院和广西农科院协作，近年评价了 871 份包括 13 个野生稻种材料对白叶枯病的抗性。发现 4 个野生稻种中的 32 份材料高抗 1 个广致病谱菌株 P6，其中可能存在栽培稻所缺乏的新基因。我国有丰富的地方和野生稻种质资源，许多单位已做了大量抗性评价工作。值得从病理、遗传育种、生物技术等各个侧面进一步研究。

原载：植物保护学报，1995，22（3）：421～426.

How to Evaluate and Identify Resistance Gene to Bacterial Blight in Wild Rice Species

Q Zhang[1] S C Lin[1] B Y Zhao[1] C L Wang[1] W C Yang[1] Y L Zhou[1] Y L Fan[1] H J Pan K J Zhao[1] D Y Li[2] C B Chen[2] L H Zhu[3] S X Tang[4]

(1. Institute of Crop Breeding and Cultivation, CAAS, Beijing 100081, China;
2. Institute of Germplasm Resources, GAAS, Naining 530007, China;
3. Institute of Genetics, Academia Sinica, Beijing 100101, China;
4. China National Rice Research Institute, Hangzhou, China)

Abstract Wild *Oryza* species are widely distributed in 77 countries. A total of 19 690 accessions of wild rice resources were collected in the main rice production countries and the International Rice Research Institutions (IRRI). The genus *Oryza* consists of 21 wild and 2 cultivated species representing AA, BB, CC, BBCC, CCDD, EE, GG, HHJJ, FF, and HHKK 10 genomes. Wild *Oryza* species own rich resources of resistance to diseases. Resistance to bacterial blight (*Xanthomonas oryzae* pv. *oryzae*, *Xoo*) exists in 15 out of 21 wild *Oryza* species. Some notable disease resistance genes, like *Gs* and *Xa21* were transferred successfully from wild *Oryza* species into the cultivated rice. Recently, a completely dominant, overall resistance gene with broad-spectrum *Xa23* was explored from *Oryza rufipogon* by systematic analysis. Two pairs of NILs, CBB23 (JG30 as recurrent parent, RP) and CBB23 (B) (IR24 as RP) and two improved cultivars with high yielding and resistance to blast, Zhongye 5112 and Zhongye 5114 carrying *Xa23* were developed. Some approaches to inter-specific hybridization between *O. sativa* and wild rice species are mentioned.

Wild *Oryza* species are widely distributed in 77 countries of Africa, Asia, Australia, and Latin America. Twenty one wild species together with two cultivated species, *O. sativa*, and *O. glaberrima* are recognized. The wild taxa have either 2n = 24 or 2n = 48 chromosomes representing AA (including A^bA^b, A^gA^g), BB, CC, BBCC, CCDD, EE, FF, GG, HHJJ, and HHKK genomes (Jena et al., 1990). Up to 2001, 19 690 collections of wild *Oryza* species were conserved involving 4 447 accessions at IRRI, 2 263 acc. in Japan, 1 591 acc. in India, and 7 322 acc. in China (Tang et al., 2004).

Some notable genes have been explored and transferred successfully from wild *Oryza* species into the cultivated rice: The well-known commercial exploitation of heterosis in rice improvement has been successfully demonstrated in China since the development of cytoplasmic male-sterile (CMS) line from *O. rufipogon* in 1973. Since then, more than 95% Chinese CMS rice lines used for hybrid rice development bear *O. rufipogon* cytoplasm (Jiang et al., 2004). The single dominant resistance gene *Gs*, identified from the one accession of *O. nivara* among 6 723 rice accessions is the only known source of resistance to grassy stunt (Khush, 1977). The first broad spectrum resistance gene to bacterial blight (BB) *Xa21* was identified (Khush et al., 1990) and cloned from *O. longistiminata* (Song et al., 1995) In addtion, *O. officinalisis* and *O. rufipogon* are good sources of resistance to brown planthopper.

Evaluation of BB resistance in wild *Oryza* species

As the genetic variability for resistance to bacterial blight (BB) is limited in the cultivated rice, plant scientists have made efforts to explore resistance resources from wild *Oryza* species since 1980s. Fifteen out of 21 wild *Oryza* species, i. e. *O. australiensis* (Multani et al., 1994), *O. eichingeri* and *O. brachyantha* (Brar et al., 1997), *O. rhizomatis* and *O. officinalis*, (Yang, et al., 1999), *O. meyeriana*, *O. punctata*, *O. latifolia*, *O. minuta*, *O. longistaminata*, *O. alta*, *O. rufipogon*, *O. gramulata*, *O. grandiglumis* and *O. glaberrima* (Zhang et al., 1994) were reported to have resistance to the disease (Table 1 and 2). In China, three wild *Oryza* species, *O. rufipogon*, *O. officinalis* and *O. meyeriana* were reported and are distributed in the southern provinces of Guangdong, Guangxi, Hainan, Yunnan and Jiangxi, *O. rufipogon* being the predominant one. Peng et al. (1992) observed high resistance to BB in *O. meyeriana* and *O. officinalisis*. Xu et al. (1990) after a large scale screening of *O. rufipogon* germplasm resources from Guangdong province found that 4 out of 2 064 accessions of showed high resistance to BB. Sun et al. (1992) reported that 9 out of 1 600 *O. rufipogon* accessions of Guangxi province were highly resistant to BB. In a collaborative project between CAAS and GXAAS, 871 accessions of wild rice involving 13 species were evaluated for resistance to 3 Chinese BB strains in 1987 - 1992; All the 12 accessions of *O. meyeriana* were highly resistant to the 3 strains and many accessions of *O. punctata*, *O. latifolia*, *O. minuta*, *O. officinalis*, *O. longistaminata*, *O. alta*, *O. rufipogon*, and *O. glaberrima* showed high resistance also (Table 1) (Zhang et al., 1994).

Table 1 Resistance to bacterial blight in wild-rice species (Zhang et al., 1994)

Species	Genome	Reaction of resistance to three strains					
		ZHE173		GD1358		HB84 - 17	
		MTS	BS	MTS	BS	MTS	BS
O. meyeriana	GG	HR	HR -	HR	HR -	HR	HR -
O. punctata	BBCC, BB	R	HR	R	R	R	R
O. latifolia	CCDD	R	R	R	R	HR	R

(continued)

Species	Genome	Reaction of resistance to three strains					
		ZHE173		GD1358		HB84 - 17	
		MTS	BS	MTS	BS	MTS	BS
O. minuta	BBCC	HR	HR	R	R	R	R
O. officinalis	CC	R	R	R	R	R	R
O. longistaminata	A^bA^b	R	R	R	R	R	R
O. rufipogon	AA	R	R	R	R	R	R
O. alta	CCDD	R	R	R	R	R	R
O. glaberrima	A^gA^g	R	R	R	R	R	R
O. granulata	GG	MR	—	MR	—	MR	—
O. nivara	AA	R	—	S	—	S	—
O. grandiglumis	CCDD	MR	MR	MS	MS)	MR	MR
O. australiensis	EE	S	—	S	—	S	—
Jingang30 (CK_1)	AA	S	S	S	S	S	S
Shenong1033 (CK_2)	AA	S	S	S	S	S	S

MTS=Maximum tillering stage, BS = Booting stage, R=Resistant, MR=Moderately resistant, S=Susceptible, MS=Moderately susceptible。

Transfer of bacterial blight resistance from wild rice to cultivated rice

Bacterial blight resistance in eight of the 12 wild rice species was successfully transferred into the cultivated rice by various approaches. *O. rufipogon*, *O. nivara* and *O. longistaminata*, the closely related wild species to *O. sativa*, can be crossed with latter by conventional hybridization supplemented with anther-culture or embryo rescue. Tissue culture, protoplast technique, somatic hybridization and other biotechnologies provide powerful means for producing interspecific hybrids between *O. sativa* and some distantly related wild species. A series of monosomic alien addition lines (MAALs) were developed through embryo rescue following hybridization between elite rice breeding lines and several distantly related *Oryza* species, and genes resistant to GS, BB, blast, BPH and WBPH respectively from 7 wild *Oryza* species (Amante et al., 1992; Brar et al., 1991; 1997; Yasuyuki et al., 1988; Jena et al., 1989; 1990; Zhu et al., 2004; Multani et al., 1994; *Huang al.*, 2001) (*Table* 2) *were separately transferred to the cultivated rice*.

New BPH resistance gene was identified from O. meyeriana *and inter-specific hybrid between* O. sativa *and* O. meyeriana *was obtained through cryopreservation and protoplasm regeneration of* O. meyeriana (*Huang et al.*, 2001). *Somatic hybrids were developed between the susceptible cultivated rice and* O. meyeriana *by asymmetric somatic hybridization* (*Zhu et al.*, 2004). *Some of the somatic hybrids showed resistance to BB* (*Table* 2).

Table 2 Methods for producing hybrids between *O. sativa* and some distantly related wild species

* Wild species	Resistance*	Method	Reference
O. minuta (BB, BBCC)	BB, blast	Embryo rescue	Amante-Bordeos et al., 1992
O. rhizomatis (CC) &	BB, BPH	Embryo rescue	Brar & Khushe, 1991, 1997
O. eichingeri (CC)		Cell fusion	Yasuyuki et al., 1988.
O. officinalis (CC)	BB, BPH, WBPH	Embryo rescue	Jena & Khush, 1989, 1990
O. meyeriana (GG)	BB, BPH	Asymmetric somatic hybridization	Zhu et al., 2004
		Cryopreservation, protoplasm regeneration	Huang & He et al., 2001
O. australiensis (EE)	BB, BPH	Embryo rescue	Multani et al., 1994,

* BB = bacterial blight, BPH = Brown planthopper, WBPH = Brown planthopper, WBPH = Whitebacked planthopper).

Performance of bacterial blight resistance in *O. rufipogon*

Collections with high resistance to BB in *O. rufipogon* are very few, however, this is a closely related species of the cultivated rice and its resistance can be easily transferred to the latter by conventional hybridization (Zhang et al., 1994) Therefore, it is necessary to well characterize the resistance type in *O. rufipogon*. According to our studies, various expressions of resistance to BB in *O. rufipogon* were observed: a) different resistance reactions (R/S or R/MR) appeared on different plants within the same acc. and even on the same plant at same growth stage ; b) both overall resistance and adult plant resistance existed; c) reversal reaction between host and pathogen might occur (Zhang et al., 1994). Sun et al. (1992) reported germplasm resources of *O. rufipogon* from the same origins used have different levels of resistance to BB. All these phenomena indicated that resistance to BB in *O. rufipogon* population is genetically heterogeneous and heterozygous since this is a predominately open-pollinated species. Oka and Morishima (1967) reported that the *O. rufipogon* and *O. nivara* usually introgressed with the cultivated rice where they were grown nearby the surrounding paddy fields. They retained a high rate of natural crossing (5%~40%) which enabled the populations to maintain in heterozygous condition. Li *et al.* (2001) compard the flower morphology between *O. rufipogon* and cultivated rice, indicated that extruded stigma and long anther in *O. rufipogon* were two main factors which influenced the introgression with the cultivated rice.

Species	Extruded stigma (%)	Anther length (mm)
O. rufipogon	80.3	5.56
O. sativa	0	2.44

Exploitation of resistance stock RBB16 and the development of NILs

In late 1980s, the Philippine race 6 (PXO99) was found to be virulent to all known dominante resistance genes from the cultivated rice. So six accessions of *O. rufipogon* with high resistance to 5 Chinese *Xoo* strains were re-evaruated using PXO99 as the test pathogen. Fortunately, one acc. named RBB16 showed high resistance to PXO99, and was supposed to have a new resistance gene. Then a doubled haploid line homozygous for BB resistance was developed by anther culture. An interspecific cross was made between RBB16 and JG30, a susceptible indica cultivar. The BB resistance of F_1s segregated as HR, MR and S in various proportions. F_1 plants with high resistance were then selected and anther-cultured. H_2 progenies showing high resistance to PXO99 were advanced to H_4 generation. All the H_4 plants were evaluated and confirmed for homozygosity in BB resistance. Subsequently, two pairs of near-isogenic lines carrying BB resistance gene from RBB16, CBB23 (with JG30 as recurrent parent, RP) and CBB23 (B) (with IR24 as RP) were developed by successive backcrossing and selfing to the generation BC_5F_4. They all showed high stable resistance to BB with the same agronomic characters of their respective recurrent parents (Zhang et al., 1998; 2002).

Resistance spectrum of CBB23

CBB23 is highly resistant to the 9 Philippine *Xoo* races at all growth stages (Table 3). IRBB21 (carrying *Xa21* from *O. longistaminata*) was susceptible to all the 9 races at seedling stage but turned resistance at late tillering stage. The other 5 dominant resistance genes *Xa3*, *Xa4*, *Xa7*, *Xa10*, and *Xa14* were all susceptible to the Philippine race 6, and showed differential reactions to some of the remaining races. Reaction types of the resistance gene in CBB23 to the 9 Philippine races were different from those of *Xa21* and other tested genes (Table 3).

Table 3 Resistance reaction to the 9 Philippine races of BB between CBB23 and the 6 genes tested (Zhang et al., 1998)

Material	Resistance gene	Resistance reaction to 9 Philippine races									Resistance type
		1	2	3	4	5	6	7	8	9	
CBB23	*Xa23*[t]	R	R	R	R	R	R	R	R	R	overall
IRBB21	*Xa21*	R	R	R	R	R	R	R	R	R	tillering
IRBB3	*Xa3*	R	R	R	R	R	S	S	S	R	adult
IRBB4	*Xa4*	R	S	M	M	R	S	R	R	S	overall
IRBB7	*Xa7*	R	R	R	S	R	S	R	R	R	adult
IRBB10	*Xa10*	S	R	S	S	R	S	R	S	S	overall
IRBB14	*Xa14*	S	S	S	S	R	S	R	S	R	overall
JG30 (CK)	Non	S	S	S	S	S	S	S	S	S	—

R=Resistant, M=Moderately resistant, S=Susceptible。

In 2001, IRBB21 (carrying *Xa21*) and CBB23 were re-evaluated for resistance at the tillering stage using 10 Philippine races, 7 Chinese pathotypes and 3 Japanese races. CBB23 was resistant to all the 20 races tested, while IRBB21 was resistant to 20 races except the Philippine race 10. CBB23 was also resistant to 8 strains from Korea tested in 2003. So far the gene in CBB23 has the broadest resistance spectrum to *Xoo* among the BB resistance genes designated.

Inheritance analysis of CBB23

In a previous study, CBB23 showed to be carrying a dominant resistance gene, but none of the known dominant resistance genes including *Xa1*, *Xa2*, *Xa3*, *Xa4* and *Xa7* were allelic to that CBB23 (Lin et al., 1992). Allelism test of the gene in CBB23 against *Xa21* showed significant segregation, indicating they are non-allelic (Table 4).

Table 4 Segregation of resistance reaction to PXO99 at seedling stage in F_1 and F_2 populations from reciprocal crosses of CBB23 and IRBB21

Cross	Reaction					Total	χ^2 (expected ratio)	*P*
	F_1	F_2		BC_1F_1				
		R	S	R	S			
JG30/CBB23	R	183	52			235	0.887 (3 : 1)	0.25～0.50
JG30/CBB23//JG30				35	32	67	0.060 (1 : 1)	0.25～0.50
IRBB21/CBB23	R	136	28			164	0.171 (13 : 3)	0.25～0.50
CBB23/IRBB21	R	164	31			195	0.159 (13 : 3)	0.25～0.50

R=Resistant, S=Susceptible。

Sequence Target Sites (STS) analysis of CBB23 with *Xa21* specific primers

Among all the known dominant BB resistance genes, only *Xa21* and the gene in CBB23 are resistant to the strain PXO99. It is important to differentiate *Xa21* and the gene in CBB23. A pair of STS primersU1/I1 was found able specifically to detect *Xa21*, in which U1 (5′-CGA TCG GTA TAA CAG CAA AAC-3′) and I1 (5′-ATA GCA ACT GAT TGC TTG G-3′) were designed accordingly to the core sequence of *Xa21* (Wang et al., 1995). The STS analysis showed that *Xa21* locus gave a 1.4 kb fragment while the gene in CBB23 amplified a 1.3 kb fragment (Fig. 1). This result confirmed at molecular level that the gene in CBB23 is different from *Xa21*. We designated this new gene as *Xa23*ᵗ (Zhang et al., 1998) which was registered as *Xa23* by the Committee on Gene Symbolization in 2001 (Yushimura, 2002).

Molecular mapping of the new gene *Xa23*

The gene *Xa23* was mapped on rice chromosome 11 with two tightly linked markers,

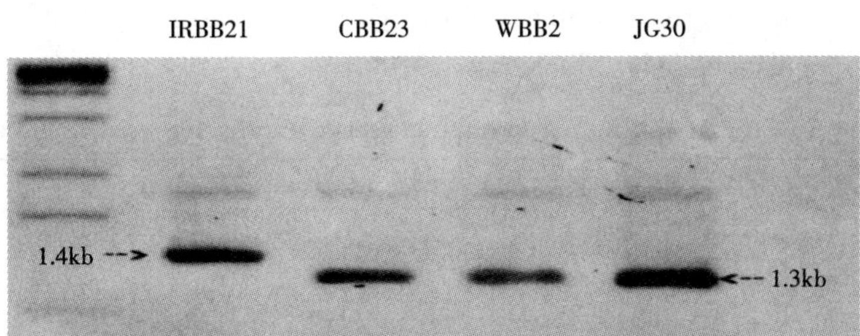

Fig. 1 Polymorphism was detected between IRBB21 and CBB23 using STS primers U1/I1 specific for *Xa21* locus (Zhao BY, Zhang et al., 1998)

1.0 cM to APKj23 and 1.9 cM to RM206 that flanking the *Xa23* locus, by SSR, AFLP, RFLP and RAPD methods

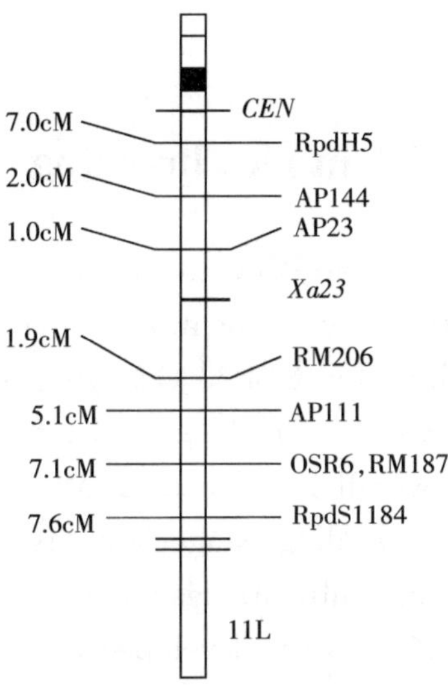

Fig. 2 Chromosome location of the new gene *Xa23* for resistance to bacterial blight (Zhang et al., 1998; Fan et al., 2006)

Effectiveness of *Xa23* and *Xa21* in genetic transfer of resistance to BB

Five susceptible cultivars, JG30, Zhenzhuai, NJ 11, 02428 and R917 crossed with CBB23 (carrying *Xa23*), respectively. The F_1 populations derived from five crosses, namely JG30/CBB23, Zhenzhuai/CBB23, N11/CBB23, 02428/CBB23, and R917/CBB23, were highly resistant to PXO99 at seedling stage, and most of the lesions were less than 0.5 cm in length (about 3% of leaf area). The five F_2 populations segregated essentially either highly resistant or susceptible. 66%- 74% showed highly resistant plants of the F_2 in the 5 populations. It would be easier and more convenient to develop highly resistant lines with desirable agronomy. The JG30 and 02428 crossed with IRBB21 (carrying *Xa21*), the F_1 populations derived from the crosses JG30/IRBB21 and 02428/IRBB21 showed moderate resistance to PXO99 at the late tillering stage (about10%- 15% of lesion area), and the two F_2 populations segregated also either moderately resistant (MR) or moderately susceptible (MS). There were 11.1%–15.4% plants with high resistancein the F_2 populations only (Table 5).

Table 5 Efficiency of *Xa23* and *Xa21* in genetic-transfer for resistance to BB

Cross	Reaction in F_1	Resistance type	Percentage of R/S plants in F_2 (%)			S/HS
			HR	MR	MS	
JG30/CBB23	HR	Seedling	74.3	0.2	1.1	24.4
Zhenzuai/CBB23	HR	Seedling	71.3	1.8	7.2	19.7
N11/CBB23/	HR	Seedling	72.7	2.8	7.5	14.0
02428/CBB23	HR	Seedling	66.2	6.3	3.5	24.0
R917/CBB23	HR	Seedling	73.4	1.1	0.7	24.8
JG30/RBB21	MR	Tillering	15.4	54.8	21.9	7.9
02428/IRBB21	MR	Tillering	11.5	65.7	15.8	7.0

HR=Highly resistant, MR=Moderately resistant, MS=Moderately susceptible, S=Susceptible, HS=Highly susceptible.

Utilization of the gene *Xa23* in rice breeding

Xa23 is an overall resistance gene to BB with complete dominance and broad spectrum. It is especially useful in hybrid rice development. In 48 crosses between susceptible dual-purpose genic male sterile rice lines and CBB23 (carrying *Xa23*), all the F_1s showed high resistance to the strain PXO99 with a small lesion length 0.3–1.0 cm (Zhang et al., 2002). Moreover, *Xa23* bears overall resistance to BB. The segregating hybrid progenies with *Xa23* can be screened at seedling stage. This is important in conventional rice breeding. In 2001, two new japonica cultivars designated as Zhongye 5112 and Zhongye 5114 (Wuxiangnuo//9520/CBB23/^{3}/ 99-1) were developed by our institute in collaborative with Wujin Rice Breeding Station, Jiangsu province and grown for demonstration in 2003 on some fields in Anhui and Zhejiang provinces. They grew robustly and were free the disease, while the adjacent commercial cultivar JG30 was susceptible to BB and lodged heavily. They are ready to release for commercial production.

ACKNOWLEDGEMENTS

We thank Professor Zhuang Q S for reviewing and revising the manuscript. We acknowledge the support of the Rockefeller Foundation (Z. Q, # 586to CAAS), and National Natural Sciences Foundation of China (#AA222151 to CAAS).

We thank P. C., Ronald T. W. Mew, Zhu L. H. for helpful suggestions and discussion of the gene location, resistance analysis; Vera Craz for supported the Philippine *Xoo* strains by Material Transfer Agreement. Chio S for giving 8 Korea strains kindly.

In: Proceeding of the 1st International Conference of Bacterial Blight (ICBB) held in Japan on March 16–19, 2004.

水稻白叶枯病抗性基因鉴定进展及其利用

章　琦

（中国农业科学院作物育种栽培研究所，北京，100081）

Highlights in Identification and Application of Resistance Genes to Bacterial Blight

Abstract The international differential system for resistance to bacterial blight (BB) was established using a common base of research methods by collaborative studies between Tropical Agricultural Research Center（TARC）, Japan, and International Rice Research Institute（IRRI）in 1982—1987. Nomenclature was unified and duplications were removed（including *Xa4*^{b}, *Xa6*, and *Xa6*）. Up to 2005, there are a total of 30 bacterial blight resistance genes registered internationally or reported in journals. Among them, the genes *Xa22t*, *Xa26t*, *xa26t*, *Xa27t*, *xa28t*, *Xa29t*, and three *Xa25t* were tentatively designated and subjected to further be re-arranged. Twenty-one and 9 out of the 30 *Xa* genes are dominant and recessive, respectively. Ten genes exhibit all-growth-stage resistance, two genes, *Xa21* and *Xa27* (from *O. minuta*) express resistance at the later tillering stage, and the other 18 genes confer only adult-stage resistance. Eighteen *Xa* genes were mapped: *Xa29t* on chromosome 1; 4 genes, *Xa1*, *Xa2*, *Xa12*, and *Xa14* on chromosome 4; *xa5* on chromosome 5; 2 genes, *Xa7* and *Xa27t* on chromosome 6; *xa8* on chromosome 7; *xa13* on chromosome 8; 7 genes, *Xa3*, *Xa4*, *Xa10*, *Xa21*, *Xa22t*, *Xa23*, and *Xa26t* on chromosome 11; and *Xa25t* on chromosome 12. Eight genes, *Xa11*、*xa15*、*Xa16*、*Xa17*、*Xa18*、*xa26t* (Nep Bha Bong To)、*Xa27t* (Arai Raj) and *xa28t* have not been mapped. No BB resistance gene was found on chromosomes 2, 3, 9, and 10. Six *Xa* genes were cloned, including *Xa1*, *xa5*, *Xa21*, *xa13*, *Xa26t*（from Minghui 63) and *Xa27*（from *O. minuta*,）. Nineteen *Xa* genes were registered by the International Committee on Nomenclation of New Rice Genes. The nomenclation of six genes were found to be duplicated: two *Xa25t* genes, one from Minghui 63 with overall resistance to Philippine race 9 and the other was a somatic mutation line（HX3）derived from tissue cultures of Minghui 63; two *Xa27t* genes were independently named in *O. minuta*, and Bangladsh cultivar Arai Raj, respectively; the dominant gene *Xa26t* is from Minghui 63, and a recessive gene *xa26t* is from Vietnam cultivar Nep Bha Bong To. Application of resistance to bacterial blight was also discussed.

Key words rice; bacterial blight; resistance; gene identification

摘要 1982—1987 年，日本热带农业研究中心（TARC）和国际水稻研究所（IRRI）合作采取统一研究方案，创建了国际水稻白叶枯病抗性鉴别系统。统一了命名，删去了与 *Xa3* 相同的 $Xa4^b$、*Xa6* 和 *xa9* 三个基因。截至 2005 年，已统一鉴定出 30 个水稻白叶枯病抗性基因，包括 21 个显性基因和 9 个隐性基因。其中 12 个是全生育期（包括 2 个分蘖期）抗性基因、18 个是成株期抗性基因。已定位的抗性基因有 18 个，即第 1 染色体上有 *Xa29t*；第 4 染色体上有 4 个，*Xa1*、*Xa2*、*Xa12* 和 *Xa14*；第 5 染色体上有 *xa5*；第 6 染色体上有 2 个，*Xa7* 和 *Xa27t*；第 7 染色上有 *xa8*；第 8 染色体上有 *xa13*；第 11 染色体上有 7 个，*Xa3*、*Xa4*、*Xa10*、*Xa21*、*Xa22t*、*Xa23* 和 *Xa26 t*；第 12 染色体上有 *Xa25t*。还未定位的有 *Xa11*、*xa15*、*Xa16*、*Xa17*、*Xa18*、*xa26t*（*Nep Bha Bong To*）、*Xa27t*（Arai Raj）和 *xa28t*。第 2、3、9、10 四个染色体上还未发现有白叶枯病抗性基因。已克隆的抗性基因有 6 个，即 *Xa1*、*xa5*、*xa13*、*Xa21*、*Xa26t*（全抗，明恢 63）和 *Xa27 t*（*O. minuta*）。已在国际水稻新基因命名委员会正式注册的有 19 个。编号重复的基因有 6 个：2 个 *Xa25t* 分别来自全生育期抗菲律宾小种 9 的明恢 63 和源于明恢 63 经组培后的体细胞无性突变系 HX3；2 个 *Xa27t*，分别来自小粒野生稻和孟加拉品种 Arai Raj；以及 1 个显性的来自明恢 63 的 *Xa26t* 和 1 个隐性的 *xa26t* 来自越南品种 Nep Bha Bong To。

在已命名的抗性基因中，*Xa23*、*Xa21* 和 *Xa7* 抗谱最广、抗性最强，*xa5*、*xa13*、和 *xa 24t* 的抗谱也较广；*Xa3* 和 *Xa4* 对我国白叶枯病致病型的抗谱较广，但要避免在中国Ⅳ型和Ⅴ型菌流行的地区应用。本文并讨论了合理利用抗性基因等问题。

关键词 水稻；白叶枯病；抗性；基因鉴定

在水稻主要病害中，利用寄主的抗性来控制稻白叶枯病（*Xanthomonas oryzae* pv. *oryzae*，*Xoo*）的效果最为明显。但是水稻白叶枯病（Bacterial blight，BB）病原菌随着抗病基因的利用和布局，其群体遗传结构即随之发生变化，促使新、老毒性基因不断交替。面对这种永世长存的生物间生存竞争和协同进化，除了改弦易辙或双管齐下探索持久抗性的可行性外，还有相当一段时期要采用基因轮换、多基因屏障或基因累加以及创造田间生物多样性等合理利用抗性基因的策略，而这都需要在不断发掘和鉴定抗性基因的基础上付诸实现。

20 世纪 60 年代末以来，水稻白叶枯病抗性基因的发掘和利用一直得到科学家们的关注。分子技术加速了这一领域的发展，优异抗性基因 *Xa21* 的问世，掀起了发掘抗白叶枯病新基因的高潮。新基因的发现、鉴定定位和克隆的报道屡见不鲜。本文作一简要回顾，聚焦于当前广泛用于育种的抗性基因信息，并进行有效利用抗性基因的探讨。

1 水稻白叶枯病抗性基因鉴定的历史回顾

水稻白叶枯病抗性基因的鉴定，经历了采用本国菌株和建立国际鉴别系统两个阶段。1982年以前，各国采用本国的代表菌株进行抗性遗传研究和基因鉴定，其结果缺少国与国之间的可比性。1982—1987年，日本热带农业研究中心（Tropical Agricultural Research Center，TARC）和国际水稻研究所（IRRI）合作创建了国际水稻白叶枯病抗性鉴别系统，统一采用日本和菲律宾两套病菌鉴别小种和研究方案，开展抗性基因鉴定，删去了3个与*Xa3*重复的*Xa4*^{b}、*Xa6*和*xa9*基因，统一命名了下列基因：*Xa1*、*Xa2*、*Xa3*、*Xa4*、*xa5*、*Xa7*、*xa8*、*Xa10*和*Xa12*（Ogawa，1993）。此后，日本和菲律宾的两套小种及与其相应的抗性基因遂成为国际公认的水稻白叶枯病鉴别系统。在IRRI继续进行双方的合作研究，陆续发现了一批抗病新基因。

2 水稻白叶枯病抗性基因统一鉴定后的进展

2.1 *xa13*、*xa24t*和*Xa14*基因鉴定

Mew等（1982）发现了菲律宾小种5和6之后，TARC与IRRI开始采用多个小种的代表菌株进行品种抗性基因的鉴定，才能同时鉴别抗性基因的异同。在发现了对小种6具有稳定抗性的BJ1、Aus 247、AC19-1-1、Chinsurah BoroⅡ和Kalimekri 7725等5个BJ1群品种之后，Ogawa（1987）采用菲律宾小种1、2、3、4和6共5个代表菌株鉴定BJ1群品种的抗性基因。结果表明这5个品种对菲律宾小种1、2、3、4和6的抗性由两对隐性基因控制，一对为*xa5*，另一对是控制小种6的新基因，被命名为*xa13*。Zhang等（1996）将*xa13*定位于第8染色体上，与RFLP标记RZ28的图距为5.1cM，与G136的图距为3.8cM。Chu等（2006a；2006b）进一步将*xa13*基因定位到一个9.2 kb的片段上。该片段只包含一个候选基因，采用将显性候选基因导入携带隐性基因*xa13*的IRBB13中，转基因植株便恢复了感病表现，完成了功能互补测试，成功地克隆了*xa13*基因。

Sidhu等（1978，1979）曾分析了DV85、DV86和DZ78三个孟加拉品种对菲律宾小种1的代表菌株PXO61的抗性后，认为其抗性受两对基因控制，一对是控制全生育期抗性的隐性基因*xa5*，另一对是成株期抗病的显性基因*Xa7*。之后，Mir等（1990）用菲律宾小种1～6鉴定出DV86、DV85、Aus 295等品种在全生育期对小种1、2、3、4和6五个小种都抗病。Khush等（1999）进一步用小种6将这3个品种与携有*xa13*的品种BJ1进行等位性测试，F_2群体呈7抗∶9感的分离比，与*xa13*不等位，并独立遗传，该新基因被命名为*xa24t*。

TN1是菲律宾白叶枯病鉴别品种之一，感染小种1、2、3、4和6，但在全生育期抗菲律宾小种5。Taura等（1987）分析了它对菲律宾小种5的抗性，表明由1对显性基因控制，被命名为*Xa14*。谭振波等（1998）将其定位于第4连锁群的RFLP分子标记RG620和G282之间。

2.2 *Xa16*、*Xa17*和*Xa18*基因鉴定

当日本的稻白叶枯病菌被划分为Ⅰ、Ⅱ和Ⅲ三个菌群后，Sakaguchi（1967）分析了兰

泰艾玛斯（Rantai Emas）群品种对日本小种的抗性是由 *Xa1* 和 *Xa2* 两对显性基因控制的。在日本白叶枯病菌群增加到 5 个小种后，Noda 等（1989）认为 Rantai Emas 群的代表品种 Tetep 对日本小种Ⅴ（菌株 H8584 和 H8581）的抗性由另一对显性基因控制，该基因被命名为 *Xa16*。因此，Tetep 携有的 *Xa1*、*Xa2* 和 *Xa16* 三对基因控制了对日本小种Ⅰ、Ⅱ和Ⅴ的抗性。

Ogawa 等（1989）分析了日本品种阿苏稔（Asaminnori）对日本小种Ⅱ（代表菌株 H8313）的成株抗性，结果表明其抗性由一对新的显性基因控制，被命名为 *Xa17*。

Yamamoto 和 Ogawa（1990）发现 IR24、Milyang 23 和丰锦（Toyonishiki）在孕穗期对两个缅甸白叶枯病菌株 BM8417 和 BM8429 具有抗性，其抗性由一对新的显性基因控制，命名为 *Xa18*。

2.3 *xa26t*、*Xa27 t* 和 *xa28t* 基因鉴定

Lee 等（2003）用菲律宾小种 1～6 评价了 IRRI 水稻种质中心的 21 个品种。根据这些品种对 6 个小种的抗性反应型被划分为两群：抗小种 1 和小种 5 的品种为一群，感小种 1、抗小种 2 和小种 5 的品种为另一群。将所有 21 个品种分别与 TN1 杂交，并与已知抗性基因品种进行等位性测定，经遗传分析确定了各品种的抗性基因，其中有 3 个被鉴定为新基因；1）1 对隐性基因控制了越南品种 Nep Bha Bong 成株期对小种 5 的抗性和对小种 1、2 和 3 的中等抗性，暂被定名为 *xa26t*；2）孟加拉品种 Arai Raj 对小种 2 和 5 的抗性由 1 对显性基因控制，与仅对小种 2 具有抗性的显性基因 *Xa10* 不等位，被定名为 *Xa27t*；3）孟加拉品种 Lota Sail 对小种 2 的抗性是由一对新的隐性基因控制的，被定名为 *xa28t*。

2.4 源于诱变品种和组织培养的抗病基因

日本遗传学家用日本小种从诱变品种中鉴定出 3 对新的隐性基因 *xa15 t*、*xa19* 和 *xa20*。Nakai 等（1988）通过热中子照射感病品种 Harebare 的种子，获得抗病诱变体 M41，其抗性由一对隐性基因控制，命名为 *xa15t*。由 Ikeda 等（1995）将 *xa15t* 与另外两个来自诱变体的隐性基因 *xa19* 和 *xa20* 进行了等位性测试，证明它们之间互不等位，将 *xa15t* 确认为新基因。

Taura 等（1991，1992）用甲基亚硝酸脲烷（MNU）处理品种 IR24 在胚胎形成单细胞期的受精卵，获得了 2739 个诱变系，发现 XM5 和 XM6 两个诱变系抗菲律宾小种 1～6。XM5 对这些小种的抗性由一对隐性基因控制，该基因并与 *xa5*、*xa8* 和 *xa13* 三对隐性基因都不等位，被命名为 *xa19*。XM6 也由一对隐性基因所控制，该基因与 *xa19*、*xa5*、*xa8* 和 *xa13* 都不等位，命名为 *xa20*。

高东迎等（2001）取我国杂交稻恢复系明恢 63 的成熟胚进行组培，在继代培养中以白叶枯病菌液作为选择压进行离体筛选，获得了抗病的体细胞无性变异系 HX3。HX3 有较广的抗谱，与 *Xa1*、*Xa2*、*Xa3*、*Xa4*、*Xa7*、*Xa10*、*Xa14*、*Xa21*、*Xa23* 等 13 个白叶枯病主要显性抗病基因的抗性反应不同，用中国病原型代表菌株（浙 173）与其反应型相似的 3 个抗性基因 *Xa4*、*Xa7* 和 *Xa21* 进行了等位性测定，表明 HX3 携有一对新的显性基因，暂命名为 *Xa25t*。

2.5 源于野生稻的抗病基因

截至 2004 年，已从野生稻中鉴定出以下 4 个白叶枯病抗性基因。

Xa21 是第一个从野生稻中被克隆出来的重要功能基因，是当时唯一含有两个结构域的新一类抗性基因（Song et al.，1995），加上其具有广谱抗性，因而 *Xa21* 一经问世，即备受国际同行关注。Khush 等（1990）报道其抗性供体为西非长药野生稻（*O. longistaminata*），在水稻分蘖盛期抗菲律宾当时全部 6 个小种。经杂交导入 IR24 后，通过回交、自交获得 BC_4F_2群体，分析其中两个 F_2群体对菲律宾小种 1、2、4 和 6 的抗性遗传，表明其广谱抗性由一对显性基因控制，有别于已鉴定的 17 个抗性基因，被命名为 *Xa21*。后来进一步选育成携有 *Xa21* 的近等基因系 IRBB21。Ronald 等（1992）将 *Xa21* 定位于第 11 染色体，随后该基因由 Song 等（1995）克隆，并迅速广泛应用于国内外水稻抗白叶枯病转基因育种。

Xa23 由章琦等（1998）鉴定。1987 年中国农科院与广西农科院合作，从普通野生稻（*O. rufipogon*）中筛选出一个全生育期高抗能侵染所有栽培稻白叶枯病抗性基因的菲律宾小种 6（P6）的新抗源 RBB16。采用野栽杂交、花培、回交、自交、目标基因鉴定等手段，育成了纯合抗病系 H4，继而育成近等基因系 CBB23（轮回亲本为 JG30）。并在 2000 年育成了第 2 对携 *Xa23* 的近等基因系 CBB23（B）（轮回亲本为具有恢复基因的 IR24），便于杂交水稻的抗性改良。在当时命名的所有白叶枯病显性抗性基因中，仅 *Xa21* 和 *Xa23* 能抗御广致病菌 P6，遂重点进行这两者之间的抗谱比较、STS 检测、等位性测定、抗性遗传传递效应等研究，确定为新基因 *Xa23*，并被定位于第 11 染色体（Zhang，et al.，2002；潘海军等，2003）。2001 年，经国际水稻新基因命名委员会正式命名为 *Xa23*。其特点是，1）抗谱广，抗菲律宾小种 1～10、中国致病型小种 1～7 和日本小种 1～3 共 20 个国内外白叶枯病鉴别菌株，*Xa21* 抗其中的 19 个菌株，不抗菲律宾小种 10。2）全生育期抗病，可在苗期鉴定，省地、高效、低耗，便于利用，*Xa21* 则须在分蘖末期转抗。3）完全显性，抗性遗传传递力强，便于育种选择（Zhang et al.，2001）。

Xa27t 是 IRRI Amante-Bordeos 等（1992）通过种间杂交、胚拯救和回交等手段，将小粒野生稻（*O. minuta*，BBCC，Acc. 101141）的白叶枯病抗性导入栽培稻 IR31917－45－3－2 并选育出 78－1－5 BC_2F_3株系，该株系抗菲律宾小种 2、3、4、5 和 6。此后，新加坡国立大学 Temasek 生命科学实验室与美国 Ohio 州立大学植物病理系合作，以 78－1－5 品系为供体，IR24 为轮回亲本，育成了近等基因系 IRBB27。IRBB27 的抗谱较广，在分蘖盛期高抗菲律宾小种 2、3、5 和 6，中抗/中感小种 4，高抗中国致病型Ⅰ、Ⅲ、Ⅴ、Ⅵ、Ⅶ，但是，其抗谱与抗菲律宾小种 6 的 *Xa21* 和 *Xa23* 不相同。此外，IRBB27 还高抗菌株 T7174（日本小种Ⅰ），A3842、A3857（印度），IXO56（印尼），NXO260（尼泊尔），R－7（泰国）以及 Aust-2131、Aust-R3（澳大利亚）等菌株，但高感中国Ⅳ型菌（ZHE173），中感中国Ⅱ型菌（HB17）以及韩国、泰国、非洲等一些菌株。遗传分析表明，IRBB27 对菲律宾小种 6 的抗性由一个不完全显性的新基因控制，暂被命名为 *Xa27t*。该基因被定位于第 6 染色体的两个 RFLP 标记 M964 和 M1197 之间，与这两个标记的遗传距离均为 0.052 cM；随后，该研究组进一步分离克隆了基因 *Xa27*（Gu et al.，2004；2005）。

Xa29t 由谭光轩等（2004）鉴定。他们从药用野生稻（*O. officinalis*）渗入栽培稻的后

代中，选育了抗菲律宾小种 1（PXO61）的水稻株系 B5，经与籼稻品种明恢 63 杂交，通过单粒传（single seed descent）获得了一个含有 187 个稳定纯合株系的重组自交系（recombinant inbred lines，RILs）群体。在植株孕穗期的抗白叶枯病鉴定和遗传分析表明，B5 对 PXO61 的抗性受控于 1 对显性基因。根据 RILs 群体的抗病性表型鉴定数据及分子标记连锁图谱，将 B5 携有的抗白叶枯病基因定位于第 1 染色体短臂的 C904 和 R596 之间，其遗传距离为 1.3cM。其基因座不同于已报道的所有抗白叶枯病基因，是 1 对新的显性抗病基因，暂被命名为 *Xa29t*。

2.6 源于中国品种的抗病基因

Xa22t。Lin 等（1996）研究了云南品种扎昌龙的抗性遗传，该品种在成株期对我国致病型Ⅰ、Ⅱ、Ⅳ和Ⅶ，菲律宾小种 1、3、4、5 和 6，日本小种Ⅰ、Ⅱ和Ⅲ，共 12 个代表菌株具有抗至中等抗性，由 1 对显性基因控制，分别与 *Xa1*、*Xa2* 和 *Xa14* 独立遗传，与 *Xa4* 呈连锁遗传，其重组率为（13.3±4.5）%，为 1 对新的显性抗病基因，该基因被定位于第 11 染色体短臂末端，与最近的 RFLP 分子标记 R543 的图距为 7.1cM，被定名为 *Xa22t*。

Xa25t。Chen 等（2002）报道了明恢 63 在全生育期高抗菲律宾小种 9。遗传分析表明它对小种 9 的抗性由 1 对显性基因控制。通过抗/感组合单粒传获得 241 个重组自交系（RILs）群体，进行了 RFLP 分子标记定位，该基因被定位于第 12 染色体上的两个分子标记 R887 和 G1314 之间，与两者的图距分别为 2.5cM 和 7.3cM。迄今在第 12 染色体上还未发现任何 *Xa* 抗性基因，应为 1 对新的显性基因，暂命名为 *Xa25t*。

Xa26t。Yang 等（2003）报道明恢 63 除携有 1 对全生育期控制菲律宾小种 9 的新基因 *Xa25t* 之外，还有 1 对在苗期和孕穗期都抗中国湖北菌株 JL691 的显性单基因 *Xa26t*。该抗性基因座的一端与 SSR 分子标记 RM224 的图距为 0.21cM，另一端与两个共分离的 RFLP 分子标记 Y6855 RA 和 S12886 的图距为 1.47cM，被定位于第 11 染色体上，与 *Xa4* 紧密连锁。该基因已由 Sun 等（2004）克隆。

截至 2005 年，已报道鉴定出 30 个水稻白叶枯病抗性基因，包括 21 个显性基因和 9 个隐性基因。其中 12 个是全生育期（包括 2 个分蘖期）抗性基因、18 个是成株期抗性基因。已定位的抗性基因有 18 个，即第 1 染色体上有 *Xa29t*；第 4 染色体上有 4 个，*Xa1*、*Xa2*、*Xa12* 和 *Xa14*；第 5 染色体上有 *xa5*；第 6 染色体上有 2 个，*Xa7* 和 *Xa27t*；第 7 染色上有 *xa8*；第 8 染色体上有 *xa13*；第 11 染色体上有 7 个，*Xa3*、*Xa4*、*Xa10*、*Xa21*、*Xa22t*、*Xa23* 和 *Xa26t*；第 12 染色体上有 *Xa25t*。还未定位的有 *Xa11*、*xa15*、*Xa16*、*Xa17*、*Xa18*、*xa26t*（*Nep Bha Bong To*）、*Xa27t*（Arai Raj）和 *xa28t*。第 2、3、9、10 四个染色体上还未发现有白叶枯病抗性基因。已被克隆的抗性基因有 6 个，即 *Xa1*、*xa5*（Iyer et al.，2004）、*xa13*、*Xa21*、*Xa26t*（全抗，明恢 63）和 *Xa27t*（*O. minuta*）。已在国际水稻新基因命名委员会正式注册的有 19 个。编号重复的基因有 6 个：2 个 *Xa25t* 分别来自全生育期抗菲律宾小种 9 的明恢 63 和源于明恢 63 经组培后的体细胞无性突变系 HX3；2 个 *Xa27t*，分别来自小粒野生稻和孟加拉品种 Arai Raj；以及 1 个显性的来自明恢 63 的 *Xa26t* 和 1 个隐性的 *xa26t* 来自越南品种 Nep Bha Bong To。现将 30 个基因的代表品种、用以鉴定的菌株、染色体定位、分子标记等基本信息汇总于表 1。

3　合理应用抗性基因

人们从半个多世纪的抗白叶枯病育种中积累了宝贵经验，也一直面临着对付病菌群体迅速变异和提供足以抵御多个病菌致病型的抗病品种的挑战。有效利用抗病基因和保持抗性基因的持久抗性应该值得关注。

3.1　抗病基因与病原菌的互动信息

一些基因曾经成功地控制了白叶枯病的为害。例如在亚洲许多国家大面积应用的携有 *Xa4* 基因的品种，*Xa1* 和 *Xa3* 在日本和韩国的应用都有效地控制了此病的为害。但是后来都因为遇到其毒性菌的侵袭而丧失了抗性。众所周知，一个抗病基因在大面积推广应用几年之后，因毒性菌频率的上升和侵袭可能失去抗性。人为混杂或自然退化也是品种抗性水平下降的一个原因，使集中应用的抗性基因的使用寿命大大缩短。历史教训的代价是深重的，但却在不断重演。当全世界广泛利用的 *Xa4* 基因在亚洲国家丧失抗性的时候，广谱抗性基因 *Xa21* 一经问世自然被竞相应用。遗憾的是有些国家又已经出现了能侵袭 *Xa21* 的菌群。Marella 等（2001）报道曾高抗印度 17 个州和全部菲律宾 6 个小种的携带 *Xa21* 的品种，已对东印度 3 个地区 50％的菌株感病。Lee 等（1999）报道携带 *Xa21* 品种在韩国刚投放时，就对 1987—1989 年采集的 15％的菌株感染，几年之后，在 1994—1996 年采集的菌株中竟有 96％能侵袭 *Xa21*。*Xa21* 在韩国的遭遇显然是源于该国原来就存在侵染 *Xa21* 的毒性菌株，随着不断扩大携有 *Xa21* 基因品种的推广面积，毒性菌株的频率便迅速增长。曾列先等（2002）报道广东省致病型Ⅳ和Ⅴ也能侵染 *Xa21*，这些信息足以表明寄主抗性基因与病菌致病基因之间连续互动的激烈较量，抗性水平越高的基因，对病原菌的选择压越强，以致加速诱发新的毒性菌群（如印度），或使原来处于微小群体的毒性菌群迅速发展（如韩国）。事实充分证明，病菌群体结构变化的监测工作是应用抗病基因前后不可缺少的信息渠道，是合理利用抗性基因的重要依据。

3.2　抗性基因的轮换利用

育种家们在理论与实践中早已重视抗病基因的轮换利用或聚合抗性基因以加强抗性力度。早在 20 世纪 70 年代末，周毓珍、朱立宏（1985）就开始采用 DV85、DV86 和 DZ78 等携有 *Xa7*＋*xa5* 的抗源，育成一批抗病品系。近年 Zai 等（2000）、吴家道等（2001）和曹立勇等（2003）通过常规杂交结合分子标记辅助选择，将 *Xa21* 导入杂交稻或常规稻，育成一批改良型抗病品系。携有 *Xa23* 基因的品系 CBB23（B）或中野 5112 等都成为育种计划的供体（Zhang et al.，2001）。这些广谱抗性基因的利用，将拓宽我国白叶枯病抗性遗传基础。随着研究的深入，基因定位、克隆，抗病新基因的不断发掘，日益丰富的遗传资源，将提供更宽的选择空间，实现基因合理轮换。

3.3　抗病基因的合理部署

国内外大量的研究证实了水稻白叶枯病原菌群体遗传结构的亚群之间存在明显的地区差别。通过植病学家提供白叶枯病菌群体变化的连续观察信息，掌握本国或地区致病型（小种）优势菌群的分布和消长情况，才能进行抗病基因合理部署和交叉利用，或采用基因累加系，创造品种抗性遗传多样性或基因屏障，实现合理利用和延长抗病基因的寿命。至今，已鉴定了 30 个类型丰富的抗白叶枯病基因，拓宽了抗性遗传改良基因源，其中，抗谱较宽的优异基因 *Xa21* 正广泛地热用于我国，其抗性尚处在高峰状态，应密切注意推广地区的菌群变化情况。*Xa23* 是个十分优异的抗性基因，正因为它具有高度抗性，不可避免也能遭遇毒性菌的袭击，不宜集中利用。还有来自特定生态环境的抗性基因，因为长期与当地白叶枯病菌群协同进化，形成了优异的抗性，如源于云南的地方稻扎昌龙（携基因 *Xa22t*），不仅抗我国致病型Ⅰ、Ⅱ、Ⅳ和Ⅶ，并且抗云南省全部 7 个白叶枯病致病型，是个适应性好的地区优异抗病基因。与植物病理学家密切合作，针对性地将 *Xa21* 与 *Xa23*、*Xa22t*、*Xa7*、*xa5*、*xa13* 等抗病基因合理轮换利用，正处在高峰期的 *Xa23*、*Xa7* 和 *Xa21* 等优异的抗病基因，应及时、因地制宜地逐步累加其他新的抗病基因，延长其有效利用寿命。

全国年度小麦条锈菌生理小种监测结果面向育种家的做法很值得借鉴。育种家可根据各地病菌致病型（小种）变化情况，抢在病菌小种积微见著之前，用携有不同抗病基因的品种有计划地进行轮换种植，做到地区内外品种布局合理，从时间和空间上切断病菌毒力新小种的定向适应、选择、繁衍和发展，也是我国抗水稻白叶枯病育种所期望实现的目标。

表 1　迄今已鉴定的水稻白叶枯病抗性基因[a]

Table 1　Summary of resistance genes to *Xoo* and their resistance preformance up to 2005

基因 Gene identified	抗性类型 Resistance type	供体品种 Donor cultivar	所用的小种或菌株表现抗[b] Resistant to *Xoo* race/strain tested	染色体 Chromosome	标记与图距 Marker & distance	参考文献 Reference
Xa1[a]（1984）	全生育期[c]	Kogyoku，Java14	X-17 T7174（1），印尼小种Ⅴ（Xo7306）	4	1.5 cM $U08_{750}$（RAPD）， 0 cM to XNpb235（RFLP）	Sakaguchi，1967； Yoshimura et al.，1995，1992
Xa2[a]（1984）	全生育期	Rantai Emas 2， Tetep	X-17，X-14 TⅠ，TⅡ；Ⅳ（Xo7306）、Ⅴ（Xo7306）	4	XNpb264、$YO3_{700}$	Sakaguchi，1967； Yoshimura et al.，1995
Xa3（1984） *Xa-w*	成株[d]	Wase Aikoku 3	TⅠ，TⅡ，TⅢ 印尼小种Ⅴ（Xo7306）； P1，P2，P3，P5	11	2.3 cM to XNbp181	Ezukaet al.，1975； Yoshimura et al.，1992，1995
Xa4[b]（同 *Xa3*）		Semora Mangga	P1			Librogo et al.，1976；
Xa6（同 *Xa3*）		Zenith	P1			Sidhu et al.，1978；
xa9（同 *Xa3*）		Sateng	T1			Singh et al.，1983
Xa4 （*Xa4*[a]）（1984）	全生育期	Sigadis， TKM-6，IR20	P1，P5，T1～T7；CⅠ，CⅡ，CⅢ，CⅣ，CⅥ，CⅦ	11	1.7 cM to XNpb78， XNpb181	Sidhu et al.，1978 Yoshimura et al.，1995；
xa5[a]（1984）	全生育期	DZ192， IR1545-339， DV85，BJ1	P1，P2，P3，P5 TⅠ～TⅦ；CⅠ～CⅦ	5	Close to XNpb396 &XNpb208 <1.0 cM to RG 207	Petpisit et al.，1977； Yoshimura et al.，1992b Mc Couch et al.，1991； Iyer et al，.2004.

（续）

基因 Gene identified	抗性类型 Resistance type	供体品种 Donor cultivar	所用的小种或菌株表现抗[b] Resistant to *Xoo* race/ strain tested	染色体 Chromosome	标记与图距 Marker & distance	参考文献 Reference
Xa7（1984）	成株	DV85，DV86，DZ78	P1，P2，P3，P5；TⅠ～TⅧ；CⅠ～CⅧ	6	6.0 cM to G1091. 3.0 cM to AFLP31－10	Kaji et al.，1995 Borineset al.，2003
xa8（1984）	成株	PI231128	P1，MR to P1，P2，P3，P5；TⅠ～TⅦ	7	19.9 cM to RM214 (STMS)	Sidhu et al.，1978； Singh et al.，2002
Xa10（1984）	全生育期	Cas209	P2，P5；TⅤ	11	5.3 cM to $O07_{2000}$ (RAPD)	Yoshimure et al.，1983； Yoshimure et al.，1995
Xa11（*Xa-pt*）（1984）	成株	IR8，944－102－2－3	TⅠ，TⅡ，TⅢA，TⅣ，TⅤ	—		Ogawa et al.，1986
Xa12（*Xa－kg*）（1984）	成株	Kogyoku，Java 14	Xo－7306 Indonesian strain (Jap. raceⅤ)	4		Ogawa et al.，1978
xa13[a]（1988）	成株	BJ1、Chinsurah Boro ll，Aus247、AC19－1－1，Kalimekri 77－5	Philippine races 1，2，4，6	8	5.1 cM to RZ28， 3.8 cM to G136	Ogawa et al.，1987； Zhang et al.，1996
Xa14（1988）	全生育期	TN1	Philippine races 5	4	RG620～G282 (RFLP)	Taura et al.，1987 谭震波等，1998
xa15（1984）	成株	M41 Harebare mutant	TⅠ、TⅡ、TⅢ (H75373)	—		Nakai et al.，1988
Xa16（1989）	成株	Tetep	Japanese raceⅤ (H8581，H8584)	—		Noda et al.，1989
Xa17（1989）	成株	Asominori	Japanese raceⅡ (H 8313)	—		Ogawa et al.，1989
Xa18（1989）	成株	IR24，Miyang23，Toyonishiki	Burmese strain BM8417、BM8429	—		Yamamoto et al.，1990
xa19（1990）	成株	IR24 的诱变体 XM5 (mutant of IR24)	P 1～6	—		Taura et al.，1991b
xa20（1990）	成株	IR24 的诱变体 XM6 (mutant of IR24)	P 1～6	—		Taura et al.，1992b
Xa21[a]	全生育期（分蘖后期开始）	长药野生稻 *o. longistaminata* (IRBB21)	P1～9；CⅠ，CⅡ，CⅢ，CⅣ，CⅥ，CⅦ；TⅠ～Ⅲ	11	0 cM to RG103 <1.0 cM to PTA 248	Ikeda et al.，1991 Ronald et al.，.1992； Chunwongse et al.，1993
Xa22t	成株	扎昌龙 Zhachanglong	P1、4、5、6；CⅠCⅡ、CⅦ、JL691 和云南若干菌株；TⅠ～Ⅲ；	11	7.1 cM from CR 543， 10.7 cM to RZ536	Lin et al.，1996

（续）

基因 Gene identified	抗性类型 Resistance type	供体品种 Donor cultivar	所用的小种或菌株表现抗[b] Resistant to *Xoo* race/ strain tested	染色体 Chromosome	标记与图距 Marker & distance	参考文献 Reference
Xa23 (2001)	全生育期	普通野生稻 *O. rufipogon* (CBB23)	P1～10；CⅠ～Ⅶ； TⅠ～Ⅲ；8 Korean strains	11	5.4 to PSR6； 1.9 cM to RM206， 0.4 cM to C1003 A	Zhang，et al.，.1998； 樊颖伦等，2006
xa24 t	全生育期	DV85，DV86， Aus295	P 1，P2 P4，P6	—		Khush et al.，2003
Xa25 t	全生育期	明恢 63	P 9 JL691	12	2.5 cM to NBSv09， 7.3 cM to G1314	Chen & Wang et al.，2002
Xa25 t	成株	明恢 63 体细胞 无性系突变体 HX3 Somaclonal mutant of Minghuei 63，HX3	CⅣ (ZHE 173)	—		Gao et al.，2001
Xa26 t [a]	全生育期	明恢 63	JL 691	11	0.21 cM to RM224. 1.47 cM to Y6855R (RFLP)	Yang et al.，2003； Sun et al，2004
xa26 t	成株	Nep Bha Bong To	P 5，MR to P1～3	—		Lee & Khush et al.，2003
Xa27 [a]	全生育期 分蘖盛期	小粒野生稻衍生系 Line (IRBB27) drived from *O. minuta*	P2，P3，P5，P6；CⅠ， CⅢ，CⅤ，CⅥ，CⅦ； some strains from Japan， India，& Indonesia	6	0 cM to C964	Amanta-Bordeos et al.，1992； Gu et al.，2004，2005
Xa27 t	成株	Arai Raj	P 2，P 5	—		Lee et al.，2003
xa28 t	成株	Lota sail	P2，P5	—		Lee，et al.，2003
Xa29 t	成株	药用野生稻 *O. officinalis* (B5)	P1	1	1.3 cM C904 - R596 (RFLP)	谭光轩等，2004

资料来源（Source of data *Xa1*～*Xa21* mainly from Ogawa，1993，and the remining from Rice Genetics Newsletter Vol. 1，5，6，7，18～21）.

[a] 已克隆（Cloned）.

[b] C，中国致病型（Chinese pathotype）；P，菲律宾小种（Philippine race）；T，日本小种（Japanese race Ⅰ，Ⅱ，Ⅲ）.

[c] 全生育期抗性（Overall resistance）.

[d] 成株期抗性（Adult plant resistance）.

国家自然科学基金资助项目（39470438，39970452）；
美国洛克菲勒基金资助项目（RF97001，* 586）；
原载：中国水稻科学（Chinese J Rice Sci.），2005，19（5）：453～459.

Studies on Genetics and Improvement of Resistance to Bacterial Blight in China (abstract)

Zhang Qi[1] **Lam Hon-Ming**[2]

(1. Chinese Academy of Agricultural Sciences, Beijing; 2. The Chinese University of Hong Kong)

Bacterial blight (*Xanthomonas oryzae* pv. *oryzae*, BB) is one of the most devastating rice diseases just next to blast in China. It was first reported in China in 1930s and has become widely spread nationwide since late 1950s. The late renowned Chinese plant pathologist Prof. Z. D. Fang initiated the study of BB in China in 1950s, by examining its causal organism, symptom, physiology, disease cycle and chemical control. Building on this foundation, extensive, comprehensive, and integrated investigations of rice resistance to BB have been conducted for more than half a century. Since the Ministry of Agriculture (PRC) launched the rice breeding program targeting at high yield, disease resistance, and good quality in early 1970s, the nationwide studies on the evaluation and identification of germplasms for BB resistance were commenced, and the researches of BB resistance rice breeding have been rapidly and extensively progressing since then. There were 911 indica and 351 japonica new cultivars developed in the period 1960—1986, representing an annual acreage of more than 6 667 hm^2. Among those, 192 (21.1%) out of 911 indica, and 90 (25.6%) out of 351 japonica cultivars were resistant to BB. Before 1980s, 31 (35.6%) out of 87 indica three-line hybrid rice combinations were developed with moderate resistance toward BB. The combinations of Weiyou 64, Weiyou 6, and Shanyou 6 had reached an annual acreage of 1.333 M hm^2. Up to 2005, some combinations exhibiting high yield, good quality and resistance to both BB and fungal blast have accumulated a large acreage, such as Shanyou 77 (2.497 M hm^2 in 1996—2005), Ⅱ You 084 (0.757 M hm^2 in 2001 - 2005), etc. For two-line hybrid rice, up to 2006, 6 out of 21 combinations with high yielding, good quality were resistant to both BB and fungal blast, such as Liangyou-peijiou, Peiza-shuangqi, and 70 you 9 (japonica). Up to 2005, the accumulated acreage of Liangyou-peijiou had reached 4.681 M hm^2, representing 50% of the total acreage of two-line hybrid rice in China.

The idea of a multidisiplinary genetic evaluation and utilization (GEU) advocated by IRRI was introduced to China by the late Chinese rice breeder Prof. S. C. Lin. This effort brought China into the international collaborative network to share information and materials on BB resistance. In the rice cultivation field of China, the genetic sources of BB resistance are very narrow. Most indica and japonica rice depend on the resistance conferred by *Xa4* and

Xa3. There is a urgent need to identify new and broad spectrum *Xa* genes. To perform extensive screening of BB resistance rice germplasms, Chinese scientists first established a standardized and stable evaluation system for BB. Until early 1990s, scientists evaluated approximately 70 424 rice accessions (from local cultivars of different ecological regions, wild *Oryza* species, improved cultivars, and foreign introductions) for BB resistance. For example, Huazhong Agricultural University and Yunnan Academy of Agricultural Sciences collaboratively identified *Xa22*[t] by investigated local BB resistant cultivars in Yunnan province. *Xa22*[t], which mapped on chromosome 11 causes resistance to Chinese pathotypes Ⅰ, Ⅱ, Ⅳ, and Ⅶ, most of the pathotypes of Yunnan province, Philippine races 1～6, as well as Japan races at the booting stage. Wild *Oryza* species is an important source for BB resistance genes. Two wild *Oryza* species including: *O. rufipogon and O. officinalis* were evaluated and characterized in these germplasms and led to the discovery of *Xa23* and *Xa29*[t], respectively in China, in which *Xa23* confers the broadest spectrum BB resistance among all *Xa* gene identified thus far. Elite germplasms and cultivars with broad-spectrum resistance to BB have been adopted for rice cultivation and breeding programs. By the end of 1990s, the broad-spectrum resistance genes, *Xa21*, *Xa7*, *xa5* and *Xa23*, have been widely used to overcome the dependence of limited genetic resources. The cloning of the dominant broad-spectrum R gene *Xa21* and the identification of the complete dominance R gene *Xa23* identified provided more reliable and durable resources of genes for improvement of BB resistance in hybrid rice in China. Classical and molecular studies of Mendelian and quantitative genetics of BB resistance in rice were also performed. The genetic behaviors and resistance phenotypes in hybrid progenies were investigated and used as a guidance for breeding. For instance, the genotypes of the R parents, such as the number of genes, dominant and recessive relationships, resistance in different growth stages might differ, depending on the interaction between host (R gene) and pathogen races (virulence gene). In addition, 6 japonica rice NILs were constructed using SN1033 as the recurrent susceptible parent, named CBB- (carrying *Xa2*, *Xa3*, *Xa4*, *Xa12*, *Xa14* and one unnamed *Xa* gene, respectively). Two other pairs NILs (CBB23 and CBB23B) were also built for *Xa23* using JG30 (indica) and IR24 (indica) as the recurrent susceptible parents in China. In the future, a combined strategy composed of traditional breeding, molecular marker-assisted selection, and transgenic technology should bring a new erato the BB resistance rice breeding program in China.

The 2nd International Conference on Bacterial Blight of Rice, 2007, October, 1－3. Nanjing.

中国杂交水稻白叶枯病抗性的遗传改良

章　琦

（中国农业科学院作物科学研究所，北京，100081）

Genetics and Improvement of Resistance to Bacterial Blight in Hybrid Rice

Abstract　Since the 1980s，hybrid rice for resistance to bacterial blight in China has been rapidly progressing. The gene *Xa4* was mainly used in indica three-line hybrid and two-line hybrid rice grown in a large area. The disease has been seen "quiet" for 20 years in China，yet in recent years it has gradually emerged prevalently in fields planted with newly released rice cultivars in the Yangtze River Valley. This situation for scientists and inevitably raised several questions：what causes the resurgence? Is the resistance of cultivars still effective? And/or is resistance breeding still one of the main objectives of rice improvement? What approach do we take on resistance breeding so that the resistance will be more durable，and the R-gene will be used more efficiently? A combined strategy composed of traditional method，molecular marker-assisted selection，and transgenic technology should bring a new era to the bacterial blight resistance hybrid rice breeding program. The review also briefly discusses and deliberates on issues related to the broadening of bacterial blight resistance，and suitable utilization of resistance genes，alternation planting of available resistance genes；and introduces the virulent populations of the bacterial blight pathogen in China even in Asia.

Key words　hybrid rice；bacterial blight；resistance breeding；genetics

摘要　20世纪80年代，中国杂交水稻抗白叶枯病育种进展迅速。在大面积种植的籼型三系和两系杂交稻中主要应用白叶枯病抗性基因 *Xa4*，白叶枯病在生产中几乎“销声匿迹”了20年。但近年来，此病却在长江流域一些种植新组合的稻田里又爆发了。什么原因引起它再度流行？品种的抗性是否还有效？抗病性遗传改良是否还是水稻育种的主要目标之一？通过哪些途径才能更有效地利用抗病基因培育出更为持久的抗性品种？为回答这些问题，就我国杂交水稻对白叶枯病抗性的改良、寄主与病原菌群体的互动演变效应进行了分析，讨论了拓宽抗性遗传基础、白叶枯病菌毒性群体结构、基因轮换和合理利用基因等问题。在采用了包括传统方法、分子标记辅助选择和转基因技术的综合策略后，杂交水稻抗白叶枯病育种研究将出现新的局面。

关键词 杂交水稻；白叶枯病；抗性育种；遗传学

水稻白叶枯病（*Xanthomonas oryzae* pv. *oryzae*，*Xoo*）是我国水稻生产上的一个严重病害。20 世纪 30 年代在苏、浙就有发现（方中达，1963）。目前，除新疆、西藏和东北的北部以外，全国都有不同程度的发生（章琦等，2007）。长江以南（包括华南和云南）和江淮平原籼稻产区是常发区。

水稻白叶枯病在 20 世纪 50 年代末至 60 年代在长江下游稻区的江苏、70～80 年代在淮南、皖中、鄂中丘陵、鄂东、江汉平原和华南等稻区多次流行（纪桐云，1989；许红，1993；钟藏文等，1998）。1976 年农业部设置了“全国抗稻瘟病、白叶枯病研究”项目，随后在“全国水稻高产、优质、抗病育种攻关”项目中，明确了抗稻瘟病和白叶枯病为品种改良的主攻目标之一。此后，大量优良、抗病组合的育成并推广，基本上控制了白叶枯病，该病害差不多“沉寂”了 20 年。然而，2005 年夏、秋两季，皖、苏又传出病情，连云港市 2.8 万 hm^2 稻田（占该市水稻总面积 16.8%）白叶枯病严重爆发，这是 20 世纪 90 年代以来病情最重年份（连云港市种子站，2005），东海县安峰乡一块 26.7 hm^2 稻田中的常规稻、杂交稻及制种田竟是枯槁一片，农民估计至多一半收成（图 1）。白叶枯病的抬头值得关注。什么原因引起它再度流行？品种的抗性是否还有效？抗病性改良是否还是水稻育种的主要目标之一？通过哪些途径才能更有效地利用抗病基因培育出更为持久的抗性品种？为回答这些问题，有必要对我国杂交水稻白叶枯病抗性遗传改良进行回顾，以确定未来的研究焦点。

图 1　江苏省连云港市东海县安峰乡白叶枯病稻田（徐大勇摄，2005）（另见彩版 9）

Fig. 1　Rice affected by bacterial blight in Anfeng，Donghai County Jiangsu Province (Xu，2005)

1 中国杂交水稻对白叶枯病的抗性改良

1.1 三系杂交籼稻

三系杂交籼稻中的野败型、矮败型、岗型、D型、印水型和红莲型等主体胞质不育系都是不抗白叶枯病的。但是，IR26、IR28、IR30、IR32、IR36、IR50、IR54以及具有IR系统亲缘的测64-7和桂99等，都是对上述不育系具有恢复力的恢复系，并且都携有一对不完全显性抗白叶枯病基因*Xa4*，因此，所配组合都有不同程度的抗性（章琦等，1997）。在2003—2005年通过省级审定的不育系中，29A、中1A和H28A也都有中等抗性（姚立生等，2003；何发青，2005）。罗彦长等（2005）结合常规杂交与分子标记辅助选择方法，聚合了基因*Xa21*和*Xa23*，育成全生育期广抗谱不育系R106A。

在1973—1986年育成的87个组合中，有31个中抗白叶枯病，占35.6%，年推广面积达0.267 M hm^2（400万亩）以上的有17个，其中6个兼抗其他病（虫），如：汕优6号、汕优桂33、汕优桂34和汕优30选兼抗稻瘟病，威优6号兼抗褐飞虱，四优6号兼抗黄矮病，威优64中抗多种病虫。在20世纪80～90年代，威优64、汕优6号、威优6号和汕优桂33的年最大推广面积都曾达到66.7万hm^2以上，威优64曾达到136.7万hm^2（1988年），在1983—2003年累计种植面积达1 227万hm^2，种植时间长达20年之久；汕优6号1981—1994年累计9 331万hm^2；威优6号的年推广面积曾达到1 739万hm^2（1985年），在1981—1995年累计6 434万hm^2，汕优桂33在1984—1999年累计4 662万hm^2（全国农业技术推广服务中心统计资料，1982—2006）。

1996年后，全国掀起“超级稻”研究，涌现出大量杂交稻组合。至2005年，在通过国家审定的169个组合中，有18个组合抗白叶枯病，占10.7%（杨仕华等，2004，2005，2006）。其中在生产上迅速达到20万hm^2（300万亩）以上的，有汕优77（1992—2005，累计249.7万hm^2）、Ⅱ优084（2001—2005，75.7万hm^2）、特优多系1号（1997—2005，25.6万hm^2）、协优9308（1998—2005，24.7万hm^2）、中9优838选（2002—2005，21万hm^2），多数组合兼抗稻瘟病（全国农业技术推广服务中心统计资料，1982—2006；杨仕华等，2004，2005，2006；盛生兰等，2005；程式华等，2004）。

在2002—2008年通过省级审定的组合中，皖稻135、Ⅱ优118、抗优98、Ⅱ优205、富优108、农华优808和Ⅱ优1511都兼抗稻瘟病，D优986、新香优906、29优559、Ⅱ优559以及天优1120还兼抗稻瘟和纹枯病或条纹叶枯病（章琦等，2007；马铮等，2007；姚立生等，2008；周明镜等，2008）。优异的双抗白叶枯和稻瘟病的超级稻中浙优1号至2006年的累计面积已达6.7万hm^2（章善庆等，2008）。表1列出80年代以来大面积生产应用及近期通过国家和省级审定的主要抗病高产、优质抗病组合。

表 1 我国主要抗病（虫）籼型三系杂交稻组合

Table 1 Main combinations of three-line indica hybrid rice with resistance to diseases and pest in China

杂交稻[1] Hybrid rice	组 合 Combination	抗病（虫）性反应[2] Reaction to						审定或育成年份 Year approved or bred	种植面积（公顷）[3] Acreage（year）[3)]/（$\times 10^4 hm^2$）
		BB	BL	SB	YD	BPH	RS		
威优 64[a]	V20A/测 64－7（IR36 衍生）	MR	MR		MR	MR		1981	1227.0（1983—2003）
汕优 6[a]	珍汕 97A/IR26	MR	MR					1976	933.1（1981—1994）
威优 6 号[a]	V20A/IR26	MR				MR		1976	752.6（1981—1995）
汕优桂 33[ab]	珍汕 97A/桂 33（IR36/IR20）	MR	MR					1983	466.2（1984—1999）
汕优桂 34	珍汕 97A/桂 34	MR	MR					1984	151.9（1986—1994）
汕优 30 选	珍汕 97A/30 选	MR	MR					1981	44.3（1984—1989）
四优 6 号	V41A/IR26	R			MR			1977	15.3（1983—1986）
汕优 77[#ab]	珍汕 97A/明恢 77	MR						1998	249.7（1992—2005）
Ⅱ优 084[#ab]	Ⅱ－32A/镇恢 084	HR	HR					2003	75.7（2001—2005）
特优多系 1 号[#a]	龙特甫 A/多系 1 号	MR	R					2001	25.6（1997—2005）
中 9 优 838 选[#]	中 9A/838 选	MR	R					2001	21.0（2002—2005）
丰优香占[#ab]	粤丰 A/R6547	HR	HR					2003	10.3（2002—2004）
K 优 047[#ab]	K17A/成恢 047	HR	MR					2001	10.3（1996—2003）
优优 128[#a]	优 IA/广恢 128	MR	R					1999	3.7（1997—2002）
K 优 402[#a]	K17A/402	MR	MR					1998	3.1（1998—2004）
中浙优 1 号[#ab]	中浙 A/抗恢 570	MR	MR					2004	7.6（2006）
中优 218[#ab]	中 9A/R218－51	MR						2004	1.3（2005）
中优 85[#ab]	中 9A/蒲恢 85	HR						2005	
宜香 9 号[#ab]	宜香Ⅰ A/宜恢 9 号	MR						2005	0.8（2005）
协优 9308[ab]	协青早 A /C57//300/ IR26	R	MR					1999	24.7（1998—2005）
Ⅱ优 128[ab]	Ⅱ－32A/恢 128	MR	HR					1999	53.2（1998—2005）
抗优 63[a]	珍汕 97A/抗恢 63	HR	HR					2002	
丰优 205[ab]	丰 7A/D205	R	R					2005	
协优 084[ab]	协青早 A/镇恢 084	MR	HR					2004	7.9（2004—2005）
皖稻 135[b]	丰源 A/YR909	HR	R					2004	
天优 372[ab]	天丰 A/广恢 372	MR	R					2005	
天优 290[ab]	天丰 A/广恢 290	MR	R					2005	
Ⅱ优 118[ab]	Ⅱ－32A/华恢 118	R	HR					2003	3.1（2004—2005）
Ⅱ优 559[a]	Ⅱ－32A/盐恢 559	MR	R	MR				2002	
29 优 559[a]	29A/盐恢 559	MR	HR	MR				2003	
新香优 906	新香优 A/蓉恢 906	R	MR	MR				2005	
K 优 507[ab]	K17A/R507	MR	HR					2002	
抗优 98[ab]	Ⅱ－32A/抗恢 98	HR	HR					2003	
Ⅱ优 205[ab]	丰 7A/D205	R/HR	R					2004	1.2（2005）

（续）

杂交稻[1] Hybrid rice	组 合 Combination	抗病（虫）性反应[2] Reaction to						审定或育成年份 Year approved or bred	种植面积（公顷）[3] Acreage (year)[3)] / (×10^4 hm^2)
		BB	BL	SB	YD	BPH	RS		
富优 108[b]	冈 46A/嘉恢 108	R	R					2005	
农华优 808[a]	M98A/MR0208	R	R					2005	
中优 2506[b]	中 9A/R2596	MR	MR					2005	
Ⅱ优 1511[# ab]	Ⅱ-32A/新恢 1511	MR	HR					2007	
D 优 986	D62A/池恢 986	MR	R	MR				2007	
天优 1120[ab]	天丰 A/盐恢 1120	HR	HR				R	2008	
冈香 707[ab]	冈香 1A/蜀恢 707	MR	HR					2008	

[1] #通过国家审定的组合，其余为省级审定的组合；[a]高产；[b]优质。[2] BB—白叶枯病；BL—稻瘟病；YD—黄矮病；BPH—褐飞虱；RS—水稻条纹叶枯病。R—抗；MR—中抗；HR—高抗；[3]种植面积根据全国农技推广中心资料整理(1982—2006)。

[1] #Registered at national level, and the others at provincial level). [a] High yield; [b] Good quality; [2] BB, Bacterial blight; BL, Rice blast ; SB, Sheath blight; YD, Rice yellow dwarf; BPH, Brown planthopper; RS, Rice stripe virus disease. R, Resistant; MR, Moderately resistant; HR, Highly resistant. [3] Planted area based on statistical data from National Agricultural Extention Service Center China (1982—2006) .

1.2 三系杂交粳稻

80 年代后，育成的 43 个组合中有 60%中抗白叶枯病。在苏、浙、皖、湘、辽、京等省（市）推广的 15 个组合中有 9 个抗病，黎明 A、丰锦 A、秋光 A、六千辛 A、当选晚 2 号 A、秀玲 A 等主要不育系也有中等抗性（章琦等，2007）。东北稻区种植最广的中抗组合黎优 57，1981—1992 年累计种植 65.7 万 hm^2（全国农技推广中心统计资料，1982—2006）。在 21 世纪审定的南方稻区的组合中，嘉优 2 号兼抗稻瘟和褐飞虱（杨伟国等，2008）；86 优 8 号、86 优 242、嘉乐优 2 号、甬优 6 号、甬优 5 号、皖稻 76、榆杂 34、徐优 201、常优 2 号、苏优 22 等均双抗白叶枯和稻瘟病；69 优 8 号、嘉优 1 号、甬优 2 号等组合兼抗纹枯病或细菌性条斑病或褐飞虱（表 2）。

表 2 2001 年来通过省级审定的部分双抗（多抗）优质或高产粳型三系杂交稻

Table 2 Some three-line japonica hybrid rice with multiple-resistance to diseases and pest, high-yield and good quality registered at provincial level in China since 2001

杂交稻 Hybrid rice	组合 Combination	抗病（虫）性反应 Reaction to					审定年份 Year approved	育成单位 bred by Inst.
		BB	BL	SB	BS	BPH		
甬优 4 号[# ab]	甬粳 2 号 A/K2001	MR					2004	宁波市农业科学院
86 优 8 号	863A/宁恢 8 号	MR	HR				2001	江苏省农业科学院
69 优 8 号[ab]	69A/R11238	R	HR	MR			2001	徐州市农业科学院
86 优 242	863A/R242	MR	HR				2002	太湖地区农业科学研究所

（续）

杂交稻 Hybrid rice	组合 Combination	抗病（虫）性反应 Reaction to					审定年份 Year approved	育成单位 bred by Inst.
		BB	BL	SB	BS	BPH		
嘉优 1 号[a]	嘉 60A/夹恢 40	MR	R		MR		2004	嘉兴市农业科学院
嘉优 2 号	嘉 60A/嘉恢 30	MR	R			MR	2007	嘉兴市农业科学院
嘉乐优 2 号[a]	151A/DH32	R	R				2004	嘉兴市农业科学院
甬优 6 号[b]	甬粳 2 号 A/K4806	R	R				2004	宁波市农业科学院
甬优 5 号[b]	甬糯 2 号 A/K6926	MR	MR				2004	宁波市农业科学院
甬优 2 号[ab]	甬粳 2 号 A/K1722	MR	MR		MR		2001	宁波市农业科学院
皖稻 76[b]	爱知香 A/MC20518	MR	MR				2004	合肥新隆水稻研究所
榆杂 34 [b]	滇榆 1 号 A/南 34	R	R				2004	云南省杂交水稻研究中心
常优 2 号[b]	武运粳 7A/C53	HR	R				2005	常熟市农业科学研究所
苏优 22[b]	武运粳 7A/R16189	R	MR				2005	江苏中江种业股份有限公司
徐优 201[ab]	徐 9320A/徐恢 201	MR	R	MR			2005	徐州市农业科学研究所
秀优 5 号 [#] [ab]	秀水 110/Yo 4	MR	R				2006	嘉兴市农业科学院

1.3 两系杂交稻

两系杂交稻的白叶枯病抗性总体上强于三系杂交稻。因为在同一亚种内常规稻的优异品种（系）都能正常恢复光、温敏核不育系的育性。这样，在选配组合时可利用的父、母本就大为拓宽，筛选产量潜力大、优质、抗性强的组合也就容易了。如特青、山青 11、余红、扬稻 6 号、双七占、D100、9311、双九、皖恢 9 号、R288、明恢 63 等已被广泛用来配制两系杂交稻，其中除明恢 63、R288 中感白叶枯病外，其他均有中等抗性。随着性状水平的提高和制种技术的不断完善，两系法选配杂交稻的相对优越性将日见彰显。

在 1994—2000 年通过省审定的 32 个组合中，17 个中抗白叶枯病，占 53.1%，有些兼抗稻瘟或其他病害（袁潜华，2001）。在杂交籼稻中，1994 年首批通过省级审定的培两优特青至 2000 年，累计种植 19.3 万 hm^2；培杂山青，培杂茂选、田两优 402、培两优 275、培杂 67、培两优余红等组合还兼抗稻瘟、稻曲病、细菌性条斑病。1994 年通过省级审定的粳型组合中，70 优 9 号、鄂粳杂 1 号和云光 8 号都双抗白叶枯和稻瘟病，鄂粳杂 1 号累计种植 38.9 万 hm^2（至 2005 年）（全国农技推广中心统计资料，1982—2006).

2000 年后，通过省级审定的双抗白叶枯和稻瘟病（虫）的组合有皖稻 119、赣业 1 号、皖稻 147、皖稻 129、两优 108 和两优培粳等（章琦等，2007）。

2006 年，在通过国家审定的 21 个组合中，两优培九、培杂双七、培杂茂三、陆两优 28、扬两优 6 号、70 优 9 号 6 个双抗白叶枯和稻瘟病，在生产上大面积应用。其中优质、高产、兼抗 1～2 种病的籼型组合两优培九从 2000 年开始，在江、皖、浙、闽、粤、桂、滇、黔、豫、鄂、湘、赣、蜀、陕、琼、渝等 16 个省（自治区、市）示范推广，2005 年的年种植面积达 0.655M hm^2，累计面积达 4.681M hm^2（2005 年）（全国农技推广中心统计资料，1982—2006）。

2007年通过国家审定的抗病优良组合有丰两优4号、新两优6380、培杂35和两优6号，前两个被农业部定名为2007年超级稻（中稻宣，2008；张从合等，2007）（表3）。

表3 两系杂交稻中主要的抗白叶枯病组合（部分内容据袁潜华报道整理）

Table 3 Main two-line hybrid rice combinations with bacterial blight resistance in China

(Adopted partly from the report of Yuan，2001)

杂交稻 Hybrid rice	类型 Type	组合 Combination	抗病（虫）性反应 Reaction to					审定年份 Year approved	审定单位 bred by Inst.
			BB	BL	BS	FS	BPH		
两优培九[# ab]	籼	培矮64S/9311	MR	HR				2001	江苏省农业科学院
培杂双七[# ab]	籼	培矮64S/双七占	MR	HR				2001	广东省农业科学院
培杂茂三[# ab]	籼	培矮64S/茂三	MR	R				2001	广东茂名市杂交稻研究中心
70优9号[# ab]	粳	7001S/皖恢9号	MR	HR				2001	安徽省农业科学院
培两优特青[a]	籼	培矮64S/特青	MR	MR				1994	湖南杂交水稻研究中心
鄂粳杂1号[ab]	粳	N5088S/R187	MR	MR				1995	湖北省农业科学院
云光8号	粳	N5088S/云恢11	MR	MR				2000	云南省农业科学院
培杂山青[a]	籼	培矮64S/山青11	MR	MR	MR			1997	广东茂名市杂优协作组
培两优余红[b]	籼	培矮64S/余红1号	MR	MR		MR		1997	湖南农业大学
培杂67[ab]	籼	培矮64S/G67	MR	MR				2000	华南农业大学
田两优402[a]	籼	田丰S/R402	HR	MR				1998	江西赣州地区农业科学研究所
培两优275	籼	培矮64S/275	R	MR				1999	广西农业科学院
培杂茂选[b]	籼	培矮64S/茂选	R	R				2000	广东茂名市杂交稻中心
赣亚1号	籼	培矮64S/Hb-01	MR	MR			MR	2002	江西滨湖农业科学研究所
皖稻129[b]	籼	培矮64S/红98	MR	HR				2004	安徽宣城市农科所
陆两优28[#a]	籼	陆18S×华28	MR	MR				2005	湖南亚华种业科学研究所
扬两优6号[#ab]	籼	广占63-4S/扬稻6号	MR	MR				2005	江苏省里下河农业科学研究所
两优培粳	粳	培矮64S/粳稻94205	MR	R				2005	河南信阳市农科所
皖稻147[ab]	籼	新安S/安选6号	R	MR				2005	安徽荃银农科所
两优108[b]	籼	培矮64S/宁恢108	MR	MR				2005	江苏省农科院
丰两优4号[# ab]	籼	丰39S/盐稻4号	HR					2006	合肥丰乐种业公司
新两优6号[# ab]	籼	新安S/安选6号	HR					2006	安徽荃银种业公司
新两优6380[# ab]	籼	03S/D208	R	R			R	2006	南京农业大学
培杂35[# ab]	籼	培矮64S/特华占35	MR	MR				2006	华南农业大学

BS=细菌性条斑病（Bacterial streak）；FS=稻曲病（False smut）.

中国水稻育种家成功地实现了预期目标，为农民提供了大量优良的抗病组合，20年来在全国有效地控制了白叶枯病的危害。但是，其“沉寂”多年后开始“复苏”，出现了一个老问题新爆发的危机。这不免会引起人们的关注和质疑，为寻找解决途径，弄清问题发生的原因很有必要。

2 寄主与病菌群体的互动演变效应

植物的抗病性是由寄主和病菌共同参与下形成的，病原物具有适应新品种的天性。老病害的再现和毒性病菌群体发生的频率，取决于病原菌对投放组合的适应力。生物界存在着这种周而复始的“沉浮”循环，才能协同演化。因此，抗性基因能否有效控制病菌小种，以及病原菌的变异动态，是有效利用抗病基因的重要依据。

2.1 水稻对白叶枯病抗性遗传变异狭窄的潜在危险

根据对 287 个抗病籼稻品种和抗源的系谱分析，IRRI 系列品种（系）、斯里兰卡品种 BG90－2 和竹占、竹印 2 号是我国利用的 3 大抗源。对其中近百个品种的抗性遗传、基因推论和 52 个抗源的基因等位性测试表明，我国三系和两系杂交稻的白叶枯病抗性均由 *Xa4* 基因控制（图 2）（伍尚忠，1990；章琦等，1988，1990，1991，1994；沙学延等，1994；施爱农等，1993；杨文才等，1995；熊卫等，2000；谢岳峰等，1990）。仅用单一抗性基因造成了寄主抗性遗传基础脆弱，在选择压的驱动下，病原菌演变出新的毒性菌株来克服该基因。不出所料，80 年代中期，广东出现了能侵袭 IR26（携 *Xa4* 基因）的毒性菌株 GD1358（中国 V 型菌），并扩展到闽、桂、浙、苏、滇和辽等省（徐羡明等，1987，1994；曾列先等，2002）。

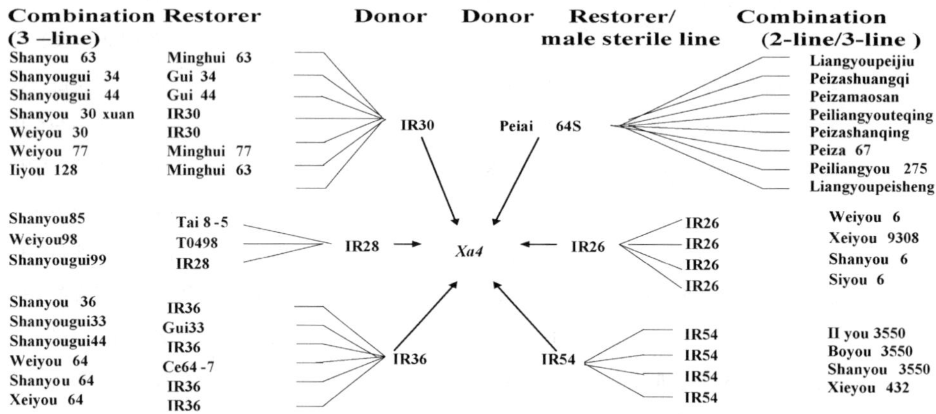

图 2 我国三系和两系杂交稻主要白叶枯病抗源（均由 *Xa4* 基因所控制）

Fig. 2 Sources of bacterial blight resistance for main hybrid rice in China (all depend on *Xa4*)

菲律宾在推广现代品种之前，其优势小种 1 占菌株采样总数的 82%。20 世纪 70 年代后，大面积推广携有基因 *Xa4* 的品种，与其亲和的小种 3 的频率为 26%，小种 1 已降至 9%；至 1988 年，*Xa4* 基因品种面积增至水稻总面积的 90%，小种 3 的频率上升到 61%。显然，*Xoo* 群体结构的变化取决于生产所用抗病品种的替换状况，品种与病菌定向选择（directional selection）的互作表露无遗。印度、印度尼西亚、韩国、尼泊尔等国家，长期大面积种植携 *Xa4* 基因的品种也都丧失了抗性（Mew et al.，1985，1992；Mew，1987）。

鉴于我国主要恢复系是都携有 *Xa4* 基因的 IR 品种，覆盖面遍及全国，随时可能遇到毒性菌的侵袭，也将无法逃脱抗性丧失的厄运。

20 世纪 90 年代在全球迅速应用的广抗谱基因 *Xa21* 先后在韩国、印度、印尼等国已遭受侵袭，我国广东Ⅳ、Ⅴ及新型菌，和云南 14 个中的 8 个菌群也都能侵染携带 *Xa21* 的品种（组合）(Lee et al.，1999；Adhikari et al.，1999；Marella et al.，2001；Muralidharan et al.，2001；曾列先等，2005；姬广海等，2003)。

2.2 中国水稻白叶枯病原菌遗传群体结构的变化特点

Zhang 等（1997）对 1984—1999 年采集的 272 个包括中国白叶枯病 7 个病原型菌株的基因组 DNA 的 RFLP 指纹进行测定，继而对 143 个代表菌株的 RFLP 和 rep-PCR DNA 指纹复合数据聚类分析，形成 3 簇（cluster）。绝大多数菌株聚于簇 1，少数在簇 2、簇 3，广东和云南的特殊单元型（haplotype）在 3 簇之外（outgroup）（图 3），说明，1）多数菌株间系统发育的亲缘关系相近，群体结构简单，这是因抗病基因利用单一所致。亚洲 7 国合作试验的结果类似，308 个菌株形成 5 簇（cluster），76 个中国菌株中的 74 个都聚于簇 1，广东、云南菌株聚于簇 5（图 4）（Adhikari et al.，1995）；2）分子型与致病型有一定对应关系，多数致病型表现遗传异质性，预示群体潜藏着演变（Zhang et al.，1997）。

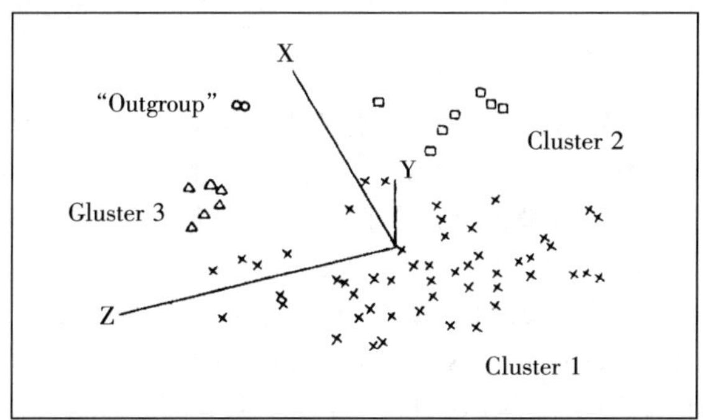

图 3　用 rep-PCR（ERIC，BOX，IS1112，IS1113 引物）和 RFLP（探针 pBSavr*Xa10*）分析中国 43 个 *Xoo* 菌株复合数据，通过三维传感聚类形成 3 簇，菌株在立体坐标图的位置根据复对应分析确定（章琦等，1997）

Fig. 3　Clustering analysis of Chinese 143 *X. o.* pv. *oryzae* strains as derived from DNA band data obtained with rep-PCR (ERIC, BOX, IS1112 and IS1113 primers) and RFLP (pBS*avrXa10* probe); The number of clusters which best fit the data was determined by the consensus of three clustering statistics. Positions of the strain on the three dimensional graph were determined by multiple correspondence analysis (Zhang et al., 1997).

正如预期所述，云南、广东菌株的毒性是特殊的，在姬广海等（2003）划分的 14 个"小种"中，有 9 个和 8 个分别能侵染 *Xa4* 和 *Xa21*。曾列先等（2005）报道了广东发现了另一个新毒性菌能侵染 *Xa1*、*Xa2*、*Xa3*、*Xa4*、*xa8*、*Xa11*、*Xa12*、*xa13*、*Xa14* 和 *Xa21* 诸多基因。

这些研究为鉴定和分析寄主抗性提供了重要信息，往往由于以下原因不能发挥寄主抗性的预期效果：1）已出现毒力更强的群体，选用的抗性基因不抗种植所在地的菌群；2）因投放的抗病品种所产生的选择压使病菌毒性已发生变化。我们可以根据对病菌群体生物学的认知，来解决这些问题并提出对策。

据许志刚等（2004）报道我国水稻白叶枯病致病型的分布是：华南籼稻区以Ⅳ型为主，

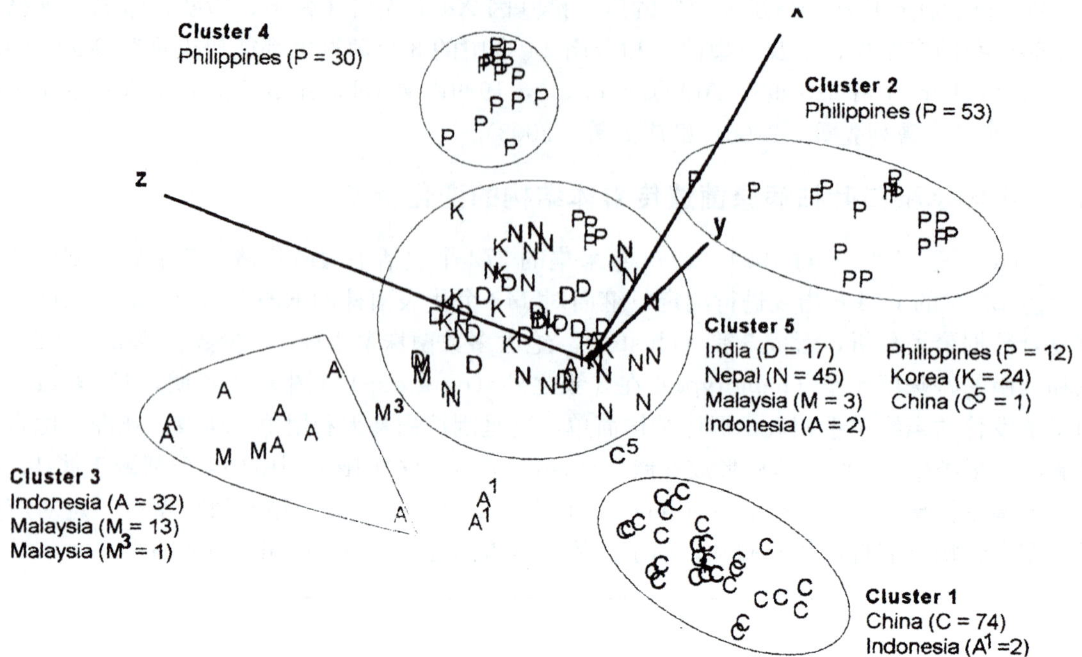

图 4　用 pBS*avrXa10* 和 pJEl 101 两个探针检测所得 RFLP 指纹复合数值的三维传感聚类分析，将中（C）、印尼（A）、印度（D）、韩（K）、马（M）、尼（N）和菲（P）7 国的 *Xoo* 菌株基因组分子型聚集为 5 簇（Adhikari et al.，1995）。

Fig. 4　Cluster analysis showing the relationships of *Xoo* strains as derives from DNA band data obtained with probes pBS*avrXa10* and pJEL101. Numbers of strains from China (C), Indonesia (A), India (D), Korea (K), Malaysia (M), Nepal (N), and the Philippines (P) are shown. The number of cluster that best fit the data was determined by the consensus of three clustering statistics (Adhikari et al.，1995).

对 *Xa4* 有毒性的Ⅴ型菌已从华南扩展至长江流域，在长江下游籼粳兼种稻区，以Ⅳ、Ⅱ型为主；北方粳稻区以Ⅱ和Ⅰ型居多，云南省分布强毒性的新致病型。这是个粗略的概况，各地需根据本区病菌群体结构，决定抗性基因的部署。

2.3　亚洲国家水稻白叶枯病菌群及有效抗性基因

水稻白叶枯病是亚洲国家十分严重的病害。在菲律宾，*Xa7*、*xa5* 和 *Xa21* 抗其 10 个小种中的大多数小种（Vera Cruz，私人通信）。在日本的 8 个小种中，小种Ⅰ和Ⅱ居于优势，有效的抗病基因是 *Xa3*、*Xa4*、*Xa12*（Noda et al.，1990）。尼泊尔的 5 个小种的毒性很强，对携带 *Xa4*、*xa5*、*Xa7*、*xa8*、*Xa14* 和 *Xa21* 的品种都能侵染，只有 2～4 个抗性基因的聚合系才是有效的（Adhikari et al.，1999）。韩国鉴定的 5 个小种中，K1～K3 遍布全国，2003 年以后，能侵染 *Xa3* 的新小种 K3a 为优势小种，韩国西南部稻田遭到严重侵袭，该小种还能侵染携带 *Xa7* 和 *Xa21* 的品种（组合）（Noh et al.，2004；Jeung et al.，2006）；韩国在 90 年代中期测定的全国菌株中的 96% 都能侵染 *Xa21*（Lee et al.，1999）。Muralidharan 等（2004）在印度调查的 11 个地区中，*xa5*、*xa13* 和（*xa13*＋*Xa21*）、（*xa5*＋*Xa21*）聚合系在 5～8 个地区是抗的。总体来看，*xa5* 和 *Xa7* 在中国、菲律宾、印尼、韩国、越南都有效。

Noda 等（1996）用日本和菲律宾的两套白叶枯病鉴别品种，测定 1982—1994 年孟加拉、印度、尼泊尔、菲律宾、马来西亚、缅甸、泰国、韩国、中国南方稻区的 397 个 *Xoo* 菌株，结果表明，孟加拉、印度、尼泊尔和我国南方稻区菌株的毒力较强。能侵染广抗谱基因 *xa5*、*Xa7*（表 4）。*Xoo* 病菌群体分布的区域差异，源于病原菌具有适应寄主抗性的遗传潜能，因此要选择适宜的抗性基因来控制地区优势菌群。

表 4　亚洲国家 BB 菌群能侵染的以及有效的主要抗性基因

［根据 Noda 等（1996）测定结果整理］

Table 4　The main attacked and effective resistance genes to BB in Asian countries

（resummerized based on Noda et al.，1996）

国家 Country	小种 Race	测试菌株 strain tested	能侵染的抗病基因 Attacked main R gene	有效抗病基因 Effective R gene
孟加拉 Bangladesh	12	74	*Xa1*、*Xa2*、*Xa3*、*Xa4*、*Xa7*、*Xa10*、*Xa11*、*xa5*、*xa8*	*Xa3*、*Xa12*、聚合系 Pyramiding line
印度 India	5	58	*Xa1*、*Xa2*、*Xa3*、*Xa4*、*Xa10*、*Xa11*、*Xa12*、*xa5*	基因聚合系
尼泊尔 Nepal	16	45	*Xa1*、*Xa2*、*Xa3*、*Xa4*、*Xa10*、*Xa11*、*Xa12*、*xa5*	*xa5*、*Xa7*、基因聚合系
缅甸 Myanmar	7	27	*Xa1*、*Xa4*、*Xa10*、*Xa11*、*xa5*	
泰国 Tailand	5	35	*Xa1*、*Xa2*、*Xa10*、*Xa11*、*xa5*	
印度尼西亚 Indonesia	8	78	*Xa1*、*Xa2*、*Xa3*、*Xa10*、*Xa11*、*xa8*	*xa5*、*Xa7*、基因聚合系
菲律宾 Philippines	5	61	*Xa1*、*Xa2*、*Xa4*、*Xa11*	*xa5*、*Xa7*、*Xa12*、基因聚合系
马来西亚 Malaysia	2	11	*Xa1*、*Xa2*、*Xa10*、*Xa11*、*xa8*	*Xa4*、*Xa7*、*Xa12*、*xa5*
中国（南方） China South	6	24	*Xa1*、*Xa2*、*Xa3*、*Xa4*、*Xa7*、*Xa10*、*Xa11*、*Xa12*、*xa8*	*xa5*、*Xa7*、*Xa22*、*Xa23*

3　杂交稻白叶枯病抗性改良途径

3.1　拓宽抗性遗传基础

在 20 世纪 70 年代末，我国育种家已着手抗性拓宽工作。朱立宏（1991）、周毓珍等（1993）和丁伦友等（2005）将 DV85 导入 TN1，育成一批携广抗谱基因 *Xa7* 的衍生抗源 TD 品系，以明恢 63 为父本，育成了携有 *Xa7* 的抗恢 63、抗恢 98 及 D205 等抗病恢复系，配制了抗白叶枯和稻瘟病的组合，抗优 63、抗优 98、Ⅱ优 205 及丰优 205 等，先后在江、浙、皖、豫、滇、鄂、赣、闽、湘等省疫情严重的地区示范和应用，迄今累计推广面积约 2.5 万 hm^2。

截止 2006 年，已鉴定的抗抗白叶枯病基因约有 30 个（章琦，2005），通过分子标记定位，鉴别出与基因紧密连锁分子标记的有 18 个，主要利用的基因 *Xa4*、*Xa7*、*Xa21*、*Xa23*、*xa5* 和 *xa13* 等都具备了紧密连锁的 PCR 标记。已克隆的抗性基因有 *Xa1*、*xa5*、*xa13*、*Xa21*、*Xa26*（明恢 63）和 *Xa27*（*O. minuta*）6 个。在我国，*Xa21* 已在我国广泛用于分子标记辅助改良杂交稻的抗性。育种家以携有 *Xa21* 的品系 IRBB21 为供体，利用与该基因紧密连锁的 PCR 标记 pTA248，经回交将它转育到明恢 63、93－11、中恢 218、蜀恢 527、R8006、R1176 和辐恢 838 等恢复系中，选育出一批抗病恢复系，测配出抗病新组合（表 5）（徐建龙，2007）。

表 5　分子标记辅助选择育成的携有 *Xa21* 的抗病籼型杂交稻组合

（徐建龙，2007）

Table 5　Resistant three-line indica hybrid rice with bacterial blight resistance gene *Xa21* bred by molecular marker assisted selection

（Xu，2007）

抗病组合或恢复系 Resistant cross or restorer	待改良恢复系 Restorer to be improved	供体（基因） Donor (gene)	分子标记 Molecular marker	育成单位 Bred by Inst.	审定年份 Year approved
协优 218	辐恢 838	IRBB21（*Xa21*）	pTA248	中国水稻所 CRRI	2002
中优 218	中恢 218	IRBB21（*Xa21*）	pTA248		2003
国稻 1 号[a]	R8006	IRBB60（*Xa21*）	pTA248		2004
国稻 3 号	R8006	IRBB60（*Xa21*）	pTA248		2004
中优 1176	R1176	IRBB60（*Xa21*）	pTA248		2005
改良蜀恢 527	蜀恢 527	IRBB60（*Xa21*）	pTA248、MP12	四川农大水稻所	
明恢 63（携 *Xa21*）	明恢 63	IRBB21（*Xa21*）	pTA 248、PTA 21	华中农业大学作物遗传改良国家重点实验室	
93－11（*Xa21*）	93－11	Minghui 63（*Bt*/*Xa21*）	pTA 248		
93－11（*Bt*/*Xa21*）	93－11	Minghui 63（*Bt*/*Xa21*）	pTA 248		
Ⅱ优 8220	密阳 46/明恢 63	IRBB21（*Xa21*）	pTA 248	浙江农业大学	2003
T71、T81	明恢 63	IRBB21（*Xa21*）	pTA 248		

[a] 通过国家审定，其余为省级审定（Registered at national level，and the others at provincial level）.

3.2　综合策略

丰富的基因源和成熟的技术，为水稻白叶枯病抗性改良奠定了基础。然而，20 世纪 90 年代以来，三系杂交稻新组合的抗性百分率分别从 35.6%（籼）和 60%（粳）下降至 10.0%和 7.8%、两系组合从 53.1%降至 28.6%（表 6）。不由让人联想到白叶枯病“沉寂”多年后“复苏”的原因是，1）新组合不携有抗病基因；2）投放前没有经过充分的抗白叶枯病性筛选；3）组合的抗性是否还生效？在战略上，对育种工作者提出了以下问题：改良白叶枯病抗性是否还是水稻抗病的主要目标之一？以哪些新的手段和途径才能更有效地利用抗病基因培育出更为持久的抗病组合？而面临的挑战是：怎样解决病原菌的快速变化？如何设计和培育具有较持久的抗性品种?

上述问题并不难解决，充分利用已有的一批优异抗病基因。结合传统方法、分子标记辅助选择和转基因技术，必然能使杂交水稻抗白叶枯病育种进入一个新的时代。在水稻白叶枯病抗性育种中，应该采取以下策略：1）把改良白叶枯病抗性作为水稻抗性育种的目标之一。2）根据优势菌群的分布，建议选用的抗病基因。长江中、下游Ⅳ型和Ⅱ型菌为优势菌，选用 *Xa21*、*Xa23*、*Xa7*、*xa5* 和 *xa13* 等；华南稻区针对Ⅴ型、Ⅳ型和新菌型，选用 *Xa23*、*Xa7* 和 *xa5* 等；云南稻区，选用 *Xa22*、*Xa23* 、*Xa7* 和 *xa5* 等；北方粳稻区，选用 *Xa23*、*Xa7*、*Xa21*、*Xa4*、*xa5* 和 *xa13* 等，制订基因轮换计划。3）培育携带多个基因的聚合系，筑成抗病基因屏障；主效与微效基因相结合，延长优异抗性基因使用寿命。

与水稻病害的斗争是无止境的。为了防患于未然，我们不仅需要培育携有不同抗性基因的品种，还要了解怎样在育种中科学地利用它们，继续检测病菌群体对投放的新品种的反应，选择当地优势菌株用于鉴定和筛选材料，并用来指导调整品种的推广。病菌毒性变异的丰富知识对于基于寄主抗性制定治理病害的策略是至关重要的，为此，国际之间、遗传育种病理等方面的密切合作是实现目标的保证。

表 6 1973—2006 年推广面积 6 667hm² （10 万亩）以上杂交稻组合的白叶枯病抗性情况

Table 6 Bacterial blight resistance combinations with annual planted area over 6 667hm² during 1973—2006

时期 Period（year）	组合数 Total no. of combinations	抗病组合 No. of resistant combinations	抗病百分率（%） Percentage of resistance（%）
三系杂交稻 Three-line hybrid rice			
1973—1986	87	31	35.6（籼）(indica)
	43	26	60.0（粳）(japonica)
1996—2005	169	18	10.7（籼）(indica)
2006	6	4	66.7（南方、粳） (South，japonica)
	64	5	7.8（北方、粳） (North，japonica)
两系杂交稻 Two-line hybrid rice			
1994—2000	32	17	53.1
2001—2006	21	6	28.6

原载：中国水稻科学，2009，23（2）：111～119.

Genetics and Improvement of Bacterial Blight Resistance of Hybrid Rice in China

Zhang Qi

Abstract Since 1980s, rice breeding for resistance to bacterial blight has been rapidly progressing in China. The gene *Xa4* was mainly used in three-line indica hybrid and two-line hybrid rice. The disease has been seen 'quiet' for 20 years in China, yet in recent years it has gradually emerged prevalently in fields planted with newly released rice cultivars in the Changjiang River valley. Under the circumstances, scientists inevitably raised several questions: what causes the resurgence and what should we do next? And/or is resistance breeding still one of the main objectives in rice improvement? What approach do we take on resistance breeding so that the resistance will be more durable, and the resistance gene will be used more efficiently? A combined strategy involving traditional method, molecular marker-assisted selection, and transgenic technology should bring a new era to the bacterial blight resistance hybrid rice breeding program. This review also briefly discusses and deliberates on issues related to the broadening of bacterial blight resistance, and suitable utilization of resistance genes, alternation planting of available resistance genes; and understands the virulent populations of the bacterial pathogen in China even in Asia.

Key words hybrid rice; bacterial blight, *Xanthomonas oryzae* pv. *oryzae*; resistance breeding; genetics; resistance gene

Bacterial blight (*Xanthomonas oryzae* pv. *oryzae*, *Xoo*) is one of the most serious diseases of rice in China. In 1930s, the disease mainly occurred in Jiangsu and Zhejiang provinces of China (Fang, 1963) . To date, it has widely spread nationwide, except in Xinjiang Uygur Autonomous Region, Tibet Autonomous Region and northern part of the northeast region of China (Zhang et al. , 2007).

The disease is most destructive in the southern part of China, the major indica rice cultivation region. In the Ezhong rice cultivation area, the middle region of the Changjiang River valley, bacterial blight epidemics occurred in 1974-1975, which affected about 292 260 hm^2 of rice fields and caused rice yield loss of 67 700t; in 1980、1983 and 1987, 141 800 t rice yield loss was caused by bacterial blight, which attacked nearly 584, 100 hm^2 paddy fields (Xu, 1993). In Anhui Province of China, the disease outbreak in 1982 a yield losses of 205 400 t (Ji, 1989).

In 1976, the Ministry of Agriculture of China established a national research target: "Resistances Toward Rice Fungal Blast and Bacterial Blight" . In the subsequent prioritized research program "Breeding Rice for High Yield, Good Quality, and Disease Resistance", resistances toward rice fungal blast and bacterial blight were enlisted as main objectives of rice improvement. Since 1980s, bacterial blight has been controlled in China for almost 20 years by popularizing rice cultivars exhibiting high yield, good quality, and resistance toward rice bacterial blight. However, in summer and autumn of 2005, epidemic of rice bacterial blight was observed again in the northern part of Jiangsu Province, China, severely affecting about 28 000 hm^2 (occupying 16.8% of the total rice cultivation area) in the Lianyungang city. In a 26.7 hm^2 paddy field in the Anfeng county, Jiangsu province, all rice (hybrid rice and other cultivars) was damaged and withered, with a greater than 50% loss of yield (Fig. 1). The recurrence of rice bacterial blight is alarming. It is a critical time to thoroughly review progress in the past to set new focuses for future researches.

Rice breeding for resistance to bacterial blight in China

In early 1970s, the Ministry of Agriculture of China established a program named 'Rice Breeding for High Yield, Good Quality, and Disease Resistance'. The bacterial blight resistance breeding research in China has been rapidly and extensively progressed since then.

Three-line indica hybrid rice

Actually, the main male sterile lines such as wild abortive (WA), dwarf abortive (DA), Gang, D, Yinshui and Honglian types developed in China are susceptible to bacterial blight, however, all the restorer lines IR26, IR28, IR30, IR32, IR36, IR50, IR54 and Ce 64-7 (IR9761-19-1) with restoring ability for those sterile lines are resistant, therefore, the hybrid rice combinations possessed various degrees of resistance to the disease (Zhang, 2007) .

From 1973 to 1986, 31 (35.6%) out of 87 three-line indica hybrid rice combinations developed were moderately resistant to bacterial blight. There were 17 combinations having an annual acreage of more than2.67 × 10^5 hm^2 . Among them, Weiyou 64, Weiyou 6, Shanyou 6, Shanyougui 33, Shanyougui 34, Shanyou 30 Xuan and Siyou 6 were also resistant to rice blast or brown planthopper. In addition, the combinations including Weiyou 64, Weiyou 6 and Shanyou 6had reached an annual acreage of 1.333×10^6 hm^2. Moreover, Weiyou 64 was grown as long as for 20 years. The acreage for other combinations were grown until middle of 1990s, and gradually cut down (Zhang, 2007) .

Since 1996, China National Rice Research Institute and its counterparts established an idea for development of 'Super Rice', it boosted the rice breeding program nationwide, and a numerous hybrid rice combinations with high-yield, good quality were developed.

Up to 2005, 169 indica combinations passed the national certification requirements, 18 (10.7%) out of these were resistant to bacterial blight or both bacterial blight and rice blast. Combinations including Ⅱ You 084, Shanyou 77, Guofeng 1 and Teyouduoxi 1 had accumulated a large acreage (Yang et al., 2004, 2006; Cheng et al., 2004; Sheng et al., 2005). The acreage of an outstanding super hybrid rice, Zhongzheyou 1, with resistance to both bacterial blight and rice blast, reached $6.7 \times 10^4 hm^2$ in 2006 (Zhang S et al., 2008).

During 2002—2008, among combinations with high yield or good quality passed the provincial certification requirements, Wandao 135, Ⅱ You 118, Kangyou 98, Ⅱ You 205, Fuyou 108, Nonghuayou 808 and Ⅱ You 1511 were resistant to both bacterial blight and rice blast, and D You 986, 29 You 559, Ⅱ You 559, Xinxiangyou 906 and Tianyou 1120 exhibited multiple resistances to bacterial blight, rice blast and sheath blight or rice stripe virus (Zhang et al., 2007; Ma et al., 2007; Yao et al., 2008; Zhou et al., 2008). Main combinations with disease (s) /insect (s) resistance, as well as high yield or good quality that has been bred since 1980s and approved and grown in a large acreage in China are listed in Table 1.

Additionally, several male sterile lines with bacterial blight resistance were developed and passed the provincial certification requirements during 2003 to 2005, such as 29A、Zhong 1A and H28A (Zhang et al., 2007). The sterile line R106A exhibiting high resistance to bacterial blight at all growth stage, which polymerized both genes *Xa21* and *Xa23* by molecular marker-assisted selection (Luo et al., 2005).

Table 1 Main combinations of three-line indica hybrid rice with resistance to diseases and insects in China.

Hybrid rice	Combination	Reaction to					Year approved or bred	Acreage (Year) / ($\times 10^4 hm^2$) [c]
		BB	BL	SB	YD	BPH		
Weiyou 64[a]	V20A/Ce 64-7 (IR36)	MR	MR		MR	MR	1981	1 227.0 (1983—2003)
Shanyou 6[a]	Zhenshan 97A/IR26	MR	MR				1976	933.1 (1981—1994)
Weiyou 6 [a]	V20A/IR26	MR				MR	1976	752.6 (1981—1995)
Siyou 6	V41A/IR26	R			MR		1977	15.3 (1983—1986)
Shanyou 30 Xuan	Zhenshan 97A/30 Xuan	MR	MR				1981	44.3 (1984—1989)
Shanyougui 33 [ab]	Zhenshan 97A/Gui 33 (IR36/IR20)	MR	MR				1983	466.2 (1984—1999)
Shanyougui 34	Zhenshan 97A/Gui 34	MR	MR				1984	151.9 (1986—1994)
Shanyou 77 #[ab]	Zhenshan 97A/Minghui 77	MR					1998	249.7 (1992—2005)
K You 402 #[a]	K17A/402	MR	MR				1998	3.1 (1998—2004)
Youyou 128 #[a]	You Ⅰ A/Guanghui 128	MR	R				1999	3.7 (1997—2002)
Xieyou 9308[ab]	Xieqingzao A /C57//300 /IR26	R	MR				1999	24.7 (1998—2005)
Ⅱ You 128 [ab]	Ⅱ-32A/Hui 128 (7 Guizao 25/Ce 64//Minghui 63)	MR	HR				1999	53.2 (1998—2005)

(continued)

Hybrid rice	Combination	Reaction to					Year approved or bred	Acreage (Year) / (×10^4 hm^2) [c]
		BB	BL	SB	YD	BPH		
Teyouduoxi 1 [#a]	Longtefu A/Duoxi 1	MR	R				2001	25.6 (1997 - 2005)
Zhong 9 You 838 Xuan [#]	Zhong 9A/838 Xuan	MR	R				2001	21.0 (2002 - 2005)
K You 047 [#ab]	K17A/Chenghui 047	HR	MR				2001	10.3 (1996 - 2003)
Kangyou 63[a]	Zhenshan 97A/Kanghui 63	HR	HR				2002	
Ⅱ You 559[a]	Ⅱ - 32A/Yanhui 559	MR	R	MR			2002	
K You 507 [ab]	K17A/R507	MR	HR				2002	
Ⅱ You 084[#ab]	Ⅱ - 32A/Zhenhui 084	HR	HR				2003	75.7 (2001 - 2005)
Fengyouxiangzhan [#ab]	Yuefeng A/R6547	HR	HR				2003	10.3 (2002 - 2004)
Ⅱ You 118 [ab]	Ⅱ - 32A/Huahui 118	R	HR				2003	3.1 (2004 - 2005)
Kangyou 98 [ab]	Ⅱ - 32A/Kanghui 98	HR	HR				2003	
29 You 559[a]	29A/Yanhui 559	MR	HR	MR			2003	
Xieyou 084 [ab]	Xieqingzao A/Zhenhui 084	MR	HR				2004	7.9 (2004 - 2005)
Zhongzheyou 1 [#ab]	Zhongzhe A/ Hanghui 570	MR	MR				2004	7.6 (2004 - 2006)
Zhongyou 218 [#ab]	Zhong 9A/R218 - 51	MR					2004	1.3 (2005)
Ⅱ You 205[ab]	Feng 7A/D205	HR	R				2004	1.2 (2005)
Wandao 135 [b]	Fengyuan A/YR909	HR	R				2004	
Zhongyou 85 [#ab]	Zhong 9A/Puhui 85	HR					2005	
Yixiang 9[#ab]	YixiangIA/Yihui 9	MR					2005	0.8 (2005)
Fengyou 205[ab]	Feng 7A/D205	R	R				2005	
Tianyou 372 [ab]	TianfengA/Guanghui 372	MR	R				2005	
Tianyou 290 [ab]	TianfengA/Guanghui 290	MR	MR				2005	
Xinxiangyou 906	Xinxiangyou A/Ronghui 906	R	MR	MR			2005	
Fuyou 108 [b]	Gang 46A/Jiahui 108	R	R				2005	
Nonghuayou 808 [a]	M98A/MR0208	R	R				2005	
Zhongyou 2506[b]	Zhong 9A/R2596	MR	MR				2005	
Ⅱ You 1511[#ab]	Ⅱ - 32A/Xinhui 1511	MR	HR				2007	
D You 986	D62A/Chihui 986	MR	R	MR			2007	
Tianyou 1120[ab]	Tianfeng A/Yanhui 1120	HR	HR				2008	
Gangxiang 707[ab]	Gangxiang A/Shuhui 707	MR	HR				2008	

[#] Registered at national level, the others at provincial level; [a] High yield; [b] Good quality; [c] Planted area based on the statistical data from the National Agricultural Extention Service Center, China.

BB, Bacterial blight; BL, Rice blast; SB, Sheath blight; YD, Rice yellow dwarf; BPH, Brown planthopper. R, Resistant; MR, Moderately resistant; HR, Highly resistant.

Three-line japonica hybrid rice

Up to late 1970s, 43 out of 72 three-line japonica hybrid rice combinations developed were moderately resistant to bacterial blight. Since then, most of the japonica combinations bred also exhibited bacterial blight resistance. The main sterile lines, such as Reimei A, Toyonishiki A, Akihikari A, Liuqianxin A, Dangxuanwan 2A and Shurei A are also resistant to the disease.

During 1980s to 1990s, 3 out of 5 japonica combinations cultivated in Northern China were moderately resistant to bacterial blight. One of them, Liyou 57 has an accumulated acreage exceeding 6.57×10^5 hm^2. Since 2001, some japonica combinations exhibiting multiple resistances to pests had passed the national and provincial certification requirements, including Yongyou 4, 69 You 8, Jialeyou 2, Jiayou 2 and so on (Table 2).

Table 2 Some three-line Chinese japonica hybrid rice with multiple-resistance to diseases and insects, as well as high-yield and good quality registered at national or provincial level since 2001 in China

Hybrid rice	Combination	Reaction to					Year approved	Breeder
		BB	BL	SB	RS	BPH		
Yongyou 4 [#ab]	Yongjing 2A/K2001	MR					2004	Ningbo Academy of Agricultural Sciences, China
86 You 8	863A/Ninghui 8	MR	HR				2001,	Jiangsu Academy of Agricultural Sciences, China
69 You 8 [ab]	69A/R11238	R	HR	MR			2001	Xuzhou Academy of Agricultural Sciences, China
Yongyou 2 [ab]	Yongjing 2A /K1722	MR	MR		MR		2001	Ningbo Academy of Agricultural Sciences, China
86 You 242	863A/R242	MR	HR				2002	Taihu Agricultural Research Institute, China
Jiayou 1 [a]	Jia 60A/Jiahui 40	MR	R			MR	2004	Jiaxing Academy of Agricultural Sciences, China
Jialeyou 2 [a]	151A/DH 32	R	R				2004	Jiaxing Academy of Agricultural Sciences, China
Yongyou 6 [b]	Yongjing 2A/K4806	R	R				2004	Ningbo Academy of Agricultural Sciences, China
Yongyou 5 [b]	Yongnuo 2A/K6926	MR		R			2004	Ningbo Academy of Agricultural Sciences, China
Wandao 76 [b]	Aizhixiang A/MC20518	MR		MR			2004	Hefei Xilong Rice Research Institute, China
Yuza 34 [b]	Dianyu 1 A/Nan 34	R	R				2004	Yunnan Hybrid Rice Research Center, China
Changyou 2 [b]	Wuyunjing 7A/C53	HR	R				2005	Changshu Agricultural Research Institute, China
Suyou 22 [b]	Wuyunjing 7A/R16189	R	MR				2005	Zhongjiang Seed Company, China
Xuyou 201 [ab]	Xu 9320A/Xuhui 201	MR	R	MR			2005	Xuzhou Academy of Agricultural Sciences, China
Xiuyou 5 [#ab]	Xiushui 110/Xiuhui 69	MR	R				2006	Jiaxing Academy of Agricultural Sciences, China

[#] Registered at national level, the others at provincial level; [a] High yield; [b] Good quality.

BB, Bacterial blight; BL, Rice blast; SB, Sheath blight; RS, Rice stripe virus; BPH, Brown planthopper. R, Resistant; MR, Moderately resistant; HR, Highly resistant.

Two-line hybrid rice

In general, the bacterial blight resistance of two-line hybrid rice is better than that of three-line hybrid rice, due to that the parents of two-line hybrid rice are widely selected and used, such as Teqing, Shanqing 1, Yuhong, Yangdao 6, Shuangqizhan, D100, 9311, Shuang 9 and Wanhui 9, which exhibit moderately resistance to bacterial blight. Twenty sterile lines released during 1996 – 2000 were evaluated with six *Xoo* strains, CⅡ (HB17), CⅣ (Zhe173), CⅤ (GD1358), GⅣ, GⅤ, and P6 (PXO99), the result showed that most of the sterile lines were resistant to CⅣ only, and susceptible to the others (Zhang et al., 2007).

Some combinations exhibiting high yield, good quality and multiple resistances to bacterial blight, rice blast or rice bacterial leaf streak, or brown planthopper had passed the national certification requirements after the year 2000. For example, the combinations of Peizashanqing, Peizamaoxuan, Peiliangyouteqing and Yunguang 8. Up to 2006, 6 out of 21 two-line hybrid rice combinations with high yield and good quality that passed the national certification requirements were resistant to both bacterial blight and rice blast, such as Liangyoupeijiu, Peizashuangqi, and 70 You 9 (japonica). Up to 2005, the accumulated acreage of Liangyoupeijiu had reached 4.681×10^6 hm^2 (NATESC, 1982 – 2006), representing 50% of the total acreage of two-line hybrid rice in China. Recently, many new two-line hybrid combinations with bacterial blight resistance, high yield, and good quality were registered at the provincial level in 2006 – 2007, such as Fengliangyou 4, Xinliangyou 6380, Peiza 35, Xinliangyou 6 and Xinliangyou 98 (Table 3), the former two were confirmed as super-rice by the Ministry of Agricultureof China in 2007 (China Rice, 2008).

Table 3 Main two-line hybrid rice combinations with bacterial blight resistance (Adopted partly from Yuan (Yuan 2001))

Hybrid rice	Type	Combination	Reaction to					Year approved	Breeder
			BB	BL	BS	FS	BPH		
Liangyoupeijiu #[ab]	indica	Pei'ai 64S/9311	MR	HR				2001	Jiangsu Academy of Agricultural Sciences
Peizashuangqi #[ab]	indica	Pei'ai 64S/Shuangqizhan	MR	HR				2001	Guangdong Academy of Agricultural Sciences
Peizamaosan #[ab]	indica	Pei'ai 64S/Maosan	MR	R				2001	Maoming Hybrid Rice Research Center, Guangdong Province
Luliangyou 28 #[a]	indica	Lu 18S/Hua 28	MR	MR				2005	Yahua Seed Research Institute
Yangliangyou 6 #[ab]	indica	Guangzhan 63 – 4S/ Yangdao 6	MR	MR				2005	Lixiahe Agricultural Research Institute, Jiangsu Province
Wandao 24#[ab]	japonica	7001S/Wanhui 9	MR	HR				2001	Anhui Academy of Agricultural Sciences
Peiliangyouteqing [a]	indica	Pei'ai 64S/Teqing	MR	MR				1994	Hunan Hybrid Rice Research Center
Ejingza 1[ab]	japonica	N5088 S/R187	MR	MR				1995	Hubei Academy of Agricultural Sciences
Yunguang 8	japonica	N5088 S/Yunhui 11	MR	MR				2000	Yunnan Academy of Agricultural Sciences
Peizashanqing[a]	indica	Pei'ai 64S/Shanqing 11	MR	MR	MR			1997	Maoming Hybrid Rice Research Center, Guangdong Province
Peiliangyouyuhong [b]	indica	Pei'ai 64S/Yuhong 1	MR	MR		MR		1997	Hunan Agricultural University
Peiza 67[ab]	indica	Pei'ai 64S/G67	MR	MR				2000	South China Agricultural University
Tianliangyou 402[a]	indica	Tianfeng S/R402	HR	MR				1998	Gangzhou Agricultural Research Institute, Jiangxi Province
Peiliangyou 275	indica	Pei'ai 64S/275	R	MR				1999	Guangxi Academy of Agricultural Sciences
Peizamaoxuan [b]	indica	Pei'ai 64S/Maoxuan	R	R				2000	Maoming Hybrid Rice Research Center, Guangdong Province

(continued)

Hybrid rice	Type	Combination	Reaction to					Year approved	Breeder
			BB	BL	BS	FS	BPH		
Ganya 1	indica	Pei'ai 64S/HB-01	MR	MR			MR	2002	Jiangxi Binghu Agricultural Research Institute
Wandao 129 [b]	indica	Pei'ai 64S/Hong 98	MR	HR				2004	Xuancheng Agricultural Research Institute, Anhui Province
Wandao 119 [b]	indica	Xuan 69S/WH26	MR					2004	Anqing Agricultural Research Institute, Anhui Province
Liangyoupeijing	japonica	Pei'ai 64S/94205	MR	R				2005	Xinyang Agricultural Research Institute, Hena Province
Liangyou 108[b]	indica	Pei'ai 64S/Ninghui 108	MR	MR				2005	Jiangsu Academy of Agricultural Sciences
Wandao 147 [ab]	indica	Xin'an S/Anxuan 6	R	MR				2005	Quanyin Agricultural Research Institute, Anhui Province
Fengliangyou 4 #[ab]	indica	Feng 39S/Yandao 4	HR					2006	Hefei Fule Seed Company
Xinliangyou 6 #[ab]	indica	Xin'an S/Anxuan 6	HR					2006	Anhui QYHF Seed Company
Xinliangyou 6 380 #[ab]	indica	03S/D208	R	R			R	2006	Nanjing Agricultural University
Peiza 35 #[ab]	indica	Pei'ai 64S/Tehuazhan 35	MR	MR				2006	South China Agricultural University
Xinliangyou 98 #[ab]	indica	Xin'an S/Hong 98	MR	MR				2007	Anqing Agricultural Research Institute, Anhui Province

Registered at national level, the others at provincial level; [a]High yield; [b]Good quality.

BB, Bacterial blight; BL, Rice blast; BS, Bacterial streak; FS, Rice false smut; BPH, Brown planthopper. R, Resistant; MR, Moderately resistant; HR, Highly resistant.

Rice breeders succeeded in achieving this objective as proven by the low incidence of bacterial blight throughout China for about two decades, and hybrid rice combinations grown by farmers possessed various degrees of bacterial blight resistance. But in recent years, the disease has gradually emerged prevalently in fields planted with some rice cultivars and hybrid rice combinations in the Changjiang River valley (Fig. 1). The resurgence of the disease after 20 years of 'quietness' indicated the potential risk in China.

Fig. 1 Rice cultivar infected by bacterial blight in Donghai Anfeng, Jiangsu province (Xu, 2005)

Discussion on the actual and potential damage situation

Reduced genetic variability of rice for resistance to bacterial blight

A smart use of the resistance genes should be based on the information of host-pathogen interactions, which has been prepared by research on the bacterial blight of rice in past 20 years. According to the studies on the pedigrees, inheritance and gene deduction of bacterial blight resistance in 287 resistant cultivars, the gene *Xa4* has been widely adopted in the development of both indica hybrid and conventional cultivars in China, it became the only source of bacterial blight resistance (Fig. 2) (Zhang et al., 1986, 1988, 1990, 1994; Xie et al., 1990; Shi et al., 1993; Sha et al., 1994; Yang et al., 1994). Such genetic uniformity has made the rice crop vulnerable to disease epidemics that cause great damage. It is well known that the frequency of a reemerging disease and virulent the emergence of pathogen population depend on the inherent ability of the pathogen to be adapted by the hosts. Introduction of a single resistance gene into cultivars often results in the shifts of virulence in *Xoo* population. Xu et al (1981) first reported that a new *Xoo* strain (Chinese pathotypeⅤ, CⅤ) was virulent to IR26 (carrying *Xa4*). It warned that the host (carrying *Xa4*) has been cultivated extensively at least for 20 years. The rapid increase in genetic diversity of the CⅤ pathotype indicated the breakdown of host resistance gene of *Xa4* (Zhang et al., 1997). Now, the pathogen has been extending from Guangdong to Fujian, Guangxi, Zhejiang, Jiangsu, Yunnan, and Jilin provinces (Xu et al., 1994). The rice cultivation area has been reduced for some combinations, for example, the area of the popular combination Shanyou 6 was declined from 3.333×10^6 hm^2 in middle of 1980s to 2.947×10^5 hm^2 in 2004 (Zhang et al., 2007).

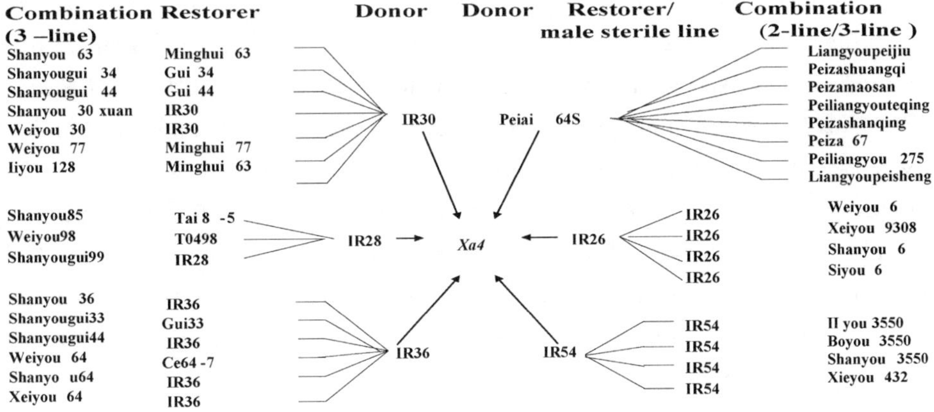

Fig. 2 Sources of BB resistance for main hybrid rice (all depend on *Xa4*)

Many cultivars with *Xa4* gene have been cultivated widely in the Philippines and in Asia, the large-scale and long-term cultivation of cultivars carrying *Xa4* has resulted in significant shifts in the race frequency of *Xoo*, also in many areas of India and Indonesia, rice cultivars with only *Xa4* have become susceptible to bacterial blight (Mew et al., 1985, 1987, 1992).

The *Xa21* gene, originally reported to be a broad spectrum resistance gene to the South Asian *Xoo* strains, has been susceptible to some strains in India, Korea, Nepal and China (Lee et al., 1999; Adhikari et al., 1999; Marella et al., 2001; Muralidharan et al., 2004; Zeng et al., 2002; Ji et al., 2003). This information suggest sthat we need to look into the pathogenic variability not only nationwide but also worldwide.

Population genetic diversity and pathogenicity of *Xoo*

Information of pathogenic variability is essential to formulate the strategy for disease management based on the resistance deployment. Plant pathogens have the intrinsic mechanism to adjust to new cultivars and to new crop production environments. Due to selection pressure, new strains of disease pathogens overcome the resistance developed, dominating the disease pathogen population and shortening the resistance life of these cultivars. This 'boom and bust cycle' continues (Mew et al., 2004). Therefore, it is necessary to continue monitoring the pathogen population in response to the newly released cultivars in order to be 'a step ahead' of the disease problem.

Zhang et al. (1997) and Wang et al. (2001) analyzed the population genetic structure of *Xoo* strains collected from 1984 to 1999 in China. The results showed that a simple population genetic structure of the pathogen presented in China. Most of the strains were grouped in clusters1 and 2, only strains from Guangdong (pathotype CⅤ) were grouped in the outgroup (Fig. 3), which was due to the host rice plants carrying a single resistance gene *Xa4* in China as mentioned above; besides, the molecular type of the pathotype Ⅴ and Yunnan strains were distinct from the other collections. As expected, the *Xoo* population of Yunnan province contained several new pathotypes which differed from the 7 national pathotypes in China (Fang et al., 1990), 9 and 8 out of the 14 *Xoo* races are virulent to cultivars carrying *Xa4* and *Xa21* genes, respectively (Ji *et al.*, 2003). Zeng et al. (2005) also found another new *Xoo* strain which has virulence to *Xa4*, *Xa21*, *Xa1*, *Xa2*, *Xa3*, *xa8*, *Xa10*, *Xa11*, *xa13*, and *Xa14*, only *Xa23* is resistant and *xa5* and *Xa7* are moderately resistant to the strain in Guangdong province, China.

In China, since 1980s, *Xoo* pathotype distribution in the major rice cultivation areas have been stable: the pathotypes Ⅱ and Ⅰ dominate in japonica rice in Northern China; the pathotypes Ⅳ and Ⅱ are major *Xoo* found in the Changjiang River valley. It is important to note that the spreading threshold value of the pathotype CⅤ (infecting *Xa4*) is alarming and now approaching 20% (Xu et al., 2004).

Distribution of *Xoo* and the available resistance genes in Asian countries

In the Philippines, 10 races have been identified, the *Xa7*, *xa5* and *Xa21* were effective to most of the races (Vera Cruz, personal communication). In Japan, the predominant races Ⅰ (59.5%) and Ⅱ (33.5%) distribute in all of the locations in the country, the effective resistance genes are *Xa3*, *Xa4* and *Xa12* (Nota et al., 1990).

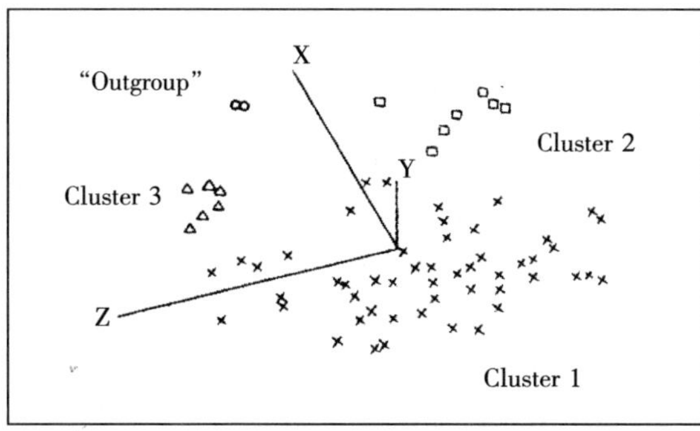

Fig. 3 Clustering analysis of Chinese 143 *X. o.* pv. *oryzae* strains as derived from DNA band data obtained with rep-PCR (ERIC, BOX IS1112 and IS1113 primers) and RFLP (pBSavrXa10 probe). The number of clusters which best fit the data was determined by the consensus of three clustering statistics. Positions of the strain on the three dimensional graph were determined by multiple correspondence analysis (Zhang et al., 1997).

Adhikari et al. (1999) reported that five virulence groups are classified in Nepal. The virulence of *Xoo* is very strong, all *Xoo* isolate collections are virulent to genes *Xa4*, *xa5*, *Xa7*, *xa8*, *Xa14* and *Xa21*. Only the pyramided lines possessing (*Xa4*+*xa5*), (*Xa4*+*xa5*+*xa13*), (*Xa4*+*xa5*+*Xa21*) and (*Xa4*+*xa5*+*xa13*+*Xa21*) are resistant to 2, 3, 3, and 5 virulence groups, respectively. Noh et al. (2004) reported that five races of *Xoo* were classified in Korea in 1980s. A new race, K3a was found to be virulent to cultivars carrying *Xa3* gene. The distribution frequency of K3a has been increasing in whole South Korea since 2003. Besides, the cultivars carrying *Xa21* gene were susceptible to 15% of *Xoo* strain collections (1987－1989) a few years after cultivation, and 96% of strain collections were virulent to *Xa21* gene in middle of 1990s in Korea (Lee et al., 1999).

Muralidharan et al. (2004) reported the field performance of genotypes to bacterial blight in 11 locations in India. All the lines with single R gene *xa5* or *xa13* recorded resistant reaction in 5－7 locations, those lines with *Xa21* were found resistant at only 3－5 locations, the pyramided lines possessing (*xa5*+*xa13*), (*xa13*+*Xa21*), (*xa5*+*Xa21*) showed resistance at more sites, the genotypes possessing (*xa5*+*xa13*+*Xa21*) showed resistance at 5－8 sites. Noda et al. (1996) examined the variation in the pathogenicity of *Xoo* from 9 countries in South and East Asia by testing on the Japan and Philippine differentials. The data were resorted based on the results in Table 4.

Table 4 Main ineffective and effective resistance genes to bacterial blight in different countries (Zhang et al., 2007; Noda et al., 1996)

Country	Race	No. of strains tested	Main ineffective resistance gene	Effective resistance gene
Bangladesh	12	74	*Xa1*, *Xa2*, *Xa3*, *Xa4*, *Xa7*, *Xa10*, *Xa11*, *xa5*, *xa8*	*Xa3*, *Xa12*, Pyramided line
India	5	58	*Xa1*, *Xa2*, *Xa3*, *Xa4*, *Xa10*, *Xa11*, *Xa12*, *xa5*	Pyramided line

(continued)

Country	Race	No. of strains tested	Main ineffective resistance gene	Effective resistance gene
Nepal	16	45	*Xa1*, *Xa2*, *Xa3*, *Xa4*, *Xa10*, *Xa11*, *Xa12*, *xa5*	*xa5*, *Xa7*, Pyramided line
Myanmar	7	27	*Xa1*, *Xa2*, *Xa4*, *Xa10*, *Xa11*, *xa5*	
Thailand	5	35	*Xa1*, *Xa2*, *Xa10*, *Xa11*, *xa5*	
Indonesia	8	78	*Xa1*, *Xa2*, *Xa3*, *Xa10*, *Xa11*, *xa8*	*xa5*, *Xa7*, Pyramided line
Philippines	5	61	*Xa1*, *XaXa2*, *Xa4*, and *Xa11*	*xa5*, *Xa7*, *Xa12*, Pyramided line
Malaysia	2	11	*Xa1*, *Xa2*, *Xa10*, *Xa11*, *xa8*	*Xa4*, *Xa7*, *Xa12*, *xa5*
South China	27	24	*Xa1*, *Xa2*, *Xa3*, *Xa4*, *Xa7*, *Xa10*, *Xa11*, *Xa12*, *xa8*	*xa5*, *Xa7*, *Xa22*, *Xa23*

Approaches to improve bacterial blight resistance in breeding hybrid rice

Broaden genetic basis for bacterial blight resistance

Actually, since early 1980s, rice breeders have been making efforts to use more genes to broaden the genetic basis for bacterial blight resistance. Zhu et al. (1990) developed 'TD' series rice lines carrying resistance gene *Xa7* derived from TN1/DV85. The resistant restorer lines, Kanghui 63, Kanghui 98 and D205 were developed using TD as female and Minghui 63 as male parents; the hybrid rice combinations, Kangyou 63, Kangyou 98, Ⅱ You 205 and Fengyou 205 exhibiting resistance to both bacterial blight and rice blast were developed using these restorer lines with the sterile lines Zhenshan97A or Ⅱ-32 A. These combinations were popularized extensively in bacterial blight serious fields of Jiangsu, Zhejiang, Anhui, Henan, Yunnan, Hubei, Hunan, Jiangxi and Fujian provinces of China and exhibited bacterial blight resistance, high yield and wide adaptability. To date, the accumulated acreage of those combinations has reached 2.50×10^6 hm^2 (Ding, 2005).

Up to 2006, a total of 30 bacterial blight resistance genes have been identified. Twenty-one and nine out of the 30 bacterial blight genes are dominant and recessive, respectively (Zhang, 2005). Eighteen genes have been mapped by molecular markers. To facilitate reliable marker-assisted selection and application, tightly linked PCR-based markers have been identified for those common resistance genes such as *Xa4*, *Xa7*, *Xa21*, *Xa23*, *xa5*, and *xa13*. Six bacterial blight genes have been cloned, including *Xa1*, *xa5*, *Xa21*, *xa13*, *Xa26(t)* (from Minghui 63) and *Xa27(t)* (from *Oryza minuta*). Since the cloning of the dominant gene *Xa21* (Song et al., 1995), there has been a major progress in applying marker-assisted back-crossing in the crop improvement programs on bacterial blight resistance for both conventional and hybrid rice (Table 5) (Xu, 2007).

Among all designated dominant bacterial blight genes, *Xa23* (*Oryza rufipogon*) (Zhang et al., 1998), *Xa22* (Lin et al., 1996), *Xa21* (Song et al., 1995) and *Xa7* (Sidhu et al., 1978) exhibit the broadest resistance spectrum and the strongest resistance,

Xa3 and *Xa4* show broad resistance spectrum against China bacterial blight pathotypes, but cannot be used in regions where the prevailing pathotypes are Ⅳ and Ⅴ.

Table 5 Three-line indica hybrid rice with bacterial blight resistance gene *Xa21* bred by molecular marker assisted selection (Xu JL, 2007)

Resistantcombination or restorer line[a]	Restorer to be improved	Donor (gene)	Molecular marker	Breeder	Year approved
Xieyou 218	Fuhui 838	IRBB21 (*Xa21*)	pTA248	China National Rice Research Institute	2002
Zhongyou 218	Zhonghui 218	IRBB21 (*Xa21*)	pTA248	China National Rice Research Institute	2003
Guodao 1 [a]	R8006	IRBB60 (*Xa21*)	pTA248	China National Rice Research Institute	2004
Guodao 3	R8006	IRBB60 (*Xa21*)	pTA248	China National Rice Research Institute	2004
Zhongyou 1176	R1176	IRBB60 (*Xa21*)	pTA248	China National Rice Research Institute	2005
Shuhui 527 improved	Shuhui 527	IRBB60 (*Xa21*)	pTA248, MP12	Sichuan Agricultural University	
Minghui 63 (carrying *Xa21*)	Minghui 63	IRBB21 (*Xa21*)	pTA 248, PTA 21	Huazhong Agricultural University	
93-11 (*Xa21*)	93-11	Minghui 63 (*Bt*/*Xa21*)	pTA248	Huazhong Agricultural University	
93-11 (*Bt*/*Xa21*)	93-11	Minghui 63 (*Bt*/*Xa21*)	pTA 248	Huazhong Agricultural University	
Ⅱ You 8220	Milyang 46/Minghui 63	IRBB21 (*Xa21*)	pTA 248	Zhejiang University	2003
T71, T81	Minghui 63	IRBB21 (*Xa21*)	pTA248	Zhejiang University	

[a] Registered at national level, and the others at provincial level.

Combined strategy for improving bacterial blight resistance

Currently, we must notice that the number of new resistant hybrid combinations released has gradually reduced since later 1990s (Table 6). Were the combinations not adequately screened prior to their release? The resistance breeding must still be one of the main objectives of rice improvement. Thus, we are faced with two key challenges: a) how to deal with the rapidly changing pathogens, b) how to design and develop durable resistant combinations.

Table 6 Bacterial blight resistant hybrid rice combinations developed during 1973-2006 [a]

Period (Year)	Total no. of combinations	No. of resistant combinations	Percentage of resistant combinations (%)
3-line hybrid rice			
1973-1986	87	31	35.6 (indica)
	43	26	60.0 (japonica)
1996-2005	169	18	10.7 (indica)
2005	34	12	35.3
2006	6	4	66.7 (South, japonica)
	64	5	7.8 (North, japonica)
2-line hybrid rice			
1994-2000	32	17	53.1
2001-2006	21	6	28.6

[a] Combinations with an annual acreage of more than 6.667 hm^2

A combined strategy composed of traditional method, molecular marker-assisted selection, and transgenic technology should bring a new era to the bacterial blight resistant hybrid rice breeding program. For prolongation of lifespan of major resistance genes, it is necessary to develop pyramided lines carrying multiple resistance genes for broadening of bacterial blight resistance, suitably utilize broad-spectrum resistance genes, such as *Xa23* (from *O. rufipogon*), *Xa22* (from Yunnan landrace, Zhachanglong), *Xa7* and *Xa21*, refine strategies for alternation planting of these available resistance genes, and adoption of 'horizontal resistance'. These measures of a truism will bring to success if seriously carried out.

Presentation paper for 5th International Hybrid Rice Symposium 2008, September, 11－15. Changsha, Hunan.

原载：水稻科学，2009，16（2）：83～92.

Types of Resistance in Rice to Bacterial Blight

Zhang Qi[1] T. W. Mew[2]

(1. CAAS; 2. IRRI)

Resistance to bacterial blight has been identified in many rice cultivars. Several types of resistance comparable to those against fungal pathogens have been recognized. We studied seedling and adult plant resistance to clarify the types of resistance. IR944 and IR1695 resistance to Philippine 4 races of *Xanthomonas campestris* pv. *oryzae* appeared at the 6th to 10th leaf position. Zenith and Malagkit Sungsong showed resistance at one or two leaf positions earlier than their progenies. IR1545 was differentially resistant to the Philippine races 1 – 3 at all growth stages, where as TN1 was susceptible. Expressed as the plants matured, the resistance of Zenith and Malagkit Sungsong was not affected by temperature change, but the lesions of IR944 and IR1695 were longer at higher temperature. A strong differential interaction was noted on selected japonica cultivars and lines from China. The differential resistance appears moderate at maximum tillering and increases at maturity; it may not be fully expressed at booting. However, seedling resistance to incompatible races was stable.

Bacterial blight (BB) of rice caused by *Xanthomonas campestris* pv. *oryzae* (*Xco*) has been recognized as an important rice disease since the introduction of improved but susceptible cultivars. These modern cultivars respond to fertilizer with luxurious growth and thus have modified the rice microenvironment to make it highly conducive to disease development. Until cultivars with genetic resistance to the disease were bred and introduced, BB posed a serious threat to rice production in the major rice-growing countries of Asia.

Many attempts have been made to control the disease, including chemical treatment and modified cultural practices, chiefly reducing the rate of N fertilizer; but the impact of such management tactics has been limited in scale and locality. Host plant resistance has proven more effective and economic. Since the release of IR20, the first modern rice cultivar in tropical Asia with genetic resistance to BB, breeding for BB resistance has become an important component of varietal improvement programs. Many cultivars and breeding lines have been developed with certain levels of resistance, but have later been discovered to be resistant only to specific isolates in one locality. Cultivar Asakaze in Japan is a good example (Ezuka and Sakaguchi, 1978). IR20, after its release in the Philippines in early 1970, was found susceptible in Isabela province (Ou, 1985). These setbacks suggest not only the existence of different pathogen races in different regions, but also that the types of resistance to BB in rice must be assessed.

The subject has been reviewed recently (Mew, 1987); in this paper we provide additional information on the topic.

General concept of host resistance

According to Van der Plank (1968), host plant resistance can be classified into vertical resistance (VR) and horizontal resistance (HR), considering both genetic and epidemiological concepts. VR is monogenic and is effective against specific races. It delays the disease by reducing the amount of effective inoculum. HR is polygenic and is effective against the most of pathogen races. When a disease has started, HR lowers its severity by reducing the apparent infection rate. HR can be studied either by an analysis of variance or by ranking the isolates in order of virulence. It is believed that a reduction of the apparent infection rate is associated with the lasting resistance of a cultivar, and that it can be measured quantitatively.

Parlevliet (1979) assessed the resistance of barley to leaf rust caused by *Puccinia hordei* Orth. at different crop growth stages. He noted several types of resistance:

- Seedling resistance is against low-infection types (0 – 2) at all growth stages.
- Intermediate resistance is against infection types 2 – 3 at all growth stages.
- Adult plant resistance is characterized by susceptible infection types (4^{-} – 4^{+}) at the seedling stage and a resistant reaction at the adult plant stage (0 – 2).
- Partial resistance shows a susceptible reaction at all growth stages (4 in seedlings, 3 – 4 in adult plants).

On analyzing the resistance of cabbage to the yellows caused by *Fusarium oxysporum* f. sp. *conglutinans* and of potato to late blight caused by *Phytophthora infestans*, Fry (1982) established that different forms of resistance have different characteristics. The magnitude of the effect may be very high, moderate, or relatively low. Its inheritance may be conditioned by one or several genes. It may be effective against some but not all pathogen isolates; that is, it may be differential. If there are at least two possibilities for each of the three characteristics, there are eight categories of resistance.

To assess host plant resistance, the growth and development of the pathogen should be measured. Equally important is the stage of host plant development. Response to infection varies with the age of the host. In fungal diseases, we can measure the growth and development of the pathogen, e. g. , its sporulation, latent period, and infection frequency。

Rsistance to bacterial blight

Resistance of plants to bacteria shares some similar characteristics with resistance to fungal pathogens. Ercolani (1972) proposed to measure the generation time in host tissues as an effective means of measuring resistance to bacterial diseases; this may be accurate, but it is tedious. Mew et al. (1981) used incubation period as a component for varietal resistance to BB in rice; this appears feasible, but it is highly influenced by the inoculum dose deposited on the infection court. The initial inoculum dose is difficult to control, and it affects the incubation period, especially for moderate resistance levels. In dicots, lesion type has been used to indicate resistance type. On BB of cotton caused by *X. campestris* pv. *malvacearum*, two distinct types of lesion are observed: watersoaked and dry (Brinkerhoff, 1963).

Bacterial blight resistance has been detected in many rice cultivars. In some cases, it is conditioned by a single gene, in other cases by many genes (Mew, 1987; Ogawa et al. , 1989, this volume). In an inventory of rice germplasm for BB resistance, Mew et al. (1981) recognized several types of resistance comparable to some of those reported by Parlevliet (1976) for barley leaf rust; this has been reviewed recently (Mew, 1987). To further clarify the types of resistance relative to adult plant resistance and seedling resistance, the following experiments were done.

When 6 rice cultivars were inoculated on the 6th leaf to the flag leaf of the same plant, TN1 was susceptible to the 4 Philippine pathogen races at all leaf positions, and IR1545 was differentially resistant to races 1 - 3 and susceptible to race 4 at all growth stages. IR944 and IR1695 were susceptible to the 4 races at the 6th to the 10th leaf positions and became resistant from either the 11th or 12th leaf to the flag leaf. Zenith and Malagkit Sungsong reacted similarly, but resistance appeared one or two leaf positions earlier than in the two progeny lines (Table 1). The lesions on leaf positions other than the flag leaf expanded with time (Fig. 1). Both experiments showed the same results.

Table 1 Disease reactions of rice leaves at different leaf positions in the main culm to 4 races of *Xanthomonas campestris* pv. *oryzae*, PXO61, PXO86, PXO71 & PXO79

Leaf position	TN1	IR1545	IR1695	IR944	TN1	IR1545	IR1695	IR944
		* PXO61				* PXO71		
6	8.5 a	3.0 a	7.0 a	5.7ab	8.5a	7.0a	7.0a	6.0b
7	7.0b	3.0a	6.0a	6.0a	7.5b	6.5b	7.0a	6.5a

(continued)

Leaf position	TN1	IR1545	IR1695	IR944	TN1	IR1545	IR1695	IR944
8	7.0b	3.0a	7.0a	5.0b	7.5b	6.5b	7.0a	5.0c
9	7.0b	2.5b	7.0a	5.5ab	7.0b	5.0d	5.0b	5.0c
10	7.0b	3.0a	5.0b	5.0b	5.0c	7.0a	5.0b	5.0c
11	7.0b	1.5c	5.0b	5.0b	7.0b	5.0d	5.5b	4.5d
12	7.0b	1.0d	3.0c	1.5f	6.5b	5.0d	1.0d	1.0f
13	6.5c	1.0d	1.5d	2.0e	6.0b	5.0d	1.0d	1.0f
14		1.0d	1.0d	2.5d		6.0c	1.0d	1.0f
15			1.5d	4.0c			2.0c	2.0e
		* PXO86				* PXO79		
6	7.0a	3.0a	6.5a	6.5a	7.0b	2.0b		
7	6.0b	3.0a	5.5b	6.0b	8.5a	2.5a		
8	5.0c	3.0a	6.5a	5.0c	7.5b	2.0b		
9	5.0c	2.5b	5.5b	5.0c	7.0b	1.5		
10	5.0c	3.0a	5.0b	5.0c	7.0b	2.0b		
11	5.0c	1.8c	5.0b	5.0c	7.0b	1.8bc		
12	5.0c	1.0d	1.0c	1.0e	5.0c	1.0d	1.0d	1.5f
13	5.0c	1.0d	1.0c	1.0e	6.5c	1.0d	1.0d	1.0g
14		1.0d	1.0c	1.0e		1.0d	1.0d	1.0g
15			1.0c	2.0d		1.0d	1.0d	3.0e

* Disease scale: 1 = <1% leaf area infected, 9 = >75%. In a column, figures followed by a common letter are not significantly different at the 5% level by Duncan's multiple range test.

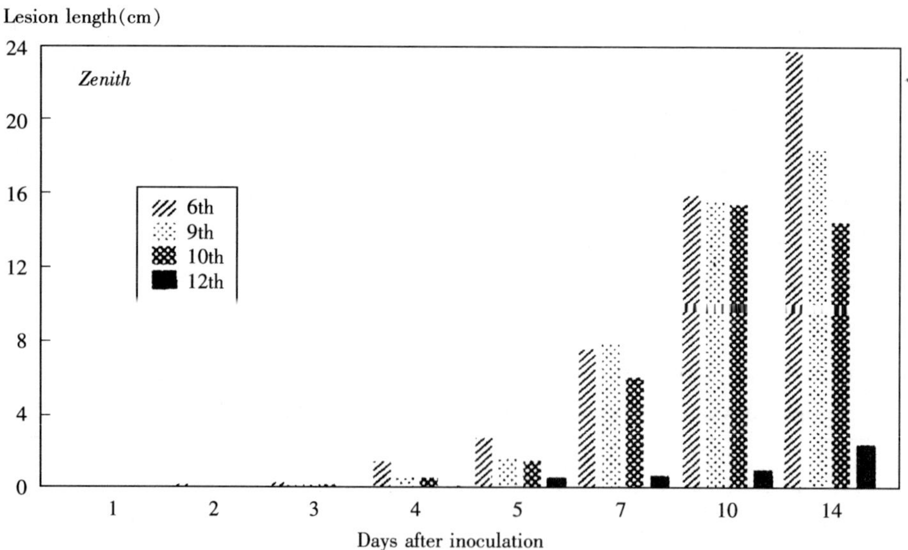

Fig. 1 Lesion development of bacterial blight at different leaf positions in cultivar Zenith in the greenhouse.

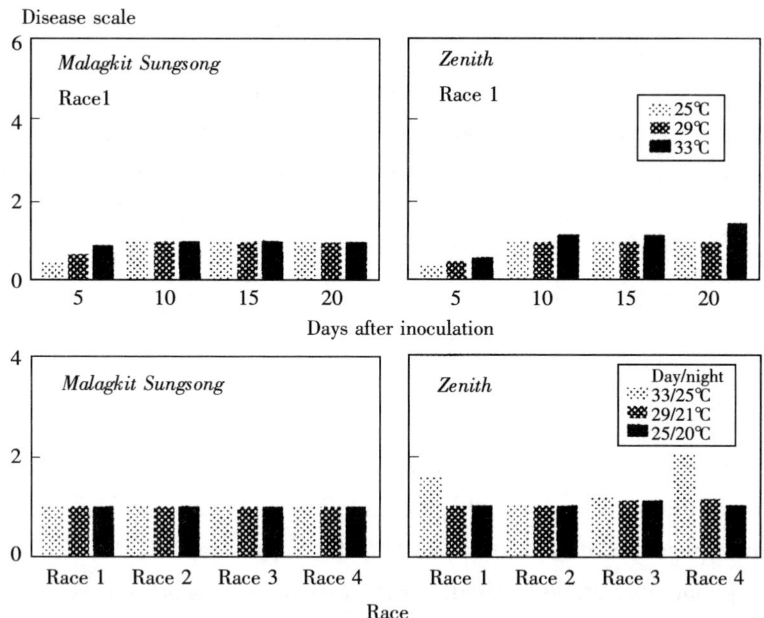

Fig. 2 Effect of temperature on disease reaction caused by 4 races of *Xoo* on Malagkit Sungsong and Zenith.

When the resistance of Malagkit Sungsong, Zenith, IR944, and IR1695 was evaluated against the 4 pathogen races at 3 temperature ranges (33/25, 29/21 and 25/20℃) in a growth chamber, the disease reaction of Malagkit Sungsong and Zenith was not affected by temperature change (Fig. 2). The adult plant resistance of these cultivars was expressed as the plants matured. The response to infection when the susceptible reaction was expressed according to leaf position, however, was not consistent in different environments, i. e. , greenhouse vs growth chamber. The resistance of Malagkit Sungsong and Zenith was demonstrated well at booting with fully expanded flag leaf. BB lesions on the two cultivars developed very slowly gradually became yellowish brown, and then ceased to progress. At low temperatures (25/20℃), these cultivars showed resistance several days earlier than that at 29/21℃.

The adult plant resistance of Malagkit Sungsong and Zenith was also fully expressed on lines derived from them-IR944 and IR1695, respectively. The lesions, however, appeared to be longer than those on their parents at higher temperature. At lower leaf positions at booting, the difference was not obvious.

Adult plant resistance can be differential (Zhang and Mew, 1985), but differential resistance at maximum tillering may not be fully expressed at booting if seedling resistance is moderate. Such resistance may be differential, yet it may not be expressed in later plant development.

Under the same temperature range in response to incompatible races, seedling resistance was stable. Lesion development was not altered by any temperature tested, but in compatible races, lesion development was faster at higher (33/25℃) than at lower (25/20℃) temperatures (Fig. 3).

The growth of *Xco* in leaf tissues was also assessed. Bacterial number was estimated by direct plating of leaf extracts on peptone sucrose agar (PSA) plates and by bacteriophage method. An aliquot (0. 1ml) of the leaf extract was uniformly spread on the surface of a petri dish containing 20ml of PSA medium. The rest was mixed with a standard phage suspension obtained by culturing P80 in PXO61 suspension for 20 – 24 h. Ten minutes after incubation at 28℃ in aeration, the mixture was centrifuged to remove excess phage. The procedure was repeated once. The precipitate was then resuspended and incubated for 20 min to estimate phage adsorption by the bacterial cells. Final phage plating was 10h later by serial dilution with PXO61 as the indicator strain. The phage count, as an indication of bacterial number, was expressed as plaque-forming units (pfu)/cm^2 of leaf area. There was a good correlation (Fig. 4) between bacterial count by dilution plating and phage count for the bacteria in the leaf extract. Based on these results, multiplication of *Xco* in leaves 6th, 9th, 10th, and $12^{th}$of Zenith was established (Fig. 5). When phage was added to the leaf maceration, the phage counts were all increased at 10h compared to 20 min after incubation, which indicated the presence of numerous bacterial cells in the leaf extracts. Similarly, the increase in plaque counts from leaf tissues 24h after inoculation indicated rapid bacterial multiplication at all 4 leaf positions, regardless of their BB resistance. Based on this indirect method of bacterial enumeration, the bacteria seem to have increased rapidly, achieved a peak population at 72 h, and then slightly declined in all 4 leaves. At 6d and 7d (120 and 158 h), the sample leaf specimens were totally infected, yet the bacterial numbers in the leaf tissues were even lower than those at 2 or 3d (24 – 72 h) after inoculation. Lesion development or expansion was significantly different on leaves 6th, 9th, and 10th from that on leaf 12th (Fig. 1), yet bacterial number and rate of multiplication followed the same trend; but the lag phase of bacterial multiplication in leaf 12th was longer, and the number was less and declined faster than in the other 3 leaves. The resistance of leaf $12^{th}$is likely controlled by mechanisms that have no effect on bacterial multiplication, e. g. , inhibition of pathogen growth.

Although the phage method provides an indirect method for counting bacteria, the number of bacterial cells can be estimated with reasonable confidence from the l-step growth curve of the phage, with 4 – 6 phages per bacterial cell. The advantage of the phage method in such a study with *Xco* is its rapid assay and minimal contamination. *Xco* is a slow-growing bacterium, and no selective medium is available to reduce contamination, especially to assay bacterial multiplication in a host plant that is grown in humid environments.

Finally, 20 selected japonica cultivars from China were tested at both maximum tillering and booting stages against Chinese isolates differing in virulence. Planting was staggered to synchronize growth stage and inoculation. Differential responses of the cultivars were recorded (Table 2), indicating strong interaction at maximum tillering. The difference in lesion length was rather small. The differential interaction was not shown at booting (Table 3). It appears

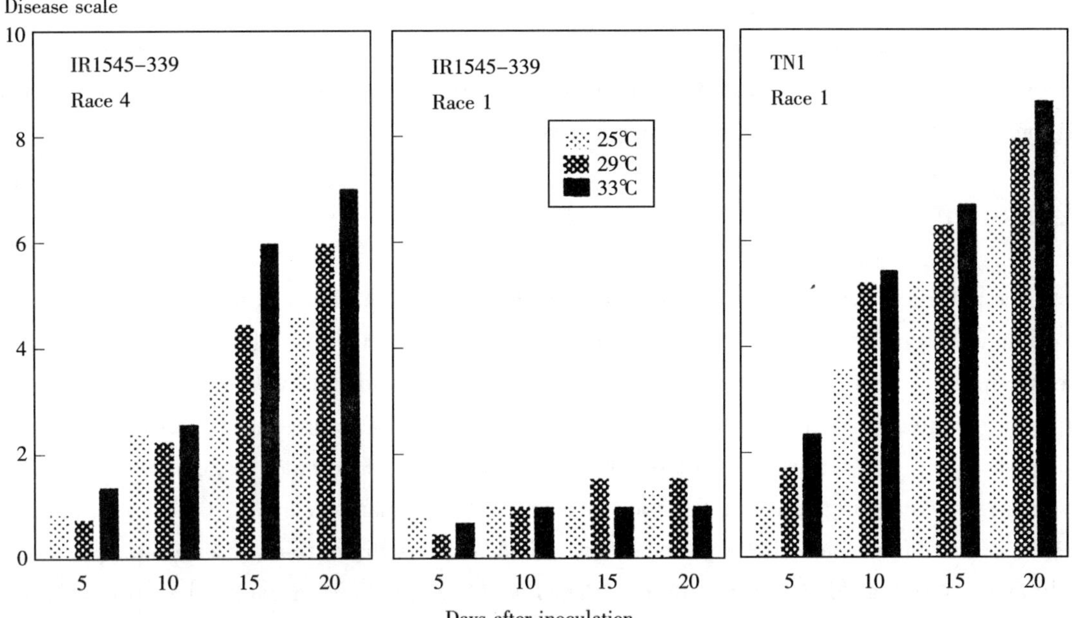

Fig. 3 Effect of temperature on lesion development inoculated by races 1 and 4 of *Xoo* on IR1545 - 339 and TN1 at booting stage.

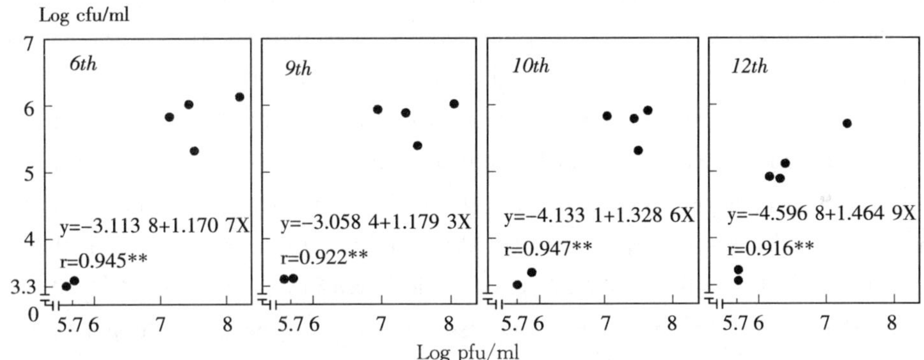

Fig. 4 Relationship between plaque-forming units (pfu) /cm^2 and colony-forming units (pfu) / cm^2 in leaf samples with bacterial blight lesions at different positions in cultivar Zenith

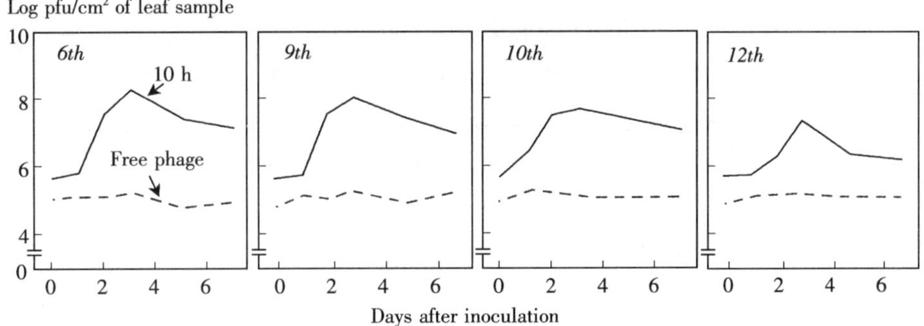

Fig. 5 Population trends of *Xoo* in the 6th, 9th, 10th and 12th leaves of Zenith at different days after inoculation

that the differential response is no longer significant in some cultivars that show strong interactions at maximum tillering with some isolates, and the resistance increases as the plants approach maturity. In this group of cultivars, resistance appears moderate at maximum tillering and increases to a high level of resistance at maturity.

Discussion

Rice pathologists have just begun to understand the types of resistance in rice to *Xco*.

The issues of quantitative resistance and HR in relation to host plant development and to disease measurement scales and indices have been reviewed recently (Mew, 1987); in response to infection, the resistance of rice plants may be expressed as different levels of resistance, or the resistance may be expressed at different stages of host plant development. Analyzing some of the data presented by Yamamoto et al. (1977) and Horino (1985), both Parlevliet (1981) and Mew (1987) recognized that differential resistance is likely at the moderate level of resistance. Baw and Mew (1988) measured lesion development on rice cultivars at different growth stages with a leaf area meter. Neglect of either category will not facilitate a good understanding of host-pathogen interactions. This applies to genetic analysis of resistance, and to the use of host plant resistance in disease management.

In response to infection by a single pathogen race, rice cultivars were classified into three or four groups. In general, susceptible cultivars such as TN1 and IR8 showed very severe symptoms at 20 d after sowing (DAS), and kresek often developed. But from 30 to 50 DAS, the lesion areas gradually decreased and became constant from 50 to 80 DAS. In cultivars with adult plant resistance, lesion areas decreased from 20 to 40 DAS, but the reaction remained susceptible. Lesions decreased abruptly at 50 DAS, and became less than 5% of the leaf area at 80 DAS. In plants with seedling resistance, the lesion area never exceeded 10% of the leaf area even at 20 DAS; differential resistance was also evident. However, seedling resistance varied from the 3-to the 6-leaf stage. Moderate resistance may increase as the plants mature. Hence, even though a differential response is present, the reaction may change when plants mature.

Resistance of rice to BB has many characteristics in common with resistance to other diseases. Types of resistance can be distinguished, but whether the resistance is stable or not depends very much on the host-pathogen interaction, which warrants further study.

Table 2 Strong interaction of selected Chinese cultivar X isolates at maximum tillering and booting stages in China, CAAS, 1986.

Cultivar	Lesion area (%) with isolate[a]				
	Zhe173	HN84 - 31	KS6 - 6	HB82 - 21	HB84 - 21
Yangeng 2	30	13	10	—	—
Yunyu 1	12	42	—	—	—
Chaitang	12	—	55	5	—

(continued)

Cultivar	Lesion area (%) with isolate[a]				
	Zhe173	HN84 - 31	KS6 - 6	HB82 - 21	HB84 - 21
Aoyu 24	9	40	—	—	—
Wei 84 - 894	—	—	5	28	—
H84 - 213	8[b]	—	—	—	50[b]
Zhong Xi 8409	35[b]	—	—	—	50[b]

[a] Mean of 3 replications. (—) = no interaction with any of the 5 isolates.

[b] Interaction at booting.

Table 3 Differential resistance of selected cultivars to some *Xco* isolates in China, CAAS, 1986.

Cultivar	Lesion area (%) at								
	Maximum tillering			Booting					
	LN44	LN49	KS6 - 6	HN37	NX40	JL65	HL69	BJ3	HLJ72
Tetep	—	—	—	31	14	24	10	—	—
Jingang 30	—	—	—	16	25	—	—	—	—
Chugoku 45	—	—	—	—	—	15	41	—	—
Cas 209	—	—	—	—	—	—	—	22	15
Zaiye Qing	—	—	—	—	—	—	—	2	26
Cheongcheongbyeo	60	1	—	—	—	—	—	—	—
Pungsenbyeo	—	1	35	—	—	—	—	—	—
Milyan	3	55	1	—	—	—	—	—	—

Published in: Pages 125 - 134 *in* book Bacterial blight of rice. 1989. International Rice Research Institute, P. O. Box 933, Manila, Philippines.

水稻白叶枯病概述

章　琦[1]　张红生[2]

（1. 中国农业科学院作物科学研究所；2. 南京农业大学）

An Overview of Rice Bacterial Blight

(1. Institute of Crop Sciences, CAAS; 2. Nanjing Agriculture University)

摘要　人们知道水稻白叶枯病已有 120 多年的历史，1911 年才被确认为细菌性病害。20 世纪 60 年代以前，白叶枯病仅在东亚发生，后来逐渐遍及东南亚、南亚等稻作国家。现今澳大利亚、非洲、南欧、美洲等都有此病发生的报道。

水稻白叶枯病是我国仅次于稻瘟病的重要水稻病害。20 世纪 30 年代江苏、浙江就有发生此病的报道。至今，除新疆、西藏自治区和东北北部以外，全国各稻区均有此病发生，尤以南方籼稻地区最为严重。1974—1975 年长江中游鄂中稻区连续暴发此病，受害面积达 292 260 hm^2，损失稻谷 67 700 t；80 年代，安徽省皖中稻区曾 4 次流行白叶枯病，仅 1982 年就减产稻谷 205 400 t。

水稻白叶枯病的症状有 4 种类型，最典型的是黄色波纹病斑的叶枯型和通常在苗期发生的整株枯萎的凋萎型，以及苍白黄叶型和淡黄色中脉型。本文简述了该病症状与气候条件、生育期、水稻类型、品种与病原菌互作以及田间诊断此病的主要依据。

水稻白叶枯病原菌的学名曾数次更易，目前是黄单胞菌属稻黄单胞杆菌白叶枯病致病变种（*Xanthomonas oryzae* pv. *oryzae*）。对病原细菌的形态、培养、保存以及白叶枯病噬菌体的特性与应用也作了简要介绍。

病菌的侵染循环一般包括病原菌的越冬场所、初侵染源、侵染过程和田间传播的再侵染。水稻白叶枯病的初侵染源主要有稻种、杂草、稻草、稻桩等，其传病作用的大小因地理环境、气候条件、农作制度等而异。在我国南方早、中稻或早晚稻交叉种植区域只要有足够数量的病原菌，就能通过灌溉水或病株提供再侵染源。一些野生稻种也可能是传染源之一。

气候条件、品种抗性和田间管理是造成白叶枯病流行的主要条件。高温、高湿、暴风雨、台风和寡日照是白叶枯病暴发流行的主要气候因素。最适气温是 25 ～30℃，最适相对湿度在 80%～90%以上，气温在 20℃以下或 33℃以上，相对湿度低于 80%时，病害则很少发生。

品种的抗病性对白叶枯病的发生与流行起决定性作用。科学肥水管理，适量适时施用氮肥、适度烤田，都有利于控制病害的发展。根据种子带菌、田间始病期观测、病菌小种组成、品种抗病性布局和气候条件等因素定期预测预报白叶枯病的发生和流行程度，可为生产管理部门制订防病决策提供依据。

白叶枯病发生的早晚和病害的严重度决定产量损失的大小。分蘖盛期至孕穗期发病严重影响籽粒的饱满度、千粒重和产量；叶发病面积率和发病叶片率直接影响有效穗数和每穗粒数，是调查病情程度的主要指标。

Abstract Rice bacterial blight (BB) has been known for more than 120 years and was confirmed as a bacterial disease in 1911. Prior to the 1960s, this disease was limited to East Asia but then gradually spread to South and Southeast Asia. In recent years, the disease has also been reported in Australia, Africa, South Europe and America.

In China, BB is one of the most devastating rice diseases just next to rice fungal blast. In the 1930s, BB mainly occurred in Jiangsu and Zhejiang provinces. To date, the disease has widely spread nationwide, except in the Xinjiang and Tibet Autonomous Regions and northern part of the northeast region of China. BB is most destructive in the southern part of China, the major indica rice cultivation region. For example, about 292 260 hm^2 of rice field was affected by the disease in the Ezhong rice cultivation area (middle region of Yangtse River), leading to a yield loss of 67 700 tons from 1974 to 1975. In 1982, a destructive disease outbreak caused a yield loss of 205 400 tons in the middle region of the Anhui province.

BB exhibits four types of disease symptoms. Two most typical types include (1) lesions found at the margin of leaf blades forming a wavy infection and yellowing pattern; and (2) the "kresek" symptom with the entire leaves turned grayish-green, rolled up and withered (usually occurred at the seedling stage). The other two types can be distinguished by the appearance of pale yellow leaves or light-yellow mid-rib. This paper summarized how the BB disease symptoms are affected by climatic conditions, growth stages, rice types and host-pathogen interactions. The basis for BB disease diagnosis in the field was also discussed.

The nomenclature for the BB pathogen had been revised several times and finally settled as *Xanthomonas oryzae* pv. *oryzae* (*Xoo*). The morphology, culturing methods, preservation techniques, and the characteristics and application of bacteriophage of *Xoo* were also introduced in this paper.

The disease cycle of *Xoo* generally involved over-wintering, primary infection, transmission in the fields and re-infection. The primary infection sources for the *Xoo* pathogen include stored rice seeds, weed hosts, infected strews, plant debris, and rice stubbles. Their relative importance in the infection process is dependent on the geographic environment, climatic conditions, agricultural system, and the rice cultivars. In South China, the re-infection source of BB is based on the agronomic system as well as adequate pathogen population in rice fields transferred by irrigation water or infected plants. Several wild rice species are also infection source of BB.

Climatic conditions, resistance of rice cultivars, and field management are major criteria governing BB epidemic. The most favorable conditions for BB development require a temperature range of 25－30℃ and the RH should be higher than 80％－90％. BB disease seldom occurs when the temperature is lower than 20℃ or higher than 33℃ or the RH is less than 80％.

The most critical determinant controlling BB occurrence and epidemic is the resistance of rice cultivars. Regional cultivation measures including regulation of nitrogen fertilizer applications and appropriate drainage system to reduce humidity in the fields could also help to reduce the disease development.

BB occurrence and the possible extent of epidemic spread can be forecasted and predicted by the amount of pathogens in seeds, the date for first appearance of disease symptoms, composition of pathogenic races, resistance of cultivars, and regional climatic changes. These data provides important information to agricultural management offices to design appropriate BB prevention measures accordingly.

The actual yield loss caused by BB depends on the growth stage of rice and the extent of the disease outbreak. If BB occurs at the maximum tillering or at the booting stage, the kernel plumpness, thousand-grain weight and the final yield will be severely hampered. The percentage of infected area in each leaf and the percentage of leaf infected per plant will respectively affect the number of grains per panicle and the total number of panicles developed. These are important parameters to estimate yield loss caused by BB.

水稻白叶枯病（bacterial blight，*Xanthomonas oryzae* pv. *oryzae*，简称 *Xoo*）自从1884年日本农民首次在九州岛福冈（Fukuoka）地区的稻田发现至今（Yamanuki et al.，1962）已有120年的历史。一个世纪以来，各国科学家在白叶枯病的发生与传播、病原学、生态学、病菌变异、综合治理、品种抗性及其改良等诸多方面作了大量的研究，积累了丰富的科学资料，在认识上也有很大的发展。近20年，随着分子生物学的兴起和生物技术的发展，国际上不少植物病理、遗传和分子生物学家在深入探索植物抗性机理这一极具挑战性的科学问题时，都把水稻白叶枯病作为模式寄主和病原菌来研究，并已成为世界研究热点课题之一（Mew，2004）。抗性遗传与致病性遗传相互依存关系的确立，使人们对抗性机理的实质有了更深层次的理解，抗性主基因的发掘与鉴定和数量抗性位点的分析与确定，为结合两类抗性的科学利用提供了有力支撑，而分子标记辅助选择也为常规育种技术升级创造了有利条件。抗病基因、无毒基因的克隆、结构、功能表达及其调节等研究也都取得了一些新的进展（David et al.，2006），为转基因育种和分子设计育种的实施做准备。本文简略介绍水稻白叶枯病的发生与传播、流行预测和病原学等一般基础知识，作为读者转入抗性研究等其他问题探讨和思考时的先导。

1 水稻白叶枯病的发现与分布

1.1 发生历史与分布

在 1911 年以前，水稻白叶枯病一直被认为是由酸性土壤引起的一个生理性病害，其主要依据是病叶上的露滴比健康叶上的露滴酸度高（Nishida，1909）；Maruyama（1908；1909）也认为酸性土壤和施用饼肥是促使白叶枯病发生的主要原因之一。这种观点直至 Takaishi（1909）从病叶上的混浊露珠中分离出细菌并将其回接到健康叶上又形成了同样的病斑后，才开始认为这可能是一种由细菌引起的病害。之后，Bokura（1911）连续 3 篇文章报道了其对白叶枯病菌的分离、病原菌在琼脂培养基上生长出黄色菌落、病菌的形态特征、生理特性以及致病性等的系统观察和研究，确认了水稻白叶枯病是由细菌引起的。

1908—1910 年水稻白叶枯病在日本西南部地区常有发生，1926 年后逐渐扩展到东北部，1962 年连北海道也有此病，之后，便在全国普遍发生（Tagami et al.，1963）。从 20 世纪 60 年代后，日本每年都有 30 万～50 万 hm^2 稻田受白叶枯病危害，中等发生田减产 20%～30%，重病田 50%以上（Mizukami et al.，1969）。作为一种细菌性病害，白叶枯病菌虽然不同于真菌孢子或稻飞虱等害虫那样常常借助大气环流远程传播，但它可以随风、雨、水流或带病稻种等途径扩散至异国他乡。20 世纪 20 年代末，朝鲜（包括今韩国）开始有白叶枯病发生（Takeuchi，1930）。1950 年前后已扩展到东亚和东南亚各主要产稻国家。印度孟买地区 1951 年开始发生此病并迅速蔓延至其他稻区，每年有数以百万公顷计的稻田罹病，重病田的产量损失高达 60%（Bhapkar et al.，1960；Srivastava，1967）。60 年代末以后，菲律宾、印度尼西亚、马来西亚、印度、巴基斯坦、孟加拉、泰国、斯里兰卡、尼泊尔、缅甸、越南、柬埔寨、韩国等由于大面积种植由国际水稻研究所（International Rice Research Institute，IRRI）育成的 IR8、IR5 等第一代高产、半矮秆、集约型改良品种并采用相应栽培措施后，水稻白叶枯病很快便成为这些国家的一个重要病害（Ou，1985；伍尚忠，1983）。1979—1980 年，印度旁遮普邦（Punjab）、北方邦（Uttar Pradesh）连续两年遭受白叶枯病危害，毁种的重病田以千公顷计（Durgapal，1985）。

现今白叶枯病除在亚洲流行外，非洲、南欧、南美、北美和澳大利亚也都有发生。1969—1971 年间，澳大利亚北部首次发现 IR8 品种、普通野生稻（*O. rufipogon*）和澳洲野生稻（*O. australiensis*）在自然条件下有此病害（Aldrick et al.，1973）。随后，塞内加尔、马里、尼日尔、上沃尔特等撒哈拉非洲国家（Awoden et al.，1984）和南美洲（Lozano，1977）都有此病的报道。进入 21 世纪后，西非马里、尼日尔的当地水稻品种以及从亚洲引进的高产品种也都曾因白叶枯病严重发生而减产（Gonzalez et al.，2004）。

1.2 中国水稻白叶枯病的发生与危害

水稻白叶枯病在中国的发生已有 70 年以上的历史，是仅次于稻瘟病的第二大水稻病害。各地农民对该病有许多叫法：浙江称“茅草瘟”，安徽叫“地火”、“游火”，湖北称为“发瘘”，江西称“黄毛瘴”，福建叫“水瘴”，两广则有“霎苗”、“着风”、“筛禾”、“燎禾”等名称，说明白叶枯病由来已久，农民对其症状都有形象化的认识。

水稻白叶枯病早在 20 世纪 30 年代在江苏、浙江就有发生（方中达，1963），到了 50 年代，陕西、安徽、云南、广东、福建、湖北等省区也有零星发生的报道（张效忠，1959；纪桐云，1989；段永嘉等，1989；伍尚忠，1983；钟藏文等，1998；许红，1993）。60年代以后，由于半矮秆、高产、不抗病品种的大范围推广和相应集约化栽培措施的普及，白叶枯病很快扩展蔓延开来，其危害也逐渐加重。目前，除新疆、西藏和东北的北部以外，南起海南岛，北至吉林，西至云贵高原，东临大海之滨以及台湾省都有不同程度的发生（图1）。但是它的主要危害地区是长江以南的主产稻区（包括华南和云贵高原）以及江淮平原，这是我国稻田分布和气候条件所决定的。在高温、多湿、雨日多、有台风袭击的年份和一定数量菌源的情况下，南方各产稻省区（除四川、广西相对较轻外）都会受到相当严重的损害。一般说来，籼稻发病较重，粳稻较轻；发病早的稻田影响较大，灌浆以后发病的影响较小。

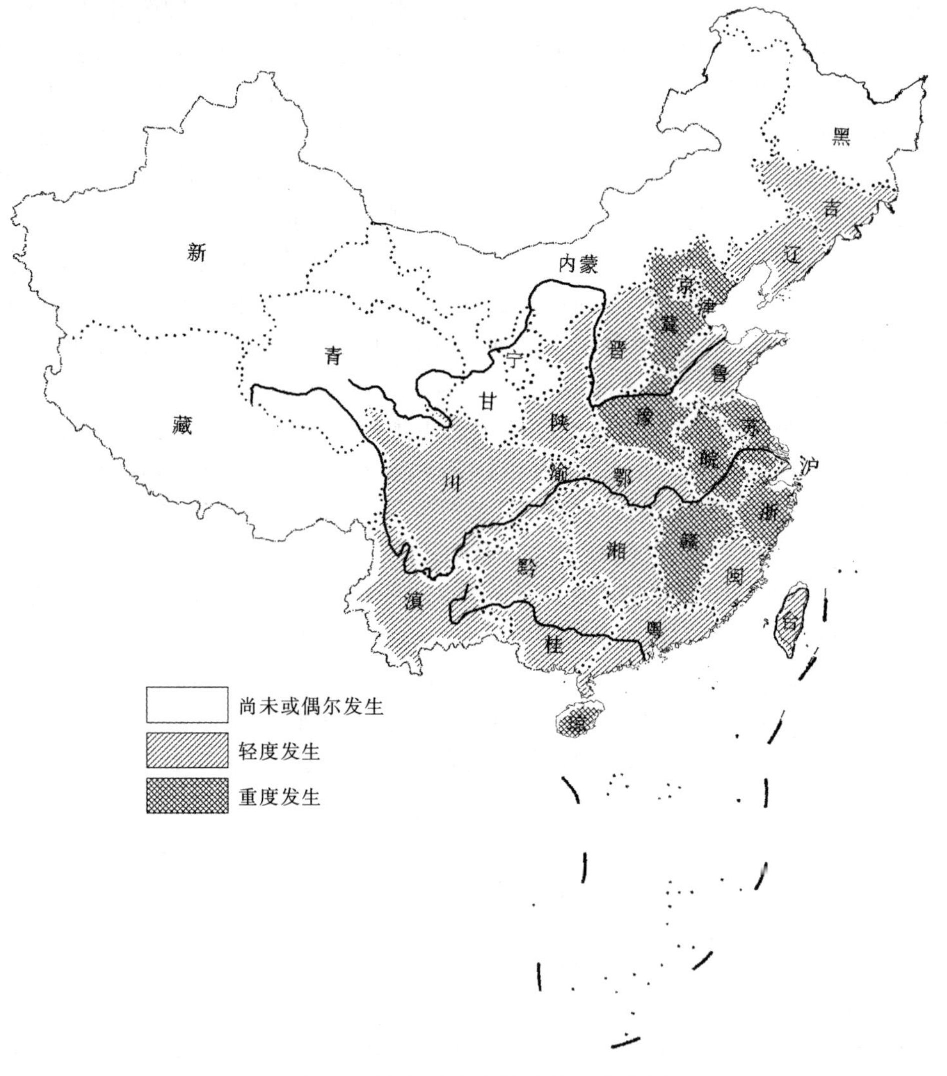

图1 中国水稻白叶枯病分布示意图

Fig. 1 Distribution of rice bacterial blight in China

白叶枯病在不同时期都有严重流行的记载。长江下游稻区的江苏省在1954年、1956年、1957年和1962年曾发生过四次严重流行。1962年该省丹阳市练湖农场一个耕作队的中稻田全部罹病，重病田减产达30%～40%。随着抗病品种的扩大推广，在相当长的一段时间基本控制了病情。然而，2005年夏、秋两季遇到了台风暴雨频繁袭击的气候，苏北地区连云港市有2.8万hm^2（合42万亩）发生了白叶枯病，占水稻总面积的16.8%，是该市20世纪90年代以来，病情最重的一年（连云港市种子站统计资料，2005)。笔者在当年9月末调查了东海县安峰乡一块26.7hm^2的稻田，远望一片枯槁、不见绿色，据当地农民估产最多只有一半收成（图2，另见彩版9)。安徽省白叶枯病严重的淮南地区的滁县、六安县等是白叶枯病常发地区，70年代发生过三次大流行，其中1974年减产稻谷165 000t；80年代以来，在皖中大部分稻区由于种植的当家品种都是感病的，80年代曾四次流行此病，其中仅1982年就减产稻谷205 400t，灾情最重的巢湖地区许多稻田甚至颗粒无收（纪桐云，1989)。后来全省推广了BG90-2、扬稻1号、2号和3号等高抗白叶枯病品种，一度基本上控制了危害。但是，扩大推广汕优63等高产、感病的杂交稻组合后病情又有回升。进入21世纪，由于扩大种植携有抗谱较广的*Xa7*、*Xa21*抗性基因的Ⅱ优129和Ⅱ优084等杂交组合，白叶枯病危害又得到了控制（王文湘个人通讯，2004)。

图2 江苏连云港市东海安峰乡严重发生白叶枯病的稻田（徐大勇摄，2005)

Fig. 2 Rice cultivar infected by bacterial blight in Donghai Anfeng Jiangsu province (Xu, 2005)

1973年，长江中游鄂中丘陵岗地稻区有61 700 hm^2（925 500亩）稻田暴发白叶枯病，损失稻谷33 100t；随后两年又连续流行危害，发病面积计292 260 hm^2，估计损失稻谷67 700 t，一度出现了“人缺粮、畜无草”的灾害年景。鄂东和江汉平原稻区在1980、1983、1987年三个白叶枯病流行年份发病面积共584 100 hm^2，估计损失稻谷141 800t（许红，1993)。

华南稻区的广东省常年遭受台风和寒露风频繁侵袭，更有利于白叶枯病的暴发流行，造成晚稻产量一直不高、不稳。福建省也是一个白叶枯病的常发区，1975年由于大面积推广种植感病品种珍珠矮而暴发了白叶枯病，全省沿江、沿海各县（市）的大片稻田远望一片枯槁，减产高达40%～50%（钟藏文等，1998）。

2 水稻白叶枯病的症状

水稻白叶枯病在全生育期都能发生，主要危害叶片，也可侵染叶鞘。其病症表现与品种的抗性、发育时期、侵染部位以及所处的环境条件等都有关系。

2.1 典型症状

白叶枯病菌主要是从叶片的水孔侵入的，因此最初的症状是在水孔分布较多的叶尖或叶缘开始出现暗绿色或黄绿色水渍状斑点，以后沿着叶缘或中脉扩展成为黄色长波纹病斑，叶面上的病、健界限分明，病斑扩及病组织，枯死后成灰白色，叶片开始萎蔫并卷曲，远看一片枯槁状，所以被称为白叶枯病。进入雨季后，空气湿度较高，在雨后傍晚和清晨露水未干时，新鲜病斑表面或病症未显的叶缘会出现蜜黄色露珠，即是由水孔溢出的菌脓（图3，另见彩版8）。

典型病症因水稻类型的不同而异。籼稻的病斑多为黄色和黄绿色，粳稻呈灰绿和灰白色。抗病品种只在叶片边缘出现黄色条状病斑，感病品种的病斑能一直延伸至叶鞘以至植株基部，由黄白色转为灰白色而后迅速干枯。

图3　白叶枯病症状

A. 叶缘伸长的浅色的水孔；B. 叶缘受 *Xoo* 自然侵染的病斑；

C. 扩展的病斑（Mew et al.，1984）；D. 病叶上的 *Xoo* 菌脓（浙江农业大学，1978）

Fig. 3　Symptom of infected leaves

A. The hydathode is near edge of blade　B. Natural infection of *Xoo* at the edge of blade

C. Expanded lesion (Mew et al.，1984) . D. *Xoo* ooze on the infected leaves (Zhejiang Agricultural University，1978)

白叶枯病的病斑形成和发展速度与水稻生育期及温度、湿度等气候条件有关。水稻在苗期就开始感染。在南方，早稻生长初期的温度较低，秧苗或分蘖期的病症不很明显，直至幼穗分化、进入孕穗期，气温逐渐升高有利于发病时才出现典型症状；连作晚稻或一季中稻的苗期也能看到典型的病症。带病秧苗被移栽到大田后半个月，稻株上就出现典型病斑。晚稻抽穗时受寒潮影响症状也不易显露，至水稻生长后期，特别在暴风雨之后的植株和叶片受到损伤，细菌便通过任何部位的伤口侵入，使布满全叶的波纹病斑迅速扩展连片，开始萎蔫卷曲干枯。在重病稻株的颖壳上会出现水渍状边缘的变色病斑，其幼嫩青绿谷粒的症状也比较明显，成熟后谷粒呈灰白或黄白色，病斑色泽不显，形成青谷或秕粒。

2.2 凋萎型症状（kresek）

1950 年 Reitsma 和 Schure 首先报道印度尼西亚出现了一种极为严重的细菌性病害，使插秧后一个月的稻苗全部枯死，当地称为“kresek”。Schure 曾鉴定其病原细菌为一个新种，命名为 *Xanthomonas kresek* Schure。直至 1964 年，Goto 在 IRRI 进行合作研究时考察了印度尼西亚等几个东南亚国家以及该所的水稻，发现都有这种类似螟虫蛀茎致伤的枯心苗。经研究，证实印度尼西亚所谓的“kresek”病其实就是白叶枯病在热带地区严重为害时的一种症状，亚洲热带许多国家都有此现象发生。凋萎型白叶枯病不局限于侵染叶部，还引起全株性凋萎，因此印度及其他各国将原来的英文病名“bacterial leaf blight”改称为 bacterial blight（Ou，1985）。

凋萎型症状主要发生在苗期。病秧移栽 1～4 周后的短期内，病情继续向其他分蘖扩展。最初受侵害的部位在植株基部的叶鞘和叶节之间，使病叶失水、青枯、卷曲、凋萎，剥开病组织的心叶，挤压其褐变部位可溢出大量黄色菌脓，而枯死病株则漂浮于水面。

在亚洲热带地区发现的这种凋萎型白叶枯病后来在中国、越南、韩国和日本等国也都有发生。如 1974 年中国台湾省嘉南地区推广了感病品种嘉农籼 11 和嘉农籼 14 之后，曾发生严重的凋萎型白叶枯病，最后只得严格限制这两个品种的种植以控制其发展（谢式坪钰，1978）。同年，中国海南岛陵水县和崖县首次爆发了凋萎型白叶枯病。1977 年春，11 个省（市）杂交稻育种队在海南进行南繁的南优 2、矮优 2、二九南等感病组合及恢复系 IR24 等约 133hm^2 秧田都发生了急性凋萎型白叶枯病，个别田块全部枯死。随后这些带病稻种又将其传播至广东、广西、湖南等省（广东省农业科学院植保所，1977；华南农学院植保系，1978）。

研究表明，凋萎的起因一般是病菌从拔秧伤根、伤茎、移栽、伤叶造成的伤口入侵，随后移向稻株的输导组织沿维管束等部位分泌毒素并急剧增殖而造成堵塞，使秧苗失水引起急性凋萎。亚洲热带地区的秧苗生长太长，稻农便掐叶插秧，所以容易出现 kresek 症状。

关于菌株毒性不同与引起急性凋萎型症状的能力是否有关，研究者所得的结果有正反两种，认为强毒菌株或无论强、弱菌株都可能引起（Devadath et al.，1970；Choi et al.，1978；Watanabe，1975；Goto，1964）。更多看法是菌株毒性的强弱取决于与不同品种的互作，该症状的出现与品种抗性应该有更直接的关系。谢式坪钰指导张胜祺的研究（1978）表明，用两个台湾本地菌株接种高雄选 1 号、台南 5 号、台中 65 等抗白叶枯病的粳稻品种都未出现急性凋萎；而用相同菌株接种嘉农籼 11、台中籼 2 号和 TN1 等感病品种 2～3 周后，都发生了不同程度的急性凋萎，以嘉农籼 11 为最重。伍尚忠等（1979）用白叶枯病凋萎型分离菌，通过刺茎、伤茎、剪根、剪叶和浸根等方法，对抗病杂交稻汕优 6 号和感病杂交稻南优 2 号进行苗期接种，结果是前者只在接种部位形成叶枯型病斑，而后者则迅速展现为凋萎型症状。上述研究充分表明品种的抗性起着决定性作用。

凋萎型症状的出现也与生育期有关，一般多在秧苗期。秧龄越小，潜伏期越短，田间凋萎型症状大都出现在秧苗移栽后 1～4 周。在成株期一般不发生急性凋萎。但是高感品种或杂交稻组合如在生长后期进行割叶或深水灌溉，有时会发生“穗凋萎”症状，表现剑叶凋萎、稻穗不结实或穗苞枯死。

2.3 苍白黄叶型症状

除“kresek”外，热带地区白叶枯病还有另一种症状，即黄白叶型（pale-yellow leaf）白叶枯病，多见于水稻成株期抽出的新叶上。病株上有 1、2 张正常大小的叶片，染病部分或整张叶片转为青黄色或淡黄白色，而其他健康叶片则保持正常绿色，其下部有少数叶片枯死。在黄白叶枯型的病叶上难以分离到细菌，但在病叶下方的节间和褐色茎秆部位却大量带菌。此后不再抽出正常新叶，能严重影响产量。

2.4 中脉型症状

在水稻分蘖或孕穗时，剑叶或剑叶下第 1～2 叶片的中脉中部开始呈现淡黄色症状，病叶两边互相折叠，从中脉向上延伸至叶尖，向下扩展至叶鞘，或由整株扩展至周围稻株，形成发病中心。取病部的分离菌进行再侵染测试时，受病菌侵染的任何部位皆可蔓延至全株，表现为全株性系统侵染。

上述症状中，以典型的和凋萎型症状对产量的危害最重。

2.5 水稻白叶枯病的诊断

水稻白叶枯病与其他寄生性或生理性病害的叶部症状很容易混淆。Ou（1985）在《Rice Diseases》一书中曾提到 1918 年 Reinking 所报道的菲律宾一种细菌性病害（未经病原物鉴定）在很长时间内被误认为“白叶枯病”，其实是水稻细菌性条斑病（bacterial leaf streak）。无巧不成书，将近 40 年后，中国广东珠江三角洲的水稻上也发现一种条斑状细菌病，当时也被认为可能是水稻白叶枯病在当地特殊条件下（包括气候、品种、栽培等综合因素）表现的条斑型症状。后经方中达等（1957）比较研究了这两种病害的症状和病原学后，才将该水稻细菌性条斑病［*Xanthomonas oryzicola* Fang et al.，*Xanthomonas oryzae* pv. *oryzicola* (Ishiyama) Swings et al.］与水稻白叶枯病区分开来，证明了这是不同病症、不同病原菌的两种水稻细菌性病害。这一研究同时也将 Reinking 于 1918 年在菲律宾发现、一直被误认为“白叶枯病”的病害正名为水稻细菌性条斑病。

水稻白叶枯病和由同属于黄单胞菌属（*Xanthomonas*）的细菌引起的水稻细菌性条斑病都能呈现叶斑和叶枯，但是前者是维管束病害（vascular disease），而后者是叶斑病害（leaf spot disease），其病原细菌限于薄壁组织危害。搞清了两者的基本区别，就容易了解水稻细菌性条斑病的特征是病斑不局限于叶尖和叶缘，在叶片的任何部位都能发病，形成水渍状细条斑，有大量细菌流胶，最后众多条斑发展为叶枯。而白叶枯病的病斑首先出现在叶尖和叶缘，沿叶脉或主脉扩展，概括其症状的三个特点是：叶面病、健界限明显；病斑边缘呈波纹状；叶面有少量菌脓。表 1 是方中达（1963）对这两种细菌病害症状特点的比较，可作为鉴别依据。

田间目测考查有疑问时，可以①剪取一小块带有健部的病叶，置于载玻片上的水滴中，加盖玻片，在低倍显微镜下检查。在病组织切口处涌出大量浅黄或无色云雾状细菌溢，就是在叶片维管束中繁殖的白叶枯病细菌。一般真菌病害或生理性枯黄的叶片则不出现此等细菌溢出现象。②剪取小块病组织放在洁净玻片上，加清水一滴，覆盖玻片，半分钟后，迎光或用放大镜观察，如切口两端的叶脉处流出乳白色菌液即为白叶枯病。③用指型管或试管加水或湿沙，剪取1～2寸长、带有健部的病叶，插入水（沙）中，一端露出，1～2h后检查，如切口处排出黄色珠状菌脓，即为白叶枯病（伍尚忠，1983）。

表1　水稻白叶枯病与水稻细菌性条斑病症状的比较（方中达，1963）

Table 1　Comparison between symptoms of bacterial blight and bacterial leaf streak（Fang，1963）

特征 Characters	水稻白叶枯病 Bacterial blight（*Xanthomonas oryzae* pv. *oryzae*）	细菌性条斑病 Bacterial leaf streak（*Xanthomonas oryzicola*）(Fang et al.)
发生部位	病菌主要从水孔或伤口侵入，从叶尖或近叶尖边缘开始发病，病斑沿叶脉或主脉扩展	病菌由气孔和机动细胞侵入，病斑不限于叶尖或叶缘，在叶片任何部分发生，病菌在维管束间的组织定殖
病斑形状	黄绿色或灰白色，向长、宽延伸长成波纹状斑，病组织不透明，呈非水渍状枯死斑，病、健界限明显	断续条斑、半透明、水渍状，连接后病斑成片，病组织大片枯死，此时与白叶枯病相似，但仔细观察可见条纹
细菌菌胶（菌脓）	在空气相对湿度很高时叶面发生数量不多的蜜黄色菌溢（菌脓）	在任何气候条件下均产生大量深黄色细菌溢脓，干燥后形成深黄色小珠
感病品种上的严重侵染状态	病斑延伸至叶鞘以至基部，最终叶片干枯萎蔫	病斑有黄色晕圈，在后期病叶全部转褐、死亡

3　水稻白叶枯病的病原细菌

3.1　白叶枯病原细菌的命名

水稻白叶枯病菌目前的学名全文是 *Xanthomonas oryzae*（Ishiyama）pv. *oryzae* Swings（黄单胞菌属稻黄单胞杆菌白叶枯病致病变种）。首先由Ishiyama在1922年定名为 *Bacterium oryzae* Ishiyama。Dowson在1939年建立黄单胞菌属后，改名为 *Xanthomonas oryzae*（Ishiyama）Dowson。Dye等在1978年把它归入油菜黄单胞菌种内作为一个变种，即 *X. campestris* pv. *oryzae*（Ishiyama）Dye。Swings等在1990年又恢复为 *Xanthomonas oryzae*。根据Swings等的鉴定，在稻黄单胞菌这个种内含有两个致病变种，一是稻白叶枯病菌致病变种（*Xanthomonas. oryzae* pv. *oryzae*，*Xoo*），这就是目前普遍采用的学名。另一个是水稻细菌性条斑病菌致病变种（*Xanthomonas oryzae* pv. *oryzicola*）。

3.2　病原细菌的形态特征

水稻白叶枯病的病原细菌为单胞，短杆状，两端钝圆，在培养基上的大小为（0.55～

0.75）μm×（1.35～2.17）μm，单鞭毛极生或亚极生（flaglium），长6～8.7μm，革兰氏染色反应阴性，无芽孢和荚膜，菌体外部有一层黏胶质的胞外多糖（Ishiyama，1922；Yushimura，1963），其化学成分是一种多糖体，能保护病菌细胞在干旱或其他不良条件下生存。培养基上的菌落为圆形、隆起、表面光滑、有光泽、黄色，病菌适宜的生长温度为26～30℃。

3.3 分离和培养

（1）分离 *Xoo*细菌在培养基上生长缓慢，较同属其他细菌分离更为困难，尤其是从稻种或土壤进行分离时难度更大。一般选择新鲜病叶样本在合适的培养基和适宜的培养条件下进行分离。培养基调节至中性和微酸性（pH6.5～7.0）为最好，培养的适宜温度为28℃左右。传统的病原细菌分离方法可参阅《植病研究方法》第三版（方中达，1998）。Zhang等（1987）在上述传统方法的基础上，曾以微型离心管代替培养皿，采用琼脂平板划线分离法，较为简便易行。琼脂平板培养基应有足够的硬度和无冷凝水，或在使用前一周制作好琼脂平板培养基，将其保存在4℃冰箱中备用。一般3天左右即可显现菌落（图4，另见彩版8）。

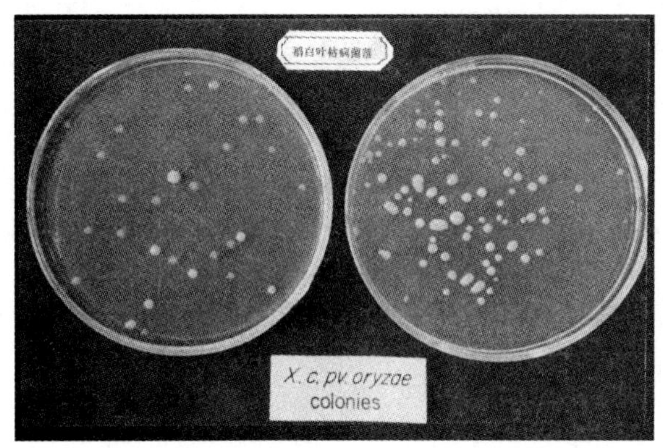

图4 *Xoo*菌单菌落（章琦等，1985）
Fig. 4 *Xoo* colonies (Zhang et al., 1985)

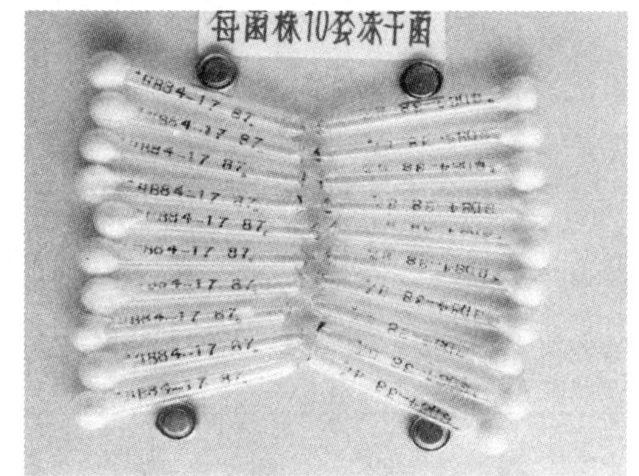

图5 冻干菌种（章琦等，1985）
Fig. 5 *Xoo* lyophilized stock (Zhang et al., 1987)

（2）培养基 马铃薯半合成培养基（PSA）和牛肉胨（NA）培养基最为适用。培养基的碳源以蔗糖取代葡萄糖为好（方中达等，1957；Watanabe，1963；1966）；最佳氮源是谷氨酸和胱氨酸（Hsu et al.，1966）。胁本哲氏（Wakimoto）马铃薯半合成培养基能加快细菌的繁殖速度，一般培养48～72h即可使用，最适于扩繁菌种。常用在对水稻种质、育种材料以及抗性与遗传研究。去除Wakimoto（WF-P）培养基中的马铃薯，按0.05g/L加入$FeSO_4$更有利于该细菌生长，因为$FeSO_4$可以克服Na_2HPO_4对菌株生长的抑制作用（Karganilla et al.，1973）。如果需要通过液体培养大量细菌，则可采用WF-P和SB（Silver and Buddenhagen）培养基。

3.4 菌种的保存

菌种保存的要点是不论短期或长期保存都要保持其专化毒性不变。

①供当年接种应用时，将长期保存的原菌种（stock）在培养基上划线恢复生长 2 代，并在鉴别品种上接种确认其专化毒性无误，然后转管数支斜面菌种在低温冰箱（－10℃）保存，约可维持半年，在此期间可随时取出解冻后再转管用于接种。

②作为病原菌种长期保存时，一般采用 15%～30%无菌甘油或脱脂牛乳冻干保存法（图 5）。

4 水稻白叶枯病细菌噬菌体

噬菌体（phage 或 bacteriophage）是以细菌和放线菌为寄主的一种病毒，分别称为细菌噬菌体和放线菌噬菌体。因其能将被感染的细菌裂解（lysis），故称为“噬细菌者”，亦即噬菌体（Okabe et al.，1963）。它对寄主细菌的杀伤力很强，一个噬菌体可在几秒内使大于其体积 200 倍的细菌不活化，并在几十分钟至几小时内使整个细菌体裂解。

4.1 噬菌体的形态和生活圈

水稻白叶枯病噬菌体多为蝌蚪状，有一个略呈多角形的头部和管状尾巴，其外壳由蛋白质构成，内部缠绕着一条细长的脱氧核糖核酸，尾部末端具有附盘和纤毛。当噬菌体接触到与其相敏感的细菌菌株时，以其尾部末端吸附在细菌细胞壁上，产生一种能破坏细胞壁化学结构的酶，将细胞壁溶成一个小孔，把 DNA 注入细菌细胞后，其蛋白质外壳留在细胞体外。注入寄主细胞的噬菌体 DNA 直接利用寄主体内现成的新陈代谢系统制造新的酶来复制噬菌体所需的 DNA、蛋白质壳和尾巴，形成新的噬菌体。其后寄主细菌就溃裂并释放出新噬菌体，即在平板培养基上所见到的透明溶菌斑（plaque）（图 6，另见彩版 8）。从噬菌体侵入到释放新噬菌体之前，称为潜伏期，待细胞裂解后即称为裂解周期（lytic cycle）。潜伏期的长短及繁殖量多少因噬菌体的种类、寄主细菌的菌株及培养条件的不同而异。白叶枯病菌噬菌体的裂解周期较短，一般约 6h 便呈现溶菌斑，10～15h 后溶菌斑在培养基上更为清晰而显现为无菌空斑，即常见噬菌斑（郭宗德，1971）。

4.2 水稻白叶枯病菌噬菌体的应用

噬菌体必须生存在与其相对应（敏感）的寄主细菌中，因此科学家常利用这种寄主专化性进行一些白叶枯病菌生态学研究。下面是几个实例。

（1）诊断 *Xoo* 菌的存在 噬菌体对寄主细菌具有很强的特异性。利用水稻白叶枯病噬菌体可以在不侵害其他细菌的情况下确定是否存在 *Xoo* 细菌。例如检测 *Xoo* 菌可能越冬的场所，诸如稻种、储藏的稻草、残存的病株、杂草、土壤中的病原体等。南京农业大学方中达实验室从 20 世纪 50 年代末以来，采用噬菌体检测来自全国稻田种子样本，发现从病田上收集的稻种都能检测到噬菌体，而从无病区和病区内无病田里收集的稻种则分离不到噬菌体。这就成为证明带病稻种是白叶枯病初侵来源的一个重要依据。

（2）测报 *Xoo* 发生的消长动态 水稻白叶枯病发病田和经过病田的灌溉水都存在噬菌体，其数量是随病情的发展而增多的。测定噬菌体的量就标示了 *Xoo* 菌的量，可以用来测报白叶枯病发生发展的动态。南京农业大学在安徽滁州地区连续进行十多年的研究，在当时水稻品种和耕作制度都较为复杂多样的情况下，所得的预测预报结果仍能看出稻田水中噬菌

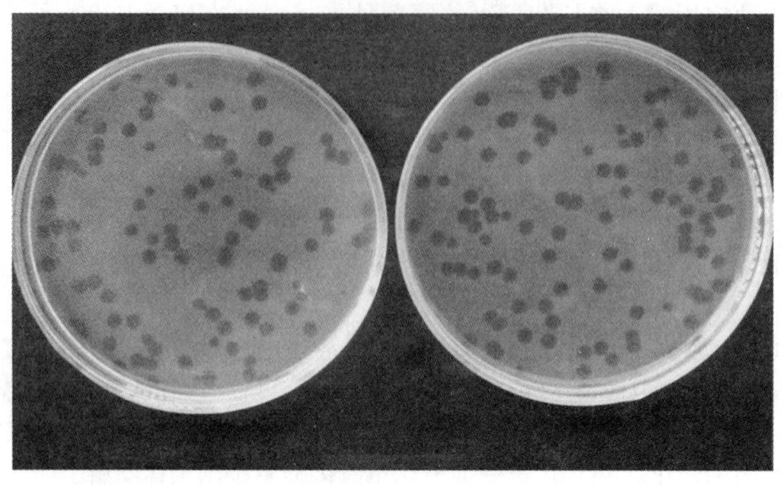

图 6　水稻白叶枯病菌的噬菌斑（章琦等，1958）

Fig. 6　Plaques of *Xoo*（Zhang et al. , 1985）

体的出现比田间病株发生提早 10～12d，可以用来确定白叶枯病药剂防治的适期（方中达，1998）。该实验室在较长时间内就是沿用这套方法向全国发布水稻白叶枯病发生和消长预报的。

(3) 测定水稻白叶枯病原细菌量　估测 *Xoo* 菌的数量也是分析该病抗性机理的一个内容。利用噬菌体的特异性及其增殖方式可以间接计算病原细菌的数量。采用这一方法需要进行一些预备试验，包括与接种 *Xoo* 菌株有特异性的噬菌体菌株的筛选，两者适宜的稀释液及最终的适宜混合浓度，噬菌体感染细菌细胞的潜伏期，噬菌体从细菌细胞开始释放至结束的时间以及每个细菌细胞所释放的平均噬菌体数量（繁殖量）。例如 Zhang 等（1985；1987）在 IRRI 合作研究中曾应用噬菌体数量这一指标来比较水稻白叶枯病成株抗性品种 Zenith 在第 6、9、10 叶位（前期感病）和第 12 叶位（后期抗病）两个生育阶段细菌增殖的差异。每隔 10 天播种一次供试品种以便同期接种，用菲律宾菌株 PXO61 接种，筛选出对 PXO61 敏感的噬菌体菌株 P80，在接种 1h（为细菌基数）、24h、72h 直至 7d 后分别测定上述各叶位单位叶面积中 *Xoo* 细菌群体的增殖动态。通过琼脂平板上形成的噬菌斑数（phage forming unit，pfu），确定每毫升含有 10^6 个能形成噬菌斑的噬菌体（10^6 pfu/ml）。根据各期数据绘制噬菌体所形成的增长曲线，从而了解该成株抗性品种从感病的营养生长期转至抗病的成株期感染叶片中 *Xoo* 菌的增殖趋势和差异（图 7，另见彩版 8）。

5　水稻白叶枯病的侵染循环

水稻白叶枯病的侵染来源包括初侵染源与再侵染源。由于不同稻区的地理环境、气候条件、农作制度和水稻品种复杂多样，各国学者对该病的初侵染来源诸如病稻种、病稻草、杂草的带菌情况，病菌残存及其传染作用所得的研究结果也不尽相同。

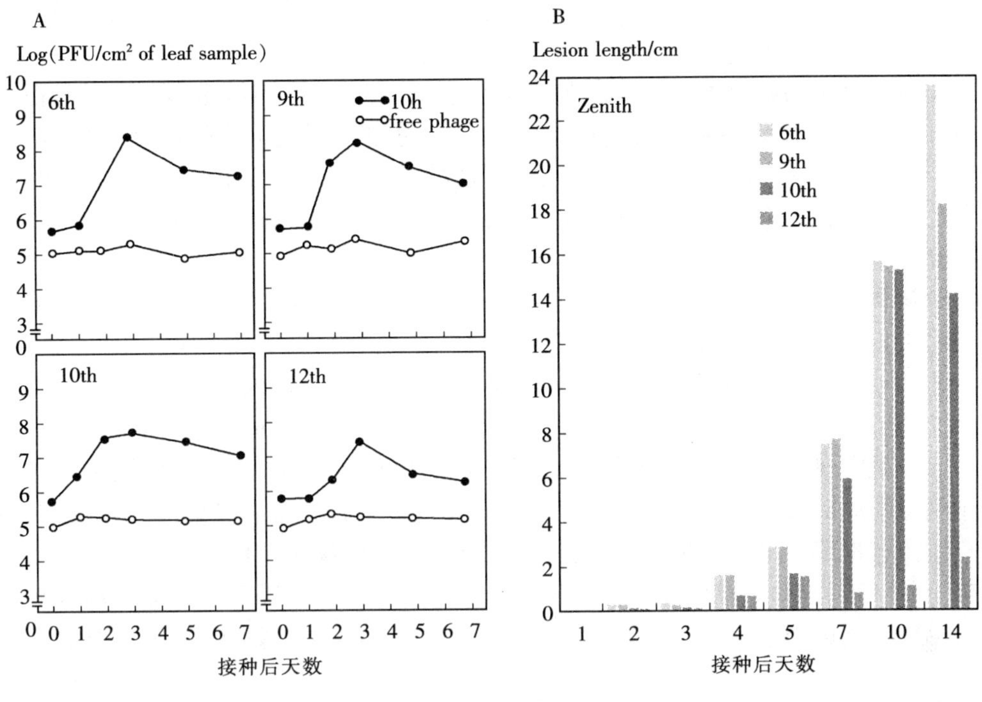

图 7

A. 用噬菌体法测定成株抗性品种 Zenith，其第 6、9、10 叶片（感病期）和第 12 叶片（抗病期）接种 *Xoo* 菌株 PXO61 后定期测定细菌增殖趋势；B. Zenith 被接种后的病斑延伸长度（章琦等，1987）

Fig. 7

A. Population trends of *Xoo* in the 6th, 9th, 10th and 12th leaves of adult plant resistance variety Zenith at different days after inoculation by using phage test method. B. Development of BB lesion at different leaf positions in Zenith (Zhang et al., 1987)

5.1 病原菌初侵染源

（1）稻种传染 带菌稻种能否传病及其重要性一直有不同看法。我国科学家从 20 世纪 50 年代开始，进行了大量研究和全国性调查，都肯定带菌稻种是白叶枯病初次侵染的主要来源。1950 至 1978 年间从南至北大量新稻区种植了来自病区的稻种都不同程度地发病。南京农业大学从全国 20 个省（市、自治区）采集了 625 个稻种样本，通过专化噬菌体检测，带菌和不带菌的稻种播种在培养皿内或温室盆栽，前者发病，后者则不发病等结果，进一步证明了稻种传病的作用。方中达等（1978）根据更换种植无病田稻种的防治效果以及将病田稻种在隔离区种植可以发病等现象，提出了稻种传病的观点，并确定稻种检疫作为防病措施之一。江苏和广东省植物病理学家曾在发生白叶枯病的杂交稻制种田观察到水稻病株在抽穗扬花期有大量细菌在花器中生长、繁殖并潜伏在成熟种子的谷颖中成为带菌稻种，其长出的幼苗根部、叶尖、芽鞘上都感染有病菌，而作为对照的无病种子则无。在干燥条件下，病菌在种子中一般可以存活半年以上（伍尚忠，1983）。不过，大多数病菌在种子贮藏过程中能被病菌上的噬菌体消解，剩下的病菌能否成功地侵染秧苗还受到温度、种子处理等多种因素的影响。据方中达等（1956）检查，稻种颖壳中大部分病菌被噬菌体寄生后，存活的病菌不到原菌量的5%，但这些病菌可存活到翌年播种期。育秧后存活的病菌可从幼苗根部、叶尖和芽鞘等处侵染水稻组织成为初侵源。

日本学者在日本的研究表明，藏存在稻种上的病原细菌到翌年生长季节（6月份）数量已大量减少，再经播种前的几天浸种也有死亡，认为稻种不是重要的传染源（Mizukami，1961；Tagami et al.，1963）。至于热带地区，稻种的传病作用很小，很少见到秧田发病，带病种子的侵染率比来自稻草、稻茬、再生稻或落粒自播稻株等的侵染率都低（Mizukami et al.，1969）。

（2）杂草传染 日本科学家的大量研究认为一些禾本科杂草是*Xoo*的越冬寄主和初侵来源。稻田和田边杂草的根和根茎是*Xoo*的重要越冬场所。Goto等（1953）以及Inoue等（1957）用*Xoo*菌接种鞘糠草（*Leersia oryzoides* var. *japonica*）、李氏禾（*Leersia oryzoides*）、茭白（*Zizania Latifolia*）、芦苇（*Phramytis communis*）、柳叶箬（*Isachus globasa*）等都能严重发病，并从上述大部分杂草上可分离到*Xoo*菌，其中鞘糠草能在自然条件下感病（Yoshimura et al.，1959）。Watanabe等（1958）根据噬菌体测定的结果认为，越冬杂草寄主鞘糠草的根际和地下茎都是*Xoo*的主要越冬场所。伊坂实人（1974）综合了在已经确定的和可疑的23种传病杂草及冬季作物中，以日本李氏禾的根际带菌越冬传病的菌源最为重要，因此把清除田间杂草、清洁灌溉水源作为消灭越冬病菌传染源的一项主要措施。

在热带地区，印度Chattopadhyay等（1968）报道莎草（*Cyperus rotundus*）和异型莎草（*Cyperus deformis*）能自然发生白叶枯病。但是Pandey等（1970）在所测试的32种杂草中没有发现任何有效寄主。Rao等（1970）和Reddy等（1974）报道*Xoo*菌的重要中间寄主李氏禾、游草（*Leersia hexandra*）雀稗（*Paspalum serobiculatum*）、稃草（*Hymenachne pseuonterrupra*）以及尼瓦拉（*Oryza nivara*）等多种野生稻在接种后能感染白叶枯病，但是在印度的上述杂草都没有自然发病，认为杂草传病的作用微不足道。

中国科学家也进行了大量的杂草传病研究。方中达等（1959，1960）首先在华南的李氏禾属植物游草（*Leersia hexandra*）上发现一种细菌病害，并曾怀疑其为白叶枯病，但是通过对菌株培养特性、生理生化反应、血清反应和致病性等研究，证明该游草上的菌株与*Xoo*菌和水稻细菌性条斑病菌都不相同，便将其鉴定为一种新的植物病原细菌（*Xanthomonas leersiae* Fang et al.）。方中达还发现在我国都感染细菌性条斑病的李氏禾属的三种植物中，除上述华南游草上的一种新病原细菌外，在华中和华东分布最广的*Leersia japonica*和浙江的*Leersia hackeleli*上也都有细菌性条斑病，经鉴定后证明，后两种李氏禾属植物上的条斑病细菌与华南游草上的细菌是相同的。由此证实了华中和华东地区的李氏禾属植物与水稻白叶枯病并没有关系。据此，方中达认为，对于日本学者提出禾本科杂草是水稻白叶枯菌的越冬场所这一结论，与中国学者的研究结果不同，由于没有看到对不同寄主来源的菌株较全面的病原学研究报道，所采用的接种方法也不相同，目前还无法做出比较和判断，有待继续探究。

关于有些野生稻种能自然发生白叶枯病的报道很多。Kulkarni等（1969）认为印度沿海地区水稻田周围经常分布有感染白叶枯病的普通野生稻（*Oryza sativa* L. *spontanea* Rosch，后统一为*O. rufipogon*——笔者注）和多年生野生稻（*O. perennis* Moench），可能是水稻白叶枯病传染源之一。Buddenhagen（1987）在澳洲、非洲、拉丁美洲和亚洲考察时，发现水稻白叶枯病能够在包括野生稻在内的一些植物上越冬，并使种植在附近的水稻发病，推测在世界各个水稻产区都有一些白叶枯病的原发中心，其中包括了非洲光身稻（*O. glaberrima*）、宿根野生稻（*O. barthii*）、长药野生稻（*O. longistaminata*）、普通野生稻（*O. rufipogon*）和澳洲野生稻（*O. australiensis*）等。

（3）病稻株和病稻茬 病菌能否在病稻株和病稻茬上越冬，时间和气候条件是这个问题的焦点。Goto 等（1953）和 Wakimoto（1954）认为 *Xoo* 在病稻草和其他残存组织上可以越冬。Inoue 等（1957）报道 *Xoo* 菌可在稻草上存活 5 个月以上，而 Tagami（1958）则说带病稻草上的病菌在潮湿条件下两个月内就全部死亡。Hsieh 等（1975）研究提出，*Xoo* 病菌在相对湿度（RH）与温度都低的条件下存活时间长，病稻草在高温、高湿时很快腐烂，病菌也迅速死亡；在温带地区，温度低于 10℃、RH 为 0%时，病菌在病叶中可存活 800d，RH 为 100%的病叶中只能存活 110d；气温 30℃，土壤湿度达 40%，在 10d 内就难以检测到存活的病菌。Nwigwe（1973）观察到病稻草中的病菌在低温、干燥条件下以凝胶包裹状态存在于叶片维管束组织的导管壁周围及孔纹导管内；翌年，越冬存活的病菌在农事作业中随着病稻草被带入田间便成为侵染源而使水稻秧苗感病。

在我国南方稻区，*Xoo* 病原细菌在单季和双季稻的病稻桩上都能存活而成为初侵染源。江苏和广东省农业科学院采取分期接种的方法测试了越冬病稻草上 *Xoo* 菌脓病菌的存活力，结果表明部分病菌可存活 6 个月以上，而残存在土表与埋入土壤中的带病残体经日晒雨淋后大多腐烂，病菌在春播前便失去了存活的可能。在南方这些温暖地区，受病菌侵染的越冬再生稻和自生稻均可成为初侵染源（伍尚忠，1983）。过崇俭等（1980）的研究表明，温带地区的杂交稻稻桩上的病菌可存活越冬直至翌年春播，并能引起发病。王法明等（1980）研究认为凋萎型白叶枯病的稻桩里的病菌可存活到翌年 5 月以后，成为初侵染源。

5.2 病原菌再侵染源

在大部分稻区只要存有足够数量的病原菌，就能使水稻遭到侵染。环境条件合适时，南方稻作区的早稻发病田可以对中稻或连作晚稻的秧苗直接再传染，在早、中稻或早晚稻交叉田则更为普遍，病田的稻草、稻桩上的病原菌可随灌溉水流或风雨传播，向其他田块扩散而再次侵染，从最初的发病中心逐渐向周围扩展。

6 水稻白叶枯病的流行和预测

水稻白叶枯病发生后能否在田间流行造成危害，主要受气候条件、品种抗病性和栽培措施三方面的影响。

6.1 气候条件

高温、高湿、多雨、风暴和寡日照是诱发白叶枯病暴发流行的主要气候因素。白叶枯病流行的最适气温是 25～30℃，最适相对湿度为 80%～90%以上，气温在 20℃以下或 33℃以上，相对湿度低于 80%时，病害发生就会受到不同程度的抑制（Horino et al.，1982）。气温的高低影响病菌的生长和潜育期的长短，一般在 22℃时潜育期为 13d，24℃时 8d，25～30℃时仅 3d。早稻秧苗期气温较低及晚稻抽穗期受寒潮影响，染病稻株一般不表现病症；适温、多雨、日照不足、露大或高温、高湿、相对湿度高时，稻株叶片水孔张开，叶面覆有水膜，有利于病菌的入侵生长。风暴造成稻株大量伤口也有利于病菌的入侵与传播。洪涝灾害淹没了稻田，降低了稻株抗病力，在地势低洼、排水不良或沿江、河、湖一带稻区更容易暴发病害流行。夏、秋季的中稻、早稻的后期和晚稻的前中期，病害往往在台风暴雨以后迅速发展。在炎热、干旱年份一般不发生白叶枯病或发病较轻。

中国稻区辽阔，农作制度和品种多样，加上地理气候环境因素的影响，不同纬度、不同地区的白叶枯病流行条件和特点不一。华南热带及亚热带籼稻区包括海南省和广东省雷州半岛南部，1 月平均温度都在 15℃以上，可全年发生白叶枯病；南方双季连作籼稻区包括广东大部、广西和福建南部区，发病季节为每年的 5～11 月，在 5～10 月的温度适宜时，早稻年度间病害流行与否与 4～6 月的月雨量和雨日的分布均匀与否有关。长江流域早、中、晚稻混栽区，早稻在 6～7 月，中稻在 7～8 月，晚稻在 8 月中旬至 9 月中旬的适温季节里，病害的流行程度则依当年温度高低、雨日长短、雨量大小以及台风的强度与频率而定。北方单季稻区包括京、津、辽宁省等粳稻区病害能否流行，与 7～8 月的温度高低、雨日长短和雨量大小密切相关，一般年份白叶枯病发病较轻或仅在局部地区发生。

6.2 品种抗病性

在有足够的菌源和有利于发病的环境条件下，品种的抗病性对能否形成白叶枯病大流行起决定性作用。品种的抗病性包括遗传的和生理的两个方面，前者的影响尤其重要，因为后者也是受遗传控制的。一般而言，糯稻的抗性最强，粳稻次之，籼稻最弱。从形态上看，株形紧凑、叶片较窄而挺直、张开角度小、水孔少而小、叶面茸毛多的品种，有抗性较强的趋势。同一个品种在不同生育阶段的代谢活动也不相同，其体内各类酶的活动和化学成分都有差异，可能是影响植株在不同生育期发生抗病性变化的原因。多数品种具有在苗期或营养生长期抗病性较弱、成株期或生殖生长期抗病性较强的现象，故有苗期抗性与成株期抗性之分，但也有不少品种是全生育抗病或感病，这些都是由其基因型决定的。

Nwigwe（1973）认为寄主植物的抗（感）性常取决于其体内新陈代谢状况是促进还是抑制寄生物生长这两者之间的平衡作用，即寄主植物细胞中的新陈代谢产物能否为病菌入侵后的生长繁殖提供所必需的营养条件，因此提出“寄生物的营养平衡假说”予以解释。方中达等（1963）在研究水稻植株体内各种氨基酸和碳源物质对 *Xoo* 病菌生长的影响时观察到，感病品种或感病状态下的植株叶色是浓绿的，其体内游离氨基酸和酰胺类化合物包括 *Xoo* 菌最嗜好的谷氨酸、天门冬氨酸和胱氨酸等的含量高，而还原糖和多元酚的含量较低。Watanabe 等（1963）发现谷氨酸、胱氨酸、天门冬氨酸、丝氨酸等都是促进 *Xoo* 病菌生长的良好有机氮源。Hsu（1967）研究发现感白叶枯病品种维管束中渗出的甘氨酸、蛋氨酸、天门冬氨酸、丝氨酸等游离氨基酸的含量比抗病品种多得多；病菌在感病品种茎秆汁液中培养时也比在抗病品种的汁液中生长得好。关于寄主植物的营养代谢与病原菌生长繁衍的关系，现有的报道都是很初步的情况，有待更多的积累和更加深入的研究。

然而，从防止病害发生的角度看，起决定性作用的还是水稻品种所携有的抗病基因。许多研究认为，多数品种对白叶枯病都具有某种程度的抗性，只是抗性水平不一、抗谱宽窄不同而已。例如 IR8、IR24 和 TN1 等感染白叶枯病的品种，一般都被认为无抗性基因，其实是当时还未找到其所能抗御的菌株，现已鉴定出它们分别具有 *Xa11*、*Xa18* 和 *Xa14* 基因，分别控制了对日本小种 1 和 2 等、缅甸菌株 BM8417、BM8429 和菲律宾小种的抗性。品种

“丧失”抗性是由于出现了能够克服其原有抗性基因的新毒性菌株。由 IRRI 育成的许多抗性基因聚合系和具有 *xa5*、*xa24* 和 *Xa7* 等多个抗性基因的孟加拉品种 DV85 等，不仅对 *Xoo* 菌具有较高的抗性，而且能抗多个小种或菌株，从而拓宽了抗谱。如能及时将这些基因转育到农艺性状好的品种中，培育出综合性状优良、抗谱广的品种在生产上种植，就可以大大延长品种的使用寿命。为了减缓病菌小种的变异和形成新优势毒性小种的速度，在强调全部种植高抗品种的同时，要注意抗病基因多样化，不能单一化，并有计划地进行抗性基因轮换，让病菌始终处在较小的选择压力下，使病害的发生维持在低水平状况，这样就能有效控制白叶枯病的危害。

此外，促进寄主抵抗力和抑制病原菌生长繁殖的科学肥水管理措施，也可在控制白叶枯病发生流行中发挥一定的作用。氮肥施用过多或过迟，绿肥压青量过大而磷钾肥不足，都能使稻株地上部分生长茂密，叶片披垂，株、行间通风透光不好，影响根系发育，促使呼吸消耗增大，这些都会显著削弱稻株的抵抗力，而稻田相对湿度增加，植株碰撞频繁所造成的创伤，则更有利于病菌的侵染和病害的发展。地势低洼的稻田应合理灌溉，适时适度烤田，以利稻株健壮生长发育，提高抗病力，并营造适宜的小气候，以阻滞病菌的传播和入侵。

6.3 病害流行的预测

在水稻生长季节采用科学的方法对当年会否发生白叶枯病流行进行预测，从长远看可为制订该病的防治决策提供依据。主要是以服务区域内的种子带菌、稻田水层中初始菌量及其消长，作为预测发病基数，应用噬菌体法或染色法检测早、中、晚稻秧苗的带菌状况，按病情测报要求，在检查菌量和田间发病中心的基础上，结合品种抗（感）性及其布局和气象条件进行预测，综合整理分析后发出预报。

（1）种子带菌测定 种子带菌是我国水稻白叶枯病发生的主要侵染源。因此在一个服务地区范围内种子带菌和稻田水中初始的菌量及其消长变化，是预测病害发生发展的基本依据（陈利锋等，2001）。检验种子带菌有直接分离、血清学和噬菌体等多种方法，以噬菌体检验法较为准确。其操作如下：在培养 3d 的试管斜面 *Xoo* 菌种中，加几毫升无菌水洗下菌苔，配置成 10^8CFU 细菌悬浮液作为指示菌备用。待测稻种充分混匀后，随机取样 5～10g，把剥脱下的稻谷颖壳剪碎，置于灭菌研钵或试管中，加无菌水 10～20ml 浸泡 5～10min，用粗滤纸过滤并吸取滤液待测。取上述浸出液 3ml，分别置于 3 个灭菌培养皿中（1ml/皿），各加入指示悬浮菌液 1ml 和经过融化并冷却至 45℃的蔗糖马铃薯培养基（PSA）或胁本哲氏（Wakimoto）培养基快速摇匀。培养 12h 后记载各培养平板中的噬菌斑数，再换算成每克种子含有的噬菌体数。在测定中应选用与本地区 *Xoo* 菌种相对应的噬菌体菌株作指示菌，才能保证测定的准确性。一般重病田中每克种子上的噬菌体可达 3 000 个以上，由于种子上的 *Xoo* 病菌及其噬菌体的存活率会随贮存时间的延长而不断降低，建议用收获后 3 个月内的种子进行测定为妥。

（2）设立预测圃观测始病期 在历年发病地区选择低洼肥沃稻田，种植当地代表性感病品种和推广品种，作为预测圃。适当多施一些氮肥，密植和深灌，或在分蘖期淹苗 1～2 次诱发病害发生。开始发病前勤加检查，一旦出现发病中心株丛，则每隔 5d 检查一次病情扩展情况。同时将若干疑似有病菌潜伏在内的病叶带回室内浸于水中后剪下，立即将其基部插入红色墨水与清水各半的混合液中，在 25～30℃下 15～30min 后观察其是否变色。如为健叶则全部染成红色，而潜育有病斑的叶片仍为绿色，可根据潜育病斑数量及不同气温条件下的潜育期来预测病害可能发生与发展的情况。如出现急性型病斑或病叶出现较多菌脓，应结合雨量、雨日、暴风雨等气象情况发出病情警报。南京农业大学许志刚等（1998）曾采用噬菌体法在秧苗移栽后每隔 3～5d 监测一次本田灌溉水中噬菌体的消长动态，用以间接了解 *Xoo* 菌的存活与增殖状况，预测该病在近期发生流行的可能性。在长江流域及北方稻区检测结果，当早、中稻田灌溉水中的噬菌体达 500 个/ml 以上，环境条件适宜，1～2 周后田间即可能出现病株；达到或超过 1 000 个/ml 以上时，3～5d 内即可出现病株。例如在浙江，当一天早、中、晚监测本田灌溉水中的噬菌体数量保持在 100 个/ml 水平时，多数田块不发病或发病很轻；当噬菌体数量达到 500 个/ml 时，则大多数田块都能发病。当然，此数值指标还应结合当地品种的抗性程度和气象因素进行分析。还要指出，施药和排水烤田等措施能使噬菌体发生钝化作用而导致监测的数值偏低，所以应在喷药或排水烤田之前采样。

(3) 品种抗病性测定 当预测圃内已开始发病时，开始对地区内生产种植的抗（感）病品种分期分类（按生育期、田间长势等）进行抗性普查。并将感病和早熟品种、长势好及偏施氮肥或低洼积水稻田列为优先检查田块。水稻品种对 *Xoo* 菌不同致病型的抗（感）性有较大差异，根据这种差别，以每个品种可抗御的致病型数作为分子，不抗的致病型数为分母，列出抗谱式以示该品种对白叶枯病抗谱的宽窄。如 DV85 对我国病菌 7 个致病型的抗病谱式为 C1～C7/0（C1 指致病型Ⅰ，余类推），即能抗所有 7 个致病型；汕优 63 的抗病谱式为 C1/C2～C7，即除抗Ⅰ型外，对 C2～C7 型都缺乏抗性；桂朝 2 号、南优 2 号的抗病谱式为 0/C1～C7，即对所有致病型都不具有抗性，对这类品种要多加注意调查其发病情况。抗谱宽的品种其田间抗性较为稳定持久，可在较大范围内使用；抗谱窄的品种应安排在非致病型流行的地区种植；对所有致病型都不具有抗性的品种，只能在无病稻区种植。

(4) 根据气候条件的发病趋势预测 在品种感病并存在较多菌源的前提下，白叶枯病流行与否取决于气候条件。在华南稻区，4～10 月份期间的月平均气温达到或超过 22℃、月雨量 250～300mm、雨日 15d 以上，均适宜发病。5 月份以前病害能否提前发生、10 月份以后病害能否延迟终止则主要看温度；5～9 月份能否暴发流行，主要看雨量、雨日及台风频率。在长江中下游地区，6 月下旬雨日数达 8d 左右，早稻可能严重发病；7 月至 8 月中旬，阴雨达 20d 以上、平均气温在 30℃以下，中稻有大发生的危险；7～8 月间阴雨日多，台风、暴雨交加的天数达 6d 以上时，晚稻严重发病。在广东省，月平均温度 26℃以上、月降雨量 200 mm 以上，是白叶枯病盛发期，可以此为主要根据再参照观察预测圃所出现的情况来预测病害的流行程度（赖传雅，2003）。

7 发病后对产量损失的估测

水稻白叶枯病导致产量损失的大小取决于病害发生的早晚和病害的严重程度。Ho 等（1976）在水稻不同生育期接种白叶枯菌的试验表明，分蘖盛期至孕穗期对病害最为敏感，这一时期发病对籽粒饱满度、千粒重和产量影响很大。Sharma 等（1984）的类似试验表明，植株最高分蘖期受害减产 54%～57%，孕穗期为 30%～48%，开花期为 29%～38%。

发病后造成有效穗数、每穗粒数及千粒重降低的直接原因是发病叶片率及每叶发病面积率，二者有密切联系，以后者的影响为主。IRRI（1972）的科学家研究认为，叶面积的发病率与产量损失高度相关，主要是空瘪粒增加、千粒重降低。Mohiuddin 等（1977）分析了接种后叶片发病面积率对产量的影响，结果表明，当 10％的叶片面积发病后，产量损失仅为 5％～6％，当 25％～100％的叶片面积发病后，产量损失为 17％～46％。

对于病害危害程度的估算，需要在田间现场进行调查统计，IRRI（1990）推荐的计算方法如下：

发病株百分率（发病率）＝ 发病样本数/调查样本总数

病情指数（感染指数）＝∑（各级病情指数×各级样本数）/（最高病情级数×调查总样本数）

病情指数是典型白叶枯病的重要指标。发病率则是调查凋萎型白叶枯病发病程度的重要指标。田间调查可以参考人工接种发病后的分级指标（表 2）进行记载，最后换算成发病率和病情指数来预测产量损失。

表 2 人工接种后叶片发病的分级标准

Table 2 Classification of leaf of infection after artificial inoculation

（IRRI，1990）

级别 Score	以病斑长度为基础 Based on lesion length	以病斑面积为基础 Based on lesion area
0 级	无病	无病
1 级	病斑长度 2～3cm	病斑面积占叶面积的 1/5 以下
3 级	病斑长度占叶片长度的 1/4 以下	病斑面积占叶面积的 1/4 左右
5 级	病斑长度占叶片长度的 1/2 以下	病斑面积占叶面积的 1/2 左右
7 级	病斑长度占叶片长度的 3/4 左右	病斑面积占叶面积的 3/4 左右
9 级	全叶发病，少数扩展到叶鞘	全叶发病

产量损失的估算　估算田间产量损失通常是用发病田（小区、株）的产量与健康田的产量相比较，以损失率表示：

损失率（％）＝［健株（区）产量－病株（区）产量］/健株产量

估算产量损失相对准确的方法是先通过小区试验找到发病程度与损失率的关系，然后以此作为依据，进行损失估计（Chiarappa，1981；Seem，1984）。即使如此，在实际应用时还可能有 10％～15％或更多的误差。试验的方法大致有以下几种：①接种对不接种，用接种的方法使植株生病，然后与不接种的健株比较，以小区为比较单位，单株计算较费时也不准确；②用防治与不防治的对比，使一部分植株不生病，然后与自然发病的植株比较；③在田间分别找有病的和无病的植株，或者发病轻重不同的植株进行比较；④用抗病和感病的品种，接种后测定其对产量的影响。

许多研究表明，叶片受害面积是估算病害损失比较好的指标。通过长期观察试验和实践，研究者可以积累水稻白叶枯病发病程度与产量损失的大量资料，从中总结形成估算产量损失的经验公式。IRRI（1990）提出如下的经验公式可供参考：

Y（％）＝1Ⅰ＋3Ⅱ＋4Ⅲ＋5Ⅳ＋7Ⅴ

式中，Y 为产量损失百分率；1，3，4，5，7 为受害指数；Ⅰ为＜20％叶面受害；Ⅱ为30％～40％叶面受害；Ⅲ为 50％叶面受害；Ⅳ为 60％叶面受害；Ⅴ为 70％及以上叶面受害。

水稻白叶枯病菌的致病性变异

章　琦

（中国农业科学院作物科学研究所）

Variability in Pathogenicity of *Xanthomonas oryzae* pv. *oryzae*

Zhang Qi

（Institute of Crop Sciences，CAAS）

摘要　水稻白叶枯病菌致病力变异研究经历了漫长的探索过程。水稻白叶枯病菌存在生理小种的发现，是该病研究史上的一件划时代的大事。

自 Kuhara 等（1958）发现 *Xoo* 菌存在致病力专化性以来，亚洲 *Xoo* 菌毒性差异的研究结果多有争议，病菌-品种之间没有呈现明显的特异性互作。直至 20 世纪 70～80 年代，Yamamoto 等发现印度尼西亚菌株与日本鉴别品种的特异性互作，迈出了亚洲 *Xoo* 菌可能存在小种的第一步。Mew 等发现鉴别品种 Cas 209 之后，菲律宾鉴别品种与 *Xoo* 菌株之间呈现了 Zadoks 提出的互作交叉反应模式，取得 *Xoo* 菌具有生理小种的佐证，明确了 *Xoo* 的生理专化作用与真菌稻瘟病及棉花角斑病等其他细菌病相似，结束了长期争议、概念混乱的局面。

20 世纪 80 年代初，Mew 等建立的首套菲律宾 *Xoo* 菌小种-品种鉴别系统包括 IR20（*Xa4*）、Cas209（*Xa10*）、IR1545 - 339（*xa5*）、DV85（*xa5*，*Xa7*）、IR24 五个鉴别品种和 4 个生理小种。至 80 年代末，菲律宾鉴定了 6 个 *Xoo* 菌小种。这套清晰的寄主-病菌间基因对基因关系的鉴别系统，便成为全球水稻白叶枯病抗性育种、新基因的发掘、鉴定，分子定位、克隆、抗性机理、病菌群体遗传学等方面研究的基础。

1986 年后，IRRI 转向 *Xoo* 菌群体遗传结构和生物学研究。在 *Xoo* 系统的生理专化性与寄主特异性的研究基础上，采用分子技术结合病菌毒性分析，揭示了田间水稻携有的不同抗性基因与 *Xoo* 群体结构的协同演变关系。Nelson 等从菲律宾 *Xoo* 病菌 DNA 指纹所显示的三个不同谱系的异质小种群中，鉴别出三个新小种，使菲律宾的 *Xoo* 小种由 6 个增至 9 个，为抗病育种提供了重要信息，为亚洲其他国家和生态区研究 *Xoo* 毒性基因-水稻抗性基因协同进化树立了典范。随后发现了对 *Xa4* 无毒力，却能侵染基因 *Xa21* 的小种 10。菲律宾白叶枯病鉴别系统包括了遗传背景一致的携有不同抗性单基因的近等基因系：IR24、IRBB4、IRBB10、IRBB5、IRBB7、IRBB14、IRBB21 及 10 个小种，各小种 1～10 的代表菌株依次为 PXO61、PXO86、PXO79、PXO71、PXO112、PXO99、PXO145、PXO280、PXO339、PXO341。

日本在 20 世纪 70 年代中期，Sato 等分析九州地区 *Xoo* 菌群体分布时鉴定到两群菌，一群能侵染当时所有 4 个鉴别品种，另一群只侵染 Kinmaza 和 Wase Aikoku 3，不侵染另两个品种，他们初步计划命名为“小种”。Yamamoto 等（1977）在日本 Kozaka 所确定的鉴别品种上测试了 71 个印度尼西亚菌株，发现其中两个菌群的毒性相似于 Sato 鉴定的菌群，与其相应品种群呈现强互作反应，1979 年声称日本 *Xoo* 菌存在小种，划分为 5 个小种和 7 个品种群。至 80 年代末，日本白叶枯病鉴别系统包括了 Kinmaze、Kogyoku、Rantai-Emas、Wase Aikoku、Java、Elwee、Heen Dikwee 和 IR8 共 8 个鉴别品种，及ⅠA、ⅠB、Ⅱ、ⅢA、ⅢB、Ⅳ、Ⅴ、Ⅵ和Ⅶ共 9 个小种。

20 世纪 70 年代，中国全面展开 *Xoo* 菌致病性分化研究。通过约 4 000 多个（次）菌株在 200 多个国内外鉴别品种上的测试表明，绝大多数病菌-品种之间表现弱互作反应。70 年代中期，段永嘉等初步发现了云南省 *Xoo* 菌株与品种之间具有质的互作反应，21 世纪前后，研究者在云南省发现了更多病菌-品种的强互作交叉反应组合，提出以小种取代致病型。

中国 90 年代建立的全国 *Xoo* 病鉴别系统包括 IR26、Java14、南粳 15、Tetep 和金刚 30 五个鉴别品种，及 CⅠ～C Ⅶ 7 个致病型。品种-菌株已呈现有特异性互作，但仍属于强和弱互作反应。

自 80 年代以来，各稻区菌群分布的变化不大。北方粳稻区仍以Ⅱ和Ⅰ型为主；南方籼稻区以Ⅳ型为主，Ⅱ型次之，并有少量Ⅴ型菌存在；长江流域籼、粳混栽稻区则以Ⅳ、Ⅱ型为主，能侵染 *Xa*4 基因品种的Ⅴ型菌已有所增加，将达到 20%的流行阈值。

随着水稻生产发展、抗病基因利用的拓宽，进一步完善全国水稻白叶枯病鉴别系统的工作是：①组成包括云南、广东等华南地区、长江中、下游及华北等稻区在内的全国协作研究组，制订一个统一的研究方案；②从携有相同抗性基因的若干品种中，选择鉴别性能最佳的品种，各稻区添加地方辅助鉴别品种；③确定全国（主要稻区）小种的代表菌株；④建立一套携有单基因抗性鉴别品种和毒性分明的病菌小种鉴别系统。

Abstract Studies on the variability of pathogenicity of *Xanthomonas oryzae* pv. *oryzae* (*Xoo*) has been a long searching process. The identification of different races in *Xoo* signified a historical milestone.

Since Kuhara and his associates discovered the existence of pathogenicity specialization in *Xoo* in 1958, similar studies in Asia often led to controversial results. The differences in *Xoo* virulence were not sufficient to demonstrate pathogenic specialization. It was until the 1970s to 1980s when Yamamoto and his associates discovered the specific interaction between the Indonesia *Xoo* strains and Japanese differentiating rice cultivars, and the presence of races within the population of *Xoo* in Asia was then first demonstrated. After Mew and his associates discovered the differentiating rice cultivar Cas 209, the differential pathogenicity response originally proposed by Zadoks was observed using differentiating rice cultivars of Philippines and different isolates of *Xoo*. This result further supported the existence of races within the population of *Xoo*, which is similar to the cases of *Magnaporthe grisea* in rice fungal blast disease and *Xanthomonas malvacearum* in cotton bacterial blight disease. The prolonged debate and confusion were then ended.

In the early 1980s, Mew and his associates established the first set of the Philippines *Xoo* race-rice cultivar differentiation system, including 4 *Xoo* races and 5 rice cultivars: IR20 (*Xa4*), Cas209 (*Xa10*), IR1545 - 339 (*xa5*), DV85 (*xa5*, *Xa7*), and IR24. By the end of 1980s, six races were identified in the Philippines. This unambiguous host-pathogen and gene-for-gene identification system has become the basis of *Xoo* resistance investigations, including breeding, gene discovery, identification, mapping and cloning, study of resistance mechanism, and pathogen genetic structures.

IRRI switched her research focus to population genetics and biology of *Xoo* after 1986. Based on the studies on *Xoo* specialization and host specificity, together with the analysis of *Xoo* virulence by molecular techniques, the co-evolution between R genes in rice cultivars and *Xoo* genetic structures was demonstrated. Using DNA fingerprinting analysis, Nelson and her associates identified 3 "new" *Xoo* races which were originated from hetrogeneous races of 3 distinct lineages. The number of *Xoo* races thus increased from 6 to 9 in Philippines. This provided important information for *Xoo* resistance breeding and set an example for the co-evolution study between *Xoo* virulence genes and rice R genes for other Asian countries and ecosystems. Subsequently, the Philippine *Xoo* race 10 which is avirulent to *Xa4* but virulent to *Xa21* was discovered. The Philippine *Xoo* race-rice cultivar differentiation system now includes a series of near isogenic lines carrying different genes: IR24, IRBB4, IRBB10, IRBB5, IRBB7, IRBB14, IRBB21 as well as 10 *Xoo* races: races 1 to 10 are exemplified by the strains PXO61, PXO86, PXO79, PXO71, PXO112, PXO99, PXO145, PXO280, PXO339, and PXO341 respectively.

In the mid-1970s, Sato and his associates analyzed the distribution of the *Xoo* population in Kyushu, Japan, and identified two bacterial groups: one group could infect all 4 differentiating rice cultivars while the other group could only infect the cultivars Kinmaze and Wase Aikoku 3. They intended to designate these bacterial groups as "races" . In 1977, Yamamoto and his associates tested 71 *Xoo* isolates from Indonesia on Kozaka's differentiating rice cultivars and observed a strong differential response from 2 *Xoo* populations that resembled those identified by Sato's group. In 1979, Japan declared the existence of races in the Japan *Xoo* population and reported 5 *Xoo* races and 7 differentiating rice cultivars. By the end of 1980s, the *Xoo* race-rice cultivar identification system established in Japan included 8 differentiating rice cultivars: Kinmaze, Kogyoku, Rantai-Emas, Wase Aikoku, Java, Elwee, Heen Dikwee, and IR8 and 9 *Xoo* races: ⅠA, ⅠB, Ⅱ, ⅢA, ⅢB, Ⅳ, Ⅴ, Ⅵ, and Ⅶ.

The nationwide studies on the variability of *Xoo* pathogenicity were commenced in China in the 1970s. Weak pathogen-cultivar associations were observed after testing more than 4 000 *Xoo* isolates on more than 200 rice cultivars. By the mid 1970s, DUAN Yongjia and his associates observed qualitative differences in the differential responses between *Xoo* pathogens and rice cultivars in Yunnan province. At the beginning of the 21th century, strong and reciprocal reactions were found in pathogen-cultivar combinations in Yunnan province. The nomenclature of *Xoo* pathotypes in China will soon be replaced by *Xoo* races.

The *Xoo* -rice cultivar identification system in China was established in the 1990s which including 5 differentiating cultivars: IR26, Java14, Nangeng 15, Tetep, Jingang 30 together with 7 *Xoo* pathotypes, CⅠ to CⅦ. Some of the *Xoo* pathotype-rice cultivar combinations exhibited certain degree of specificity. However, the reaction was still categorized quantitatively into weak and strong responses.

Since the 1980s, *Xoo* pathotypes distribution in the major rice growing areas in China have been quite stable: the pathotypes Ⅱ and Ⅰ dominate in japonica rice in North China; the pathotype Ⅳ, followed by the pathotype Ⅱ and to a lesser extent the pathotype Ⅴ have been found in indica rice in South China; the pathotypes Ⅳ and Ⅱ are major *Xoo* found in Yangtze valley where a mixed cultivation of japonica and indica rice is common. It is important to note that the spreading threshold value of pathotype Ⅴ (which can infect rice carrying *Xa4*) is alarming and now approaching 20%.

Subsequent to the development of rice production and the extended utilization of R genes, the major tasks to optimize the national *Xoo* race-rice cultivar differentiation system including: (1) to formulate a unified research plan via a nationwide cooperative efforts of Yunnan and Guangdong provinces, Yangtze valley, and the North China; (2) to identify common differentiating rice cultivars containing non-overlapping R genes, supplemented with local cultivars; (3) to establish nationwide representative strains for different pathotypes; (4) to establish a set of differentiation system containing differentiating rice cultivars carrying a single R gene and a collection of *Xoo* pathotypes of unambiguous pathogenicity.

植物病原菌致病性变化是普遍存在的现象，微生物的演变也正源于这种变化。就植物病害而言，病原菌致病性变化是植物品种抗性基因压力所驱动的。为了适应寄主品种的更换，病原菌毒性型遗传结构产生相应变化以求生存。病原菌在经受不同品种的选择压力下，一般有连续性和非连续性两种变异类型，亦即寄主-病菌之间弱或强的互作效应。了解病原菌毒性的变异动态，可为品种抗性遗传改良提供重要科学依据。病原菌小种或毒性型的地区分布、优势菌群的消长、小种毒性分化等，都是制定抗病育种计划或策略所不可缺少的信息。

Stakman 等（1917）发现小麦杆锈菌毒性的快速变化严重干扰了抗锈育种工作。经过深入的试验研究，首次提出小麦杆锈菌存在生理小种。这是植物病理学上一项最早的重大研究成果，它推动了不同植物病原菌生理小种的研究，先后发现了小麦叶锈菌（1926）、小麦条锈菌（1934）、马铃薯晚疫病菌（1926）、小麦白粉病菌（20 世纪 30 年代）、稻瘟病菌（1963）以及许多其他植物病原菌存在的生理分化现象（李振歧，1995）。

水稻与白叶枯病菌（*Xoo*）之间是否存在鉴别互作反应的问题有很长一段时期不很明确。最初科学家都认为 *Xoo* 菌的毒性属于连续性差异。直至 1957 年，日本抗病品种朝风突然丧失了抗性，Kuhara 等（1958）通过病理学分析，证实抗病品种朝风的感病化是遭遇了当地毒性菌株的侵袭，提出了 *Xoo* 菌具有致病性专化作用（pathogenicity specialization）。但是，此后对亚洲白叶枯病菌致病力差异的研究结果多有争议，有些品种与病菌组合的互作表现较强的特异性，而另一些则为弱的互作反应，认为品种与病菌互作的差异性小，不足以说明两者之间明显的特异性互作（IRRI，1966；Devadath et al.，1969；Ou et al.，1971；Buddenhagen et al.，1972）。20 世纪 70～80 年代，IRRI 进行了大规模水稻品种与 *Xoo* 菌的互作反应测试，其结果仍然介于弱互作与强互作之间。之后，Mew 以生物变异的概念为理论依据进行了系统研究，在发现鉴别品种 Cas 209 后，品种与 *Xoo* 菌株的寄主-病菌之间的强互作交叉反应，显示了热带水稻生产环境的 *Xoo* 明显的生理专化作用，明确了水稻白叶枯病菌的生理专化作用与稻瘟病、锈病及晚疫病等真菌病原菌是相似的。Mew 对 *Xoo* 生理专化性的系统研究和有理有据的论述，结束了这一研究领域长期争议和概念混乱的局面（Mew et al.，1979；1980；1981；1982；1985；Mew，1987）。

日本也获得了菌株与相应抗性基因品种之间垂直抗性的强互作证据，宣称白叶枯病菌存在生理小种现象（Horino et al.，1981）。

我国 20 世纪 70 年代以来，方中达等（1981；1991）首先开展全国水稻白叶枯病菌毒性变异研究，并于 1985—1988 年组织全国协作组，初步筛选出一套鉴别品种。中国主要稻作省（自治区）农业科研单位和院、校科学家先后对 *Xoo* 菌的毒性变异进行了大量研究，取得了翔实的数据和信息，促进了水稻抗白叶枯病育种和应用基础研究。

本文主要叙述国内外水稻白叶枯病菌毒性变异研究进展、理论依据以及研究方法，探讨我国标准鉴别系统的建立及其在品种抗性遗传改良中的应用。

1　植物病原菌生理小种

1.1　生理小种概念

生理小种（physiological race）是种（species）、变种（variety 缩写 var.）或专化型（formae specilis，缩写 f. sp.）内由生物型组成的群体。病菌小种之间没有形态上的差异，而是根据其对携有不同抗性基因品种的致病力差异来划分的，可以理解为与主效抗性基因有特异性互作的毒性类型，或称为致病型（pathotype）。“小种”既是客观存在的，也是人为划分的病原菌类群，因为划分多少个小种取决于选用多少个鉴别品种。鉴别寄主中具有的垂直抗病基因越多，所划分的小种数就越多。归在一个小种中的菌株对供试鉴别品种的反应型应该是一致的，但是对这套鉴别品种以外的品种，则可能有不同的抗性表现型。这表明将病原菌划分为小种，不过是暂时、人为地根据菌株在鉴别品种上的毒性反应型的归类，是有时间性的，不是固定不变的，但是，小种的变异和多样性确是客观存在的。

致病性（pathogenicity）是指一种病原菌对不同种类寄主植物的致病能力。例如 *Xoo* 菌对水稻能致病（pathogenic），就是水稻的病原物（pathogen），*Xoo* 菌对小麦不能致病（nonpathogenic），就不是小麦的病原物，而小麦也就成了水稻白叶枯病菌的非寄主（nonhost）。作为某种植物的病原物，能克服该植物的抗病性，两者发生亲和反应（compatible reaction），寄主植物表现感病；若不能克服抗病性，则两者发生非亲和反应（incompatible reaction），寄主植物表现抗病。由此可知，致病性是指病原物在“种”的水平上对寄主植物的专化性致病力。

毒性（毒力，virulence）指某种植物的病原物虽然在种的水平上可以发生亲和性关系，但同一种病原物内的不同小种（菌系、菌株）对其寄主植物的不同品种的致病力不相同，有强、弱、无之别，这种在小种-品种水平上的病原菌致病力称之为毒性。例如菲律宾 *Xoo* 菌小种 1 对 Cas 209 品种能够致病，即称小种 1 对 Cas 209 品种有毒力（virulent），而小种 1 对 DV85 品种不能致病，即称小种 1 对 DV85 品种无毒力（avirulent）。根据 Heath（1991）提出的基本亲和模型，推测基因对基因互作产生了病原菌对寄主选择压力的一种反应。病原物的某个小种对寄主植物某品种有没有毒性，主要取决于该小种的毒性基因能否克服该品种的抗病基因，如能克服，两者产生亲和反应，表明该小种对此品种有毒性；如不能克服，则双方产生非亲和反应，表明该小种对此品种无毒性。所以毒性是指病原物小种对其寄主品种所具备的特异致病能力。

1.2 生理小种的鉴定

Zadokes 和 Schien 在其专著 *Epidemiology and plant disease management*（Oxford University Press，1979）中论述了病原菌与寄主品种之间的互作反应模式，其描述具体而形象。病原菌和寄主品种原本是两个独立的可变因子，但当两者建立了互作关系后，即会产生不同的应答反应（response reaction）。以病菌的两个菌株（P1、P2）对两个品种（C1、C2）的自由组合为例，可能产生 16 种互作结果，可被归纳为五种互作反应模式。

①反应一致（uniform）：品种 1 和 2 与病原菌 1 和 2 的组合都不亲和（抗 R）或都亲和（感 S），即这两个品种对两个菌株全抗，或两个菌株对两个品种都有毒力；

②病菌鉴别型（P-differential）：菌株 1 与品种 1、2 都亲和，或都不亲和；菌株 2 对品种 1 和 2 的毒性正相反，都不亲和或都亲和；

③品种鉴别型（C-differential）：品种 1 与菌株 1、2 都亲和，或都不亲和，品种 2 与菌株 1、2 的反应则正相反，都不亲和或都亲和；

④强互作反应（strongly interactive）：菌株 1 对品种 1 侵染（S），对品种 2 不侵染（R）；而菌株 2 对品种 1 不侵染（R），对品种 2 侵染（S）。即菌株 1 只侵染品种 1，不侵染

品种2；菌株2不侵染品种1，只侵染品种2，表明菌株1、菌株2是两个毒力不同的菌株，这两个菌株与相同品种的致病反应是截然不同的。当出现了这种菌株间对品种间的特异性反应时。表明病菌已存在生理小种。即两对品种与病原菌之间表现出如同模式4的强互作交叉反应，菌株中就存在小种。可见，生理小种的确定既需要具备毒力差异明显的菌株，又要有辨别力分明的鉴别品种，通过这种鉴别系统才能鉴定出病菌生理小种；

⑤弱互作反应（weakly interactive）：1对菌株（P）与品种（C）组合表现非亲和反应，其他3对菌株（C）与品种（P）组合均为亲和反应，或一个品种感（S）或抗（R）两个菌株，另一个品种只感或抗一个菌株。两对菌株与品种之间的互作表现为第5种这样8个弱互作反应模式，表明菌株间对品种间不存在特异性反应（图1）。

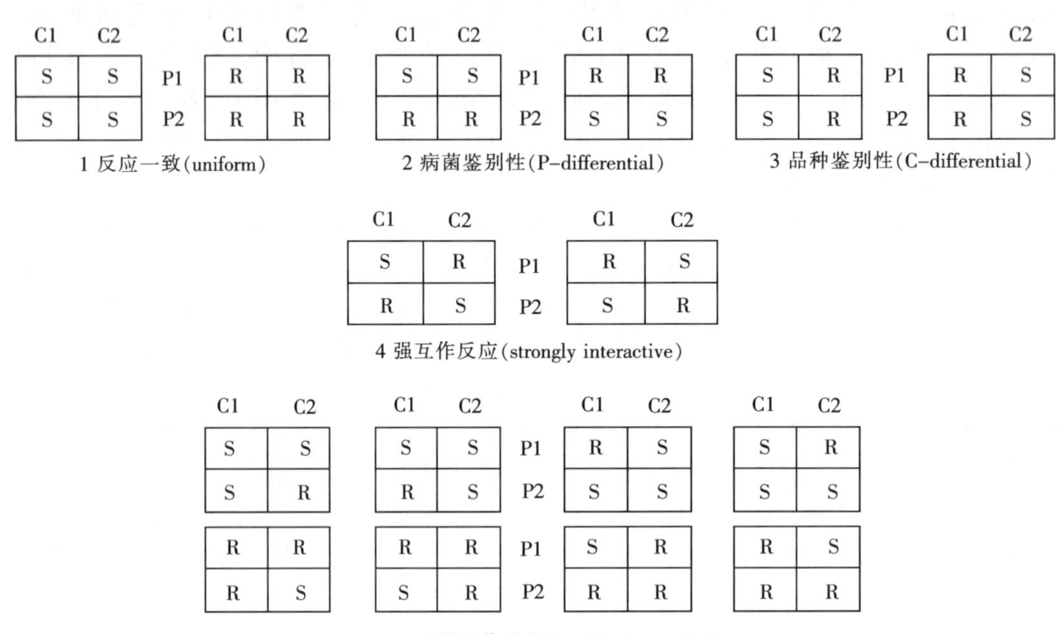

图1 Zadoks的2对病菌-品种互作反应模式（改自 Zadoks et al.，1979）

Fig. 1 Reaction models of two pairs of pathogen-cultivar interactions (Modified form Zadoks et al.，1979)

2 水稻白叶枯病菌致病性的演变

2.1 水稻白叶枯病菌致病性专化作用的发现

1957年，日本源于黄玉的抗白叶枯病品种“朝风”在育成当年，种植于Kyushu Fukuoka（福冈），Okawa和Beniya的157 hm^2 稻田突然严重罹病。Kuhara等（1958）首先以“丧失抗性（break-down of resistance）”的术语报道了这一事件。随后，测试了采自“朝风”和另外两个稻区的*Xoo*菌株对黄玉、朝风和感病品种十石（Jukkoku）的毒性。发现来自Beniya稻区的朝风品种的菌株能侵染黄玉、朝风和十石3个品种，而另外两地稻叶上的菌株只能侵染感病品种十石，不侵染黄玉和朝风，可见朝风抗性的丧失是遇到了不同地区的毒性菌株，不同地区分布着不同的毒性菌群。差不多就在同一时期，Kusaba 等（1958）在抗病品种黄玉和感病品种 Asahi 上测试了采自日本不同地区的更多菌株，结果显示，有些菌株能侵染这两个品种，另一些则不能侵染黄玉，也提出了 *Xoo* 菌株具有致病力特异作用。使研究者们开始认同 *Xoo* 菌毒性存在差异的事实。这一认识改变了日本和亚洲此前 20 多年白叶枯病的研究方向。

2.2 早期关于水稻白叶枯病菌生理专化性的争议

在 Kuhara 等提出水稻白叶枯病菌具有致病性专化作用之后，有相当长的一段时期亚洲所报道的有关水稻和 *Xoo* 菌株之间的鉴别互作存有争议。一些研究显示品种与病菌之间是垂直的寄主-病菌关系，表现小种专化性差异（Ezuka et al.，1974；Reddy et al.，1976；Mew et al.，1979），另一些研究则认为 *Xoo* 的毒性差异只是量的不同，病菌毒性由东亚、东南亚到南亚逐渐增强（Buddenhagen et al.，1972；Ou et al.，1971；Rao et al.，1971）。Buddenhagen 等认为，亚洲热带白叶枯病分化的表现是复杂的，一个国家和另一个国家的 *Xoo* 毒力不同，但是这些菌株在鉴别品种上的毒力差异太小，不足以说明存在明显的致病性专化作用。因此，认为增强品种对白叶枯病的抗性，可能不至于产生导致水稻品种与菌株之间出现垂直互作关系的新毒性小种。这个结论的问题之一是怎样来评价不同病害的抗性。稻白叶枯病是一种维管束病害，即使是抗病品种，人工接种或自然侵染后都能表现病斑，因此水稻对白叶枯病的抗性需要根据病斑的发展进行评价。它与稻瘟病、锈病等真菌病害不同，人们可以根据目测寄主上有没有病斑来评价寄主对这些真菌病害的抗性，而评价白叶枯病的抗性则是以病斑大小（面积/长度）及侵染形态为依据的。想当然地期望在白叶枯病侵染系统中寻找有病斑或无病斑反应是不大可能的。这就是他们将水稻对白叶枯病的抗性完全视为数量差异的症结所在。这一偏颇结论源于他们认为病原菌都应该按真菌病害的模式变异。不合理的分析方法自然得出不正确的结论，因此才认为 *Xoo* 病原菌“不存在”致病性专化性、“也不会”发生变异、品种抗性“始终不会”受到侵袭而能够持久。然而，从 20 世纪 70 年代以后，这个论点就被携有已知抗性基因的品种陆续遭受 *Xoo* 菌侵袭的事实所否定。

1972 年 IRRI 育成的抗白叶枯病品种 IR20 在菲律宾北部 Isabeia 省突然罹病（Ou，1985）。1975 年又发现与 IR20 一样都携有 *Xa4* 抗性基因的 IR30 在菲律宾 Iloilo 省 Panpy 岛和 Davao 省 Mindanao 感病。测试表明这两个品种上的分离菌与 1963 年从 Davao 采集的菌株 B56 的毒性相似（Mew et al.，1979）。说明 *Xoo* 菌存在毒性专化性及群体毒性的异质性，病菌毒性基因与某些寄主抗性基因是亲和的，也是“朝风感病化”在热带亚洲地区的再现，这一结果促使对白叶枯病原菌致病力分化的研究更为深入。

Kauffman 等（1972）用 7 个水稻品种测定了 161 个印度 *Xoo* 菌株，多数菌株在品种上的差异较小，虽然其中 H100 和 H161 两个菌株对 Malagkit Sungsong 和 BJ1 两个品种的毒性表现一定程度的抗、感之弱互作反应，但从品种鉴别力角度看，7 个品种的抗性差异都很小，它们对多数菌株缺乏鉴别功能。

Ezuka 等（1974）用携有不同抗性基因的一套鉴别品种测试菲律宾菌株，表现出清晰的毒性差异，但在品种-菌株之间都是弱的互作反应；然而，用这套鉴别品种测试日本菌株时，却发现了两者间的特异性互作，不过，在任一个抗病品种-毒性菌株组合之间还没有出现交叉反应（reversal reaction）（表 1）。直至 Mew 等（1980）发现了品种 Cas209 并作为菲律宾鉴别品种后，水稻-病菌之间才真正出现了强互作交叉反应。Cas209 感菌株 PXO61、高抗 PXO86，而 IR20 抗 PXO61、感 PXO86，该现象符合 Zadoks 等的寄主-病原菌的特异性交叉互作反应模式，提供了在水稻- *Xoo* 互作系统中存在生理小种的有力佐证，从而结束了水稻白叶枯病菌有、无生理专化性的争议。

表 1　日本和菲律宾水稻白叶枯病原型与其鉴别品种之间的互作反应[a]

Table 1　Interaction between Japanese and Philippine pathotypes of *X. compestris* pv. *oryzae* and differential rice cultivars of Japan and IRRI

(Horino et al., 1980)

鉴别品种 Differential cultivar	抗性基因[b] R gene	日本致病型[c] Japanese pathotype					菲律宾小种[d] Philippine race			
		T7174	T7147	T7133	H75303	H75304	PXO61	PXO86	PXO79	PXO71
		Ⅰ	Ⅱ	Ⅲ	Ⅳ	Ⅴ	Ⅰ	Ⅱ	Ⅲ	Ⅳ
Japanese differential										
Kinmaze	non	7.0[e]	7.0	6.8	7.0	6.6	6.7	6.4	6.7	6.9
Kogyoku	*Xa1*, *Xakg*	0.3	6.7	7.0	6.6	0.1	6.9	6.6	6.4	6.9
Te-tep	*Xa1*, *Xa2*	0.0	0.2	6.6	6.0	0.0	4.8	6.3	6.0	4.9
Wase-Aikoku 3	*Xa3*	1.0	0.7	0.2	6.7	5.1	0.9	0.5	0.8	0.3
Java 14a	*Xa1*, *Xa3*, *Xakg*	0.0	2.0	1.3	6.1	0.1	0.7	0.4	2.0	1.2
IRRI differential										
IR8	?	6.3	1.2	1.9	6.5	1.1	6.6	6.4	6.2	5.9
IR20	*Xa4*	0.0	2.3	3.9	4.5	0.0	1.9	6.3	7.0	3.0
IR1545 - 339	*xa5*	1.5	2.0	2.0	1.4	1.8	1.9	2.0	1.7	5.6
DV85	*xa5*, *Xa7*	0.0	2.0	1.8	1.9	0.0	1.5	1.5	1.8	1.8
Cas209	?	6.5	6.8	7.0	7.0	3.1	6.0	0.0	6.8	6.8
Cempo selak	?	1.7	1.2	0.1	6.0	6.1	1.6	1.2	2.6	3.1

[a] 抽穗期针刺接种（Inoculation by the pin-prick method at heading stage）。

[b] 当时尚未鉴定（Not yet identified）。

[c] 日本记录数据（Data recorded in Japan）。

[d] 菲律宾记录数据（Data recorded in Philippines）。

[e] 3 次重复（3 张叶片）的病情数据平均值（Represent average values of disease rating indices of 3 leaves/replications）。

3　日本水稻白叶枯病菌致病性演变

3.1　日本水稻白叶枯病菌生理专化作用研究

日本科学家发现了水稻白叶枯病致病性专化作用之后，根据品种对菌株的抗性类型划分了品种群，又据菌株对品种的毒性划分了菌群。由于各自所用的分群方法、鉴定系统或定名的不同，结果是零星或重叠的（Yoshimura et al., 1960；Kuhara et al., 1965；Kusaba et al., 1966；Washio et al., 1966）。Sakagushi 等（1968）根据 863 个品种对日本Ⅰ、Ⅱ、Ⅲ三个菌群的不同抗性反应型，划分为 3 个品种群。Kozaka（1969）在 Sakagushi 分群方法的基础上，根据病菌在一套水稻品种上的鉴别反应，将其分为 3 个菌群，并按这些品种对三个菌群的抗性和感病性分为金南风，黄玉，Rantai-Emas 和早生爱国群 4 个抗性品种群，大家接受了这个应用较为方便的系统。随后，Ezuka 等（1974）又用品种 Tetep 替换了 Kozaka 系统中的 Rantai-Emas 测定了 120 个日本菌株。结果是以菌群Ⅰ为主，依次是菌群Ⅱ和Ⅲ。

菌群Ⅰ仅侵染对日本菌株没有抗性功能的金南风品种群，菌群Ⅱ侵染金南风品种群和携有 *Xa1* 基因的黄玉品种群；菌群Ⅲ侵染金南风、黄玉、携有 *Xa1*、*Xa2* 基因的 Rantai-Emas（Tetep）群，没有菌株能侵染来自 Wase Aikoku 3（*Xa3*）抗性的 Chugoku 45 品种，表明早生爱国品种群抗所有三个菌群。这些结果已表明了品种与菌株之间存在鉴别互作关系，但是在 Kozaka 的数据中，鉴别品种与菌株之间还是没有强互作交叉反应（表 2）。当时，日本科学家确信菌株具有不同毒性，但是由于这三个菌群对品种群的致病力只有数量差异（quantitative difference）或侵袭性（aggressiveness）的表现，没有发现两者之间有特异互作反应，觉得还不足以提出生理小种。

1976 年 Sato 等分析 Kyushu 地区 *Xoo* 病菌群体分布时鉴别出两个菌群，一群侵袭当时包括 Wase Aikoku 3 在内的所有 4 个鉴别品种，另一群侵袭 Kinmaze 和 Wase Aikoku 3，但不侵袭黄玉和 Rantai-Emas，他们拟将这些菌群定名为“小种”。差不多就在同时，Yamamoto 等（1977）在 Kozaka 确定的鉴别品种上测试了 71 个印度尼西亚菌株，鉴别出三个菌群，一群菌株的毒性类似于能侵袭 Rantai-Emas 的日本菌群Ⅲ，另外两群菌株的毒性与 Sato 等鉴定的毒性反应类似，并与其相应品种群呈现出强互作反应，将其分别定名为Ⅳ群和Ⅴ群。这套菌群中的群Ⅲ与群Ⅴ，群Ⅱ与群Ⅴ，分别与其相应品种群之间的互作呈现出交叉反应。上述结果可以表明 *Xoo* 菌株在适宜的鉴别品种上进行测试，就有可能发现菌株致病性专化作用及品种与菌株之间的强互作反应，但是他们所采用的 *Xoo* 菌株不是本国的，是跨国界的，因此这应是迈出了第一步。Horino（1978）报道曾在一套携有不同抗性基因的鉴别品种上测试了 1973 年和 1975 年的日本菌株，也鉴定出Ⅵ和Ⅴ群。诚然，如果在一套合适的鉴别品种上测试更多菌株样品就可能发现毒性专化作用。

日本科学家在这期间进行了大量研究，虽然认为 *Xoo* 菌存在小种，初步也划分了 5 个“小种”，但是认为还需要通过研究，发现更多的病菌/品种专化反应组合来证实 *Xoo* 菌小种的存在。

表 2 水稻品种与白叶枯病菌株之间的鉴别反应［根据 Kozaka（1969）整理］

Table 2 Differential reactions between rice cultivars and *Xanthomonas oryzae* pv. *oryzae* strains

（Summarized from Kozaka，1969）

品种群 Varietal group	抗性基因 Resistance gene	代表品种 Representative cultivar	对菌群的抗性反应[a] Reaction to strain group		
			Ⅰ （T7174）	Ⅱ （T7147）	Ⅲ （T7133）
Kinmaze	non	金南风（Kinmaze）、霜月（Shimotsuki）、十旭（Jikkyoku），农林 37（Norin 37），旭（Asahi）	S	S	S
Kogyoku	*Xa1*	黄玉（Kogyoku）、全胜 17（Zensho 17），农林 37（Norin 27）	R	S	S
Rantai-Emas	*Xa2*	Rantai-Emas、Tetep、Nigeria 5	R	R	S
Wase Aikoku	*Xa3*	早生爱国、中新 120（Nakashin 120）、中国 45（Chugoku45）领先稻（Lead Rice）	R	R	R

[a] R，抗（Resistant）；S，感（Susceptible）。

在 Yamada 等（1979）发现籼稻 Elwee 和 Heen Dikwee 两个新品种群后，增加了品种-病菌交叉反应的组合，对日本菌群Ⅰ～Ⅴ的反应型依次分别为 SRRSR 和 SRRSS。初步确认了 *Xoo* 存在小种，遂将其划分为 5 个小种和 7 个品种群。

Horino 等（1978）用日本鉴别品种也从测试的印度尼西亚菌株中，鉴定出一个能侵染金南风和 Rantai-Emas 群，但不侵染黄玉和早生爱国群的新小种，定名为小种Ⅵ。同时发现早生爱国品种群中，存在对小种Ⅴ抗感反应不同的品种，一些品种保持原来的感病反应，另一些品种是抗小种Ⅴ的，便从早生爱国品种群中划分出爪哇品种群，增加了交叉互作反应分明的病菌-品种组合。

Ogawa（1983）发现 IR8 品种对日本小种Ⅰ和Ⅲ中的菌株表现鉴别反应，根据这些菌株对 IR8 的不同毒性反应型，将小种Ⅰ和小种Ⅲ的菌株分别划分为ⅠA、ⅠB 和ⅢA、ⅢB。

Noda 等（1987，1989b）1985 年采集了 162 个来自日本不同地区的 *Xoo* 菌株，在日本原来 5 个鉴别品种上测试，发现来自宫崎（Miyazaki）省的 H8581 和 H8584 两个菌株，除了 Tetep 之外，侵染金南风、黄玉、Chugoku 45 和 Java14 四个品种。在重复测定中，又发现它们只侵染金南风品种群中的金南风，不侵染该品种群中的密阳 23 和 IR24，侵染黄玉品种群的黄玉，不侵染该品种群另一些品种和 IR 品种，遂将这些菌株划分为小种Ⅶ。至 20 世纪 80 年代末，在日本白叶枯病鉴别系统中，包括 Kinmaze、Kogyoku、Rantai-Emas、Wase Aikoku、Java、Elwee、Heen Dikwee 和 IR8 八个鉴别品种及ⅠA、ⅠB、Ⅱ、ⅢA、ⅢB、Ⅳ、Ⅴ、Ⅵ和Ⅶ九个小种（表 3）。日本科学家在 20 世纪 50～60 年代进行了大量水稻白叶枯病研究工作，成果卓著，但是在病菌致病性分化方面未获得预期结果（Wakimoto，1967；Goto，1965；Ezuka et al.，1974；Nikai et al.，1975）。后来除日本菌株外，还用印度尼西亚、菲律宾、斯里兰卡等外国菌株在籼稻品种或日本鉴别品种上进行测试，Yamamoto 等（1977）发现印度尼西亚 *Xoo* 菌株与日本鉴别品种出现了寄主-病菌之间的特异性互作，表明白叶枯病菌的毒性也存在着质量特异性，这在探索 *Xoo* 菌存在生理小种上跨出了重要一步。但是，令人费解的是日本病理学家却先后将印度尼西亚病菌作为本国白叶枯病鉴别系统中小种Ⅳ、Ⅴ、Ⅵ的代表菌株。这种做法很容易造成概念上的混乱。根据病原菌-寄主协同演化规律，以外国菌株作为本国的代表小种，怎样反映本国水稻生产上主栽品种与白叶枯病菌之间的互作演变呢？事实上根据 Sato 等在 Yamamoto 等（1977）之前的报道，已经在日本 Kyushu 地区鉴定出新菌群，1973 年和 1975 年 Horino 测试菌株，也鉴定出Ⅵ和Ⅴ群。是什么原因没有采用这些本国分离菌作为日本小种群代表菌株，至今尚无任何合理的解释。这里还需要澄清的一个问题是，根据 R 基因和毒性基因之间存在协同进化关系的一般规律，日本稻作生产上应该采用过国外籼稻种植区的抗性基因，但据日本报道仅发现了“小种Ⅳ和Ⅴ”后才开始引用细菌的一个广谱抗性籼稻品种 Lead rice 和 IR28 等 IR 品种作为抗性供体，转向水平抗性育种。那么，Sato 及 Horino 等先后测试的分离菌何以与亚热带环境的印度尼西亚菌株相类似？唯一的解释是这些菌株来自日本温带环境种植的主栽粳稻品种，而这些品种或许具有南亚籼稻抗病基因。这一解释的理由似嫌不足。我们只能从另一个角度来看，即怎样研究和综合分析寄主、病原物及生态环境三者之间的关系，才能清晰地阐明所得的结果并进行合理的诠释。

表 3 日本水稻白叶枯病品种-病菌互作反应［据 Ogawa 等（1983）和 Noda 等（1989）数据归纳］

Table 3 Cultivar-pathogen interaction of *X. oryzae* pv. *oryzae* in Japen

［Summarized from Ogawa et al.，1983；Noda et al.，1989］

品种群/品种 Cultivar group or cultivar	代表品种 Representative cultivar	对小种的反应[a] Reaction to race								
		ⅠA T7174	ⅠB T7156	Ⅱ T7147	ⅢA T7133	ⅢB Q6803	Ⅳ H75303	Ⅴ H75304	Ⅵ	Ⅶ H8584
Kinmaze group	Kinmaze，Koshihikari Milyang 23，IR24	S	S	S	S	S	S	S	S	S R
Kogyoku group	Kogyoku，Tokai 12 Daiyoshi，Calorina	R	R	S	S	S	S	R	R	S R
Rantai-Emas group	Te-tep，Rantai-Emas 2	R	R	R	S	S	S	R	S	R
Wase Aikoku group	Wase Aikoku3，Hosokawa	R	R	R	R	R	S	S	R	S
Java group	Jave 14，Jamica	R	R	R	R	R	S	R	R	S
Elwee group	Wlwee，Dickwee－1	S	—	R	R	—	S	R	—	—
Heen Dikwee group	Heen Dikwee－1，M104	S	—	R	R	—	S	S	—	—
IR8	IR8	S	R	R	R	S	S	R	—	R

[a] R，抗（Resistant）；S，感（Susceptible）；—，未测试（Un-tested）。

3.2 日本水稻白叶枯病菌生理小种分布及其时空变化

日本农民种植的抗白叶枯病新品种朝风，在福冈地区种植当年发生严重罹病事件，应该不是由于长期种植抗病品种的选择压引起的病菌群体变化，而是朝风遇到了早已潜伏于当地的毒性菌群的侵袭，属于突发疫情。这是个群体生物学问题，自然界中白叶枯病原菌通常多存在毒性杂合群体，优势菌群的建立有一个或长或短的时间段，菌群的分布也有地域差异，这一事件为制订地区性抗病育种计划提供了一个重要信息，表明侦察寄主与病原菌互动变化在抗病育种中的重要性。

日本科学家在确定了 *Xoo* 菌具有专化作用后，继续进行病菌群体变化动向的监测，将侦察全国病情的发生、小种分布频率及其消长测报作为常规工作。20 世纪 70 年代，Ezuka 等（1974a）用鉴别品种测试了日本东海近畿（Tokai-Kinki）地区 120 个菌株，其中小种Ⅰ的菌株占 75.8%，为当地优势菌；小种Ⅱ和小种Ⅲ分别占 17.5%和 6.7%。后来，Horino 等（1978）侦察了 1973—1975 年的水稻白叶枯病小种全国分布状况。所测试的 427 个菌株中，小种Ⅰ占 60.7%，小种Ⅱ约占 30%，小种Ⅰ、Ⅱ几乎遍布全国；小种Ⅲ和小种Ⅳ的频率分别为 8.4%和 0.26%，只发现 1 个菌株属小种Ⅴ，小种Ⅲ、Ⅳ、Ⅴ分布在不同地区。Morinaka 等（1978b）侦察近畿（Kinki）、中国（Chugogu）、四国（Shikoku）小种分布的结果类同，以小种Ⅰ和Ⅱ为主，其次是小种Ⅲ，没有测到小种Ⅳ和Ⅴ。

Horino 等（1981；1981c）和 Noda 等（1987；1990；1996b）在 20 世纪 80～90 年代侦察日本 *Xoo* 小种分布的结果：①仍然是优势菌群的小种Ⅰ和Ⅱ几乎遍布日本所有地区；②除1983、1985 年外，小种Ⅲ主要分布在日本中部和西部；③在日本南部的九州和冲绳，除小种Ⅳ外，其他 6 个小种都有发生。日本全国小种Ⅱ频率的上升，是因为其主栽品种都是不抗小种Ⅱ的携有 *Xa1* 的黄玉群品种，其北部地区广泛种植了感染日本所有菌群的金南风群品种。高频率的菌群Ⅱ和群Ⅲ，也都分布在广泛种植感病的金南风品种群地区。Tanaka 等（2004）测试了 1999—2002 年在 Hiroshima 省的内陆山地 8 个点的 449 个菌株，结果表明该地区分布了Ⅰ、Ⅱ、Ⅴ、Ⅶ 四个小种群，小种Ⅴ和Ⅶ是新小种，该省各地每年的小种构成多有变化，小种Ⅰ和Ⅱ的频率逐年增长。在 Hiroshima 受到侵染的品种多属于不携有对日本小种有任何抗病基因的金南风群。显然，白叶枯病菌小种的分布，主要受品种基因型布局的影响。日本病理学家发表的上述结果表明：①日本至今仍然以侵染粳稻的小种Ⅰ、Ⅱ和Ⅲ为主，在 80～90 年代分布于全国 *Xoo* 的小种中，没有发现相同于来自印度尼西亚的日本小种Ⅳ；②日本小种Ⅴ也只是 Hiroshima 省的新菌群。显然，日本至今仍然是粳稻生产国家。

4　菲律宾水稻白叶枯病菌致病性演变

4.1　菲律宾水稻白叶枯病菌生理专化作用研究

IRRI 最初将 20 世纪 60～70 年代期间的菲律宾 *Xoo* 菌株在 IR8（对菲律宾菌株无抗性基因）、IR20（*Xa4*）、IR1545 - 339（*xa5*）3 个品种上测试，虽然已鉴定出少数品种-菌株组合的互作反应，但是两者的互作不是清晰的抗病或感病反应，初步定名为 4 个“毒性群”（virulence group）（Mew et al.，1979）。随后通过不同设计和反复测试，从 500 个品种中选出 6 个鉴别品种，即对菲律宾菌株无抗性功能的 IR8 为感病对照品种、用于生产的主要品种 IR20（*Xa4*）、IRRI 抗白叶枯病育种主体抗源品种 IR1545 - 339（*xa5*）、RP291（*xa5*）、DV85 和 ARC5756（*xa5*、*Xa7*）。根据 1963 年和 1970 年以后从菲律宾不同地区采集的 80 个 *Xoo* 菌株在这套鉴别品种上的反应，划分为 3 个毒性群。各菌群在品种上的病斑长度均值没有很大差异，但可以区分出毒性群之间的不同，并且在病菌与品种之间表现了垂直互作关系。IR8 对 3 个 *Xoo* 菌群都感病；携有 *xa5*、*Xa7* 两对基因的 DV85 和 ACR5756 抗所有菌群；携有 *xa5* 的 IR1545 和 RP291 与携有 *Xa4* 的 IR20 对群Ⅲ都是感的，群Ⅲ对 *xa5* 的毒性强于 *Xa4* 基因的品种，而群Ⅱ对 *Xa4* 的毒性较强。方差分析显示品种-菌株（C×P）组合间的互作反应为显著差异，虽然菲律宾 *Xoo* 菌在这套鉴别品种上出现了某种程度的致病性专化作用，但是测试的任何品种与菌株组合均未出现交叉反应互作。

至 70 年代中期，Mew 等（1980）在测试的大量品种中发现了一个有明显鉴别力的塞内加尔品种 Cas209，将原来菲律宾第 2 毒性群菌株区分为截然不同的两群，即 Cas209 高抗的 PXO86、PXO63 等和高感的 PXO79、PXO87 等菌株，说明该菌群中混有不同毒性的菌株，同时也出现了品种-菌株强互作交叉反应（表 4，Mew et al.，1982）。为验证这一重要发现，他们采用剪叶、针刺两种接种法，不同菌液浓度、在植株不同生育期接种，接种后 3d、7d、14d 调查，按病斑占叶片面积百分率和测量病斑长度调查法。结果显示，Cas209 的鉴别性是清晰、稳定、一致的，植株在各个生育期对原第 2 群菌株的抗、感鉴别反应分明，验证了初试结果。品种与菲律宾 *Xoo* 菌的互作呈现了符合 Zadoks 寄主-病菌间的强互作交叉反应模式：Cas209 感 PXO61（小种 1），抗 PXO86（小种 2）；IR20 抗 PXO61，感 PXO86。根据 IR8、IR20、Cas209、IR1545 - 339 和 DV85 五个品种的鉴别反应，将菲律宾菌株区分的 4

个菌群，正式命名为小种，该结果所呈现的小种专化现象与稻瘟病等真菌病害及棉花角斑病（*Xanthomonas malvacearum*）等其他细菌病害是类似的（Mew et al.，1982）。

表 4 Gas 209 品种对菲律宾 *Xanthomonas oryzae* pv. *oryzae* 菌群Ⅱ菌株的鉴别反应

Table 4 Differential response of rice cultivar Cas 209 to virulence group Ⅱ strains of *Xanthomonas compestris* pv. *oryzae*

(Mew et al.，1982)

菌株[a] Strain	病情级[b] Disease index		
	IR8	IR20	Cas209
PXO79（Ⅱ）	9	7	9
PXO87（Ⅱ）	9	7	9
PXO88（Ⅱ）	9	7	9
PXO18（Ⅱ）	9	7	9
PXO22（Ⅱ）	9	7	9
PXO86（Ⅱ）	9	7	1
PXO63（Ⅱ）	9	7	1
PXO78（Ⅱ）	9	7	1
PXO82（Ⅱ）	9	7	1
PXO83（Ⅱ）	9	7	1
PXO61（Ⅰ）	7	3	7
PXO71（Ⅲ）	8	4	9

[a] 括号内表明毒性群（Virulence groups of isolate indicated in parentheses）。

[b] 根据 1～9 的病情分级标准：1 表示侵染面积为叶片面积的 1%，3、5、7、9 分别为侵染面积占叶片面积的 5%、6%～25%、26%～50%和>50%（Based on disease scale of 1～9，where 1=1% of lesion area affected，and 3，5，7，9 = 5%，6%～25%，26%～50% and >50% of lesion area affected）。

IRRI 植物病理学家确定了菲律宾水稻白叶枯病存在小种并建立了鉴别系统后，随着生产的发展，继续侦察病菌致病性的演变，陆续发现了新小种。Mew 等（1982；1985；1987；1992）将 1972—1988 年采集的 960 个菌株在 IR20（*Xa4*）、Cas209（*Xa10*）、IR1545 - 339（*xa5*）、DV85（*xa5*，*Xa7*）和 IR24（感病）5 个鉴别品种上测定，发现了小种 5 和小种 6。小种 5 只分布于菲律宾吕宋岛北部海拔 5 000 英尺高原山区的 Benaue，这里每年只种植一季传统品种，小种 5 的毒性谱很窄，只侵染 IR24 和 IR8（*Xa11*）。小种 6 仅分布在吕宋岛南部的 Laguna 省，其毒性谱很宽，能侵染当时分别携有 *Xa4*、*Xa10*、*xa5*、*Xa7*、*Xa11* 基因的各个鉴别品种。其他 4 个小种的毒性谱是，小种 1 侵染携有 *Xa10*、*Xa11* 基因的品种，小种 2 侵染 *Xa4*、*Xa11* 基因，小种 3 侵染 *Xa4*、*Xa10*、*Xa11* 基因，小种 4 侵染 *xa5*、*Xa7*、*Xa10*、*Xa11* 基因。这套鉴别品种与鉴别菌株的基因互作表现分明而稳定（表 5）。该研究为水稻白叶枯病抗性基因发掘、鉴定等遗传学研究，品种抗性遗传改良、分子定位、基因克隆、抗性机制等研究奠定了基础。

表 5 菲律宾水稻白叶枯病 6 个小种与鉴别品种的互作反应［据 Mew 等（1985，1987）资料整理］

Table 5 Interaction between differential cultivars and *Xoo* in the Philippines

（Summarized from Mew et al.，1985；1987）

鉴别品种[a] Differential cultivar (R gene)	对小种的抗性反应[b] Reaction to race					
	小种 1 Race1 (PXO61)	小种 2 Race2 (PXO86)	小种 3 Race3 (PXO79)	小种 4 Race4 (PXO71)	小种 5 Race5 (PXO112)	小种 6 Race6 (PXO99)
R24（0）	S（40～40）[c]	S（30～35）	S（35～40）	S（40～45）	S（40～45）	S（45～50）
R20（*Xa4*）	R（3～5）	S（30～35）	S（30～35）	MR（15～20）	R（1～3）	S（40～45）
Cas209（*Xa10*）	S（40～45）	R（1～3）	S（35～40）	S（40～45）	R（1～3）	S（45～50）
R1545－339（*xa5*）	R（3～5）	R（5～10）	R（3～5）	S（30～35）	R（3～5）	S（40～45）
DV85（*xa5*，*Xa7*）	R（1～3）	R（1～3）	R（1～3）	R（3～5）	R（1～3）	MR（25～30）

[a] 水稻株龄：播种后 40d (Plant age：40d after sowing)。

[b] R，抗（Resistant）；MR，中抗（Moderately resistant）；MS，中感（Moderately susceptible）；S，感病 (Susceptible)。

[c] 括号中的数字是病斑占叶片面积的百分率 (Numbers in paretheses are ranges of % lesion area)。

IRRI 的植物病理学家连续 20 多年系统侦察了菲律宾不同地区抗病品种的 *Xoo* 菌株之毒性群的变化。通过所采集菌株样品的数据分析，表明菲律宾 *Xoo* 群体遗传结构的变化取决于生产上所用的抗病品种的替换状况。明确了专化性病菌（毒性基因）与田间抗病品种（基因）互作的密切关系（Mew et al.，1992）。

1986 年洛克菲勒基金会设立了水稻生物技术研究项目后，IRRI 转向 *Xoo* 病原菌群体遗传结构和生物学研究。该项研究的宗旨是了解水稻品种的抗性特异性与 *Xoo* 的系统发育关系。Mew 的深层次研究目的是期望通过综合分析两种生物间的互动信息，揭示田间水稻品种携有的不同抗性基因与 *Xoo* 群体结构的协同演变关系，明确抗病品种（基因）被投放之后病菌群体将发生的变化，为实现抗病性的相对稳定和抗病育种设计的更新提供重要信息和科学依据。Mew 等的 *Xoo* 生理专化性与寄主特异性研究及其理论，成为分子生物学家们以某抗病基因为标准来观察 *Xoo* 群体变化及其理论诠释的研究基础。IRRI 的分子遗传学家 Nelson 和 H. Leung 与 Mew 的紧密合作，进一步揭示并阐明了菲律宾 *Xoo* 群体变异动态。

Nelson 等（1994）以 Mew 等建立的这套鉴别系统为依据，通过分子技术结合病菌的毒性分析，观察了菲律宾 *Xoo* 小种的内在遗传变异，发现在 6 个小种中，小种 1、3、5 三个菌群均分别含有几个差异明显的分子型，表现为多元演化杂合型，这些小种很可能潜匿着不同的毒性型。在原有鉴别品种的基础上，再增加几个已知的抗病基因品种来测定那些聚类于不同簇群的各个小种之代表菌株后，发现了 3 个与原小种毒性截然不同的新菌群：①小种 5 菌株在 IRBB14（*Xa14*）品种上被划分为两个毒性群，一群抗 IRBB14，其毒性与原小种 5 相同；另一群感 IRBB14，其毒性不同于小种 5，被命名为小种 7；②根据小种 1 参试代表菌株对 IRBB4（*Xa4*）、IRBB5（*xa5*）、IRBB10（*Xa10*）、IRBB14（*Xa14*）的不同毒性反应型，也分成两群，一群的系列反应依次为 RRSS，对 IRBB14 有毒力，其毒性与小种 1 相同；

另一群为 RRSR，对 IRBB14 无毒力，其毒性与小种 1 不同，被定名为小种 8；③小种 3 中的一群菌株对 IRBB7 无毒力，其毒性与小种 3 相同，另一群则能侵染 IRBB7，不同于小种 3，被命名为小种 9。这样，从毒性异质型的小种 1、3、5 中，分别鉴别出与原菌群毒性不同的新小种。自此，菲律宾白叶枯病菌小种由 6 个增至 9 个。这一研究洞察了田间水稻品种的抗性特异性与 *Xoo* 系统发育的关系。Relson 等根据菲律宾 *Xoo* 小种研究结果，进一步揭示了热带地区 *Xoo* 的演化，为水稻抗白叶枯病育种设计提供了新的依据。这一趋势可能同样存在于其他地区或生态区。假如我们掌握其他国家或生态区域类似的研究资料，则可能形成亚洲 *Xoo* 毒性基因-水稻抗性基因协同进化的一个极有理论价值的假说（Mew，2006，个人通讯）。

20 世纪 90 年代中期以来，随着新基因 *Xa21* 的问世与迅速应用，在菲律宾发现了能侵染该基因的新小种 10，小种 10 虽然能侵染 *Xa21*，却对 *Xa4* 无毒力。IRRI 建立的菲律宾白叶枯病鉴别品种逐渐采用了遗传背景一致的携有抗病单基因的近等基因系。迄今，菲律宾水稻白叶枯病鉴别系统包括 10 个 *Xoo* 小种和 7 个鉴别品种（表 6）。

表 6　菲律宾水稻白叶枯病 10 个小种对鉴别品种的互作反应

Table 6　Interaction between rice and 10 races of *Xanthomonas oryzae* pv. *oryzae* in the Philippines

品系[a] Line	Race 1 PXO61	Race 2 PXO86	Race 3 PXO79	Race 4 PXO71	Race 5 PXO112	Race 6 PXO99	Race 7 PXO145	Race 8 PXO280	Race 9 PXO330	Race 10 PXO341
IR24	S[b]	S	S	S	S	S	S	S	S	S
IRBB4	R	S	S	I	R	S	R	R	S	R
IRBB10	S	R	S	S	R	S	R	S	S	S
IRBB5	R	R	R	S	R	S	R	R	R	R
IRBB7	I	R	R	(S)	R	S	R	R	(S)	I
IRBB14	S	S	S	S	R	S	S	R	S	S
IRBB21	R	R	R	R	R	R	R	R	I	S

a　以 IR24 为遗传背景的近等基因系（Near-isogenic line with IR24 background）。

b　以病斑长度表达抗感反应（R and S reactions are expressed in lesion length）。R，抗（Resistant），<5cm；I，中抗/中感（Moderately resistant/ moderately susceptible），5～10cm；S，感（Susceptible），>10cm。

4.2　菲律宾水稻白叶枯病菌生理小种分布及其时空变化

IRRI 的植物病理学家从 20 世纪 70 年代开始，即对菲律宾白叶枯病菌毒性群（小种）分布、抗病品种布局等进行了定点侦察。菲律宾病菌小种的发生频率随抗病品种的更换而变动。1972—1974 年，在低地稻区推广现代品种之前，当时的优势菌群小种 1 占菌株采样总数的 82%。1975—1979 年、1980—1984 年，在低地稻区及一些山区大面积推广了遗传相对均质的携有抗性基因 *Xa4* 的品种，为与其亲和的小种 2 和小种 3 提供了扩展条件，小种 2 和 3 分别以样本总数 43%和 26%的频率高于已降至 9%的小种 1。1985—1988 年，携有 *Xa4* 基因品种的种植面积已增至菲律宾水稻总面积的 85%～90%。至 90 年代中期，在吕宋岛中部 Nueva Ecija、Tarlac 和 Bulacan 三个省所采集的菌株样本中，小种 3 的频率上升到 61%，小种 2 则降至 39%，品种与病菌定向选择（directional selection）的互作关系表露无遗，病菌群体结构也随之发生了明显的变化（Mew et al.，1982，1992）。

显然，寄主品种的更换是引起病菌群体变化的主导因子，环境条件也发挥了一定作用。

例如小种5只分布在平均气温21℃的吕宋岛北部的高原山区，那里每年只种植一季传统品种，这种特殊分布结构的形成，也与传统品种对当地环境的适应性有关，在这种生态条件下形成了寄主与病菌的共存关系。菲律宾有些来自地方品种的菌群表现出广谱毒性，如小种6对携有 *Xa7* 或 *xa5* 和兼有 *Xa7*、*xa5* 广谱抗性基因的品种都有毒性，不过其发生的频率很低，只出现在吕宋岛南部的 Laguna 省。能侵染 *Xa7* 或 *xa5* 的小种4也仅在吕宋岛西南 Palawan 岛屿的一个地区被发现。

IRRI 的病理学家通过长期的系统侦察，明确了菲律宾水稻白叶枯病菌小种消长及其群体变化的主导因子是品种基因型的分布，特殊生态环境条件的参与也有一定的影响。阐明了水稻白叶枯病专化性病菌与抗病品种互作，寄主-病菌协同进化（co-evolution）的密切关系，是寄主-病原菌-环境三者互相作用的典型。病菌群体已发生及正在发生的变化，为 IRRI 育种家提供了适时调整抗病育种方案以及进行抗性基因布局和培育基因聚合系的重要依据。

5　亚洲国家水稻白叶枯病菌致病性分化

了解各国 *Xoo* 病菌-品种互作研究信息，对水稻遗传材料的选育和国际交流十分重要。一些科学家用菲律宾和日本鉴别系统中的鉴别品种测试亚洲有些国家 *Xoo* 菌株的毒力，多数国家的白叶枯病鉴别系统还有待完善。

Noda 等（1996）曾用日本和菲律宾的两套共12个鉴别品种，包括金南风、黄玉、中国（Chugoku）45、IR8、IR20、IR1545－339、DV86、Java14、Te-tep、Cas209、PI231129 和 IR24，测定了1982—1994年从孟加拉、印度尼西亚、尼泊尔、菲律宾、马来西亚、缅甸、泰国、韩国、中国南方稻区9个国家或地区收集的397个 *Xoo* 菌株，比较了不同国家白叶枯病菌群的毒性差异及其分布。从所划分的27个菌群中（表7），可大致了解亚洲国家的 *Xoo* 毒性状况。

①菌群1、2和3集中了毒性最强、毒力谱最广的菌株，这些菌株能侵染包括抗谱较宽的 DV85（*xa5*、*xa13*、*Xa7*）或 Java 14（*Xa1*、*Xa2*、*Xa12*）或 IR1545（*xa5*）等所有12个或11个品种，这3群菌株主要来自孟加拉、缅甸，其中2个来自中国云南。

②菌群4～7菌株的毒力较强，能侵染包括上述广抗谱品种在内的10个鉴别品种，来自印度、尼泊尔、孟加拉、印度尼西亚，其中4个来自中国华南菌株。

③菌群8～17的菌株能侵染8或9个鉴别品种，来自印度、尼泊尔、孟加拉、印度尼西亚、泰国、缅甸和中国南部等地。所测试的鉴别品种对这些菌群的抗、感反应差异很大，多数表现特异性互作的病菌-品种组合也都集中在这些菌群。如菌群10和11中的孟加拉（BD）菌株与品种 IR8 和 DV85；菌群7和16中的印度菌株与品种 IR1545－339 和 Java 14；菌群8和12中的尼泊尔（NP）菌株与 IR20 和 IR1545－339 等组合，都表现强的互作交叉反应。

④菌群18～25的菌株能侵染6或7个鉴别品种，主要来自东南亚的菲律宾、印度尼西亚、缅甸、泰国、马来西亚，其中17个来自中国南部。印度尼西亚所有87个参试菌株无一能够侵染 IR1545 和 DV85，表明印度尼西亚可以种植或利用携有 *xa5*、*xa7* 基因的品种；鉴别品种 Cas209 所抗的菌群18、21和22仅分布在菲律宾，所有这些菌株都不能侵染中国

表7 亚洲国家水稻白叶枯病菌群与品种的互作反应及其分布（Noda et al.，1996）

Table 7 Interaction between differential rice cultivars and races of *X. o.* pv. *oryeae* and their distribution in Asian countries（Noda et al.，1996）

小种[a] Race	鉴别品种 Differential cultivar												分布[b] Distribution								
	Kinmaze	Kongyoku	Chugoku45	IR8	IR20	IR1545 - 339	DV85	Java14	Tetep	Cas209	PI231129	IR24									
1	S[c]	S	S	S	S	S	S	S	S	S	S	S	BD	—	—	—	—	—	—	—	—
2	S	S	S	S	S	S	S	S	S	S	R	S	BD	—	—	MY	—	—	—	—	—
3	S	S	S	S	R	S	S	S	S	S	S	S	BD	—	—	—	—	—	—	—	CN
4	S	S	S	S	S	S	R	S	S	S	R	S	BD	ID	NP	—	—	—	—	—	—
5	S	S	S	S	S	R	R	S	S	S	S	S	—	—	—	—	—	IN	—	—	—
6	S	S	S	S	R	R	S	S	S	S	S	S	—	—	—	—	—	—	—	—	CN
7	S	S	R	S	S	S	S	R	S	S	S	S	BD	ID	—	—	—	—	—	—	—
8	S	S	S	S	S	R	R	S	S	S	R	S	—	ID	NP	MY	—	—	—	—	—
9	S	S	R	S	S	S	S	R	S	S	R	S	BD	—	—	—	—	—	—	—	—
10	S	S	R	S	S	S	R	R	S	S	S	S	BD	—	—	—	—	—	—	—	—
11	S	S	R	R	S	S	S	R	S	S	S	S	BD	—	—	—	TL	—	—	—	—
12	S	S	S	S	R	S	R	S	S	S	R	S	BD	ID	NP	—	—	—	—	—	—
13	S	S	S	S	R	R	R	S	S	S	S	S	—	—	—	—	—	IN	—	—	CN
14	S	S	R	S	R	S	S	R	S	S	S	S	BD	—	—	—	—	—	—	—	—
15	S	S	R	S	S	R	R	R	S	S	S	S	—	—	—	—	—	IN	—	—	—
16	S	S	S	S	R	R	R	S	S	S	R	S	BD	ID	NP	MY	—	IN	—	—	CN
17	S	S	S	R	R	R	R	S	S	S	S	S	—	—	—	—	—	IN	—	—	—
18	S	S	R	S	S	R	R	R	S	R	S	S	—	—	—	—	—	—	PP	—	—
19	S	S	R	S	S	R	R	R	S	S	R	S	—	—	—	—	—	IN	PP	—	—
20	S	S	R	S	R	R	R	R	S	S	S	S	—	—	—	MY	TL	IN	—	MS	CN
21	S	S	R	S	S	R	R	R	S	R	R	S	—	—	—	—	—	—	PP	—	—
22	S	S	R	S	R	R	R	R	S	R	S	S	—	—	—	—	—	—	PP	—	—
23	S	S	R	S	R	R	R	R	S	S	R	S	BD	—	NP	MY	TL	—	PP	MS	CN
24	S	S	S	R	R	R	R	S	R	S	R	S	—	—	—	—	TL	IN	—	—	—
25	S	S	R	S	S	R	R	R	R	S	R	S	—	—	—	—	TL	—	—	—	—
26	R	R	R	R	S	R	R	R	S	R	R	R	—	—	—	MY	—	—	—	—	—
27	R	R	R	R	R	S	R	R	S	S	R	R	—	—	—	MY	—	—	—	—	—

[a] 暂定小种群（Tentative name）。

[b] BD，孟加拉（Bangladesh）；ID，印度（India）；NP，尼泊尔（Nepal）；MY，缅甸（Myanmar）；TL，泰国（Thailand）；IN，印度尼西亚（Indonesia）；PP，菲律宾（Philippines）；MS，马来西亚（Malaysia）；CN，中国（China）。

[c] R，（抗 Resistant）；S，感（Susceptible）。

45、IR1545－339、DV85 和 Java 14 品种；大部分鉴别品种包括中国 45、IR20、IR1545－339、DV85 和 Java14 抗马来西亚的多数菌株，也抗泰国的优势菌群 20，只有菌群 10 唯一的菌株 TL8203 不仅侵染 IR20、IR1545－339，而且侵染 DV85，表明泰国菌群的毒性已开始分化。

⑤菌群 26、27 的菌株对感病品种金南风和 IR24 无毒力。Yamamoto 等（1990）曾鉴定出 IR24 对其中的缅甸菌株 MY8417 和 MY8429 具有抗性，根据其遗传分析结果，将 IR24 对这 2 个菌株的抗性基因定名为 *Xa18*。缅甸也是至今种植地方品种较多的国家，该国病菌的毒力谱较窄。

上述研究提供了进行种质交流的详细信息。可根据病菌的毒性情况，在育种和生产中来选择适宜所在国的抗病基因。资料表明，南亚的孟加拉、印度、尼泊尔的 *Xoo* 菌的毒力较强，以孟加拉为最，该国的菌群 1 能侵染全部 12 个鉴别品种，它也是遍布孟加拉的优势菌。南亚比东南亚国家的 *Xoo* 菌株之毒力谱更宽。我国云南省的菌株也有相当强的毒性。

鉴于不同国家 *Xoo* 菌与当地品种基因型的长期互作关系，其菌群毒性分化各异，因此，各自鉴定和发展适合于本国的携有不同抗病基因的鉴别品种，以明确其病菌小种群的毒性特点是十分必要的。

韩国科学家十分重视水稻白叶枯病鉴别系统的研究。虽然还没有明确其鉴别品种的抗性基因，但是已鉴定了 K1（KXO599）、K2（KXO579、KXO649）、K3、K4 和 K5 五个小种，并且连续侦察全国的小种分布及其消长变化。K1、K2、K3 几乎遍布全国，20 世纪 80 年代 K1 的分布频率远远高于 K2 和 K3，分布在全国 60%以上的水稻种植面积上，不过 K1 的毒力谱比其他 4 个小种都窄。K4 在 1980、1981、1985 年发生，但 1986 年则几乎没有发现。1977—1979 年连续发现小种 K5，1980 年以后已不多见。90 年代推广了携有 *Xa1* 和 *Xa3* 基因品种之后，与这两个抗病基因不亲和小种 K1 减少了，而 K2、K3 则日益增多。2001 年 Noh 等（2004）用韩国鉴别品种分析了来自 Jeonnam 省的 15 个菌株，因为这些菌株能够侵染 K3 所抗的品种而被区分了出来，其毒性强于 K3，弱于 K4 和 K5，便作为一个新小种 K3a。2003 年 K3a 的频率在韩国所有南方稻区迅速上升。另外一个在韩国发生的特别现象是携有广谱抗性基因 *Xa21* 的品种在该国推广后不久，就对 1987—1989 年采集的 15%的菌株感病，至 90 年代中期，采集菌株中的 96%都能侵袭 *Xa21*（Lee et al.，1999）。*Xa21* 遭遇印度（Goel et al.，1998）、尼泊尔（Adhikari et al.，1999）和中国广东、云南以及其他亚洲国家或地区毒性菌株的侵袭，是在当地生产中应用数年之后才发生的，而 *Xa21* 在韩国所发生的情况不同，该基因遇到了潜在于当地的毒性菌株，这些原地菌株的存在及其演变是个值得探索的问题。

6　中国水稻白叶枯病菌致病性分化研究

6.1　中国水稻抗白叶枯病致病型研究概况

20 世纪 70 年代，中国开始进行 *Xoo* 菌致病性分化研究。南京农业大学、云南农业大学，粤、桂、苏、滇、浙、湘、鄂、闽、皖、蜀、宁等省（自治区）农业科学院、大专院校的植物病理学家进行了 *Xoo* 病菌毒性变异研究工作。据不完全统计，测试了将近 4 000 多个（次）菌株的毒性，收集、筛选了包括日本与 IRRI 白叶枯病鉴别寄主在内的 200 多个水稻品种。由于病菌-品种互作组合不同，采用方法不一，所获结果难以进行比较。但是大量的研究反映了当时当地菌株的毒性变化及其分布状况，为抗病育种提供了重要信息。按照病菌-品种互作反应表现，有弱互作、特异性互作和交叉强互作三种模式（表 8）。

表 8 中国水稻主要稻作区水稻白叶枯病菌致病型分化简况（据孙恢鸿 2004 年资料增补）

Table 8 Summary of pathogenicity variability of *Xanthomonas orizae* pv *oryzae* in main rice growing regions（provinces）**in China**（Supplemented from Sun，2004）

地区 Region	时期 Duration	菌株数 No. of strains	鉴别品种 Differential cultivar	特异性 Specificity	致病型及参考文献 Pathotype & reference
华南、华北（东北）、长江流域	1975—1988	835	IR26、南粳 15、Tetep、金刚 30、Java14	有	CⅠ～Ⅶ 7 型（方中达等，1990）
全国鉴测	1999—2000	100	金刚 30、IRBB14、IRBB3、IRBB4、IRBB5、Java14	有	C1～C Ⅶ（许志刚等，2004）
长江以南稻区	1996—1999	69	金刚 30、Tetep、南粳 15、Java14、IR26	有	0～C Ⅶ、Ⅳ（王春莲等，2001）
云南	1976—1987	412	IR26、秋二矮、珍白 18、窄叶青 8 号、Tetep、成都矮、金南风、74－105、南粳 33、农林 27	有强互作交叉反应	9 个（段永嘉等，1987）
	1986—1987	73	IR26、Heen-Dikwee-1、珍白 18、早生爱国 3 号、Tetep、窄叶青 8 号、IR8、南粳 33	有强互作交叉反应	10 个（肖火根等，1989）
	1994—1996	138	IRBB－1～5、IRBB－7、－8、－10、－11、－13、－14、－21	有强互作交叉反应	10 个（李成云，2002）
高原粳稻区	2001—2003	36	毫糯扬、TN1、黄玉、珍珠矮、IR26、南粳 33、金南风	有强互作交叉反应	9 个（余腾琼等，2005）
	2001	51	丰锦、IRBB1、IRBB14、IRBB3、IRBB4、IRBB5、IRBB21	有强互作交叉反应	14 个（姬广海等，2003）
广东	1979—1980	211	IR26、农垦 57、窄叶青 8 号、包胎矮、金刚 30	无	Ⅰ～Ⅴ（Ⅵ）（徐羡明等，1981；1994）
	1988—1993	450	同上		
广西	1977—1983	236	IR26、闽香糯 3 号/南粳 15、窄叶青 8 号、团结 1 号/江宁糯、金刚 30	无	Ⅰ～Ⅴ（Ⅵ）（孙恢鸿等，1987）
	1986—1989	72	金刚 30、Tetep、南粳 15、Java14、IR26	无	Ⅰ～Ⅴ（Ⅵ）（孙恢鸿等，1993）
福建	80 年代	298	IR26、南粳 15、农垦 57、江宁糯、金刚 30	强弱差异	Ⅰ～Ⅳ（钟藏文等，1998）
江苏	1976—1980	49	金刚 30、江宁糯、矢租、金南风、南粳 15、大治 1 号、IR26	有	11 群（江苏农科院，1981）
浙江	1985—1987	115	金刚 30、窄叶青 8 号、南粳 15、农垦 57、IR26	有	Ⅰ～Ⅴ（Ⅳ）（谢关林，1988）
	1990—1994	203	金刚 30、Tetep、南粳 15、Java14、IR26	有	Ⅰ～Ⅶ（Ⅳ）（王汉荣等，1995）

（续）

地区 Region	时期 Duration	菌株数 No. of strains	鉴别品种 Differential cultivar	特异性 Specificity	致病型及参考文献 Pathotype & reference
湖南	1981—1983	102	金刚 30、IR2、江宁糯、窄叶青 8 号、74-105、南粳 15 日本和菲律宾两套鉴别品种	有	9 个（陈寘等，1984）
安徽	1977—1978	35	IR26、鄂 42、农垦 57、矮育粳、棒槌稻、金刚 30	无	Ⅰ～Ⅲ（承河元，1979；1999）
	1997—1998	30	IR26、Java14、Tetep、金刚 30＼Java14	无	
湖北	1989—1991 1992—1994	208 108	金刚 30、Tetep、南粳 15、Java14、IR26	无	0～Ⅳ（成国英等，1994；钱汉良等，1997）
四川	1976—1989	160	IR26、DV85、金南风、Tetep、TN1、早生爱国 3 号、Java14、南粳 15、黄玉、金刚 30、八四矮 63、南京 11	强弱差异	0～Ⅳ（何明等，1993）
宁夏	1985—1987	77	IR26、南粳 15、矢租、金南风、金刚 30	有	8 群（肖思心等，1988）
合计		4 068			

6.2 全国水稻白叶枯病菌致病性分化研究

方中达等（1981）首先从 32 个国内、外抗性强弱不等的品种中，选择了 IR26、南粳 15、金南风、江宁糯、金刚 30 五个鉴别品种，测试了 1975—1979 年间从全国采集的 258 个菌株。结果表明，不同菌株的毒力虽有明显差异，但未发现特异性反应。根据菌株在这 5 个品种上的反应型，划分为 4 个毒性强弱不等的菌系群。

许志刚等（1987）将上述研究中测试的多数菌株，加上 1983—1985 年采自全国的 253 个菌株，在 DV85、矢租、南粳 15、金南风、金刚 30 五个品种上测试，划分为 5 个菌系群，出现了少数表现较强的特异性反应组合，但多数组合为弱互作关系。

为了探明中国稻区病菌的毒性分化，1985—1988 年间，由南京农业大学、江苏省农业科学院、广东省农业科学院和中国农业科学院组成了全国白叶枯病菌致病性研究组，制订了统一试验方案，收集了全国 835 个菌株，分别在北京、南京、扬州、广州等地，以 30 个鉴别品种进行预备试验，确定了 IR26、Java14、南粳 15、Tetep 和金刚 30 五个鉴别品种。据参试菌株在这套鉴别品种成株期接种的抗性反应，划分为 7 个致病型（pathotype）。在 IR26、Java14、南粳 15、Tetep 和金刚 30 上的系列反应型分别依次为 RRRRS（CⅠ）、RRRSS（CⅡ）、RRSSS（CⅢ）、RSSSS（CⅣ）、SRRSS（CⅤ）、RRSRS（CⅥ）和 RSSRS（CⅦ）。其中有些菌株与品种组合已表现出特异性互作关系，如南粳 15 与广东菌株 GD1358 组合表现高抗并出现褐斑，而对 OS-225 和 JS49-6 高度感病；相反，IR26 对 OS-225 和 JS49-6 表现高抗，对 GD1358 为中度感病。但是多数病菌-品种组合表现弱互作反应模式，或者仅一方有鉴别性，病菌-品种间没有呈现强互作交叉反应。

全国在此阶段的主要致病型，在北方粳稻区多为Ⅱ和Ⅰ型；南方籼稻区以Ⅳ型为主，Ⅱ型次之，并有少量Ⅴ型菌存在，长江流域籼、粳混栽稻区则以Ⅱ、Ⅳ型为主。致病型-品种互作及不同农业生态区的代表菌株见表 9。至今，这套鉴别系统一直应用于我国抗白叶枯病育种或抗性遗传研究（方中达等，1991）。

在我国水稻主要生态区中，品种布局、生态环境特殊的华南和云贵高原等稻区，其 *Xoo* 致病型的变化与全国多数地区不同，是我国水稻白叶枯病鉴别系统逐步完善中的重要组成部分。

表 9　中国水稻白叶枯病菌致病型及代表菌株（方中达等，1990）

Table 9　Pathotypes and representative strains of *Xanthomonas oryzae* pv. *oryzae* in China (Fang et al.，1990)

致病型 Pathotype	在鉴别品种上的抗性反应[a] Reaction on differential cultivar					代表菌株及来源 Representative strain and origin		
	金刚 30	Tetep	南粳 15	Java14	IR26	南方稻区 South rice region	长江流域 Yangtze River	北方稻区 North rice region
Ⅰ	S	R	R	R	R	GD1329（粤）	OS200（云），JS97－2（泸）	HLJ85－72（黑）
Ⅱ	S	S	R	R	R	GZ1008（黔）	OS40，KS－6－6（苏）	HB85－17（冀）
Ⅲ	S	S	S	R	R	FJ856（闽）	JS158－2（浙）	NX85－42（宁夏）
Ⅳ	S	S	S	S	R	GX878（桂）	浙 173（浙）	JL86－76（吉）
Ⅴ	S	S	R	R	S	GD1358（粤）	—	—
Ⅵ	S	R	S	R	R	—	OS198（桂）	LN85－57（辽）
Ⅶ	S	R	S	S	R	—	JS49－6（湘），OS225（浙）	—

[a] R，抗 (Resistant)；S，感 (Susceptible)。

6.3　华南水稻白叶枯病菌致病性分化

6.3.1　中国水稻白叶枯病菌Ⅴ型致病型的发现　20 世纪 70 年代，我国从 IRRI 引进了一批携带抗白叶枯病 *Xa4* 基因的品种，在各地表现高度抗病，被直接用于大面积生产，并作为主要抗白叶枯病育种的抗源。

徐羡明等（1981）用广东省 5 个白叶枯病鉴别品种 IR26、农垦 57、窄叶青 8 号、包胎矮和金刚 30，测试了 1978—1980 年间在全省 62 个县（市）采集的 211 个 *Xoo* 菌株，首先发现湛江和梅县的 8 个菌株能中度侵染携有 *Xa4* 基因的品种 IR26（表 10），该报道曾引起业内同行的疑义。但是在 80 年代中期，通过全国 *Xoo* 菌致病性变异的协作研究，终于肯定了这一事实，并将能侵染 IR26 的该菌群划分为中国Ⅴ型菌。

广东省病理学家继续侦察发现，1981—1994 年间，除Ⅳ型菌仍然为该省优势菌群外，Ⅴ群菌的出现频率已从 1983 年的 0.7%上升至 1994 年的 12.2%（徐羡明等，1987；1994；曾列先等，2002）。

表 10 广东省水稻白叶枯病 V 型菌在鉴别品种上的反应（徐羡明等，1981）

Table 10 Reaction of Guangdong pathotype V on the Guangdong differntial cultivars (Xu et al., 1981)

菌株 Strain	来源 Origin	IR26 病级[a,b] Score	农垦 57 病级 Score	窄叶青 8 号 病级 Score	包胎矮 病级 Score	金刚 30 病级 Score
242	佛山坑仔口	4. 15 (MS)	6.00 (HS)	6. 00 (HS)	6. 00 (HS)	6. 00 (HS)
172	高州木头塘	4. 15 (MS)	5. 60 (HS)	5. 87 (HS)	5. 95 (HS)	6. 00 (HS)
105	阳江西岸	3. 95 (MS)	5. 29 (S)	5. 78 (S)	6. 00 (HS)	6. 00 (HS)
234	大埔古城	3. 80 (MS)	5. 95 (HS)	5. 95 (HS)	6. 00 (HS)	6. 00 (HS)
262	从化旗杆	3. 70 (MS)	5.10 (S)	5.93 (HS)	5. 85 (HS)	5. 83 (HS)
238	五华农科所	3. 65 (MS)	5. 93 (HS)	6. 00 (HS)	6. 00 (HS)	6. 00 (HS)
158	信宜农科所	3. 60 (MS)	4. 50 (S)	5. 55 (HS)	5. 60 (HS)	5.95 (HS)
217	阳春城南	3. 56 (MS)	5. 07 (S)	5.93 (HS)	6. 00 (HS)	6. 00 (HS)

[a] 调查标准——病斑长度（lesion length）：1 级，0～10 mm；2 级，11～20 mm；3 级，21～30 mm；4 级，31～40mm；5 级，41～80mm；6 级，81 mm 以上。

[b] HR，高抗（Highly resistant），1～1.4；R，抗（Resistant），1.5～2.4；MR，中抗（Moderately resistant），2.5～3.4；MS，中感（Moderately susceptible），3.5～4.4；S，感（Susceptible），4.5～5.4；HS，高感（Highly susceptible），5. 5～6。

6.3.2 华南与亚洲主要稻作国家白叶枯病菌毒性比较 伍尚忠 1979 年在国际水稻白叶枯病联合鉴定现场考察中，发现南亚地区的强毒菌株能侵染携有 *Xa4* 甚至 *xa5* 和 *Xa7* 基因的品种。随后，伍尚忠（1983；1985）在 IRRI 进行了 3 项合作试验，比较了我国华南与南亚、东南亚国家白叶枯病菌群之间的异同与渊源，结果如下。

①中国 5 个鉴别品种与华南及印度、越南、泰国、孟加拉、尼泊尔、菲律宾等国家代表菌株间的互作反应结果表明，各国菌株的毒性存在明显差异。在Ⅰ、Ⅱ、Ⅲ、Ⅳ、Ⅴ五个菌群中，Ⅰ群菌的毒力最弱，Ⅱ群菌有中等毒力，Ⅲ、Ⅳ为强毒菌群，Ⅴ群菌的毒性最强，该群菌中的菌株能不同程度地侵染 IR26。其中东南亚参试菌株的毒力很强，分别属于Ⅲ、Ⅳ和Ⅴ群。印度、尼泊尔和孟加拉的菌株与广东 4 个菌株都属毒性最强的Ⅴ群菌（表 11）。

表 11 华南及东南亚水稻白叶枯病菌在中国鉴别品种上的毒性比较（摘自伍尚忠等，1983）

Table 11 Comparison of virulence reaction of strains of *Xanthomonas compestris* pv. *oryzae* between South China and Southeast Asia on the Chinese differential cultivars (Based on Wu et al., 1983)

菌株 Strain	地点 Origin	菌群 Group	在鉴别品种上的病级及相应反应型[a,b] Reaction on the Chinese differential cultivar				
			JG30	Baotaiai	Zaiyeqing 8	Nongken 57	IR26
RB003	广东番禺 (GD)	Ⅰ	7.0 S	3.1 R	1.1 R	1.2 R	0.5 R
RB16-7	广东崖县 (GD)	Ⅱ	9.0 S	4.0 S	2.5 R	1.3 R	1.0 R
RB129	印度 (India)	Ⅲ	7.0 S	5.5 S	4.2 S	2.3 R	1.4 R
RB054-5	广东陵水 (GD)	Ⅲ	7.5 S	5.1 S	3.4 R	3.2 R	1.2 R

（续）

菌株 Strain	地点 Origin	菌群 Group	在鉴别品种上的病级及相应反应型[a,b] Reaction on the Chinese differential cultivar				
			JG30	Baotaiai	Zaiyeqing 8	Nongken 57	IR26
RB135	印度（India）	Ⅲ	8.8 S	6.7 S	5.6 S	2.8 R	1.9 R
RB113	泰国（Thailand）	Ⅲ	9.6 S	6.7 S	6.2 S	2.3 R	1.9 R
RB068－5	越南（Vietnam）	Ⅳ	21.2 S	10.7 S	10.2 S	6.7 S	1.7 R
RB139	菲律宾（Philippines）	Ⅳ	18.2 S	18.5 S	13.8 S	4.6 S	1.3 R
RB068	越南（Vietnam）	Ⅳ	23.9 S	23.4 S	10.7 S	1.5 R	0.2 R
RB111	泰国（Thailand）	Ⅳ	22.6 S	16.0 S	17.4 S	12.5 S	2.1 R
RB064－2	广东陵水（GD）	Ⅳ	28.1 S	12.7 S	18.2 S	13.5 S	0.6 R
RB131	尼泊尔（Nepal）	Ⅴ	18.0 S	15.2 S	10.7 S	4.3 S	5.7 S
RB172	广东高州（GD）	Ⅴ	20.3 S	14.6 S	18.3 S	11.1 S	4.6 S
RB105	广东阳江（GD）	Ⅴ	22. S	15.5 S	15.9 S	11.5 S	4.3 S
RB133	尼泊尔（Nepal）	Ⅴ	20.1 S	19.3 S	18.1 S	7.4 S	5.4 S
RB234	广东大埔（GD）	Ⅴ	24.1 S	16.4 S	19.7 S	13.6 S	4.0 S
RB242	广东佛山（GD）	Ⅴ	19.3 S	23.2 S	20.9 S	14.6 S	4.9 S
RB137	孟加拉（Bangladesh）	Ⅴ	28.6 S	21.3 S	21.5 S	10.0 S	4.4 S
RB125	印度（India）	Ⅴ	24.2 S	19.1 S	25.4 S	14.3 S	6.0 S

a 3次重复（Three replications），病斑长度（Lesion length）（cm）。

b R，抗（Resistant），0.1～3.5cm；S，感（Susceptible），＞3.5cm。

②用日本、IRRI及中国共三套鉴别品种，测试和比较华南与菲律宾 *Xoo* 的毒性。结果表明，在中国5个鉴别品种上划分为5个菌群的华南菌株，在日本、IRRI鉴别寄主上只能划分成3个菌群。华南菌群Ⅲ和Ⅳ的毒力与菲律宾菌群Ⅰ相似，菌群Ⅴ的毒性与菲律宾菌群Ⅳ相似。

③用日本和菲律宾两套鉴别寄主比较华南与印度尼西亚、泰国、菲律宾各群代表菌株的毒力。结果表明，华南的Ⅲ、Ⅳ菌群与印度尼西亚、泰国、菲律宾的某些菌群的毒力是类似的（表12）。

总之，在国与国、地区与地区之间的白叶枯病菌毒性类型，既有相同亦有相异之处；华南地区当时的新毒性菌群Ⅴ及其优势菌Ⅳ型与南亚和东南亚国家菌群的毒性是类似的。

6.3.3 中国水稻白叶枯病Ⅴ型菌的形成背景 我国Ⅴ型菌的发生频率不仅在广东省逐渐上升，并且不断扩大蔓延，由广东省原发地西部扩散至中部，继之广西的南宁、玉林、梧州、柳州和桂林等市，以及福建、江苏、浙江、云南甚至宁夏等省（徐羡明等，1994；林声和等，1986；肖火根等，1989；曾列先等，2002；孙恢鸿等，1993）。Ⅴ型菌具有极大的威胁性，2003年广东台山一块种植携有 *Xa4* 基因的新品种，因遭遇Ⅴ型菌的袭击基本毁种（周少川，个人通讯，2004），因此，Ⅴ型菌已成为当前广东省水稻抗白叶枯病育种必须抗御的菌群。

中国Ⅴ型菌的产生与其原发地广东省大面积种植的品种基因型及其地理位置有密切关系。该省位于我国大陆最南端，南临热带海洋，地跨中亚热带和北热带，属南亚湿暖及边缘热带湿热季风气候。从地理、气候等环境条件到栽培品种的亲缘关系以及 *Xoo* 菌毒力等情况都与东南亚相近。菲律宾、印度、印度尼西亚等国家引进的 IRRI 品种遭遇毒性菌的侵袭早于我国广东省，随着品种的交流，增加了病菌迁移的机会，Ⅴ型菌在广东省的出现应该是必然的。

表 12 华南与菲律宾、泰国和印度尼西亚的对应菌群毒性比较（伍尚忠等，1985）

Table 12 Comparison of virulence reaction of *Xanthomonas compestris* pv. *oryzae* strains in Philippines, Thailand , Indonesia and South China on the International differentials（Wu et al. , 1985）

菌株 Strain[a]	日本鉴别寄主 Japanese differentials					菲律宾鉴别寄主 Philippine differentials				
	金南风 Kinmaze	黄玉 Kogyoku	Tetep	Wase-aikoku	Java14	IR8	IR20	IR1545	DV85	Cas209
P-Ⅰ	S[b]	S	S	R	R	S	R	R	R	S
I-B	S	S	S	R	R	S	R	R	R	S
T-Ⅰ	S	S	S	R	R	S	R	R	R	S
C-Ⅲ	S	S	S	R	R	S	R	R	R	S
C-Ⅳ	S	S	S	R	R	S	R	R	R	S
P-2	S	S	S	R	R	S	S	R	R	R
T-Ⅲ						S	S	R	R	S
P-3	S	S	S	R	R	S	S	R	R	S
I-A	S	S	S	R	R	S	S	R	R	S
P-4	S	S	S	R	R	S	R/S	S	R	S
I-C	S	S	S	R	R	S	R	S	R	S
C-Ⅴ	S	S	S	R	R	S	S	S	R	S

[a] P：菲律宾小种（Philippine races），1（PXO61），2（PXO86），3（PXO79），4（PXO71）。I：印度尼西亚菌群（Indonesian strain group）。T：泰国菌群（Thailand strain group）。C：中国华南菌群（South China strain group）。

[b] R，抗（Resistant）；S，感（Susceptible）。

6.4 云南省水稻白叶枯病菌致病性分化

6.4.1 云南省白叶枯病菌-水稻特异性互作 云南省地形复杂，生态环境特殊，地方品种多样。20 世纪 50 年代末开始流行水稻白叶枯病，到 80 年代已成为重要病害（段永嘉等，1985）。该省植病学家自 70 年代中期开始研究全省 *Xoo* 菌致病性分化。首先对不同海拔、不同生态地区 94 个县（市）的不同品种类型上采集的 710 个（次）菌株进行了测试。直至 21 世纪初的研究结果都表明，其 *Xoo* 菌与品种之间是强互作特异性关系，发现了很多交叉互作反应的品种-菌株组合。

段永嘉等（1983；1985；1987）自 1976 至 1987 年，先后用抗、感反应明显、稳定，生育期一致，适于本省的籼稻、粳稻 74-105、IR26、秋二矮、珍白 18、窄叶青 8 号、Tetep、成都矮、金南风、南粳 33、农林 27 共 10 个鉴别品种，加上金南风、黄玉、Rantai-Emas、Java14、Heen-Dikwee-1 日本鉴别品种和一个有鉴别作用的品系 IR2071-656-5-5（携有 *Xa4*），测定了全省 94 个县（市）采集、分离的 412 个菌株的毒性。结果初步表明云南菌株与品种的互作属于质量差异。首先，弥渡、祥云、龙陵等县的菌株在鉴别品种上的系列反应型，分别与日本小种Ⅱ、Ⅲ、Ⅳ相同，显示了品种-菌株为质量差异的互作关系，如龙陵和晋宁的菌株都能侵染早生爱国、Java14、Heen 和 Dikwee 品种群，与日本小种Ⅳ的毒性反应是相同的。其次，用 IR2071－636－5－5 测定的云南菌株也有类似 Ogawa 用 IR8 测定日本小种Ⅰ、Ⅲ所呈现的抗、感交叉反应现象。

肖火根等（1989）在段永嘉等的研究基础上，用云南、日本和 IRRI 三套鉴别品种测定了包括云南不同地区新采集的共 73 个菌株。结果表明，在这三套鉴别品种上都出现了 Zadokes 的两对菌株-品种之间强互作交叉反应模式。这些菌株在一套云南新鉴别品种上，就有 21 对组合呈现交叉反应，其中 86－64－4 和 86－64－3 两个菌株分别与 IR8 和窄叶青 8 号两个品种，86－35－1 和 86－52－1 菌株分别与 Heen-Dikwee－1 和窄叶青 8 号的互作更为清晰分明（表 13）。

表 13 选出的鉴别品种与云南白叶枯病菌群的特异性反应（肖火根等，1989）

Table 13 Specific reactions between the selected differential cultivars and pathogen groups in Yunnan (Xiao et al.，1989)

品种 Cultivar	菌株[a] Strain																					
	Ⅰ	Ⅱ	Ⅲ	Ⅳ	Ⅴ	Ⅵ	Ⅶ	Ⅷ	Ⅸ	Ⅹ	Ⅺ	Ⅻ	XⅢ	XⅣ	XⅤ	XⅥ	XⅦ	XⅧ	XⅨ	XX	XXⅠ	XXⅡ
IR_{26}	**R**	S	R	R	R	R	R	R	R	R	R	R	R	R	R	R	R	R	R	R	R	S
Heen-Dikwee-1	**R**	S	R	R	R	R	R	R	R	R	R	S	R	S	R	S	S	S	R	R	S	S
珍白 18	**R**	S	R	R	R	R	R	R	R	R	S	R	S	R	R	S	R	S	R	R	S	S
早生爱国 3 号	**R**	S	R	R	R	S	S	S	S	R	S	S	S	R	S	R	S	R	R	R	S	S
Tetep	**R**	S	R	R	R	S	S	R	R	S	R	R	S	S	S	S	S	S	R	S	S	S
窄叶青 8 号	**R**	S	R	S	R	S	R	S	R	S	S	R	S	S	S	S	S	R	S	R	S	S
IR_8	**R**	R	S	R	S	R	S	R	S	S	R	S	S	S	S	S	S	S	S	R	S	S
南粳 33	**R**	S	R	S	S	S	S	S	S	S	S	S	S	S	S	S	S	S	S	S	S	S

[a] R，抗（Resistant）；S，感（Susceptible）。

此外，云南不同地区的毒性菌群差异也已反映在生产中，1985 年育成的抗病品种滇新 10 号在白叶枯病流行的新平和元江两地，表现一抗一感，经研究明确了它们分别属于云南Ⅶ和Ⅸ两个不同毒性群。认为以小种替代致病型的条件可能成熟了。

进入 21 世纪以来，云南高原粳稻区的水稻白叶枯病日益严重。余腾琼等（2003；2005）研究了该稻区病菌的毒性分化。用毫糯扬、TN1、黄玉、珍珠矮、IR26、南粳 33 和金南风共 7 个云南高原粳稻作为鉴别品种，将 36 个菌株划分为 9 个致病型，其反应型为Ⅰ型，RRRRRRS；Ⅱ型，RRRRSRS；Ⅲ型，RSRRRSS；Ⅳ型，RRRRRSS；Ⅴ型，RRSSSSS；Ⅵ型，SSRSSSS；Ⅶ型，SSSSSSS；Ⅷ型，RRSRRSS；Ⅸ型，RRRSSSS，显示了菌株与品种之间的非连续性反应。一些菌株鉴别品种组合表现的交叉互作是清晰的。如致病型Ⅴ和Ⅵ与鉴别品种 TN1 和黄玉，以及致病型Ⅷ和Ⅸ与鉴别品种黄玉和珍珠矮的交叉互作反应非常明显，其显著性测试结果见表 14。

表 14　云南高原粳稻区白叶枯病菌致病型与品种间交叉互作反应显著性测验（余腾琼等，2005）

Table 14　Significance test of reversible reaction between pathotypes and cultivars（Yu et al.，2005）

品种 Cultivar	致病型 Pathotype	病斑长度[a]/cm Lesion length	*t* 值 t value	双尾 *P* 值 Sig.（2 tailed）	*F* 值	*P* 值
TN1	Ⅴ	1.78±0.71（R）	−8.649	8.07×10⁻⁴**		
	Ⅵ	13.74±3.06（S）				
					180.733	3.1×10⁻¹⁴*
黄玉（Kogyoku）	Ⅴ	9.52±1.98（S）	6.573	8.83×10⁻⁶*		
	Ⅵ	2.14±2.42（R）				
黄玉（Kogyoku）	Ⅷ	8.5±1.78（S）	5.034	0.001*		
	Ⅺ	2.93±0.85（R）				
					85.492	8.08×10⁻⁸*
珍珠矮（Zhenzhuai）	Ⅷ	2.81±0.57（R）	−13.739	7.6×10⁻⁷*		
	Ⅺ	8.07±0.5（S）				

a　R，抗（Resistant）；S，感（Susceptible）。

*　0.01 水平的显著性差异（Significantly different at probability level of 1%）。

**　极显著（Highly significant）。

6.4.2　云南省水稻白叶枯病菌毒性特点　云南省 *Xoo* 菌的毒性较强，致病型结构复杂，与全国致病型有很大差异。姬广海等（2003）以丰锦、IRBB1、IRBB14、IRBB3、IRBB4、IRBB5 和 IRBB21 作为鉴别品种，测试了云南各区域的 51 个代表菌株。根据其互作反应，划分为 14 个“小种”，其中竟有 8 个菌群能侵染或中度侵染 *Xa21*。这些菌株在全国鉴别品种上测试的结果表明，在 14 个“小种”中，除 1～8 外，另 6 个小种与全国 7 个致病型的毒性反应型不同。余腾琼等（2005）对云南高原粳稻区菌株的毒性研究也表明，在所划分的 9 个致病型中，就有 5 个（Ⅱ、Ⅴ、Ⅵ、Ⅶ、Ⅸ）能侵染 IR26，而在全国 7 个致病型中，只有Ⅴ型菌能中度侵染 IR26，即使其中有一群与全国Ⅴ型菌相似，另外 4 个则有别于全国其他菌型。Noda 等（1996）用日本和菲律宾两套共 12 个鉴别品种测试 1987—1994 年我国南方的 24 个菌株，发现 2 个云南菌株除 IR20 外，还能侵染 DV85、IR1545 - 339 等 11 个鉴别品种。Noda 等（2004）又用这 12 个鉴别品种，测试了 1994—1998 年间采集的 38 个云南菌株，将其划分为三群，当时来自云南北部和中部高原粳稻区的群Ⅰ和群Ⅱ菌株的毒力较弱，分别只对 2 个和 4～7 个品种有毒性，来自南部籼型常规稻和杂交稻种植地区的菌群Ⅲ的毒力较强，对 9～12 个品种都有毒力。而余腾琼等（2005）近年的研究表明，云南高原粳稻区菌株毒性已经趋强、致病型多样。前后两个研究间隔不到 10 年，表明云南高原粳稻区菌株毒性分化的速度很快，病菌群体结构的变化复杂。这种状况在梁斌等（2003）的研究中也得到了确认，他们采用 rep-PCR、IS-PCR 标记技术，将云南高原粳稻区 17 个菌株和本省其他 7 个地区的菌株聚集成 4 个簇（cluster），高原粳稻区的菌群竟然在所有 4 簇中都有分布。Zhang 等（1995，1997）在采用 RFLP、rep-PCR 等分子技术分析全国 *Xoo* 菌遗传多样性中，就发现云南菌株的分子型始终与众不同。后来，王春连等（2001）用全国鉴别品种对

1996—1999 年长江以南 *Xoo* 菌进行测试，结果表明，这些特殊分子型的云南菌株是一群有别于中国 7 个致病型的新菌群。以上研究充分表明云南白叶枯病菌小种构成的复杂性及其特殊性，颇有继续研究的价值。

6.5　中国水稻白叶枯病菌致病型分布

南京农业大学每 10 年进行一次全国 *Xoo* 菌致病力分化监测，指出全国病菌群体构成基本稳定，只有个别菌群有所变化。目前仍以Ⅱ型和Ⅰ型为优势群，但能侵染 *Xa4* 基因品种的Ⅴ型菌的频率有所增加，并即将达到 20%的流行阈值（许志刚等，2004）。

各稻区白叶枯病致病型的分布，在南方籼稻区以Ⅳ型最多。对 *Xa4* 和 *Xa21* 有毒性的Ⅴ型菌频率，不仅在原发地广东省有所上升，并已从华南扩展到长江流域，甚至西北的宁夏。长江中下游籼、粳稻区，仍以Ⅳ型、Ⅱ型为主。北方粳稻区则以Ⅱ和Ⅰ型居多。

云南省是我国 *Xoo* 菌群体结构最具多样性的地区，无论在该省西南部籼型杂交稻和常规稻区或高原粳稻区，病菌毒性类群都比较复杂。该省 *Xoo* 菌群体结构的特点，反映了其与寄主品种布局和生态环境的密切关系。姬广海等（2003）报道在全省 14 个“小种”中，YN3 和 YN8 为优势小种，YN3 主要分布于红河、德宏地区，YN8 则在大理、玉溪等地。只有聚合 2～4 个抗白叶枯病基因的品系，才能抗御分布在该省的毒性菌群，这些菌群能侵袭 *Xa21*、*xa5*、*Xa7* 等广抗谱基因。但是，在籼稻区表现感病的携有 *Xa21*、*xa5* 基因品种，却能抵抗高原粳稻区菌株；云南品种扎昌隆 *Xa22t* 能抗御高原稻区的全部菌株，表现广谱抗性。云南 *Xoo* 菌“小种”的分布明显受地区间隔影响，与寄主品种基因型的对应关系也很清晰，该省科学家的研究结果为云南地区水稻品种白叶枯病抗性遗传改良提供了重要依据。

7　水稻白叶枯病鉴别系统的建立

7.1　鉴别品种的选择

病原菌生理专化（physiologic specialization）划分的基本单位是生理小种，这是根据病菌在一套标准寄主鉴别品种上引起的一系列有毒或无毒反应型（reaction pattern）来划分的。因此能够引起病菌致病力表达强弱反应分明的一套鉴别品种，是测定病菌变异的基本条件之一。

日本病理学家山田昌雄（1990）述及日本初期选择稻瘟病鉴别品种时，只根据各菌株反应型来确定。随着遗传学的研究，发现有以下缺陷：①携有两个或两个以上抗性基因品种，基因型构成太复杂，不适于用作鉴别品种；②缺乏正在育种中利用的抗性基因，因此不能鉴别是否存在着与这些抗性基因对应的毒性小种；③有些鉴别品种携有生产应用品种中不具备的未知抗性基因，按其所细分的毒性小种无实际意义；④一套鉴别品种中有相同抗性基因，其鉴别力没有差异而成为多余的品种。我们不妨借鉴其某些经验，参考下列条件选择鉴别品种：①鉴别力强，对病菌小种敏感，非抗即感，界限分明；②稳定性，在不同条件下的鉴定结果一致；③代表性，生产推广的代表品种或主要育种抗源；④用已知单基因抗性品种，既能避免只采用国际鉴别寄主的局限性，也可弥补只用本国（地区）品种的不足，便于国际交流。如果有一套遗传背景相同、携有与小种相应的单基因抗性近等基因系，则更为理想。以上条件①和②是鉴别品种的基本选择要求，条件③表明鉴别品种不是一成不变的，生产上的代表品种和采用的抗源亲本（基因），必然随着生产发展不断变更，原有鉴别品种的代表性会越来越小，根据各地生产品种的抗性变化情况选择鉴别品种，同时采用若干可供变换的后备鉴别寄主。

如果只用一套国际鉴别品种，虽然便于国际交流，但所测菌群难以反应本国生产的实际状况，或者根本缺乏鉴别性。正如 Stakman 等（1962）指出，所有植物病原菌小种不可能归于一览表内，因为：①难于从全世界数千个寄主品种或野生生物型上获得适宜样本；②难于确定在任何时间采用的鉴定方法都是可信赖的。对于广泛发生在全球稻作国家的水稻白叶枯病而言，很难用一套鉴别品种就能测定复杂的农业生态系统中如此多样的病菌毒性。20 世纪 80 年代 Mew 与伍尚忠和 Horino 在 IRRI 的合作研究中，发现不同国家或地区之间病菌毒性类型的差异很大，不同国家所用的鉴别品种只适用于本国（地区），用日本的鉴别体系不能区分菲律宾小种群，反之亦然。方中达等（1991），孙恢鸿（2004）和谢关林（1989）的研究结果表明，无论是日本或 IRRI 的鉴别品种都只能将中国华北、广西的白叶枯病菌划分为有毒或无毒两群；将浙江菌株划分为三群。但是，当这些菌株在中国推广品种上测试时，则能表现鉴别反应。用日本和 IRRI 两套鉴别品种测试过广泛分布于印度的菌株，也仅分为两个小种，毫无疑问，如果增加了携有不同抗性基因的印度推广品种测试，必然可以区分出更多小种。大量研究表明，各国白叶枯病菌致病型分化都有其自身特点。因此，采用本国主要推广的品种和抗源作为基本鉴别品种，根据不同农业生态区的情况，增加地区辅助品种，再选用若干代表性国际鉴别寄主，可以及时测定新致病型和掌握小种消长信息，这也是多数国家和我国一些省（区）所采用的做法。

7.2 中国及主要稻区（省）的鉴别品种的选用

表 8 中已列出我国主要科研、教学院校不同时期进行白叶枯病菌毒性分化研究所采用的鉴别品种。多数省（自治区）所采用的基本鉴别品种大同小异。

（1）金刚 30，中熟籼稻，高感，至今未发现与其相应的无毒基因。多数菌株形成的病斑长度都达到叶长的 50%左右；南粳 15（*Xa3*）是我国主要抗源亲本，一直作为抗中国Ⅰ、Ⅱ群菌的粳稻代表品种；IR26（*Xa4*）是中国Ⅴ群菌的鉴别品种；Java14（*Xa1*、*Xa3*、*Xa12*）高抗中国多数菌株，对广东、广西、江西等地部分菌株表现稳定感病。

在具有 IR24 遗传背景的近等基因系中，IRBB3（*Xa3*）有广谱抗性；IRBB4（*Xa4*）与 IR26 携有相同的抗性基因，一直被用作中国Ⅴ型菌的鉴别品种；IRBB5（*xa5*），抗所有中国菌株；IRBB14（TN1，*Xa14*）对菲律宾小种 5、小种 8 以及我国有些菌株具有特异抗性，是菲律宾鉴别系统中新增加的品种。IRBB21（*Xa21*）是近年广泛用于国内外育种的白叶枯病抗源供体。

（2）云南省采用的主要鉴别品种有珍珠矮（无）、金南风（无）、黄玉（*Xa1*、*Xa12*）、IRBB1（*Xa1*）、IRBB3（*Xa3*）、早生爱国 3 号（*Xa3*）、IRBB4（*Xa4*）、IR26（*Xa4*）、IRBB5（*xa5*）、IRBB14（*Xa14*）、TN1（*Xa14*）、IRBB21（*Xa21*）、IR8（*Xa11*）、Tetep（*Xa1*、*Xa2*、*Xa16*）、丰锦（*Xa18*）以及未知抗病基因品种 Heen-Dikwee－1、珍白 18、窄叶青 8 号、南粳 33、毫糯扬。上述鉴别品种基本上可归纳为三类：①已明确抗性基因，单基因，或携有 2～3 个基因。②不同品种（系）携有相同基因，如 IRBB3、早生爱国 3 号、

南粳15等品种都携有 *Xa3*，*Xa4*（IRBB4、IR26），*xa5*（IRBB5、IR1545－229、矢租），*Xa14*（IRBB14、TN1），*Xa1*（IRBB1、黄玉）等基因。这些品种的遗传背景不一，有时即使与相同菌株的互作结果也未必一致。如余腾琼等（2003）用黄玉测试云南高原粳稻区的第8致病型菌株为抗，而以黄玉为供体育成的IRBB1却表现感病；又如，携有相同白叶枯病抗性基因的两套不同遗传背景的近等基因系，CBB系统品系（轮回亲本沈农1033）和IRBB系统品系（轮回亲本IR24）对相同的对应菌株进行比较，CBB系统的抗性水平高于IRBB系统（许志刚等，2004）。

（3）未知抗性基因。根据选择鉴别品种的原则，提出如下建议：①对病菌小种具有抗、感分明的抗性基因，如 *Xa4*、*Xa14* 等；②我国主栽杂交稻组合和栽培稻品种所携有的抗病基因，仍以 *Xa4*、*Xa3* 为主；已经和正在育种中利用的抗性基因有 *Xa21*、*Xa7*、*xa5* 和 *Xa23* 等；*Xa22t* 是有潜在利用价值的抗性基因；③避免采用相同抗性基因的品种，不宜采用携有一个以上抗性基因的品种，根据地区菌株情况，增加辅助鉴别品种；④籼稻、粳稻兼备；⑤明确所采用的鉴别品种之抗性基因，包括那些已经或正在利用的地区特异鉴别品种。在携有相同抗性基因的品种中，经过统一鉴定、比较，确定一个代表品种。

7.3 水稻白叶枯病鉴别系统研究的若干问题

水稻白叶枯病菌致病性分化研究中的一个普遍问题是菌株毒性没有明显差异。Mew（1987）曾对此作了两种解释：①有些菌株对某些品种的毒性范围可能很宽，而对另一些品种较窄，因此菌株在一系列测试品种上产生的病斑长度的均值经过制表计算成毒性指数后，可能出现一个连续增加或递减的数据，就使菌株和品种的个体之间显示不出清晰的互作反应；②采用了只有相似的或中等抗性的、缺乏鉴别力的品种。

对于水稻白叶枯病菌是否存在小种的质疑，人们从IRRI对菲律宾 *Xoo* 的系统研究中已经有了答案。但是这里还要稍微介绍一下值得借鉴之处。Mew（1987）剖析了产生怀疑的焦点在于观察和分析 *Xoo* 鉴别毒性的方法——人们在测试两对不同的品种和菌株后，在寄主-病菌组合中通常只得到“发病”或“不发病”的表象，并没有去仔细分析是菌株的毒性特异性还是寄主的抗性特异性作用，把水稻对 *Xoo* 的抗性看作绝对的而不是相对的性状。事实上，不同寄主-病菌组合之特异作用的程度是不相同的，实验数据经过仔细比较分析后，才能发现逐步研究的焦点所在。与所有科学研究一样，任何项目首先应基于某一理论依据，提出理论假设后，再从不同环节来制定试验方案。Mew首先确定 *Xoo* 与任何生物一样都存在变异，那么是什么样的变异？用什么方法来描述其变异？Mew在寻找 *Xoo* 病菌有没有毒性专化性的答案的过程中，首先进行了大规模的品种、菌株测试，在菌株被划分为“毒性群”的基础上，曾通过如下设计进一步观察菌群内菌株的毒性差异，从菲律宾菌群1、菌群2中各选5个共10个菌株，在IR8、IR20和IR1545－339三个品种上进行测试。10个菌株与IR8品种之间都表现亲和互作，就在与IR1545－339都不亲和的菌株中，出现了与众不同的个别菌株，通过比较菌株清晰可辨的侵染性后，菌株之间的毒性差异就区别了出来。观察与IR20品种亲和的菌群2所形成的病斑要比不亲和的菌群1的长得多，不过，归于同一菌群中的菌株所产生的病斑长度也不是一致的，这样，就可以从不同病斑长度结合病斑的侵染型，甚至从亲和组合中分辨出不亲和组合；或从不亲和组合中区分出亲和反应的菌株。通过不同试验设计和重复测试来分析这些表型结果，去发现亲和与不亲和组合之间的那些病斑长度显著不同的病菌-品种组合。结果表明，水稻-病菌系统是存在毒性专化作用的，尽管这些表现还只属于弱互作，但从中了解到有些“毒性混杂”的异质型菌群。当 Mew 等发现了鉴别品种 Cas 209 后，终于将“毒性混杂”的菌群区分为清晰的两个小种群，出现了寄主-病菌间的强的互作交叉反应，取得了水稻白叶枯病菌存在着生理专化作用的佐证，它们与稻瘟病、锈病等真菌病菌生理专化作用是相似的，从而形成了一套鉴别性能强的水稻- *Xoo* 系统。

7.4 建立和完善我国水稻白叶枯病鉴别系统的商榷

我国水稻白叶枯病菌毒性分化研究的历史悠久，种植水稻的省（自治区）的植物病理学家几乎都进行了这项研究工作。随着水稻生产快速发展，品种不断更换，所利用的抗白叶枯病基因由单一化而逐渐增多，无论是全国性的或不同农业生态区的白叶枯病鉴别系统的建立或完善都有迫切的需要。对照国内外经验和存在的共性问题，提出以下商榷意见。

（1）被广泛用于国际交流和研究的菲律宾 *Xoo* 鉴别系统所包括的鉴别品种与小种之间表现的寄主-病菌的垂直互作反应清晰（clear-cut），充分呈现寄主-病菌基因对基因关系。显然，筛选鉴别力强的品种和毒性差异分明的菌株是建立鉴别系统的关键。我国白叶枯病鉴别系统的完善，首先要进一步筛选理想的鉴别品种和代表菌株。其中还涉及调查和选择的标准。

由于白叶枯病为维管束病害，细菌的增殖受制于维管束，一般形成不整齐花斑，不易进行准确的测量。大多数白叶枯病寄主-病菌反应不总是很分明的。在抗病育种或遗传研究中抗性和毒性是两个相对术语，从各世代筛选综合性状较好的抗或中抗植株，或根据分离群体内抗、感个体所自然形成的峰槽进行数据分析时，则可按照一般调查标准。然而，作为鉴别品种与鉴别菌株，则应强调选择那些抗/感反应和毒性清晰的个体。将抗感病斑长度的交界处作为区分抗、感的界限是不合理的。例如准备入选的“抗病”鉴别品种上的病斑长度甚至已经超过了 5 cm，这种在“抗”与中抗之间的抗性品种，一般由多基因或由主效基因与修饰基因协同作用控制，这种呈现侵袭性差异（aggressiveness difference）的病菌与品种互作双方，一旦被选为鉴别品种或菌株，其稳定性差，经不起环境条件的考验，显然不是理想的。必须选择互作反应清晰、稳定的品种和菌株。

比较记载方法的结果表明，对于鉴别性明显的品种，测量病斑长度和目测病斑面积都能表明菌株与品种互作的实际反应。那些包括了籼、粳及爪哇型等不同亚种的鉴别品种，其植株叶片的大小、长短等形态各异，当调查其抗病或感病反应时，几种调查方法的结果并没有明显差异，但是对于中抗至中感反应，测量病斑与叶片长度的比值，往往高于病斑面积占叶片面积之比，两种调查结果可能有明显差异。病斑占叶片面积的百分率则比较合理。如果采用遗传背景一致的近等基因系作为鉴别品种，其生育期、形态等性状基本一致，则可以避免上述干扰，可以采用任何记载方法。

（2）进一步完善我国鉴别系统。方中达等（1990）所建立的中国 *Xoo* 菌鉴别系统，是在我国水稻生产中长期集中利用 *Xa4* 或 *Xa3* 抗病基因的阶段进行的。因此，病菌群体组成相对简单，病菌与品种间的互作基本上表现为强与弱的数量差异。10 多年来，都用其测试全国主要稻区优势菌群的消长，并出现了能侵染 *Xa4* 的第Ⅴ致病型。这套鉴别系统基本上客观地反映了我国当时水稻白叶枯病菌毒性分化状况，一直是全国抗白叶枯病育种和抗性研究的主要工具，也为我国水稻白叶枯病鉴别系统的进一步完善奠定了基础。

随着我国水稻生产的发展及抗病新基因的发掘、引进和利用，我国水稻主栽品种的白叶枯病抗性基因范围正在拓宽。在国内病理学家的不断研究中，白叶枯病菌与品种之间的互作关系已出现了特异性互作组合，甚至发现了较多垂直的、强互作交叉反应的品种-菌株组合。虽然各自采用的鉴别品种、接种方法、调查方法等有所不同，但通过对照和比较，结果大体上是类似的，已经存在协调的基础。云南省具有的特殊生态条件和遗传多样性的品种，提供了深入研究的条件。进一步的工作是：①各地区以本地水稻生产为基础，建立区域鉴别系统。在此基础上，组成以云南、广东省等华南、长江中下游、华北等稻区在内的全国协作研究组，制订统一的研究方案，进行比较，寻找共性结果；②从携有相同抗性基因的若干品种中，选择鉴别性能最佳的品种，并确定适合特殊稻区测试的地方辅助鉴别品种；③选择全国包括主要稻区的小种代表菌株；④形成包括遗传背景清楚、携有单基因抗性品种和毒性分明的鉴别菌株的鉴别系统。

根据我国水稻生产快速发展的需要，完善全国水稻白叶枯病鉴别系统的目标，应该在今后的共同合作研究中得以实现。

水稻白叶枯病菌的群体结构和遗传多样性*

章　琦[1]　林汉明[2]

（1. 中国农业科学院作物科学研究所；2. 香港中文大学生物系）

Population Structure and Genetic Diversity of *Xanthomonas oryzae* pv. *oryzae*

Zhang Qi[1]　Han-Ming Lam[2]

（1. Institute of Crop Sciences, CAAS; 2. Department of Biology, The Chinese University of Hong Kong）

摘要　20 世纪 80 年代中期，洛克菲勒基金会设立了生物技术项目，促进了 *Xoo* 菌群体遗传结构研究。Leach 等（1988）首先从 *Xoo* 基因组中检测到两个 DNA 重复序列，用 pJEL101（IS1112）为探针进行菲律宾水稻白叶枯病菌 RFLP 分析。全球开始广泛采用分子技术探究 *Xoo* 病原菌群体生物学和遗传学及寄主-寄生物关系的系统发育。

在测试水稻白叶枯病菌群体遗传结构的分子标记中，采用插入因子 IS1112、IS1113 和 pBS*avrXa*10（1 个来自 *Xoo* 的无毒基因家族）等 RFLP 探针，可获得清晰、稳定、数据一致的指纹图谱。用 rep-PCR 引物 IS1112（J3）、ERIC、REP 等基于聚合酶链式反应标记的测试结果相同于 RFLP，且快速、简易、费用低廉。

90 年代，IRRI、美国、中国、菲律宾、韩国、尼泊尔、印度、印度尼西亚和马来西亚等国科学家协作，采用分子技术结合病菌毒性分析，研究 *Xoo* 病菌群体遗传结构演变，结果如下：

（1）病菌分子型与致病型之间存在一定对应关系。菲律宾病菌小种或中国致病型的 RFLP 型，在两个小种或致病型中都没有出现重叠。

（2）菲律宾 *Xoo* 群体遗传结构系统分析形成的立体图像，清晰地展示了病菌群体遗传演变的内在联系，表明：①分子型多元演化的异质性小种孕育了新毒性型，并分离出 3 个新小种；②寄主基因型是引起病菌遗传适应性演变的主因。通过病菌分子型遗传多样性差异图像，推演了菲律宾 *Xoo* 菌群系统发育及其历史渊源，诠

* 林汉明博士参与撰写的工作部分由香港研究资助局资助项目 CUHK4273/02M 及香港教资会卓越学科领域植物及农业生物科技项目 AoE－B－07/09 资助。

释了品种-病菌-环境间的互动关系。

（3）*Xoo* 菌群体结构差异存在于国与国或一国的不同农业生态地区。种质交换促成了病菌迁移，并可突破地域屏障。寄主选择和病菌迁移，导致 *Xoo* 菌群体遗传多样性的地区差异。

中国 *Xoo* 病菌群体遗传结构分析表明：①其结构简单，明显受制于寄主抗病基因的单一利用，侵袭 *Xa4* 基因的广东Ⅴ型遗传多样性值上升最快，传递了品种丧失抗性的信息；②分子型与致病型有一定对应关系，多数致病型具遗传异质性，中国Ⅴ型菌及云南菌群不同于其他地区，云南菌群含有中国 7 群菌之外的多个新毒性型，具备发现特异鉴别品种和小种的研究条件，对于完善全国水稻白叶枯病鉴别系统具有重要作用；③不同稻区病菌分子型有明显区别，提供了因地制宜利用抗病基因的信息。

（4）Levy 等首先提出以谱系代表菌株取代病菌小种进行抗性评价的设想，以分子技术分析所传递的新毒性菌信息为基础，为抗性遗传改良计划提供超前性科学依据，改变寄主品种基因追逐病菌快速变异的被动局面。该设想能否实现，尚需要更多研究和生产实践的验证。

Abstract In the mid 1980s, Rockefeller Foundation established a biotechnology program to promote studies on the genetic structure of *Xoo* populations. Leach et al. (1988) first detected two repetitive DNA sequences in *Xoo*. Using pJEL101 (IS1112) as the RFLP probes, they successfully analyzed the *Xoo* population in the Philippines. Since then, molecular techniques were adopted worldwide to study the population biology, genetics, and phylogenetics of *Xoo* with respect to host-pathogen interaction.

Among the RFLP probes used in the study of the genetic structure of *Xoo* populations, the insertional elements IS1112, IS1113, and pBSavrXa10 (an *avr* gene from *Xoo*) exhibited clear, stable, and reproducible DNA fingerprinting patterns. Primers for rep-PCR such as IS1112 (J3), ERIC, and REP also gave results consistent to those of the RFLP experiments. PCR based markers have additional advantages over RFLP since the reaction is faster, simpler, and cheaper.

A combined molecular and pathologenic study related to the evolution of *Xoo* population genetics was launched in the 1990s. The research was led by IRRI, with collaborative efforts by scientists from USA, Philippines, China, Korea, Nepel, India, Indonesia, and Malaysia. The major findings included:

1 Some correlations exist between molecular types and pathotypes of *Xoo*. No overlapping RFLP types were found in the *Xoo* races from Philippines or the *Xoo* pathotypes from China.

2 The 3-D clustering tree resulted from genetic structure analysis of *Xoo* races in Philippines clearly indicated the internal connection among *Xoo* population and genetics during the evolution process: (1) polyphyletic evolution of molecular

types in a heterogenous population led to new pathotypes, resulting in the isolation of 3 new races; (2) the genotype of host plants was the main factor inducing an adaptive evolution in *Xoo* genetics to achieve the diversity in the molecular types of *Xoo*. From these results, the evolutionary and historical relationship among *Xoo* races in Philippines were deduced. The dynamic relationship among host plant genotypes, *Xoo* pathotypes and environmental factors was explained.

3 There were variations in the *Xoo* population structure among different countries or among different agroecological regions in the same country. The geographical barrier was broken due to the pathogen migration as well as the activity of germplasm exchanges. Host selection and pathogen migration led to genetic diversity of *Xoo* in different regions.

Analysis of *Xoo* population genetic structure in China led to the following findings. First, the simple population genetic structure was due to the cultivation of host plants carrying a single R (*Xa4*) gene. The rapid increase in genetic diversity of the *Xa4* insensitive Guangdong V indicated the breakdown of host resistance. Second, some correlations were found between molecular types and pathotypes. Most pathotypes were heterogenous populations. The Chinese *Xoo* pathotype V's molecular type and *Xoo* population from Yunan were distinct from those from other regions. The *Xoo* population from Yunan included new pathotypes differed from the 7 known *Xoo* pathotypes in China. New and unique races and pathotypes may be identified therein. It may help in refining the national *Xoo* evalution system. Third, different molecular types were found in different rice cultivation areas. This provides information for the utilization of disease resistance genes in different environments.

4 Levy et al. first suggested to use representative strains from phylogenetic analysis to replace the traditional races for evaluation of host resistance. This harvests on the new information from molecular technologies, providing scientific basis for proactive plan to improve disease resistance. It may end the current passive situation that the application of host genotypes is solely a reaction to the rapid evolution of pathogens. This advocation awaits further verifications by experiments and actual cultivation practices.

在作物抗病育种中最受关注的是寄主抗性的丧失。品种丧失抗性的主要原因是病菌群体遗传结构发生了变化（出现新小种或增加侵袭性），即病原菌的毒性发生了变异。群体遗传结构的变异是指组成一个生物群体的个体之间存在的遗传变异所含的异质性或多样性，以及这些属性的时空演变。用传统病理学手段测试病原菌致病力的表现可以观察到它们能够侵袭哪些寄主基因型，但是不能洞悉病菌群体结构变异及其毒性分化的遗传实质，因此很难深入地分析病菌群体遗传多样性变化和生理小种的系统发育及其演化。比如菌株在一套鉴别品种上测定后，虽然反应型类似的菌株被划分在某一小种内，而事实上有些小种的遗传组分可能已经或正在发生不同程度的变化，只是传统的检测技术无法确定而已。用分子标记技术结合多维数据分析，可以在病菌基因组 DNA 水平上得到对相应寄主基因型毒性的遗传信息，假如该愿望能够实现，只需将异质型菌群的代表菌株在生产中的主要抗性基因品种上进行测定，就可能揭示病原菌毒性演变的遗传实质。根据病菌群体结构与有关毒性变化的情况，为制定深层次育种策略、品种抗性的持续改良提供重要的前瞻性信息。

20 世纪 80 年代，科学家开始用分子标记技术研究寄主-病原菌互作关系的群体分子遗传学。Cook 等（1989）采用 RFLP 技术检测了香蕉细菌萎蔫病菌（*Pseudomonas solanaceanum*）三个小种、五个变种（pathovars）的遗传多样性，结果表明病菌 RFLP 型与其毒性和寄主的过敏性反应有较明显的对应关系。Levy 等（1991）研究南美洲稻瘟病菌（*Pyricularia grisea*）RFLP 型与致病型的关系，发现哥伦比亚稻瘟病菌谱系（lineage，相同 RFLP 型中的代表型）的毒性与对品种的抗性是相应的，其结果也比较简单，来自一个谱系的几个或所有菌株侵染抗病品种的选择性十分明显，病菌谱系对品种的侵染范围表现出某种遗传约束力。Chen 等（1995）用 RFLP 探针 MGR586 分析菲律宾 Laguna 省高地稻区 1 516 个稻瘟病菌株（*Pyricularia grisea*）群体结构，发现病菌遗传谱系毒性与寄主基因型的关系相当复杂，但是各谱系菌株都显示出一定的毒性谱，不同谱系只侵染不同抗性基因品种，寄主对病菌群体结构有较强的选择作用，品种的地区分布及其种植季节都能影响菌株“亚群”的分布。因此，认为病原菌群体结构与其毒性特异性之间的相关性或许能为制定育种策略、合理采用寄主基因提供信息。20 世纪 80 年代在洛克菲勒基金（The Rockefeller Foundation）立项资助下，国际水稻研究所（IRRI）、美国、菲律宾、中国、尼泊尔、韩国、印度、印度尼西亚等国家植物病理学家和分子遗传学家先后开展了水稻白叶枯病菌（*Xoo*）群体结构和遗传多样性研究。20 多年来，应用多种分子标记进行了亚洲主要稻作国家 *Xoo* 群体结构的研究，诠释了 *Xoo* 的群体生物学和遗传学寄主-寄生物关系的系统发育（phylogeny）。研究取得很大进展，使水稻白叶枯病成为植物病害研究中广泛采用分子标记技术结合传统方法进行病菌群体演化和抗病育种研究的一个模式病害。本文将叙述有关基本概念、用于分析 *Xoo* 菌群体结构的方法、研究进展，以及今后值得探索的问题。

1　用分子技术研究水稻白叶枯病菌群体结构和遗传多样性

了解病原菌群体遗传结构变化动态基于病菌基因组 DNA 指纹图谱分析，即主要利用随机分散的、具重复序列的遗传因子，形成能进行数值分析的指纹型。采用的主要技术为 DNA 杂交（如 RFLP 分子标记法）和聚合酶链反应（PCR）扩增（如 RAPD、rep－PCR 等）。

1.1　分析白叶枯病菌的 RFLP 分子标记

利用 RFLP（restriction fragment length polymorphism）方法产生 DNA 指纹的基本原理是用限制性内切核酸酶将病菌基因组切割成一系列不连续的 DNA 片段，通过凝胶电泳把这些片段按分子质量大小逐一分开，再以随机重复分布于基因组内的核酸标记作为探针进行 DNA 杂交和分析，形成指纹形态，来描述病菌基因组中多个基因位点上呈现的群体遗传结构的动态差异，便于较为全面和精确地反映其遗传结构实质。

重复插入序列（repetitive insertional sequence）家族所具备的特性提供了观察 *Xoo* 基因

组多样性的分子标记（Leach et al.，1995）。Leach 等（1988，1990）首先从 *Xoo* 基因组中检测到两个 DNA 重复序列。其中一个 2.4kb 的 *Eco*R1－*Hin*dⅢ片段克隆于 pUC18 质粒内，定名为 pJEL101，该片段携有插入序列 IS1112。在已测试的 100 多个 *Xoo* 菌株基因组中，都含有 80 多个 IS1112 拷贝，pJEL101 因此可以作为 RFLP 探针。她们最初用以测试小范围菌株样本的指纹时，可以观察到菌株 DNA 型与小种之间的某种联系，但是大范围样本的测试结果却不太清晰。

为了进一步有效地观察病菌 DNA 谱系与毒性之间的关系，Nelson 等（1994）采用了不同插入序列作为探针，根据其 *Eco*RⅠ酶切 *Xoo* 基因组 DNA 所呈现的杂交谱型，选择了 4 个转座因子（transposable element），包括来自 *Xoo* 菌株 PXO112 的 TNX1、TNX6、TNX7 和来自菌株 PXO285 的 TNX8。TNX8 与重复序列杂交形成的质粒具有原核的 IS，将其定名为 IS1112（等同于 Leach 等报道的 IS1112）。TNX1 也有典型的插入序列特点，被定名为 IS1113。经引导式再取样分析（Bootstrap Analysis），IS1112 和 IS1113 的指纹图谱比较清晰。其中以 IS1112 所提供的信息最为详细，指纹带也很多，但带数过多，难以准确读数。IS1113 同样能形成明显的谱系组群，杂交带较 IS1112 减少一半，易于读数，两个探针都可应用。

尽管移动因子（如转座子）在显示微生物遗传分化上能起一定作用，但是在提高分子识别或选择效率上，则不会像无毒基因那样能够激发病菌群体产生较强的选择压。Hopkins 等（1992）从 *Xoo* 中克隆了一个 3.1kb *Bam*H1 片段的无毒基因 pBS*avr Xa*10 探针。Nelson 等（1996）比较了无毒基因 pBS*avr Xa*10 和 4 个转座因子 IS1112、IS1113、TNX6、TNX7 共 5 个探针，分析了 155 个 *Xoo* 菌株的遗传分化效果，其中 pBS*avr Xa*10、IS1112 和 IS1113 的 RFLP 指纹图谱相对稳定、清晰，数据一致，便于分析。研究表明，虽然用 RFLP 分析是研究 *Xoo* 菌株群体遗传结构多样性及其演变可信度较高的技术，但却费时费工、费用高昂。其后利用较简易的 PCR 指纹测定，可以快速、大规模地进行病菌标样的分析，使深入检测广阔地区的病原菌群体多样性变为可行。

1.2　分析白叶枯病菌的 PCR 分子标记

基于聚合酶链式反应（polymerase chain reaction；PCR）的 *Xoo* 菌群分子标记技术，包括随机扩增多态 DNA（randomly amplified polymorphic DNA，RAPD）及连接介导 PCR（ligation-mediated PCR）等（George，1995；Nelson，1996）。美国堪萨斯州立大学和 IRRI 的实验室以 IS1112 因子末端来命名的 IS－PCR 引物（后定名 J3），具有上述 RFLP 探针的优点。用这些引物鉴定 *Xoo* 菌群的结果与 RFLP 分析相类似。

基于散布重复序列的 PCR（rep－PCR）是快速简易、费用低廉，无需进行 Southern 印迹检测的分子标记技术。在原核生物内，rep－PCR 技术的引物序列来自三种主要基因标志：①基于基因外重复回纹 REP（repetitive extragenic palindromic）序列（Versalovic et al.，1991）；②基于大肠杆菌重复基因间共有序列 ERIC（enterobacterial repetitive intergenic consensus）（Hulton et al.，1991）；③基于 BOX 序列（Martin et al.，1992）。De Bruijn（1992）较早采用了 rep－PCR DNA 引物 REP 和 ERIC，对土壤革兰氏阴性细菌分离菌基因组进行了分类，发现各分离菌 DNA 的 PCR 产物指纹图谱具有特异性，与参试细菌种之间的系统发育数据高度相关。Eardly 等（1990）根据 14 个酶基因位点等位基因检测，发现土

壤固氮菌 *Rhizobium meliloci* 分离菌中 REP 和 ERIC PCR 产物的一系列分子型明显地按系统发育树状排列，结果都表明 rep－PCR 可用于鉴定细菌菌株的遗传分化并进行分类。Louws 等（1995）发现 rep－PCR 可以产生与细菌 *Xanthomonas* 及 *Pseudomonas* 不同致病变种（pathovar）之特异基因组指纹。后来又发现可以对 *X. campestris* pv. *vesicatoria* 基因组结构进行分类。

rep－PCR 技术能否取代 RFLP 方法分析白叶枯病菌群？Vera Cruz 等（1996）通过两个实验比较了 rep－PCR 与 RFLP 分子标记两种技术检测 *Xoo* 的结果。一个实验是从 Leach（1992）和 Nelson 等（1994）用 RFLP 测试的菲律宾 *Xoo* 所形成的 4 个遗传谱系中选出了 30 个代表菌株，将 rep－PCR 的 REP 和 ERIC 两个引物扩增产物指纹型的组合数据，进行三维聚类统计分析，结果也形成了类似 RFLP 的 4 个谱系，并且也测出属于小种 1 中的一个独特菌株 PXO35，完全符合 RFLP 所测的结果。另一个实验是用 RFLP 探针 IS1113 与 rep－PCR 引物 ERIC 和 REP 分析了一块连续三年种植的感病品种上的 *Xoo* 群体遗传多样性，两种技术的结果也相一致。

用 ERIC 和 REP 等长度为 18～22 个碱基的 rep－PCR 引物分析 *Xoo* 菌群体遗传结构的结果比只有 10 个碱基的 RAPD 引物更为稳定和可信，与 RFLP 的分析结果也类似，并且可用来分析大规模样品。

至今可用来分析 *Xoo* 病菌群体结构的分子标记有 4 个转座因子、无毒基因家族以及一套基于 PCR 反应的标记。

1.3　水稻白叶枯病菌群体结构的研究方法

分子技术作为研究植物病原菌群体遗传结构的一种手段，其基本方法是相同的。但对分布在不同国家的 *Xoo* 菌而言，所选择的分子标记、测试致病力所采用的鉴别品种和菌株样本的规模、统计分析等总体试验设计，则需要因材、因地制宜。

1.3.1　分子标记的选择　用不同分子标记分析菌株基因组 DNA 所产生的单元型（Haplotype）称为分子型、RFLP 型、PCR 型，这些指纹类型及被识别的多态性有所不同。适合的分子标记与所测菌株 DNA 会形成分辨率高、清晰和稳定的指纹，所呈现的谱系与小种存在某种程度的对应关系。用其较准确的数据可以构建详细的立体树状谱系图，具有提高识别病菌内在遗传差异的可行性。可以进行微小变化、病菌（致病型/小种）内部所发生的细微遗传差异、系统发育动态的剖析，洞察深层次区域间病菌演变的遗传实质。这正是传统植物病理学方法所不能及的。

在上述 RFLP 探针中，用 pBS*avr Xa*10、IS1112 和 IS1113 分析菲律宾、中国等亚洲主要产稻国家 *Xoo* 的结果均比较理想。相比之下，pBS*avrXa*10 的指纹型较为清晰，便于读数，但需要在恒温 4℃的冷室内进行 70 多个小时的凝胶电泳；采用 IS1112 和 IS1113 只需在室温电泳 20～22h，但指纹较多，清晰度稍差。

在 rep－PCR 的 4 个引物中，ERIC 扩增产物显示的带位较多，用其分析中国病菌群体结构的结果与用 RFLP 的区别不大；IS1112 带型的鉴别力较大；REP 形成的指纹型比较单一，但这 3 个引物所显示的表型图都是可信的，类同于用 RFLP 探针的结果，可以应用。用 BOX 分析中国菌株的指纹型过于单一，鉴别力很低。

1.3.2　抗性基因品种的选择　用分子技术测出病菌基因组 DNA 的异质性小种后，要选择合适的抗病基因品种测试异质性小种代表菌株的毒性，来确定病菌群体遗传结构的变异。对菲律宾 *Xoo* 群体以 IS1113 为探针的 RFLP 分析结果表明，一些小种的菌株呈现几个单元型，如小种 1、2、3 和 5 分别有 7、5、8 和 4 个单元型（RFLP 代表型），表明它们都是处于遗传异质状态的小种群，如以 80%相似系数水平进行聚类分析，上述各小种的菌株都会被聚集于不同的簇（cluster）。如小种 1、小种 3 和小种 5 的参试菌株分别分布在 3 个、2 个和 3 个簇内。病菌分子型这些细微的遗传变化传递了其毒性型可能已经发生演变的潜在信息。下一步就是病菌的毒性是否发生了变化的求证。具体的做法是选择小种不同单元型的代表菌株，采用除标准鉴别品种之外的若干已知抗性基因品种，通过鉴定和比较这些特殊单元型菌株与其原有小种的毒力，便有可能发现不同毒性的菌株，从而鉴别出新的小种。Leach 等（1992），Nelson 等（1994）和 Vera Cruz 等（1996）连续分析菲律宾 *Xoo* 群体遗传结构后，从原有 6 个小种中，发现并鉴定出小种 7、8、9 三个新小种。

可选择如下品种用于致病力的测试。

①具有鉴别性的携抗病新基因 *Xa21*（IRBB21）、*Xa23*（CBB23）、*Xa22t*（扎昌隆）等品种。

②携有水稻白叶枯病抗性基因的近等基因系（NIL）。以 IR24 为轮回亲本的籼稻 NILs IR-BB1（*Xa1*）～IRBB14（*Xa14*）、IRBB 21（*Xa21*）、IRBB 27（*Xa27*）、CBB23（B，*Xa23*）；以沈农 1033 为轮回亲本的粳稻 NIL CBB2（*Xa2*）、CBB3（*Xa3*）、CBB4（*Xa4*）、CBB12（*Xa12*）和 CBB14（*Xa14*）。

③携有已知抗性基因的供体品种，籼稻有 Tetep（*Xa1*、*Xa2*、*Xa12*）、DV85（*xa5*、*Xa7*）、BJ1（*xa5*、*xa13*）、IR1545（*xa5*）、CAS209（*Xa10*）和 TN1（*Xa14*）。粳稻有黄玉（*Xa1*，*Xa12*）、早生爱国 3 号（*Xa3*）、矢租（*xa5*）和南粳 15（*Xa3*）等。

④具有鉴别性的其他已知抗性基因品种

1.3.3 菌株取样 根据研究目的和试验假设来确定取样的规模、时间、地点、寄主品种，进行系统的随机取样。只有采集的样本代表研究地区病菌群体的整体遗传变异，才能明确遗传多样性的分布与全貌。因此要求掌握所在地区的环境以及农业生态系统（包括寄主植物、品种演变历史）、病菌迁移、突变、遗传重组等基本情况。

样本的大小和范围取决于发生频率最低或很低的珍稀目标菌株能够按预定概率出现而不至于遗漏。一般可按下列公式计算：

$$N \geqslant \log \left[1-P/\log(1-F)\right]$$

N 是样本数，F 为已知其发生频率最低或很低的目标菌株频率，P 是取样时设定的概率（保险系数一般为 0.99 或 0.95）。其原理是，设已知最低或很低的目标菌株频率为 1/100，则其他菌株的频率总和为 1－1/100＝99/100。要使所有非“目标”菌株在全部样本中的概率小于 0.01，即 $(99/100)^N$，就可以保证频率最低或很低“目标”菌株的出现达到预期概率 0.99，即所取的全部样本中至少有一个是“目标”菌株。用对数计算结果，$N \geqslant 454.5$ 个样本。

菌株概率（F）	取样数（*N*）	
	P ＝ 0.99	P ＝ 0.95
0.01	454.5	295.7

0.05	83.7	58.3
0.10	43.7	28.4

［P ＝ 保险系数（概率）］

总之，取样策略就是以最小菌株的样本数代表病菌群体的最大遗传多样性。IRRI 和美国学者的合作研究提供了一个值得参考的例子。Ardales 等（1993；1996）在研究菲律宾吕宋岛各种不同农业生态系统 *Xoo* 群体多样性时，沿着全岛 310km“横切面”，选择山区地方品种和低地现代品种共 78 个点，系统采集了 1 200 多个样本，通过分析这样规模的样本，其结果能清晰地显示出不同生态系统、地点、田块之间病菌群体遗传差异，甚至获得在一个采集点上的菌株指纹谱系与小种群的符合率高达 99%的完美结果。当然，这是属于一个较为完整的大规模试验。一般研究应进行必要的指导性预备实验以减少取样误差，在了解病菌群体多样性的大致轮廓之后，再设计取样方案，包括时间、范围（空间）、地点、寄主基因型等因子和取样数量。

1.3.4 数据分析

（1）遗传变异估测 生物群体结构的演变是由突变、选择、迁移、遗传漂变等引起基因频率变化而造成的。因此，病菌群体基因频率是用于测量群体遗传变异度的基本单位。Nei（1981；1987）用基因多样性概念来描述生物的有性和无性群体遗传变异。基因多样性（H）是来自一个群体标本的两个单元型个体在一个基因位点获得两个不同等位基因的概率。一个基因的多样性值为 1 时，意指来自一个群体标本基因位点的两个等位基因是不相同的；如果所测标本的基因位点没有等位基因变异，则其多样性值为 0，即在遗传上是一致的。由于不了解许多无性群体中表现型的遗传控制，通常称这种情况为单元型。因此用不同单元型组成的复合基因型频率可以衡量其遗传多样性。例如，病菌菌株基因组 DNA 杂交分析中所呈现的各种各样指纹型经处理后就可当作单元型对待，用单元型频率代替基因频率以相似的方式即可计算出遗传多样性，其公式如下：

$H=[n/(n-1)](1-\sum X_i^2)$，$X_i$ 为群体中第 i 个样品的单元型出现的频率；n 是所测菌株数。

（2）系统发育分析（phylogenetic analysis） 用基因和单元型频率数据构建系统发育树状图可以描绘个体之间的遗传谱系关系。Li 等（1991）将树状分析法分为距离矩阵法（distance matrix）和最大节约法（maximum parsimony）。前者用原始数据获得相似性或距离矩阵来构建系统发育树。后者直接用核苷酸序列、内切酶位点、指纹杂交带的有或缺等数据，按计算程序取其最小值逐步演算，最后形成系统发育树。一般情况下，数据被输入后即形成基因序列或呈现表型的特点。以二进制数值编码，分别以 1 或 0 表达各菌株基因组 DNA 指纹在同一位置上有带或无带，并被指派至相应的各等位基因位点，这时并不知道具体的等位基因位点，将这些数据输入电脑进行相似性系数分析（Dice's coefficient of similarity）。采用简单的非加权配对算术平均法（UPGMA；unweighted pair group method of arithmetic means），将被输入数据转化成相似性矩阵或遗传距离矩阵。距离矩阵法根据相似系数水平构成树状图（phenogram）；最大节约法则分析描绘为进化树（cladogram），用以表明祖先-后代的谱系关系。图 1 显示了 DNA 杂交数据如何转化为便于描述菌株间亲缘关系的表型图解。

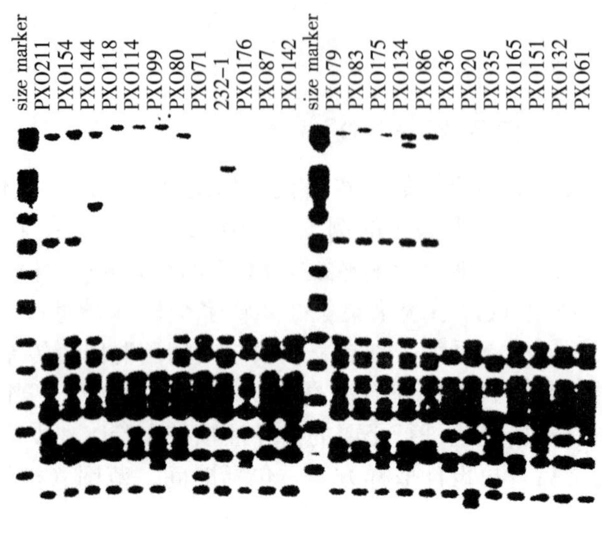

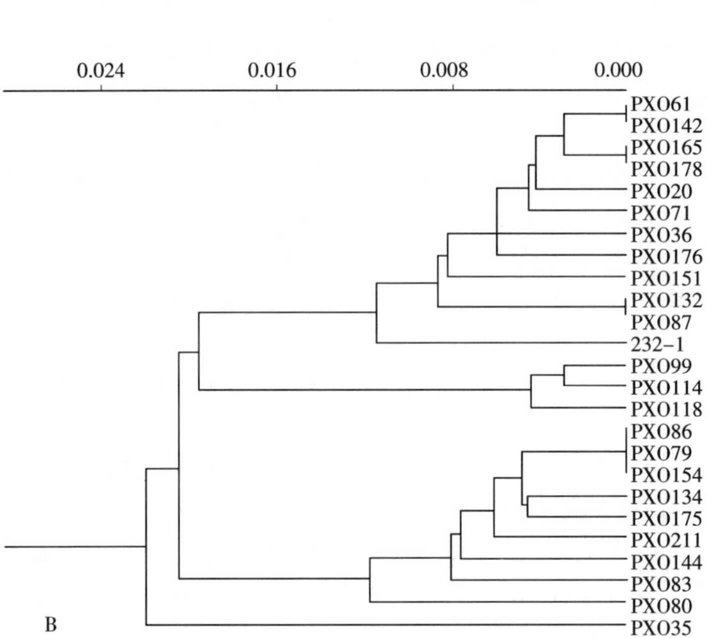

图 1　DNA 杂交数据转化为菌株遗传关系（Leung et al.，1993）。用无毒基因 avrXa10 为杂交探针的 *Xoo* 菌株 RFLP 分析（A）。以 1 或 0 代表同一位置上有带或无带数据，通过 UPGMA 法转化为遗传距离所形成的表型图（B）。数据引自 Vera Cruz 等（1992）的文献

Fig. 1　DNA hybridization data translated into genetic relationships (Leung et al., 1993). RFLP analysis of *Xanthomonas oryzae* pv. *oryzae* by using of cloned avirulence gene avrXa10 as a hybridization probe (A). The presence and absence of a band at a position was recorded as 1 and 0. The data were transformed into genetic distance to produce a phenogram by the unweighted pair group method (UPGMA) (B). Data provided by Vera Cruz et al., (1992)

2 病菌群体遗传结构系统发育与致病型谱系的对应

随机分子遗传标记所区分的病菌谱系能否与按其毒性划分的小种相对应是首先要明确的问题。Levy 等（1991a）用重复因子（repetitive element）MGR586 作为 RFLP 探针分析美国稻瘟菌，发现存在 8 个可以区别的谱系，这些谱系与致病型分群有密切联系，其结果相当完美。他们进一步用 DNA 指纹分析了哥伦比亚水稻种质筛选圃的稻瘟菌，发现可将 50 多个致病型（pathotype）划分为 7 个谱系（lineage），各谱系与一个特异致病型有关，同时观察到在一个谱系中的几个致病型毒性之间表现出逐步变化的相互关系，表明用分子标记能够快速显示致病型遗传谱系的演化状况（Levy et al.，1991b）。尽管有些研究表现出很多致病型与谱系之间的对应关系相当复杂，但是 Levy 等的研究发出了可以详细分析这两者之间存在着某种关系的信息。

利用病菌基因组的特异重复序列对水稻白叶枯病菌基因组 DNA 指纹分析的研究，不仅观察到了分子型不同的谱系的菌群与其毒性的对应关系，并且揭示了病菌群体遗传结构的形成及其演变的某些内在联系。

2.1 水稻白叶枯病菌分子型与致病型演变的关系

在 Mew 等（1992）对菲律宾 *Xoo* 菌小种及其系统发育研究的基础上，Leach 等（1990；1992）用 pJEL101 作 RFLP 杂交探针分析了菲律宾不同地区包括小种 1～6 六个小种的 98 个菌株，结果是：①在所发现的 27 个 RFLP 型中，没有一个 RFLP 型在两个小种中重叠出现，表明菲律宾 *Xoo* 小种的划分是清晰的，这就具备了从交叉繁杂的分子型中剖析病菌群体结构变化的基础；②各小种群的菌株基因组 DNA 所形成的 RFLP 型多寡不一，小种 1、2、3 和 5 分别有 7、5、8、4 个 RFLP 型，这些小种在遗传上显然处于异质（heterogeneous）状态，表现出多元演化（polyphyletic）；而小种 6 只有两个 RFLP 型，小种 4 仅一个 RFLP 型，是遗传同质（homogenous）菌群（表 1）；③各小种菌株的 RFLP 型的差异不一，小种 5 与小种 2 的多数菌株的 RFLP 型非常相似，只有一条带不同；但有些小种内菌株间的 RFLP 型的差异很大，如小种 1 的 RFLP 型 1 与型 3 之间竟有 12 条带不同，小种 1 的一个菌株 PXO35 在聚类分析中单独形成一个特殊谱系。在初步揭示了菲律宾 *Xoo* 系统发育及其演变的内在联系之后，下一步的研究焦点便集中于多元型演化小种群的成因及其演变。

Nelson 等（1994）采用 4 个转位因子以及从 *Xoo* 分离的 1 个无毒基因，进一步分析一套较大的菌株群体，在检测的 155 个菲律宾 *Xoo* 菌株 RFLP 型数据中，发现了更为详细的群体结构和清晰的谱系关系，虽然 RFLP 数据取决于所用的探针，但是形成的小种谱系进化图却是一致的（图 2）。结果表明两个主要分子谱系都含有几个小种，如图 2 B 所示，小种 1、2、3、5 被聚集于 B 簇，小种 1、2、3、4 被聚集于 C 簇；而有些小种的菌株却又被分散于不同谱系中，如小种 1 菌株分布于 B、F、C 簇，上述提到的小种 1 菌株 PXO35 仍然被单独划分于 B 簇；小种 5 菌株分散于 B、D 和 A 簇；小种 2、3 菌株却始终都在几个相同的簇群中；上述结果确认了在 6 个菲律宾 *Xoo* 小种中，除小种 4 和小种 6 外，小种 1、2、3、5 是表现多元演化（polyphyletic）的 4 个异质菌群。将几个分子标记 RFLP 带型组合

表 1　菲律宾 6 个 *Xoo* 小种 RFLP 型的频率及各群估测的遗传多样性（Leach et al.，1992）

Table 1　Frequency of RFLP types of *Xoo* according to race grouping and the estimated genetic diversity of each group（Leach et al.，1992）

RFLP 型[a] RFLP type	菌株 strain	总群体数 Total Population		各小种 RFLP 型的频率[b] Frequency of RFLP type in each race grouping											
				1		2		3		4		5		6	
		菌株数 No. of strains	小种% % of race	菌株数 No. of strains	小种% % of race	菌株数 No. of strains	小种% % of race	菌株数 No. of strains	小种% % of race	菌株数 No. of strains	小种% % of race	菌株数 No. of strains	小种% % of race	菌株数 No. of strains	小种% % of race
1	PXO61	9	9.2	9	47.3										
2	PXO132	5	5.1	5	26.3										
3	PXO35	1	1.0	1	5.3										
4	PXO36	1	1.0	1	5.3										
5	PXO68	1	1.0	1	5.3										
6	PXO157	1	1.0	1	5.3										
7	PXO170	1	6.1	1	5.3										
8	PXO63	6	10.2			6	31.5								
9	PXO78	10	1.0			10	52.6								
10	PXO137	1	1.0			1	5.3								
11	PXO83	1	1.0			1	5.3								
12	PXO171	1	1.0			1	5.3								
13	PXO79	1	1.0					1	6.7						
14	PXO142	1	1.0					1	6.7						
15	PXO141	6	6.1					6	40						
16	PXO143	1	1.0					1	6.7						
17	PXO87	2	2.0					2	13.3						
18	PXO167	2	2.0					2	13.3						
19	PXO176	1	1.0					1	6.7						
20	PXO177	1	1.0					1	6.7						
21	PXO125	6	6.1							6	100				
22	PXO80	17	17.3									17	65.4		
23	PXO154	4	4.1									4	15.4		
24	PXO145	3	3.1									3	11.5		
25	PXO183	2	2.0									2	7.7		
26	PXO114	5	5.1											5	38.5
27	PXO115	8	6.1											8	61.5
菌珠总数 Total		98		19		19		15		6		26		13	
遗传多样性值 Genetic diversity		0.93		0.73		0.65		0.84		0		0.55		0.51	

[a] 用探针 pJEL101、以 *Eco*R1 酶切 *Xoo* 菌株总基因组 DNA 、Southern 杂交的 RFLP 型（RFLP type was determined from Southern blot analysis of *Eco*R1 - digested, total DNA from strains of *X. oryzae* pv. *oryzae* by using pJEL101 as probe）。

[b] 用水稻鉴别品种 IR8（*Xa11*）、IR20（*Xa4*）、IR1545（*xa5*）、DV85（*xa5*，*Xa7*）和 Cas209（*Xa10*），接种参试菌株后，根据其抗性反应型确定的小种［Race was determined by inoculation to the rice differentials IR8（*Xa11*），IR20（*Xa4*），IR1545（*xa5*），DV85（*xa5*，*Xa7*），and Cas209（*Xa10*）］。

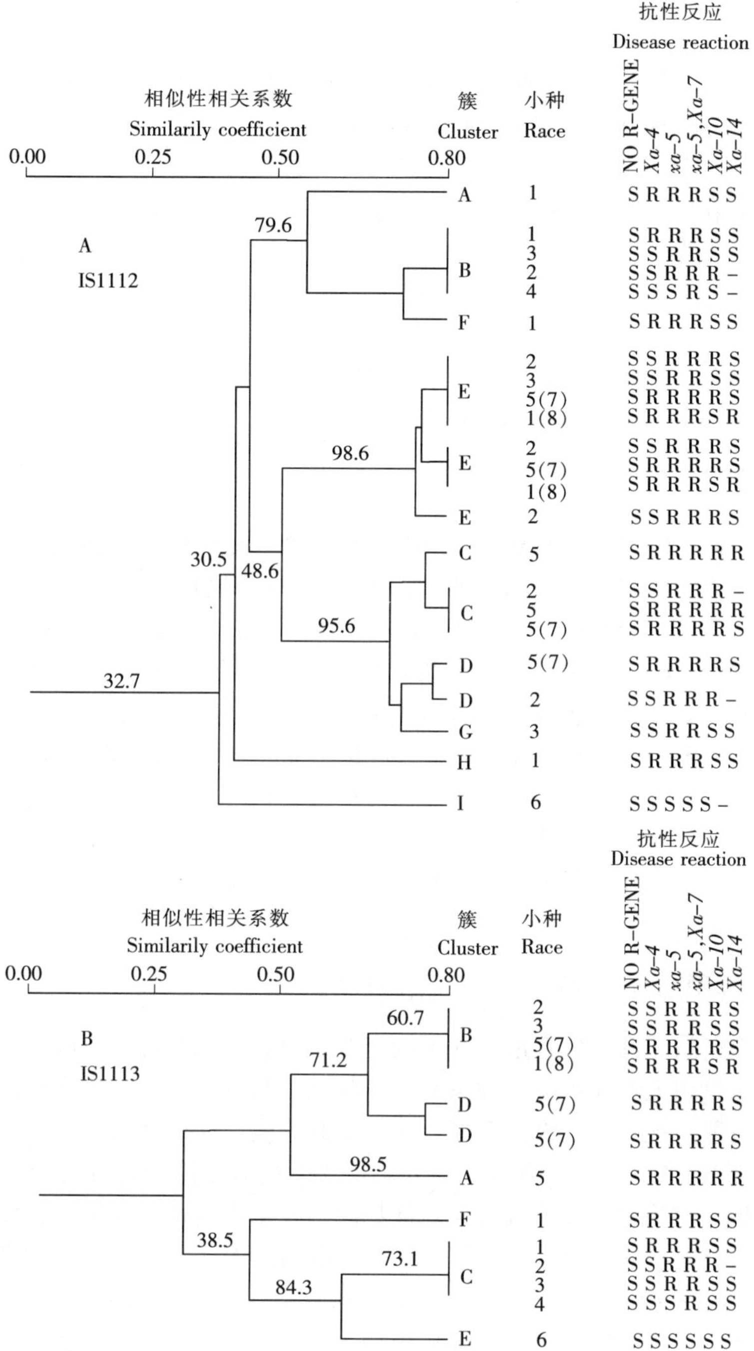

图 2 根据非加权配对算术平均法（UPGMA）构建以探针 IS1112（A）和 IS1113（B）分析 155 个 *X. oryzae* pv. *oryzae* 菌株的 RFIP 数据为依据的小种、谱系演变树状图（Nelson，1994）病菌小种群与 RFLP 谱系群相对应，括号中的数字为命名的小种，图中每一个小种对一套携有已知抗病基因的质量抗性反应。S，感；R，抗

Fig. 2 Dendrograms constructed with UPGMA on the basis of RFLP data with the probes IS1112 (A) and IS1113 (B), for a collection of 155 strains of *X. oryzae* pv. *oryzae* (Nelson, 1994). Numbers shown in parentheses represent the race designations. The qualitative disease reaction of each race is shown for a set of rice lines carrying known resistance genes, S, susceptible; R, resistant

数据聚类统计分析所形成的树状系谱图，显示了详细的病菌系统发育及彼此间的亲缘关系。当用 20 个携有不同抗病基因品种和近等基因系测试这些菌株的致病力时，便从小种 1 和小种 5 的菌株中，分别鉴定出其毒性型有别于原小种的两个新小种：①将分散于 3 个簇群的小种 1 的 10 个代表菌株在对其有鉴别性的抗病基因品种 IRBB1（*Xa1*）、IRBB2（*Xa2*）、IRBB14（*Xa14*）、Tetep（*Xa1*、*Xa2*、*Xa12*）上测试后，从小种 1 中鉴别出毒性与其完全不同的一个新的菌型（菌株 PXO35），定名为小种 8；②分别出现在 3 个簇群中的小种 5 的代表菌株，在鉴别品种 TN1（*Xa14*）上被划分为两个毒性型，一个对 *Xa14* 不亲和，保持了原小种 5 的抗性，另一个与 *Xa14* 亲和，能侵染 *Xa14*，从小种 5 鉴别出一个新毒性型，定名为小种 7。

该研究采用了遗传标记分子探针，把这些遗传证据结合传统病理学侦察结果，构建了菲律宾整体 *Xoo* 病菌群体遗传结构的立体图像，清晰地呈现出病菌群体及其亚群体遗传多样性的演变，初步诠释了品种-病菌-环境的互动关系。为继续揭示和分析表现多元演化的菌群提供了重要信息。

Vera Cruz 等（1996）系统地分析了另一个表现多元演化的菲律宾小种 3 的演变动态。能侵袭 *Xa4* 基因的小种 3 在菲律宾一些地区繁衍生息了较长时间，具备了积累遗传多样性的条件。他们用 RFLP 和 rep－PCR 技术，对菲律宾吕宋岛一块常年重病稻田进行了连续三年的小生境自然侵染的 *Xoo* 群体结构分析，结果表明有两个簇群与 Ardals 等（1996）分析的吕宋岛 *Xoo* 群体的 4 个簇群中的两个是对应的。这两个簇分别由 4 个和 3 个 RFLP 型组成。毒性测定结果表明 IRBB7（*Xa7*）品系将这 7 个 RFLP 型的 32 个菌株区分为毒性截然不同的两个群，一群不侵染 IRBB7，其毒性与小种 3 相同；另一群则能侵染 IRBB7，从而自小种 3 中鉴别出一个新菌群，被命名为小种 9。分子技术结合菌株毒性分析，从处于杂合状态的小种 1、3、5 中，分别鉴别出 3 个新小种，由此，菲律宾 *Xoo* 小种从原来的 6 个增至 9 个。

2.2 白叶枯病菌分子型与致病型系统发育关系

普遍认为病菌从无毒性到有毒性的自然突变是植物病原菌群体适应寄主群体变化的一种手段。这样可以设想无毒性可能是病菌的原始状态。与毒性谱广的致病型相比，对寄主毒性谱窄的致病型可能与祖先的亲缘关系更近些。由此推测，菲律宾高原稻田优势菌小种 5 应是该地区的原始菌群，它对 *xa5*、*Xa10*、*Xa14* 等多个抗病基因都无毒力；小种 1 则应是低地稻田的原始菌群，因为它是该地区最早的优势菌，其毒性谱也比同簇的其他小种窄。

Nelson 与 Mew 等（1994）根据分子标记技术揭示的菲律宾 *Xoo* 群体谱系演变树状图，对菲律宾 *Xoo* 菌的系统发育及其演变作了如下推论：①因为能侵袭 *Xa4* 基因的小种 2 取代了早先广泛分布在吕宋岛低地稻区的优势小种 1 的变化，所以小种 2 有可能是小种 1 的突变菌株，然而小种 2 与小种 1 的 RFLP 带型却有很大差异，也不属同一个簇，从而排除了这种可能。出乎意料的是，后来成为低地稻区优势菌群的小种 2 却与高地优势菌小种 5 的 RFLP 带型非常相似，两者因仅一带之差而被聚于一簇，表明小种 2 很可能由小种 5 突变而来，或随着引种从他处迁移过来；②小种 3 能侵袭 *Xa4* 基因，因而在种植携有 *Xa4* 品种的吕宋岛地区增长得很快，小种 3 与小种 1 的 RFLP 型只有一带之差，表明它们的亲缘关系很近。根据菌群在抗性基因长时间的选择压下能获得毒性而产生新小种的一般规律，从图 2（IS1112

的簇 E、C、D、G = IS1113 的簇 A、B 和 D）中可推测，由小种 5 产生了新小种 7，继而产生了小种 2，并且形成了小种 3 的一个单独谱系；另一个主要谱系，是由小种 1 产生了小种 3 的第二个谱系。

分子探针的精细分析结合毒性测定，形成了病菌群体遗传结构演变的清晰图像，就可能在寄主呈现明显选择作用之前获得病菌系统发育的潜在演变信息。因为中性遗传标记可能反映与病菌小种演变的相关程度，即如果小种变化快，能促使分子表现型有相应的快速变化，因此可以预期在一个特殊分子谱系中就会聚集几个小种；相反，如果小种变化慢，随机的分子多样性则会在各个小种中出现，导致小种与谱系之间的相关性很小。Leung 等（1993）结合致病型和分子分析描绘了菲律宾 *Xoo* 致病型进化图（图 3），归纳了菲律宾 *Xoo* 病菌系统发育谱系和致病型之间的关系。以寄主的鉴别作用阐明了两者之间不同的相关程度：①完全相关。遗传谱系与致病型之间是一对一的关系，表明致病型在分子谱系中是相对稳定的，寄主的选择性弱或致病型变异（分化）低；②部分相关。部分致病型在不同的分子谱系中；③没有相关，随机出现。两者没有系统关系，由于寄主明显强烈的鉴别选择作用，而导致一个分子谱系中存在几个致病型。当然，除了主要受到寄主品种和病菌基因型的制约之外，还会受到环境等其他因素的影响。

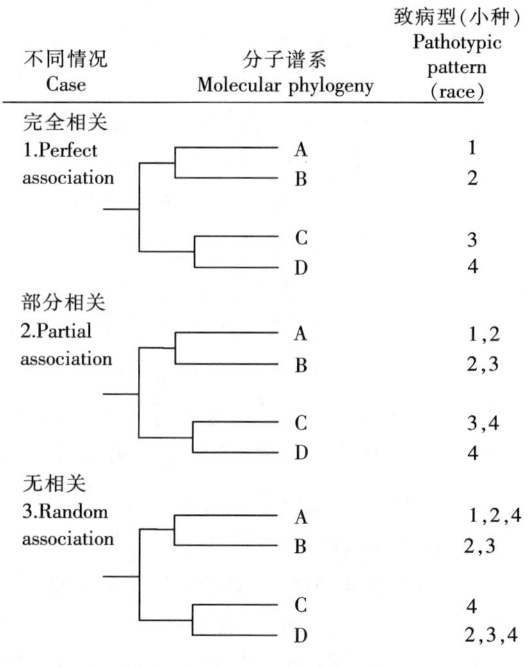

不同情况 Case	分子谱系 Molecular phylogeny	致病型(小种) Pathotypic pattern (race)
完全相关 1.Perfect association	A	1
	B	2
	C	3
	D	4
部分相关 2.Partial association	A	1,2
	B	2,3
	C	3,4
	D	4
无相关 3.Random association	A	1,2,4
	B	2,3
	C	4
	D	2,3,4

图 3　不同情况下病菌系统发育的分子谱系与致病型之间的相关模式（Leung et al.，1993）。A、B、C 和 D 是中性分子标记分析形成的谱系；1、2、3 和 4 是同一套病菌标样经鉴别寄主测试的致病型（小种）

Fig. 3　Different levels of associations between molecular lineages and pathotypic patterns（Leung et al. ，1993）. Lineages A，B，C，and D are defined by neutral molecular markers whereas pathotypic patterns 1，2，3，and 4 are defined by disease interactions with differential hosts

3 水稻白叶枯病菌在不同时空的群体遗传多样性

3.1 时空对白叶枯病菌群体遗传多样性的影响

长期以来，在传统病理学水平上进行病菌群体系统侦察，为抗病基因的合理调配提供了重要信息。Mew 等（1992）从 20 世纪 70 年代开始进行菲律宾白叶枯病的系统监测，定点侦察主产稻区的品种基因型布局、病菌小种频率消长动态，阐明了品种与病菌之间定向选择（directional selection）的密切关系，病菌群体结构随着品种交替，发生了明显变化。

Leach 等（1992）在 Mew 的研究基础上，选择了菲律宾 1972—1976 年、1977—1981 年和 1986—1988 年三个时段采集的 *Xoo* 菌株标样，用 RFLP 探针 pJEL101 估测的群体遗传多样性分别为 0.89、0.90 和 0.92，表明病菌群体变异在这 15 年间均保持较高的水平。连续 3～5 年的 3 个时间段的病菌群体遗传分化系数（coefficient of genetic differentiation；Gr）仅为 0.01，表明时间对遗传差异的影响较小。

病菌群体遗传结构的地区多样性，充分展示在 Ardals 等（1996）对跨越菲律宾不同海拔农业生态区 *Xoo* 菌群体结构的系统研究中。他们分析了菲律宾的 310 km 横切面、覆盖种植传统品种和现代品种的稻田病菌群体结构。菌株样本采自一年一季种植地方品种的高地稻区，兼种地方品种和改良品种的高、低地衔接区和种植双季改良品种的低地灌溉区三个不同农业生态区。在 13 个地区采集了 1 247 个菌株样本，用 RFLP 探针 IS1113 和限制性内切酶 *Pst* Ⅰ分析结合致病力测定，进行了全部菌株基因组 DNA 指纹聚类分析，形成群体和亚群体结构。结果表明不同农业生态区、不同地区病菌亚群之间的差异相当明显。13 个地区间的遗传差异很大（差别系数 Gst ＝ 0.64），形成了 4 簇，山区和低地稻区菌株各成一簇，介于山区和低地间的菌株形成另外两簇。在 78 个采样点中有 70 个（采用 IS1113）和 77 个（采用 *Pst* Ⅰ）点的菌群遗传多样性有明显差异。这种地区、地点，甚至田块之间的小生境差别，表明耕作环境和品种是其主要的影响因子。病菌迁徙现象也是明显的，如有 5 个 RFLP 型菌株都出现在几个采集点中，其中 1 个竟分布在全岛所有 6 个省的 10 个点中。

吕宋岛病菌分子型的地区性与病菌小种群的分布也是相对应的。在 13 个地区中，吕宋岛北部 6 个地区共形成了 19 个 RFLP 型，其中两个地区就有 11 个单元型，它们只分布在适于种植地方品种的高地稻区，这里仅分布小种 5，其中的 1 号区存在小种 5 及由其演化的新小种 7。在吕宋岛中部至南部高低地缓冲区至低地的其余 7 个地区中，小种 2 分布于一个区；小种 3 则集中在中部普遍种植携有 *Xa4* 基因品种的 5 个地区，并与小种 2 共存；在地方品种和现代品种兼有的第 6 和第 12 区，能发现小种 1 和小种 2 的菌株（表 2）。

该研究在 Mew 等（1992）建立的菲律宾 *Xoo* 菌的系统侦察基础上，采用分子技术获得了相当完美的结果，进一步演绎了品种与病菌系统发育、毒性遗传变异在不同生态环境条件下的密切互动关系。

亚洲主要稻作国家 *Xoo* 菌群体遗传结构地区间隔相当清晰。Adhikari 等（1995）用 pJEL101 和 pB*Savr Xa10* 两个 RFLP 探针，对由亚洲 7 个水稻生产国家菲律宾、中国、韩国、印度、尼泊尔、印度尼西亚和马来西亚 1980—1990 年间采集的 308 个 *Xoo* 菌株基因组 DNA 进行指纹聚类分析，构成了区域分明的 5 个簇群（图 4）。其中菲律宾的 95 个菌株组成了清晰的簇 2、3 和 4 三个簇，簇 3 聚集的多数菌株来自东南亚地区的马来西亚和印度尼西亚；中国的

表 2 用 RFLP 探针 IS1113 检测菲律宾吕宋岛 13 个采样点的水稻白叶枯病菌株单元型及小种分布（据 Ardales，1996 年资料整理）

Table 2 Distribution of haplotypes derived from RFLP probe IS1113 and races of *Xanthomonas oryzae* pv. *oryzae* at 13 sites sampled in the Philippines (Summarized from Ardales, 1996)

地区（省） Region (province)	农业生态型[a] Type of agroecosystem	小种 Race	单元型数 No. of haplotype	种植品种 Cultivar planted
吕宋岛北部 **Northern Luson**				地方品种
1 Banaue，Ifugao	TH	5、7	7	Traditional
2 Banaue，Ifugao	TH	5	4	Traditional
3 amulditan，lfgao	TT	8	1	Traditional
4 Lamut，lfugao	IT	2	2	IR60[b]，IR64[b]
5 Nueva Vizcaya	IT	2	2	IR20[b]，IR60[b]
6 Nueva Vizcaya	I/TT	1、2	3	C4[c]
吕宋岛中部 **Central Luson**				
7 Nueva Ecija	IIL	2、3	3	IR64[b]，IR72[b]
8 Tarlac	IIL	2、3	2	IR72[b]
9 Tarlac	IIL	2、3	4	IR72[b]
10 Tarlac	IIL	2、3	2	IR72[b]
11 Tarlac	IIL	2、3	2	IR72[b]
12 Bulacan	IIL	2	5	R10，IR66[b]
吕宋岛南部 **Southern Luson**				
13 Laguna	I/TIL	1、2	5	Malagkit[d]，IR74[b]
Laguna[e]	I/TIL	6	2	
Pelauan[e]		4	1	

[a] 依据海拔和种植的品种划分的农业生态型（Sites were categorized into different agroecosystems based on cultivars planted and the elevation of the sides）：TH，种植地方品种的高地稻区（Traditional highland，planted a traditional rice cultivar）；TT，种植传统品种的高、低过渡地带（Traditional transition zone site planted a traditional cultivar）；IT，种植改良品种的过渡地带（Transition zone site planted an improved cultivar）；I/TT，种植改良和传统品种的过渡地带（Transition zone site planted improved and traditional cultivars）；IIL，种植改良品种的低地灌溉田（Irrigated lowland site planted an improved cultivar）；I/TIL，种植传统品种和改良品种的低地灌溉田（Irrigated lowland site planted improved and traditional cultivars）。

[b] 携有 *Xa4* 基因的改良品种（Improved cultivar with the *Xa*4 gene）。

[c] 不携有 *Xa4* 基因的改良品种（Improved cultivar without the *Xa*4 gene）。

[d] 携有 *Xa3* 基因的传统糯稻品种（Traditional glutinous cultivar with gene *Xa*3）。

[e] 信息来源：Mew 等（1992）的研究资料［Information based on study from Mew et al.，(1992)］。

绝大多数菌株（74/76 个）自成一簇（簇 1）；印度、尼泊尔、韩国参试的所有菌株则聚集于簇 5，簇 5 还包括了中国的强致病力Ⅴ型菌以及另外 3 个国家的部分菌株。研究表明：①地区差异明显地限定了病菌群体分布。在一个地区、一个国家和国家之间的遗传区别是明显的，寄主基因型起了主要分隔作用，病菌的缓慢迁移（扩散）也产生了一定影响。由于一些地区偏于集中种植某些品种或者有些品种只适合在某些国家（地区）种植，从而更加凸显了地域性差异；②一些病菌群体的分布突破了地域屏障，如簇 5 就聚集了亚洲 7 个国家的菌株（图 4）。有些 DNA 指纹图单元型的分布非常广泛，甚至在不同国家出现相同单元型，显然，这是病菌随着国际种质交流活动而出现的迁移现象。

有些地区群体显示了密切的遗传关系。例如，印度尼西亚与菲律宾之间既有明显不同的菌株分子型谱系，也有相似的分子型。印度尼西亚 Java 西部 125 个菌株样本的 75%是优势菌，在其形成的一个谱系中，发现十分类似于菲律宾小种 3（侵染 *Xa4*）单元型的菌群，该地区多年来广泛种植携有 *Xa4* 基因品种 IR64，以致该菌群广泛分布在这里，历史上的迁徙也造成这种情况（George et al.，1997）。

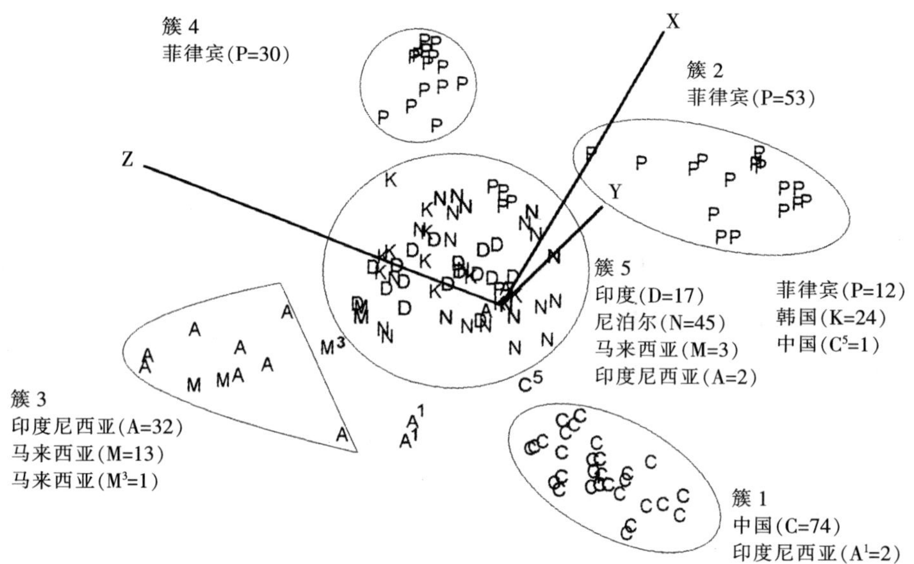

图 4　用 pBS*avr Xa10* 和 pJEl 101 两个探针检测所得 RFLP 指纹复合数值进行三维传感聚类分析，将中国（C)、印度尼西亚（A)、印度（D)、韩国（K)、马来西亚（M)、尼泊尔（N）和菲律宾（P）7 个国家的 *Xoo* 菌株基因组分子型聚集为 5 簇；菌株在立体坐标图上的位置由复合对应分析法确定（Adhikari et al.，1995）

Fig. 4　Cluster analysis showing the relationships of *X. oryzae* pv. *oryzae* strains as derives from DNA band data obtained with probes pBS*avr Xa10* and pJEL101. Numbers of strains from the China (C), Indonesia (A), India (D), Korea (K), Malaysia (M), Nepal (N), and the Philippines (P) are shown. The number of cluster that best fit the data was determined by the consensus of three clustering statistics; position of strains on a three-dimensional graph were determined by multiple correspondence analysis (Adhikari et al.，1995)

Xoo 病菌群体结构表现的区域差异，是因为病原菌具有独特的适应寄主抗性的遗传潜能。因此不同区域需要选择适宜的寄主基因型来控制地区优势菌群。Adhikari 等（1997）

报道聚集于同一簇群的尼泊尔和印度多数菌株的毒性较强、毒谱也很宽，能侵染 *Xa2*、*Xa3* 和 *Xa4* 甚至广谱抗性基因 *xa5* 和 *Xa7*，只有 *Xa21* 才能够抵御尼泊尔多数菌株。但是 *xa5* 和 *Xa7* 在中国、菲律宾、印度尼西亚、韩国却都是有效的，而 *Xa21* 却已经在这些国家遇到了毒性菌的侵染。

印度学者根据病菌群体结构遗传多样性表现的地域差异制订了相应的育种计划（Shanti et al.，2001）。他们测试了 *Xoo* 常年严重发病的东印度库塔克（Cuttack，Orissa）、中央（Madhya Pradesh）和北方（UttarPradesh）3 个邦 8 个地点共 450 个菌株群体结构分子型及代表菌株致病型。结果表明病菌分布的区域差异分明，普遍分布于库塔克 Orissa 邦和中央邦的两个单元型代表菌的毒性最强，都能侵袭 *Xa3*、*Xa4*、*xa5* 等抗性基因，其中一个单元型菌株甚至能侵袭抗谱很宽的 *Xa21* 和 *xa13*，只有累加几个基因的聚合系才能抵御它们。因此培育基因聚合系就成为印度制订区域抗性育种基因部署计划的一个重要部分。

大量研究足以证明水稻白叶枯病抗性基因在不同地区的应用效果之所以不同，是存在病菌群体结构的差异所致。根据病菌亚群空间（地区）分布，选择适合的地区性抗病资源，或进行基因轮换应用，在某些地区丧失了抗性的基因，可能抗御另一些地区的病菌群体，可以最大限度地发挥抗病基因的作用。

3.2 影响水稻白叶枯病菌群体结构及遗传演变的其他因素

了解病原菌群体遗传结构演变的主要因素及时空影响，便于调控两种生物间的协同进化趋于有利方向。

品种选择在诸多因素中对病菌群体遗传结构变化起着最为重要的作用。人为导向的抗病育种和为了适应某种农业生态条件而采用的品种都是病原菌毒性选择的原始驱动因素。历史上积累了大量长期侦察病菌致病力变化的数据，足以阐明人为品种选择对病菌群体结构演变的影响。Leonard（1987）列举的玉米小斑病菌小种的消长就是个典型例子。1970 年，对玉米小斑病菌（*Bipolaris maydis*）T 小种高度感病的雄性不育细胞质杂交玉米在美国的种植面积达到全国玉米总面积的 75%以上时，突发了玉米小斑病大流行，造成严重经济损失。尽管 T 小种群体在那段时期剧增，量大面广，但是并没有完全取代当地的 O 小种，当又改用了抗 T 小种的玉米细胞质雄性不育系后，O 小种频率很快又上升到即将流行的水平。水稻抗病基因对 *Xoo* 群体结构变化的影响也很大。菲律宾水稻品种的布局左右其 *Xoo* 小种的起伏、群体遗传结构的变化就很典型。20 世纪 70 年代以来，菲律宾有 65%的水稻面积种植携 *Xa4* 基因品种，能侵染该基因的小种 2 的发生频率迅速上升至 80%，取代了原来的优势小种 1，成为最大的菌群，但是那些不种植携有 *Xa4* 基因品种的地区，仍然存在小种 1。显然，寄主选择对病菌群体结构的变化发挥了主导作用。

寄主生态环境对于病菌群体结构的形成也有一定影响。寄主选择对病菌群体遗传发生关键作用的同时，环境因子也发挥了一定作用。例如，菲律宾 *Xoo* 小种 5 只在海拔 1 524m、平均气温 21℃的吕宋岛北部高原山区才能发现，两千年以来这里每年只种植一季传统品种。毒性谱很窄的小种 5 除了侵染山区的传统品种外，只侵染对多数小种感病的 IR24 和 IR8（*Xa11*）品种。这正是自然界病原菌与其特异寄主在相同生态环境中进行自然选择的典型协同进化现象。

此外，在一定农业生态系统中，当某种或若干种因素突然发生变化时，病菌群体也可能受到冲击。比如在耕作季节的间歇时期，病菌经常遭受诸如作物轮作换茬、农药使用等各种农业措施的干扰，其后果十分明显，因此也常作为一种抑制病害发展的措施。品种更替、环境条件、耕作改制、农业措施、病原菌基因流动等都可能影响病原菌的演化，但其中作用最大的当属品种，它不仅参与了其他因素的影响，并且直接从遗传上制约病原菌的变化。了解病菌群体遗传多样性在不同时空变化的影响、寄主基因型、种植强度（包括时间）对病菌变异的冲击等信息，为合理选择和调度寄主抗性提供了不可缺少的依据。

4 病原菌群体遗传结构变异信息在抗性遗传改良中的应用

人们从不同方面和不同层次进行病原物和寄主两种生物的各自不断变异又相互选择和协同进化的研究，目的都是在保护环境和可持续发展的前提下有效地控制病害。

病原菌群体遗传结构信息对于鉴定和描述寄主抗性都很重要。往往由于以下原因不能发挥寄主抗性的预期效果：①筛选的抗性资源不抗种植该品种所在地的菌群，以致育成的抗病品种到农民地里后表现感病；②病菌因投放的抗病品种所产生的选择压发生了变化；③病菌迁移。对病菌群体生物学的认知可用来解决这些问题和提出对策。事实上人们早已倡导以群体遗传学途径来控制病害，并且积累了经验。在广大地区控制小麦秆锈病就是个成功例子，很少有工作能够像对小麦秆锈菌的长期消长侦察分析做得这样深入（Wolfe，1985）。实践证明采用分阶段、分地区进行病菌群体结构图解是很可取的，它为育种家传递了较为精确、有预见性的抗病育种指导性信息。

以往抗病育种一直采用病菌小种或当地“优势”菌株进行人工接种评价抗病种质品种或育种材料的传统方法。事实上科学家早已觉察到有些菌株不大可能代表自然界以及包括演变前后的毒性病菌群体，因此所育成的抗病新品种往往赶不上病菌早已开始演变的速度，始终处于追逐病菌变化的被动局面。怎样解决？应该选择那些已经包含在田间的多样性菌株，来接种评价种质或育种材料，只有这样筛选的抗性品种在农民地里才会是抗病的。哪些是真正的代表菌株？选择几个？病原菌群体遗传结构分析或许能提供以下信息：①群体中已经发生了遗传变异及其特异型的个体；②品种整个种植过程中的不同地理区域谱系多样性及其形成的渠道。分子标记技术可以较精确地分析病菌毒性变异，观察到病菌有关个体的遗传属性，确定与寄主相互适应（亲和）的那些病菌的遗传谱系，结合毒性测试，即可用各谱系代表菌株进行抗性评价，为育种家提供前瞻性的科学信息。通过选择能抵御当地或不同生态系统专化病菌群体的抗性基因，就可能找到有效控制当地病菌群体的寄主基因型。采用谱系代表菌株取代小种进行抗性评价的设想首先是由 Levy 等（1991）提出的。他发现哥伦比亚稻瘟病菌谱系与品种之间有很强的联系，某些特殊谱系有选择地侵袭一些品种，既然寄主的选择是促使病菌群体结构变异的主要因子，病菌群体结构的系统发育组分则有可能呈现与寄主抗性基因型相对应的分子特异型，在分子水平上传递新毒性菌的信息。据此，从抗病育种角度衡量，了解病菌谱系和小种都有重要作用。哥伦比亚国际农业热带中心（CIAT）田间稻瘟病数据提供了一个持久抗病因子组合控制病菌谱系的初步设想。根据谱系-毒性型相关分析，了解病原菌由于遗传适应性而产生的演变，为抗性遗传改良计划提供超前性科学依据（Leung et al.，1993）。这一设想和观点，是其对病原菌因遗传适应性产生演变的研究、观察、分析的结果，也许是科学理论和实践的一种探索。它是否能够立即有效地用于生产，还需要经过广泛的实践和进一步研究，在实践中得到证实。

部署基因型的重要问题是一个优良抗病品种在多大范围内能被稳定利用。各国有关 *Xoo* 菌群体遗传结构研究所显示的国与国、一个国家不同地区之间病菌亚群所表现的差异，都与寄主基因型的利用有密切关系，自然也成为抗性基因选择和合理布局的信息来源。准确观察病菌亚群空间分布，选择能够控制该地区病菌群体的广抗谱品种，也表明在某地发生感病化的品种，可在另一些地区种植。系统观察的任何空间病菌群体变化都是进行抗性基因部署的依据。

5 中国水稻白叶枯病菌群体结构

在国家自然科学基金、洛克菲勒基金和“948”等项目资助下，中国农业科学院作物所1989年开始与IRRI及美国堪萨斯州立大学合作，采用分子标记技术结合病菌致病力测试探究中国 *Xoo* 群体遗传结构及其多样性时空变迁与寄主抗性基因利用的关系。分析了1984—1999年中国北自黑龙江，南至海南，西至宁夏、云南共20个省、市、区272个 *Xoo* 菌株群体和亚群体结构变化与致病型分布及水稻抗性基因型利用的关系。

5.1 白叶枯病菌分子型群体结构特点

Zhang 等（1995）用 pJEL101 和 pB*Savr Xa10* 两个 RFLP 探针分析78个主要来自华北、东北和少数来自华中、华南18个地区1981—1985年白叶枯病菌株的群体遗传结构，结果各形成16个单元型。Zhang 等（1997）后来采用 rep-PCR 技术分析了143个1984—1994年间包括中国7个致病型代表菌株，增加了华中、华南地区共20个省的标样。所用的4个 rep-PCR 引物 ERIC、IS1112、IS1113、BOX 分别形成了17、14、9、12个单元型，将4个引物所形成的指纹数据进行组合分析，共形成54型。两种方法的分析结果相似，多数单元型之间显示1个或3～4个带位不同，唯有广东、云南一些菌株稳定地显示特殊指纹型；绝大多数菌株间的差异较小，都被聚合于1簇，特殊单元型则自成1簇。用 pJEL101 为探针杂交的16种 RFLP 型，在相似系数为85%时勉强能分为6簇，其中簇1、簇2、簇3十分相近；如果相似系数水平为80%时，这3簇也只能聚为一簇，第4、5、6簇分别是广东菌株 GD1358、云南菌株 YUN5 和辽宁菌株 LN43。该结果似乎与中国复杂的自然环境、不同稻作区域以及多种水稻类型的水稻生产情况难以对应。

Zhang 等（1997）进一步同时采用 pB*Savr Xa10* 探针及 ERIC、IS1112、IS1113、BOX 4个引物检测和获得了来自我国4个稻区、7个致病型共143个 *Xoo* 代表菌株基因组 DNA 的 RFLP 和 rep-PCR DNA 指纹型复合数据，通过三维综合聚类，形成了清晰的3簇。绝大多数菌株为簇1，少数菌株（包括 LN43）为簇2、簇3，另有广东和云南的两个特殊单元型独立于3个簇群之外（图5）。

同期，又选出76个中国 *Xoo* 代表菌株加入美国堪萨斯州立大学进行的亚洲7国 *Xoo* 菌群体遗传多样性联合试验。308个参试菌株的 RFLP 型形成了5个簇，76个中国菌株中有74个都被聚集于簇1，只有伍尚忠等最早发现于广东的中国致病型Ⅴ代表菌株 GD1358 被聚于簇5，簇5聚集了所有7个国家的若干菌株（图4）。

多次分析结果表明中国 *Xoo* 菌群体遗传结构非常单一，全然有别于亚洲其他国家（图

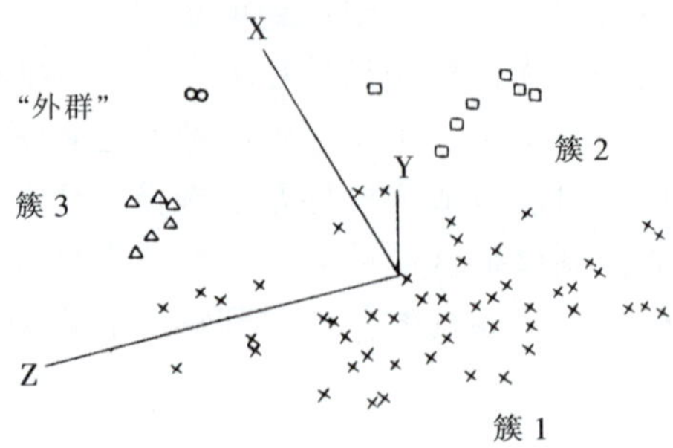

图 5　用 rep－PCR（ERIC，BOX，IS1112，IS1113 引物）和 RFLP（探针 pBS*avr Xa10*）分析中国 143 个 *Xoo* 菌株复合数据（章琦等，1997）。通过三维同感聚类形成 3 簇，菌株在立体坐标图的位置根据复对应分析确定

Fig. 5　Clustering analysis of Chinese 143 *X. o.* pv. *oryzae* strains as derived from DNA band data obtained with rep－PCR（ERIC，BOX，IS1112 and IS1113 primers）and RFLP（pBS*avr Xa*10 probe）（Zhang et al.，1997）. The number of clusters which best fit the data was determined by the consensus of three clustering statistics. Positions of the strain on the three dimensional graph were determined by multiple correspondence analysis

6）。分析其原因还是与品种的布局有密切关系。尽管中国水稻分布面广、面积大、类型多，但是长期以来种植的籼型和粳型的常规稻或杂交品种（组合）主要携有 *Xa4* 或 *Xa3* 两个抗性基因，寄主抗性基因型十分单一，这可能是中国白叶枯病菌之间的系统发育亲缘关系很近，病原菌群体遗传结构比较单一的原因，从寄主-病原菌共同演变的规律初步揭示了中国长期局限利用单一抗性基因的问题。

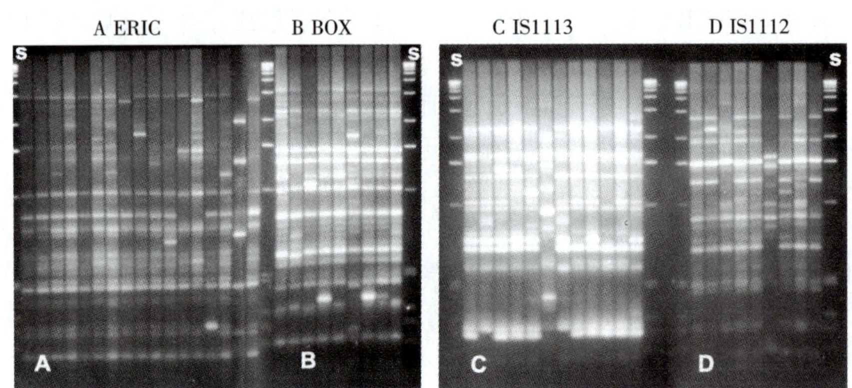

图 6　中国水稻白叶枯病菌株基因组 DNA 用 rep－PCR 不同引物扩增产物指纹型，ERIC（A）、BOX（B）、IS1113（C）和 IS1112（D）（章琦等，1997）。左、右 泳道 "S" 为 1kb DNA 分子质量标准大小

Fig. 6　Rep－PCR fingerprinting patterns from genomic DNA of Chinese *X. o.* pv. *oryzae* strains（Zhang et al.，1997）. The representative patterns of ERIC，BOX and IS1113－PCR，IS1112－PCR are shown in panels A，B，C，and D. The left and right lanes labeled "S" also show the DNA molecular marker（1 kb ladder）；the sizes are indicated in base pairs

5.2 不同稻区及水稻类型白叶枯病菌群体遗传多样性

章琦等用 pB*Savr Xa*10 为探针分析 143 个中国 *Xoo* 菌株基因组 DNA，产生了 30 个单元型。其中华南、长江流域、华北、东北稻区分别有 6、11、3 和 2 个专有单元型，其余 8 个型跨越两个稻区。在介于华南、华北两个稻区的长江流域的 15 个单元型中，除 11 个专有型外，其余 4 个与华南、华北稻区相重叠；华北与东北稻区也有 5 个重叠；而单元型 16 则分布于所有 4 个稻区，表明中国不同稻区之间 *Xoo* 菌群体遗传多样性是有差异的，由于水稻种质或品种在地区之间频繁交流，促成了菌株迁移和单元型的跨区现象（Zhang et al.，1997）。后来用 rep－PCR 的 EREC、IS1112、IS1113、BOX 四个引物分析的同一套 143 个 *Xoo* 菌株所获得的 DNA 指纹数据，则形成了 54 个单元型，确认了由 4 个 rep－PCR 引物组合数据分析所得的中国各稻区 *Xoo* 菌群体遗传多样性特点与用 RFLP 所分析的结果相同（表 3）。

表 3 华南、长江流域、华北和东北稻区 *X. o.* pv *oryzae* 菌单元型（4 个 rep－PCR 引物）**的频率及其遗传多样性**（章琦等，1997a）

Table 3 Frequency and genetic diversity of haplotype（with 4 rep－PCR primers）**from genomic DNA of *X. o.* pv *oryzae* distributed in the South，Yangtze River region，North，and Northeast China**（Zhang et al.，1997a）

单元型 Haplotype	代表菌株 Representative strain	总群体数 Total population		不同稻作区 *Xoo* 单元型的频率 Frequency of haplotype in different rice-growing regions							
				华南 South China		长江流域 Yangtze River Region		华北 North China		东北 Northeast China	
		菌株数 No. of strain	菌株% % of strain	菌株数 No. of strain	菌株% % of strain	菌株数 No. of strain	菌株% % of strain	菌株数 No. of strain	菌株% % of strain	菌株数 No. of strain	菌株% % of strain
1	GD2270	2	1.4	2	10.5	0		0		0	
2	GD2340	1	0.7	1	5.3	0		0		0	
3	GD2400	2	1.4	2	10.5	0		0		0	
4	GD2090	1	0.7	1	5.3	0		0		0	
5	GD2370	3	2.1	3	15.8	0		0		0	
6	ZHE16	1	0.7	0		1	1.7	0		0	
7	GD2283	6	4.2	2	10.5	4	6.9	0		0	
8	ZHE48	5	3.5	0		5	8.6	0		0	
9	ZHE46	1	0.7	0		1	1.7	0		0	
10	JLN691	5	3.5	0		5	8.6	0		0	
11	GD2409	2	1.4	1	5.3	1	1.7	0		0	
12	ZHE34	8	5.6	0		8	13.8	0		0	
13	ZHE18	1	0.7	0		1	1.7	0		0	
14	FJ1	12	8.4	2	10.5	10	17.2	0		0	
15	AH6	3	2.1	0		3	5.2	0		0	
16	KS1－21	2	1.4	0		2	3.4	0		0	
17	HUN3	23	16.1	0		1	1.7	22	55	0	
18	HUN7	1	0.7	0		1	1.7	0		0	
19	HUN8	1	0.7	0		1	1.7	0		0	

（续）

单元型 Haplotype	代表菌株 Representative strain	总群体数 Total population 菌株数 No. of strain	总群体数 Total population 菌株% % of strain	华南 South China 菌株数 No. of strain	华南 South China 菌株% % of strain	长江流域 Yangtze River Region 菌株数 No. of strain	长江流域 Yangtze River Region 菌株% % of strain	华北 North China 菌株数 No. of strain	华北 North China 菌株% % of strain	东北 Northeast China 菌株数 No. of strain	东北 Northeast China 菌株% % of strain
				不同稻作区 *Xoo* 单元型的频率 Frequency of haplotype in different rice-growing regions							
20	HUN10	7	4.9	0		6	10.3	1	2.5	0	
21	HB30	1	0.7	0		0		1	2.5	0	
22	BJ1	2	1.4	0		0		2	5.0	0	
23	NX40	1	0.7	0		0		1	2.5	0	
24	NX42	1	0.7	0		0		1	2.5	0	
25	SX73	1	0.7	0		0		1	2.5	0	
26	HLJ72	5	3.5	1	5.3	0		0		4	15.0
27	YUN5	1	0.7	1	5.3	0		0		0	
28	GX325	1	0.7	1	5.3	0		0		0	
29	LN47	1	0.7	0		0		0		1	3.8
30	LN48	1	0.7	0		0		0		1	3.8
31	JL76	12	8.4	0		0		0		12	46.1
32	LN52	1	0.7	0		0		0		1	3.8
33	JL60	1	0.7	0		0		0		1	3.8
34	HLJ68	1	0.7	0		0		0		1	3.8
35	HLJ71	1	0.7	0		0		0		1	3.8
36	BJ74	3	2.1	0		0		3	7.5	0	
37	NX1	1	0.7	0		0		1	2.5	0	
38	NX39	1	0.7	0		0		1	2.5	0	
39	HB25	3	2.1	0		0		3	7.5	0	
40	HB16	1	0.7	0		0		1	2.5	0	
41	JL62	3	2.1	0		0		0		3	11.5
42	HUN1	1	0.7	0		1	1.7	0		0	
43	HUN9	1	0.7	0		1	1.7	0		0	
44	HUN2	1	0.7	0		1	1.7	0		0	
45	HUN5	1	0.7	0		1	1.7	0		0	
46	HUN4	1	0.7	0		1	1.7	0		0	
47	HB15	1	0.7	0		0		1	2.5	0	
48	GX1	1	0.7	1	5.3	0		0		0	
49	LN43	1	0.7	0		0		0		1	3.8
50	GD1358	1	0.7	1	5.3	0		0		0	
51	AH1	1	0.7	0		1	1.7	0		0	
52	KS3-1	1	0.7	0		1	1.7	0		0	
53	HUN6	1	0.7	0		1	1.7	0		0	
54	BJ3	1	0.7	0		0		1	2.5	0	
菌株总数 (Total No. of strains)		143		19		53		40		26	
遗传多样性值 (Genetic diversity)			0.95		0.93		0.96		0.69		0.77

综合两类分子标记分析结果，来自不同水稻类型上的菌株单元型多样性值为：籼型杂交稻0.85，籼型常规稻0.77，粳型常规稻0.71，反映了中国南方籼型杂交稻和常规稻白叶枯病菌群体结构的变异性一般大于华北和东北粳型稻区。两类分子标记分析单元型的地区差异都是明显的。长江流域衔接华南和华北稻区，兼具籼稻、粳稻、常规稻和杂交水稻，长江以北只种植粳稻，表明稻区之间、水稻类型之间病菌群体结构差异都与品种选择有着密切关系，与中国稻作生态区的水稻分布类型的生产情况相当一致（Zhang et al.，1997a；1997b）。

5.3 白叶枯病菌分子型群体结构与致病型关系

方中达等（1990）用5个水稻白叶枯病鉴别品种将中国*Xoo*菌划分为7个致病型，基本反映了当时全国白叶枯病菌的致病力分化状况。章琦实验组先后用RFLP、rep-PCR和IS-PCR技术分析了1981—1985年、1989—1999年和1996—1999年不同稻区采集的*Xoo*标样，研究其群体结构与致病型演变关系（Zhang et al.，1995，1996，1997；王春连等，2001），结果如下：

（1）多数致病型群体为遗传异质性 综合分析两个分子探针pJEL101和pBS*avr Xa10*检测病菌群体结构所得数据总计有29个单元型，各致病型含有的单元型多寡不同。致病型Ⅰ、Ⅱ、Ⅲ、Ⅳ、Ⅴ、Ⅵ和Ⅶ分别含有11、13、2、3、1、4和1个单元型，致病型Ⅰ、Ⅱ所含分子型远远多于其他致病型。唯致病型Ⅴ和型Ⅶ各含1个单元型。该数值虽然与参试菌株标样的数量有关，但足以表明我国*Xoo*菌群遗传多样性和多数致病型存在分子型异质性。

（2）病菌分子型与致病型之间的相应关系 多数*Xoo*致病型菌株基因组DNA都有1～2个主要单元型，彼此间的重叠单元型并不多，如致病型Ⅰ以单元型1和2为主；致病型Ⅱ以单元型9居多；致病型Ⅳ以单元型22为主，致病型Ⅵ为单元型1、型26各半，另外，JL691（湖北）和YN5（云南）这两个不属于这7个致病型的菌株分别为单元型28和单元型29。每次测试始终呈现特殊型的广东菌株GD1358（致病型Ⅴ）和云南YN5（未知型）的毒性也有别于其他致病型。在1996—1999年中国长江以南地区69个菌株分子型及其致病型中，鉴别出以云南菌株为主包括YN5在内的一个新菌群。

余腾琼等（2003，2005）报道云南高原粳稻白叶枯病菌在该地区鉴别品种上发现了特异性交叉互作关系，所划分的9个致病型中，有5个群能侵染携有*Xa4*基因的品种IR26，假设其中一群与全国Ⅴ型菌相似，其余4群毒性类型则可能有别于全国菌群，具有进一步测试价值。尽管方中达等（1990）当时分析的中国*Xoo*菌与鉴别品种之间还没有出现交叉反应分明的互作关系，但是根据病菌致病型与分子型之间必然存在的相应关系，表明方中达等所建立的鉴别系统反映了当时中国*Xoo*菌的毒性分化状况，他们的研究为未来完善中国鉴别系统奠定了基础。随着分子技术的应用和病菌毒性研究工作的深入，实现发掘分辨性强的鉴别品种，完善中国水稻白叶枯病鉴别系统的目标应指日可待。

（3）中国水稻白叶枯菌致病型遗传多样性不同时空的变化 20世纪80年代和90年代对我国*Xoo*菌群遗传多样性估值分别为0.83和0.84，时间段之间的变化不大。但是有些致病型遗传多样性有较大的变化，致病型Ⅰ、Ⅱ、Ⅲ、Ⅳ、Ⅴ、Ⅵ和Ⅶ遗传多样性值在80年代依次为0.83、0.52、1.00、0.40、0.00、0.75、0.00；90年代依次为0.79、0.77、0.48、0.53、0.79、0.86和0.00（表4）。在10年期间，多数致病型的多样性保持原状，型Ⅲ下降较多，从1.00降至0.48；而侵染*Xa4*的Ⅴ型的升值很大，由0增至0.79。近15

年来病理学检测的结果也有相似反映，Ⅴ型菌在80年代发生频率仅为0.5%，至1999年上升为16%（表5）。中国Ⅴ型菌时空变化的分子技术遗传结构分析与病理学致病力检测结果是对应的，从理论上证实了中国水稻大面积长期种植携有单一抗病基因 *Xa*4 品种，致使能侵袭 *Xa*4 基因的Ⅴ型菌群体急剧发展，从原发地广东省扩展至闽、浙、苏、皖等省。病原菌群体结构随寄主基因型分布而变化的情况与菲律宾、尼泊尔等亚洲其他国家的情况十分类似。根据中国Ⅴ型菌被聚集在亚洲7国 *Xoo* 菌簇5的状况，可以推测首先出现Ⅴ型菌的广东省与东南亚国家相邻，品种交流频繁，Ⅴ型菌由国外迁移而来，抑或潜伏于华南本地的弱势菌群因得势而上升，是个值得深入研究的问题。

表4　中国水稻白叶枯病7个致病型菌株基因组DNA分子型遗传多样性变化[a]
（章琦等，1995）

Table 4　Variation of genetic diversity of molecular types in 7 Chinese *Xoo* pathotypes
（Zhang et al.，1995）

菌株采集年代 Year of strain collection	总值 Total	中国7个致病型遗传多样性 Genetic diversity in 7 Chinese pathotypes						
		Ⅰ	Ⅱ	Ⅲ	Ⅳ	Ⅴ	Ⅵ	Ⅶ
80年代（1981—1984）	0.83	0.83	0.52	1.00	0.40	0.00	0.75	0.00
90年代（1989—1994）	0.84	0.79	0.77	0.48	0.53	0.79	0.86	0.00

[a] 根据RFLP（探针 pB*Savr Xa10*）和 rep-PCR（引物 ERIC）对病菌基因组总DNA分析数据摘录［Summarised from data of RFLP（pB *Savr Xa10*）and rep-PCR（ERIC）analysis from genomic DNA of *Xoo*］。

表5　中国水稻白叶枯病致病型Ⅴ型菌发生频率及其分布

Table 5　Frequency and distribution of *Xoo* strains belonging to Chinese pathotype Ⅴ

年代 Year	发生频率% Frequency	省份 Province	地点 Site	参考文献 Reference
1980	0.5	广东	零星地块	伍尚忠等，1984
1989—1992	1.0～12.0	广东、福建	中山、佛冈、福山、番禺、电白、信宜、化州、电白、徐闻等	徐羡明等，1994
1990—1994	4.8～12.0	浙江	嘉兴、台州、宁波、舟山	王汉荣等，1995
1999	16.0	江苏、安徽	南京、连云港、六安	个人通信，2000

用分子标记技术对中国 *Xoo* 菌群体遗传结构的分析初步明确了：①病菌分子型与致病型存在一定对应关系；②中国多数致病型的分子型具遗传异质性；③病菌分子型存在明显地区差异；④病菌分子型群体结构单一，反映品种抗性基因利用集中，遗传基础狭窄。进一步的工作是：①在现有作物病（虫）害预测预报的基础上，探索采用分子标记技术结合传统病理学手段检测全国稻区主要病害病菌变化，建立积累中、长期病菌消长资料网；②依据病菌群体遗传结构的致病型和谱系变异以及品种基因布局信息，进一步分析寄主-病菌互作协同进化规律，为水稻常规杂交育种结合分子技术进行持久抗性育种提供预见性理论依据；关切全球性病菌群体结构变化，实现高产、优质、抗病水稻品种进入国际市场；③云南、贵州等西南地区病菌群体结构及其系统发育具有重要探索价值，加强这些地区的研究将有助于完善我国 *Xoo* 鉴别系统。制订延长品种抗性寿命策略，逐渐实现全国不同区域水稻品种和致病型群体相互选择、相互适应和协同进展的目标。

稻种资源的白叶枯病抗性评价

章　琦

（中国农业科学院作物科学研究所）

Evaluation of Rice Germplasm Resources for Resistance to Bacterial Blight

摘要　稻属植物有 23 个种，包括 21 个野生种和 2 个栽培种。其中普通野生稻（*O. rufipogon*，AA）、长药（*O. longistaminata*，$A^b A^b$）、非洲光身稻（*O. glaberriama*，$A^g A^g$）、尼瓦拉（*O. nivara*，AA）、斑点（*O. punctata*，BBCC，BB）、小粒（*O. minuta*，BBCC）、药用（*O. officinalis*，CC）、紧穗（*O. eichingeri*，CC）、宽叶（*O. latifolia*，CCDD）、高秆（*O. alta*，CCDD）、大颖（*O. grandiglumis*，CCDD）、澳洲（*O. australiensis*，EE）、短药（*O. brachyantha*，FF）、疣粒（*O. meyeriana*，GG）、颗粒（*O. granulata*，GG）等 15 个野生稻种都含有抗白叶枯病基因。已鉴定的 4 个抗白叶枯病基因如 *Xa21*、*Xa23*、*Xa27t* 和 *Xa29t* 分别源自 *O. longistaminata*、*O. rufipogon*、*O. minuta* 和 *O. officinalis* 野生稻种。截至 2004 年已先后将 6 个远缘种的抗白叶枯病基因转移到栽培稻，并育成了单体异附加系（MAAL）或种间杂交种。

20 世纪 50 年代至 90 年代初，我国植物病理学家与育种家共评价了约 70 424 份次稻种资源对白叶枯病的抗性，从不同生态区地方品种、野生稻的育成品种（系）和国外引进种质中评选出 2 929 份中抗以上的种质和品种，其中不乏对白叶枯病抗谱宽、兼抗其他病虫、优质、适应性广的材料，为进一步选育抗病品种、发掘新基因奠定了良好基础。

抗病的地方品种资源中表现较为突出的有：①云贵高原的扎昌龙、中白米、鲁士谷、毫梅、云南大粒，贵州麻谷、西南 175、孟关大麻谷等品种表现抗性稳定、抗谱宽、适应性广，除已鉴定出扎昌龙的新基因 *Xa22t* 外，尚有一些值得进一步挖掘的抗病新基因；②太湖流域的小爱 2 号、风景稻、大绿种、矮绿种、铁粳青等 11 个抗病优质品种；③两广地区的插仔、龙虾种、南海白壳、玻璃占、大子谷、大糯和大黄谷等 35 个兼抗稻瘟和褐飞虱、白背飞虱中 1～2 个病虫；④洞庭湖平原的双抗红、工人糯、紫谷、背子谷和冬香糯等 29 个兼抗稻瘟、白背飞虱、褐飞虱中 1～3 个病虫并且优质的抗源品种。

1981 年开始，我国参加 IRRI 的国际水稻遗传评价项目，引进了大量籼稻材料。在南方，粤、桂、湘等省（区）农业科学院和中国水稻所从 2 000 多份材料中评价出一大批对白叶枯病抗性稳定、抗谱广的品种材料广泛用于抗病育种。如广西壮族自治区鉴定出 BG90－2、GE456、BR51－282－8、IR20、IR52 等 54 个抗源品种。其中 GE456（桂引 1 号）于 1981 年被育成了抗病、高产、优质的早籼品种梅桂 1 号；又以由 IR36 育成的抗病恢复系桂 33 与珍汕 97A 配组了抗病、优质、适应性广的杂交稻汕优桂 33，后者在 1987 年在南方稻区就推广了 86.7 万 hm^2。据不完全统计，全国这批外引材料中的 35 个高产、优质、抗病、适应性广的优良品系，已由各地自行定名为新品种直接在生产上推广种植，其中 BG90－2、IR20、IR26、IR28、IR29、IR36 和 IR54 等 18 个高抗品种已被广泛用作抗源，同时也作为选育杂交稻的主体恢复基因源。

1977—1989 年，中国农业科学院从 5 330 份次引进的粳型稻种资源中，鉴定出秋光、丰锦、中新 120、中国 45、早生爱国 3 号、阿苏稔、黎明、银河、喜峰和日本晴等 36 个成株抗性稳定的品种，其中高产优质抗病的秋光、丰锦等是东北稻区直接在生产上种植的主要品种，秋光在 1982—2005 年间的累计面积达 210.9 万 hm^2（3 163 万亩）。还鉴定出稀有的两个全生育期高抗我国 6 个代表菌株和菲律宾 1～4 四个小种的粳稻品种矢租和爪哇稻印度诺。

在 28 个引进的和国内的广亲和材料中，有 6 个品种抗谱较广，抗 20 个国内外小种代表菌株中的 18/19 个。如 Dular、Pecos、Aus373、华师大 7 009 四个品种在全生育期抗 19 个菌株，69232－2、鉴 12 两个品种在成株期抗 18 个。广泛被利用的培矮 64 在成株期抗 15 个菌株，但不同年份和季节所表现的抗性不稳。

日本在 20 世纪 20 年代就开始了抗白叶枯病稻种资源的筛选工作。由病区自然发病评价品种抗性发展到人工接种，其主要的经验是根据全国白叶枯病菌群的变化，来确定采用的小种数进行接种，以及按参试品种对各小种的反应型划分为抗性品种群，使育种家对每份种质资源的抗性水平和抗谱信息一目了然。

在日本地方稻种资源中，抗白叶枯病的品种不多、抗谱也较窄、多数为成株抗性品种。如 1962—1965 年间，在苗期接种的 400 份地方品种中，只有 11 个抗病品种，并只抗日本小种Ⅰ。在剑叶期接种的 800 个品种中，有 88 个品种（11.1%）抗小种Ⅰ，如农林 27、霜月、爱优、大淀、不知火、福锦等品种；34 个品种（4.3%）抗Ⅰ、Ⅱ、Ⅲ三个小种，如早生爱国 3 号、中新 120、中国 45、撰一等；没有抗日本小种Ⅱ的地方品种。20 世纪 70 年代中期，Uchiyamada 等（1977；1978）用小种Ⅰ～Ⅳ的四个代表菌株评价 1 194 个日本地方品种的抗性，有 318 个品种抗病，其中 241 个品种只抗小种Ⅰ（RSSS），2 个品种抗小种Ⅰ、Ⅱ（RRSS），24 个品种抗小种Ⅰ、Ⅱ、Ⅲ（RRRS），51 个品种对 4 个小种为交叉反应（RSMS、SRMS、SRRS、SSRS、SSMS、SRSS 等），没有发现抗小种Ⅳ的地方品种。

60 年代中期至 70 年代末期，日本为了改变本国资源贫乏、白叶枯病抗性类型单一的情况，十分重视从不同国家引进稻种资源，评价和筛选到大批优良、广抗谱资源。Sakagushi 等（1968）和 Uchiyamada 等（1977；1978）先后用日本 3 个菌群或 4 个小种测试，从约 2 800 份引自国外的稻种资源中筛选到大批抗性强、抗谱宽的籼稻品种。如抗日本白叶枯病小种Ⅰ和Ⅱ的品种 Rantai Emas 2、Tetep 、Tadukan、Nigeria、RP9－3 等；抗小种Ⅰ、Ⅱ和Ⅲ的品种 Lead Rice、Java14 、Zenith、RP291－7、IR2153－159－1－4 等；抗小种Ⅰ、Ⅱ、Ⅲ和Ⅳ的 Sigadis、TKM6、Pelita I/1 等携有 *Xa*4 基因的品种。这些抗谱较宽的品种主要来自 IRRI 和印度尼西亚、印度、巴基斯坦、孟加拉、斯里兰卡等亚洲热带国家。

IRRI 是 1975 年就倡导开展“稻种资源遗传评价与利用”的国际农业研究机构，该项目促进了国际间自由交换、综合评价、迅速利用优异水稻种质，带动提高作物育种水平和发展粮食生产。从 1981 年开始，IRRI 组织了 62 个国家参加由该所主持的抗两病（白叶枯病、稻瘟病）、三虫（褐飞虱、白背飞虱、稻瘿蚊）、耐旱、耐盐碱、耐冷等多抗性鉴定与筛选协作网。在国际水稻界和遗传育种领域产生了划时代的影响。

IRRI 对稻种资源的抗病评价的特点是依据菲律宾白叶枯病菌群体结构的变化，采用各阶段的优势小种筛选鉴定新抗性基因源，进行基因轮换。

20 世纪 60 年代，IRRI 最初用 10 个菲律宾白叶枯病菌株从几千份稻种资源中评价出 Sigadis、TKM6、PelitaI/1、ARC5756、DL5、Dhariai、DV85 和 DV86 等 300 多个抗病品种，1969 年后，IRRI 用已育成了抗白叶枯病的 IR20、IR26、IR28、IR29、IR30、IR34 和 IR36 等品种，在亚洲产稻国家广泛种植或作抗源利用。

当 IRRI 病理学家发现了抗菲律宾代表菌株的 Sigadis、TKM6、PelitaI/1 等骨干抗源品种及其衍生的 IR 品种都感染菲律宾北部的一个菌株“Isabella”后，便扩大采用抗“Isabella”的广抗谱品种 BJ1、Dular、AZ192、DV85、W1263 和 Zenith 等抗源品种在育种中利用，育成 IR9623、IR14093、IR4475B 等一批新品种。

1975 年用菲律宾白叶枯病小种 1 的代表菌株 PXO61 开始进行大规模抗性鉴定。从 40 000 多份稻种资源中鉴定出 2 274 份抗病材料，再用菲律宾 4 个小种测定其中的 1 441 个抗病品种，被划分为 Java14 群（*Xa3*）、TKM6 群（*Xa4*）、DZ192 群（*xa5*）、CAS209 群（*Xa10*）、Mond Ba 群（*Xa4*＋*Xa10*）、DV85（*xa5*＋*Xa7*）、Makhmal Mehi（*xa5*＋*Xa4*）、BJ1（*xa5*＋*xa13*）共 8 个抗性品种群，这些抗性品种群的品种主要分布在印度、孟加拉、斯里兰卡、印度尼西亚、菲律宾和尼泊尔等国家。IRRI 保存的各抗病种质编号都有清楚明了的抗性和其他性状档案，很值得借鉴。本文对稻种资源抗性接种鉴定技术也作了简要说明。

Abstract Twenty one wild species and 2 cultivated species were included in the genus *Oryza*. Bacterial blight (BB) resistance genes were found in 15 wild *Oryza* species, including *O. rufipogon* (AA), *O. longistaminata* ($A^b A^b$), *O. glaberriama* ($A^g A^g$), *O. nivara* (AA), *O. punctata* (BBCC, BB), (*O. minuta*, BBCC), *O. officinalis* (CC), *O. eichingeri* (CC), *O. latifolia* (CCDD), *O. alta* (CCDD), *O. grandiglumis* (CCDD), *O. australiensis* (EE), *O. brachyantha*

(FF), *O. meyeriana* (GG), *O. granulata* (GG) . The 4 BB resistance genes, *Xa21*, *Xa23*, *Xa27t* and *Xa29t*, were derived from wild *Oryza* species *O. longistaminata*, *O. rufipogon*, *O. minuta*, and *O. officinalis*, respectively. Up to 2004, BB resistance genes in 6 distantly related wild species were introgressed into the cultivated rice and developed as monosomic alien addition lines (MAALs) or inter-specific hybrids.

From the 1970s to the early 1990s, plant pathologists and rice breeders in China evaluated approximately 70 424 rice accessions (from local cultivars of different ecological regions, wild *Oryza* species, improved cultivars, and foreign introductions) for BB resistance, among them, 2 919 were identified as resistant lines, and subsequently many elite materials with broad-specturm resistance to BB and other diseases or pests, high grain quality, and wide adaptation. Such work has laid foundation for further development of more desirable BB resistant cultivars and exploration for new resistance genes.

Some excellent local resistant germplasm resources include the followings:

(1) Germplasms originated from the plateaus of Yunnan and Guizhou that exhibit broad - specturm BB resistance, wide adaptability, such as Zhachanglong, Zhonbami, Lushigu, Haomei, Yunnandali, Gueizhoumagu, Xinan175, and Mengguandamagu. These germplasms may provide new sources of BB resistance genes as exemplified by the recently identified *Xa22t* from Zhachanglong.

(2) Eleven local cultivars from Taihu Valley such as Xiaoai 2, Fengjingdao, Daluzhong, Ailuzhong, and Tieganqing exhibit stable broad-spectrum BB resistance and high grain quality.

(3) Thirty-five BB resistance local cultivars screened from the Guangdong province are also resistant to 1～2 diseases or pests among fungal blast, brown planthopper and whitebacked planthopper, such as Chazai, Longxiazhong, Nanhaibaike, Bolizhan, Dazigu, Danuo, and Dahuanggu.

(4) In the region of Dongting Lake Plain, 29 cultivars such as Shuangkanghong, Gongrenou, Zigu, Beizigu, and Dongxiangnou exhibit high grain quality and resistance to 1～3 diseases or pests among BB, fungal blast, whitebacked planthopper and brown planthopper.

From 1980s, a large number of indica cultivars were introduced into China via the "International Rice Genetic Evaluation Program" . Collaborative efforts from the Academies of Agricultural Sciences of Guangdong, Guangxi, and Hunan and the Chinese Rice Research Institute obtained a large amount of stable, broad-spectrum BB resistance genetic materials which widely adopted in BB resistance breeding programs from 2 000 accessions. For example, the Guangxi Autonomous Regional Academy of Agricultural Sciences evaluated 54 cultivars such as BG90 - 2, GE456, BR51 - 282 - 8, IR20 and IR52, including the cultivar GE456 which was used as R donor and developed the indica cultivar Meiguei No. 1 that exhibits BB resistance, high-yield, and good grain quality in 1981. At the same time the BB resistance restorer line Gui 33 developed from IR 36 hybridized with the male sterile line Zhenshan 97A and generate the hybrid rice Shanyougui 33 which exhibits BB resistance, wide adaptability, and high grain quality. By 1987, the accumulated growth acreage of Shanyouguei 33 had reached 867 000hm^2 in the southern part of China. According to incomplete figures, there are at least 35 new BB resistant lines named and used in the agricultural production, which also exhibits BB resistance, high-yield, good grain quality and wide adaptability. Eighteen out of the 35 highly resistant cultivars, such as BG90－2, IR20, IR26, IR28, IR29, IR36 and IR54 are widely used as sources of resistance genes and are also employed as restorer lines in hybrid rice production.

From 1977 to 1989, The Chinese Academy of Agricultural Sciences identified 36 japonica cultivars such as Akihikari, Toyonishiki, Nakashin 120, Wasa Aikoku 3, Asominori, Reimei, Kiho, and Nipponbare with stable BB adult plant resistance from 5 330 accessions foreign introductions. Two out of the 36 cultivars, Akihikari and Toyonishiki with high-yield, good grain quality have been planting wildly as the main cultivars in the Northeast China up to now, the accumulated growth acreage of Akihikari had reached 2 100 000 hm^2 from 1983 to 2005 in the Northeast China. Two unique javanica cultivars, Shizu and Yindunou showed high resistance to 6 representative BB strains of China and races 1 to 4 from Philippines at all growth stages.

Twenty eight rice cultivars with wide compatibility originated from China locally or foreign introductions were evaluated with 20 representative strains from Philippines, Japan and China. Six out of the 28 cultivars showed broad spectrum BB resistance. Dular, Pecos, Aus373, and Huashida 7009 were resistant to 19 strains at all growth stages. 69232－2 and Jian 12 were resistant to 18 strains at the adult stage. Peiai 64, which is used widely, is resistant to 15 strains at the adult stage and the resistance showed unstable based on years or seasons.

From the 1920s, the screening for BB resistant rice germplasms were commenced in Japan by natural evaluation in the BB epidemic fields in the beginning, and then by artificial inoculation method. The key experiences are: definition of the races used for inoculation was dependent on the variation of the *Xoo* groups in the country, and the rice germplasms were classified into resistance cultivar groups based on the reaction patterns against the all races tested. So, each acc. of the germplasms has surveyability information of BB resistance type and R spectrum.

Among the local rice germplasms in Japan, the resistant materials are not so much; most of them were resistant at the adult stage and exhibit narrow resistance spectrum. from 1962 to 1965, only 11 out of the 400 cultivars inoculated were resistant to Japanese race I at the seedling stage. Among the 800 cultivars tested at the adult stage , 88 (11.1%) cultivars were resistant to Japanese race I, such as Norin 27, Shimotsuki, Aimasari, Oyodo, Shiranui, and Fukunishiki; 34 cultivars (4.3%) were resistant to Japanese races Ⅰ, Ⅱ, Ⅲ, such as Wase Aikoku, Nakashin 120, Chugoku 45, and Sen-ichi, and none was resistant to race Ⅱ.

In the middle 1970s, Uchiyamada et al. (1977; 1978) evaluated 1 194 local rice germplasms with four Japanese races, Ⅰ, Ⅱ, Ⅲ, Ⅳ. Three hundred and eighteen out of 1 194 cultivars screened were resistant to BB, in which 241 cultivars were only resistant to race I (RSSS); 24 cultivars were resistant to races Ⅰ to Ⅲ (RRRS); 2 cultivars were resistant to race Ⅰ, Ⅱ (RRSS); the resistance pattern of 51 cultivars to the 4 races were RSMS、SRMS、SRRS、SSRS、SSMS、SRSS; and none was resistant to race Ⅳ.

From the middle 1960s to the late 1970s the scientists had a regard for introducing BB resistance resources from different countries into Japan. Sakagushi et al. (1968), Uchiyamada et al. (1977; 1978) successively evaluated about 2 800rice accessions of foreign introductions with 3 or 4 Japanese *Xoo* races. A number of cultivars with high and broad spectrum resistance to BB were screened. some cultivars were resistant to races Ⅰ and Ⅱ, e.g. Rantai Emas 2, Tadukan, Tetep, Nigeria, RP9-3; cultivars were resistant to races Ⅰ～Ⅲ, e.g. Lead Rice, Java14, Zenith, RP291-7, IR2153-159-1-4; cultivars were resistant to races Ⅰ～Ⅳ, e.g. Sigadis, TKM6, Pelita I/1 carrying gene *Xa4*. Most of the BB resistance cultivars were introduced from IRRI and tropical Asia countries, e.g Indonesia, India, Sri Lanka, Bangladesh, the Philippines, Myanmer, and Pakistan.

IRRI is the International agricultural research institution for providing an innovative approach to develop "Genetic Evaluation and Utilization" of rice germplasms. The program has facilitated the international free exchange, comprehensive evaluation, and rapid utilization of the elite rice materials and improved the crop breeding level and developed the cereals production. From 1981, 62 countries have taken part in the symbiotic network program for evaluation multiple resistance to 2 diseases, bacterial blight and rice blast, 3 pests, brown planthopper (*Nilaparvata lugens*), whitebacked planthopper (*Orseolia oryzeae*), and gall midge *Sogatella furcifera*; and tolerance to drought, salt-alkali, and cold. etc. The program evoked great repercussion in the rice and genetics breeding fields.

The experience for evaluation the BB resistance in rice germplasms at IRRI is that according to the diversification of the Philippine *Xoo* structure population in the different phase, the predominant BB races were used to inoculate the tested materials, then, genes rotation was carried out with those new resistance cultivars.

In the 1960s, IRRI evaluated more then 300 resistant cultivars such as Sigadis,

TKM6, PelitaI/1, ARC5756, DL5, Dhariai, DV85 and DV86 from thousands rice accessions using 10 Philippine BB strains. Since 1969, BB resistance cultivars, IR20, IR26, IR28, IR29, IR30, IR34 and IR36 were developed and widely planted or used as R donors for rice breeding in Asian countries.

The principal resistance donors Sigadis, TKM6, Pelita I/1 and their derived lines IR20 and IR22 were susceptible to the strain, "Isabella" from north Philippines. Since then, the cultivars IR9623, IR14093, IR4475B were developed using BJ1, Dular, AZ192, DV85, W1263, Zenith which were resistant to all 10 BB strains .

In 1975, a large scale evaluation of more than 40 000 rice germplasms was preformed to identify 2 274 resistance cultivars using PX061 (the representative strain of Philippine race 1) . Further examination using 4 Philippine races divided 1 441 (out of the 2 274 resistant cultivars) germplasms into 8 resistance cultivar groups: Java14 (*Xa3*), TKM6 (*Xa4*), DZ192 (*xa5*), CAS209 (*Xa10*), Mond Ba (*Xa4*+*Xa10*), DV85 (*xa5*+*Xa7*), Makhmal Mehi (*xa5*+*Xa4*), and BJ1 (*xa5*+*xa13*) . These BB resistance groups of rice germplasms distributed mostly in India, Bangladesh, Sri Lanka, Indonesia, Nepal and Philippines.

The characterizations of diseases and pests resistance and other traits were described clearly and perspicuously in all of the rice germplasms at IRRI and it is good for reference.

Finally, summarized the technology for evaluation and identification of BB resistance.

品种资源又称种质资源或遗传资源。水稻品种资源在其生养栖息的万千环境中经历漫长岁月的自然演化和人工培育与选择，形成了丰富多彩、各具特点的遗传属性，蕴藏有各色各样、功能特异的宝贵基因，是人们选育高产、优质、多抗新品种和开展遗传、进化、栽培、生理以及相关学科研究的材料基础。

稻种资源包括野生稻、地方农家品种、现代改良品种、国外引进品种以及各种遗传研究材料。野生稻和地方品种资源蕴藏有丰富的抗病虫、耐逆境、广适应性以及优良品质等优异基因。系统评价和研究这些种质资源，了解其各方面特性，明确其所携带的优异基因，可以为品种改良、生产发展以及相关基础理论研究提供各种各样有价值的信息。因此国际农业研究机构和主要产稻国家都十分重视稻种资源的收集、整理、评价和利用研究。

据国际农业研究磋商小组（Consultative Group of International Agricultural Research, CGIAR）统计，截至20世纪90年代中期，全球稻属遗传资源约有130 000多份不重复的种质，分别保存在国际水稻研究所（IRRI）及水稻主要生产国。我国从20世纪10年代就开始进行水稻选种工作，南方稻区的一些稻作、稻麦改良场或农事试验场也同时着手收集农家品种和近邻国家的优良品种。据报道，从20世纪初期至1990年底，通过4个阶段的稻种资源征集与考察活动，我国已完成了第二次全国稻种资源统一编目，列入编目的国内外品种资源共61 358份，其中地方品种46 885份，野生稻资源4 655份，选育品种2 396份，杂交稻三系资源1 039份，外国（系）6 383份（应存山，1996）。

1956年，方中达等（1957）在国内首先研究水稻品种对白叶枯病抗性鉴定的方法，并鉴定了145个水稻品种资源的白叶枯病抗性。1957年2月，全国水稻白叶枯病座谈会把品种抗白叶枯病鉴定作为重要研究内容之一，全国植病研究者开始进行水稻抗白叶枯病的研究和评价工作。进入70年代，国家将评价、研究和利用抗白叶枯病资源作为全国水稻品种改良研究的一个重要组成部分。1976—1979年，由农业部委托浙江省农业科学院主持全国稻瘟、白叶枯等主要病害抗性鉴定协作网，南北稻区多数省（市）农业科学研究单位和大专院校参加了这一协作研究，其中部分单位又继续参加了80年代的全国水稻育种攻关项目，承担新品种（系）对稻瘟、白叶枯病的抗性鉴定工作；1980—1984年中国农业科学院作物品种资源所组织了国外引进稻种资源抗三病两虫的联合鉴定。据1993年的不完全统计，从70 424份（次）国内外品种材料中评选出2 929份白叶枯病中抗以上的水稻种质和品种（表1）。其中包括一大批抗性强的野生稻种、地方品种、国外引进种和国内育成品种，有些品种直接在生产上种植，有些用作抗病育种的骨干供体，并且从中鉴定出若干优异的抗白叶枯病新基因。

表1　我国水稻品种及种质资源的白叶枯病抗性评价（1957—1993）[a]

Table 1　Evaluation of bacterial blight（BB）resistance in rice cultivars and germplasms in China（1957—1993）

单位 Institute	总数 Total	抗病 No. of R Acc.	主要抗病品种和种质资源 Main resistant cultivar and germplasm	参考文献 Reference
南京农业大学	1 209	279	瘦田赤、矮脚齐子、湖南早、一线红、红节川、太和粘、红米子、小冬稻、叶里培、陈家种、老大稻等	方中达等，植物病理学报，1963
湖北省农业科学院	2 992	59	鄂晚5号、嘉农选优6号、BW196，IR系统品种	全国水稻白叶枯病防治研究资料第二辑，1976
福建省农业科学院	8 000	220	引自IRRI品种	全国水稻白叶枯病防治研究资料第二辑，1976
安徽省农业科学院	3 452	44	紫色粳、富光16、大野中稻和IR系统品种（系）	全国稻病抗性鉴定会议资料，1976—1987
浙江省农业科学院	1 350	39	农试4号、厦引11、辐包矮22、福锦、神乐糯及IR系统品种	全国稻病抗性鉴定会议资料，1977
江苏省农业科学院	2 382	738	奎稻、75-47鉴选、7432-15、贵州麻谷、蚌珠芒、南京罗汉黄、金银粳、徐辐4号、贵农育等	江苏省农科院植保所、粮食所，江苏农业科学，1980
福建农学院	5 000	59	梅菲早2号、植科1号、十石、龙城12、D32、C104、矢租等	夏怡厚等，福建农学院学报，1985
江苏省农业科学院	4 819	20	南粳15、南粳11、邳早15、盐粳2、农垦58、紫金粳、筑紫晴、扬稻1等	江苏省农业科学院植保所、粮食所，江苏农业科学，1980
江苏省农业科学院	1 389	11	小爱2号、风景稻、大绿种、矮绿种、铁粳青、红芒糯、大稻头糯、红壳稻、长秸糯、早糯稻等	过崇俭等，江苏农业学报，1986
中国水稻所	970	32	惠谷占、玻美占、齐双占、二华早、青叶青丰、浙湖2号等	黄世文等，1986

（续）

单位 Institute	总数 Total	抗病 No. of R Acc.	主要抗病品种和种质资源 Main resistant cultivar and germplasm	参考文献 Reference
浙江省农业科学院	366	11	丰阳矮 217、单 32、青封占 35、玻齐占、晚六早、齐双占等	"七五"全国水稻育种组新品种抗性鉴定，1987
中国农业科学院作物所	3 234	54	秋光、秀峰、早生爱国 3 号、中新 120、筑紫晴、京系酒、关东 60、银河、喜峰、黎明、银河、矢租、印度诺等	全国稻病抗性鉴定会议资料 1976—1987"75"全国水稻育种组新品种抗性鉴定 1987，章琦等，1986
中国农业科学院植保所	62	5	*O. laftifolia*（W14）、*O. minuta*（W17、W44）、*O. officinalis*（W46）、*O. punctata*（W50）	韩玉梅，赵荣杰，植物保护学报，1988
湖南省农业科学院	4 615	457	假蚂蚱谷、背子谷、紫谷、二百号、双抗红、江水香糯、冬香糯、香稻、浏阳大糯、千禾糯、白麻谷、早禾糯、麻糯谷、六十早、红其苏等	侯小华等，湖南农业科学，2001
华中农业大学	6 184	50	扎昌龙、中白米、黄牛尾香糯、安宁早、鲁士谷、毫梅、雾露谷、云南大粒、云香糯 1 号、八月糯、马罗糯谷、矮脚糯（1），长毛糯、中云香糯和毫双 7 号等	陈勇，谢岳峰等，1990
广东省农业科学院	5 361	162	早仔种、宫牌种、马尾齐、红梅糯、牛团糯、红壳糯、早水玻璃占、麻谷	徐羡明等，广东农业科学，1990
	2 064（野生稻）	10	*O. rufipogon*（S1008，S1115，S2165，S2178，S3116，S3375）等	广东省农业科学院水稻所，植保所，1989
广西壮族自治区农业科学院	7 964	61（MR）	小颗油占、黑毛粳、大糯、香糯、白旱谷、白皮山糯谷等	孙恢鸿等，广西农业科学，1992
	1 752（野生稻）	57	YD2-0363、YD2-0601、YD2-0795（RBB16）、YD2-0684 等	孙恢鸿等，植物保护学报，1991
贵州省农业科学院	3 759	103	西南 175，大秆早、贵州麻谷、伊势景、孟关大麻谷、麻油占、香糯、红占谷、寸谷糯、三百棒等	张再兴等，中国稻种资源，应存山主编，1993
中国农业科学院品种资源所	3 500	458	IRRI 品种为主	袁世传等，1993
合计（Total）	70 424	2 929		

[a] 章琦整理，不包括台湾省资料（2005）（Summarized by Zhang，and barring data of Taiwan，2005）。

1　野生稻种资源的白叶枯病抗性评价

1.1　稻属野生种资源概述

自 1753 年林奈（Linnaeus）将普通栽培稻定名为 *Oryza sativa* L. 以来，稻属植物的分类研究一直没有间断。20 世纪 20 年代以前主要依据植物形态学进行分类，30 年代以来结合细胞学、细胞遗传学知识，着重考察体细胞染色体数、倍数性、杂交亲和力、杂种 F_1 在减数分裂时的染色体配对行为、后代育性以及农艺性状等予以修订补充和完善，一直到近 20 年才开始应用 RFLP、AFLP、SSR、SNP 等分子标记和种子储藏蛋白分析等分子生物学技术进行深化研究。Khush 等（2001）综合了前人在稻属植物分类研究上的结果和意见，提出稻属植物可归结为 23 个种（species），包括亚洲栽培稻（*O. sativa*）和非洲栽培稻（*O. glaberrima*）两个栽培种和 21 个野生种，分别含有 AA、BB、CC、BBCC、CCDD、EE、GG、FF、HHJJ 和 HHKK 等 10 个染色体组的分类体系，其中二倍体种 15 个，四倍体种 7 个，二倍体/四倍体种 1 个（表 2）。普通野生稻种群（*O. rufipogon* complex）中有 6 个携有 AA 染色体组的野生种是栽培稻的近缘种，其中多年生普通野生稻（*O. rufipogon*）是亚洲栽培稻的祖先种，长药野生稻（*O. longistaminata*）是非洲栽培稻的祖先种，至于其他染色体组的野生种则与栽培稻的亲缘关系相距较远（吴强等，1998；何予卿等，2001）。

表 2　稻属的种名、染色体组和地域分布（Khush and Brar，2001）

Table 2　Chromosome number, genomic composition and geographic distribution of *Oryza* species.
（Khush and Brar，2001）

种 Species	2n	染色体组 Genome	地域分布 Distribution
普通栽培稻种群复合体 （*O. sativa* complex）			
普通栽培稻（*O. sativa*）	24	AA	全球（Worldwide）
普通野生稻（*O. rufipogon*）	24	AA	亚洲热带、亚热带（Tropical & Subtropical Asia）、热带澳大利亚（Tropical Australia）
尼瓦拉野生稻（*O. nivara*）	24	AA	亚洲热带、亚热带
非洲光身稻（*O. glaberrima*）	24	AA	西非（West Africa）
短舌野生稻（*O. breviligulata*）	24	AA	非洲（Africa）
长药野生稻（*O. longistaminata*）	24	AA	非洲
南方野生稻（*O. meridionalis*）	24	AA	热带澳大利亚
长粒野生稻（*O. Glumaepatula*）	24	AA	南美、中美（South & Central America）
药用野生稻种群复合体 （*O. officinalis* complex）			
斑点野生稻（*O. punctata*）	24，48	BB，BBCC	非洲
小粒野生稻（*O. minuta*）	48	BBCC	菲律宾、巴布亚新几内亚（Philippines & Papua New Guinea）
药用野生稻（*O. officinalis*）	24	CC	亚洲热带、亚热带 热带澳大利亚
根茎野生稻（*O. rhizomatis*）	24	CC	斯里兰卡（Sri Lanka）
紧穗野生稻（*O. eichingeri*）	24	CC	南亚、东非（South Asia & East Africa）
宽叶野生稻（*O. latifolia*）	48	CCDD	南美、中美（South & Central America）
高秆野生稻（*O. alta*）	48	CCDD	南美、中美
大颖野生稻（*O. grandiglumis*）	48	CCDD	南美、中美

（续）

种 Species	2n	染色体组 Genome	地域分布 Distribution
澳洲野生稻（*O. australiensis*）	24	EE	热带澳大利亚
疣粒野生稻种群复合体 （*O. meyeriana* complex）			
颗粒野生稻（*O. granulata*）	24	GG	南亚、东南亚（South & Southeast Asia）
疣粒野生稻（*O. meyeriana*）	24	GG	东南亚（Southeast Asia）
马来野生稻种群复合体 （*O. ridleyi* complex）			
长护颖野生稻（*O. longiglumis*）	48	HHJJ	印度尼西亚、巴布亚新几内亚（Indonesia & Papua New Guinea）
马来野生稻（*O. ridleyi*）	48	HHJJ	南亚（South Asia）
尚未划分的野生稻群 （Unclassified）			
短药野生稻（*O. brachyantha*）	24	FF	非洲
极短粒野生稻（*O. schlechteri*）	48	HHKK	巴布亚新几内亚

野生稻在各种生态环境下经受了近百万年的进化历程，形成了各具其自身特点的遗传多样性，蕴含有栽培稻所缺乏的诸多有利性状（抗病、虫，抗逆境和雄性不育等）基因。国内外利用野生稻种质改良栽培稻品种的成功例子很多。例如，至今唯一存在于尼瓦拉野生稻中控制水稻草状矮缩病毒病（grassy stunt）的显性抗性基因*Gs*，是Lin等（林克治等，1972）从IRRI的6 723份稻种资源中通过大规模筛选与研究后发现的，随即被广泛应用于IRRI的育种计划，并育成了携带*Gs*基因的IR28、IR29和IR30等IR品种（Khush，1977）。又如野生稻中含有一些细胞质雄性不育（*CMS*）基因，中国三系杂交籼稻的突破就是由于1970年李必湖在海南省崖县南红农场发现一株花粉败育的中国普通野生稻（简称野败）（李丁民等，1991），之后，通过与栽培稻杂交、回交转育成不育系和保持系才得以实现三系配套，并于1976年第一次在大面积生产上推广应用（袁隆平，2002）。再如*Xa21*、*Xa23*和*Xa27t*三个对白叶枯病具有广谱抗性的显性基因分别来源于长药野生稻、普通野生稻和小粒野生稻（*O.minuta*）。迄今，只有这3个基因能够抗御具广谱毒性的菲律宾白叶枯病小种6（PXO99），该小种能侵袭现有栽培稻在此前定名的所有的白叶枯病显性抗性基因（章琦，2005）。最近，又从药用野生稻（*O. officinalis*）中发现了一个抗菲律宾白叶枯病小种1的新基因*Xa29t*（谭光轩等，2004）。由此可见，野生稻种质确是改良栽培稻某些特性的优异源泉，应该大力发掘其潜在的利用价值。

野生稻广泛分布于亚洲、非洲、拉丁美洲和澳大利亚，共计77个国家。20世纪60～80年代是收集野生稻资源的高峰时期。国际水稻研究所、西非水稻发展协会（WARDA）和国际热带农业研究所（CIAT）等曾在世界范围内组织多次稻属遗传资源培训和收集活动，从而推动了主要产稻国家野生稻资源收集和研究工作。截至2001年，国际农业研究机构和主要产稻国家共保存野生稻种质约17 605份（汤圣祥等，2004）。

1978—1982年中国农业科学院作物品资所主持了全国野生稻考察和收集工作。通过对306个县、3 531个乡的考察，收集到原产于我国的3个野生稻种即普通野生稻、药用野生稻和疣粒野生稻（*O. meyeriana*）的种茎、种子共 3 238 份（全国野生稻资源考察协作组，1990）。在进行全国野生稻收集的同时，还从菲律宾、印度、马来西亚、法国等 43 个国家引进野生稻种质 558 份。至 2000 年，编入全国野生稻资源目录的种质计有国内外 19 个野生稻种共 7 322 份（张万霞等，2004）。

1.2 野生稻的抗性评价

从野生稻中发掘优异的白叶枯病抗性基因一直是国内外研究者的殷切期望。Sakaguchi 等（1968）测定了 11 个野生稻种 112 份材料对日本白叶枯病菌的抗性，发现宽叶野生稻（*O. latifolia*，CCDD）和颗粒野生稻（*O . granulata*，GG）中的有些材料对日本小种Ⅱ具有抗性，在日本地方品种中很少有抗日本小种Ⅱ的。彭绍裘等（1982）1978—1981 年曾多次用湖南白叶枯病菌对云南省 3 个野生稻种进行抗性鉴定，结果表明疣粒野生稻高度抗病至接近免疫，药用野生稻的抗性较好，而普通野生稻则感病。

野生稻是喜高温多湿植物，在我国广东、广西、云南、海南、福建、江西、湖南和台湾等省（自治区）都有分布，其中海南和两广地区集中分布了普通野生稻和药用野生稻。1982 年广东省农业科学院（1989）开始用本省白叶枯病优势菌株鉴定了 2 064 份普通野生稻材料的抗性，按Ⅰ～Ⅴ级调查标准评价出中抗以上材料 387 份，占总数的 18.8%，其中 4 份高抗（HR，Ⅰ级），56 份抗（R，Ⅱ级），余为中抗（MR，Ⅲ级）。在这些抗病种质中，有不少为兼抗稻瘟病、优质或耐寒的材料，如兼抗稻瘟病和优质的种质有 S1011、S1042、S1045、S1108、S1117、S1118、S1131、S1136、S1142、S2126、S6165、S7072 等；兼抗稻瘟病、优质和苗期耐寒的种质有 S2035 等；兼具优质、耐旱和耐涝的种质有 S3045、S3305、S6023 等；兼具优质的种质有 S1008、S1115、S2165、S2178、S3116、S3375、S6013、S6161、S7304、S7591 等。孙恢鸿等（1992）从 1981 年开始就对广西壮族自治区农业科学院保存的普通和药用两个野生稻种共 1 752 份进行了系统的抗白叶枯病鉴定。按抗病级别Ⅰ～Ⅴ区分，从 1553 份普通野生稻材料中筛选出 46 份抗至高抗材料，占总数的 2.96%，其中高抗（HR，Ⅰ级）材料 2 份，抗（R，Ⅱ级）的材料 44 份；在 199 份药用野生稻中，抗（R，Ⅱ级）的材料 11 份，多数材料属中抗，少数为中感，没有发现感的编号。他们在用广西优势菌（致病型Ⅳ）进行抗性初筛的基础上，又用全国致病型Ⅰ～Ⅴ共 5 个代表菌株对中抗以上的材料进行抗性稳定性鉴定，发现普通野生稻中有 22 份材料抗谱宽而稳定，其中的 YD2－0795（全国编号）就是后来被鉴定为优异抗性基因 *Xa23* 的原始材料。章琦等（1994）1987—1992 年曾用菲律宾小种Ⅰ、日本小种Ⅲ和中国致病型Ⅱ（HB17）的 3 个代表菌株接种鉴定了国内外 13 个野生稻种 871 份材料的抗性，发现除澳洲野生稻（*O. australiensis*）外，其余 12 个野生稻种都有抗病材料。其中，全部 12 份疣粒野生稻（*O. meyeriana*）对 3 个小种都是高度抗病的；在普通野生稻（*O. rufipogon*，AA）、非洲栽培稻（*O. glaberriama*，AA）、长药（*O. longistaminata*，AA）、斑点（*O. punctata*，BB，BBCC）、小粒（*O. minuta*，BBCC）、药用（*O. officinalis*，CC）、宽叶（*O. latifolia*，CCDD）以及高秆（*O. alta*，CCDD）等 8 个种中也都有抗性上等的材料；大颖（*O. grandiglumis*，CCDD）、颗粒（*O. granulata*，GG）、尼瓦拉野生稻（*O. nivara*，AA）中也有些中抗的材料（表 3）。此外，有些野生稻种对白叶枯病的抗扩展性远远强于栽培稻，例如，药用野生稻在接种 60d 后其病斑长度仍小于 1 cm，植株青枝绿叶，而携有 *Xa4* 基因的栽培稻 IR20 的病斑则已由 1 cm 扩展至 3～4 cm。又据报道，紧穗野生稻（*O. eichingeri*，CC)、澳洲野生稻（*O. australiensis*，EE）和短药野生稻（*O. brachyantha*，FF）对白叶枯病的抗性已经被成功地转移到栽培稻（Yasuyuki et al.，1988；Multane et al.，1994；Brar et al.，1997)。由此可见，至少有 15 个野生稻种含有抗白叶枯病种质，说明野生稻种是改良水稻白叶枯病抗性的重要资源。

表 3 13 个野生稻种对 3 个白叶枯病小种（致病型）代表菌株的抗性
（章琦等，1994）

Table 3 Resistance to 3 representative strains of *Xoo* races (pathotypes) **in 13 wild *Oryza* species**
(Zhang et al.，1994)

种名 Species	染色体组 Genome	对 3 个 *Xoo* 菌株的抗性反应[a,b] Reaction of resistance to 3 *Xoo* strains					
		HB17（中国Ⅱ型） (Chinese pathotype Ⅱ)		PXO61（菲小种 1） (Phi. race 1)		T7174（日本小种 1） (Jap. race 1)	
		MTS	BS	MTS	BS	MTS	BS
O. meyeriana	GG	HR	HR—	HR	HR—	HR	HR—
O. minuta	BBCC	HR	HR	R	R	R	R
O. punctata	BBCC，BB	R	HR	R	R	R	R
O. latifolia	CCDD	R	R	R	R	HR	R
O. officinalis	CC	R	R	R	R	R	R
O. longistaminata	A^bA^b	R	R	R	R	R	R
O. rufipogon	AA	R	R	R	R	R	R
O. alta	CCDD	R	R	R	R	R	R
O. glaberrima	A^gA^g	*R*	*R*	*R*	*R*	*R*	*R*
O. granulata	*GG*	*MR*	—	*MR*	—	*MR*	—
O. grandiglumis	*CCDD*	*MR*	*MR*	*MS*	*MS*	*MR*	*MR*
O. nivara	*AA*	*R*	—	*S*	—	*S*	—
O. australiensis	*EE*	*S*	—	*S*	—	*S*	—
Jingang 30 (*CK*)	*AA*	*S*	*S*	*S*	*S*	*S*	*S*

[a] *MTS*，最高分蘖期（*Maximum tillering stage*)；*BS*，孕穗期（*Booting stage*）。

[b] *R*，抗（*Resistant*)；*MR*，中抗（*Moderately resistant*)；*S*，感（*Susceptible*)；*MS*，中感（*Moderately susceptible*)。

1.3 普通野生稻白叶枯病抗性的遗传异质性和多样性

虽然普通野生稻中高抗白叶枯病的材料很少，但它是栽培稻的近缘种，并且还具有抗早衰（*early senescence*)、耐寒和优质等特性，只要用常规杂交就可以将其有利基因导入栽培稻。但是普通野生稻有开放授粉的习性，所以自然群体内对白叶枯病的抗性存在着遗传异质性。章琦等（1994）在鉴定广西壮族自治区 585 份普通野生稻种质的白叶枯病抗性时，发现多数抗病材料在遗传上是异质（*heterogeneous*）的，表现为同一编号的植株间有抗/感的不同反应以及处于相同生育期的同一植株上的不同叶片，对相同菌株的抗性反应不一。另外，孙恢鸿等（1992）也观察到从同一植株上收下的种子，在来年检测其所属植株的白叶枯病抗性时仍然存在抗/感的分离现象。

Hamamatsu 等（1992）比较了泰国普通野生稻 4 个自然群体和孟加拉国普通野生稻 2 个自然群体对日本白叶枯病菌小种Ⅰ、Ⅱ、Ⅲ和Ⅳ的抗性差异，并与中国、孟加拉、日本栽培品种的抗性反应对比，其结果为：①在泰国的 4 个普通野生稻自然群体中，2 个深水多年生野生稻群体的绝大多数植株对日本 4 个小种的反应型是抗的（*RRRR*），2 个浅水一年生野生稻的抗性反应差异则很大，对日本 4 个小种表现抗（*RRRR*）或感（*SSSS*）；②孟加拉国的 2 个深水稻田中杂草型普通野生稻对 4 个日本小种的反应型有 *RRSR*、*RRRS*、*SRRR*、*RSRR*、*SRRS*、*RSSR*、*SRSR*、*SSSS* 和 *RRRR* 等 6～9 种表现型；③中国云南稻、*IR*8、*BR*3、*BR*14 等栽培品种对 4 个日本小种的反应型简单，只有 1～4 种抗性反应型；④*Pajam*、*BR*3 和*BR*14 等栽培品种则表现全抗日本 4 个小种（*RRRR*）。该研究还表明远离栽培稻田的普通野生稻群体其抗性遗传异质性较小，而生长在栽培稻田中的普通野生稻群体其抗性遗传异质性较大。这是由于野生稻的花器发育特性（花粉发育不良、开颖时间长、柱头伸长外露等）有利于接受外来花粉，因而出现了距离栽培稻较近的野生稻群体其种间自然异交率较高，而与栽培稻相距较远的普通野生稻群体则种间自然异交率相对较低的现象。

1.4 利用野生稻的途径

利用种间杂交将野生稻的有利基因导入栽培稻，要根据其亲缘关系的远近采取不同的方法。对于含有 *AA* 染色体组的栽培稻近缘野生种，一般可采用常规有性杂交获得 F_1，然后再回交、自交若干世代，即可育成携有目标基因的栽培稻新种质。例如携有长药野生稻的 Xa*21* 基因的 *IRBB*21 和携有普通野生稻的 Xa*23* 基因的*CBB*23 就是通过常规杂交育成的抗白叶枯病的栽培稻品系（*Zhang et al.*，2004）。至于与栽培稻亲缘关系较远的野生稻种，如疣粒野生稻等，很难采用有性杂交将其优良抗性渗入到栽培稻，采用原生质体培养和体细胞杂交技术，则有可能将不同染色体组的有利基因聚合在一起。近年这方面的尝试已有若干成功的事例。何光存等（2004）针对疣粒野生稻的愈伤组织缺乏分化能力，先将其于液氮中进行超低温处理，取出后令其解冻恢复生长并培育成胚性细胞系（*embryonic cell line*）。然后，采用一系列原生质体培养技术，再生出原生质体植株，其外形与原始植株没有差异并具有与原始供体相同的抗白叶枯病能力。至于体细胞杂交或融合，有对称体细胞杂交（融合）与不对称体细胞杂交（*asymmetric somatic hybridization*）之分。前者指渗入双方的原生质体直接进行融合，但在导入目标基因的同时，也带入许多不利基因。因此，有关研究者采取各种可能的物理或化学处理，打破亲本一方完整的染色体结构后再用于体细胞杂交，这种只有部分染色体转入杂种细胞中的融合方式称为不对称体细胞杂交。近年来采取不对称体细胞杂交已成为产生体细胞核杂种或胞质杂种的常用手段。例如以药用野生稻、紧穗野生稻和短药野生稻作为基因供体亲本，分别与栽培稻进行体细胞杂交，都已获得了种间体细胞杂种（*Hayashi et al.*，1988；*Brar et al.*，1997）。朱永生等（2004）也报道了栽培稻（*AA*）与疣粒野生稻（*GG*）两个亲缘关系很远的种间不对称体细胞杂交的成功实例。他们以软 *X* 射线处理的疣粒野生稻原生质体作供体，用粳稻品种 02428 原生质体经碘乙酰胺（*IOA*）失活后作受体，采取 *PEG*（*polyethylene glycol*，聚乙二醇）法诱导原生质体融合，由于代谢功能互补，成功地从其融合产物中获得了愈伤组织并分化出 72 个再生植株。这些融合植株在形态上已接近栽培稻，通过微卫星分子标记和细胞遗传学分析，确认是栽培稻和疣粒野生稻的体细胞杂种。由它产生的体细胞杂种系在成株期接种菲律宾小种 1（*PXO*61）和小种 3（*PXO*79）时，约 50%的植株表现为高度抗病，其抗性与作为供体的药用野生稻相近，但仍处在分离状态，尚需采用回交结合胚拯救方法才能获得有利用价值的育种材料。

在远缘杂交，特别是亲缘关系稍远或较远的种属间杂交中，时常出现杂种胚不能正常成活或夭折的问题，胚拯救是广泛用于获得远缘杂交种子的技术。*Amante-Bordeos* 等（1992）将四倍体小粒野生稻（O. minuta，*BBCC*）与栽培稻 *IR*31917 杂交后获得的幼胚，用赤霉酸和萘乙酸等植物生长剂在 28℃下进行幼胚暗处理，生出幼芽后置于荧光、水培等适宜生长的条件下离体培养，以拯救其蛋白质合成，待植株正常成长后，再按照常规的连续回交、自交方法将小粒野生稻抗白叶枯病基因导入栽培稻，育成了抗菲律宾小种 2、3、5、6 的广谱抗性新品系。随后由 *Gu* 等（2004）对其进行抗性遗传、基因鉴定和定位研究，认为其携有一个新基因并被命名为 Xa*27*t。谭光轩等（2004）从药用野生稻渗入栽培稻的后代中发现并定位了一个抗白叶枯病新基因 Xa*29*t。*Jena* 等（1990）通过种间杂交也将药用野生稻的抗病（虫）性导入栽培稻，获得了抗白叶枯病、褐飞虱或白背飞虱的一系列单体异附加系（*monosomic alien addition lines*，*MAAL*）。

截至 2004 年的报道，已有 6 个远缘野生稻种的白叶枯病抗性导入了栽培稻并获得种间杂交种或单体外源附加系（表 4）

表 4 获得栽培稻与若干远缘种间杂交种的方法[a]

Table 4 Methods for producing interspecific hybrids between *O. sativa* and some distantly related wild *Oryza* species

野生稻种 Wild rice species	抗病（虫）性[b] Disease and pest resistance	方法 Method	参考文献 Reference
小粒野生稻 (*O. minuta*) (BB, BBCC)	BB，BL	种间杂交、胚拯救 (Interspecies hybridization, embryo rescue)	Amante-Bordeos et al. , 1992
短药野生稻 (*O. brachyantha*) (FF)	BB	种间杂交、胚拯救	Brar et al. , 1996
紧穗野生稻 (*O. eichingeri*) (CC)	BB	体细胞杂交 (Somatic hybridization)	Yasuyuki et al. , 1988
药用野生稻 (*O. officinalis*) (CC)	BB，BPH，WBPH	种间杂交、胚拯救	Jena and Khush，1989；1990
疣粒野生稻 (*O. meyeriana*) (GG)	BB，BPH	不对称体细胞杂交 (Asymmetric somatic hybridization)	Zhu et al. , 2004
疣粒野生稻 (*O. meyeriana*) (GG)	BB	超低温冷藏、原生质再生 (Cryopreservation, protoplasm regeneration)	何光存等，2004
澳洲野生稻 (*O. australiensis*) (EE)	BB，BPH	胚拯救	Multani et al. , 1994

[a] 根据有关文献资料整理（Summariesd from references）。

[b] BB，白叶枯病（Bacterial blight）；BL，稻瘟病（Rice blast）；BPH，褐飞虱（Brown planthopper）；WBPH，白背飞虱（Whitebacked planthopper）。

2　中国地方稻种和引进稻种资源的白叶枯病抗性评价

我国的地方稻种资源主要分布在云贵高原、华南、长江下游太湖流域和长江中游洞庭湖平原等地区。这些地区时常发生白叶枯病，有些年份还较为严重，因此所在省（市、自治区）农业科学院都很重视对其所搜集到的品种资源进行抗病性评价和研究。

2.1　地方稻种资源的抗性评价

2.1.1　云南稻种资源　1956 年，西南农业科学研究所联合云南省农业厅和省试验站开始对全省稻种资源进行考察和收集。至 1980 年云南省农业科学院已保存有 6 000 余份稻种资源，并集中在元江县农业科学研究所进行繁种、性状调查和整理编目，已编入《云南省稻种资源目录》的种质计有 5 128 份（陈勇等，1993）。

云南省农业科学院、元江县农业科学研究所和华中农业大学合作，在 1986—1987 年三季用致病性稳定的湖北白叶枯病菌株江陵（JL）691 对 6 184 份次云南地方稻品种进行人工接种鉴定，表现抗的品种（病斑 0～3 cm）有 345 个，占总数的 5.6%，其中抗性稳定的约 50 份；中抗品种（病斑 3.1～6 cm）824 个，占总数的 13.3%；没有发现免疫品种。抗病品种的数量和频率与水稻类型、海拔高度及地区分布有一定关系，抗病品种的频率方面，粳稻高于籼稻，陆稻高于水稻，糯稻高于黏稻；海拔低于 1 400 m 时，抗病品种频率随高度降低而增高，海拔在1 800 m 以上时，抗病品种频率则随高度增高而提高；按地区来看，滇西北的抗病品种频率最高，依次为滇西南、滇东南和滇东北（陈勇等，1990）。

同一时期，华中农业大学与云南省农业科学院合作，采用菌株 JL691 以及若干国际白叶枯病菌鉴别小种研究了 21 个抗白叶枯病性状稳定的云南地方稻种的抗性遗传，发掘出一个新基因 *Xa22t*（Lin et al.，1996）和一批新抗源（谢岳峰等，1990）。其抗性遗传行为可归纳为：①12 个品种的抗性各受一对显性基因控制，但这些品种所抗御的菌株来源及菌株数则多寡不一：除扎昌龙携有一对新基因 *Xa22t* 外，中白米、黄牛尾香糯、安宁早、鲁士谷和毫梅 5 个品种抗湖北菌株 JL691、菲律宾小种 1、5、6 和日本小种Ⅰ等共 5 个菌株（熊卫等，1994）；堆金子、毫糯扬和干炸谷 3 个品种抗菲律宾小种 1，其中干炸谷的抗性基因与 *Xa3* 等位或相同（高和平等，1996）；云香糯、毫双 7 号和长毛糯 3 个品种抗 JL691 和菲律宾小种Ⅰ，其中云香糯和毫双 7 号的抗性基因可能与 *Xa4* 等位，长毛糯的抗性基因与 *Xa3* 等位（谢岳峰等，1990）。②受一对隐性基因控制，如冬糯对 JL691 的抗性遗传（林兴华等，1992）。③受两对隐性基因控制，如云南大粒对 JL691 的抗性。④同一品种与不同菌株互作后分别有显性或隐性遗传，如八月糯、冷水谷、87304、矮脚糯 1 和矮脚糯 0 五个品种对 JL691 的抗性由一对显性基因控制；对菲律宾小种 1 的抗性则由一对隐性基因制约。⑤品种 87305 对 JL691 的抗性由两对显性基因控制，对菲律宾小种 1 的抗性则由两对隐性基因制约。此外，还发现若干个携有与已知基因不等位的新抗源，尚需进一步鉴定研究（表 5）。

2.1.2　太湖流域稻种资源　早在 1957 年和 1958 年，方中达等（1963）用南京白叶枯病菌株 OS－2 和 OS－6 对 1209 份江苏、浙江、安徽和江西 4 省保存的部分地方水稻品种进行了抗性测定，筛选出高抗～抗（1～Ⅱ级）品种 26 份，占 2.2 %，其中籼稻有瘦田赤、湖南白、矮脚齐子、湖南早、一线红、红节川、太和黏、红米子、小冬稻，粳稻品种有叶里培、

表 5 云南地方稻种的白叶枯病抗性遗传[a]

Table 5 Inheritance of resistance to bacterial blight in Yunnan local rice resources[a]

品种 Cultivar	类型 Type	所用菌株 Strain used[b]	遗传模式 Mode of inheritance	鉴定基因 Gene identified	参考文献 Reference
扎昌龙	粳（japonica）	P6	显性单基因 （1dominant gene）	*Xa22t*	Lin et al. ,1996.
中白米	籼（indica）	P6，P1，P5， T1、JL691	显性单基因	—	熊卫等，1994
黄牛尾香糯	籼	P6，P1，P5， T1、JL691	显性单基因	—	同上
安宁早	偏籼	P6，P1，P5， T1、JL691	显性单基因	—	同上
鲁士谷	偏籼	P6，P1，P5， T1、JL691	显性单基因	—	同上
毫梅	偏粳	P6，P1，P5， T1、JL691	显性单基因	—	
堆金子	粳	P1	显性单基因	—	高和平等，1996
毫糯扬	籼	P1	显性单基因	—	同上
干炸谷	籼	P1	显性单基因	*Xa3*	同上
云香糯	籼	JL691、P1	显性单基因	*Xa4*	谢岳峰等，1990
毫双七号	籼	JL691、P1	显性单基因	*Xa4*	同上
长毛糯	粳糯(japonica waxy)	JL691、P1	显性单基因	*Xa3*	同上
风情 2 号	粳糯	JL691	显性单基因	*Xa－j*	万焕桥等，2000
冬糯	籼	JL691	隐性单基因 （1recessive gene）	*xa－k*	林兴华等，1992
云南大粒	粳	JL691	两对隐性基因 （2recessive genes）	—	罗利军等，1991
八月糯	粳糯	JL691 P1	显性单基因 隐性单基因	—	谢岳峰等，1990
冷水谷	粳黏	JL691 P1	显性单基因 隐性单基因	—	同上
矮脚糯（1）	粳糯	JL691 P1	显性单基因 隐性单基因	—	同上
矮脚糯（2）	籼	JL691 P1	显性单基因 隐性单基因	—	同上
87304	籼	JL691 P1	显性单基因 隐性单基因	—	同上
87305	籼	JL691 P1	两对显性基因 （2recessive genes） 两对隐性基因 （2dominant genes）	—	同上

[a] 根据谢岳峰论文选集整理（2004）(Summarized from Xie Academic Volume，2000)。

[b] P1，菲律宾小种 1（Philippine race 1）；P5 小种（Race 5），P6 小种（Race 6），T1 日本小种 1（Japanese race 1）。

陈家种、老大稻、中齐家青、白壳小稻、菊花黄、黑种、齐家青野稻、矮萁黄、矮子晓稻、好土粳、笔杆青、霜降乌、湖足糯、红壳糯、晚长糯和农林 16 号。中抗品种 253 个，约占 21.0%。

太湖流域种植粳稻的历史悠久，晚粳地方品种资源极为丰富。过崇俭等（1986）1982—1984 年用苏、浙白叶枯病菌株对 1 389 份太湖流域和苏北地区地方粳稻品种资源进行人工接种和田间自然诱发鉴定，筛选出抗白叶枯病品种 281 份，占总数的 20.2%，其中小爱 2 号、风景稻、大绿种、矮绿种、铁粳青、红芒糯、大稻头糯、红壳稻、长秸糯、早糯稻、白壳晚稻等 11 个优质晚粳品种对江苏、浙江两省 6 个白叶枯病菌株 KS8－4、KS6－6、KS1－12、KS1－21、OS30 和 ZHE 173 都表现稳定的抗性，其抗谱较广（表 6）。万建民等（1990）用 KS6－6、KS1－21、JS 49－6、ZHE 173 四个苏、浙白叶枯病菌株进一步分析了小爱 2 号、风景稻和大绿种 3 个品种的抗性遗传，这些品种对不同菌株的抗性表现了 1～2 对显性或隐性主基因控制或受多基因控制的不同遗传行为。

表 6　太湖流域粳稻品种对江苏、浙江 6 个白叶枯病菌株的抗性[a]

Table 6　Resistance to 6 strains of bacterial blight in japonica rice cultivars from Tai Lake Valley

品种 Cultivar	来源 Origin	类型 Type	米质 Grain quality	抗性反应（按 5 级调查标准） Resistance reaction
铁粳青	桐乡	粳（japonica）	高（High）	HR-R（0.5～1.1）
风景稻	吴县	粳	高	HR-R（0.8～1.9）
大绿种	青浦	粳	高	HR-R（0.9～1.8）
白壳晚稻	嘉善	粳	高	R（1.2～2.0）
小爱 2 号	昆山	粳	高	HR-R（0.8～1.8）
红壳稻	吴县	粳	高	HR-R（0.9～1.5）
矮绿种	嘉兴	粳	高	HR-R（0.7～2.0）
红芒糯	昆山	糯（glutinous）	高	HR-R（1.0～16）
大稻头糯	吴县	糯	高	HR-R（0.9～1.7）
长秸糯	武进	糯	高	HR-R（0.8～1.9）
早糯稻	镇江	糯	高	HR-R（0.9～1.9）
昆农选 16（CK）	昆山	粳	—	S-SH（3.1～5.0）

[a] 供试的 6 个白叶枯病菌株 KS 8－4、KS 6－6、KS 1－12、KS 1－21、OS30 和 ZHE173（Six strains of *Xoo* used KS 8－4，KS 6－6，KS1－12，KS－1－21，OS30，and ZHE173）。

2.1.3　两广地区稻种资源

（1）广东省稻种资源　早在 20 世纪 30 年代丁颖就开始调查、收集全省地方稻种资源的工作，但大规模对稻种资源进行白叶枯病抗性评价的工作是在“七五”（1986—1990）期间。广东省农业科学院将 5 361 份包括海南岛在内的广东地方品种进行了白叶枯病、稻瘟病和稻飞虱的抗性鉴定，筛选出 162 份对白叶枯病具有中抗以上的品种材料，其中水稻 150 份、陆稻 13 份。水稻方面按抗病等级分，高抗（HR）品种有早仔种和宫牌种 2 个；抗的（R）品种有马尾齐、红梅糯、牛团糯、红壳糯、早水玻璃占、麻谷、麻占、大麻谷等 25 个；中抗（MR）品种有玻璃占、牛尾早、黄占、银占、包胎江、长尾、白壳赤和毛占谷等 88 个。兼抗白叶枯和稻瘟病或褐/白背飞虱的有 35 个，其中兼抗两病的有大黄谷、惠州占、黑谷、君仔、高脚早熟、普宁马龙牙等 12 个；兼抗白叶枯、褐飞虱或白背飞虱的有 16 个，如玻璃占、大子谷、大糯、大雨糯、丝花白、迟赤、蒲水莲和立狗糯等；三抗白叶枯病、稻瘟病和褐飞虱或白背飞虱的有 7 个，即插仔、龙虾种、南海白壳、齐眉、早白谷、八月迁和大熟。陆稻方面，抗白叶枯病的有山兰、旱稻、山禾、山谷糯、地禾糯等 13 个品种（徐羡明等，1990，表 7）。上述具有中度抗性的玻璃占曾作为抗源与六二占杂交，育成了抗病、矮秆、优质的华南地区白叶枯病骨干抗源品种玻璃矮，后者又被用作抗性供体育成了玻惠占、玻齐占、珍占等一系列抗病优质品种。此外，广东省有若干栽培历史悠久、抗病虫能力强的优质地方品种曾在育种中发挥了十分重要的作用。如抗病品种竹占，20 世纪 30 年代就被原中山大学用其与印度 2 号杂交育成了竹印 2，由其衍生的华竹矮，一度成为华南稻区抗白叶枯病育种的骨干抗源品种，先后衍生出青华矮 6 号、晚华矮 1 号和青叶青丰等一批高抗白叶枯病的优质、高产品种。又如 1952 年广东朝安晚籼地方品种绞盘矮从中系选出的抗病品种溪南矮，曾于 1962—1973 年间累计推广了 107 万 hm^2（1 605 万亩）。

表 7　广东省抗白叶枯病和兼抗其他病虫的水稻品种资源（徐羡明等，1990）

Table 7　Rice varietal resources resistant to bacterial blight and other diseases or pests in Guangdong province (Xu et al., 1990)

品种 Cultivar	类型 Type	所抗病虫[a] Resistance to disease & pest	品种 Cultivar	类型 Type	所抗病虫[a] Resistance to disease & pest	品种 Cultivar	类型 Type	所抗病虫[a] Resistance to disease & pest
早仔种	籼 (indica)	BB (HR)	麻占	籼 (indica)	BB (R)	插仔	粳 (japonica)	BB BL BPH
宫牌处	粳 (japonica)	BB (HR)	柏粒占	籼	BB (R)	香花糯	粳	BB WPH
牛团糯	粳	BB (R)	花苗仔	籼	BB (R)	立狗糯	粳	BB WPH
红壳糯	粳	BB (R)	白粳占	籼	BB (R)	丝花白	籼 (indica)	BB WPH
三粒寸	粳	BB (R)	大麻谷	籼	BB (R)	龙芽仔	籼	BB WPH
株早	籼	BB (R)	惠州占	籼	BB BL	广西种	籼	BB WPH
大须大川谷	籼	BB (R)	黑君仔	籼	BB BL	12 号	籼	BB WPH
红谷仔	籼	BB (R)	黑谷	籼	BB BL	增城银占	籼	BB WPH
花头白	籼	BB (R)	高脚早熟	籼	BB BL	迟赤	籼	BB WPH
马尾齐	籼	BB (R)	普宁马龙牙	籼	BB BL	蒲水莲	籼	BB WPH
红脚糯	籼	BB (R)	乌督占	籼	BB BL	指天占	籼	BB WPH
早水玻璃占	籼	BB (R)	园身早	籼	BB BL	暹种	籼	BB WPH
红六十占	籼	BB (R)	大黄谷	籼	BB BL	有须糯	籼	BB WPH
麻谷	籼	BB (R)	黄壳赤	籼	BB BL	龙虾种	籼	BB BL WPH
白花占	籼	BB (R)	慢种快	籼	BB BL	南海白壳	籼	BB BL WPH
十月仔	籼	BB (R)	邓源糯	粳	BB BL	齐眉	籼	BB BL WPH
黄花	籼	BB (R)	大糯	粳	BB BPH	早白谷	籼	BB BPH WPH
矮脚白壳 178	籼	BB (R)	500 粒江米	粳	BB BPH	八月迁	籼	BB BPH WPH
苏仔	籼	BB (R)	大南糯	粳	BB BPH	大熟	籼	BB BPH WPH
细黄谷 6	籼	BB (R)	大子谷	籼	BB BPH			
大杀谷	籼	BB (R)	玻璃占	籼	BB BPH			

[a] BB，白叶枯病（Bacterial blight）；BL，稻瘟病（Rice blast）；BPH，褐飞虱（Brown planthopper）；WPH，白背飞虱（Whitebacked planthopper）；HR，高抗（Highly resistant）；R，抗（Resistant）。

（2）广西稻种资源 广西壮族自治区农业科学院从 20 世纪 30 年代初就开始进行稻种资源的收集、整理、保存和研究，至 90 年代初已有 8 790 份种质被编入全国目录。该院又在 70～90 年代连续鉴定了 7 964 份地方稻种资源和育成品种对本省白叶枯病优势菌群Ⅳ型的抗性，没有发现免疫和高抗的种质，表现中抗的品种材料有 61 份，占总数的 0.8%，主要是粳型糯稻材料，如大糯、早糯、黄糯、香糯、江州糯、东糯、黑芝糯、腰伍糯、无芒糯、侯线糯等；中感的材料有 516 份，占 6.5%；感病的有 7 369 份，占 92.7%。不同水稻类型的抗病品种情况是粳稻多于籼稻，糯稻多于黏稻，中晚稻多于早稻，陆稻多于水稻；来自不同农业生态区的种质，其抗性程度轻重不一，一般是分布在山地的品种抗病率高，在博白、崇左、宁明等山地少的县，其品种几乎全部感病（孙恢鸿等，1991）。

2.1.4 湖南省稻种资源 湖南省品种资源丰富。20 世纪 70 年代中期，该省植病和昆虫学家开始了水稻品种对稻瘟病、白叶枯病和稻飞虱等病虫抗性研究，并且较早就与抗病虫育种密切结合。1986—1990 年湖南省农业科学院水稻所用该省白叶枯病强致病力菌株 HA33 对 4 615 份水稻地方稻种资源和引进品种进行了抗性鉴定，筛选出中抗以上品种材料 457 份。其中籼稻 375 份，占籼稻总数的 9.4%，按抗性 1～9 级划分，高抗（HR，1 级）材料 11 份，有假蚂蚱谷、背子谷、紫谷、ARC6231、ARC6565、ARC7090、IET3096 等；抗（R）的材料 14 份，如二百号、牛屎麻、百粒糯、香稻、双抗红等；其余 350 份为中抗（MR）材料。在 46 份中抗以上的粳稻（占粳稻总数的 15.4%）中，只有背子谷是高抗的，抗（R）的材料也只有江永香稻，其余 44 份为中抗。在 36 个中抗以上的糯稻中（占糯稻总数的 13.2%），只有工人糯、冬香糯和百粒糯 3 份抗（R）的材料，其余 33 份为中抗。有些抗病材料兼抗稻瘟病或白背飞虱或褐飞虱并具有中等以上米质，如双抗红、工人糯、紫谷、背子谷、冬香糯等（表 8）（侯小华等，2001）。

表 8 湖南省抗白叶枯病及其他病虫的水稻种质（侯小华等，2001）

Table 8 Rice germplasms with multiple resistance to bacterial blight and other diseases or pests in Hunan province（Hou et al.，2001）

品种 Cultivar	类型 Type	米质 Grain quality	所抗病虫[a,b] Resistance to pests			
			BB	BL	BPH	WBPH
双抗红	籼（indica）	优（high）	R	R	R	—
工人糯	糯（glutinous）	优	R	R	R	—
黄毛黏	籼	中等（medium）	MR	MR	MR	—
迟麻谷苏	籼	优	MR		HR	MR
暗嘴红	籼	中	MR	—	HR	R
红禾糯	糯	优	MR	—	HR	R
冬毛黏	籼	中	MR	MR	MR	—
坳冬黏	籼	中	MR	MR	MR	—
大叶毛	籼	优	MR	—	MR	MR
麻壳糯	粳（japonica）	优	MR	MR	MR	—

（续）

品种 Cultivar	类型 Type	米质 Grain quality	所抗病虫[a,b] Resistance to pests BB	BL	BPH	WBPH
紫谷	籼	优	HR	HR	—	—
背子谷	粳	优	HR	HR	—	—
假蚂蚱谷	籼	中	HR	HR	—	—
金钱糯	糯	优	MR	HR	—	—
响壳糯	糯	优	MR	HR	—	—
二百号	籼	中	R	—	R	—
早麻谷	籼	中	MR	—	HR	—
麻三百粒	粳	优	MR	—	HR	—
东升高脚	籼	优	MR	—	HR	—
毛黏	籼	中	MR	—	HR	—
黄麻黏	籼	优	MR	—	HR	—
红米糯	籼	优	MR	—	HR	—
长子三十黏	籼	优	MR	—	HR	—
江永香稻	粳	优	R	MR	—	—
冬香糯	糯	优	R	—	—	R
白壳糯	糯	优	MR	—	—	R
大叶香麻黏	籼	中	MR	—	—	R
洋禾种	籼	中	MR	—	—	R
香稻	粳糯(japonica glutinous)	优	R	MR	—	—

[a] BB，白叶枯病（Bacterial blight）；BL，稻瘟病（Rice blast）；BPH，褐飞虱（Brown planthopper）；WBPH，白背飞虱（Whitebacked planthopper）。

[b] R，抗（Resistant）；MR，中抗（Moderately resistant）。

2.2 引进稻种资源的抗性评价

20 世纪 50 年代以来，南方各省（市、区）都在大力加强水稻品种资源的收集与研究。70 年代初我国水稻工作者开始注意引进 IRRI 育成的籼稻品种（系），并从中评价出可以直接在南方稻区生产利用和用作杂交水稻恢复系的 IR 品种。1981 年中国农业科学院和国际水稻研究所达成了互相交换水稻种质资源的协议。也就是在这一年，我国南北稻区的主要农业研究机构（包括院校）开始参加由 IRRI 组织的国际水稻遗传评价项目，从而显著扩大了国际种质资源的拥有量和推动了对这些资源的病（虫）抗性评价和利用工作。

现将全国引进稻种资源抗白叶枯病鉴定的主要工作情况和结果扼要叙述如下：

①1979—1982 年，中国农业科学院作物品种资源所组织全国有关省（市、区）农业科学院植保所，对引自 82 个国家以及 IRRI 的共 3 500 多份籼、粳稻种质进行了稻瘟病、白叶枯病、黄矮病、稻飞虱和稻纵卷叶螟的抗性鉴定。其中白叶枯病抗性鉴定工作由品资所（北京）、广东、江苏、四川三省植保所分别采用华北和上述 3 省的水稻白叶枯病各一个代表性菌株对数以

千计水稻种质进行人工接种。结果表明，按抗性1～9级划分，有459份材料分别对这4个白叶枯病菌株具有中等以上抗性；在籼稻中，抗全部4个代表菌株（RRRR，1级）的有IRRI的IR17494-32-1-1-3-2和IR19670-57-1-1-3两个品种；抗3个、中抗1个菌株（RRR MR，3级）的品种有10个，包括IR54、IR19660-274—3-3-1-3、IR17494-32-1-1-3-2和IR2823-399四个IRRI品系，印度尼西亚的M61B-15-3-1和B2360-8-5-LR4-3，斯里兰卡的BG90-2和BW242-5-5，以及印度的RP633-76-1和UPR79-23。在粳稻品种中，没有全抗4个菌株的品种，表现抗3个菌株、中抗1个菌株（RRR MR）的有来自朝鲜的龙城30和赤标以及巴基斯坦的CTG1118共三个品种。全部结果已汇编于《引入稻种资源主要农艺性状和抗三病两虫鉴定结果》(中国农业科学院作物品种资源所编，1983)。

②我国参加IRRI组织的国际水稻遗传评价项目是从1981年开始的，至今仍继续进行。参加单位有14个，组成一个协作网，由中国水稻所主持。根据该所负责国际水稻遗传评价组织工作的汤圣祥（2002）的20年总结报告，各单位根据本生态区的特点和工作条件，分别从IRRI引进12类试验圃的各种材料共7 989份进行了试种、鉴定和遗传评价。其中白叶枯病鉴定圃（IRBBN）工作由中国水稻研究所、广西、广东和湖南等省（区）为主的农业科学院植保或水稻所承担，挑选其中的2 000多份国外种质，种植在当地白叶枯病疫区，采取自然发病和用本省（区）优势菌群或强毒性菌株人工接种相结合的方法进行抗性鉴定。经初评和复评，选择历年鉴定表现抗性好而稳定的材料，推荐给生产直接利用或作为抗源亲本在育种计划中应用。根据汤圣祥等（2002）报道，已有35个高产、优质、抗病、适应性广的国外品种材料由各地自行定名作为新品种直接在生产上推广种植，其中高抗白叶枯病的18个品种，如BG90-2、IR22、IR20、IR26、IR28、IR29、IR30、IR36、IR50、IR54、IR64等已被广泛用作抗源，这批品种同时也是我国籼型杂交稻的主体恢复基因源，可参阅表9和表10。

表9　直接用于我国生产的INGER抗白叶枯病种质（摘自汤圣祥等，2002）

Table 9　INGER entries with BB resistance released as cultivars for production in China

(Summarized from Tang et al., 2002)

INGER编号 INGER entry	品种命名 Cultivar in China	原产地 Origin	推广年份 Year released	主要种植省（区） Area planted
BG90-2	扬稻1号	Srilanka	1983	江苏、安徽
BG90-2	扬稻2号	Srilanka	1985	江苏、安徽
Chianung Sen Yu 13	E108	Taiwan, China	1980	湖南
Chianung Sen Yu 26	川米1号	Taiwan, China	1990	四川
IR26	IR26	IRRI	1980	安徽、四川
IR29	HA糯15	IRRI	1990	湖南
IR36	IR36	IRRI	1984	湖南、广西
IR54	七桂早25	IRRI	1989	广东
IR64	IR 64	IRRI	1989	广西、湖北
IR1561-228-3-3	32选5 32	IRRI	1980	湖南、广西
IR15853-89-7E-P3	N90	IRRI	1981	广西

（续）

INGER 编号 INGER entry	品种命名 Cultivar in China	原产地 Origin	推广年份 Year released	主要种植省（区） Area planted
IR19274-26-2-31-2	湘晚籼 2	IRRI	1992	湖南、江西
IR21015-80-3-3-1-2	N304	IRRI	1983	广西
IR21929-12-3-3	闽抗 108	IRRI	1986	福建
IR28125-79-3-3-2	桂 713	IRRI	1989	广西
IR9129-102-2	国际优占	IRRI	1985	广东
IR9965-48-2	晚选 35	IRRI	1981	广东

表 10　我国用作杂交水稻恢复系的 INGER 抗白叶枯病种质（摘自汤圣祥等，2002）

Table 10　INGER entries with BB resistance used as restorer parents in hybrid rice production in China (Summarized from Tang et al.，2002)

组合 Combination	恢复系 Restorer line	原产地 Origin	亲本（A/R） Parentage
白优 63	IR30	IRRI	白 A/（IR30/圭 630）
博优 3550	IR54	IRRI	博 A/（IR54/青四矮 16）
Ⅱ优 3550	IR54	IRRI	IIA/（IR54/青四矮 16）
金优 F6	IR24，IR26，IR30	IRRI	金 A/（10-35//IR36/IR24//圭 630/IR30）
汕优 34	IR30	IRRI	珍汕 97A/（IR30/IR24）
汕优 36	IR36	IRRI	珍汕 97A/IR36
汕优 63	IR30	IRRI	珍汕 97A（IR30/圭 630）
汕优 64	IR9761-19-1	IRRI	珍汕 97A/IR9761-19-1
汕优 85	IR28	IRRI	珍汕 97A/（台雄 2 号/IR28）
汕优 3550	IR54	IRRI	珍汕 97A/（青四占/IR50）
汕优 5111	IR36	IRRI	珍汕 97A（IR20913-B-60/IR36）
汕优桂 3	IR36	IRRI	珍汕 97A（IR36/IR24）
汕优桂 8	IR2061	IRRI	珍汕 97A（IR2061/IR661）
汕优桂 33	IR36	IRRI	珍汕 97A（IR36/IR24）
汕优桂 44	IR36	IRRI	珍汕 97A（IR36/Tai24）
汕优 96	IR50	IRRI	珍汕 97A（测 64//丛圭 3/IR50）
汕优桂 99	IR2061	IRRI	珍汕 97A/（*O. rufipogon*//IR2061/IR661）
威优 77	IR30	IRRI	V20A/（IR9761-19-1/IR30/圭 630）
威优 111	IR30	IRRI，韩国（Korea）	V20A/（IR30/圭 630//密阳 46）
香优 63	IR30	IRRI	湘香 2A（IR30/圭 630）
协优 432	IR54	IRRI	协青早 A/（测 64-7/IR56）
协优 9308	IR26	IRRI	协青早 A（C57//300 号/IR26）

以上是总的概括，下面再按籼、粳稻分别将广西、广东、湖南三省和中国农业科学院作物所抗性评价的主要情况与结果作些补充，因为它们在承担国际水稻白叶枯病鉴定圃（IRBBN）的任务之前，就进行了较大量的品种抗性评价和抗病育种工作。

(1) 籼稻 广西壮族自治区农业科学院植保所1975—1983年间，采用本区7～9个菌株人工接种与自然诱发相结合，反复多次对2 179份（包括国际白叶枯病鉴定圃在内）稻种资源按1～9级的标准评价了白叶枯病的抗性，结果表明高抗（HR）的材料75份，占总数的3.4%，抗/中抗（R/MR）的材料1 110份，占50.5%，其中抗性稳定、抗谱较广的（1～2级）材料有54个，如BJ1、DV85、TKM6、BG90-2、GE456、BR51-282-8、IR20、IR22、IR26、IE34、IR36、IR42、IR50、IR54、IR1529-680-3-2等（表11，孙恢鸿等，1985）；已用于该院和全自治区各地育种计划的有120多个抗病材料，如深受本地区农民欢迎的高产、优质、抗病的早籼品种梅桂1号就是用桂引1号（GE456）与感病的当家品种红梅早杂交后于1981年育成的；用IR36与IR24杂交，于1981年育成新抗病恢复系桂33，随后又用它与珍汕97A配组，于1983年育成高产、优质、抗白叶枯病、适应性广的杂交稻汕优桂33，1987年在南方稻区累计推广了80.67万hm^2（1 210万亩）（李丁民等，1991）。

表11 1974—1983年广西壮族自治区评价白叶枯病抗性稳定的引进稻种资源
（孙恢鸿等，1985）

Table 11 Introduced rice germplasms showing high and stable resistance to bacterial blight in Guangxi Autonomus Region as evaluated during 1974—1983 (Sun et al., 1985)

品种 Cultivar	抗性级别 Score			品种 Cultivar	抗性级别 Score		
	测定次数 No. of testing	变幅 Range	平均病级 Mean		测定次数 No. of testing	变幅 Range	平均病级 Mean
IR20	8	1.0～2.0	1.7	IR1833-208-1-3	6	1.2～2.5	1.9
IR22	7	1.0～2.1	1.8	IR11418-64-2	8	1.0～1.5	1.2
IR28	20	1.0～2.5	1.9	JPR70-30-7	10	1.5～2.0	1.8
IR30	17	1.0～2.5	1.9	IR1545-339-2-2	10	1.1～2.0	1.8
IR32	8	1.0～2.0	1.8	IR9828-91-2-3	11	1.0～2.5	1.9
IR36	39	1.5～2.5	2	IR9129-209-2-2-2-1	6	1.5～2.0	1.9
IR42	16	1.0～2.0	1.6	IR13429-196-1	8	1.5～2.5	2
IR50	7	1.4～2.5	1.9	IR13429-287-3	7	1.0～2.5	1.9
IR52	5	1.4～2.5	1.9	IR2153-338-3	19	1.3～2.5	2
IR54	4	1.1～2.0	1.8	IR3536-31-1	11	1.0～2.0	1.3
IR34	23	1.0～2.5	1.7	G. E. 456（桂引一号）	29	1.0～1.5	1.4
BG90-2	5	1.3～2.0	1.9	IR11297-139-2-2	3	1.1～2.0	1.7
DV85	7	1.0～2.0	1.7	BR109-74-2-2-2	3	1.2～2.0	1.6
BJ1	7	1.0～2.0	1.7	IR9703-41-3-3-1	3	1.0～2.0	1.3
UPR-4-25-1	12	1.0～2.5	1.8	IR9201-301-1-3-1-3	4	1.0～2.0	1.5
TKM-6	5	1.5～2.1	1.9	IR4442-46-3-3-3	3	1.0～2.0	1.3

（续）

品种 Cultivar	抗性级别 Score			品种 Cultivar	抗性级别 Score		
	测定次数 No. of testing	变幅 Range	平均病级 Mean		测定次数 No. of testing	变幅 Range	平均病级 Mean
Rasht507	4	1.1～2.6	1.8	IR868-231-2-2-3-2	3	1.0～2.0	1.4
UCP28	2	1.2～2.0	1.6	BR51-282-8	11	1.0～2.0	1.3
ASD7	9	1.4～2.6	1.7	IR8608-189-2-2-1-3	6	1.0～2.0	1.7
DZ192	2	1.5～2.6	2	IR2798-88-3-2	4	1.2～2.0	1.8
PI215936	2	1.6～1.6	1.6	RP633-76-1	3	1.0～2.0	1.3
IR1529-680-3-2	15	1.0～2.5	1.5	IR13423-17-1-2-1	3	1.0～2.0	1.3
IR2006-Pn-12-2-2	9	1.1～2.5	1.8	BR171-2B-8	3	1.5～2	1.8
BR319-1-HR12	2	2.0	2.0	CK IR26			1.3
UPR79-14	3	1.0～2.0	1.7	CK 金刚 30			3.7

广东省农业科学院植保所1980—1984年间在研究全省白叶枯病菌致病性变异的基础上，选用优势致病型Ⅳ的代表菌株鉴定了中国农业科学院品资所和国际白叶枯病鉴定圃引进的共2 012份籼稻种质的抗性。按1～5级的标准评价，其中高抗（1级）的种质29份，抗的（2级）187份，中抗的（3级）285份，又从这些抗病种质中挑选出57份经历年重复鉴定，认为抗性表现和经济性状都好的材料为重点种质，再进一步对本省白叶枯病Ⅰ～Ⅴ五个致病型及稻瘟病和褐飞虱进行抗谱评价，其中三抗的有IR36、IR42和IR4563-52-1-3-6三个品种；双抗白叶枯和稻瘟病的有IR28、Edzna A79（墨西哥）、IR4442-46-3-3-3、IR29、BR-161-2B-59和DZ192等13个品种；双抗白叶枯和褐飞虱的有IR17525-114-2-1-1、UPR82-1-7、IR54、IR9129-441-3-3-1和IR9769-67-3五个品种；抗本省白叶枯病Ⅰ～Ⅴ五个致病型的有Lal Ahu（巴基斯坦）、CNBP220-226（IRRI）、DZ78（孟加拉）、IR1545、RP633-76-1（印度）等16个品种；抗Ⅰ～Ⅳ四个致病型的有IR20、IR30、IR26、BR319-1和IR2307-117-2-1等8个品种。这些优良抗病材料都及时地用于该院和全省各地育种计划（徐羡明等，1986）。

湖南省农业科学院水稻所从20世纪70年代中期开始，长期在白叶枯病常发区安仁县驻点，进行品种抗性遗传改良工作。每年采用当地强毒性菌株HA33于分蘖期接种。至1992年，从累计1 190份（次）鉴定材料中，按1～5级的标准，评价出1级高抗品种37个，其中重复鉴定2次以上的高抗品种有IR9828-91-2-2、IR1561-228-3-3、IR9830-26-3-3、IR-BB、RP2151-173-1-8、RP2151-192-1、RP2151-33-2、DV85、BR161-2B-53、BR-161-2B-58和B4075D-PN-13-1计11个；抗（3级）的品种材料675个，又从中挑选出一批兼抗其他病害并且品质优良的材料作为亲本，在全省早籼新品种选育中起到了骨干作用（李友荣等，1993）。

(2) 北方粳稻　1977—1989年，中国农业科学院作物所评价了5 330份（次）引进粳稻资源和育种材料对崖雕1-1（海南凋萎型菌株）、江苏强毒性菌株KS8-4和辽宁菌株E2-2的抗性反应，共鉴定出具1～3级（按1～9级制）抗性稳定的粳稻品种38个，如引自日本的秋光、丰锦、黎明、秀峰、中新120、早生爱国3号、筑紫晴、京系酒、稻叶早生、关东60、银河、喜峰、农林35（关东30）、农垦58（世界）、农垦51和日本晴等，多数为成株抗性。其中1970年和1976年引进的丰锦和秋光已在东北粳稻区直接生产应用，丰锦在1978年就种植了19.53万hm^2（293万亩）。秋光于1987年在辽宁省的种植面积也达21.53万hm^2（323万亩），二者都表现高产、优质、抗白叶枯病。秋光的熟期略早，至今已有30年的种植历史。据全国农业技术推广服务中心统计，1983—2005年间的累计面积有210.87万hm^2（3 163万亩），这在东北粳稻区是不多见的。顺便指出，粳稻品种中具全生育期抗性的很少，只有来自朝鲜的矢租（1.5～1.6级）和印度的印度诺（爪哇型，1.2～1.4级）这两个品种在全生育期高抗我国南北稻区6个白叶枯病代表菌株和菲律宾1～4四个小种，它们对菲律宾小种1的抗性均受隐性的*xa5*所控制，而印度诺除*xa5*之外还携有另一对隐性的未知基因控制其成株抗性，值得注意利用（章琦等，1986；1988）。

2.3 广亲和品种材料的抗性评价

籼粳亚种间杂种优势的利用是水稻育种扩大遗传变异的重要途径，但是水稻亚种间杂交常有结实困难和后代的不育性障碍而限制其被广为利用。日本学者池桥宏等（Ikehashi et al.，1984）首先提出了广亲和性（wide compatibility）概念。它是指有一些水稻品种与籼稻或粳稻杂交时，当代结实率较高，胚和胚乳发育较正常，其杂种后代都表现出正常的育性，具有这一特性的品种称为广亲和性品种。此后，广亲和品种便成为遗传育种学家克服籼粳杂种育性障碍的重要桥梁材料。广亲和基因资源的发掘、鉴定、研究和利用等工作也就受到作物科学界的广泛关注。根据池桥宏的研究，广亲和品种主要出现在印尼爪哇稻的婆罗（buru）稻生态型品种和印度、孟加拉的一些生态型品种中（顾铭洪等，2001）。我国曾陆续引进一些广亲和水稻品种，并已在杂交水稻育种中发挥了重要作用。如湖南杂交水稻研究中心罗孝和等（1989）以爪哇稻培迪作母本，育成了广亲和品种轮回422、PC311、培矮64和光温敏核不育系培矮64S等，这些都可能选配出一系列优势组合。杨文才等（1995）1991—1993年间用6个菲律宾小种（1～6）、4个日本小种（Ⅰ～Ⅳ）、7个中国致病型，以及JL691（鄂）、LN44（辽）、KS1-21（苏）共20个白叶枯病菌株评价了28个引进的和在国内广泛应用的广亲和品种的抗性。结果表明有6个品种具有广谱抗性，如Dular（印度秋稻）和Pecos（美国）抗19个菌株（仅感染菲律宾小种6），品种Aus373（印度秋稻）抗18个菌株（感染菲律宾小种6和3），华师大7009（中国）抗18个菌株（感染菲律宾小种6和3），上述4个都是全生育期抗性品种；69232-2和鉴12在成株期抗18个菌株（感菲律宾小种6和日本小种Ⅳ）；广泛利用的广亲和品种培矮64在成株期抗15个菌株，但是受年份、季节和生育期的影响而对少数菌株的抗性反应不很稳定；029在成株期除了抗8个菌株外，感染其他菌株或抗性不稳定。（表12）。其余22个品种如培迪、Bellement、CA537、02428、早AB78、迟AB78、轮回422、测09、CA537、CA544、Calotoc、CP231、CPSL0-1、E164、JW-9、Ketan Nangka、Nangka、Lemont、Mcp231-4、Mcp231-7、PC311和Pelde感染所有20个菌株。用菲律宾小种1（PXO61）、中国致病型Ⅱ（HB17）和湖北菌株JL691三个菌株进行白叶枯病抗性遗传分析和基因推导表明，Dular除携有一对隐性基因*xa5*控制其全生育期对这3个菌株的抗性外，尚有一对与*xa8*不等位的未知隐性成株抗病基因；69232-2和鉴12的成株抗性是由*Xa3*控制的。培矮64的亲本之一是测64，后者是IRRI抗白叶枯病品系IR9761-19-1的衍生系；029的亲本之一是IR36，推测这两个品种的抗性也是由

*Xa*4 控制的，但其抗性表达时期与其供体相比，已有所延后。

表 12 8 个广亲和品种对 20 个水稻白叶枯病菌株的抗性反应（杨文才等，1995）

Table 12 Reactions of resistance to 20 strains of *Xoo* in 8 wide compatibilitg rice cultivars（Yang et al.，1995）

品种 Cultivar	来源 Origin	所用接种菌株 Strains inoculated												
		P1[c]	P2 C5	P3	P4	T4	P5 C2 C4 KS-1-21	P6	C1 JI691	C3	C6	C7	T1 LN44	T2 T3
Dular 籼	印度（India）	R[b]	R	R	R	R	R	S	R	R	R	R	R	R
Pecos 粳	美国（USA）	MR	R	MR	R	R	R	S	R	R	R	R	R	R
Aus373 籼	印度	R	R	R	S	R	R	S	R	R	R	R	R	R
华师大 7009 粳	中国（China）	R	R	S	R	R	R	S	R	R	R	R	R	R
69232-2[a] 籼	中国	R	R	R	R	S	R	S	R	R	R	R	R	R
鉴 12[a] 籼	美国	R	R	R	R	S	R	S	R	R	R	R	R	R
培矮 64[a] 籼	中国	R	R	S	S	R	R	S	S	R/S	R	R/S	R	R
029[a] 粳	中国	R	R	S	R/S	S	R	S	R	R/S	R/S	S/R	S/R	S

[a] 为成株期抗病品种，其他为全生育期抗病品种（Cultivars with adult plant resistance，and the others with overall resistance）。

[b] R，抗（Resistant）；MR，中抗（Moderately resistant）；S，感（Susceptible）。

[c] P1～P6，菲律宾小种（Philippine races）；T1～T4，日本小种（Japanese races）；C1～C7，中国致病型（Chinese pathotypes）。

3 日本稻种资源的白叶枯病抗性评价

日本育种家早在 1904 年就开始了水稻抗稻瘟病育种工作（Matsuo，1967），为 20 年后开展抗白叶枯病育种研究提供了丰富的材料基础和技术储备。工作初期着重在自然发病严重的疫区进行田间选择无病或发病较轻的单株，直至 1957 年发现常年抗病的品种朝风突然丧失了抗性之后，意识到需要注意研究病菌群体结构变异的问题，开始采用相应的人工接种方法对稻种资源进行较系统地抗性评价和筛选。Ogawa 等（1979）在回顾 1962—1965 年间采用针刺法鉴定 1 200 份本国地方稻种资源对白叶枯病的抗性表现时，就提到当时日本的水稻白叶枯病菌已被区分为Ⅰ、Ⅱ、Ⅲ三个菌群（70 年代后期开始命名为小种），并将其结果归纳如下：①苗期（8～9 叶期）接种的 400 个品种中只有 11 个抗菌群Ⅰ；②成株期接种的 800 个品种中有 4.3%抗Ⅰ、Ⅱ、Ⅲ三个菌群，如早生爱国 3 号、中新 120、中国 45、撰一（Sen-ichi、）等 34 个品种；有 11.1%抗菌群Ⅰ，如农林 27（Norin 27）、朝风（Asakaze）、霜月（Shimotsuki）、爱优（Aimasari）、大淀（Oyodo）、不知火（Shiranui）、福锦（Fukunishiki）等 88 个品种；③多数抗病地方品种，包括最早选出的滋贺关取 11、庄兵卫、高农 35 三个抗源品种及其衍生品种只抗菌群（小种）Ⅰ，少数抗菌群（小种）Ⅲ。没有发现抗小种Ⅱ的地方品种。

发现了小种Ⅳ以后，Uchiyamada 等（1977，1978）使用日本的Ⅰ、Ⅱ、Ⅲ、Ⅳ四个小种评价了 1 194 个日本地方水（陆）稻品种对白叶枯病的抗性。所得结果与 Ogawa 等的相似：在318个抗病品种中以陆稻居多，占有242个，水稻品种有76个。根据其习惯上按Ⅰ、Ⅱ、Ⅲ、Ⅳ四个小种顺序排列的反应型是：在318个抗病品种中，只抗小种Ⅰ（RSSS）的品种有241个，抗Ⅰ和Ⅱ（RRSS）的品种只有2个，抗小种Ⅰ、Ⅱ、Ⅲ（RRRS）的品种有24个，有51个品种是抗此感彼或抗彼感此的交叉反应（RSMS、SRMS、SRRS、SSRS、SSMS、SRSS等），没有发现抗小种Ⅳ的品种。表明日本地方品种对白叶枯病的抗性类型比较单一，抗谱较窄。

日本为了改变本国资源贫乏、抗性类型单一的情况，从不同国家引进了大量抗谱广、抗性类型多的籼稻品种。Sakagushi等（1968）最早测试了704份日本引进的稻种资源。它们对日本白叶枯病小种Ⅰ、Ⅱ、Ⅲ的抗性结果是：①来自不同国家的绝大多数品种抗日本小种Ⅰ；②亚洲热带国家的多数品种如Rantai-Emas 2（印度尼西亚）、Tadukan（菲律宾）、Tetep（孟加拉）、Nigeria（斯里兰卡）、RP9－3（印度）等抗日本小种Ⅱ，不抗小种Ⅲ，而日本小种Ⅱ几乎能侵染所有的日本品种；③Lead Rice（缅甸）、Java14（印尼）和Zenith（USA）等品种抗日本小种Ⅲ，但这类的品种还不够多。

Uchiyamada等（1977）评价了1437份引自世界各地的稻种资源对日本白叶枯病小种Ⅰ～Ⅲ的抗性，有421个（29.3%）抗病品种，根据对3个小种按顺序排列的反应型结果是：290个（20.2%）只抗小种Ⅰ（RSS或MSS）；46个品种（3.2%）抗小种Ⅰ和Ⅱ（RRS或RMS、MRS、MMS）；52个品种（3.6%）抗/中抗小种Ⅰ～Ⅲ（RRR或RRM、RMR、MRR、MMR、MMM）；33个品种（2.3%）对3个小种表现抗此感彼或抗彼感此的交叉反应（SRS、SSR、SSM、RSM、MSM）。Uchiyamada等（1978）后来又用日本白叶枯病小种Ⅰ～Ⅳ评价了另外648份引自世界各地的稻种资源的抗性，有142个（21.9%）抗病品种，其中93个品种（14.4%）抗小种Ⅰ；11个品种（1.7%）抗小种Ⅰ和Ⅱ；8个品种（1.2%）抗小种Ⅰ～Ⅲ，如RP291－7、IR2153－159－1－4等；15个品种（2.3%）抗小种Ⅰ～Ⅳ，如Sigadis、TKM6、Pelita I/1等携有*Xa4*基因的品种；15个品种（2.3%）对3或4个小种表现抗此感彼或抗彼感此的交叉反应。总的来说，日本从不同国家引进了抗谱广、抗性类型多的籼稻品种，丰富了国家抗病基因库。这些对白叶枯病抗谱较广的品种主要来自IRRI和印度尼西亚、印度、巴基斯坦、孟加拉、斯里兰卡等国家。

4 IRRI稻种资源的白叶枯病抗性评价

IRRI 1960年成立后，十分重视全球稻种资源包括野生稻种的收集、保存、评价和研究。IRRI也是最早倡导开展“稻种资源遗传评价与利用”的国际农业研究机构。早在1975年IRRI组织和协调了由各国水稻机构和国际农业研究中心组成的全球性国际水稻遗传评价研究项目（International Network of Genetic Evaluation for Rice，INGER，前称IRTP），旨在促进国际间自由交换、综合评价、迅速利用优异水稻种质，以带动提高作物育种水平和发展粮食生产。截至1987年，IRRI的国际水稻种质中心（IRGC）已保存有85 000份栽培稻种资源（应存山，1989），其中多数材料是通过该国际项目收集到的。据汤圣祥等（2002）报道，该项目从1991年开始设置了抗两病（白叶枯病、稻瘟病）、三虫［褐飞虱（*Nilaparvata lugens*）、白背飞虱（*Sogatella furci-fera*）、稻瘿蚊（*Orseolia oryzeae*）］、耐旱、耐盐碱、耐冷等多抗性鉴定与筛选、产量观察、产量试验、杂交稻、粳稻、旱稻、耐冷和优质等 12 种试验圃，曾组织 62 个国家参加，使全球稻种遗传资源得以进行深入研究与高效利用，至今仍在发挥其积极作用，推动了各国水稻育种工作不断向纵深发展，在国际水稻界和遗传育种领域产生了划时代的影响。

4.1 IRRI 稻种资源的白叶枯病抗性评价和抗性品种群

从 20 世纪 60 年代起 IRRI 植物病理学家就开始用当时菲律宾的 10 个代表性白叶枯病菌株（划分小种之前）评价了国际水稻种质中心初期保存的数千份稻种资源的抗性（IRRI's Annual Reports for 1964～1969；Ou et al.，1971），鉴定出 Sigadis、TKM6、PelitaI/1、ARC5756、DL5、Dhariai、DV85 和 DV86 等 300 多个抗病品种。1969 年，IRRI 育种家开始采用上述一些抗病种质育成了抗白叶枯病的 IR20、IR26、IR28、IR29、IR30、IR34 和 IR36 等品种，在亚洲产稻国家广泛种植或作抗源利用。70 年代中期，IRRI 病理学家发现了其中抗菲律宾代表菌株的 Sigadis、TKM6、PelitaI/1 等骨干抗源品种及其衍生的 IR 品种都感染菲律宾北部名叫“Isabella”的一个菌株，便扩大采用能抗“Isabella”侵染且抗谱较广的品种如 BJ1、Dular、AZ192、DV85 和 W1263 Zenith 等作为新抗源品种在育种中利用，育成 IR9623、IR14093、IR4475B 等一批新品种。

1975 年 IRRI 科学家又开展了更大规模的白叶枯病抗性鉴定，先用菲律宾白叶枯病主要小种 1 的代表菌株 PXO61 评价了 40 000 多份 IRRI 稻种资源的抗性，从中选出 3 000 多份抗病材料种植在白叶枯病筛选圃（BBSN），又用菲律宾 1～4 小种的代表菌株在分蘖期和孕穗期重复进行人工剪叶接种鉴定。至 1980 年，按病斑占叶面积百分率分 1～9 级的调查标准，鉴定出 2 274 份抗性为 1～3 级的抗病品种，其中包括全生育期高抗 PXO61（小种 1）的 DZ192、DV85、DV86 等 100 多个品种，在 IRRI 的抗性育种和遗传研究中广为利用（表 13）（Mew et al.，1981）。

表 13 用于 IRRI 育种方案中的水稻白叶枯病抗性供体亲本（Mew et al.，1981）

Table 13 Parental donors for resistance to bacterial blight of rice used in the IRRI crossing program (Mew et al.，1981)

品种 Cultivar	来源 Origin	品种 Cultivar	来源 Origin
Xa 3		*xa 5*	
Wase Aikoku	日本（Japan）	Dular	印度
Xa 4		ARC 5756	印度
CO10	印度（India）	Chinsurah Boro Ⅱ	印度
CO22	印度	BJ 1	尼泊尔（Nepal）
TKM 6	印度	Dharial	苏里南（Surinam ）
Taothabi	印度	Dhariai	苏里南
Pelita I/1	印度尼西亚（Indonesia）	DL 5	巴基斯坦（Pakistan）
Pelita I/2	印度尼西亚	DV29	孟加拉（Bangladesh）
Sigadis	印度尼西亚	Dv139	孟加拉
Gui Do	韩国（Korea）	DV 192	孟加拉

（续）

品种 Cultivar	来源 Origin	品种 Cultivar	来源 Origin
Pakheng	老挝（Laos）	Aus32	孟加拉
Xa 3（$Xa4^b$）[a]		Aus251	孟加拉
Rathu Heenati	斯里兰卡（Sri Lanka）	Aus449	孟加拉
Sintawati	印度尼西亚	Bageri	孟加拉
Remadja	印度尼西亚	DD48	孟加拉
Rengesi	印度尼西亚	Kele	孟加拉
Rante	印度尼西亚	Kaliboro 600	孟加拉
Pulut Bongo	印度尼西亚	*xa5 Xa7*	
Pulut kemandi	印度尼西亚	DV85	孟加拉
Bajang	印度尼西亚	DV86	孟加拉
Bulu Putih	印度尼西亚	*Xa3*（*Xa6*）[a]	
Kentjana	印度尼西亚	Zenith	美国（U. S. A）
Camor	印度尼西亚	M. Sungsong	菲律宾（Philippines）
Ketan Pandan	印度尼西亚	*xa8*	
Ketan Temon	印度尼西亚	Pl 231129	美国
Ketan Gondel	印度尼西亚	*Xa3*（*xa9*）[a]	
Pare Djerah	印度尼西亚	Sateng	老挝

[a] $Xa4^b$、*Xa6* 和 *xa9* 均已确认为 *Xa3*（the three genes $Xa4^b$、*Xa6* and *xa9* were confirmed as same as the *Xa3*）。

1982—1987 年 IRRI 同日本热带农业研究中心合作期间，曾用当时菲律宾白叶枯菌小种 1～4 的代表菌株分别对筛选出来的抗病品种进行了复评，根据这些品种的反应型划分为抗性品种群，对于那些在抗病型上呈现异质的品种，按照其抗/感和抗/抗杂交组合 F_1 的抗性反应型再分为抗性品种亚群（Ogawa et al.，1991）。比较这些抗性品种亚群和已知抗病基因供体的抗性反应型，就可以推知其所携带的白叶枯病抗性基因。据此将 1 441 个品种材料划分为 8 个抗性品种群，其中 Java14、TKM6、DZ192 和 CAS209 四个抗性品种群分别是由 *Xa3*、*Xa4*、*xa5* 和 *Xa10* 四个单基因控制的，其余的 Mond Ba、DV85、Makhmal Mehi 和 BJ1 四个抗性品种群分别携有 *Xa4*＋*Xa10*、*xa5*＋*Xa7*、*xa5*＋*Xa4* 和 *xa5*＋*xa13* 两对抗性基因。各个抗性品种群的抗谱、抗性水平、病斑型及抗性基因列于表 14。据该表所示，Java14 抗性品种群计有 389 个品种，抗菲律宾小种 1～4（RRRR），其病斑具褐色边缘，推测携有 *Xa3*；TKM6 抗性品种群的品种 272 个，对小种 1、2、3 和 4 的抗性分别为抗、感、感和中抗（RSS MR），携有 *Xa4* 基因；DZ192 抗性品种群的品种 356 个，抗小种 1、2、3，中感小种 4，（RRR MS），携有 *xa5* 基因；Cas209 抗性品种群的品种 16 个，高抗小种 2，感小种 1、3 和 4（S HR SS），是典型的 *Xa10* 抗性反应型。携有两对抗性基因的 4 个抗性品种群显示了拓宽抗谱和基因互补的优势——Mond Ba 抗性品种群携有 *Xa4*＋*Xa10*，*Xa10* 抗小种 2，*Xa4* 抗小种 1 和 4，该品种群对菲律宾小种 1、2、3 和 4 的表现为抗、高抗、感和中抗（R HR S MR）；DV85（*xa5*＋*Xa7*）、Makhmal Mehi（*xa5*＋*Xa4*）和 BJ1（*xa5*＋*xa13*）三个抗性品种群都抗菲律宾 1～4 四个小种——DV85 抗性品种群的 *xa5* 控制全生育期对菲律宾小种 1、2 和 3 的抗性，中感小种 4，而其 *Xa7* 则在成株期抗菲律宾 4 个小种，所以抗谱较广；Makhmal Mehi 抗性品种群的 *xa5* 与 *Xa4* 抗性效应互补，全抗菲律宾 4 个小种；BJ1 抗性品种群中的 *xa13* 对小种 4 的高度抗性弥补了 *xa5* 中度感染小种 4 的不足，也表现为全抗菲律宾 4 个小种。这是一个比较系统的水稻白叶枯病抗性品种群划分的研究结果。

表 14 根据对菲律宾 4 个白叶枯病小种反应型划分的水稻抗性品种群及其基因型
(Ogawa et al., 1991)

Table 14 Classified rice resistance cultivar groups based on reactions to the 4 Philippine *Xoo* races
(Ogawa et al., 1991)

品种群 Cultivar Group	品种数 No. of cultivars	对 *Xoo* 菌小种反应型[a] Reaction to *Xoo* races				推测基因 Presumed gene
		1	2	3	4	
Java 14	389	RB	RB	RB	RB	*Xa3* (R R R R)
TKM 6	272	R	S	S	MR	*Xa4* (R S S MR)
DZ 192	356	R	R	R	MS	*xa5* (R R R MS)
CAS 209	16	S	HR	S	S	*Xa10* (S HR S S)
Mond Ba	22	R	HR	S	MR	*Xa4*+*Xa10* (R HR S MR)
DV 85	8	HR	HR	HR	R	*xa5*+*Xa7* (R R R R)
Makhmal Mehi	20	R	R	R	R	*xa5*+*Xa4* (R R R R)
BJ 1	21	R	R	R	R	*xa5*+*xa13* (R R R R)

[a] HR，高抗 (Highly resistant)；RB，抗性病斑有褐色边缘 (Resistance reaction with browning margin around lesions)；R，抗 (Resistant)；MR，中抗 (Moderately resistant)；MS，中感 (Moderately susceptible)；S，感 (Susceptible)。

4.2 全球稻种资源的白叶枯病抗性类群及其地理分布

IRRI 所鉴定的抗白叶枯病水稻种质主要来自印度、孟加拉、尼泊尔、印度尼西亚、菲律宾、斯里兰卡等亚洲国家。这些国家明显受季风变化影响的气候、不同的生态环境和长期的稻作栽培与人工选择，造就了其稻种资源的遗传多样性。20 世纪 70 年代，这些国家与 IRRI 合作，用本国白叶枯病菌对当地大量水稻种质进行了接种鉴定，筛选出一批批抗病的稻种资源在国际间进行互相交换。菲律宾具有丰富的野生稻种质，曾将 1962—1976 年间征集的 607 份疣粒野生稻和 580 份小粒野生稻种质送至 IRRI 保存。IRRI 面向全球收集水稻遗传资源的主要来源区域也是东南亚和南亚的产稻国家。IRRI 的植物病理学家、遗传育种学家和访问学者在合作进行全球稻种资源白叶枯病抗性品种群及其基因型分析时，也就对这些国家稻种资源的白叶枯病抗性类群的地理分布有了一定程度的了解。现将所得结果归纳在表 15 和表 16 以便查考。主要情况如下：①多数抗病的稻种资源分布在印度、孟加拉、斯里兰卡、印度尼西亚和菲律宾等国；②Java 14 抗性品种群（*Xa3*）分布最为广泛，遍及东南亚、南亚、东亚、非洲以至拉丁美洲，但比较集中分布在印度尼西亚和孟加拉；③TKM6 抗性品种群（*Xa4*）主要分布在孟加拉湾沿海地区以及印度、斯里兰卡和印度尼西亚等国；④DZ192 抗性品种群的品种（*xa5*）多集中在孟加拉、印度和尼泊尔；⑤Cas209 抗性群的品种（*Xa10*）分布在越南穿越缅甸直至沿着喜马拉雅山脉南部的尼泊尔；Mond Ba 抗性品种群的品种（*Xa10*＋*Xa4*）主要分布在孟加拉和印度；⑥BJ1 抗性品种群的品种（*xa5*＋*xa13*）集中分布于印度。另外，TN1 抗性品种群的品种（*Xa14*）呈现独特渐变的地理分布，从亚洲东部至喜马拉雅山脉的丘陵地带都有发现，在马来西亚、越南和中国也比较普遍（Sidhu et al.，1978；1979；Endo et al.，2004）。当然，上述抗性品种群的划分主要是根据对菲律宾一些小种的反应型确定的，有一定的局限性，但在目前国际间还不能进行菌种交流的情况下，上述工作所提供的信息对于水稻抗白叶枯病育种和相关研究还是很有价值的。

表 15　水稻品种与菲律宾白叶枯病 4 个小种互作反应型类群的世界分布

（Summarized from Ogawa et al.，1991）

Table 15　Distribution of rice groups according to reaction patterns of cultivars to 4 Philippine *Xoo* races in the world（Summarized from Ogawa et al.，1991）

原产地 Origin	群 Group			
	Java14 *Xa3*	TKM 6 *Xa4*	DZ 192 *xa5*	CAS 209/Mond Ba，*Xa10*/（*Xa10*＋*Xa4*）
中国（China）	5			
中国台湾（Taiwan，China）	5	1		
韩国（Korea）	5			
越南（Vietnam）	2	2	2	
柬埔寨（Cambodia）		1	1	
老挝（Laos）	9	9	9	
泰国（Thailand）	8	4		
马来西亚（Malaysia）	1	1		
缅甸（Myanmar）	1	9	3	1
菲律宾（Philippines）	11			
孟加拉（Bangladesh）	72	46	360	24
尼泊尔（Nepal）		1	17	
印度（India）		61	74	8
巴基斯坦（Pakistan）		1	10	
斯里兰卡（Sri Lanka）	1	59	2	
印度尼西亚（Indonesia）	204	42	4	2
埃及（Egypt）	3			
象牙海岸（Ivory Coast）	9			
塞内加尔（Senegal）		?		
塞拉里昂（Sierra Leone）	1			
利比里亚（Liberia）	13	9		1
马尔加什（Malagasy）	4			
苏里南（Surinam）			2	
伊朗（Iran）	2			
美国（USA）	3			

印度、孟加拉、斯里兰卡、印度尼西亚、尼泊尔五个国家的水稻白叶枯病主要抗性种质

资源及其主导抗性基因见表 16。

表 16 南亚、东南亚国家水稻白叶枯病主要抗性种质资源［根据 Ogawa 等（1991）资料整理］

Table 16 Main bacterial blight resistance germplasm in South and Southeast Asian countries

（Summarized from Ogawa et al.，1991）

印度[a] India	孟加拉 Bangladesh	斯里兰卡 Sri Lanka	印度尼西亚 Indonesia	尼泊尔 Nepal
Xa 4[b]	*xa5*	*Xa3*	*Xa3*	*xa5*
ADT3,[c] ADT4，ADT9，ADT10，ADT14，ADT17，ADT18～19，ADT21～23，ASD 4，ASD5，ASD8，ASD5，ASD8，CO10，CO2,CO4,CO5，CO7，CO9，CO22,CO11，CO17，CO19,CO27，CO29，CO30，HR12，Hema，KH998，Kolong Bao，Poongar，Samba Polan，TKM 6，TKM 3，Taothbi，WC1263 等	Aus 32，Aus 251，Aus 449，Aus 70 ，Bazail975，Begeri，DZ192，Birmadla，Bolun，Boteswae 2，Chakila，Cia,Karia，Katachikon，DD48，DB3，DD49，DD100，DF1，DV29，DF1，DL5，DV32，DNJ142，DV52，DV139，DZ 192,Jhur，Karia，Katachikon，Kele，Kaliboro 600，Lohargura，Lakami Laksmilota，Loroi，Pankhirai，Sofahar 等	Halsudu Heenati，Halsuduwee，Mahamawee，Malalwariyan,Murunga Balawee,Patchaiperumal，Peurm Karumal,Perum Karuppan，Race，Rathu Heenati，Sudumalawee，Yakadawee 等	Andel,Andel Welut，Bajang,Bulu，Beak Ganggap,Butu Putih，Butu Samping,Camor，Dete,Dewi,Djambu，Gondel,Gundul,Jedish1，Ketan Kengkung，Lumbu，Pulut Bongo，Nangka，Rante，Remadja，Roder，Sintawati 等	Bageri，Bangaluwa，Devarasi,Dudhi,Lal Ahu，Lalaka Gadur,Lal Sar，Maturi,Matury,Nakhi，Rerm Bilash,Sajani，Sokan Dhan,Tally 等
xa5		*Xa4*	*Xa4*	
ARC 7098，Taothabi，ARC 7128，ARC 7043，ARC10376，ARC5756，Dular，Koalar ata，ARC5756,PI 180060－1 等		Bathkiriel,Murungawee，No. 20,Wanni Dahanala 等	Cisadane，Pelital/1，Pelital/2,Sigadis 等	
Xa3	*xa5*，*Xa7*，*xa 24t*		*xa5*	
CO21，Perunel 0. 69－18，Taothabi 等	DV85，DV86，Aus 295 等		Latu	
xa5＋*xa13*	*Xa10*			
BJ 1，Chinsurah Boro Ⅱ，AC 19－1－1，Tepal (Bolo)，Kalimekri 77－5,Long Grain 等	Aghanisail，Bhua Balam，Ganga Sagar,Huson Dhaw,Kanchanchari，Khar Mao,SumaSail 等			

[a] 国家 (Country)。

[b] 抗性基因 (Resistance gene)。

[c] 主要抗病种质 (Main resistant germplasm)。

5 稻种资源的白叶枯病抗性评价方法

作物品种抗病性的鉴定或研究，因对象和问题而有群体、个体、组织、细胞乃至分子水平之分，其所采取的方法与技术也各不相同。对于一般的抗病性鉴定来说，有田间鉴定和室内鉴定，鉴定时期分苗期和成株期，鉴定方式有自然发病（田间）和人工接种。接种菌株一个或几个，接种方法也各色各样。田间自然发病有其方便和可取之处，但受气候、土壤、管理，特别是年度间病情流行情况和发病均匀度等的影响较大，不大可靠和稳定。人工接种需要自己繁殖或有单位提供所需的纯真菌种，略为费事，但可通过菌株选择、接种技术改进、接种物浓度控制等措施使其发病适时、轻重得体、分布均匀，这样就可以获得比较可靠的结果。采取人工接种与自然鉴定相结合的方式则可以更全面地反映供试品种材料在生产实际中的抗性表现，值得提倡。

5.1 接种鉴定方法

水稻白叶枯病的接种方法要根据研究目标来确定。病原细菌是通过植株的伤口入侵的。叶片边缘的水孔和叶面、根、茎的伤口都是病菌侵入稻株的途径，所以人工接种总是与适时适量地人为产生伤口有关联。实际操作时可以选择以下几种接种方法。

（1）喷雾法 用喷雾器将菌液直接均匀喷洒在移栽前1～2d的育苗箱（盘）的秧苗上，接种后注意保湿，特别在空气相对湿度较低的地区。此法接近自然侵染状况，其潜伏期一般长于剪叶接种法，常用来判别品种材料抗侵入或抗扩展的特性，也适于比较品种材料的抗性类型和田间抗性水平。

（2）浸灌法 此法接近自然侵染状况。首先由方中达等（1956）提出，将混有菌液的水浸灌或漫灌苗床，观察白叶枯病侵染循环的途径，一般在育苗箱或小面积试验田中进行。

（3）浸根法 拔出秧苗时已然伤了根，或在接种前再剪去部分根系，移栽前浸在菌液中3～6h，观察稻苗受侵染后所呈现的kresek（凋萎型）症状的发生发展情况，用以比较品种材料的感染程度或病菌致病力强弱（Yoshimura et al.，1965）。后来Watanabe（1975）对此作了改进，将干稻种直播于混有菌土的育苗盒中，使其在发芽过程中有直接受到细菌感染的均等机会，能获得较高的“kresek”诱发率。

（4）针刺法 Reitsma等（1950）采用针刺、喷雾、菌液浸苗3种方法来查证最初在印度尼西亚发生稻苗白叶枯病“kresek”症状的侵染途径，针刺和菌液浸苗的方法都可以引起较高的发病率。Mukoo等（1951）和Yoshida等（1961）改进了针刺接种法，所用针数由多枚改为单针或双针，将待接种的叶片平展于浸有接种菌液的海绵上（置于小盒中），当针束刺进叶片时，也插入了浸有菌液的海绵，拔出针束时就完成了接种操作。

（5）剪叶法 Kauffman等（1973）根据水稻白叶枯病为细菌性维管束病害的特点，采用灭菌手术剪沾了试管中的菌液后剪去叶尖1～3cm（视稻株生育期而定），使菌液直接进入切伤口（图1）。剪叶法与针刺法所产生的抗性反应效果相似，相关系数为 $r=0.728\sim0.753$（伍尚忠，1983）。IRRI进行大面积接种时，采用改进的接种剪，即在果园专用修剪刀的交叉支点上安装一个可盛菌液的塑料小瓶，在剪叶的同时就自然流出菌液侵染剪口，操作简单，大大提高了接种效率，已被广泛用于育种试验田大规模接种、种质资源评价和抗性遗传研究（图2）。

5.2 菌株选择和接种时期

（1）菌株选择 进行稻种资源白叶枯病抗性鉴定，应以当地有代表性的优势菌株为首

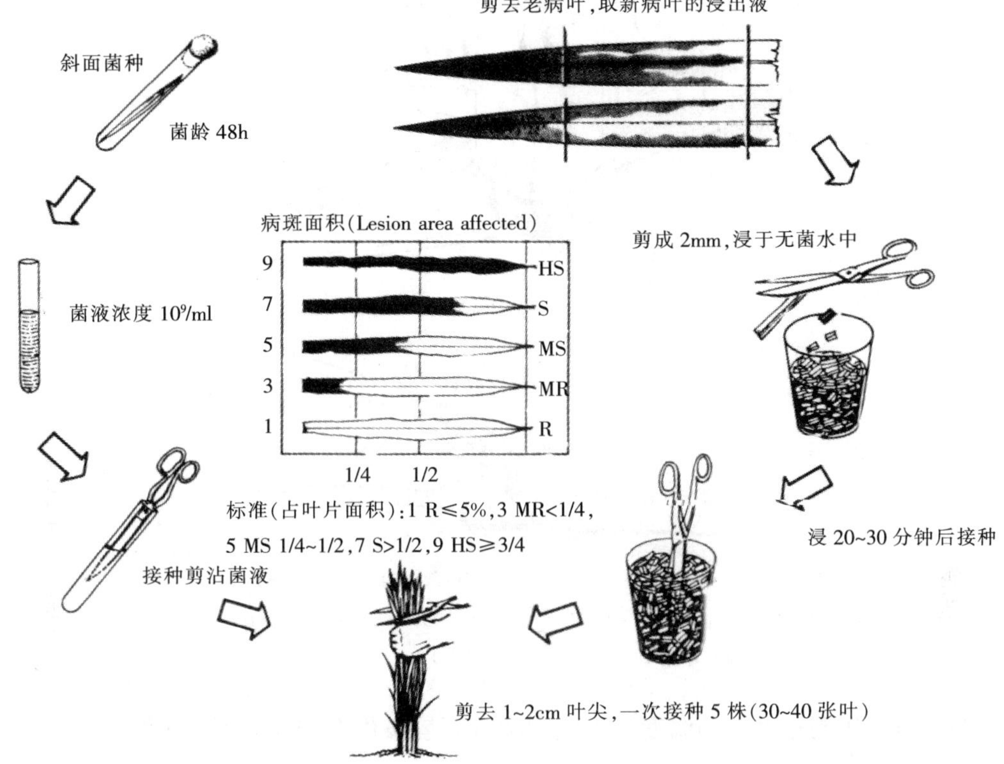

图 1　IRRI 田间剪叶接种白叶枯病及调查标准图示（Kauffman et al.，1973）

Fig. 1　Clipping method for BB inoculation in field at IRRI (Kauffman et al.，1973)

选，再根据其他研究目的来补充选用相应的菌株。菌株在培养基上培养 48～72h，接种菌液浓度调整到 10^8～10^9 细胞/ml。由于病原细菌之间一般会产生拮抗作用而使致病力下降，因此避免用几个菌株的混合液接种，如需要用多个小种的代表菌株时，应该采用同株分蘖分别接种法。

（2）接种时期　水稻对白叶枯病抗性的表达时期主要取决于品种基因型，但与毒性不同的病菌互作之后也可能有所变化。许多日本研究者认为水稻品种的白叶枯病抗性一般在苗期表现不稳定，建议在成株期接种，这可能与他们当年所鉴定的材料是日本地方粳稻种质资源有关，因为多数抗病粳稻品种到成株期才表达其抗性。IRRI 则多在分蘖盛期（播种后 45～50d）进行接种。在不了解待测品种资源抗性类型的情况下，应该首先进行苗期接种，避免漏掉宝贵的全生育期抗性材料。一般许多苗期抗病的水稻品种多为全生育期抗性，并且比较稳定；在孕穗期再接种一次，则具有成株期抗性的材料也不会丢失。所以苗期和孕穗期都进行接种鉴定更为合理。要强调的一点是不论室内或室外鉴定，每次（包括每批）必须设置对照品种。

5.3　抗性调查

（1）调查时间　无论是苗期或是孕穗期鉴定，都应在感病对照品种充分发病、病情发展相对稳定时进行调查。一般苗期在接种后 10～14d 进行，孕穗期接种和调查间隔时间为14～

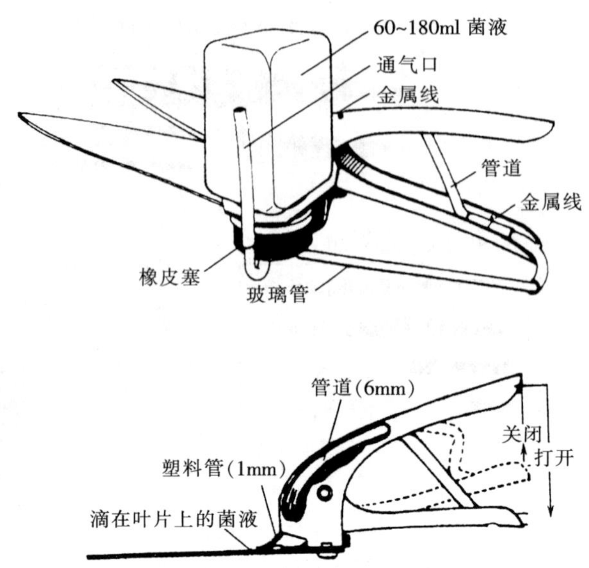

图 2 安装了菌液瓶的白叶枯病剪叶接种剪刀图（IRRI 提供）

Fig. 2 Grass clipper and an attached bottle of bacterial suspension for bacterial blight inoculation (Courtesy of IRRI)

21d。因为病程长短与接种后的温度、天气阴晴等环境条件有密切关系，所以应参考对照品种的病情进程确定调查时间。

(2) 调查标准 水稻品种对白叶枯病的抗性以抗扩展为主，所以主要根据病情表现作为评定标准。可以按病斑反应型进行分级，依据侵染部位反应的强弱、形状、色泽、病斑大小等特征来划分级别。但是病斑反应型的这些特征表现常与所采用的接种方法有关，所以不同接种方法的调查标准也不完全一样。

现举 3 种分级标准实例作为参考：

①IRRI 根据 Kauffman（1973）剪叶接种法提出的 SES（Standarlized evaluation system）目测病情调查标准

a 叶枯型症状分级标准（上部 3 张接种叶片的平均病斑面积%，即病斑面积/叶面积）：

0 无病斑

1 少于 1%（病斑面积/叶面积）

3 1%～5%

5 6%～25%

7 26%～50%

9 51%～100%

b 凋萎型症状分级标准（调查 100 穴以上稻苗的平均凋萎株百分率%）：

0 无病

1 少于 1%

3 15%

5 6%～25%

7 26%～50%

9 51%～100%

以上 1、3 级为抗病，5、7、9 级为感病。

②Ezuka 和 Horino（1974a）针刺法的病情调查标准（图 3，另见彩版 8）：

0 无病斑

1 近针刺口有褪色痕迹

2 病斑长度短于 15mm

3 病斑长度短于叶长的 1/4（自接种针口向上量至叶尖，下同）

4 病斑长度在叶长的 1/4～1/2 之间

5 病斑长度在叶长的 1/2～1 之间

6 病斑占全叶长，但仍保留少许绿色部分

7 全张叶片枯萎或死亡，无绿色部分

以上 1、2 和 3 级为抗病，4、5、6 和 7 级为感病。

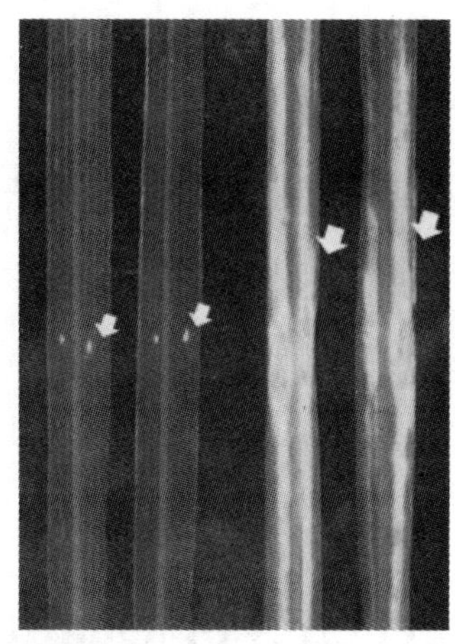

图 3　双针接种法将日本 *Xoo* 小种Ⅰ接种在品种黄玉（抗病，左）和金南风（感病，右）的反应型（Ezuka and Horino，1974）

Fig. 3　Reactions of Kogyoku（left，resistant）Kinmaze（right，susceptible）to Japanese *Xoo* race Ⅰ by double needle-pricking method（Ezuka and Horino，1974）

③中国水稻白叶枯病情调查标准［Chinese Evaluation System（CES），由全国稻白叶枯病协作组制定］

病　级	剪 叶 法	针 刺 法
0（免疫）	剪口处无病斑	刺口周围无病斑
1（抗）	剪口处有小病斑，长度不超过 2 cm	刺口周围有小病斑，超过 2 cm
2（中抗）	病斑下延伸 2 cm 以上，不超过叶长的 1/4	病斑纵向扩展超过 2 cm，不超过叶长的 1/4
3（中感）	病斑占叶面积的 1/4～1/2	病斑纵向扩展，上半叶枯死
4（感）	病斑长度占叶片的 1/2～3/4	病斑长度占叶片的 1/2～3/4
5（高感）	病斑长度超过叶片的 3/4	病斑长度占叶片的 1/2～3/4

日本和中国的植物病理学家对上述 3 种调查标准作了比较。Ezuka 等（1974）分级标准与 IRRI 分级标准的相关系数 r ＝ 0.753 0**，中国分级标准与 IRRI 分级标准的相关系数 r ＝0.892 2**（伍尚忠等，1983），都达到极显著正相关，表明这几种分级标准都是适用的。

此外，接种后所呈现的病斑侵染型也是目测水稻对白叶枯病抗感发展的重要依据。感病品种的叶片在接种后 3～4d 的剪口处就会出现水浸状暗绿色病斑，随后病斑迅速延伸，粳稻的病斑转为灰白，籼稻转为黄色，病健交界处均呈现波纹状；高抗品种叶片剪口处为白色线状，多数抗病的爪哇型品种呈现褐色边缘病斑，这些都是重要的抗病型特征。

应该指出，品种材料间抗病性强弱的划分是相对的，它受接种方法、部位、接种菌液浓度和试验的环境条件以及评定病级尺度的掌握等自然与人为因素的影响，因此，抗性鉴定的方法、调查标准应该相对固定。

水稻白叶枯病质量抗性遗传和抗病主基因鉴定

章　琦

（中国农业科学院作物科学研究所）

Genetics of Quality Resistance and Identification of Major Resistance Genes to Rice Bacterial Blight

摘要　水稻白叶枯病抗性遗传研究是在1957年日本抗病品种朝风种植不久就遭遇到当地的毒性菌株侵袭，突然丧失了抗性之后才引起重视的。日本科学家Nishimura（1961）最早发现黄金丸（Koganemaru）和Pi 1这两个品种对白叶枯病的抗性受控于一对显性抗病基因。Sakaguchi等（1968）首先提出日本抗性品种群与日本菌群的概念以及二者之间的对应关系，成为研究作物特异性病害及其遗传的基本准则。IRRI成立后不久开创性地建立了“水稻种质资源遗传评价与利用研究”项目。亚洲主要产稻国家及美国在各自独立开展水稻品种改良的同时，先后参加了这个具有前瞻性的大协作。众多不同专业的科学家共同围绕着水稻白叶枯病抗性基因的挖掘、鉴定与利用这个核心问题，相继开展了抗病性鉴定、抗性遗传、抗性基因鉴定与定位、分子标记、基因克隆以及与抗性相关的分子生物学基础等研究，取得了令人瞩目的进展。本文着重将半个世纪以来在水稻白叶枯病主效抗性基因鉴定方面的进展和成果作一回顾性阐述。为便于读者理解，先从经典遗传学角度概括一下水稻品种白叶枯病抗性遗传的一般规律：水稻白叶枯病是小种专化性病害，品种抗性与小种毒性存在着典型的对应关系。品种抗性基本上是受核基因组中的主效基因控制的，其遗传行为或表现取决于双亲各自携有的主效抗性基因的多少、异同及其属性，包括抗性基因的显隐性、基因互作、独立遗传或连锁遗传以及抗性类型（全抗或成抗）等。一般而言，品种间的抗性是一对或两对基因的差异，表现为典型的单基因或双基因遗传，也有两对基因以上差异的遗传，但较少见。这些都是属于质量性状遗传。

水稻白叶枯病抗性基因的鉴定经历了采用本国菌株和建立国际鉴别系统两个阶段：1982年以前，主要由开展抗性基因鉴定工作较早的日本和IRRI分别采用日本和菲律宾的白叶枯病代表菌株进行测试。其他国家也是采用本国的菌株自行鉴定的。这样所得的结果缺乏可比性，难以进行概括。1982年以后，IRRI和日本TARC合作研究，采取统一的鉴别小种、鉴别品种和实验方案开展白叶枯病抗性基因鉴定，删去了重复（包括与*Xa4*相同的$Xa4^b$，与*Xa3*相同的*Xa6*和*xa9*），统一了命名，并创建了国际水稻白叶枯病抗性鉴别系统。这样就使各国同行在进行本国白叶枯病抗性鉴定时，可以通过与国际统一鉴别系统（特别是鉴别品种）所测试的结果相比较后，与国际接轨、鉴别异同、“对号入座”。

截至 2006 年，已统一鉴定出 30 个水稻白叶枯病抗性基因，包括 21 个显性基因和 9 个隐性基因。其中 12 个是全生育期（包括 2 个分蘖盛期开始抗病）抗性基因、18 个是成株期抗性基因。已定位的抗性基因有 18 个，其中第 1 染色体上有 *Xa29t*（成抗）；第 4 染色体上有 4 个，*Xa1*（全抗）、*Xa2*（全抗）、*Xa12*（成抗）和 *Xa14*（全抗）；第 5 染色体上有 *xa5*（全抗）；第 6 染色体上有 2 个，*Xa7*（成抗）和 *Xa27t*（全抗）；第 7 染色上有 *xa8*（成抗）；第 8 染色体上有 *xa13*（成抗）；第 11 染色体上有 7 个，*Xa3*（成抗）、*Xa4*（全抗）、*Xa10*（全抗）、*Xa21*（全抗）、*Xa22t*（成抗）、*Xa23*（全抗）和 *Xa26t*（全抗）；第 12 染色体上有 *Xa25t*（全抗）。还未定位的有 *Xa11*、*xa15*、*Xa16*、*Xa17*、*Xa18*、*Xa26t*（*Nep Bha Bong To*）、*Xa27t*（Arai Raj）和 *xa28t*。第 2、3、9、10 四个染色体上还未发现有白叶枯病抗性基因。已克隆的抗性基因有 6 个，即 *Xa1*（全抗）、*xa5*（全抗）、*xa13*（全抗）、*Xa21*（全抗）、*Xa26t*（全抗，明恢 63）和 *Xa27t*（全抗，*O. minuta*）。已在国际水稻新基因命名委员会正式注册的有 19 个。编号重复的基因有 6 个：2 个 *Xa25t* 分别来自全生育期抗菲律宾小种 9 的明恢 63 和源于明恢 63 经组培后的体细胞无性突变系 HX3；2 个 *Xa27t*，分别来自小粒野生稻和孟加拉品种 Arai Raj；以及 1 个显性的来自明恢 63 的 *Xa26t* 和 1 个隐性的 *Xa26t*（来自越南品种 Nep Bha Bong To）。

在已命名的抗性基因中，*Xa23*、*Xa21* 和 *Xa7* 的抗谱最广、抗性最强，*xa5*、*xa13* 和 *xa24t* 的抗谱也较广；*Xa3* 和 *Xa4* 对我国白叶枯病致病型的抗谱较广，但要避免在中国Ⅳ型和Ⅴ型菌流行的地区应用。

近等基因系是鉴别基因异同及其抗性强弱或其他属性的重要遗传工具材料。日本 TARC 和 IRRI 合作构建了分别以 IR24（籼稻）（IRBB1－11 和 IRBB21）、Milyang23（籼粳交）（IRBB101－111）和 Toyonishiki（粳稻）（IRBB201－211）为轮回亲本的 3 套近等基因系，分别携有 *Xa1*、*Xa2*、*Xa3*、*Xa4*、*xa5*、*Xa7*、*xa8*、*Xa10*、*Xa11*、*Xa14* 和 *Xa21* 等 11 个基因；新加坡国立大学构建了 IR24 为轮回亲本、携有 *Xa27*ᵗ的 IRBB27 一个近等基因系。中国也构建了以沈农 1033（粳稻）为轮回亲本的 6 个近等基因系（CBB 系统），分别携有 *Xa2*、*Xa3*、*Xa4*、*Xa12*、*Xa14* 和一个有别于 *Xa7* 但尚未定名的新基因共 6 个基因；以及以 JG30（籼稻，CBB23）和 IR24（籼稻 CBB23B）为轮回亲本的两对携有 *Xa23* 的近等基因系。

Abstract The genetic study of BB resistance in rice germplasm did not draw much attention until a sudden outbreak of BB epidemic in 1957, caused by a new *Xoo* virulent strain that was pathogenic to the then commonly used *Xoo* resistant rice cultivar Asakaze. The Japanese scientist Nishimura (1961) was the first one to discover that BB resistance of two cultivars, Koganemaru and Pi 1 is controlled by a single dominant gene. Sakaguchi et al. (1968) brought forth the concept of BB resistance rice variety group and *Xoo* bacteria pathotype group as well as their corresponding relationship, which became the basic guideline for studying specific pathogens and resistance genetics in crops. Right after the establishment of IRRI, the "Genetic Evaluation and Utilization" (GEU) program was launched. Despite that major rice cultivating Asian countries and USA had independently launched similar programs, they all decided to join this IRRI initiative. Combined efforts from different expertises addressed core questions of BB resistance including discovery, evaluation, cloning and utilization of important genes, development of molecular markers, as well as studies on related aspects in plant molecular biology. This paper provides a synopsis on the achievements and progresses made in the identification and evaluation of BB resistance genes over the 50 years.

Using classical genetics, the general principles of BB resistance in rice can be summarized as follows: BB is a race specific disease. Resistant rice cultivars and the corresponding races exhibit a classical gene-for-gene relationship. BB resistance is mainly conferred by major resistance genes in nucleus but not through cytoplasmic inheritance. The genetic behaviors and resistance phenotypes in hybrid progenies are dependent on the genotypes of the parents, such as the number of genes, dominant and recessive relationships, resistance in different growth stages, location of genes on chromosomes, and genetic epistasis. In general, the inheritance of BB resistance in rice is determined by 1 or 2 genes. These are genes controlling qualitative heredity. On the other hand, involvement of more than 2 genes is uncommon.

Worldwide examination and evaluation of BB resistance genes have been conducted in two phases: using national pathogen races and establishing an international evaluation system. Before 1982, evaluations of BB resistance genes were mainly performed in Japan and IRRI, using Japanese races or Philippine races respectively. Some other countries also used their local races for evaluation. The results were then incomparable and inconclusive. From 1982 onwards, IRRI and TARC in Japan collaboratively established an international evaluation system for BB resistance. Standardized collections of pathogen races, rice cultivars, and experimental protocols were established. Nomenclature was unified and duplications were removed (including $Xa4^b$, *Xa6*, and *xa9*). Through this unifying system, results from different countries can now be compared.

Up to 2006, a total of 30 BB resistance genes were identified. Twenty-one and 9 out of the 30 *Xa* genes are dominant and recessive, respectively. Twelve genes exhibit all-growth-stage resistance (allR; including 2 genes that confer BB resistance at the tillering stage). The other 18 genes confer only adult-stage resistance (adultR). Eighteen *Xa* genes were mapped: *Xa29t* (adultR) on chromosome 1; 4 genes, *Xa1* (allR), *Xa2* (allR), *Xa12* (adultR), and*Xa14* (allR) on chromosome 4; *xa5* (allR) on chromosome 5; 2 genes, *Xa7* (adultR) and *Xa27t* (allR) on chromosome 6; *xa8* (adultR) on chromosome 7; *xa13* (adultR) on chromosome 8; 7 genes, *Xa3* (adultR), *Xa4* (allR), *Xa10* (allR), *Xa21* (allR), *Xa22t* (adultR), *Xa23* (allR), and *Xa26t* (allR) on chromosome 11; and *Xa25t* (allR) on chromosome 12. Eight genes, *Xa11*, *xa15*, *Xa16*, *Xa17*, *Xa18*, *xa26 t* (Nep Bha Bong To), *Xa27t* (Arai Raj) and *xa28 t* have not been mapped. No BB resistance gene was found on chromosomes 2, 3, 9, and 10. Six *Xa* genes were cloned, including *Xa1* (allR), *xa5* (allR), *Xa21* (allR), *xa13* (adultR), *Xa26 t* (from Minghuei 63, allR) and *Xa27t* (from *O. minuta*, allR). Nineteen *Xa* genes were registered by the International Committee on Nomenclation of New Rice Genes. The nomenclation of six genes were found to be duplicated: 2 *Xa25t* genes, one from Minghuei 63 with overall resistance to Philippine race 9 and the other was a somatic mutation line (HX3) derived from tissue cultures of Minghuei 63; 2 *Xa27t* genes were independently named in *O. minuta*, and Bangladsh cultivar Arai Raj, respectively; the dominant gene *Xa26t* is from Minghuei 63, and a recessive gene *xa26 t* is from Vietnam cultivar Nep Bha Bong To cultivar.

Among all designated *Xa* genes, *Xa23*, *Xa21*, and *Xa7* exhibit the broadest resistance spectrum and strongest resistance strength; *xa5*, *xa13*, and *xa24t* also exhibit broad resistance spectrum. *Xa3* and *Xa4* show broad resistance spectrum against Chinese BB pathotypes, but cannot be used in regions where the prevailing pathotypes are Ⅳ and Ⅴ.

Near isogenic lines (NILs) are important genetic tools for comparing resistance genes in terms of their similarities and differences, resistance strengths, and other properties. TARC in Japan and IRRI together constructed 3 sets of near isogenic lines (NILs) using IR24 (indica), Milyang 23 (indica-japonica hybrid), and Toyonishiki (japonica) as the recurrent susceptible parents (for 11 *Xa* genes including *Xa1*, *Xa2*, *Xa3*, *Xa4*, *xa5*, *Xa7*, *xa8*, *Xa10*, *Xa11*, *Xa14* and *Xa21*, respectively). The IRBB27 (*Xa27t*) was built using IR24 as recurrent parent by the National University of Singapore. In China, 6 NILs were constructed using SN1033 (japonica) as the recurrent susceptible parent (carrying *Xa2*, *Xa3*, *Xa4*, *Xa12*, *Xa14* and one unnamed *Xa* gene, respectively). Two other NILs (CBB23 and CBB23B) were also built for *Xa23* using JG30 (*indica*) and IR24 (*indica*) as the recurrent susceptible parents.

植物抗病性是由遗传决定的一种生物学潜能。和其他生物学性状不同，它是在寄主和病原菌双方参与下共同形成的。当寄主受到病原菌侵染后，由于寄主与病原菌彼此亲和能力的不同，形成了不同程度的抗病或感病表型。植物抗病性遗传学实质上就是植物—病原物相互关系的遗传学。60多年前 Flor (1942) 就提出“基因对基因”假说以阐明寄主—病原物之间的相互关系，二者在相互作用、竞争、适应的情况下，各自不断变异而又相互选择，这就是两者的协同进化（coevolution）关系。水稻与白叶枯病确实存在着这种典型的关系，它们之间既有基因对基因的关系，又各自独立地受到外部环境条件的影响。所以在研究植物抗病性遗传时必须从寄主—病原物—环境三元关系的出发点着眼，才能取得接近实际的结果。

当今，对动植物、微生物和人类遗传学研究均已进入分子水平。这些研究的许多结果表明，它与经典的性状遗传分析是一致的。本文将着重讨论水稻白叶枯病抗性传统遗传学的一般规律，扼要回顾水稻白叶枯病主效抗性基因的鉴定历程与截至 2006 年已鉴定命名的质量抗病基因概况。

1　水稻白叶枯病抗性遗传的一般规律

国内外对水稻白叶枯病抗性遗传研究的报道较多，与其他病害的抗性遗传有不少共通之处。本节只对其一般规律做些概括，并选用一些国内学者的工作实例予以说明，当然也包括国外科学家的一些典型资料。这样可以和随后的主效抗性基因鉴定多以日本和 IRRI 科学家的工作为主来阐述的方式相协调。

水稻品种对白叶枯病的抗性表现依其所携带的抗病基因和所面对的病菌小种（致病型 pathotype）而异。抗病基因有主效和微效之分，主效基因的抗病效应明显，多具有小种专化性（对特定的一个或几个小种有效），而微效基因的数目较多，单个效应很小，但有累加效果，可积微见著地对一个或多个小种发挥中感至中等抗性的作用。在一般情况下，水稻品种对白叶枯病的抗性由主效基因控制，但常常还有若干正向或负向的微效基因起着补充和修饰作用。

携有某一主效基因的水稻品种，抗一个或几个白叶枯病菌小种，不抗另一个或另一些白叶枯病菌小种，这个现象十分普遍。例如，日本品种黄玉（携 *Xa1* 基因）抗日本小种 1，不抗小种 2 和 3；IRRI 品种 IR20（*Xa4*）抗菲律宾小种 1 和 5，而不抗小种 2、3、4 和 6；携有 *Xa21* 的近等基因系 IRBB21 抗菲律宾小种 1～9 九个小种，感染小种 10；携有 *Xa23* 的近等基因系 CBB23 则抗菲律宾小种 1～10 十个小种等。这都表明主效基因的抗谱有宽有窄，其抗性强度也不尽相同。

1.1　显隐性及基因互作

显隐性是同一基因座（位点）上不同等位基因之间相互作用的表现。水稻品种对白叶枯病的抗性在遗传上较多表现为显性，但也有一些由隐性抗病基因控制的抗性。所谓显性是指它的抗病效应可在杂合（即与隐性基因同时存在）状态下表达出来，而隐性基因只能在纯合状态下表达其抗性功能，在杂合状态时由于受到与其相对应的显性感病基因的抑制（基因座内互作）而不能表达出来，因而在表型上是感病的。显性抗病基因的显性程度有所差异，有完全显性和不完全显性或部分显性之分。

携有某一显性抗病基因的品种与相应的感病品种杂交时，F_1为抗病，F_2群体中抗病植株占多数，如属完全显性时，则为 3 抗∶1 感的分离比例，以后世代的抗病家系也占多数；如属不完全显性，则呈 1 抗∶2 中抗或轻感∶1 感的分离比例，其以后世代仍以抗至中抗的家系居多。而携有一对隐性抗病基因的品种与其相对应的感病品种（携有显性基因）杂交时，其 F_1为感病，F_2群体中感病的植株居多数，如感病性属完全显性，则为 1 抗∶3 感的比例；如感病性为不完全显性，则呈 1 抗∶2 中感∶1 感的比例。显然，在抗病育种中，利用显性抗病基因比隐性抗病基因有利得多。但显性抗病基因暴露在众多病菌小种或菌株毒性变异的竞争环境中，受侵袭的机会也较多，相对来说其抗性较不稳定。

水稻品种对白叶枯病抗性的显隐性关系有时也会因为病菌小种不同而变化，即它的遗传表现会因品种（抗病基因）—病菌（毒性基因）组合的不同而有所差别。例如周泰初等（1982）在研究高抗白叶枯病的朝鲜粳稻品种矢租对江苏菌株 KS1-3 的抗性遗传时，观察到泗稻 4 号（感）/矢租（抗）组合 F_1 对 KS1-3 表现抗病（按 9 级标准为 1.53±0.78），F_2 的抗、感分离观察值为 142 抗∶46 感，符合 3∶1 的比例，其 χ^2（0.064）概率（*P* 值）为 0.90～0.75，表明矢租对 KS1-3 的抗性由一对显性基因控制。而章琦等（1986）分析矢租对菲律宾小种 1（PXO61）的全生育期抗性遗传，在组合金南风（感）/矢租（抗）中 F_1 对 PXO61 表现全生育期感病（病斑面积 49.5%±5%），F_2 群体在分蘖期和孕穗期的抗、感分离观察值分别为 96 抗∶278 感和 101 抗∶266 感，是一对隐性基因的差异，符合 1∶3 的比例，其 χ^2 测验的概率即 *P* 值分别为 0.75～0.90 和 0.25～0.50；在随机取样种植的 64 个 F_3 家系中，调查其在两个生育期的抗性分离比例，其实际观察值符合 1 抗∶2 分离∶1 感的比例，χ^2 测验的 *P* 值分别为 0.90～0.95 和 0.50～0.75，证实矢租对 PXO61 的全生育期抗性是由一对隐性基因控制的。基因等位性测定表明该隐性基因与 *xa5* 等位。又如万建民等（1990）用 4 个中国菌株研究太湖地区几个抗病粳稻品种与同一个感病品种（南粳 33）杂交后的抗性遗传，观察到抗病品种小爱 2 号对 KS6-6、KS1-21 和 JS49-6 三个江苏菌株的抗性表现受两对隐性的互补基因控制（F_1 感病，F_2 分离比为 1 抗∶15 感），而对浙江菌株 ZHE173 的 F_2 分离群体则呈单峰连续分布，属多基因控制；抗病品种风景稻对 ZHE173 和 KS1-21 两个菌株的抗性由一对隐性基因控制，而对 KS6-6 和 JS49-6 两个菌株的抗性则由多基因控制。这些事例都表明水稻品种抗白叶枯病的显隐性及其遗传行为因不同毒性小种或不同品种—菌株组合而异。表 1 列举若干品种—菌株之间的互作实例。

表 1　水稻品种对白叶枯病菌不同小种（致病型）代表菌株的抗性遗传行为[a]（章琦整理）

Table 1　Genetic behaviors of rice cultivars resistance to some *Xoo* representative strains of different races（pathotypes）（Summarized by Zhang）

品种 Cultivar	小种或菌株 Race or strain	遗传表现 Genetic behavior	参考文献 Reference
矢租	菲律宾小种 1～4[a] (Philippine races 1～4)	一对隐性基因（*xa5*） (Single recessive gene)	章琦等，1986
	KS1-3	一对显性基因 (Single dominant gene)	周泰初等，1982
马罗糯谷和八月糯	江陵 691 J L691	一对显性基因	谢岳峰等，1989
	菲律宾小种 1 (Philippine race 1)	一对隐性基因	

（续）

品种 Cultivar	小种或菌株 Race or strain	遗传表现 Genetic behavior	参考文献 Reference
蚌珠芒	P1、T3、C2	一对显性基因（*Xa3*）[a]	章琦等，1990
	KS 1-3	两对显性互补基因 （Two dominant complementary genes）	周泰初等，1982
IR26	菲律宾小种 1	一对显性基因（*Xa4*）[a]	Petpisit et al.，1977
	日本 5 个小种 （Japanese races Ⅰ～Ⅴ）	多基因（Polygen）	Horino et al.，1980
小爱 2 号	KS6-6 和 KS1-21、JS49-6	两对重复显性基因 （Two dominant genes）	万建民和朱立宏，1990
	ZHE 173	多基因	
风景稻	ZHE 173 和 KS1-21	一对隐性基因	
	KS6-6 和 JS49-6	多基因	

[a] 经等位性测定（Confirmed by allelism test）。

抗病品种与感病品种杂交，就其对某一特定菌株而言，多为一对有时两对基因（差异）的遗传，当然也有两对以上的例子，但较少见。这里举一个 3 个不同抗感组合的寡基因抗性遗传实例：Yamamoto 等（1990）分析研究了 Toyonishiki、Milyang 23 和 Cas209 三个品种对缅甸（BM8429）、日本（H8584）和菲律宾（PXO86）3 个 *Xoo* 菌株的抗性遗传，供试 3 个组合的亲本、F_1 及 F_2 群体的抗性结果见表 2。

表 2　3 个组合的亲本、F_1 及 F_2 群体对 3 个菌株的抗性反应

（根据 Yamamoto 等，1990 年数据整理）

Table 2　Reaction of parents、F_1 and F_2 populations derived from 3 crosses to 3 strains of *Xoo*（Summarized on basis of Yamamoto et al.，1990）

组合 Cross	对 3 个菌株的抗性反应 Reaction to 3 strains of *Xoo* in												预期比例 Expected ratio	*P*
	P_1	P_2	F_1	F_2										
					RRR[a]	RRS	RSR	SRR	RSS	SRS	SSR	SSS		
Toyonishiki/Cas209	RSS	SRR	RRR	O[b]	125	40	36	43	10	18	15	7	27∶9∶9∶9∶3∶3∶3∶1	0.7～0.8
				E	124	41	41	41	14	14	14	5		
Milyang23/Cas209	RRS	SRR	RRR	O	160	42		57		20			9∶3∶3∶1	0.3～0.5
				E	157	52		52		17				
Toyonishiki/Milyang23	RSS	RRS	RRS	O		186			52				3∶1	0.2～0.3
				E		180			60					

[a] 反应型第 1、2、3 字母分别对缅甸菌株 BM8429、日本菌株 H8584 和菲律宾菌株 PX086 的抗性反应（The combined three capitals stand for reaction to strains BM8429，HB584 and PX086，from the left to the right，respectively）。R，抗（Resistant）；S，感（Susceptible）。

[b] O 为观察值（株数）；E 为期望值（O，Observed number of plants；E，Expected number of plants）。

由表 2 可知，在第一组合的 F_1 中，来自 Toyonishiki 对缅甸菌株的抗性和来自 Cas209 对日本菌株和菲律宾菌株的抗性都是显性，F_2 群体分别对上述 3 个菌株的抗、感分离的观察值依次为 211∶83、226∶68 和 219∶75 株，都是一对显性基因的差异，符合 3∶1 的比例，其 χ^2的 P 值分别为 0.2～0.3，0.3～0.5 和 0.8～0.9，其中 Toyonishiki 携有抗缅甸菌株 BM8429 的一对显性基因，Cas 209 携有抗日本菌株 H8584 的一对显性基因和抗菲律宾菌株 PXO86 的一对显性基因，彼此间都是独立遗传，其三抗、双抗、单抗和全感的 8 种表型符合 27∶9∶9∶9∶3∶3∶3∶1 的比例，其 χ^2的 P 值为 0.7～0.8。

在第二组合中，Milyang 23 抗缅甸菌株和日本菌株而感染菲律宾菌株。它与 Cas 209 杂交的 F_1，其抗性都是显性，F_2群体对日本菌株表现抗病，没有感病植株（279∶0），而对缅甸菌株和菲律宾菌株的抗感分离的观察值分别为 202∶77 和 217∶62 株，也都是一对显性基因的差异，符合 3∶1 的比例，其 χ^2的 P 值分别为 0.3～0.5 和 0.2～0.3，说明 Milyang 23 携有一对抗缅甸菌株的显性基因和一对抗日本菌株 H8584 的显性基因，这两对基因也是独立遗传，其双抗、两种单抗和不抗的四种表型符合 9∶3∶3∶1 的比例。其 χ^2 的 P 值为 0.3～0.5。

从第三组合得知 Toyonishiki 和 Milyang 23 是一对显性基因的差异，前者抗日本菌株，后者不抗，两者杂交的 F_1是抗的，F_2群体的抗感分离的观察值为 186∶52，符合 3∶1 的比例，其 χ^2的 P 值为 0.2～0.3；并且证实这两个品种都携有抗缅甸菌株的一对显性基因和都不携有抗菲律宾菌株的一对显性基因。后来的研究又证明 Toyonishiki 与 Milyang 23 携有相同的抗缅甸菌株（BM8429）的显性基因 *Xa18*；Milyang 23 除携有 *Xa18* 外还携有抗日本菌株 H8584 的显性基因 *Xa16*；而 Cas209 除携有 *Xa16* 外还携有一对抗菲律宾小种 2（PXO86）的显性基因 *Xa10*。

应该补充指出，当两对显性基因控制某一特定菌株的抗性时，其 F_2群体的抗性分离会因抗性基因的效应大小、显性程度及其纯合度而有不同程度的表现。在独立遗传的情况下，如果两对基因的效应一强一弱，可能出现 9 高抗∶3 抗∶3 中抗或中感∶1 感的分离比；如二者效应相等，但有累加作用，可呈 9 高抗∶6 抗∶1 感；如有一对为部分显性，可能出现 9 抗∶3 中抗∶3 轻或中感∶1 高感，或 9 抗∶3 中抗∶4 感；如果纯合显性的贡献比杂合显性的贡献大，则呈 7 高抗∶8 中抗∶1 感；如果抗病程度与显性基因数成正比，则会出现 1∶4∶6∶4（不同级别的抗）∶1 感；如果两个完全显性的基因的效应相同，可以彼此替代，携有一个完全显性基因就是抗病的，则呈 15 抗∶1 感的分离等。经典遗传学教科书常把上述最后这种分离比作为基因互作的一种形式，是由重复基因（duplicate genes）造成的，其实这并不是真正意义上的基因互作。

非等位基因间的互作在抗病性遗传上也是很普遍的。在国内的报道中，最常见的两对基因间的互作有互补作用（complementary effect）和上位性（epistasis）。互补作用是指两对基因互为条件地协同构成抗病性或感病性，因而在 F_2代分离时表现为 9 抗∶7 感（显性互补）或 1 抗∶15 感（隐性互补）的比例，前者的基因型与表型（基因型分类）是 9A－B－（抗，A－代表 AA 或 Aa）∶3A－bb（感）∶3 aa B－（感）∶1 aabb（感），后者是 9A－B－（感）∶3A-bb（感）∶3 aa-B（感）∶1 aabb（抗）。也有由一对显性基因与一对隐性基因各自构成抗病性或感病性，彼此之间有上位作用，其 F_2的表型分离比为 13 抗∶3 感或 3 抗∶13 感。上位性是指一对基因的表达对另一对基因的表达占压倒优势，可由前者的表型取代后者的表型，前者是上位基因，后者为下位基因。上位性也有显性上位与隐性上位之分，前者由显性基因控制，后者由隐性基因控制，而所控制的性状也有抗病性与感病性的区别。但是当显隐性与上位性并存时，在两对显性基因或两对隐性基因分别对同一病菌小种表现抗病性或感病性的情况下，都会出现一些因抗感并存而“无所适从”的无效基因型（如在 A -为上位基因时的 3 aaB -基因型，其中 aa 为感病而 B -为抗病，或在 B -为上位基因时的 3A-bb 基因型，其中 A -为抗病而 bb 为感病；以及在 aa 为上位基因时的 3 A-bb 基因型，其中 A -为感病而 bb 为抗病，或在 bb 为上位基因时的 3 aaB -基因型，其中 aa 为感病而 B -为抗病，都是无效基因型）。只有当一对显性基因和一对隐性基因共同控制对一个病菌小种的抗性而且显性基因和隐性基因都具有上位性时，才能使 F_2 群体中 4 类基因型都是有效的，即：9A -B -（抗）：3A-bb（A、b 上位时为抗，a、B 上位时为感）：3 aaB -（A、b 上位时为感，a、B 上位时为抗）：1 aabb（抗），从而形成了 13 抗：3 感的比例。此外，还有一种专门抑制别的非等位基因的表达而自己却没有表型效应的显性抑制基因（inhibitor），在抗病性遗传只能出现抗与不抗或感与非感两类表型的情况下，这种基因由于不能形成有效基因型而“派不上用场”，所以尚未见诸报道。当然，由于基因差异的对数不同，参与互作的对数各异（且不说两个基因以上的互作），而且各对基因互作的效应也不一样，其分离比也随之发生变化，再加上不同基因的显性度不同，因此分离比会出现各式各样的变异。

应该指出，正因为基因效应与显性程度经常混杂交织在一起，在表型上往往难以明确区分，再加上非等位基因间的互作各式各样、情况复杂，上面所说的比较简单的各类 F_2 群体分离比例，不论是由等位基因间（即基因座内）的相互作用还是由非等位基因间的相互作用引起的，都不能只凭 F_2 群体当代的表型资料作出推断，还必须根据 F_3 家系间和家系内的分离数据或 F_1 的双向回交后代的数据来验证，并联系 F_1 代和亲本世代群体的抗性表现加以综合分析，才能作出有说服力的结论。

表 3　较常见的两对抗性基因互作实例（章琦据文献整理，2006）

Table 3　Common types of resistance gene interaction as shown in dihybrid ratios（Summarized from references by Zhang，2006）

基因互作 Gene interaction	F_1抗性 F_1 reaction	F_2抗感比 F_2 R/S ratio	组合 Cross	所用菌株 *Xoo* strain used	参考文献 Reference
显性互补 (Dominant complementary for resistance)	R	9∶7	IR26（R）/陆财号（S）	大冶菌株 w-9-27	谢岳峰，1979
	Nearly R	9∶7	陆财号（S）/IR26（R） 南京 11（S）/21610，DV85 衍生系，(R)	江苏菌株 KS-6-6	邵启明等，1992
显性重复 (Dominant duplicate)	R	15∶1 15∶1 15∶1	金南风（R）/中新 120（R） 中新 120（R）/农垦 58（R） 黄牛尾香糯（R）/IR26（R） IR1529（R）/GE456（R） IR1529（R）/辛尼斯（R）	湖北菌株 Jl 691 同上 JL C1701	张端品等，1982 熊卫等，2000 曾左葵等，1983
一显一隐，显性隐性上位并存 (1 dominant and 1 recessive with both dominant & recessive epistases)	R	13∶3	TNI（S）/DV85（R）， TNI（S）/DZ78（R）， DV85（R）/广陆矮 4（S）	KS-8-4 JL 1701	高山林等，1982 曾左葵等，1983
	Nearly R	13∶3	IR1529（R）/Dular（R） 南京 11（S）/21612（R）	KS-6-6	邵启明等，1992

注：R，抗病（Resistant）；S，感病（Susceptible）。

顺便说明一下，水稻白叶枯病抗性大多受控于显性基因，所以隐性上位出现的频率相对较少，而在国内白叶枯病抗性遗传研究中，上述 13 抗∶3 感的报道常有出现。这是 20 世纪 70 年代引入 Dular、DV85、DV87 等一批携有 *xa5* 或 *xa5* ＋*Xa7* 基因的抗源材料后，因其抗性强、抗谱较宽而被广泛用于抗病育种所造成的现象。表 3 列举的是常见的一些基因互作实例。

1.2　连锁遗传

以上是两对基因在独立遗传下对一个病菌致病型的抗性分离情况。假设 A－a、B－b 两对基因分别控制 1 和 2 两个小种的抗病性，那么双抗/双感杂交后，其 F_2就会出现双抗小种 1、2（RR）、抗 1 感 2（RS）、感 1 抗 2（SR）、双感小种 1、2（SS）4 种表型。在独立遗传下，如果抗病是显性（即 A－或 B－），上述 4 种表型（或是基因型归类，即 A－B－、A－bb、aaB－和 aabb）的理论频率，依 RR（前为对小种 1，后为对小种 2 的反应，R 为抗）、RS（S 为感）、SR 和 SS 的顺序应为 9∶3∶3∶1；如果感病是显性，其频率则为 1∶3∶3∶9 的比例。如若抗病基因是一显一隐（A－和 bb），上述表型顺序的频率依次为 3∶9∶1∶3；如若感病基因是一隐一显（aa 和 B－），则上述表型顺序的频率依次为 3∶1∶9∶3 的比例。由上可知，不论对小种 1 或 2，当抗病为显性时，其 F_2分离比都是 3 抗∶1 感；当感病为显性时，其 F_2分离比都是 1 抗∶3 感。当抗病为一显一隐时，小种 1 的 F_2分离比为 3 抗∶1 感，小种 2 的 F_2分离比为 1 抗∶3 感；而当感病的为一显一隐时，小种 1 的 F_2分离比为 1 抗∶3 感，小种 2 的 F_2分离比为 3 抗∶1 感。若 F_2群体中亲本类型的频率（如 AABB 与 aabb 在相引相连锁中或 AAbb 与 aaBB在相斥相连锁中）与独立遗传时的频率相比偏多，其相应的重组类型的频率偏少，并在统计学上与预期比值差异显著，则表示这两对亲本的两个基因是连锁遗传，即这两个基因位于同一染色体上相对靠近的部位，有较多机会连带传递给下一代的倾向，其连锁强度（即紧密程度）依这两个基因的遗传距离而定，距离越近，连锁遗传越紧密。遗传距离是以重组率来表示的，重组率即重组类型的比率，也称交换值，其最大值为 50%。如果位于同一染色体上的两个基因相距很远，其重组率就会达到 50%，与独立遗传时的重组率一样，不存在连锁遗传。重组类型个体的出现概率与连锁强度相关而与连锁的相引相或相斥相无关。下面举两个连锁遗传的实例予以阐明。

曾左葵等（1983）在已知 DV85（R）/广陆矮 4 号（S）对湖北两个菌株 C0104 和 C0503 的抗性分别受一对隐性基因控制，其 F_2的分离比例符合 1 抗∶3 感之后，调查了该群体对两个菌株的 4 种抗感表型，即 RR（前者指 C0104，后者指 C0503）、RS、SR 和 SS 的分布依次为 78∶10∶8∶202 株，显著偏离了独立分离所预期的 1∶3∶3∶9 的比例（其 χ^2的 $P<0.002$），其中亲本类型明显偏多，重组类型明显偏少。经计算，重组率为 6.23%，属紧密连锁遗传。

Mir 等（1991）用 3 个抗菲律宾小种 1、2、3、4 和 6 但有不同抗性基因的品种 Long Grain、Kali Mekir 77－5 和 Aus 295 与对这 5 个小种都感病的 TN1 品种杂交，研究其抗性遗传模式。该 3 个组合的 F_1对菲律宾小种 1、2、3 和 4 为全生育期感病，说明抗病基因是隐性的。其 F_2群体在苗期和分蘖盛期分别对各个小种都呈 1R∶3S 的分离，但各单株依次对这 4 个小种的抗性反应可归纳为 SSSS、SSSR、RRRS 和 RRRR 四种表型，可见其对小种 1、2 和 3 的反应型相同。显然，这是含有两对隐性基因的遗传，可是又不符合 9（全感）∶3∶3∶1（全抗）的独立分离比例。亲本类型明显偏多，重组类型明显偏少（表 4），经计算平均重组率为 9.68%。为了验证上述结果，研究者从 TN1/Long Grain 的 F_2群体中随机取样 79 株种植 F_3家系，每家系29株，在分蘖期接种上述4个小种，调查其抗性反应的分离情况，并分为有分离的家系和不分离的家系两大类。结果发现，与预期的独立分离相比较，亲本类型的家系远远多于非亲本类型的家系，说明由这两对隐性基因控制的白叶枯病抗性是连锁遗传。通过相关研究（包括等位性测定）和有关资料比较分析，最后确认是一个与 *xa13* 等位、对菲律宾1、2、3、4和6五个小种都具有抗性的隐性基因与抗菲律宾小种1、2、3和5的 *xa5* 相连锁，而且已查明这3个抗病品种与TN1杂交的3个组合的 F_2 群体在开花期抗小种6的个体也抗小种4，而抗小种4的个体并不一定都抗小种6，表明Long Grain、Kali Mekir 77－5和Aus 295三个抗病品种对菲律宾小种1、2、3和4是由全生育期抗性基因 *xa5* 控制，对菲律宾小种1、2、3、4和6五个小种则由成株抗性基因 *xa13* 控制，*xa13* 和 *xa5* 紧密连锁。

表4 TN1/Long Grain等3个组合 F_1 和 F_2 在分蘖盛期对4个菲律宾小种的抗性反应［根据Mir等，(1991)数据整理］

Table 4 Resistance reaction to 4 Philippine races of *Xoo* in 3 crosses from TN1/Long Grain and others at the maximum tillering stage (Summarized from Mir et al., 1991)

组合 Cross	F_1 对4个小种的反应型 F_1 reaction to 4 races	F_2 对4个小种的反应型[a] F_2 plants reaction to 4 races				χ^2 (9∶3∶3∶1)	P	重组率 (%)
		SSSS	SSSR	RRRS	RRRR			
TN1/LongGrain	SSSS[a]	282	32	12	88	241.82	<0.01	9.39
TN1/Kali akri77－5	SSSS	271	27	14	87	244.10	<0.01	10.04
TN1/Aus295	SSSS	236	26	20	72	196.98	<0.01	9.48

[a] 依次对菲律宾小种1～4的反应（The combined four capitals stand for order of reaction to Philippine races 1～4，from left to right，respectively）。R，抗病（Resistant）；S，感病（Susceptible）。

1.3 成株期抗性的遗传

和小麦对锈病的抗性一样，水稻对白叶枯病的抗性也有全生育期抗性（全抗）与成株期抗性（成抗）之分，前者在整个生育期均抗病，后者在苗期感病，到孕穗或开花之后转变为抗病（图1，另见彩版8）。

Sidhu等（1978）用携有 *Xa6*（后改名为 *Xa3*）基因的两个成株期抗病的菲律宾品种Malagkit Sungsong和美国品种Zenith与感病品种TN1杂交，研究其抗性遗传时发现其 F_1 在花期之前感病，花期之后才转为抗病，F_2 在花期之前的抗感分离为1抗∶3感的隐性单基因遗传，花期以后则转为3抗∶1感的显性单基因遗传。他们将此遗传现象称为“显性转换”。张端品等（1984）和赵新平等（1986）对成株抗性品种进行遗传分析时也都观察到在植株抗性的显性反转过程中，杂合株转变为抗病的时期比纯合株稍晚些，因此认为在研究成株抗性及其遗传模式时要注意掌握适时播种和适时鉴定的时间。他们认为在华中地区苗期鉴定在播种后40d左右（6～7叶龄期），成株鉴定在70d后或11叶龄以后为宜。

成抗/感病：粳稻和爪哇稻中的成抗品种多受一对显性基因控制，在与感病品种杂交时，不论后者是籼稻或粳稻，其 F_1 在苗期感病，成株期抗病，F_2 群体在苗期均感病，到成株期都发生抗感分离，比例为3抗∶1感；回交到感病亲本的 BC_1F_1 群体，苗期均感病，成株期抗感分离为1∶1；自交的 F_3 家系在苗期全部感病，在成株期纯合抗病、抗感分离和纯合感病的家系为1∶2∶1的比例。

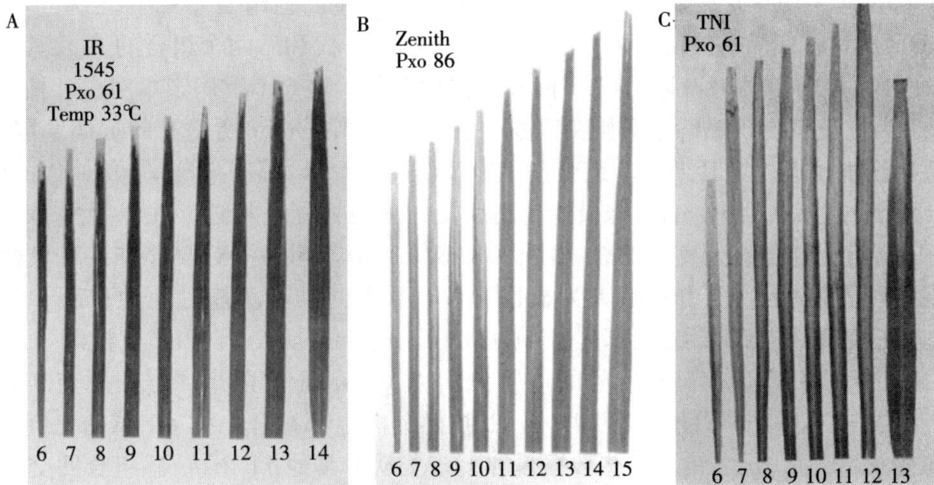

图 1　水稻对白叶枯病不同生育期抗性类型品种的病斑反应（章琦等，1985）。

A. 全生育期抗性；B. 成株抗性；C. 感病品种

Fig. 1　Lesion reactions of cultivars with different BB resistance type (Zhang et al., 1985).

A. Over-all resistance. B. Adult plant resistance. C. Susceptible cultivar

成抗/全抗：谢岳峰指导赵新平等（1985；1986）进行了较系统的水稻白叶枯病成株抗性遗传研究。结果表明当全抗品种的抗病基因为显性时，如爪哇稻 Ketan Bajong（KB）/IR28 或 Pulu Kamandi（PK）/IR28 对湖北菌株江陵 691 菌株的反应，F_1 苗期抗病，成株期也抗病；F_2 群体苗期抗感分离为 3 抗∶1 感，成株期全抗病，表明 KB 或 PK 的成抗基因与 IR28 的全抗基因是等位的，即彼此是同一基因座上的复等位基因。

当全抗品种的抗病基因为隐性时，如 KB/IR1545－339 或 KB/DV85 对江陵 691 的反应，其 F_1 苗期感病，成株期抗病；F_2 群体在苗期就出现 3 抗∶1 感的分离比，成株期也表现抗感分离，但符合 13 抗∶3 感的比例。这表明 KB 的成抗显性基因与 IR1545－339 或 DV85 的全抗隐性基因不等位，呈独立遗传，并且成抗显性基因对全抗隐性基因的显性等位基因是上位的，而隐性全抗基因对成抗显性基因的隐性等位基因也是上位的，即显性上位和隐性上位的互作关系同时存在。

成抗/成抗：假如两个品种的成抗基因相同，则其 F_1 和 F_2 群体都是苗期全感病，成株期全抗病，不分离；如果成抗基因不同，如南粳 15/Zenith 或早生爱国 3 号/Zenith 对江陵 691 菌株的反应，则 F_1 苗期感病，成株期抗病；F_2 群体的苗期全感病，成株期的抗感分离为 15 抗∶1 感，表明南粳 15 或早生爱国 3 号对江陵 691 的成抗基因与 Zenith 的成抗基因不等位，呈独立遗传，这两对显性基因的作用是重复的，只要有一对（或一个）存在就是抗病的（张端品等，1984）。

1.4　细胞质效应

长期以来，绝大多数研究者在随机选取抗病亲本作母本或父本进行抗病性遗传研究中所得的结果认为，无论在常规稻或杂交稻，是籼籼交或籼粳交，水稻对白叶枯病的抗性受控于核基因，与细胞质无关或关系很小。只有极少数的报道采用专门的设计研究其中是否存在细胞质的影响，其结果还不够肯定，但也都同意即使有些影响，起主导或决定性作用的仍是核基因。曾左葵等（1983）较详细地比较了常规稻 6 个抗/感组合（4 个抗病的 IR 品种、DV85 和 Dular 与感病品种广陆矮 4 号杂交）正反交的 F_1代和后两个组合的 F_2及其回交到广陆矮 4 号的 BC_1F_1等群体对江陵 C1701 菌株的抗性表现。这 6 个抗病亲本对白叶枯病大都为中抗，较少为抗至高抗，感病亲本则为高感。在 F_1代，除 Dular 所携有的抗病基因为隐性，其 F_1与感病亲本相近外，其余 5 个组合按 1～9 级分级标准，都有中抗（3 级）以上的抗性，高抗（1 级）和高感（9 级）的植株极少，正反交的表现相同或相近。在 F_2代，DV85/广陆矮 4 号正反交组合的抗感分离都符合 13 抗∶3 感的比例，因为 DV85 含有一对显性抗病基因和一对隐性抗病基因，前者对后者为显性上位，后者对前者为隐性上位；而 Dular/广陆矮 4 号这对正反交组合的抗感比都是 1 抗∶3 感。两个正反交组合的 χ^2 检测，P 值都在 0.25～0.90 之间。DV85/广陆矮 4 号的正反交 F_1以及 Dular/广陆矮 4 号的正交（缺反交的回交）F_1回交到广陆矮 4 号的 BC_1F_1群体，其抗感分离也都符合 1 抗∶1 感的比例，χ^2测验的 P 值都在 0.10～0.25 之间。这些数据再加上该文献中其他 11 个抗/感组合回交到感病亲本的 BC_1F_1群体的抗性分离数据，无一例外地证实了水稻白叶枯病抗性遗传与细胞质无关。

朱立宏指导王建设（2000）较系统地研究了杂交稻白叶枯病抗性与细胞质的关系。他们用 8 种不同胞质类型的不育系及其保持系，与 9 个恢复系按 P/Q 交配设计配制了各种可能的杂种组合，研究其对中国白叶枯病Ⅳ型菌代表菌株 ZHE173（致病力较强）和中国Ⅱ型菌代表菌株 KS－6－6 的抗性表现。所得结论是：杂交稻的抗性均由核基因控制，与胞质无关。因为不育系与保持系或不育系杂种与保持系杂种是同核异质系。应用 t 测验分析这两类亲本之间和这两类杂种之间的胞质效应（包括不同胞质类型间）差异（共 72 次对比），都未达到显著水平，说明无论不育系与保持系间、不育系胞质类型间或两类杂种质核互作间的效应都没有实质性差异，即不论杂交稻或常规稻，对白叶枯病的抗性受核基因控制，与胞质基因无关。

1.5 籼粳亚种间杂交的抗性遗传

籼稻、粳稻是栽培稻的两个亚种，具有不同的演化背景和遗传基础。它们在植株形态、生长发育、环境反应、生理特性等方面有明显的差异，杂交后除了受亲和力、当代结实性和后代育性的干扰外，还可能存在由其他内在因素引起的某些性状表达与分离偏离一般遗传规律的情况，值得深入研究。

迄今，籼粳杂交后代的白叶枯病抗性遗传行为大致有以下几种情况：①可能有特异性。万建民（1985 年硕士学位论文）用江苏菌株 KS 6－6 分析了爪哇稻 DV85 的白叶枯病抗性在南京 11 和 TN1 籼稻背景下的表现，是受一对显性基因和一对隐性基因共同控制的，二者分别存在显、隐性上位的互作关系；而在农垦 57 和桂花黄的粳稻背景下却表现为两对显性基因的互补作用。他还注意到粳稻品种早生爱国 3 号分别与籼、粳不同亚种的品种杂交时，其 F_2群体的抗感分离比例也不相同。这是品种之间的差异还是亚种不同的影响，尚待明确。②无特异性。张耀钢等（1991）用 3 个源于 DZ78、4 个源于 DV85 共 7 个籼型衍生抗源品种同 5 个粳稻、2 个籼稻感病品种按 NCⅡ设计配制了 49 个杂交组合，用江苏菌株 KS6－6 接种了全部 F_1和部分组合的双向回交 BC_1F_1、BC_2F_1和自交 F_2群体，结果都未发现不同来源的抗病基因在籼、粳两种遗传背景下的表达及其抗性遗传行为有明显差别。③抗性类型转换或表达的迟早有差异。张林等（1988）以早生爱国 3 号、中新 120、南粳 15 3 个具成株抗性的粳稻品种和IR28、IR1545－339两个具全生育期抗性的籼稻品种分别与感病籼稻珍珠矮和感病粳稻米泉秃芒杂交，用江陵691接种，分析了这些组合在F_1、F_2、BC_1F_1三个世代的抗性表现。结果表明成株期抗性在亚种间杂交呈现一对显性或隐性基因的遗传，与亚种内杂交一致；但在IR28和IR1545－339两个全生育期抗性的籼稻品种与感病粳稻品种杂交的F_2群体中，苗期抗病株的比例较亚种内杂交的要小得多，而苗期感病的一部分植株到成株期则转变为抗病。章琦等（1996）将籼稻的全生育期抗病基因*Xa4*转育至全生育期感病的粳稻轮回亲本中时，也观察到育成的品系CBB4所携带的*Xa4*基因的抗性表达时期由苗期延后至分蘖盛期。看来水稻白叶枯病抗性在亚种间和亚种内杂交的遗传行为还可以进行深一步的研究。

2 日本水稻白叶枯病抗性遗传研究

1957年在日本发现原来抗白叶枯病的品种"朝风"突然遭到当地毒性菌株的侵袭而表现感病（Kuhara et al.，1958）之后，人们才意识到水稻白叶枯病菌也存在着特异致病性（specific pathogenicity）的事实。这个发现对研究病菌和品种抗性的遗传与变异都有很大的促进作用。水稻白叶枯病的抗性遗传研究起步较晚，与麦类锈病，玉米大、小斑病等世界性主要病害的抗性研究相比，延迟约半个世纪。根据Nishimura（1961）的报道，日本学者在20世纪50年代后期首先开始了这项工作。他们在进行人工接种鉴定水稻品种对白叶枯病抗性的同时，形成了一套对品种抗性遗传评价与利用的例行做法，从而推动了抗性遗传研究的发展。

Nishimura（1961）最早发现黄金丸（Koganemaru）和Pi 1这两个品种对白叶枯病的抗性是由一对显性基因控制的，并指出该基因与无叶舌（lg）性状连锁，位于国际统一命名的（以下的表述相同，连锁群的命名自1990年后已与染色体编号统一）第4染色体（在Nishimura的染色体编号中编为第11号）。其后，Sakaguchi等（1968）根据日本白叶枯病菌株对日本品种的毒性反应型，将病菌分成Ⅰ、Ⅱ和Ⅲ 3个菌群，又根据日本品种对这3个菌群的抗性反应型划分为4个抗性品种群。随后，许多日本遗传学家相继运用这些抗性品种群对日本菌群的对应关系，分析国内外一些抗病品种的抗病基因时，明确了*Xa1*、*Xa2*、*Xa3*、*Xa11*和*Xa12*五个基因的一些遗传属性，现简要叙述如下。

Sakaguchi（1967）研究了黄玉和兰泰艾玛斯（Rantai Emas）两个抗病品种群的抗性基因。他将黄玉群品种和兰泰耶玛斯群品种与对日本菌群都感病的金南风群品种杂交，分析了64个抗/感组合的F_1和F_2群体对菌群Ⅰ和Ⅱ的抗性反应，表明黄玉群品种对菌群Ⅰ的抗性由一对显性基因控制，该基因被命名为*Xa1*；兰泰耶玛斯群品种对菌群Ⅱ的抗性是由两对显性基因控制，一对被命名为*Xa2*，另一对为*Xa1*。通过相互易位系法进行连锁遗传分析，将*Xa1*和*Xa2*定位在第4染色体上，*Xa1*与*Xa2*的重组率为2%～16%。Yoshimura等（1992；1995）用RAPD和RFLP技术也将*Xa1*定位于第4染色体，与RAPD分子标记$YO8_{750}$相距1.1±1.4cM，与*XNpb78*（RFLP）相距1.7 cM。这个结果与Nishimura（1961）的报道是一致的。

Ezuka等（1974；1975）分析了抗谱较广的早生爱国（Wase Aikoku）群品种对日本Ⅰ、Ⅱ和Ⅲ三个菌群的抗性，证明是由一对显性基因控制的。由于该基因在苗期表现感病，到成株期才转为抗病，是一个典型的成株抗性基因，因此将其命名为 *Xaw*。后来按照国际水稻基因命名规则将 *Xaw* 改名为 *Xa3*。成株抗性基因 *Xa3* 与全生育期抗性基因 *Xa1* 和 *Xa2*（全抗）独立遗传。Ogawa 等（1986）用日本小种（即最初的菌群）Ⅲ和 4 个菲律宾小种 1～4 分析了早生爱国 3 号、中国 45（衍生于早生爱国 3 号）和 Java 14 三个品种的抗性，表明它们对上述 5 个小种在成株期的抗性都是由 *Xa3* 控制的。后来将 *Xa3* 定位于第 11 染色体，与 RFLP 分子标记 *XNpb*181 的图距为 2.3 cM（Yoshimura et al.，1992）。

Yamamoto 等（1977）将一个对日本品种群具有明显交叉反应的印度尼西亚菌株 Xo 7306作为日本菌群Ⅴ的代表菌株之后，Ogawa 等（1978）分析了黄玉和 Java 14 这两个品种对菌群Ⅴ的抗性是由一对显性成株抗性基因控制的，该基因被命名为 *Xakg*，后改名为 *Xa12*，与 *Xa1* 紧密连锁，位于第 4 染色体。

Ogawa 等（1986）分析了来自籼稻品种 Peta 的品系 IR944 - 102 - 2 - 3、IR1529 - 680 - 3 - 2 和品种 RP9 - 3 对日本小种Ⅰ、Ⅱ、Ⅲ的抗性是由一对显性基因控制的，该基因与 *Xa1* 和 *Xa3* 独立遗传，被命名为 *Xa11*。80 年代初，日本与 IRRI 合作研究水稻白叶枯病鉴别品种的抗性基因期间，首先用日本和菲律宾的小种比较双方鉴别品种的抗性，发现对菲律宾小种都感病的鉴别品种 IR8 却是抗日本小种ⅠB、Ⅱ、ⅢA 和Ⅴ的。Ogawa 等（1991）又发现 IR8 与品种 RP 9 - 3 对日本小种的抗性反应型相似，便分析了 IR8 对日本小种ⅠA、ⅠB、ⅢA 的抗性遗传，结果表明 IR8 携有一对显性基因控制对这三个小种的抗性，该基因与 RP 9 - 3 所携有的抗性基因 *Xa11* 是等位的。

3 国际水稻研究所（IRRI）水稻白叶枯病抗性遗传研究

IRRI 在 1960 年建所之后就全力投入培育高产、优质、抗稻瘟病、白叶枯病、稻虱等多种病虫害的优良品种。随着第 1 个抗白叶枯病品种 IR20 的育成，又相继育出 IR22、IR26、IR28、IR29、IR30、IR36、IR38 等抗白叶枯病品种，这些品种很快在菲律宾和亚洲产稻国家大面积种植。不久，IR20 在菲律宾吕宋岛遭遇到新毒性菌株的侵袭而丧失了抗性。随后的抗性遗传分析表明绝大多数 IR 品种的抗性是由一对不完全显性基因 *Xa4* 控制的，该基因容易感染这个新毒性菌株。此后，IRRI 科学家便加强了对全所稻种资源和育成品种的抗性评价和抗性遗传研究，鉴定出更多的抗白叶枯病基因用于育种计划（Ou，1985）。

1975 年以后，IRRI 普遍采用菲律宾白叶枯病小种Ⅰ的代表菌株 PXO61 分期分批地对所收集的 40 000 多份水稻种质进行接种鉴定，筛选出 2 274 份抗病的品种资源（Mew et al.，1981）。之后，该所遗传学家陆续对这些抗病资源的代表品种（系）进行抗性遗传分析，先后鉴定出 *Xa4*、*xa5*、*Xa4*^{b}、*Xa6*、*Xa7*、*xa8* 和 *xa9* 七个抗性基因。现简述如下。

Petpisit 等（1977）、Olufowote 等（1977）用菲律宾小种 1 的代表菌株 PXO61 鉴定了被 IRRI 广泛用作抗源供体的 IR20（IR262/TKM6）、IR22（IR8/Tadukan）、IR1529 -680 - 3 [（Sigadis/TN1）/IR24]、IR1330 - 3 - 2（源于印度品种 W1263）和印度尼西亚品种 Pelita I/1（IR5/Syntha）的抗性，认为它们都受一对不完全显性基因控制。该基因被命名为 *Xa4*，由 Yoshimura 等（1992）定位于第 11 染色体，与 RFLP 分子标记 *XNpb78* 相距 1.7cM。由此可知，TKM6、Sigadis、W1263 和 PelitaI/1 等抗源品种的抗性都是由相同的一对基因控制的。IR1545 - 284（IR24/DZ192）、印度品种 RP - 291 - 7（BJ1/IR8）、印度地方品种 Kele

以及 Chinsuraḥ BoroⅡ的抗性则是由另一对隐性基因控制的，被命名为 *xa5*（Petpisit et al.，1977）。McCouch 等（1991）将 *xa5* 定位于第 5 连锁群（第 5 染色体）的上端，与 RFLP 标记 RG207 紧密连锁，与 *XNpb* 396 和 *XNpb* 208 相接近。之前，Yoshimura 等（1984）曾根据三体分析法将 *xa5* 定位于第 5 连锁群，与形态突变体标记 *spl* - 7 连锁，二者的遗传图距为 6.1cM；后来他又对经典遗传学研究资料与 RFLP 图谱进行相关性分析，将 *xa5* 定位于第 5 染色体的 *XNpb*396 分子标记附近，与 McCouch 的研究结果相同。*Xa4* 与 *xa5* 呈独立遗传（Yoshimura，1992）。

Librojo 等（1976）用可能携带 *Xa4* 或 *xa5* 的一些 IRRI 品种，分别与 IR22（*Xa4*）和 IR1545 - 339（*xa5*）进行基因等位性测定，发现印度尼西亚品种 Semora Mangga 在 *Xa4* 基因座上有另一个基因，它的抗性表现时期与 *Xa4* 不同，要在孕穗至开花期才能充分表达，而携有 *Xa4* 基因的品种则在全生育期都抗病，因此，该基因被定名为 $Xa4^b$。

Sidhu 等（1978；1979）研究了菲律宾抗病品种 Malagkit Sungsong 及其衍生品系 IR944 - 102 和美国品种 Zenith 及其衍生品系 IR1698 - 241 对 PXO61 的抗性遗传。它们与感病品种 TN1 杂交后，F_1植株在花期之前感病，花期之后才表现抗病，F_2群体于花期之前呈 1 抗∶3 感的隐性单基因分离比例，花期以后则转为 3 抗∶1 感的显性单基因分离比例。这个将显隐性反转的成株抗性基因被命名为 *Xa6*，与第 11 染色体上的 *Xa4* 相连锁，其重组率为 26%，但与 *xa5* 呈独立遗传。他们又分析了 DV85、DV86 和 DZ78 三个孟加拉品种对 PXO61 菌株的抗性遗传，表明是受两对基因控制的，一对是控制全生育期抗性的隐性基因 *xa5*，另一对是在孕穗期表现抗病的显性成株抗性基因 *Xa7*，后者与 *Xa4*、*xa5* 和 *Xa6* 相互独立，被定位于第 6 染色体（Kaji et al.，1995）。Sidhu 等（1978）对成株抗性的美国品种 PI231129 进行遗传分析，鉴定出一对与 *xa5* 不等位的新的隐性成株抗性基因，被命名为 *xa8*，也与 *Xa4*、*xa5* 和 *X*a6 呈独立分离。*xa8* 被初步定位于第 7 染色体上，与微卫星分子标记 RM214 的遗传距离为 19.9cM（Singh et al.，2002）。

Singh 等（1983）在用一系列抗病品种与已知抗性基因进行等位性测定研究中，发现两个老挝品种 Sateng 和 Khao Lay Nhay 携有一对隐性抗性基因，它在分蘖盛期至成株期都表现抗病，与 *Xa6* 以 5.9%的重组率相连锁，与 *Xa 3*、*xa 5*、*Xa7* 和 *xa 8* 呈独立分离，被命名为 *xa9*。

Yoshimura 等（1983；1995）分析鉴定了 IRRI 后来新增加的来自西非塞内加尔的一个关键性鉴别品种 Cas209 对菲律宾小种 1、2、3、和 4 的抗性反应，发现它在全生育期高抗小种 2，其抗性由一对显性基因控制。该基因位于第 11 染色体上，与 *Xa4* 以 27.4%的重组率相连锁，与 RAPD 分子标记 O 07_{2000} 的图距为 5.3cM，与 *xa5* 呈独立遗传，被命名为 *Xa10*。

以上两节所述的分别在日本和 IRRI 进行鉴定的水稻白叶枯病抗性基因，除 *Xa12* 是采用印度尼西亚菌株之外，其他都是用本国或所在国的代表性菌株接种的。其他国家的科学家也是采用本国的菌株对本国品种或一些引进品种进行抗性鉴定和遗传分析，因为在通常情况下微生物是不允许在国际之间流通传播的。例如 Watanabe（1976）在斯里兰卡用该国的 *Xoo* 菌株 CAR 1 分析鉴定早生爱国 3 号的抗性是受两对显性基因控制的，分别被定名为 *Xa - a* 和 *Xa - k*；Zenith 和 P1209938 的抗性受另外两对显性基因 *Xa - i* 和 *Xa - p* 控制。谢岳峰实验室 Yu 等（1986）分析 IR28 对湖北菌株 JL691 的抗性是由两对显性基因 *Xa - a* 和

*Xa-h*控制的，*Xa-h* 以17%的重组率与 *Xa-a* 相连锁等等。类似的例子还很多，由于没有和后来国际统一的抗性基因鉴定工作接轨，不清楚其是否与已经正式登录的 *Xa* 或 *xa* 基因有所区别，就不多加叙述了。

4 国际统一的水稻白叶枯病抗性基因鉴定

水稻品种的抗病基因与白叶枯病菌的毒性基因存在对应关系。采用本国的菌株对本国品种或一些引进品种进行接种鉴定，无论是为了研究或利用都是可行和必要的，但在比较来自不同国家的抗病基因的异同及其属性，特别是功能属性时，则无能为力。因为不同国家的菌株很可能不一样，更何况它们所赖以生存的原先环境和寄主品种又各不相同呢？也就是说，前面所述那种"各行其是"的做法，彼此间所得的研究资料和鉴定结果缺乏可比性，也在一定程度上影响到这一研究领域的进展和发展空间。因此，IRRI 与日本热带农业研究中心（Tropic Agriculture Research Center，TARC）通过商议和合作，在 1982—1987 年间采取统一的鉴别小种，按照统一的研究方案，建立一套国际水稻白叶枯病抗性鉴别系统，对各种类型的品种材料进行了抗性基因的重新鉴定，并结合原始文献记载加以整理、规范和核实，最后登录在案。其主要情况和结果如下所述。

4.1 日本热带农业研究中心（TARC）和 IRRI 各自命名的白叶枯病抗性基因的异同

首先，日本和 IRRI 科学家各自采用所在国水稻 *Xoo* 小种的代表菌株，测定和比较所有业已命名的抗性基因的一些代表品种对白叶枯病的抗性反应型，以确定这些基因的异同。例如，用日本小种鉴定表现为抗病的黄玉（*Xa1*、*Xa12*）和 Tetep（*Xa1*、*Xa2*、*Xa12*）都感染 4 个菲律宾小种 1～4；用菲律宾小种 2 鉴定表现抗病的品种 Cas209（*Xa10*）都感染 7 个日本小种，即：小种ⅠA（T7174）、ⅠB（T7156）、Ⅱ（T7147）、ⅢA（T7133）、ⅢB（Q6803）、Ⅳ（H75373）和Ⅴ（H75304），这表明黄玉、Tetep 对菲律宾小种和 Cas209 对日本小种都不具有抗性基因（Ogawa et al.，1987）。而用菲律宾小种鉴定的具抗性基因的一些爪哇型代表品种诸如 Zenith（*Xa6*）、Sateng（*xa9*）、Semora Mangga（$Xa4^b$）和用日本小种鉴定的具抗性基因 *Xa3* 的 Java 14、中国 45 和早生爱国 3 号对日本或菲律宾小种均表现相似的抗性反应型，病斑边缘呈褐色反应（Kaku et al.，1978）。经 Ogawa 等（1986）重新接种分析后，发现原先鉴定为 $Xa4^b$ 的品种（Semora Mangga）对 4 个菲律宾小种的抗性反应型与携有 *Xa4* 的 IR20 并不相同：IR20 对菲律宾小种 1～4 的反应型依次为 RSSM，而以 Semora Mangga 为代表的这群爪哇型品种其反应型均为 RRRR（表 5）。最后确认 Zenith、Semora Mangg、Cenpo Selak 及 Sateng 所携带的抗性基因与 *Xa3* 是等位的（表 6），因此删除了 $Xa4^b$、*Xa6*、*xa9* 这 3 个基因。章琦等（1991）用菲律宾小种 1～4、中国致病型2（HB17）和日本小种Ⅰ（T7174）共 6 个菌株测定两个中国抗源品种邳早 15 和蚌珠芒与日本早生爱国 3 号和 Zenith 所携基因的等位性，结果也表明这 4 个品种对上述 6 个菌株的成株抗性都是由 *Xa3* 基因控制的（表 7）。

表 5 若干爪哇型品种对菲律宾 4 个白叶枯病小种在孕穗期的抗性反应 [据 Ogawa 等 (1986) 数据整理]

Table 5 Reaction of some javanica rice cultivars to 4 Philippine races of *Xoo* at booting stage (Summarized from Ogawa et al., 1986)

品种 Cultivar	病斑长度 (cm) 和反应型[a] Lesion length & reaction			
	Race 1	Race 2	Race 3	Race 4
IR 24 (S check)	24.7 (S)	18.5 (S)	20.6 (S)	18.4 (S)
IR 20	1.8 (R)	13.8 (S)	14.8 (S)	7.4 (MR)
早生爱国 3 号(Wase Aikoku 3)	1.8 (R)	0.7 (R)	0.6 (R)	0.6 (R)
中国 45 (Chugoku 45)	2.0 (R)	1.5 (R)	1.1 (R)	0.7 (R)
Java 14	2.1 (R)	1.2 (R)	1.0 (R)	0.8 (R)
Semora Mangga	1.9 (R)	0.7 (R)	0.7 (R)	0.8 (R)
Zenith	3.4 (MR)	1.2 (R)	0.5 (R)	0.5 (R)
Cempo Selak	1.7 (R)	1.8 (R)	0.7 (R)	0.3 (R)
Sateng	6.9 (MR)	8.7 (MR)	7.8 (MR)	3.4 (MR)

[a] 15 张接种叶片的平均病斑长度，接种后 14 天调查 [Average lesion length in 15 inoculated leaves (3 leaves from 5 plants) at 14 d after inoculation]。R，抗 (Resistant)；MR，中抗 (Moderately resistant)；S，感 (Susceptible)。

表 6 抗/抗杂交组合 F_1 和 F_2 世代对日本和菲律宾菌株在成株期的抗性反应 [据 Ogawa 等 (1986) 研究数据整理]

Table 6 Reaction of resistance to strains of *Xoo* from Japan and the Philippines in F_1 and F_2 progenies of R/R at the booting stage (Summarized from Ogawa et al., 1986)

组合 Cross	反应型 Reaction F_1	反应型 Reaction F_2 株数 No. of plants observed		χ^2 (3 : 1)	P
在 TARC (At TARC)	对日本小种ⅢA (JapaneseⅢ A)	R	S		
金南风/ Zenith (Kinmaze/Zenith)	R	282	109	1.726	0.1～0.2
丰巾/ Zenith (Toyonishiki/Zenith)	R	306	92	0.800	0.3～0.5
中国 45/ Zenith (Chugoku 45/Zenith)	R	398	0		
Toyonishiki/Sateng	R	299	98	0.021	0.8～0.9
Chugoku 45/ Sateng	R	356	0		
Sateng/Zenith	R	302	0		
在 IRRI (At IRRI)	对菲律宾小种 1～4 的反应型 Reaction to 4 Philippine races: F_1	F_2 株数 No. of plants observed: RRRR	SSSS		
IR24/ Zenith	RRRR[a]	165	47	0.960	0.3～0.5
Chugoku 45/Zenith	RRRR	230	0		

（续）

在 IRRI (At IRRI)	对菲律宾小种 1～4 的反应型 Reaction to 4 Philippine races				
	F_1	F_2株数 No. of plants observed			
		RRRR	SSSS		
Java 14/Zenith	RRRR	277	0		
IR24/Sateng	RRRR	139	58	2.073	0.1～0.2
Toyonishiki/Sateng	RRRR	172	51	0.539	0.3～0.5
Chugoku 45/ Sateng	RRRR	253	0		
Java 14/Sateng	RRRR	339	0		
Sateng/ Zenith	RRRR	355	0		
Java14/Semora Mangga	RRRR	331	0		

[a] 依次对 4 个菲律宾小种 1～4 的反应（The combined four capitals stand for order of reaction to Philippine races 1～4，from the left to the right，respectively)。R，抗病（Resistant)；S，感病（Susceptible)。

表 7 邳早 15 和蚌珠芒等成株抗性品种之间杂交的 F_1、F_2群体对多个水稻白叶枯病菌株的抗性反应（章琦等，1991)

Table 7 Reaction to strains of BB in F_1 and F_2 populations from the crosses between Peizao15，Bangzhumang and cultivars with adult plant resistance（Zhang et al.，1991)

组合 Cross	对菌株的反应型[a,b,c] Reaction pattern			F_2对菌株的反应型 Reaction pattern of F_2					
	P_1	P_2	F_1	抗 RRRR	感 SSSS	总数 Total	抗 RR	感 SS	总数 Total
早生爱国 3 号/ Zenith (Wase Aikoku 3/Zenith)	RRRR[a]	RRRR	RRRR	282	0	282			
邳早 15/早生爱国 3 号 (Pizao15/Wase Aikoku 3)	RR[b]	RR	RR				248	0	248
邳早 15 /Zenith (Pizao 15/Zenith)	RR[c]	RR	RR				298	0	298
蚌珠芒/早生爱国 3 号 (Bangzhu-mong/Wase Aikoku 3)	RR[b]	RR	RR				238	0	238
蚌珠芒/Zenith (Bangzhu-mong/ Zenith)	RR[n]	RR	RR				175	0	175

[a] 依次对 4 个菲律宾小种 1～4 的反应（The combined four capitals stand for order of reaction to Philippine races 1～4，from the left to the right，respectively)。R，抗（Resistant)；S，感（Susceptible)。

[b] 对中国（HB17）和日本小种 1（T 1）两个菌株的反应［The two capital letters stand for order of reaction to Chinese strain HB17，and Japanese race 1（T1)，respectively］。

[c] 对中国（HB17）和菲律宾小种 1（P1）两个菌株的反应（The two capital letters stand for order of reaction to Chinese strain HB17 and Philippine race 1，respectively)。

根据 TARC 和 IRRI 联合采用日本和菲律宾两套白叶枯病菌小种统一检测的结果，Ogawa 等（1987）确认了 *Xa1*、*Xa2*、*Xa3*、*Xa4*、*xa5*、*Xa7*、*xa8*、*Xa10*、*Xa11* 和 *Xa12* 十个抗

性基因的存在，其中 *Xa3*、*Xa7*、*xa8*、*Xa11* 和 *Xa12* 都是成抗基因。此后，他们所用的鉴别小种及其相对应的抗性基因便成为国际公认的水稻白叶枯病鉴别系统，双方的合作研究也就集中在 IRRI 进行。随着菲律宾 *Xoo* 新小种的出现，许多新的抗性基因又先后被发现并依次予以命名。

4.2 统一鉴定后的水稻白叶枯病抗性基因

4.2.1 *xa13*、*xa24ʻ*和 *Xa14* 的鉴定 当 IRRI 发现了菲律宾小种 5 和 6 之后（Mew et al.，1979；1982），又鉴定出一批能抗小种 6 的品种。据 Ogawa 等（1987）报道，BJ1、Aus247、AC19-1-1、Chinsurah Boro ll 和 Kalimekri 77-5 5 个 BJ1 群品种对菲律宾小种 1、2、3、4 和 6 均具有稳定的抗性。已知携有隐性抗性基因 *xa5* 的品种 IR1545-339 表现抗菲律宾小种 1、2、3，中感小种 4，感小种 6。Ogawa 等首先验测 BJ1 群品种的抗性基因与 IR1545-339（*xa5*）是否等位。将 BJ1 群品种与 IR1545-339 杂交，其 F_1 植株抗菲律宾小种 1 和 2，中感小种 4，感小种 6（R R MS S），这种反应模式与 IR1545-339 相类似（表 8）。IR1545-339 与 BJ1 群品种杂交组合的 F_2 对小种 1 和 2 没有出现感病株，对小种 4 均为中抗/中感，没有清晰的抗感分离，初步推知 BJ1 群品种具有隐性抗性基因 *xa5*；但是对小种 6 却呈现 1 抗：3 感分离比（表 9）；又将 IR24（感病）分别与上述 BJ1 群的 5 个品种杂交，同时用对 IR1545-339 有鉴别力的菲律宾小种 1、2、4 和 6 接种。BJ1 群品种与 IR24 杂交的 F_1 植株对这 4 个菲律宾小种均为感病（SSSS），而其 F_2 群体对这 4 个小种的抗性反应型可分为 4 类：类 1 为 SSSS，全感这 4 个小种；类 2 为 RRMS，与IR1545-339 类似；类 3 为 RRRR，与 BJ1 群类似；类 4 为 SSSR，仅抗小种 6，其分离比为 9：3：3：1（表 10）。该结果表明 BJ1 群品种的抗性是由两对隐性基因控制的，一对为全生育期抗性基因 *xa5*，另一对是控制小种 6 的成株抗性新基因，被命名为 *xa13*（Ogawa et al.，1987）。将 *xa13* 定位于第 8 染色体上，与 RFLP 标记 RZ28 的图距为 5.1cM，与 G136 的图距为 3.8cM（Zhang et al,，1996）。Chu 等（2006a；2006b）进一步将 *xa13* 基因定位到一个 9.2 kb 的片段上，该片段只包含一个候选基因，通过候选基因的功能互补测试，*xa13* 基因已被成功地克隆。

表 8 亲本和 BJI 群品种与 IR1545-339 杂交 F_1 对 4 个菲律宾小种的抗性反应［据 Ogawa 等（1987）研究数据整理］

Table 8 Reaction of parents and F_1 plants from the crosses between IR1545-338 and BJ1 group cultivars to Philippine races（1，2，4 and 6）**of *Xoo* at the booting stage**（Summarized from Ogawa et al.，1987）

品种或组合 Cultivar or cross	病斑长度（cm）和反应型[a,b] Lesion length（cm）& reaction			
	小种 Race 1	小种 Race 2	小种 Race 4	小种 Race 6
IR24	17.1（S）	13.4（S）	13.7（S）	15.5（S）
IR1545-339	1.5（R）	1.9（R）	7.8（MS）	19.3（S）
BJ1	1.1（R）	1.2（R）	1.2（R）	2.6（R）
AC19-1-1	0.7（R）	0.6（R）	0.5（R）	1.5（R）

（续）

品种或组合 Cultivar or cross	病斑长度（cm）和反应型[a,b] Lesion length（cm）& reaction			
	小种 Race 1	小种 Race 2	小种 Race 4	小种 Race 6
Chinsura Boro Ⅱ	1.6（R）	1.7（R）	1.3（R）	2.3（R）
Aus 274	1.7（R）	3.2（R）	2.0（R）	2.5（R）
Kalimakri 77－5	1.6（R）	1.5（R）	1.2（R）	1.4（R）
IR1545－339/BJ1 F_1	1.3（R）	2.8（R）	9.2（MS）	15.4（S）
IR1545－339/Aus274 F_1	2.1（R）	3.3（R）	11.2（MS）	18.8（S）

[a] 接种后18天的15张叶片（5株，每株3张叶片）的平均病斑长度（cm）［Average lesion length in cm in 15 inoculated leaves (3 leaves from 5 plants) at 18d after inoculation］。

[b] R，抗（Resistant）；MS，中感（Moderately susceptible）；S，感（Susceptible）。

表9 IR1545－339与BJ1群品种杂交F_1、F_2群体在孕穗期对菲律宾4个小种的抗性反应（Ogawa et al.，1987）

Table 9 Reaction of F_1 and F_2 populations from the crosses between IR1545－339 and BJ1 group cultivars to 4 Philippine races of *Xoo* at booting stage（Ogawa et al.，1987）

组合 Combination	F_1反应型[a,b] Reaction of F_1	F_2反应型 Reaction of F_2		χ^2 (1∶3)	P
		RRRR	RRMS		
IR1545－339/BJ1	RRMS	78	184	3.2	0.05～0.1
IR1545－339/AC19－1－1	RRMS	67	192	0.1	0.7～0.8
IR1545－339/ Chinsurah Boro Ⅱ	RRMS	69	184	0.7	0.3～0.5
IR1545－339/Aus 274	RRMS	70	200	1.1	0.7～0.8
IR1543－339/ Kalimakri 77－5	RRMS	81	187	3.9	0.02～0.05

[a] 依次对4个菲律宾小种1、2、4和6的反应（The combined four capitals stand for order of reaction to Philippine races 1，2，4 & 6 from the left to the right，respectively）。

[b] R，抗（Resistant）；MS，中感（Moderately susceptible）；S，感（Susceptible）。

表10 IR24与BJ1群品种杂交组合F_1和F_2群体在孕穗期对菲律宾小种1、2、3、4的抗性反应［据Ogawa等（1987）数据整理］

Table 10 Reaction of F_1 and F_2 populations from crosses between IR24 and BJ1 group cultivars to 4 Philippine races of *Xoo* at booting stage（Summarized from Ogawa et al.，1987）

组合 Cross	F_1反应型[a,b] Reaction of F_1	F_2反应型[a,b] Reaction of F_2				χ^2 (9∶3∶3∶1)	P
		SSSS	RRMS	RRRR	SSSR		
IR24/BJ1	SSSS	164	27	31	29	34.2	＜0.01
IR24/AC19－1－1	SSSS	117	35	16	37	1.2	0.7～0.8
IR24/Chinsurah Boro Ⅱ	SSSS	173	43	23	30	14.7	＜0.01

（续）

组合 Cross	F_1反应型[a,b] Reaction of F_1	F_2反应型[a,b] Reaction of F_2				χ^2 (9∶3∶3∶1)	P
		SSSS	RRMS	RRRR	SSSR		
IR24/Aus 274	SSSS	171	40	11	47	6.9	0.05～0.1
IR24/Kalimakri 77－5	SSSS	155	39	21	39	4.1	0.2～0.3

a 依次对 4 个菲律宾小种 1～4 的反应（The combined four capitals stand for order of reactions to Philippine races 1, 2, 3 & 4 from the left to the right, respectively）。

b R，抗（Resistant）；M，中抗（Moderately resistant）；S，感（Susceptible）。

Ogawa 等（1987）根据印度的 Tepa 1 对菲律宾小种 1、2、4 和 6 的抗性反应型是 RRRR，认为它可能也属于 BJ1 群品种。

1990 年 Mir 和 Khush 用菲律宾小种 1～6 鉴定出 DV86、DV85、Aus295、Tepa 1 和 Long Grain 等品种在全生育期对小种 1、2、3、4 和 6 五个小种都抗病。Khush 等（1999）使用小种 6 分析 BJ1 群的 4 个品种 Kalimekri 77－5、Chinsurah Boro Ⅱ、Aus 247、AC19－1－1和上述 5 个品种是否也携有 *xa13* 或别的抗性基因。他们用 BJ1 分别与这 9 个品种杂交进行等位性测验，在 9 个组合中，Kalimekri 77－5、Chinsurah Boro Ⅱ、Aus247、AC19－1－1 11、Tepa 1、Long Grain 6 个品种与 BJ1 杂交的组合的 F_1植株均抗小种 6，其 6 个 F_2群体都没有分离出感病植株，表明这 6 个品种对小种 6 的抗性是由 *xa13* 控制的；而 BJ1/DV86、BJ1/DV85 和 BJ1/Aus 295 三个组合的 F_1植株感小种 6，F_2群体呈 7 抗∶9 感的分离比，说明 DV86、DV85 和 Aus 295 存在一个与 *xa13* 不等位的全生育期抗病基因。已知除了 *xa13* 外，还没有在全生育期抗菲律宾小种 6 的其他隐性基因，因此这应该是一个新基因，遂将其命名为 *xa24t*（表 11）。

表 11 BJ 1 与 9 个抗病品种杂交组合 F_1和 F_2群体对菲律宾小种 6 的抗性反应［据 Khush 等（1999）数据整理］

Table 11 Reaction of F_1 and F_2 populations from crosses between BJ 1and 9 resistance cultivars to Philippine race 6 of *Xoo* (Summarized from Khush et al., 1999)

组合 Cross	对小种的反应型[a] Reaction to race 6 (PXO99)					
	F_1植株 F_1 Plant	F_2群体 F_2 Populations				
		R	S	比例 Ratio	χ^2	P
BJ1/DV85	S	279	406	7∶9	2.54	0.10～0.20
BJ1/DV86	S	356	427	7∶9	0.94	0.30～0.50
BJ1/Aus295	S	323	444	7∶9	0.84	0.30～0.50
BJ1/Kalimekri 77－5	R	617	0	1∶0		
BJ1/Chinsurah Boro 2	R	755	0	1∶0		
BJ1/Tepa 1 (Bopro)	R	693	0	1∶0		
BJ1/Aus 274	R	732	0	1∶0		
BJ1/AC19－1－1	R	776	0	1∶0		
BJ1/Long Grain	R	745	0	1∶0		

（续）

组合 Cross	对小种的反应型[a] Reaction to race 6 (PXO99)					
	F_1植株 F_1Plant	F_2群体 F_2Populations				
		R	S	比例 Ratio	χ^2	P
DV85/DV86	R	781	0	1∶0		
DC85/Ais295	R	769	0	1∶0		

[a] R，抗病（Resistant）；S，感病（Susceptible）。

品种 TN1（台中本地 1 号）在全生育期对菲律宾 6 个小种中的小种 1、2、3、4 和 6 高度感病，而高抗小种 5。Taura 等（1987）比较了 TN1 与菲律宾和日本两套携有白叶枯病已知抗性基因的鉴别品种对菲律宾 6 个小种的抗性差异，结果表明它的抗性反应型与这些鉴别品种不同，并用菲律宾小种 1、2、4 和 5 分析了 TN1 的抗性遗传并进行了等位性测验，证明其对小种 5 的全生育期抗性是由一对新的显性基因控制的，被命名为 *Xa14*，1988 年在国际水稻新基因命名委员会注册（Kinoshita et al.，1988）。Taura 等（1992）用三体技术将其定位在第 4 染色体上。谭震波等（1998）又用 RFLP 标记将该基因定位于第 4 连锁群的分子图谱上，与 RFLP 分子标记 RG620 和 G282 分别相距 20.1cM 和 19.1cM。

4.2.2 *Xa16*、*Xa17* 和 *Xa18* 的鉴定 前面曾提到，当日本的白叶枯病菌被划分为Ⅰ、Ⅱ、Ⅲ三个菌群时，Sakakuchi（1967）已表明兰泰艾玛斯抗性品种群对日本小种Ⅰ和Ⅱ的抗性是由 *Xa1* 和 *Xa2* 两对显性基因控制的。在日本的 *Xoo* 菌群增加到 5 个小种后，Noda 等（1989）研究表明兰泰艾玛斯群的代表品种 Tetep 对日本小种Ⅴ（菌株 H8584 和 H8581）的抗性是由另一对显性基因控制的，后者被命名为 *Xa16*。因此，Tetep 携有 *Xa1*、*Xa2* 和 *Xa16* 三对基因，分别控制了对日本小种Ⅰ、Ⅱ和Ⅴ的抗性。

1989 年 Ogawa 等分析了日本粳稻品种阿苏稔（Asominori）对日本菌株 H 8313 的成株抗性，发现也是由一对新的显性基因控制的，该基因被命名为 *Xa17*。

Yamamoto 等（1990）报道了日本和 IRRI 培育的三套水稻白叶枯病抗性近等基因系的 3 个轮回亲本 IR24、Miliyang 23 和丰锦（Toyonishiki），对当时的日本 5 个小种和菲律宾 6 个小种都是感病的，但能抗 BM8417 和 BM8429 两个缅甸菌株。用感染缅甸菌株的品种 Cas209 同丰锦、Miliyang23 和 IR24 杂交，分析这 3 个组合的 F_1、F_2群体对两个缅甸菌株的抗性，结果表明它们均携有一对相同的显性基因控制这两个菌株的抗性。3 个品种之间抗性基因的等位性测验结果是：丰锦/Miliyang 23、丰锦/IR24 和 Miliyang 23/IR24 的 F_1对两个菌株都抗病，F_2也都没有分离出感病植株，表明丰锦、Miliyang 23 和 IR24 对缅甸菌株 BM8417 和 BM8429 的抗性是由一对新的显性基因控制的，该基因被命名为 *Xa18*。

Xa16，*Xa17* 和 *Xa18* 三个基因都在 1989 年国际水稻新基因命名委员会注册。

4.2.3 *xa15t*、*xa19*、*xa20* 和 *Xa25t* 的鉴定 日本遗传学家用日本小种Ⅰ、Ⅱ、Ⅲ和Ⅳ从突变品种中鉴定出 3 个新的隐性基因 *xa15 t*、*xa19* 和 *xa20*。

Nakai 等（1985，1988）通过热中子照射日本感病的粳稻品种 Harebare 的种子，获得一批对白叶枯病有不同抗性水平的突变体，其中 M41 突变体在开花期抗日本 4 个小种Ⅰ、Ⅱ、Ⅲ和Ⅳ。Nakai 等用感病品种 Harebare 与 M41 杂交，研究 M41 的抗性遗传模式。该组合的 F_1植株对这 4 个小种都感病，说明其抗病基因是隐性的。其 F_2群体对各小种的分离观察值都呈 1R：3S 的分离，其 χ^2 依次为 0.29、0.001、0.02、0.38，P 值为>0.5 或>0.9（表 12），F_3家系抗感纯合度符合 1：2：1 的比例，χ^2 为 0.025，P 值为 0.70～0.90，表明 M41 在成株期对 4 个日本小种的抗性由一对隐性基因控制。由于在粳稻中还没有抗日本小种的隐性基因，因此应该是一个新基因，暂命名为 *xa15t*。当时尚没有发现能抗日本菌株的其他隐性基因，无法进行等位性测定而没有被确认。Ikeda 等（1995）后来将 *xa15t* 与下面叙述的另外两个来自化学突变体的隐性基因 *xa19* 和 *xa20* 进行了等位性测试，证明它们之间并不等位，才被确认为 *xa15* 。

表 12 M41 与感病品种 Harebare 杂交组合 F_1 和 F_2 群体对日本 4 个小种的抗性反应［据 Nakai 等（1988）数据整理］

Table 12 Reaction of F_1 and F_2 populations from Harebare（S）/M41 to 4 Japanese races of *Xoo*（Based on Nakai et al.，1988）

Race	F_1 反应型 F_1 Reaction	F_2 反应型[a] F_2 Reaction				
		R[a]	S	总数 Total	χ^2（1：3）	*P*
Ⅰ（T7174）	S	120	381	501	0.29	>0.50
Ⅱ（T7147）	S	123	368	491	0.001	>0.90
Ⅲ（T7133）	S	133	394	527	0.02	>0.50
Ⅳ（H 75373）	S	121	387	508	0.38	>0.90

[a] R，抗病（Resistant）；S，感病（Susceptible）。

Taura 等（1987，1992b）用甲基亚硝酸脲烷（MNU）处理感病的 IR24 籼稻品种胚胎形成单细胞期的受精卵，获得了一批对白叶枯病有不同抗性水平的突变体，其中 XM5 和 XM6 两个突变体在孕穗期至开花期都中抗菲律宾 6 个（1～6）小种。抗性遗传分析表明 XM5 对这些小种的抗性均由一对隐性基因控制，与 IRBB 5（*xa5*）、IRBB 8（*xa8*）、BJ 1（*xa13*）三个品种进行基因等位性测验，结果表明与 *xa5*、*xa8* 和 *xa13* 三个隐性基因都不等位，被命名为 *xa19*（Taura et al.，1991）。XM6 对菲律宾小种 1～6 的抗性也是由一对隐性基因控制的，用菲律宾小种 1、4 和 6 对它进行等位性测试，与 *xa19*、*xa5*、*xa8* 和 *xa13* 都不等位，被命名为 *xa20*（Taura et al.，1992）。

1990 年 *xa15* 、*xa19* 和 *xa20* 三个基因都在国际水稻新基因命名委员会注册。

Gao DY 等（高东迎等，2001）用杂交稻汕优 63 的恢复系明恢 63 的成熟胚进行组培，在继代培养中以白叶枯病菌液作为选择压进行离体筛选，获得了一个抗病性稳定的体细胞无性突变系（somaclonal mutant）HX3。抗性遗传分析表明 HX3 对所用白叶枯病菌株 ZHE173 的抗性由一对显性基因控制。随后用 22 个包括菲律宾白叶枯病小种 1～5、日本小种Ⅲ（T7133）、和Ⅳ（T4）以及 15 个中国的 ZHE173、BJ84-3、KW-2、OS218、Z126、Z125、X-S-2、OS17、GX09、OS35、Hen114、4、209、OS171 和 83-69 共 22 个菌株测试并比较了 HX3 与 *Xa1*～*4*、*Xa7*、*Xa10*、*Xa11*、*Xa12*、*Xa14*、*Xa16*、*Xa17*、*Xa18*、*Xa21* 共 13 个显性抗性基因的抗性反应。结果表明，HX3 在成株期抗菲律宾小种 1、3 和 4，感菲律宾小种 2 和 5，抗日本菌株 T7133 和 T4，在 15 个中国菌株中，除了 OS171 和83-69 两个菌株外，抗其他 13 个菌株（多数未标明所属致病型），说明 HX3 与上述 13 个主要显性抗病基因的抗谱都不相同。HX3 与当时已鉴定的新基因 *Xa22* 和 *Xa23* 的抗谱也不相同

（*Xa22* 抗菲律宾小种 2、感 ZHE173；*Xa23* 抗菲律宾所有小种）。最后采用中国致病型Ⅳ的代表菌株 ZHE173 测验 HX3 和抗该菌株的 *Xa4*、*Xa7* 和 *Xa21* 三个抗病基的等位性关系，结果显示 HX3 与它们都不等位，并独立遗传，表明 HX3 携有一个新的显性成株抗性基因，也被暂命名为 *Xa25t*。

4.2.4 *Xa22t*、*Xa25t*、*Xa26* 的鉴定 *Xa22t* 谢岳峰等（1990）从云南地方品种中发现了一批新的抗白叶枯病资源，Lin 等（林兴华等，1996）分析了其中一个扎昌龙品种的抗性遗传。该品种在成株期对中国致病型Ⅰ、Ⅱ、Ⅳ和Ⅶ，菲律宾的小种 1～6，日本小种Ⅰ、Ⅱ和Ⅲ共 12 个小种（致病型）的代表菌株具有抗至中等抗性（表 13）。用菲律宾小种 1 分析了珍珠矮（感）/扎昌龙的 F_1、F_2、B_1F_1 群体的抗性反应及其分离比，表明其抗性由一对显性基因控制。与已知显性基因的等位性测验表明，该抗性基因分别与 *Xa1*、*Xa2* 和 *Xa14* 呈独立遗传，与 *Xa4* 呈连锁遗传，其重组率为 13.3±4.5%，证明这是一个新的显性成株抗性基因，将其定位于第 11 染色体的长臂末端，与最近的分子标记 R543 的图距为 7.1cM，暂定名为 *Xa22*。据余腾琼等（2005）报道携有该基因的品种扎昌龙能抗御云南高原粳稻地区所有测试过的 9 个 *Xoo* 致病型。

表 13 扎昌龙及感病对照品种珍珠矮和 IR24 对白叶枯病菌株的抗性反应［据 Lin 等（1996）资料整理］

Table 13 Reaction of ZCL, Minghui 63, and susceptible cultivars ZZA, IR24 to strains of *Xoo* at the booting stage (Summarized from Lin et al., 1996)

所用菌株[a] Strain used	IR24	扎昌龙（ZCL）[b]	珍珠矮（ZZA）	所用菌株[a] Strain used	IR24	扎昌龙（ZCL）[b]	珍珠矮（ZZA）
HLJ 72（C1）	—	R	S	PXO71（P4）	S	R	S
HB17（C2）	—	R	S	PXO112（P5）	S	R	S
ZHE173（C4）	S	MS	S	PXO99（P6）	S	R	S
JS 49-6（C7）	—	MR	S	PXO339（P9）	S	—	S
JL691	S	R	S	T7174（TⅠ）	—	R	S
PXO61（P1）	S	R	S	T7147（TⅡ）	S	R	S
PXO79（P3）	S	MR	S	T7133（TⅢ）	S	MR	S

[a] C，中国致病型（Chinese pathotype）；P，菲律宾小种（Philippine race）；T，日本小种（Japanese race）。

[b] R，抗（Resistant）；RM，中抗（Moderately resistant）；MS，中感（Moderately Susceptible）；S，感（Susceptible）。

Xa25 t 陈惠兰等（Chen，Wang et al，2002）报道了明恢 63 在全生育期高抗菲律宾小种 9（PXO399）。她们以珍汕 97（感）与明恢 63（抗）杂交，用菲律宾小种 9 分析其后代的抗性遗传，观察到其 F_1 抗病，F_2 群体（1 187 株）呈现 3 抗∶1 感的分离比率，表明明恢 63 对菲律宾小种 9 的抗性受一对显性基因控制。她们又对该组合由单粒传产生的 241 个重组自交系进行了 RFLP 分析，在所构建的全长 1 790cM、覆盖有 221 基因座的遗传图谱中，将该抗病基因定位在第 12 染色体上的两个 RFLP 分子标记 R887 和 G1314 之间，与两者的图距分别为 2.5cM 和 7.3cM。由于至今在第 12 染色体上还未发现有白叶枯病抗性基因，该全生育期显性抗性基因应该是一个新基因，暂命名为 *Xa25t*。它与来自明恢 63 的体细胞无性突变系 HX3 所携带的新显性抗病基因重名（见 4.2.3）。

Xa26t 明恢 63 除了携有上述已被定位于第 12 染色体的 *Xa25 t* 基因外，据杨之新、王石平等（2003）报道，它还携有一个在苗期和成株期都高抗 JL691 的基因。为了确定这两

个基因是否相同，他们用JL691在成株期接种由珍汕97（感）/明恢63（抗）衍生的抗病品系R202与珍汕97杂交的500株F_2群体，其抗感分离为372∶128，呈现3∶1的比例（χ^2=0.096，*P*值>0.5），表明这是一个不同于*Xa25t*的显性基因，暂命名为*Xa26 t*。将该基因定位在水稻第11号染色体上的RFLP分子标记RM224和Y6855R之间，与这两个标记分别相距0.21 cM和1.47 cM。随后该基因由孙新立等克隆（Sun，Wang et al.，2004）。

4.2.5　*xa26 t*、*Xa27 t* 和 *xa28 t* 的鉴定　Lee和Khush等（2003）用菲律宾小种1～6评价了IRRI种质中心的21个品种。这些品种对6个菲律宾小种中的小种4、5和6的抗性反应型相同，因此根据其对小种1和小种2的不同反应型划分为两个品种群——抗小种1和小种5的品种为一群，感小种1、抗小种2和小种5的品种为另一群，表14列出从两个群中各选用5个代表品种的抗性反应。将所评价的21个品种都与TN1杂交，并分别与已知其抗性基因的品种进行等位性测验。经遗传分析后确定了各品种所携有的抗性基因，发现有3个品种携有新基因：①越南品种Nep Bha Bong To对菲律宾小种5的抗性和对小种1、2和3的中等抗性由一对隐性成株抗性基因控制，该基因暂定名为*xa26 t*，与来自明恢63且已被克隆的*Xa26*的编号相同；②感菲律宾小种1的孟加拉品种Arai Raj对小种2和5的成株抗性由一对显性基因控制，该基因与已知仅抗小种2的显性基因*Xa10*不等位，暂定名为*Xa27t*，但与近期已克隆的来自小粒野生稻的分蘖盛期显性抗病基因*Xa27 t*重名（见4.2.6）；③感菲律宾小种1，也被划分在第2品种群的孟加拉品种Lota Sail虽然也在成株期抗菲律宾小种2，但是其抗性是由一对隐性基因控制的，应不同于抗菲律宾小种2的Arai Raj的显性基因，暂定名为*xa28 t*。

表14　水稻品种对菲律宾6个*Xoo*小种的抗性反应

［据Lee等（2003）的资料整理］

Table 14　Reactions of rice cultivars to the 6 Philippine races of *Xoo*

（Summarized from Lee et al.，2003）

品种 Cultivar	来源 Origin	小种1 Race 1	小种2 Race 2	小种3 Race 3	小种4 Race 4	小种5 Race 5	小种6 Race 6
群1（Group 1）							
WC1263	印度（India）	R[a]	S	S	S	R	S
Karia	斐济（Fiji）	R	R	R	S	R	S
Latu	印度尼西亚（Indonesia）	R	R	R	S	R	S
Ba Kieu	越南（Vietnam）	MR	MR	MR	S	R	S
Nep Bha Bong To	越南	MR	MR	MR	S	R	S
群2（Group 2）							
Arai Raj	孟加拉（Bangladesh）	S	R	S	S	R	S
Lota Sail	孟加拉	S	R	S	S	R	S
Khar Mao	孟加拉	S	R	S	S	R	S
Ganga Sagar	孟加拉	S	R	S	S	R	S
BhuaBalam	孟加拉	S	R	S	S	R	S

[a] R，抗（Resistant）；M，中抗（Moderately resistant）；S，感（Susceptible）。

4.2.6 *Xa21*、*Xa23*、*Xa27 t* 和 *Xa29 t* 的鉴定 野生稻中蕴藏着一些珍贵的抗病基因，截至 2004 年已鉴定出以下 4 个抗白叶枯病基因。

Xa21 Khush 等（1990）曾详细叙述它的由来。1977 年 Khush 访问印度库塔克（Cuttack）中央水稻研究所时，该所植物病理学家 S. Devadath 博士曾介绍其保存的一个源自西非马里的长药野生稻（*O. longistaminata*）能抗印度当时所有的白叶枯病小种，并已提供印度的几个研究所使用。1978 年，印度 Rajendra 农业大学植物育种学家 Chaudhary 博士将该材料馈赠给 Khush，后者便在 IRRI 用它的纯系进行了白叶枯病抗性评价和研究，发现该材料在水稻分蘖期能抗当时菲律宾白叶枯病全部 1～6 六个小种。经杂交导入 IR24 后，通过回交、自交获得了 BC_4F_2 群体。分析其中的两个 F_2 群体对菲律宾小种 1、2、4 和 6 的抗性遗传，表明是由一对显性基因控制的，并与已鉴定的 17 个抗性基因不同，遂命名其为 *Xa21*。后来 IRRI 又用全部 10 个菲律宾白叶枯病小种鉴定，*Xa21* 除感小种 10 以外抗其他 9 个小种，并且也抗日本小种Ⅰ、Ⅱ、Ⅲ，以及中国 7 个致病型中的 6 个（感Ⅴ型）（Zhang et al.，2002），是一个抗谱广的基因。不久便迅速广泛地用于国内外抗白叶枯病育种和抗性机理研究中。Ikeda 等（1991）首先将其定位于第 11 染色体，与 *Xa10* 紧密连锁。Ronald 等（1992）通过染色体步移技术确定 *Xa21* 的遗传物理图谱。Song 等（1995）采用图位克隆技术时运用同源序列法的策略加速了该基因的分离进程，*Xa21* 成为第一个从野生稻中被克隆出来的抗白叶枯病基因。Song 等（1997）和 Wang 等（1998）进一步揭示了 *Xa21* 基因的分子结构和 *Xa21* 基因家族成员及其抗性功能。但在 20～21 世纪之交前后，韩国、印度、菲律宾和我国云南、广东等地都发现有侵染 *Xa21* 的毒性菌株（小种），值得注意（Goel et al.，1998；Lee et al.，1999；曾列先等，2002；姬广海等，2003）。

Xa23 其基因供体为广西壮族自治区农业科学院野生稻资源圃中编号为 RBB16 的普通野生稻（*Oryza rufipogon*）种质。1987 年开始，中国农业科学院与广西农业科学院合作，用中国致病型Ⅱ（HB17）、日本小种Ⅲ和菲律宾小种 1 三个白叶枯菌群的代表菌株鉴定 871 份野生稻种质，筛选出 61 份高抗材料，又用能侵染所有栽培稻抗性基因的菲律宾小种 6（PXO99）进行重复鉴定，发现其中的 RBB16 在全生育期高抗小种 6 而引起了注意。将 RBB16 与感病籼稻品种金刚 30（JG30）杂交，其 F_1 植株出现野/栽杂种特有的杂合抗性，选择高抗植株进行花培，获得纯合双单倍体抗病系 H_4。再以 JG30 为轮回亲本与 H_4 杂交、回交和自交，同步进行抗性鉴定和农艺性状选择至 BC_5F_4，育成了携有该抗病基因的近等基因系 CBB23。分析金刚 30/CBB23 的 F_2 群体对菲律宾小种 6 的抗性遗传，表明是由一个显性基因控制其全生育期抗性（章琦等，1994）。2000 年又育成了以具有恢复基因的籼稻品种 IR24 为轮回亲本的第二个近等基因系 CBB23（B）（Zhang et al.，2002）。CBB23 在全育期高抗 10 个菲律宾小种（1～10）、7 个中国致病型（Ⅰ～Ⅶ）、3 个日本主要小种（Ⅰ～Ⅲ）和 8 个韩国菌株共 28 个小种（致病型）的代表菌株（章琦等，2004，图 2，另见彩版），而 *Xa21* 则不抗其中的菲律宾小种 10。应该说，CBB23 所携有的基因是迄今已定名的抗性基因中抗谱最广、抗性最强的一个。已命名的其他全生育期抗病的显性基因（除 *Xa21* 外）都不抗菲律宾小种 6。抗性遗传研究表明，CBB23 的抗性基因与包括 *Xa21* 在内的所有已命名的显性基因都不等位。采用 *Xa21* 基因的专化引物对 CBB23 进行序列靶位点测试（Sequence Target Sites Test，STS），确认了 CBB23 所携带的是一个与 *Xa21* 不同的新基因，将该基因定位于第 11 染色体上，与 RFLP 分子标记 RM206 和 C1003A 的图距分别为 1.9 cM 和 0.4

cM（Zhang，Zhao et al.，1998；樊颖伦等，2006）。2001 年，携有 *Xa23* 基因的粳稻品系中野 5114 种植在安徽省卢江县白叶枯病重病田时，表现高度抗病，与毗邻种植的感病品种有明显区别（章琦等，2004；图 3）。2001 年由国际水稻新基因命名委员会正式注册为 *Xa23*（Yoshimura et al.，2001）。

图 2　*Xa23* 对 28 个国内外白叶枯病小种代表菌株的抗性反应（章琦等，2004）

Fig. 2　Reactions of *Xa23* resistance to 28 representative strains of *Xoo* races from Philippines, Japan，Korea，and China（Zhang et al.，2004）

图 3　携有 *Xa23* 的品系中野 5114 在安徽省疫区庐江县及其感病亲本（章琦等，2004；王守海摄于 2001 年）

Fig. 3　The line Zhongye 5114 carrying *Xa23* and its susceptible parent（JG 30）grown in Anhui Province in 2001（Zhang. et al.，1998；Photo from Wang，2001）

Xa27t　IRRI 的 Amante-Bordeos 等（1992）通过种间杂交、胚拯救和回交等手段，将小粒野生稻（*O. minuta*，BBCC，Acc. 101141）的白叶枯病抗性基因导入栽培稻 IR31917 -

45-3-2并选育出78-1-5 BC_2F_3株系，该株系抗菲律宾小种2、3、4、5和6。此后，新加坡国立大学 Temasek 生命科学实验室与美国 Ohio 州立大学植物病理系合作，以78-1-5品系为供体，IR24为轮回亲本，育成了近等基因系 IRBB27。IRBB27 的抗谱较广，在分蘖盛期高抗菲律宾小种 2、3、5 和 6，中抗/中感小种 4，高抗中国致病型Ⅰ、Ⅲ、Ⅴ、Ⅵ、Ⅶ，其抗谱与抗菲律宾小种 6 的 *Xa21* 和 *Xa23* 不相同。此外，IRBB27 还高抗菌株 T7174（日本小种Ⅰ）、A3842、A3857（印度）、IXO56（印度尼西亚）、NXO260（尼泊尔）、R-7（泰国），以及 Aust-2131、Aust-R3（澳大利亚）等菌株，但高感中国Ⅳ型菌（ZHE173），中感中国Ⅱ型菌（HB17）以及韩国、泰国、非洲等一些菌株。遗传分析 IRBB27 对菲律宾小种 6 的抗性由一个不完全显性的新基因控制，暂命名为 *Xa27t*。该基因被定位于第 6 染色体的两个 RFLP 标记 M964 和 M1197 之间，与这两个标记的遗传距离均为 0.052cM；随后，该研究组进一步分离克隆了 *Xa27 t*（Gu et al.，2004；2005）。

Xa29t　谭光轩、何光存等（2004）从栽培稻与药用野生稻（*O. officinalis*）杂交后代渗入系中选育出抗菲律宾小种 1 的水稻株系 B5，将 B5 与籼稻品种明恢 63 杂交，通过单粒传（single seed descent）方法获得了一个含有 187 个稳定纯合的重组自交系（recombinant inbred lines，RILs）群体。根据这个群体在抽穗期的白叶枯病抗性鉴定和遗传分析结果，抗病株系和感病株系之比符合 1∶1（$\chi^2=0.68$，$P<0.30$），表明 B5 对菲律宾小种 1 的抗性受控于 1 对显性基因。随后采用分离集团分析法（BSA）检测与抗病基因相连锁的 RFLP 标记，根据分析所构建的分子标记连锁图谱和 RILs 群体的抗病性鉴定资料，将 B5 携有的抗白叶枯病基因定位在第 1 染色体短臂上的 C904 和 R596 两个分子标记之间，这两个标记的遗传距离为 1.3cM。该基因对 RILs 群体抗病性变异的贡献率为 52.96%，是一个效应较大的主效基因。由于其基因座不同于已报道的所有抗白叶枯病基因，是一个新的显性成株抗性基因，暂命名为 *Xa29t*。

综上所述，截至 2006 年已先后鉴定出 30 个抗水稻白叶枯病主基因，包括 21 个显性基因和 9 个隐性基因，其中有 12 个全生育期（包括 2 个分蘖盛期表达的抗性基因）和 18 个成株期抗性基因。有 18 个基因已定位在第 1（*Xa29 t*），第 4（*Xa1*、*Xa2*、*Xa12*、*Xa14*），第 5（*xa5*），第 6（*Xa7*、*Xa27 t*），第 7（*xa8*），第 8（*xa13*），第 11（*Xa3*、*Xa4*、*Xa10*、*Xa21*、*Xa22 t*、*Xa23*、*Xa26 t*）和第 12（*Xa25 t*）染色体上。目前在第 2、3、9 和 10 染色体上还没有发现白叶枯病抗性基因。已在国际水稻新基因命名委员会正式注册命名的有 20 个。已经克隆了 *Xa1*、*xa5*、*xa13*、*Xa21*、*Xa26 t* 和 *Xa27 t* 六个基因。在这 30 个基因中，有 2 个基因暂定名为 *Xa25 t*：一个来自明恢 63，全生育期抗菲律宾小种 9，在冠以 *Xa* 的显性抗病基因中，只有这一个定位于第 12 染色体；另一个是源于明恢 63 的体细胞突变系 HX3 的成株抗性基因，尚未见被定位的报道，应该是两个不同的抗性基因。有 2 个基因暂名为 *Xa27t*：一个来自小粒野生稻的衍生品系 IRBB27，在分蘖期开始抗菲律宾小种 2、3、5 和 6，定位于第 6 染色体；另一个来自孟加拉国品种 Arai Raj，抗菲律宾小种 2 和 5。还有 1 个来自越南品种 Nep Bha Bong To 的隐性基因 *xa26 t* 和一个来自明恢 63 的显性基因 *Xa26t*。上述同名重号的抗病基因都需要重新注册排名。同时经 IRRI 和日本联合鉴定后删除了与 *Xa3* 相同的 3 个基因，即 $Xa4^b$、*Xa6* 和 *xa9*。应该说，通过国际协作统一采用国际鉴别系统进行抗性鉴定分析是科学、可信的，因为它所反映的寄主与病原菌互作、对应关系比较清晰。现将这 30 个基因的基本情况和特点综合于表 15，以便查考。从总体上看，目前

表 15 迄今已鉴定的水稻白叶枯病抗性基因(章琦整理,2006)

Table 15 Summary of resistance genes to *Xoo* and their main resistance preformance in rice identified up to 2006(Summarized by Zhang,2006)

基因 Gene identified	抗性类型 Resistance type	供体品种 Donor cultivar	所抗的小种或菌株[b] Resistant to *Xoo* race/strain tested	染色体 Chromosome location	标记与图距 Marker and distance	参考文献 Reference
Xa1[a] (1984)	全生育期	Kogyoku,Java14	X-17			Sakaguchi,1967
			T7174(1),印度尼西亚小种 V(Xo7306)	4	1.5 cM $U08_{750}$(RAPD), 0 cM to XNpb235 (RFLP)	Yoshimura et al.,1995;1992
Xa2[a] (1984)	全生育期	Rantai-Emas 2,Tetep	X-17,X-14			Sakaguchi,1967
			TⅠ,TⅡ;Ⅳ(Xo7306)、Ⅴ(Xo7306)	4	XNpb264、$YO3_{700}$	Yoshimura et al.,1995
Xa3 (1984)*Xa-w*	成株	Wase Aikoku 3	TⅠ,TⅡ,TⅢ,印度尼西亚小种Ⅴ			Ezuka et al.,1975
			(Xo7306);P 1,P2,P3,P5	11	2.3 cM to XNbp181	Yoshimura et al.,1992;1995
Xa4 (同*Xa3*)		Semora Mangga	P1			Librogo et al.,1976
Xa6 (同*Xa3*)		Zenith	P1			Sidhu et al.,1978
xa9 (同*Xa3*)		Sateng	P1			Singh et al.,1983
Xa4 (*Xa4*[a])(1984)	全生育期	Sigadis,TKM-6,IR20	P1			Sidhu et al.,1978
			P1,P5,T 1~T7;CⅠ,CⅡ,CⅢ,CⅣ,CⅥ,CⅦ	11	1.7 cM to XNpb78 XNpb181	(Yoshimura et al.,1995)
xa5[a] (1984)	全生育期	DZ192	P1			Petpisit et al.,1977
		IR1545-339,DV85,BJ1	P1,P2,P3,P5TⅠ~TⅦ;CⅠ~CⅦ	5	Close to XNpb396 &XNpb208 <1.0 cM to RG 207	Yoshimura et al.,1992b Mc Couch et al.,1991 Iyer et al.,2004
Xa7 (1984)	成株	DV85,DV86,DZ78	P1,P2,P3,P5;TⅠ~TⅦ;CⅠ~CⅦ	6	6.0 cM to G1091. 3.0 cM to AFLP31-10	Kaji et al., 1995 Borines et al.,2003
xa8 (1984)	成株	PI231128	P1			Sidhu et al.,1978
			P1,MR to P1,P2,P3,P5;TⅠ~TⅦ	7	19.9 cM to RM214(STMS)	Singh et al.,2002

（续）

基因 Gene identified	抗性类型 Resistance type	供体品种 Donor cultivar	所抗的小种或菌株[b] Resistant to *Xoo* race/strain tested	染色体 Chromosome location	标记与图距 Marker and distance	参考文献 Reference
Xa10 (1984)	全生育期	Cas209	P2，P5；TⅤ	11	5.3 cM to $O07_{2000}$ (RAPD)	Yoshimure et al.，1983 Yoshimure et al.，1995
Xa11 (*Xa-pt*) (1984)	成株	IR8 944-102-2-3	TⅠ，TⅡ，TⅢA，TⅣ，TⅤ	?		Ogawa et al.，1986
Xa12 (*Xa-kg*) (1984)	成株	Kogyoku，Java 14	Xo-7306 Indonesian strain (Jap. raceⅤ)	4		Ogawa et al.，1978
xa13[a] (1988)	成株	BJ1、Chinsurah Boro Ⅱ， Aus247、AC19-1-1， Kalimekri 77-5	Philippine races 1，2，4，6	8	5.1 cM to RZ28，3.8cM to G136	Ogawa et al.，1987 Zhang et al.，1996
Xa14 (1988)	全生育期	TN1	Philippine races 5			Taura et al.，1987
				4	RG620～G282 (RFLP)	谭震波等，1998
xa15 (1984)	成株	M4：Harebare mutant	TⅠ、TⅡ、TⅢ(H75373)	—		Nakai et al.，1988
Xa16 (1989)	成株	Tetep	Japanese raceⅤ(H8581，H8584)	—		Noda et al.，1989
Xa17 (1989)	成株	Asominori	Japanese raceⅡ(H 8313)	—		Ogawa et al.，1989
Xa18 (1989)	成株	IR24，Miyang23， Toyonishiki	Burmese strain BM8417、BM8429	—		Yamamoto et al.，1990
xa19 (1990)	成株	IR24 的诱变体 XM5 (mutant of IR24)	P 1～6	—		Taura et al.，1991b
xa20 (1990)	成株	IR24 的诱变体 XM6 (mutant of IR24)	P 1～6	—		Taura et al.，1992b
Xa21[a]	全生育期 (分蘖后期开始)	长药野生稻 (*O. longistaminata*) (IRBB21)	P1～9；CⅠ，CⅡ，CⅢ，CⅣ，CⅥ，CⅦ；TⅠ～Ⅲ	11	0 cM to RG103， <1.0 cM to PTA 248.	Ikeda et al.，1991 Ronald et al.，1992 Chunwongse et al.，1993
Xa22t	成株	扎昌龙(Zhachanglong)	P1，P4～P6；CⅠ，CⅡ，CⅦ；JL691 和 云南若干菌株；TⅠ～Ⅲ；	11	7.1 cM from CR 543， 10.7 cM to RZ536	Lin et al.，1996

（续）

基因 Gene identified	抗性类型 Resistance type	供体品种 Donor cultivar	所抗的小种或菌株[b] Resistant to *Xoo* race/strain tested	染色体 Chromosome location	标记与图距 Marker and distance	参考文献 Reference
Xa23 (2001)	全生育期	普通野生稻 *O. rufipogon* (CBB23)	P1～10;CⅠ～Ⅶ;TⅠ～Ⅲ; 8 Korean strains	11	5.4cM to PSR6,1.9cM to RM206,0.4 cM to C1003 A	Zhang et al.,. 1998;樊颖伦等,2006
xa24 t	全生育期	DV85,DV86,Aus295	P1,P2,P4,P6	—		Khush et al.,2003
Xa25 t	全生育期	明恢 63	P9 JL691	12	2.5 cM to NBSv09、 7.3 cM to G1314	Chen and Wang et al.,2002
Xa25 t	成株	明恢 63 体细胞无性系突变体 HX3(Somaclonal mutant of Minghuei 63,HX3)	CⅣ(ZHE 173)	—		Gao et al.,2001
Xa26 t [a]	全生育期	明恢 63	JL 691	11	0.21 cM to RM224.	Yang et al.,2003
					1.47 cM to Y6855R (RFLP)	Sun et al.,2004
xa26 t	成株	Nep Bha Bong To	P 5,MR to P1～3	—		Lee,Khush et al.,2003
Xa27 [a]	全生育期 (在分蘖盛期开始)	小粒野生稻衍生系 [Line (IRBB27) derived from *O. minuta*]	P2,P3,P5,P6;CⅠ,CⅢ,CⅤ,CⅥ,CⅦ; some strains from Japan,India,& Indonesia	6	0 cM to C964	Amanta-Bordeos et al.,1992 Gu et al.,2004;2005
Xa27 t	成株	Arai Raj	P2,P5	—		Lee et al.,2003
xa28 t	成株	Lota sail	P2,P5	—		Lee et al.,2003
Xa29 t	成株	药用野生稻 [*O. officinalis* (B5)]	P1	1	1.3 cM C904－R596 (RFLP)	Tan et al.,2004

资料来源(Source of data):*Xa1* ～*Xa21* mainly from Ogawa,1993,and the remaining from Rice Genetics Newsletter Vol. 1,5,6,7,18～21。

[a] 已克隆(Cloned)。

[b] C,中国致病型(Chinese pathotype);P,菲律宾小种(Philippine race);T,日本小种(Japanese race)Ⅰ,Ⅱ,Ⅲ。

抗谱广的显性基因有 *Xa23*（来自 *O. rufipogon*）、*Xa21*（来自 *O. longistaminata*）、*Xa7*（IRBB7）、*Xa27t*（来自 *O. minuta*）和 *Xa22 t*（扎昌龙）；抗谱较广的隐性基因有 *xa5*（IRBB5）、*xa13*（IRBB13）和 *xa24 t*。但上文已指出 *Xa21* 在一些国家和我国广东、云南省已受到毒性菌株的侵染，利用时应予以注意。

5 传统的抗性遗传研究和基因鉴定的方法

我国水稻白叶枯病抗性鉴定与遗传研究曾经历过一段零星、分散和方法不统一的过程。1991—1994 年，中国农业科学院、南京农业大学和华中农业大学在国家自然科学基金资助下，共同制定了统一的研究方案，提出合作鉴定与分析国内品种包括地方品种和野生稻资源对白叶枯病的抗性及其相关基因，以及探讨进一步完善抗性遗传研究的方法（章琦等，1996）。鉴于研究方法的重要性，特予以扼要阐述。

5.1 试验材料的确定

5.1.1 病菌的选择 正式命名的白叶枯病抗性基因，都是按照 IRRI 和 TARC 根据鉴别寄主和病原菌小种相应原则合作建立的国际鉴别系统进行鉴定的。因此，今后仍要沿用这套国际鉴别小种的代表菌株（当然需要不断地调整和补充）来测定和比较未知抗源所携带的抗病基因，以及与国际已命名的基因及其代表品种在抗性反应和抗谱上的异同。

在分析抗病性的遗传方式和测试新老抗病基因的等位性时，均应采用与参试基因相应的专化小种。用那些能侵袭几乎所有已知抗性基因的广谱毒性和毒性谱宽窄不一但专化性强的小种进行检测，最有可能在特异反应组合中发现新抗源。应用国外菌株测试时，务必在严密的隔离条件下进行，避免造成扩散和污染。

鉴于我国采用的白叶枯病菌的毒性多数呈现连续性变异，与鉴别品种之间的互作关系不像国际鉴别系统那样清晰。为了适应不同生态区生产应用的需要，可考虑采用：①本地区或不同生态区的优势致病型，例如云南地区新的流行菌群（小种）；②引起生产上大面积种植的抗病品种丧失其抗性的毒性菌株，如能侵袭广泛分布于我国品种中的抗病基因 *Xa4* 的中国Ⅴ型菌，其代表菌株 GD1358 就颇具挑战性；③近期发生频率上升很快的菌群。

5.1.2 感病品种/亲本标准系 理想的感病品种/亲本的基本条件是：①在全生育期对绝大多数小种（致病型）高度感病，对所采用的菌群代表菌株不具有抗性功能；②农艺性状和适应性相对较好，生育期适中；③纯合的品种。鉴于自交作物品种群体内经常混有遗传异质性的个体，诸如天然杂交、突变和来源不明的微小变异的发生，世代材料过早混合以及扩大繁殖时操作不当等，都足以造成品种在遗传上的不纯。作为遗传研究用的对照感病的品种或亲本，应采取自交一至二代并严格去杂或提纯选择，以提高品种群体的同质性和个体的纯合性，育成标准系。这是必须经常做的工作。

此外，品种抗性遗传行为除受制于所用的小种（致病型）之外，很大程度上还受到水稻生态类型的影响，其中以籼稻与粳稻配制的组合对全生育期抗性之遗传行为的干扰最为明显。因此，标准系要籼、粳稻品种兼备，以便分析、比较亚种内或亚种间抗/感杂交组合的遗传行为。

5.1.3 已知抗性基因的品种 待测品种需要同抗性基因已知的品种进行基因等位性测试方

能辨明其异同。国内外构建的携有单基因抗性、遗传背景相同的近等基因系是进行等位性测试最理想的工具。目前，有些基因还没有育成近等基因系，则可用抗性基因的原始供体品种。但在选用之前，应充分了解这些携有已知基因代表品种（系）的抗谱（抗哪些小种）、抗性生育期类型（全生育期或成株期）、抗性遗传方式（显隐性、基因对数、互作）等基础资料，以确定其是否适用。

5.2 抗性鉴定

影响抗性鉴定结果准确性的主要因素有：接种物的纯度、生活力及菌株数，参试品种材料的纯度或真实度及生育期，以及外界环境条件（温度、湿度、光照等）及其寄主/病菌的相互作用。调查标准也会直接或间接影响所得数据的可信度。

5.2.1 病菌的提纯复壮 自然寄生于水稻上的白叶枯病菌群体，除其主体菌落外，常常混有不同的或有潜在毒性的群落。开展抗性鉴定和遗传分析时所用的菌株应是经过单胞分离、提纯、有正常活力和基本同质纯合的群体。每次使用前应在鉴别品种上接种，检验其毒力及专化反应的稳定性。因此，选用的菌株必须经常进行更新复壮。

5.2.2 接种菌株数的确定 选用什么菌株和菌株数是根据待测品种的抗谱、抗病专化性和地区抗病性鉴定目标等来确定的。

(1) 多个菌株 水稻白叶枯病抗性通常由一对或一对以上的基因控制一个或几个病菌小种。鉴定可能携有一对基因以上抗谱较广的抗性品种时，要采用若干个相对应小种的代表菌株同时进行接种，才可能弄清楚该品种所携有的各个抗性基因与不同病菌小种（毒性基因）的互作关系，从而分辨出相应的抗病基因。例如，BJ 1 抗性群品种全生育期抗 6 个菲律宾小种（1～6），在前人试验中已知其抗性属隐性遗传，而当时已经鉴定出的隐性抗病基因只有 *xa5* 一个，因此待测材料必须首先与 *xa5* 进行等位性测试，并且应该采用与 *xa5* 相对应的菲律宾小种 1、2 和 4 进行检测；小种 6 是用来检测 BJ 1 群品种是否含有新基因的特异菌株，因此，这需要采用小种 1、2、4 和 6 四个代表菌株进行接种分析。Ogawa 等（1987）用这 4 个小种鉴定出 BJ 1 品种群含有 *xa5* 和 *xa13* 两个隐性抗病基因就是个典型的例子。

(2) 单个菌株 采用优势小种的单个菌株分析鉴定抗病基因的例子很多，如 *Xa25t*、*Xa22t* 等基因就是分别用中国的优势菌 ZHE173 和湖北菌株江陵 691 鉴定出来的。另一种是采用特定的专化菌株进行分析，如 Lee 等（1999）在基本了解待测品种的抗性的基础上，只用对某一特定基因型具有专化性反应的菌株（如菲律宾小种 6）鉴定出 DV85 含有的新基因为 *xa24t* ；又如，Zhang 等（1998）根据当时已知的显性抗病基因中只有 *Xa21* 抗菲律宾小种 6，在测试 *Xa23* 与 *Xa21* 的等位性关系时，只需采用菲律宾小种 6，或采用唯一对 *Xa23* 和 *Xa21* 具有特异性的菲律宾小种 10 也可行。

5.2.3 接种时期 水稻对白叶枯病的抗性表达，从生育期的角度可分为全生育期抗性和成株期抗性两类。全生育期抗性的品种应在苗期和孕穗期分两次接种，对生育期不一致的分离群体，则需要在孕穗期接种两次，使生长发育滞后的植株有同等机会接受侵染。成株期抗性的品种应在孕穗期接种，如为分离群体，则根据植株陆续进入孕穗期的具体情况也需要接种1～2 次。

5.2.4 抗性调查 接种与调查的间隔时间要因时因地制宜，多在感病对照品种充分发病、病情发展相对稳定时进行。一般在接种后 14～21d 调查。每个植株被接种的各个菌株各调查

3 张叶片，以病斑占叶片面积的百分率或病斑长度（cm）作为抗感反应值，统计时取其平均值。在对由寡基因控制的分离世代群体进行遗传分析时，一般以抗、感分布曲线所形成的峰谷为界限来划分抗性类型。抗感分界的数值可参照亲本的表现来决定，一般以病斑面积小于叶面积的 20%～25%或病斑长度小于 5cm 时为抗病。

5.3 抗性遗传研究

5.3.1 配组方式 如不涉及细胞质基因，正反交的遗传效应应该相同，以何者为母本或父本都可以，可视开花情况而定。但是由于抗病供体品种多来自境外，综合性状水平一般较差，甚至存在严重缺点，习惯上以抗病供体为父本，以具有良好农艺性状的感病亲本为母本，因为多数抗病基因为显性，便于区别真伪杂种；换言之，一般用隐性基因型的品种或个体做母本比较相宜。BC_1F_1的配制可用以验证 F_1配子的分离比，回交时，如抗病亲本的抗性为显性，用 F_1与感病亲本回交时，以后者做母本更好；如抗性为隐性，则用 F_1与抗病亲本回交，正反交均可。F_3是按 F_2单株的种子分别种植为株系或家系（要求其群体大小统一），这样可以检验上下代、亲子间抗感分离比的对应关系和比较株（家）系间的抗性表现或分离的异同。

5.3.2 各世代的群体容量（大小） 两个亲本品种及 F_1群体，可酌情接种 10～20 株。对分离世代材料，要有足够大的群体，在供体品种抗性基因对数不清楚的情况下，F_2分离世代应考虑种植足以分析具两对基因独立遗传的分离群体，建议种 300 株以上，若为连锁遗传，群体应尽量增大些。BC_1F_1一般为 100～150 株；F_3家系数约占 F_2群体的 1/3 左右，每家系约种植 25～30 株。注意观察家系间和家系内的分离情况及纯合度，联系到 BC_1F_1及亲本世代的表现进行统计分析，就可以得出准确、可信的遗传模式、行为和结论。

供遗传分析的全套材料，包括双亲、F_1、BC_1F_1、自交 F_2和随机取样种植的 F_3家系共 5 个世代，除每年按世代推移进行种植并接种外，还要尽可能准备好各代备用种子，以便最后一年做到“四世同堂”，在一块地上同时接种、观察、记载，以避免因年份不同（气候及其他环境条件）而造成的一系列干扰或影响。因此要有计划地分期配置组合，分别繁殖后代和安全保存种子。

5.3.3 遗传分析 根据 F_1和亲本品种的抗感反应，一般可以对被测品种抗性的显隐性作出判断。根据 F_2和 BC_1F_1群体各个植株的抗感反应型及其频率分布，再以 F_3各家系的抗感表现及其纯合或杂合分离的情况与频率分布作为主要依据，可对控制各被测品种抗性的基因数、分离比例、基因效应、基因互作等进行系统概括，确定其遗传模式是一对基因还是两对（或两对以上）基因遗传，是独立分离或连锁遗传等，并进行有关参数估算和统计检测。此外，还要注意分析有无细胞质遗传的情况。

5.3.4 传统的抗性基因鉴定 上述抗性鉴定和抗性遗传研究都是为抗性基因鉴定做准备的。其实，抗性遗传分析基本上已为抗性基因鉴定提供了许多基础信息，抗性基因与染色体的关系可通过连锁分析来确定，需要逐个基因分析鉴定，其所采用的方法不一，留在下面叙述。抗性基因是新的还是已知的基因，需要通过基因等位性测验来确认。如待测品种只携有一对显性抗病基因，在进行基因等位性测定时可将被测品种与已知携有一对显性基因的代表品种杂交，并采用与已知显性抗病基因相应的菌株进行接种。如果 F_1为抗病，F_2群体也全部抗病，没有出现感病植株，可以推断两者的抗病基因是等位关系；若 F_1虽为抗病，而 F_2群体有抗感植株分离，则表明彼此不等位，即待测品种携有与已知显性抗病品种不同的抗病基因。如待测抗病品种为隐性遗传时，应与已知的隐性抗病基因的代表品种杂交，并采用与已知隐性抗病基因相应的菌株接种，如 F_1 为感病，表明两者不等位；如 F_1 抗病，则有可能携有相同的隐性抗病基因，须继续分析其 F_2 群体的抗性是否有分离发生，如果 F_2 群体全部抗病，则可确认这两个（对）基因是等位的。当待测品种携有两对或两对以上的抗病基因时，需经基因分离、重组与纯合后，取其在抗性遗传上表现为不同类型的代表家系，分别进行等位性测试，才能准确判定各个（对）基因的等位关系和遗传效应。

5.4 抗性基因定位

给基因定位，传统的做法有三种，即标记基因法、相互易位系法和初级三体法。随着分子标记技术的迅速发展和日臻完善，应用分子标记进行基因定位已渐普及，因为它比较直观、精确，可以一步到位，不仅能给主效基因定位，还可通过数量抗性位点（QRL）分析给微效多基因体系中遗传效应较大的 QRL 定位，这是传统方法无法做到的。此法虽然对技术条件的要求较高，费用也较高，但无须事先备有成套适用的标记基因系、相互易位系和初级三体系，而后三者的制备和使用也是很费时费力的。这里仅对三个传统方法作一简述。

5.4.1 标记基因法 标记基因（marker gene）是指控制独特的植株形态特征或生理特点并容易在表型上加以识别的主效基因。由这种基因控制的性状在遗传学上称为标记性状。遗传研究中所采用的标记性状，一般都是由显隐性分界鲜明的、由单个主效基因控制的性状，携有标记性状的品种（系）叫标记基因系，与标记性状有关的主基因所属的连锁群应是已知的。利用标记基因系与待测抗病基因供体杂交，在其早期世代分析标记性状与待测性状（抗病性）之间是否存在连锁，就可以对待测抗病基因进行染色体定位。此法要求备有在水稻12个染色体上都分别携有由一些标记基因所构成的一套标记基因系列，这些标记基因较均匀地分布在染色体上，并且对生长发育和受精结实没有不良影响。20世纪60年代，日本的Takahashi（1964）和印度的 Misro 等（1966）分别确立了粳稻和籼稻的12个连锁群。其后，经过诸多研究者的工作积累，这两个连锁图已渐趋完善，粳籼之间与各自标记基因相对应的连锁群也基本一致，可以利用标记基因法对抗病基因进行染色体定位。余功新等（1990）研究了 IR28 所携带的分别对湖北菌株江陵 691 和 OS－75 表现为抗病的基因 *Xa－a* 和 *Xa－h* 与一些标记性状的遗传关系，结果表明这两个基因与标记基因 *la*（散生，lazy growth habit）的交换值分别为 27.9%和 33.7%，按现今国际统一的染色体命名方案，应位于第 11 号染色体。*Xa－a* 和 *Xa－h* 的交换值为 17.0%，推测三个基因的线性排列是：

$$\underset{33.7\%}{\underline{la\ \overline{27.9\%}\ Xa-a\ \overline{17.0\%}\ Xa-h}}$$

5.4.2 相互易位系（reciprocal translocation line）**法** 相互易位是指非同源染色体之间部分区段相互交换而引起的染色体结构变异。带有这种染色体结构变异的品系称为相互易位系。纯合的相互易位系在外观上与正常品系基本一样，减数分裂时染色体配对正常，育性也正常。用它与正常品系杂交后，F_1 植株在减数分裂中期Ⅰ会形成一个环状或链状四价体。这个四价体在赤道板上排列有相间式（1，3）或相邻式（1，2 和 4，1）三种取向；向两极分离后，相间式的染色体组成完整，产生正常可育的配子，而相邻式的染色体组成则存在部分重复与缺失，形成无活力的配子，因而会出现半不育的植株。在一般情况下，相互易位杂合体自交后（即 F_2代）可育株与半不育株的比例为 1∶1。根据这些特点，利用相互易位系与待定位基因的供体杂交后，调查其 F_2群体中半不育性状和待测性状（即抗病性）的分离情况，并辅以与半不育性相关联的染色体构型检测，就可以推断相互易位系的易位点（即半不育性）和抗病基因是否存在连锁关系，从而对抗病基因进行染色体定位。此法要求有一套或一些已知其参与易位的相应染色体编号的相互易位系。在明确抗病基因位于哪个染色体之后，还要通过标记基因法或初级三体法进一步确定该抗病基因是在哪一号染色体上。这个方法不能一步到位，而且已知的标记基因并不多，在实际应用时受到较大限制。上文曾提到 Nishimura（1961）首先用此法将黄金丸和 Pi1 品种的抗白叶枯病显性基因定位在他编号为第 11 的染色体上（后更名为第 4 染色体）；Sakaguchi（1967）也用同样方法将 *Xa1*（与 Nishimura 1961 年报道的显性基因相似）和 *Xa2* 定位于第 4 染色体上。

5.4.3　初级三体（primary trisomics）法　初级三体是指二倍体物种的基因组中某一对同源染色体多了一条而其他染色体数目不变的一种非整倍体变异类型。在减数分裂时这个三体染色体形成一个二价体和一个单价体或一个三价体，然后分向两极，其中的单价体常因行动滞后而丢失，而多了一条染色体的雄配子往往没有生活力，不能够传宗继代。同时，三体上的基因型还有三显（AAA）、双显（AAa）、单显（Aaa）和无显（aaa）之分，在按染色体随机分离的情况下，单显式三体的雌配子频率为 1A∶2a∶2Aa∶1aa ＝ 1A∶1a，雄配子频率为 1A∶2a；双显式三体的雌配子频率为 2A∶1a∶1AA∶2Aa ＝ 5A∶1a，雄配子频率为 2A∶1a，其后代的表型分离比例都与二体的明显不同。因此，利用一套已知其外加染色体编号的初级三体系列（感病）做母本，分别与待定位抗病基因的供体（二体）杂交，F_1代通过细胞学检测筛选出三体植株，分别调查其自交 F_2群体的抗病性分离情况，就可以对该抗病基因进行染色体定位。20 世纪 70 年代，国际上利用初级三体法进行水稻白叶枯病抗性基因定位的共有 *Xa1*、*Xa2*、*Xa3*、*xa5*、*Xa10* 和 *Xa12* 六个，前 5 个基因也已采用分子标记法进行了定位，并获得与初级三体法相同的结果。

6　水稻白叶枯病抗性近等基因系的构建

作物的不同品种因为遗传背景不同彼此之间无法精确比较某个性状或基因在不同基因型间的差异或表现。近等基因系（Near-isogenic lines，NIL）是指仅某个性状（在此为抗病性）或基因的基因型不同而其遗传背景则近于相同的一系列品系，用它们来研究或检测基因效应的异同就可以克服上述干扰，也适于用作鉴定病菌小种异同的鉴别品种。近 30 年来国内外学者都在利用对重要病、虫害具单基因抗性的近等基因系来鉴定新的抗性基因和检测新的病（虫）小种（生物型）。由于近等基因系的遗传学特点，采用分子标记检测时，只要在目标基因所在区域显示不同，就可以通过与目标基因紧密连锁的 DNA 分子标记对目标基因进行快速定位。因此，近等基因系又是开展基因的分子定位、克隆和抗病性、致病性机理研究的重要工具。通常近等基因系的株型都已经过遗传改良，所以也是抗病育种的优良抗性供体。

6.1　国际合作构建的近等基因系

日本和 IRRI 各有一套水稻白叶枯病鉴别品种。建立一套国际鉴别小种就需要统一的鉴别品种。1982—1987 年间，日本 TARC 和 IRRI 合作培育了三套 IR 稻白叶枯病抗性近等基因系，其后建立了包括鉴别品种和鉴别菌株（小种）在内的国际水稻白叶枯病鉴别系统（Ogawa et al.，1991）。这三套近等基因系分别具有 IR24（籼稻）、密阳 23（Milyang 23）（籼/粳交育成）和丰锦（Toyonishiki）（粳稻）的遗传背景，共携带 10 个主效抗性基因。

构建的主要程序如下：首先选择轮回亲本和目标抗病基因。根据两套病菌小种接种检测结果，确定对日本和菲律宾小种都感病的 IR24、密阳 23 和丰锦 3 个不同类型品种作为轮回亲本。采用当时已鉴定的 10 个抗病基因及其供体亲本作为待测品种，即 *Xa1*（Kogyoku）、*Xa2*（Tetep）、*Xa3*（Chugoku 45、Java14 和 Zenith）、*Xa4*（IR20）、*xa5*（IR1545）、*Xa7*（DV85）、*xa8*（PI231129）、*Xa10*（Cas 209）、*Xa11*（IR 8）和 *Xa14*（TN1）。然后进行转育和命名。将上述 10 个抗病基因的代表品种分别与 3 个轮回亲本杂交、回交，在各世代接种鉴定，每个组合的各世代至少选择 7～14 个抗病且性状近似轮回亲本的植株进行回交，回交 4 次后自交 4～6 次至 $BC_4F_{4\text{-}6}$。此时入选后代的农艺性状已基本等同于轮回亲本而携带的抗病基因则各异，即育成了近等基因系列。

以 IR24 为轮回亲本的一套近等基因系，命名时在 IR-BB 之后连接基因号码表示之。如携有基因 *Xa1* 者命名为 IR-BB1，依次为 IR-BB2、IR-BB3、IR-BB4、IR-BB5、IR-BB7、IR-BB8、IR-BB10、IR-BB11、IR-BB14 以及后来培育的 IR-BB21。以密阳 23 和 Toyonishiki（丰锦）为轮回亲本的两套 NIL 分别为 IR-BB101～IR-BB111 和 IR-BB201～IR-BB211（Dgaua et al.，1991）。

6.2 中国构建的粳稻近等基因系

作者 20 世纪 80 年代在 IRRI 进行合作研究时，曾多次用菲律宾 *Xoo* 鉴别小种的代表菌株测定上述三套近等基因系的抗性反应。以 IR24 为遗传背景的一套 NILs 各基因之专化抗性清晰、稳定，农艺性状一致，其他两套 NILs 的专化抗性不够清晰，多数近等基因系的抗性反应呈中抗偏感，不太稳定，甚至还保留着抗性供体的某些表型性状，未达到纯合一致，不好利用（Zhang et al.，1990）。因此，在 1986—1996 年间，章琦等参考 Ogawa 的方法，在轮回亲本选择和目标基因转入后的验证方面作了一些改进，育成了一套具粳稻遗传背景的近等基因系。其选育步骤如下：

（1）轮回亲本的选择 作为选育抗病 NILs 的轮回亲本，关键是感病性要典型。由于粳稻品种对白叶枯病菌一般都具有由主效或主效加微效基因控制的成株期抗性，较难选出对国内外主要鉴别菌株都高度感病的品种。经大量筛选后，确定全生育期对绝大多数菌株高度感病、株型较好的粳稻品种沈农 1033 作为轮回亲本，避免了由于感病亲本自身具有某种程度的“抗性”而干扰目标抗病基因的导入。

（2）确定目标基因 在当时国际命名的 12 个抗病基因中，排除了对中国白叶枯病致病型没有抗性功能的显性基因 *Xa1*、*Xa10* 和 *Xa11* 以及当时育种中不常用的隐性基因 *xa5* 和 *xa8*，选择了对我国水稻生产上有利用价值的 6 个抗性基因：*Xa2*（兼抗稻瘟病）、*Xa3*、*Xa4*（我国主体抗源）、*Xa7*、*Xa12*（抗谱较广）和 *Xa14*（对中国菌株具有鉴别性能）。它们的代表品种是：Tetep（*Xa2*）、Wase Aikoku 3（*Xa3*）、IR20（*Xa4*）、DV85（*Xa7*）、Java14（*Xa12*）和 TN1（*Xa14*）。

(3) 转育 将6个抗病基因的代表品种分别与轮回亲本沈农1033杂交、回交，采用与目标基因相对应的鉴别菌株接种鉴定。根据二项式分布原理（莫惠栋，1979）按通常认为的置信概率估算每次回交时的最小群体，即在各世代选择不少于7～14株高度抗病且农艺性状逐渐类似轮回亲本的植株进行回交直至$BC_{5-6}F_1$，然后自交至$BC_{5-6}F_{5-4}$，育成了一套各个目标基因对相应小种的专化抗性清晰、稳定，农艺性状与轮回亲本基本一致的NILs。转育过程见图4。

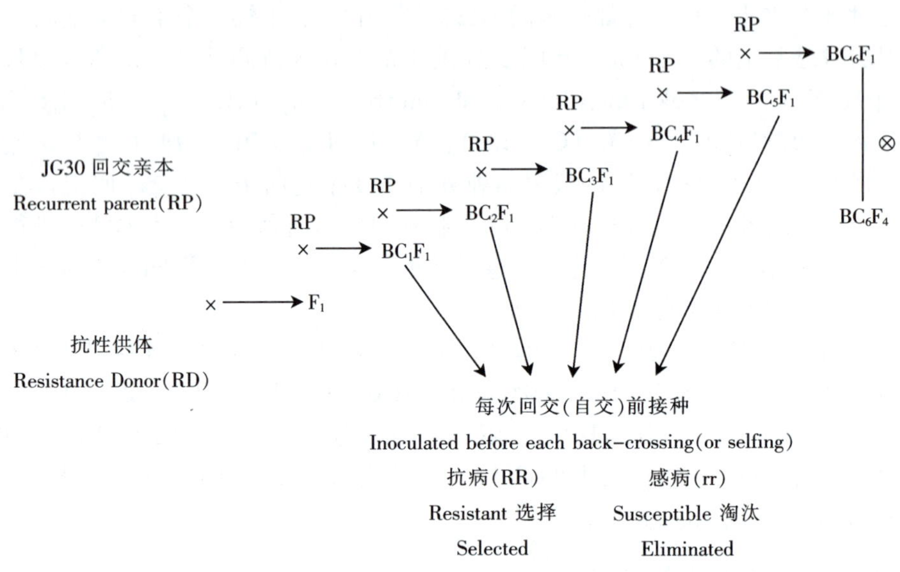

图4 水稻白叶枯病抗性近等基因系选育程序［根据Ogawa（1991）资料增补］

Fig. 4 Breeding procedure of rice near-isogenic lines for resistance to bacterial blight (Supplemented from Ogawa, 1991)

(4) 验证 这套以CBB命名的NILs经抗性遗传分析并与供体基因进行等位性测定后表明，CBB2、CBB3、CBB4、CBB7、CBB12和CBB14各携有一对抗性基因，分别与其供体的基因是等位的；但是当采用与*Xa7*基因相对应的菌株P1测定CBB7与IRBB7（携有*Xa7*）的等位性时，两者却不等位，表明CBB7携有的抗性基因与*Xa7*基因不同。证实这个结论的进一步依据有：①Lu（陆朝福，1998）将CBB7所携有的抗性基因定位于水稻第11染色体，而Kaji等（1995）将*Xa7*定位于第6染色体（表15）；②Kansas州立大学Leach植物病理实验室的测试结果是，含有对*Xa7*基因无毒力的*Xoo*菌无毒基因*avrXa7*的两个菌株能侵染CBB7（感）。这些结果表明CBB7所携有的这个显性成株抗性基因是来自DV85的另外一对未知基因，该基因的抗谱很广，至2003年仍没有发现可用于进行基因等位性测验的特异性菌株，值得继续寻找适合的鉴别菌株，揭示这个未知基因（章琦等，1996；1998；个人通讯）。

鉴于这套粳稻近等基因系中的多数基因是由籼稻导入粳稻背景的，目标基因是否会随着多次回交、继代选择而丢失或削弱，目标基因的特异抗性是否稳定、清晰，其农艺性状是否一致，需要通过病理、遗传、育种等方面的检测，我们经连续3年采用与目标基因相对应的鉴别菌株，在不同生育期测定和比较这6个近等基因系及其供体的抗性之后，结果有5个近等基因系的抗性与其供体是完全一致的（表 16），其特异抗性稳定、清晰，而 CBB4 虽也显示出清晰的特异抗性，却在分蘖期才表达，比其全生育期抗性基因 *Xa4* 的供体 IR20 滞后一段时间。

表 16 中国粳稻近等基因系与其抗性供体对白叶枯病抗性反应的 3 年比较[a]（章琦等，1998）

Table 16 Three year comparison of resistance reaction to bacterial blight between Chinese japonica rice near-isogenic lines and their donors in 1994—1996（Zhang et al.，1998）

测试材料 Material	年份 Year tested							
	1994			1995		1996		
	所用菌株							
	T2、P1 或 P5			T2、P1 或 P5		T2、P1 或 P5		
	TS[b]	BS	TS	BS		TS	BS	
轮回亲本 RP								
沈农 1033（Shennong 1033）	S[c]	S		S	S		S	S
供体（Donor）								
Tetep（*Xa 2*）	S	R		S	R		S	R
DV85（*Xa 7*，*xa5*）	R	R		R	R		R	R
TN1（*Xa14*）	R	R	R	R		R	R	
近等基因系（Near isogenic line）								
CBB2（*Xa 2*）	S	R		S	R		S	R
CBB7（*Xa*？未知）	S	R		S	R		S	R
CBB14（*Xa 14*）	R	R	R	R		R	R	

a 携有 *Xa2*、*Xa7* 和 *Xa14* 的试材分别用菌株 T2（日本小种Ⅱ）、P（菲律宾小种）1 和小种 5 接种［NILs carrying *Xa2*，*Xa7* and *Xa14* were inoculated with strains T2（Japanese raceⅡ）P1，P5（Philippine races 1 & 5），respectively］。RP，轮回亲本（Recurrent parent）。

b 接种时的生育期（Inoculation at growth stage）；TS，分蘖期（Tillering stage）；BS，孕穗期（Booting stage）。

c R，抗（Resistant）；S，感（Susceptible）。

连续 3 年田间和室内考察了这 6 个近等基因系与轮回亲本沈农 1033 的主要农艺性状，两者之间差异在统计学上均不显著，说明它们的遗传背景基本一致（表 17），田间农艺性状表现也极为一致（图 5，见彩版 9）。这套 NILs 已在国内育种及相关研究中得到有效利用。例如，浙江省农业科学院利用 CBB7 育成了 3 个全生育期抗性、抗谱广的高产晚粳品系 D601、D602 和 D603，已成为周边地区育种单位的优良抗源（徐建龙等，1996）；中国农业科学院作物所花培组用 CBB4 育成了抗病、高产粳稻品种中花 14 和中花 15。中国科学院遗传所朱立煌、陆朝福进行的分子检测表明，这套 NILs 基因组只在抗病基因区域有差异，是水稻白叶枯病抗性基因定位、克隆和抗性机理研究的理想材料。他们已在克隆 CBB7 的抗性基因的进程中获得了两个候选基因（Lu et al.，1998），值得进一步进行基因鉴定并了解其功能属性。

除构建了以沈农 1033 为轮回亲本的（粳稻）6 个抗白叶枯病近等基因系外，1993—1996 年，Zhang 等（2002）用高感籼稻品种金刚 30 为轮回亲本，育成了一对以广谱抗性基

因 *Xa23* 为供体的籼稻近等基因系 CBB23；1998—2001 年出于改良杂交稻白叶枯病抗性的需要，又育成了另一对以携有恢复基因的 IR24 籼稻品种为遗传背景的 *Xa23* 近等基因系 CBB23（B）（章琦等，2004）。这些基础材料的创制已经有了良好的开端，如果能够尽早将这三套（籼稻、粳稻、杂交稻的恢复基因）白叶枯病抗性近等基因系列"备齐"，并随着鉴别系统的演变发展而补充更新，则必将有力地推动我国水稻白叶枯病抗性改良及相关研究的发展。

图 5　中国粳稻白叶枯近等基因系（Zhang et al.，1996）

A. 6 个 CBB 系统的粳稻白叶枯近等基因系的田间表现　B. 该套近等基因系的轮回亲本（沈农 1033）及其中两个品系 CBB2 和 CBB3

Fig. 5　The Chinese near-isogenic japonica rice lines for BB resistance（Zhang et al.，1996）.

Performance of six near-isogenic japonica rice lines for BB resistance in the field. The recurrent parent（SN1033）of the NILs，and 2 out of the six lines，CBB2 and CBB3

表 17 中国抗白叶枯病粳稻近等基因系的农艺性状表现

(1994—1996 年)(章琦等，1998)

Table 17 Performance of agronomic characters in Chinese japonica rice near-isogenic lines resistant to BB in 1994—1996 (Zhang et al.，1998)

轮回亲本 近等基因系 Recurrent parent NIL	播种至抽穗期 /天 Seeding to heading (day)	株高/cm Plant height (cm)	有效分蘖数 Effective tillers/plant	穗长/cm Panicle length (cm)	结实率/% Grain setting (%)	千粒重/g 1000 - grain weight (g)
沈农 1033 (Shennong1033)	122	83.9	9.8	15.8	89.1	22.5
CBB2	124	81.6	10.8	16.2	89.5	23.1
CBB7	124	83.0	10.9	15.6	87.6	22.7
CBB14	127	85.4	11.5	15.6	87.0	21.4
CBB3	126	84.3	9.8	16.4	89.2	23.6
CBB4	119	85.7	10.1	16.4	87.6	21.9
CBB12	128	85.4	9.9	15.0	86.4	22.4
平均值 $\bar{x}$	124	84.2±2.6	10.5±2.2	15.9±0.7	87.9±1.2	22.2±1.6

水稻抗白叶枯病常规育种

章　琦　阙更生

（中国农业科学院作物科学研究所）

Traditional Rice Breeding for Resistance to Bacterial Blight

摘要　1923年日本开始水稻抗白叶枯病育种。经历了质量抗性育种和质量抗性与数量抗性并举两个阶段。1930—1972期间育成全胜、农林27、黄玉等28个携有*Xa1*基因的粳稻品种。其抗性都来自滋贺关取11、高农35和庄兵卫3个抗源品种。60年代以早生爱国3号、中新120等携有*Xa3*的品种为抗源，育成70X-46、中国45（Chugoku 45）、BL1等品种。70年代以后，采用质量抗性结合数量抗性育种，以Lead rice、R28、Pelita I-1等作为供体育成70X-11、70X-37、70X-42、日本晴、阿苏稔等兼备质量和数量抗性的品种，这些品种高抗日本小种Ⅰ和Ⅴ，中抗小种Ⅱ、Ⅲ和Ⅳ，有效地控制了白叶枯病的危害。日本科学家根据白叶枯病菌小种群的变化，按品种对日本小种群的抗性反应型，划分为金南风、黄玉、Rantai Emas、早生爱国3号和爪哇14五个以至更多的抗性品种群，为制定不同时期的育种计划和品种推广提供重要的依据。

中国水稻抗白叶枯病育种始于20世纪30年代，丁颖主持育成了抗白叶枯病品种竹印2号。60年代末开始全国性抗病育种。农业部先后设置了抗稻瘟、白叶枯病协作研究和“全国水稻高产、优质、抗病育种攻关”项目，抗白叶枯病育种研究工作全面开展，进展很大。

籼稻年推广面积在6 667hm²（10万亩）以上的品种在1960—1970年间177个，17个抗白叶枯病，占9.0%；1971—1980年间296个，53个抗病，占18.5%；1981—1986年间87个，32个抗病，占36.8%。在南方稻区大面积种植的主要高产、优质抗病的籼稻品种有双竹占、鄂早6号、二九丰、特青、湘早籼3号、扬稻2号等。截至2005年，通过国家审定的34个籼稻中的12个（35.3%）抗白叶枯病，如扬稻6号、嘉育948以及抗3种病虫害的特汕占25。

粳稻年推广面积在6 667hm²（10万亩）以上的品种在1960—1970年间123个，25个抗白叶枯病，占20.3%；1971—1980年间182个，52个抗病，占28.6%；1981—1986年间46个，13个抗病，占28.3%。在南方稻区大面积种植的主要抗病粳稻品种有农垦58、农垦57及其衍生品种农虎、鄂宜、鄂晚、矮粳、城特和扬粳等品种。北方稻区有引自日本的品种秋光、丰锦和黎明等以及我国育成的中系、中作、中花系列、冀粳 8 号、新稻 68－11 等。截至 2006 年，通过国家审定的 6 个南方粳稻中，武运粳 7 号、浙农大等 4 个品种抗病；64 个北方粳稻中，中作 93、辽粳 294 和新稻 10 号等 5 个品种抗白叶枯病。

在籼型杂交稻中，80 年代中期前育成的 87 个组合中的 35.6%中抗白叶枯病。年最高推广面积 133.33 万 hm^2（2 000 万亩）以上的有威优 64、威优 6 号和汕优 6 号。截至 2005 年，通过国家审定的高产、优质双抗白叶枯和稻瘟病、大面积推广种植的组合有Ⅱ优 084、汕优 77、国丰 1 号和特优多系 1 号等。

粳型杂交稻中，东北稻区的 5 个组合中有 3 个中抗白叶枯病，其中黎优 57 累计种植上千万亩。21 世纪审定了一批南方稻区的抗白叶枯病并兼抗其他病（虫）的粳型杂交稻组合，如甬优系统和 69 优 8 号、嘉乐优 2 号、榆杂 2 号和常优 2 号等。

2000 年后通过了大批省级审定的高产、优质组合抗多种病（虫）的两系杂交稻。如抗白叶枯病且兼抗稻瘟病、细菌性条斑病或褐飞虱的组合培杂山青、培杂茂选、培两优特青、培两优余红和云光 8 号（粳）等。在 2006 年前通过国家审定的 21 个组合中，两优培九、培杂双七、70 优 9 号（粳）等 6 个组合都双抗白叶枯和稻瘟病。优质、高产籼型组合两优培九在南方 16 个省（区、市）在 2005 年累计种植了 468.1 万 hm^2，占国内两系杂交稻总面积的 50%。

中国水稻抗白叶枯病育种的主要问题是利用的抗性基因源单一，籼稻和粳稻分别以 *Xa4* 和 *Xa3* 为主。20 世纪 90 年代以后已开始利用 *Xa7*、*xa5*、*Xa21*、*Xa23* 等广抗谱基因，已采用常规结合分子标记辅助选择及转基因技术培育抗病基因聚合系。

IRRI 在 20 世纪 60 年代末提出由多学科参加的“水稻遗传评价与利用研究”，并组建了国际水稻遗传评价协作网，实现了国际间水稻种质自由交换、综合评价及快速利用，开辟了“应用科学的探索之路”，推动了全球水稻丰产、广适性、多抗性育种取得卓著成就。该所从 IR77 之后的品种，便均由应用所在国命名，其品种遂在全球稻作国得到更充分的利用。

20 世纪 60～80 年代，IRRI 育成 IR20、IR22 至 IR66 等 27 个携有 *Xa4* 基因品种（除 IR56 外）。随后扩大利用 *Xa3*、*xa5*、*xa5*＋*Xa7* 和 *xa8* 等基因。IR36、IR38、IR42 和 IR50 等品种具有高产、优质、广适应性以及抗白叶枯、稻瘟、通戈洛、草丛矮缩病和包括褐飞虱、叶蝉、稻螟、稻摇蚊等 6～8 种病虫。到 90 年代末，IRRI 采用由生物学和传统育种深入到分子标记辅助选择培育多个抗性基因聚合系的策略，用 *Xa4*、*xa5*、*Xa7*、*Xa10*、*xa13* 和 *Xa21* 等基因累积一批携有 2～5 个基因并表现高度抗病的聚合系（IRBB50～IRBB66）。

最后就如何实现抗白叶枯病基因拓宽、合理利用、基因聚合、基因的轮换以及“水平抗性”的利用等策略进行了粗略的讨论和商榷。

Abstract Japan commenced rice breeding for bacterial blight (BB) resistance in 1923. The first phase of breeding program was limited to qualitative resistance while both qualitative and quantitative resistances were included in the second phase. During 1930－1972, 28 BB resistant japonica cultivars carrying the *Xa1* gene were obtained, including Zensho 26, Norin 27, and Kogyuku. Their resistance was derived from 3 donors: Shiga-sekitori 11, Kono 35, and Shobei. In the 1960s, cultivars carrying *Xa3*, such as 70X－46, Chugoku 45, and BL1 were bred, and their resistance was derived from Wase Aikoku and Nakashin 120. Since 1970s, both qualitative and quantitative resistances were adopted in the breeding strategy to generate cultivars such as 70X－11, 70X－42, Nipponbare, and Asominori. Their resistance was mainly derived from the cultivar "Lead rice", IR28, and Pelita I－1. These new cultivars exhibit high resistance to *Xoo* Japanese races I and V and moderate resistance to races II, III, and IV. The BB disease in Japan has been controlled effectively since then. Based on the reaction patterns to Japanese *Xoo* races I to V, rice cultivars were classified into at least five resistance groups, including Kinmaze, Kogyuku, Rantai Emas, Wase Aikoku, and Java 14. This categorization provided scientific bases for designing breeding program in different periods and formulating strategies for popularization.

Rice breeding program for BB resistance was initiated in China in the 1930s. Ding Ying's group successfully generated the first BB resistant cultivar in China named "Zhuyin No. 2" . Such breeding program has been launched at the national level since the end of 1960s. The Ministry of Agriculture (PRC) established prioritized research programs such as "Coordinated Research for Fungal Blast and BB Disease Resistance" and "Rice Breeding for High Yield, Good Quality, and Disease Resistance" . The rice BB resistance breeding researches in China have been rapidly and extensively progressing since then.

There were 177, 296, and 87 cultivars of indica rice having an annual acreage of more than 6, 667 hm^2 in the period 1960－1970, 1971－1980, and 1981－1986, respectively. Among those, 17 (9.6%) out of the 177, 53 (18.5%) out of 296 cultivars, and 32 (36.8%) out of 87 cultivars were resistant to BB. Several cultivars exhibiting high yield, good quality, and BB resistance were popularized extensively in South China, including Shuangzhuzhan, Ezao 6, Erjioufeng, Teqing, Xiangzaoxian 3, and Yangdao 2. Up to 2005, 12 (35.3%) out of 34 indica cultivars passed the national (PRC) certification requirements were resistant to BB. Some examples including Yangdao 6, Jiayu 948, and Teyouzhan 25 exhibit multiple resistances to 3 diseases and pests.

There were 123, 182, and 46 cultivars of japonica rice having an annual acreage of more than 6, 667 hm^2 in the period of 1960－1970, 1971－1980, and 1981－1986, respectively. Among those, 25 (20.3%) out of 123, 52 (28.6%) out of 182 and 13 (28.3%) out of 46 cultivars were resistant to BB. The major BB resistant japonica cultivars popularized in large areas in South China included Nongken 58, Nongken 57 and their derivatives such as Nonghu, Eyi, Ewan, Aigeng, Chengte, Yanggeng, etc. The major BB resistant japonica cultivars popularized in North China included Akihikari, Toyonishiki, and Reimei introduced from Japan and Zhongxi, Zhongzuo, Zhonghua series, Jigeng 8, and Xindao 68 - 11 developed locally. Up to 2006, 4 out of 6 Southern japonica cultivars and 5 out of 64 Northern japonica cultivars which had passed the national (PRC) certification requirements were resistant to BB. Some examples include Wuyugeng 7 and Zhenonda 454 (Southern cultivars), and Zhongzuo 93, Liaogeng 294, and Xindao 10 (Northern cultivars).

Before the 1980s, 31 (35.6%) out of 87 indica three-line hybrid rice combinations developed were moderately resistant to BB. The combinations composed of Weiyou 64, Weiyou 6, and Shanyou 6 had reached an annual acreage of 133.3 万 hm^2. Up to 2005, some combinations exhibiting high yield, good quality and resistance to both BB and fungal blast had passed the national (PRC) certification requirements. Combinations composed of Ⅱ you 084, Shanyou 77, Guofeng 1, and Teyouduoxi 1 have accumulated a large acreage.

Before the 1980s, 3 out of 5 japonica three-line hybrid rice combinations cultivated in North China were moderately resistant to BB. One of them, Liyou 57 has an accumulated acreage exceeding 65.7 万 hm^2. In 21st century, three-line hybrid rice combinations exhibiting BB resistance and multiple resistance to other pests had passed the national (PRC) certification requirements, including combinations composed of Yongyou 4, 69 you 8, Jialeyou 2, and Jleyou 2.

For two-line hybrid rice, combinations exhibiting high yield, good quality and multiple resistance to BB and blast or rice bacterial leaf streak or brown planthopper had passed the national (PRC) certification requirements after year 2000. Some examples included combinations composed of Peiza-shanqing, Peiza-maoxuan, Peiliangyou-teqing, and Yunguang 8 (japonica) . Up to 2006, 6 out of 21 combinations with high yielding and good quality that passed the national (PRC) certification requirements were resistant to both BB and fungal blast. Some examples included combinations composed of Liangyou- peijiou, Peiza-shuangqi, and 70 you 9 (japonica). Up to 2005, the accumulated acreage of Liangyou-peijiou had reached 4.681 万 hm^2, representing 50% of the total acreage of two-line hybrid rice in China.

The gene *Xa4* was mainly used in both the indica conventional cultivars and hybrid rice, while *Xa3* was used in the japonica cultivars. By the end of 1990s, the broad-spectrum resistance genes, *Xa23*, *Xa21*, *Xa7*, and *xa5* have been widely adopted to overcome the dependence of limited genetic resources. A combined strategy composed of traditional method, molecular marker-assisted selection, and transgenic technology should bring a new era to the BB resistance rice breeding program in China.

In the late 1960s, IRRI advocated the idea of a multidisiplinary genetic evaluation and utilization (GEU) program and actualized this through establishing the International Network of Genetic Evaluation for Rice (INGER) in 1973 to facilitate free exchange, integrated evaluation, and rapid utilization of rice germplasms. Such "adventure in applied science" led by IRRI boosted the rice breeding programs worldwide and attained important accomplishments in high-yield, wide adaptability, and multiple disease and insect resistances. After the generation of IR77, IRRI announced to pass the nomenclature rights to the countries adopting IRRI's new rice cultivars. This move broadened the applications of the IRRI lines worldwide.

From the 1960s to 1980s, IRRI successfully bred IR20, IR22 to IR66, a total of 27 cultivars carrying the *Xa4* gene (except IR56). Subsequently, extended use of *Xa3*, *xa5*, *xa5*＋*Xa7*, and *xa8* were practised. Cultivars developed by IRRI (such as IR36, IR38, IR40, IR42, and IR50) were high yield, of high grain quality, and exhibiting wide adaptability and resistance to BB, fungal blast, tungro, grassy stunt, and to 6 to 8 insects including brown planthopper, green leafhopper, rice stem borer, and gall midge. By the end of 1990s, IRRI's combined approach using biotechnology and conventional breeding has extended to molecular marker-assisted selection and transgenic strategies. The genes *Xa4*, *xa5*, *Xa7*, *Xa10*, *xa13*, and *Xa21* were selected to develop pyramiding lines (from IRBB50 to IRBB66) carrying 2 to 5 genes each, which showed high resistance level to BB evidently.

Finally, this paper also briefly discussed and deliberated on issues related to the broadening of BB resistance, controlled and suitable utilization of broad-spectrum resistance genes, development of pyramiding lines carrying multiple resistance genes, refined strategies for alternation planting of available resistance genes, and adoption of "horizontal resistance".

利用品种的抗病性防治农作物病害是最经济、最简便而又立竿见影的技术措施，受到国内外广泛的重视。据估计，农作物病害中有80％以上要靠或主要靠寄主本身的抗性来控制。在过去的一个世纪里，应用这种遗传改良的方法治理农作物病害有了很大的发展，它的理论基础、技术策略、工作方法以及目标要求都在不断变革、改进和完善之中。这是作物育种、遗传、病理、生理、生态、栽培以及分子生物学等诸多学科知识积累、相互沟通、深化应用的结果。本文仅就我国主要水稻抗白叶枯病常规育种工作进展情况以及日本（最早开展这方面工作）和位于菲律宾的国际水稻研究所（IRRI，当前的国际水稻育种中心）的育种工作中值得借鉴的经验予以概述。

1 中国水稻抗白叶枯病常规育种

中国水稻育种工作始于20世纪20年代，当时是调查、收集地方品种，通过品种观察、鉴定试验，挑选其中表现优良的品种就地推广种植，同时在优良品种中开展大批量穗选，进行纯系育种；30年代初开始试行品种间杂交育种，后来抗日战争爆发，半壁江山相继沦陷，只能退缩到西南后方几省在十分艰难的条件下维持工作。50年代中期以后水稻育种工作有了很大发展，其进程大体上可分为3个阶段：①60～70年代，矮化育种取得显著成效，育成和推广了一大批矮秆高产品种，实现了半矮秆品种的普及，使产量大幅度提高。②70年代中期实现杂交水稻三系配套的突破以后，育成和推广了诸如南优、汕优、威优系统以及两优培九和两优特青等一大批三系、两系高产优良组合，使我国水稻杂种优势的生产利用一直居国际领先地位。现在我国杂交稻的种植面积已超过常规品种的面积。③90年代以后，品种产量潜力的提高已进入杂种优势利用与理想株型相结合的超高产（或超级稻）育种阶段。据中国水稻研究所程式华等（2004）报道，目前小面积单产记录已超过12t/hm²，说明水稻的增产潜力还很大，大家正在向这一方向继续努力，使其在更大面积上扩展种植。

抗病育种是水稻改良的重要组成部分。由于稻瘟病在中国的发生与危害比白叶枯病更为普遍和严重，大部分抗病育种工作着重在解决稻瘟病问题。相对说来，水稻抗白叶枯病育种及相关研究起步较晚，参加单位较少，投入的力量也较小，真正有针对性地开展水稻抗白叶枯病育种是从20世纪60年代后期才开始的。但是最早涉及水稻抗白叶枯病品种的来源问题可追溯到30年代初。丁颖在主持前中山大学稻作实验场工作时，曾采用一个优质、抗病虫力强的广东地方品种竹占与印度2号杂交，育成了米质好、抗病性强的品种竹印2号（柯苇等，1991），这个品种后来被发现对白叶枯病具有较好的抗性，由它衍生了一批抗白叶枯病品种，一度在生产上种植。为了加强稻种资源抗病材料的评价，1976年农业部设置了全国抗稻瘟、白叶枯病研究协作项目，开展大规模的水稻种质资源抗性鉴定和抗源筛选。80年代初，又在“全国水稻高产、优质、抗病育种攻关”项目中，明确规定对稻瘟病和白叶枯病的抗性是品种改良的主攻目标之一，并由浙江省农业科学院负责组织全国水稻新育成品种（系）抗病虫性的评价，有力地推动了水稻品种对白叶枯病抗性改良工作的开展。1996年后，超高产和优质米成为水稻育种的主攻目标，各稻区近期育成的水稻品种其产量潜力水平不断提高，稻米品质有明显改善，抗病性也有不同程度的增强。

1.1 常规稻抗白叶枯病育种

1.1.1 籼稻 20世纪50年代初，农业部倡导和发动群众评选良种，就地在生产上扩大种植，科研单位也加强系统育种，优中选优、引进境外良种以及开展杂交育种等多种渠道并进，育出一批优良品种，收到了良好效果。如前浙江省稻麦改良场（农业改进所的前身）从宁波晚青中系选出丰产性好、株高适中、病虫害轻、抗逆稳产的浙场9号，1939年开始在浙、赣、闽、湘等省推广种植，至60年代累计推广了40万hm²（600万亩）（柯苇等，1991）。那时对水稻白叶枯病抗性鉴定尚未开始，在育种中着重选留发病轻或较轻的品种在生产上应用。

60～70年代，全国水稻品种基本实现了矮秆化，籼稻地区推广种植的矮脚南特、珍珠矮、广场矮等及其衍生品种虽然高产但都不抗白叶枯病，加上采用密植、增施化肥等集约化栽培措施，反而促使白叶枯病蔓延发展。章琦等（1988）曾对这一时期在生产上种植的32个籼稻品种进行了白叶枯病抗性鉴定，结果有25个品种（占78%）高度感染当时全国6个代表性菌株KS6－6（苏）、KS1－20（苏）、ZHE173（浙）、064－2（粤）、HN－1（豫）和HB29（冀）。这些品种遍布南方各省包括白叶枯病常发地区，在流行年份对生产造成严重危害。

根据《中国水稻品种及其系谱》（林世成，1991）记述及全国农业技术推广服务中心（简称全国农技推广中心）统计资料（1982—2006年），我国在60年代后育成的籼稻品种其白叶枯病抗性有所改进，生产上推广抗病品种的比率有所提高。1960—1970年间育成的品种年推广面积在 0.67 万 hm^2（10 万亩）以上的有 177 个，其中 17 个品种抗/中抗白叶枯病，占 9.6%，表现较为突出的品种如 1966 年育成的双抗白叶枯和稻瘟病的优质米品种双竹占，1975 年在广东推广 2.4 万 hm^2，至 1995 年南方各地累计种植面积 57.9 万 hm^2（869 万亩）。该品种被江西省引进直接在生产上种植后，1979 年由该省邓家埠良种场从中系选出江西丝苗，仍然保持抗白叶枯、稻瘟病和米质优良的特点，在全省种植了 20 万 hm^2（300 万亩），一直到 2001 年还有农民种植。

在 1971—1980 年间育成的籼稻品种年推广面积在 0.67 万 hm^2（10 万亩）以上的有 296 个，其中 53 个品种抗/中抗白叶枯病，占 18.5%，并出现了一批连续种植 10 年以上，累计面积达数百万公顷的抗白叶枯病、高产、优质品种。如抗病品种鄂早 6 号（1983—2000，累计面积 164.3 万 hm^2，2 465 万亩）；中抗白叶枯病的高产品种广二 104（1981—1992，累计 179.7 万 hm^2，2 695 万亩）；中抗白叶枯和稻瘟病的品种青秆黄（1981—1994，99.5 万 hm^2，1 492 万亩）；双抗白叶枯和稻瘟病的优质品种湘早籼 3 号（1985—1992 年，58.1 万 hm^2，871 万亩）。此外，HA70317－7（IR36/广解 8 号）继承了 IR36 的多抗性特点，除了抗白叶枯病外，还兼抗稻瘟病、叶蝉和褐飞虱，在 1986—1987 两年内就种植了 20.3 万 hm^2（305 万亩）。

在 1981—1986 年间育成的籼稻品种其白叶枯病抗性在数量和抗性水平上都有明显提高，年推广面积在 0.67 万 hm^2（10 万亩）以上的有 87 个，中抗—高抗白叶枯病的品种有 32 个，占 36.8%。如高抗品种特青（1987—1995，88.2 万 hm^2，1 323 万亩）；由高抗白叶枯病的品种华竹矮衍生了青华矮 6 号、晚华矮 1 号和青叶青丰等一批高产、优质、高抗的品种，1984—1995 年种植了约 76.9 万 hm^2（1 154 万亩）；在生产上累计种植面积千万亩以上的双抗白叶枯和稻瘟病的主要品种有二九丰（1984—1994，254.5 万 hm^2，3 817 万亩）；以扬稻 2 号为代表的扬稻 1～4 号优质品种（1985—1996，78 万 hm^2，1 170 万亩）；78130（1985—2003，103.3 万 hm^2，1 519 万亩）；79106（1985—2000，69.6 万 hm^2，1 044 万亩）以及三二矮（1986—2004，70.7 万 hm^2，1 060 万亩）。此外，这一期间育成的三抗或兼抗其他病（虫）的品种有南农籼 2 号，高抗白叶枯和稻瘟病、中抗褐飞虱；金陵 57 和南京 14 等抗白叶枯病和褐飞虱；83007 兼抗纹枯病等，详见表 1。这些高产、抗病、稳产品种的推广种植有效地控制了长江流域、华南和西南等南方稻区的白叶枯病危害。

从 20 世纪 90 年代至 2005 年，通过国家级审定的籼稻品种（组合）共 271 个，华南和长江流域稻区有 191 个，其中常规品种 34 个，抗白叶枯病的有 12 个，占 35.3%。这些品种不仅抗病，并且高产或优质或两者兼具，其中种植面积较大、兼抗稻瘟病和褐飞虱的品种有特籼占 25（1997—2005，46.5 万 hm^2，697 万亩），兼抗稻瘟病的品种有嘉育 948（1997—2005，149.7 万 hm^2，2246 万亩）、扬稻 6 号（2000—2005，32.1 万 hm^2，481 万亩）和闽岩糯（1994—2005，19.9 万 hm^2，298 万亩）；在 7 个单抗白叶枯病的品种中，有些很快就在生产中扎根，如粤丰占（2000—2005，29.3 万 hm^2，440 万亩）、丰华占（2002—2005，12.4 万 hm^2，186 万亩）和嘉早 935（1999—2005，16.4 万 hm^2，246 万亩）等；2005 年通过国家审定的高产品种鄂早 18 高抗白叶枯病（1 级），2004 年已种植 4.1 万 hm^2（62 万亩）（杨仕华等，2005；中国稻米，2005；全国农业技术推广服务中心，1982—2006）。

同一时期通过省级审定的部分抗白叶枯病品种中，兼抗稻瘟和纹枯病的，如扬稻 8 号和中早 22；双抗白叶枯和稻瘟病的品种有云超 7 号、丰丝占和德优 11 等；高抗白叶枯病的品种有皖稻 115 和皖稻 117 等，其余品种对白叶枯和稻瘟病都有中度以上的抗性（表 2）。

表 1 1960—1986 年育成大面积种植的主要籼稻抗白叶枯病品种[a]

Table 1 Main Chinese indica rice cultivars with bacterial blight resistance released during 1960 - 1986

品种[b] Cultivar	组合 Cross	育成年份/单位 Year developed & bred by Inst.	抗病(虫)性反应[c] Reaction to					种植面积/公顷 Acreage /hm²	系谱图号[d]及参考文献 Genealogy & reference
			BB	BL	SB	BPH	RB		
浙场 9 号	宁波晚青系选	1939,前浙江省稻麦改良场	R				MR	1964 年,107 000	柯韦等,1991
双竹占	中竹 2915/塘竹 1923	1966,广东省农业科学院	R	R				1975—1995,579 000	林世成等,1991;4 - 6
江西丝苗	双竹占系选	1979,江西邓家埠良种场	MR	MR				1984—2001,133 000	4 - 6
鄂早 6 号	红梅早/IR8//72 - 11/二九矮 7 号	1979,湖北省农业科学院	R					1983—2000,1 643 000	2 - 13
广二 104	广二矮 5 号系选	1976,广东省农业科学院	MR					1981—1992,1 797 000	2 - 10
青秆黄	广陆矮 4 号/龙革 16	1975,浙江省农业科学院	MR					1981—1994,995 000	2 - 9
湘早籼 3 号	IR36/广解 8 号	1979,湖南省农业科学院	R	R				1985—1992,581 000	2 - 16
HA79317	IR36/广解 8 号	1979,湖南省农业科学院	R	R			R(BPH)	1986—1987,203 000	2 - 16
特青	叶青伦/特矮	1984,广东省农业科学院	HR					1987—1995,882 000	2 - 9
青华矮 6 号[b]	晚青/青兰矮 F_4//华竹矮	1981,广东省农业科学院	HR					1984—1988,418 000	章琦等,1990; 伍尚忠等,1990;4 - 6
晚华矮 1 号	青华矮 6 号/青兰矮 32	1983,广东省农业科学院	R					1986—1990,229 000	4 - 6
青叶青丰	青华矮 6 号/青丰占	1986,广东省农业科学院	R	R				1987—1992,123 000	4 - 6
华竹矮[b]	[双华矮/(辐竹二号/ IR20/竹印 2 - C6965)]	70 年代,广东省农业科学院	HR					华南稻区 BB 骨干抗源	伍尚忠等,1990;4 - 6
二九丰[b]	IR29/原丰早	1982,嘉兴市农业科学研究所	R	R				1984—1994,2 545 000	施爱农等,1993;2 - 16
扬稻 2 号[b]	BG90 - 2 系选	1981,江苏里下河农业科学研究所	HR	R				1985—1992,474 000	沙学延等,1994;3 - 10
扬稻 3 号[b]	扬稻 2 号/桂朝 2 号	1985,江苏里下河农业科学研究所	R	R				1987 年,140 000	同上
扬稻 4 号[b]	杨稻 2 号/杰雅辐选	1985,江苏里下河农业科学研究所	R	R				1991—1996,223 000	同上

（续）

品种[b] Cultivar	组合 Cross	育成年份/单位 Year developed & bred by Inst.	抗病(虫)性反应[c] Reaction to BB	 BL	 SB	 BPH	 RB	种植面积/公顷 Acreage /hm^2	系谱图号[d]及 参考文献 Genealogy & reference
扬稻1号[b]	BG90-2系选	1981，江苏里下河农业科学研究所	HR					1986—1987，69 000	同上
78130	汕优2号/V20	1981，福建漳州市农校	R	R				1985—2003，1 030 000	2-12
79106	汕优2号/V20	1982，福建漳州市农校	R	R				1985—2000，696 000	2-12
三二矮	双二占/广二104	1983，广东龙门县农业科学研究所	R	R				1986—2004，707 000	2-10
南农籼2号[b]	IR24/广陵矮4号//BG90-2	1985，南京农业大学	HR	HR			MR(BPH)	1989年，16 000	沙学延等，1994；3-9
金陵57	Co^{60}辐射BG90-2	1985，南京市农业科学研究所	R				R(BPH)	1986—1992，104 000	3-10
南京14	SI-Pi 681032系选	1984，江苏省农业科学院	R				HR(BPH)	1986—1989，35 000	—
四喜占	IET2938/桂朝2号	1983，湖北省农业科学院	R					1988年，413 000	3-6
湘晚籼1号	IR22/红米冬占/广秋矮///ASD7	1982，湖南省农业科学院	R	R				1989—1995，263 000	4-12
83007[b]	矮南早1号/GEB451F$_3$//GE45//广陵矮6号	1981，原华中农学院	HR				MR(SB)		施爱农等，1993；2-3
三黄占[b]	团黄占/外选35	广东省农业科学院	R						施爱农等，1993；

[a] 根据《中国水稻品种及其系谱》和全国农技推广服务中心资料整理(Based on the book *Rice Varieties and Their Genealogy in China* and statistical data from National Agricultural Extention Service Center)。

[b] 经抗性遗传分析的品种(Cultivars genetically analyzed for resistance to BB)。

[c] BB，白叶枯病(Bacterial blight)；BL，稻瘟病(Rice blast)；SB，纹枯病(Sheath blight)；BPH，褐飞虱(Brown planthopper)；RB，稻螟(Rice borer)。R，抗(Resistant)；HR，高抗(Highly resistant)；MR，中抗(Moderately resistant)。

[d] 系谱图号源自《中国水稻品种及其系谱》(Genealogy figure number from the book *Rice Varieties and Their Genealogy in China*)。

表 2 1996—2005 年通过国家、省级审定的部分抗白叶枯病（多抗）优质或高产籼稻品种

Table 2 A part of Chinese indica cultivars with resistance to bacterial blight and other pests, high-yield and good quality released and registered during 1996－2005

品种[#] Cultivar	组合 Cross	抗病（虫）性反应[c] Reaction to					审定年份/单位 Year approved & bred by Inst.	参考文献 Reference
		BB	BL	SB	BPH	WPH		
特籼占 25[#,a,b]	特青 2 号/粳籼 89	MR	R		MR		2001，广东佛山市农业科学研究院	杨仕华等，2005
嘉育 948[#,a,b]	YD4－4/嘉育 293－T8	MR	MR				2001，浙江省嘉兴市农业科学院	同上
扬稻 6 号[#,b]	扬稻 4 号/3021	R	R				2001，江苏省里下河农业科学研究所	同上
闽岩糯[#,b]	闽糯 580/717	R	MR				1997，福建省龙岩地区农业科学研究所	同上
粤丰占[#,b]	粤香占/丰矮占	MR					2003，广东省农业科学院	同上
丰华占[#,b]	丰八占 1 号/华丝占	MR					2003，广东省农业科学院	同上
嘉早 935[#,a,b]	Z91－105///优 905/嘉育 293// Z91－43	MR					2000，浙江省嘉兴市农业科学院	同上
湘晚籼 12[#,a,b]	92W93/GER3	MR					2004，湖南省水稻所	同上
鄂糯 7 号[#,a,b]	R82033－41/BG90－2 突变体	R					1999，湖北省荆州市农业科学研究所	同上
嘉早 312[#,a]	嘉早 935/Z9505//Z96－10	MR					2003，浙江省嘉兴市农业科学院	同上
山溪占 11[#,a,b]	山溪占 175/籼黄占 8	R					1999，广东省佛山市农业科学研究所	同上
鄂早 18[#,a]	中早 81/嘉早 935	HR					2005，湖北省黄冈市农业科学研究所、省种子集团公司	中国稻米，2005
扬稻 8 号[a,b]	^{60}Co－γ射线辐照扬稻 6 号	MR	HR	R			2001，江苏省里下河农业科学研究所	黄年生等，2003
中早 22[a]	Z935/中选 11	R	MR			MR	2004，中国水稻研究所	马良勇等，2006
云超 7 号[a]	八宝米/滇瑞 409//DR138	R	R				2004，云南省农业科学院	李全衡，2005
丰丝占[a,b]	丰八占/珍丝占	R	R				2004，广东省农业科学院	李宏等，2005
矮秀占[b]	丰矮占 1 号//新麻占/七秀占	MR	HR				2003，广东省农业科学院	司徒志谋，2003
中健 2 号[b]	Starbonnet/IR841	R	MR				2003，中国水稻研究所	王健龙等，2003
扬辐籼 6 号[a]	^{60}Co~γ射线辐照扬稻 6 号	MR	R				2004，江苏省里下河地区农业科学研究所	何震天等，2005
中早 16[b]	嘉兴 13/中早 11（IR72/中优早 81）	MR	MR				2002，中国水稻研究所	马良勇等，2003
越糯 6 号[a,b]	越糯 96－10/Z9505	MR	MR				2004，浙江省绍兴市农业科学研究所	刘明放等，2004
临籼 21[a,b]	扰特谱/IR26	MR	MR				2005，云南省临沧地区农业科学研究所	李金衡，2005
浙农 7 号[b]	丝 3 号/浙农 947	MR	MR				2004，浙江大学，绍兴县种子公司	俞琦英，2004
皖稻 115[a]	9311 中选择的自然变异株	HR					2004，安徽农业大学	王辉，2004
皖稻 117[b]	CPS1017/毫格劳//新秀 299	HR					2004，安徽省农业科学院水稻所	王辉，2004
中佳早 6 号[b]	Z99－312/WT4	MR	R				2005，中国水稻研究所	刘鑫等，2006
粤航 1 号[b]	长丝占太空诱变	M	R				2005，广东省农业科学院	周汉钦等，2006
嘉育 21[b]	G96－29（纯）/YD951）	MR	MR				2005，浙江省嘉兴市农业科学院，余姚市种子公司	刘鑫，2006

（续）

品种# Cultivar	组合 Cross	抗病（虫）性反应[c] Reaction to					审定年份/单位 Year approved & bred by Inst.	参考文献 Reference
		BB	BL	SB	BPH	WPH		
德优11[b]	德优3号/德农211	R	R				2005，云南省德宏傣族景颇族自治州农业科学研究所	李金衡，2006
德优12[b]	德优3号/德农211	MR	MR				2005，云南省德宏傣族景颇族自治州农业科学研究所	李金衡，2006

[#] 通过国家审定的品种，其余为省级审定的品种（Cultivars marked with # were registered at national level，and the others at provincial level）。

[a] 高产（Cultivars with high yield）。

[b] 优质（Cultivars with good quality）。

[c] BB，白叶枯病（Bacterial blight）；BL，稻瘟病（Rice blast）；SB，纹枯病（Sheath blight）；BPH，褐飞虱（Brown planthopper）；白背飞虱，（Whitebacked planthopper）；R，抗（Resistant）；MR，中抗（Moderately resistant）；HR，高抗（Highly resistant）。

1.1.2 粳稻 我国粳稻品种中分早、中、晚粳三种生态类型。早粳多分布在北方稻区，中粳分布在苏、沪、浙、皖的中稻区，栽培面积不大。晚粳集中在长江中下游各省。

我国粳稻品种的白叶枯病抗性一般较好。据中国《水稻品种及其系谱》记述和全国农业技术推广服务中心统计资料，年推广面积在 0.67 万 hm^2（10 万亩）以上的品种在 1960—1970 年间有 123 个，其中 25 个抗白叶枯病，占 20.3%；1971—1980 年间的品种有 182 个，其中 52 个抗白叶枯病，占 28.6%；1981—1986 年间的品种有 46 个，其中 13 个抗白叶枯病，占 28.3%。

我国长江中下游的太湖流域稻区种植粳稻历史悠久。1958 年从日本引进的两个粳稻品种农垦 58（世界）和农垦 57（金南风）在增强我国南方稻区粳稻品种的白叶枯病抗性及其改良和提高产量潜力中发挥了重要作用。农垦 58 首先在江苏省试验、示范，该品种在白叶枯病年年严重的吴县的陆墓、长桥、保安等疫区种植，并进行了人工接种白叶枯菌测试，与全省晚粳优良品种相比，以农垦 58 的抗病力最强（王法明等，1964）。农垦 58 除了抗白叶枯病外，还具有矮秆、耐肥和高产等优良性状。在 60 年代中期至 70 年代初期，苏、浙、沪、皖、湘、鄂、川等省（市）累计种植达 946.7 万 hm^2（14 200 万亩），实现了上述地区晚粳品种的第一次大更换。与此同时许多育种家相继在农垦 58 中进行系选，选育出一大批抗白叶枯病的优良品种。

在 1960—1970 年间，用农垦 58 作亲本选育的农虎 6 号（农垦 58/老虎稻）综合了双亲的优良性状，抗白叶枯、稻瘟和小球菌核病，耐肥抗倒，丰产和适应性好，成为农垦 58 的主要接班品种，在长江中下游地区大面积推广，种植长达 15 年之久。据不完全统计，该品种 1970—1985 年间在浙、苏、沪、皖、湘、鄂、蜀、赣 8 个省（市）累计推广达 421.8 万 hm^2（6 191 万亩）。从农垦 58 中还系选出若干双抗白叶枯病和稻瘟病的品种，如高产、优质、广适性的沪选 19 在 1978 年就推广了 29.9 万 hm^2（448 万亩），此后广泛分布在长江流域的 7 省 1 市，北达河南信阳、西南至贵州；双抗白叶枯和稻瘟病的系选品种还有苏粳 2 号（1974—1986，27.5 万 hm^2，413 万亩）和苏州青（1967 年 9.3 万 hm^2，139 万亩）等。

在 1971—1986 年间，用农垦 58 及农虎 6 号系选或杂交育成了以鄂宜、鄂晚、秀水、矮粳、祥湖等编号的大批抗白叶枯病的优良品种，其中累计种植面积 60 万 hm^2（900 万亩）

以上的品种有鄂宜 105（1982—1999，330 万 hm^2，4 962 万亩）、鄂晚 5 号（1982—1999，265.3 万 hm^2，3 980 万亩）、秀水 48（1983—1992，126.1 万 hm^2，1 891 万亩）和矮粳 23（1981—1988，65.3 万 hm^2，998 万亩）（全国农业技术推广服务中心统计资料，1982—2006），这些品种都曾成为长江中下游地区的当家品种。

农垦 57 也是一个高产、优质和适应性好的中粳品种，1964—1995 年间累计种植了 67.1 万 hm^2（1 006 万亩），与此同时还被广泛用作杂交亲本（邹江石等，1991）。农垦 57 并不抗日本当时的 3 个白叶枯病小种Ⅰ～Ⅲ，但从 60 年代末以后，由其衍生的子孙两辈 25 个品种中，却有一批在成株期高抗中国白叶枯病的主要致病型菌株，如南粳 11、南粳 15、南粳 23、南粳 35、盐粳 2 号、盐粳 3 号、盐粳 20、扬粳 201 等，其中南粳编号品种截止 80 年代末的不完全统计，在江苏省累计种植了 67.8 万 hm^2（1 017 万亩）。这些品种虽然直接应用在生产上的面积不很大，但是它们从 70 年代以来却被广泛用作抗白叶枯病的供体，所衍生的高抗白叶枯病品种遍及全国，其中盐粳编号品种在江苏等省累计种植面积约在两千万亩以上（1983—1995，134.1 万 hm^2，2 011 万亩）；扬粳 201 在同一地区 1985—1990 年间也种植到 24.5 万 hm^2，合 368 万亩。章琦等（1988）曾测定包括上述品种和主要抗病供体共 46 个品种（系）对白叶枯病的抗性。结果表明，矮粳 23、浙粳 66、秀水 48、城特 232、农虎 6 号、嘉湖 5 号、邳早 15 等在成株期对我国 6 个白叶枯病代表菌株（KS 6－6、ZHE173、064－2、KS1－20、HN01 和 HB29）都具有中等以上抗性。

此外，云南省育种家从 70 年代以来，育成了一批抗白叶枯病品种，由于高原地区生态环境复杂，多数品种的生产应用面积较小，其中种植面积较大的品种有楚粳 3 号（1985—2001，38.7 万 hm^2，581 万）和云粳 136（1984—1990，13.5 万 hm^2，202 万亩）（全国农业技术推广服务中心，1982—2006）。这两个品种对稻瘟病也有中等抗性（表 3）。

表 3　20 世纪 70—80 年代育成大面积种植的粳稻抗白叶枯病品种[a]

Table 3　Main Chinese japonica rice cultivars with resistance to bacterial blight released during the 1970s－1980s

品种 Cultivar	组合 Cross	育成年份/单位 Year developed & bred by Inst.	抗病（虫）性反应[d] Reaction to			种植面积公顷 Acreage/hm^2	系谱图号[c] Genealogy
			BB	BL	SB		
农垦 58（世界）	—	约 20 世纪 50 年代末	R			1964—1988，9 656 000	7－6
农垦 57[b]（金南风）	良作/爱知中生旭	1957，引自日本				1964—1995，671 000	6－3
农虎 6 号	农垦 58/老虎稻	1965，原浙江省嘉兴地区农业科学研究所	R	R		1970—1985，4 217 000	7－7
沪选 19[b]	农垦 58 系选	1964，上海市农业科学院	R	R		1964—1984，315 000	7－6
苏州青[b]	农垦 58 系选	1965，江苏省苏州地区农业科学研究所	R	R		1967，927 000	7－6
苏粳 2 号[b]	嘉农 08 系选	1969，江苏省苏州地区农业科学研究所	MR	MR		1974—1986，275 000	7－6
南粳 11[b]	农垦 57 系选	1968，江苏省农业科学院	R			1982－1984，34 000	6－3
南粳 15	农垦 57/农垦 51	1968，江苏省农业科学院	HR			1975—1979，141 000	6－3
南粳 23	农垦 57 系选	1970，江苏省农业科学院	R			1983—1988，397 000	6－3
鄂宜 105[b]	农垦 58 系选	1978，湖北省宜昌地区农业科学研究所	MR			1982—1999，3 300 000	7－6
鄂晚 5 号[b]	鄂晚 3 号/4243	1979，湖北省农业科学院	R	R		1984—1999，2 513 000	7－6
秀水 48	辐农 709/京引 154//辐农 709	1979，原嘉兴地区农业科学研究所	MR	MR		1983—1992，1 261 000	7－7
矮粳 23	矮粳 22/桂花黄//农红 73	1974，浙江省农业科学院	R			1982—1988，653 000	7－2

（续）

品种 Cultivar	组合 Cross	育成年份/单位 Year developed & bred by Inst.	抗病（虫）性反应[d] Reaction to			种植面积公顷 Acreage/hm^2	系谱图号[c] Genealogy
			BB	BL	SB		
湘虎25	湘粳8号/农虎6号	1976，浙江农业大学	R			1983，7 000	7-5
盐粳2号[b]	南粳11系选	1979，江苏省盐城县农业科学研究所	R			1983—1995，1 019 000	6-3
盐粳3号[b]	南粳11系选	1979，江苏省盐城县农业科学研究所	HR			1987—1990，248 000	6-3
盐粳20	农垦57系选	1979，江苏省盐城县农业科学研究所	R			1986，41 000	6-3
浙粳66	矮粳23/百哥稻+花培	1979，浙江省农业科学院	R			1983—1984，15 000	7-9
祥湖47	覆农709/京引154//鉴3/嘉湖4号	1979，原浙江省嘉兴地区农业科学研究所	MR	MR		1983—1988，233 000	7-7
扬粳201[b]	南粳11/南粳33	1982，江苏省里下河地区农业科学研究所	HR			1985—1990，245 000	6-3
南粳35	农垦57/桂花黄//IR26///台中育39	1983，江苏省农业科学院	R			1982—1986，105 000	6-3
楚粳3号[b]	植生1号/假若叶	1980，云南省楚雄彝族自治州农业科学研究所	MR	MR		1985—2001，387 000	
云粳136	西南175系选	1983，云南省农业科学院	MR	HR		1984—1990，135 000	6-6
京越1号[b]	水原300粒/越路早生	1967，中国农业科学院	MR	MR		1969—1996，1 600 000	5-12
新稻68-11[b]	山法师系选	1965，河南省新乡农业科学研究所	R	R		1978—1994，439 000	5-6
冀粳1号[b]	福稔/桂花黄	1969，河北省农垦科学研究所	MR			1983—1985，99 000	5-9
鱼农1号[b]	农垦57系选	1973，山东省鱼台县良种场	R	R		1983—1986，15 000	5-9
辽盐2号[b]	丰锦系选	1979，辽宁省盐碱利用农业科学研究所	MR	MR		1988—1994，355 000	5-6
冀粳8号[b]	冀粳6号系选	1985，河北省农垦科学研究所	R	R		1985—1998，412 000	
中百4号[b]	喜峰/南粳15	1980，中国农业科学院	HR		MR	1988，10 000	5-4

[a] 根据《中国水稻品种及其系谱》和全国农业技术推广服务中心资料整理（Based on the book *Rice Varieties and Their Genealogy in China* and statistical data from National Agricultural Extention Service Center）。

[b] 优质（Cultivars with good quality）。

[c] 系统图号源自《中国水稻品种及其系谱》（Genealogy figure number No. from the book *Rice Varieties and Their Genealogy in China*）。

[d] BB，白叶枯病（Bacterial blight）；BL，稻瘟病（Rice blast）；SB，纹枯病（Sheath blight）；R，抗（Resistant）；HR，高抗（Highly resistant）；MR，中抗（Moderately resistant）。

据中国水稻所杨仕华等（2005）统计，在1996—2005年间通过国家审定的南方稻区6个粳稻品种中，高产优质、抗白叶枯病的品种有武运粳7号、浙农大454、镇稻6号、绍糯9714等4个，其中武运粳7号累计种植了数千万亩（1998—2005年，累计231.1万hm^2，3 466万亩）。又据文献报道的不完全统计，至2006年，通过苏、浙、皖、豫、滇等省审定的高产或优质粳稻新品种约有78个，其中47个（占60.3%）是抗白叶枯病的。在生产上应用面积较大的品种有兼抗稻瘟和纹枯病的武香粳14（2002—2005，76.4万hm^2，1 146万亩）以及兼抗稻瘟病的秀水994（1996—2005，43.3万hm^2，650万亩）。此外，有些品种除了抗白叶枯病外，还兼抗其他病（虫），如兼抗稻瘟和纹枯病的品种盐粳9号、华粳4号、华粳5号、郑花辐9号、嘉991和苏香粳2号；兼抗稻瘟和细菌性条斑病的云超6号、云恢188和秀水217；兼抗稻瘟和稻曲病的南粳41、风稻16号和风稻18号；兼抗稻瘟和条纹叶枯病的扬辐粳7号和岫粳11号；兼抗稻瘟病的云超7号、扬粳糯1号、华粳2号、秀水

表 4 2006 年前国家和省级以上审定的部分双抗(多抗)优质或高产粳稻品种(章琦根据有关资料整理,2006)

Table 4 Some Chinese japonica rice cultivars with multiple-resistance to diseases (including BB) **registered from later 1990s up to 2006** (Summarized by zhang,2006)

品种# Cultivar	组合 Cross	抗病(虫)性反应[c] Reaction to							审定年份/单位 Year approved & bred by Inst.	种植面积[d]/公顷 Acreage /hm^2	参考文献 Reference
		BB	BL	SB	BS	FS	BPH	RS			
武运粳 7 号[#,a,b]	加 48/香糯 9121//丙 815	MR							2000,江苏省武进市农业科学研究所	1998—2005,2 311 000	杨仕华等,2005
绍糯 9714 [#,b]	绍紫 9012/绍糯 45//绍间 9/3/绍糯 119	MR							2004,浙江省绍兴市农业科学研究所	2002—2005,47 000	同上
镇稻 6 号[#,a,b]	盐粳 2 号/秀水 04	R							1999,江苏省丘陵地区镇江农业科学研究所	—	同上
浙农大 454 [#,a,b]	丙 889/中 8029//台 202	MR							2001,浙江大学农业与生物技术学院	—	同上
中作 93[#,a,b]	中作 44/竹恢早//垦丰 10 号	MR	MR					MR	1998,中国农业科学院作物所	1988—2004,210 000	同上
辽粳 294[#,a,b]	79-227/83-326	R	R						1999,辽宁省农业科学院稻作所	1998—2005,839 000	同上
超产 1 号[#,a,b]	青系 96 号/BG90-2//下北	R	R						1999,吉林省农业科学院水稻所	1997 年,63 000	同上
连粳 2 号[#,a]	台湾稻	R							1999,江苏省连云港市农业科学所研究所	—	同上
新稻 10 号[#,a,b]	豫粳 6 号/新稻 89402	MR	MR						2004,河南省新乡市农业科学研究所	—	王书玉等,2004
武香粳 14[b]	京 58/3/248-5/254-13//武香粳	MR	HR	R					2003,常州市武进稻麦育种场	2002—2005,764 000	江祺详等,2004
秀水 994[a,b]	嘉 59 天杂/丙 9543	R	R						2003,浙江省嘉兴市农业科学院	1996—2005,433 000	边永高等,2004
南粳 41[a,b]	(R405/NJ30152)/静千子千米	MR	MR			R			2003,江苏省农业科学院	2003—2004,52 000	仲维功等,2003
嘉 991[a,b]	武运粳 7 号//嘉 62/水晶 1 号	MR	HR	MR					2005,浙江省嘉兴市农业科学院	2004—2005,115 000	李金军等,2003
华粳 2 号[a,b]	武育粳 3 号/香粳 111	R	R						2003,江苏省大华种业公司	—	王子明等,2003
华粳 5 号[a,b]	D26 系选	R	MR	MR					2005,江苏省大华种业公司	—	王子明等,2006
华粳 4 号[a,b]	镇稻 88/风尾 6 号	R	MR	MR					2005,江苏省大华种业公司	—	王子明等,2005
盐粳 9 号[b]	武育粳 3 号/武运粳 8 号	HR	MR	MR					2005,江苏省沿海地区农业科学研究所	—	唐红生等,2006
苏香粳 2 号[b]	88B225/S015	MR	MR	MR					2002,江苏省太湖地区农业科学研究所	—	周新伟等,2003
秀水 217[a]	丙 94-168/丙 95-237//丙 89-90/JF81	MR	MR	MR					2003,浙江省嘉兴市农业科学院	—	俞琦英,2003

（续）

品种# Cultivar	组合 Cross	抗病(虫)性反应[c] Reaction to							审定年份/单位 Year approved & bred by Inst.	种植面积[d]/公顷 Acreage /hm^2	参考文献 Reference
		BB	BL	SB	BS	FS	BPH	RS			
扬辐粳7号[a,b]	辐射诱变	MR	R					R	2004,江苏省里下河农业科学院研究所	—	陈秀兰等,2004
扬粳糯1号[b]	香粳糯91-21/广陵香糯	HR	R						2004,江苏省里下河农业科学研究所	—	王保和等,2004
郑花辐9号[b]	$^{60}Co \sim \gamma$射线	R	R			MR			2004,河南省农业科学院	—	王越涛等,2005
云超6号[b]	密阳23/云玉1号/玉糯2号/京黄126/3/IR64719-163-3	MR	MR			R			2005,云南省农业科学院、玉溪市农业科学研究所	—	李贵勇等,2005
皖稻84	秀水664/粳系212	MR	R						2005,安徽农业大学	—	王辉,2005
岫粳11号	合系2号/岫粳4号	MR	MR					R	2005,云南省保山市农业科学研究所	—	李金衡,2005
风稻16号[b]	合系15/风稻9号	MR	MR			MR			2005,云南省大理白族自治州农业科学研究所	—	同上
风稻18号[b]	合系40号选育	MR	MR			MR			2005,同上	—	同上
昌粳8号	预110系选	R	MR						2005,云南省昌宁县农技站	—	同上
云粳优一号[a,b]	银条粳/合系34	R	R						2004,云南省农业科学院	—	同上
云恢188[b]	BL4/晋红1号	R	MR		R				2005,云南省农业科学院	—	李金衡,2006
靖粳10号[b]	楚粳23号/合系41号	MR	R						2005,云南省曲靖市农技站	—	同上

\# 通过国家审定,其余为省级审定的品种(Cultivars marked with # were registered at national level,and the others at provincial level)。

a 高产(Cultivars with high yield)。

b 优质 (Cultivars with good quality)。

c BB,白叶枯病(Bacterial blight);BL,稻瘟病(Rice blast);SB,纹枯病(Sheath blight);BS,细菌性条斑病(Bacterial streak);FS,稻曲病(Rice false smut);RS,条纹叶枯病(Rice stripe virus);BPH,褐飞虱(Brown planthopper);R,抗(Resistant);MR,中抗(Moderately resistant);HR,高抗(Highly resistant)。

d 据全国农技推广服务中心资料整理(1982—2006)(Planted area based on statistical data from National Agricultural Extention Service Center,1982—2006)。

994、昌粳8号、云粳优1号和靖粳10号等品种（表4）。

北方粳稻地区辽阔，生态环境差异较大，白叶枯病的发生情况也不相同。西北和除辽宁省以外的东北稻区基本上不发生白叶枯病或偶尔轻微发生，而华北稻区则病情较重，后者及辽宁省较重视抗白叶枯病品种的选育。在20世纪60～70年代从日本引进的京引编号品种中，表现高产、优质、抗白叶枯病的有秋光、丰锦、早锦和黎明等品种，先后在东北、华北稻区直接在生产上应用。据全国农技推广中心统计资料（1982—2006），至2005年，秋光累计推广了216.8万hm^2（3 252万亩）；丰锦仅在1978年就推广了19.5万hm^2（293万亩），到1995年已超过80万hm^2（1 200万亩），早锦和黎明至90年代中期也分别种植到60万hm^2（900万亩）和16.7万hm^2（251万亩），这在北方稻区是不多见的。

北方粳稻的抗白叶枯病育种始于20世纪70年代。中国农业科学院作物所林世成主持该所水稻育种工作时，借鉴IRRI的经验组建了GEU小组，首先进行了抗稻瘟、白叶枯病和褐飞虱等病虫种质资源的筛选工作。以黎明、秋光、早生爱国3号、喜峰、南粳15等抗源品种作为抗性供体，对杂交育种分离世代材料同步进行抗病性鉴定和农艺性状选择，每年将高代品系种植在稻瘟病疫区接受自然考验，在70～80年代先后育成了对白叶枯病和稻瘟病具有中等抗性的中百4号、中丹、中作编号等品种，相继在生产上推广种植。该所1967年育成的中抗白叶枯和稻瘟病的京越1号品种表现优质、高产、稳产，曾在华北、东北稻区种植达27年之久（1969—1996，160万hm^2，2 400万亩），到2003年在辽宁省稻病常发的丹东地区仍有种植；优质、稳产品种中作93从1985年至今一直在天津和东北稻区种植，累计面积约26.7万hm^2（400万亩）。河北、河南、天津等省（市）农业科学院（所）也先后育成一批抗白叶枯病的冀粳、辽盐、红旗、临稻等编号品种，其中累计面积在33.3万hm^2（500万亩）以上的品种有兼抗稻瘟病的新稻68－11（1978—1994，43.9万hm^2，658万亩，一度为河南主栽品种之一），冀粳8号（1985—1998，41.2万hm^2，618万亩），辽盐2号（1988—1994，35.5万hm^2，532万亩）（表4）。

据杨仕华等（2005）报道，北方稻区在1996—2005年间通过国家审定的64个粳稻品种中，绝大多数是东北的，由于当地主要进行抗稻瘟病育种，所以有记录的抗白叶枯病的品种只有辽粳294、中作93、超产1号、连粳2号及新稻10号等5个高产或优质的品种。其中辽粳294在辽宁等省从1998—2005年已累计种植83.9万hm^2（1 259万亩）（杨仕华等，2005；中国稻米，2006；全国农业技术推广服务中心资料）。

1.2　杂交稻抗白叶枯病育种

1.2.1　籼型三系杂交稻　在我国籼型三系杂交稻中先后育成的野败型、矮败型、冈型、D型、印水型和红莲型等主体胞质不育系都不抗白叶枯病。1986—1990年间李宣铿等（1993）鉴定了湖南杂交稻中心的三系材料对白叶枯病的抗性，在175份不育系和185份保持系中，仅有1份不育系和3份保持系表现为中抗，其余材料都不抗白叶枯病。但是，在对野败型、冈型、D型、矮败型、印水型和红莲型不育系具有恢复力的恢复系中，除IR24、泰引1号和密阳46以外，从IRRI引进的一批抗白叶枯病的品种如IR26、IR28、IR30、IR32、IR36、IR50和IR54等以及具有IR系统亲缘的测64－7（IR9761－19－1测交筛选）、桂99[龙野5－3/（IR661/IR2061，即IR28）]等都携有抗性基因*Xa4*，因而在选育抗白叶枯病杂交稻组合中发挥了重要作用，也缓解了改善不育系的白叶枯病抗性难度较大的问题。尽管如此，在20、21世纪交替前后还是育成了若干抗病的三系籼稻不育系。例如在2003—2005年通过省级审定的不育系中，29A（珍汕97B/V20B）、中1A（902//^{6}PS-32A）和H28A（金23A/岳4B/福伊B//316B）对白叶枯病都有中等以上抗性（姚立生等，2003；童汉华等，2006；何发青等，2005）。罗彦长等（2005）还通过常规杂交与分子标记辅助选择相结合的方法，成功地将抗性新基因 *Xa21* 和 *Xa23* 聚合在一起，育成了全生育期抗谱广的不育系R106A，对提高三系法不育系的抗性进行了有益的尝试。

在1973—1986年间育成的87个籼型三系杂交稻中，有31个组合中抗白叶枯病，占35.6%，其中推广面积达26.7万 hm^2（400万亩）以上的有17个，包括7个除抗白叶枯病外还兼抗其他病（虫）的组合，即：兼抗稻瘟病的汕优6号、汕优桂33、汕优桂34和汕优30选，兼抗褐飞虱的威优6号，兼抗黄矮病的四优6号和对白叶枯、稻瘟、黄矮病及稻飞虱都有中度抗性的威优64。在这些组合中，威优64、汕优6号、威优6号和汕优桂33在80～90年代，年最大推广面积都曾达到66.7万 hm^2（1 000万亩）以上。威优64的年推广面积曾达到136.9万 hm^2（2 053万亩，1988年），其在1983—2003年间的累计种植面积1 227万 hm^2，合18 405万亩，汕优6号1981—1994年累计933.1万 hm^2，13 997万亩，威优6号的年推广面积曾达到173.9万 hm^2（2 608万亩，1985年），其在1981—1995年累计752.6万 hm^2（11 289万亩），汕优桂33 1984—1999年累计501.7万 hm^2（7 526万亩，表5）（全国农业技术推广服务中心统计资料，1982—2006）。威优64的种植时间长达20年之久，其他组合的种植面积直到90年代中期后才逐渐减少。

表5 中国籼型三系杂交稻的白叶枯病抗性反应

（章琦根据有关资料整理，2007）

Table 5 Bacterial blight resistance in Chinese 3-line indica hybrid rice and it's restorers（Summarized by Zhang，2007）

杂交稻 Hybrid rice	组合 Combination	抗病（虫）性反应[c,d] Reaction to					审定或育成年份 Year approved/bred	种植面积[d]/公顷 Acreage /hm^2	参考文献 Reference
		BB	BL	SB	YD	BPH			
威优64[a]	V20A/测64-7（IR36衍生）	MR	MR		MR	MR	1981	1983—2003，12 270 000	林世成等，1991；袁隆平，2002；全国农技推广中心，1982—2006
汕优6[a]	珍汕97A/IR26	MR	MR				1976	1981—1994，9 331 000	同上
威优6号[a]	V20A/IR26	MR				MR	1976	1981—1995，7 526 000	同上
汕优桂33[a,b]	珍汕97A/桂33（IR36/IR20）	MR	MR				1983	1984—1999，4 662 000	同上
汕优桂34	珍汕97A/桂34	MR	MR				1984	1986—1994，1 519 000	同上
汕优30选	珍汕97A/30选	MR	MR				1981	1984—1989，443 000	同上
四优6号	V41A/IR26	R			MR		1977	1983—1986，153 000	同上
汕优77[#,a,b]	珍汕97A/明恢77	MR					1998	1992—2005，2 497 000	杨仕华等，2005
Ⅱ优084[#,a,b]	Ⅱ-32A/镇恢084	HR	HR				2003	2001—2005，757 000	盛生兰等，2003
特优多系1号[#,a]	龙特甫A/多系1号	MR	R				2001	1997—2005，256 000	杨仕华等，2005
中9优838选[#]（国丰1号）	中9A/838选	MR	R				2001	2002—2005，21 000	杨仕华等，2005
丰优香占[#,a,b]	粤丰A/R6547	HR	HR				2003	2002—2004，103 000	徐卯林等，2004

（续）

杂交稻 Hybrid rice	组合 Combination	抗病（虫）性反应[c,d] Reaction to					审定或育成年份 Year approved/bred	种植面积[d]/公顷 Acreage /hm^2	参考文献 Reference
		BB	BL	SB	YD	BPH			
K优047[#,a,b]	K17A/成恢047	HR	MR				2001	1996—2003，103 000	杨仕华等，2005
优优128[#,a]	优IA/广恢128	MR	R				1999	1997—2002，37 000	杨仕华等，2005
K优402[#,a]	K17A/402	MR	MR				1998	1998—2004，31 000	杨仕华等，2005
中优223[#,a,b]	中A/R223	MR					2001		杨仕华等，2005
协优80[#,a]	协青早A/R80	MR					2004		杨仕华等，2005
中优218[#,a,b]（国丰2号）	中9A/R218-51	MR					2004	2005，13 000	杨仕华等，2005
协优5968[#,a]	协青早A/15968	MR					2004	2005，9 000	杨仕华等，2005
新优赣22[#,b]	新A/752	MR					1999		杨仕华等，2005
中优85[#,a,b]	中9A/蒲恢85	HR					2005		中国稻米，2006（1）
宜香9号[#,a,b]	宜香ⅠA/宜恢9号	MR					2005	2005，8 000	中国稻米，2006（1）
内香2550[#,b]	内香-2A/内恢2550	MR					2005		中国稻米，2006（1）
宜香10号[#,b]	宜香ⅠA/宜恢10号	MR					2005		中国稻米，2005（6）
Ⅱ优363[#,a]	Ⅱ-32A/蜀恢363	MR					2005		中国稻米，2005（6）
协优9308[a,b]	协青早A/C57//300号/IR26	R	MR				1999	1998—2005，247 000	程式华等，2004
Ⅱ优128[a,b]	Ⅱ-32A/恢128（七桂早25/测64//明恢63）	MR	HR				1999	1998—2005，532 000	李传国等，2000
抗优63[a]	珍汕97A/抗恢63	HR	HR				2002		丁伦友，2005
丰优58（丰优205）[a,b]	丰7A/D205	R	R				2005		丁伦友，2005
协优084[a,b]	协青早A/镇恢084	MR	HR				2004	2004—2005，79 000	盛生兰等，2005
皖稻135[b]	丰源A/YR909	HR	R				2004		王辉，2004
天优372[a,b]	天丰A/广恢372	MR	R				2005		黄德娟等，2005
天优290[a,b]	天丰A/广恢290	MR	R				2005		黄彗君等，2005
Ⅱ优118[a,b]	Ⅱ-32A/华恢118	R	HR				2003	2004—2005，31 000	王子明等，2003
Ⅱ优559[a]	Ⅱ-32A/盐恢559	MR	R	MR			2002		姚立生等，2003a
29优559[a]	29A/盐恢559	MR	HR	MR			2003		姚立生等，2003b
蓉稻10号（新香优906）	新香优A/蓉恢906	R	MR	MR			2005		李金衡，2006
K优507[a,b]	K17A/R507	MR	HR				2002		张洪熙等，2003
抗优98[a,b]	Ⅱ-32A/抗恢98	HR	HR				2003		周玉萍等，2003
Ⅱ优205[a,b]	丰7A/D205	R/HR	R				2004	2005，12 000	丁伦友，2005
红优527[a,b]	红矮A/蜀恢527	R					2004		谢崇华等，2005
富优108[b]	冈46A/嘉恢108	R	R				2005		谢崇华等，2005
农华优808[a]	M98A/MR0208	R	R				2005		庆容等，2006
中优2506[b]	中9A/R2596	MR	MR				2005		李民迪等，2006

\# 通过国家审定，其余为省级审定的品种（Cultivars marked with # were registered at national level，and the others at provincial level）。

a 高产（Cultivars with high yield）。

b 优质（Cultivars with good quality）。

c BB，白叶枯病（Bacterial blight）；BL，稻瘟病（Rice blast）；DF，黄矮病（Rice yellow dwarf）；BPH，褐飞虱（Brown planthopper）；R，抗（Resistant）；MR，中抗（Moderately resistant）；HR，高抗（Highly resistant）。

d 种植面积根据全国农业技术推广服务中心资料整理（Planted area based on statistical data from National Agricultural Extention Service Center）。

中国“超级稻”研究计划启动后，育种家都在杂交稻的基础上进一步挖掘产量潜力。在1998—2004年间参加南方籼稻品种（组合）区域试验的723个供试材料中，杂交稻就有625个，占86.4%，其中涌现出一大批高产组合，稻米品质也有明显改善，但抗病的组合不多。1996—2005年间通过国家级审定的169个籼型三系杂交稻中，只有18个组合抗白叶枯病，占10.7%（杨仕华等，2005；中国稻米，2005；2006），其中Ⅱ优084、特优多系1号、中9优838选、丰优香占、K优047、优优128和K优402等7个组合还兼抗稻瘟病。这批组合发展很快，在生产上迅速达到20万hm^2（300万亩）以上的有汕优77（1992—2005，累计249.7万hm^2，3 745万亩）、Ⅱ优084（2001—2005，75.7万hm^2，1 135万亩）、特优多系1号（1997—2005，25.6万hm^2，384万亩）、中9优838选（2002—2005，21万hm^2，315万亩）等（表5）（全国农业技术推广服务中心，1982—2006）。

与此同时育种学家们也育成了一批优良的抗白叶枯病恢复系，因为自从出现IRRI的一些IR品种（系）对国内不育系有较好的恢复力而且抗病性较强，大家都认为通过恢复系提高杂交稻的抗病性要比改良不育系的抗性容易收效。其中还利用了除*Xa4*外的其他抗白叶枯病基因。例如：①20世纪70年代末，南京农业大学朱立宏等（1990）、周毓珍等（1993）将携有抗谱较广的*Xa7*和*xa5*两个抗性基因的DV85与TN1杂交，育成一批携有*Xa7*的中间衍生抗源TD后，又以TD为母本，明恢63作父本，相继育成了携有*Xa7*的抗恢63、抗恢98及D205等抗病恢复系，再用这些恢复系分别与汕优97A、Ⅱ-32A等不育系配制组合，育成了双抗白叶枯和稻瘟病的抗优63、抗优98、Ⅱ优205及丰优205等组合。这些组合先后在苏、浙、皖、豫、滇、鄂、赣、闽、湘等省白叶枯病严重地区试验和生产应用，表现高产、抗病、适应性强，迄今累计推广面积约250万hm^2（3 750万亩），已成为安徽、江苏等省白叶枯病疫区的主栽组合（丁伦友，2005）。②江苏省丘陵地区镇江农业科学研究所在选育恢复系时，都由省植保所进行抗性鉴定，他们用其所育成的优质强恢复系镇恢084（明恢63/特青1号）与一系列不育系配制了协优084、汕优084、天丰优084、丰优084、金优084以及Ⅱ优084等组合，对长江流域白叶枯病优势菌株KS-6-6、ZHE173和JS49-6都有1-3级水平的抗性（盛生兰等，2003；2005）。③广东省农业科学院始终坚持选育抗本省白叶枯病优势菌C4和强致病菌C5的目标，成功地育成了抗白叶枯病和稻瘟病的天优372、天优290、华优153、天优998、博优368等高产、优质、抗病的系列组合，已于2004、2005年通过省级品种审定（黄德娟等，2005；黄彗君等，2005；梁世胡等，2005）。④近年利用分子标记辅助选择（MAS）手段改良恢复系抗性的报道很多。中国水稻研究所采用MAS与常规测恢相结合的手段，育成了一个优质抗病的强恢复系R218，用它育成了中抗白叶枯病的中优218、中9优218等高产、优质组合；1999年该所育成的超级杂交稻协优9308（协青早A/C57//300号/IR26）不仅具有超高产潜力，而且双抗白叶枯和稻瘟病，并且米质优良，截至2004年，已累计推广71.3万hm^2（1 070万亩）（程式华等，2004）。

此外，在2002—2005年通过省级审定的一批高产或优质组合中，皖稻135、Ⅱ优118、抗优98、Ⅱ优205、富优108和农华优80都双抗白叶枯和稻瘟病，蓉稻10号、29优559和Ⅱ优559还兼抗白叶枯、稻瘟和纹枯病三种病害。在表5中列出20世纪80年代以来若干大面积生产应用和近期通过国家和省级审定的部分抗病高产或优质的组合。

杂交稻的抗性水平主要取决于抗病基因在F_1植株上的显性表现程度。我国迄今所利用的恢复系，其对白叶枯病的抗性多半是由不完全显性基因*Xa4*控制的，与感病不育系配制组合后的抗性表现仅为中抗至中感水平，且 *Xa4* 在华南部分稻区已经遇到了毒性菌株的侵袭。因此，今后在选配父母本时，除考虑扩大抗性基因源外，应以双方或一方携有完全显性抗病基因为首选组合，从后面这一点看，解决杂交稻的抗病性问题要比常规稻更为简单和有效。

1.2.2 粳型三系杂交稻抗白叶枯病育种 20 世纪 70 年代末以前，我国育成的 43 个粳型杂交稻三系材料和组合中有 60%中抗白叶枯病。之后，育成的多数组合对白叶枯病也有中等以上抗性。在苏、浙、皖、湘、辽、京等省（市）推广种植的 15 个组合中有 9 个抗病，其不育系如黎明 A、丰锦 A、秋光 A、六千辛 A、当选晚 2 号 A、秀玲 A 等都是中抗白叶枯病的。

《杂交水稻学》（袁隆平，2002）中列举的粳型三系杂交稻 5 个主要恢复系中，有 3 个中抗白叶枯病：20 世纪 90 年代育成的 C418（轮回 422/密阳 23），恢复力强，配合力较好，兼抗稻瘟病；光身恢复系 HP121（中国 PL1/多收系 2//晚恢 9 号），恢复力强，兼抗稻瘟病；77302－1（C57/IR28//科情 3 号/京引 37）。另外两个，一是应用最广的恢复系 C57，不抗白叶枯病；一是 N138，抗性不详。

80～90 年代我国生产上种植的 5 个主要粳型三系杂交稻组合为：在东北稻区应用最早、种植面积最大的黎优 57（黎明 A/C57），1981—1992 年累计种植面积 65.7 万 hm^2（985 万亩），笔者曾用 KS－8－4（江苏）、崖凋 1－1（海南）及 LN2－2（辽宁）3 个白叶枯病菌株对它进行测试，均表现中度抗病；秀优 57（秀岭 A/C57）则中抗白叶枯和稻瘟病。长江流域种植的组合有赣化 2 号（IR24/献党 1 号），中抗白叶枯病；优质晚粳六优 1 号（六千辛 A/77302－1），具有一定的白叶枯病抗性；优质、中抗稻瘟病的泗优 422（泗稻 8 号 A/轮回 422）对白叶枯病菌株 KS－6－6、KS－1－21 和浙 173 的抗性为中抗/中感。此外，1998 年通过辽宁省审定的箨优 418（箨锦 A/C418）中抗/中感白叶枯病，高抗稻瘟病，中抗纹枯病。

朱立宏等（1990）选育的具 DV85（携 *Xa7* 抗性基因）亲缘的抗白叶枯病品系 TD－1 和 TD－2，在用于改良籼型恢复系抗性的同时，也转育成粳恢 98、珍恢 084、扬 818 等粳型恢复系。Zhang 和 Zhu（2004）用这些恢复系配制的粳优 63、粳优 98 和 K 优 818 等优质高产组合，均继承了 *Xa7* 的优良抗性。

进入 21 世纪后，长江流域一些省（市）农业科学院（所）又育成了一批抗白叶枯病、兼抗稻瘟或其他病害的新组合。截至 2005 年，通过省级审定的高产/优质粳型三系杂交稻组合中，86 优 8 号、86 优 242、嘉乐优 2 号、甬优 6 号、甬优 5 号、皖稻 76、榆杂 34、徐优 201、常优 2 号、苏优 22 等均双抗白叶枯和稻瘟病；69 优 8 号、嘉优 1 号、甬优 2 号等组合除兼抗稻瘟病外，对纹枯病或细菌性条斑病或褐飞虱也有中等抗性（表 6）。

表 6 进入 21 世纪初通过省级审定的部分双抗（多抗）优质或高产粳型三系杂交稻（章琦整理，2006）

Table 6 Some 3-line Chinese japonica hybrid rice with multiple-resistance to diseases and pest, high-yield and good quality registered at provincial level in entering the 21st century (Summarized by Zhang, 2006)

杂交稻 Hybrid rice	组合 Combination	抗病（虫）性反应[c] Reaction to					审定年份/单位 Year approved & bred by Inst.	参考文献 Reference
		BB	BL	SB	BS	BPH		
甬优 4 号[#,a,b]	甬粳 2 号 A/K2001	MR					2004，浙江省宁波市农业科学院	杨仕华等，2005
86 优 8 号	863A/宁恢 8 号	MR	HR				2001，江苏省农业科学院	谷福林等，2001

（续）

杂交稻 Hybrid rice	组合 Combination	抗病（虫）性反应[c] Reaction to					审定年份/单位 Year approved & bred by Inst.	参考文献 Reference
		BB	BL	SB	BS	BPH		
69优8号[a,b]	69A/R11238	R	HR	MR			2001，江苏省徐州市农业科学研究所	刘超等，2003
86优242	863A/R242	MR	HR				江苏太湖地区农业科学研究所	朱勇良等，2003
嘉优1号[a]	嘉60A/夹恢40	MR	R		MR		2004，浙江省嘉兴市农业科学院	刘鑫等，2005
嘉乐优2号[a]	151A/DH32	R	R				2004，同上	同上
甬优6号[b]	甬粳2号A/K4806	R	R				2004，浙江省宁波市农业科学院	同上
甬优5号[b]	甬糯2号A/K6926	MR	R				2004，同上	同上
甬优2号[a,b]	甬粳2号A/K1722	MR	MR		MR		2001，同上	陈志良等，2003
皖稻76[b]	爱知香A/MC20518	MR	MR				2004，安徽省合肥新隆水稻所	王辉，2004
榆杂34[b]	滇榆一号A/南34	R	R				2004，云南省杂交水稻研究中心	李金衡，2005
常优2号[b]	武运粳7A/C53	HR	R				2005，江苏省常熟市农业科学研究所	邓建平，2005
苏优22[b]	武运粳7A/R16189	R	MR				2005，江苏省中江种业公司	同上
徐优201[a,b]	徐9320A/徐恢201	MR	R	MR			2005，江苏省徐州农业科学研究所	王健康等，2006

[#] 通过国家审定，其余为省级审定的品种（Cultivars marked with ＃ were registered at national level，and the others at provincial level）。

[a] 高产（Cultivars with high yield）。

[b] 优质（Cultivars with good quality）。

[c] BB，白叶枯病（Bacterial blight）；BL，稻瘟病（Rice blast）；SB，纹枯病（Sheath blight）；SR，条纹叶枯病（Rice stripe virus）；BPH，褐飞虱（Brown planthopper）；R，抗（Resistant）；MR，中抗（Moderately resistant）；HR，高抗（Highly resistant）。

1.2.3 籼（粳）型两系杂交稻抗白叶枯病育种 石明松（1981；1985）首先观察到调节水稻光周期可以诱导发生雄性不育，并于1973年在晚粳农垦58大田中发现3个雄性不育株，经数年研究提出了通过改变播种期使水稻小花分化期植株处在适宜的长日条件下而不育，进而采用两系法产生杂种优势的设想。随后湖北省科学技术委员会、国家自然科学基金委员会先后将“光敏水稻核不育研究”列为重大项目，掀起了两系法杂交稻应用基础研究的高潮，很快就实现了两系杂交稻在生产上的大面积应用（卢兴桂，2001）。

1988年，湖北省农业科学院、华中农业大学和武汉大学等用农垦58S（S表示不育系）转育的粳型光敏核不育系31111S、WD1S、N5047S和籼型温敏不育系W6154S通过了国家鉴定。此后育成和通过省级审定的籼、粳稻光、温敏核不育系的不育基因也主要分别来源于农垦58S和安农S-1（程式华，1995）。由农垦58衍生的一批粳稻品种对白叶枯病具有成株抗性，但从自然突变的农垦58S选育出来的光、温敏不育系是否仍保持其原有抗性则有待明确。章琦等2001年用中国的Ⅱ（HB17）、Ⅳ（浙173）、Ⅴ型（GD1358），广东Ⅳ、广东Ⅴ型菌和菲律宾小种6共6个代表菌株鉴定了20个“九五”期间育成的两用核不育系在成株期的抗性，结果表明其中多数不育系仍具有抗性，但抗谱较窄。广为利用的两个粳型不育系7001S、N5088S和两个籼型不育系培矮64S、香125S对我国优势菌Ⅳ型均表现中抗；蜀光612S、蜀光570S、W9864S、W9593S、2301S、6442S、1290S、7975S、8011S、399S、

GD-8S 等 12 个不育系也对中国Ⅳ型菌具有中等抗性，但是这 16 个不育系都不抗上述其他 5 个菌株。苏紫 S 不育系除了抗中国Ⅳ型菌外，还抗中国Ⅱ型菌，但感染中国Ⅴ型、广东Ⅳ、广东Ⅴ型菌和菲律宾小种 6；而 T78A、XO7S、3418S 3 个不育系则感染全部 6 个菌株。这说明多数两用核不育系对我国白叶枯病优势菌群虽然具有一定的抗性，但大都不如农垦 58，需要进一步改良、提高抗谱和抗性强度。此外，培矮 64S、测 64S、133S 和 625-9S 等 4 个不育系对湖南强毒性菌表现为中抗，而安湘 S 和湘早 22S 则是感病的（李友荣等，1999）。近年罗彦长等（2003）通过常规杂交结合分子标记辅助选择转育成携有抗性基因 *Xa21* 的光敏核不育系 3418S，中抗白叶枯病，但抗谱有所拓宽，在改良两用不育系的抗性上跨出了重要的一步。

两系杂交稻对白叶枯病的抗性总体上强于三系杂交稻。因为在同一亚种内常规稻的优异品种（系）都能正常恢复光、温敏核不育系的育性，这样在选配两系杂交稻组合时，可利用的父、母本就大为拓宽，从中筛选产量潜力大、品质上佳、抗病性强、抗谱较广的优异组合也就容易了。如特青、山青 11、余红、扬稻 6 号、双七占、D100、9311、双九、皖恢 9 号、R288、明恢 63 以及新选育的 D68 等已被广泛用来配制两系杂交稻，其中除明恢 63、D68、R288 中感白叶枯病外，其他均具有中等抗性。随着两系性状水平的提高和制种技术的不断完善，两系法选配杂交稻的相对优越性将日见彰显。

在 1994—2000 年间通过省（市）级品种审定的 32 个两系杂交稻组合中，有 17 个组合中抗白叶枯病，占 53.1%，有些组合还兼抗稻瘟或其他病害（袁潜华，2001）。在籼型两系杂交稻中，1994 年首批通过省级审定的组合培两优特青由于导入特青对白叶枯和稻瘟病的抗性，在历年白叶枯病严重的河南信阳地区表现高度抗病（牟同敏，2001），1996—2000 年间累计种植 19.3 万 hm^2（合 290 万亩）；对白叶枯、稻瘟病和细菌性条斑病都有中等抗性的培杂山青，1998—2004 年间累计推广 30.7 万 hm^2（460 万亩）；培杂茂选、田两优 402、培两优 275、培杂 67 等组合双抗白叶枯和稻瘟病；培两优余红兼抗白叶枯、稻瘟病和稻曲病；两优培胜中抗白叶枯病。1994 年首批通过省级审定的粳型两系杂交稻中的双抗白叶枯和稻瘟病组合，主要有皖稻 24（70 优 9 号），以及鄂粳杂 1 号和云光 8 号，其中高产优质的鄂粳杂 1 号 1996—2005 年间累计种植 38.9 万 hm^2（584 万亩，全国农业技术推广服务中心）。

在 2000 年后通过省级审定的两系杂交稻中，抗白叶枯病的优质组合有皖稻 119，兼抗其他 1～2 个病（虫）害的高产、优质组合有赣亚 1 号，兼抗细菌性条斑病和褐飞虱，以及皖稻 147、皖稻 129、两优 108 和两优培粳等兼抗稻瘟病（杨仕华等，2005；王辉，2004；2005；高北平等，2003；邓建平，2005；2006；董家胜，2005）。在 2006 年前通过国家审定的 21 个高产、优质两系杂交稻中，两优培九、培杂双七、培杂茂三、陆两优 28、扬两优 6 号、70 优 9 号（皖稻 24）等 6 个组合都双抗白叶枯和稻瘟病，这些组合已在生产上大面积推广应用。其中籼型两系杂交稻两优培九，表现优质、高产、适应性强、抗白叶枯病、高抗稻瘟病、纹枯病也轻，2000 年开始在江、皖、浙、闽、粤、桂、滇、黔、豫、鄂、湘、赣、蜀、陕、琼、渝等 16 个省（市、自治区）示范、推广，目前已成为全国种植面积最大的两系中籼，超过两系杂交稻总面积的 50%，截至 2005 年，累计种植面积达 468.1 万 hm^2（7 022 万亩），2005 年的年种植面积达 65.5 万 hm^2（983 万亩），是全国杂交稻年推广面积最大的组合。另外，培杂双七在 1998—2005 年间累计种植 58.5 万 hm^2（877 万亩）。培杂茂三在 1999—2005 年间累计种植25.3 万 hm^2（380 万亩）。其他优良组合见表 7。

表 7　两系杂交稻中主要的抗白叶枯病组合[d]

Table 7　Main two-line hybrid rice combinations with bacterial blight resistance[d]

杂交稻[#] Hybrid rice	类型 Type	组合 Combination	抗病（虫）性反应[c] Reaction to					审定年份和单位 Year approved & bred by Inst.
			BB	BL	BS	FS	BPH	
两优培九[#,a,b]	indica	培矮 64S/9311	MR	HR				2001，江苏省农业科学院
培杂双七[#,a,b]	indica	培矮 64S/双七占	MR	HR				2001，广东省农业科学院
培杂茂三[#,a,b]	indica	培矮 64S/茂三	MR	R				2001，广东 省茂名市杂交稻研究中心
陆两优 28 [#,a]	indica	陆 18S×华 28	MR	MR				2005，湖南亚华种业科学研究所
扬两优 6 号[#,a,b]	indica	广占 63－4S×扬稻 6 号	MR	MR				2005，江苏省里下河农业科学研究所
皖稻 24 (70 优 9 号)[#,a,b]	japonica	7001S/皖恢 9 号	MR	HR				2001，安徽省农业科学院
培两优特青[a]	indica	培矮 64S/特青	MR	MR				1994，湖南杂交水稻研究中心
鄂粳杂 1 号[a,b]	japonica	N5088S/R187	MR	MR				1995，湖北省农业科学院
云光 8 号	japonica	N5088S/云恢 11	MR	MR				2000，云南省农业科学院
培杂山青[a]	indica	培矮 64S/山青 11	MR	MR	MR			1997，广东省茂名市杂优协作组
培两优余红[b]	indica	培矮 64S/余红 1	MR	MR			MR	1997，湖南农业大学
培杂 67[a,b]	indica	培矮 64S/G67	MR	MR				2000，华南农业大学
田两优 402[a]	indica	田丰 S/R402	HR	MR				1998，江西省赣州地区农业科学研究所
培两优 275	indica	培矮 64S/275	R	MR				1999，广西壮族自治区农业科学院
两优培胜[a]	indica	培矮 64S/胜优 2 号	MR					1999，湖北杂交水稻研究中心
皖稻 147[a,b]	indica	新安 S/安选 6 号	R	MR				2005，安徽荃银农业科学研究所
培杂茂选[b]	indica	培矮 64S/茂选	R	R				2000，广东省茂名市杂交稻研究中心
赣亚 1 号	indica	培矮 64S/Hb－01	MR	MR			MR	2002，江西省滨湖农业科学研究所
皖稻 129[b]	indica	培矮 64S/红 98	MR	HR				2004，安徽省宣城市农业科学研究所
皖稻 119[b]	indica	宣 69S/WH26	MR					2004，安徽省安庆市农业科学研究所
两优培粳	japonica	培矮 64S/粳稻 94205	MR	R				2005，河南省信阳市农业科学研究所
两优 108[b]	indica	培矮 64S/宁恢 108	MR	MR				2005，江苏省农业科学院

[#] 通过国家审定，其余为省级审定的品种（Cultivars marked with ＃ were registered at national level，and the others at provincial level）。

[a] 高产（Cultivars with high yield）。

[b] 优质（Cultivars with good quality）。

[c] BB，白叶枯病（Bacterial blight）；BL，稻瘟病（Rice blast）；BS，细菌性条斑病（Bacterial streak）；BPH，褐飞虱（Brown planthopper）；R，抗（Resistant）；MR，中抗（Moderately resistant）；HR，高抗（Highly resistant）。

[d] 部分内容据袁潜华（2001）报道整理［Adopted party from the report of Yuan QH（2001）］。

顺便说一下，我国育种家在 20 世纪 90 年代开始探索籼粳亚种间杂种优势的利用时发现，广亲和品种在其中显示了威力。例如以具广亲和性的爪哇稻培迪为亲本育成的不育系培矮 64S 是目前用于测配亚种间杂交组合配合力的主干亲本，也是目前应用最广泛的广亲和光温敏核不育系之一，上述的许多优良组合都是用它来配组的。据章琦实验组杨文才等（1995）用国内外 20 个白叶枯病主要鉴别小种对培矮 64S 进行的抗谱测定，培矮 64S 的抗谱与携有 *Xa4* 基因的 IRRI 品种相类似，也不抗中国Ⅴ型菌。培矮 64S 是以农垦 58S 为母本与培矮 64（IR36 衍生系 IR9761－19 系选）杂交育成，推测上述那些组合由培矮 64S 配成的组合其抗性也有可能是由 *Xa4* 控制的。广西农业大学的刘丕庆实验室李柳娟等（2005）用 SSR 分子标记 M55 检测了广西地区推广面积较大的 45 个两系和三系杂交稻组合，结果表明，两优培九、培杂双七、Ⅱ优 084 等 33 个组合（占 73.3%）均携有 *Xa4* 纯合基因；汕优

63、博优 253 等 9 个组合（占 20%）携有 *Xa4* 杂合基因，进一步证明我国杂交稻白叶枯病抗性的主体基因是 *Xa4*。

1.3 中国水稻抗白叶枯病品种的抗性渊源

籼稻品种抗白叶枯病育种的抗源主要有三种：

①竹占和竹印 2 号　竹占是广东省地方品种，其衍生品种竹印 2 号继承了竹占的优质、抗病虫力强等特点，50 年代前后在华南稻区曾大面积用于生产，并作为粤、闽、赣、湘等省水稻育种的骨干亲本。60～80 年代中期，在由竹印 2 号衍生的 8 辈 19 个推广面积较大的品种中，高抗白叶枯病的青华矮 6 号、竹包 384、晚华矮 1 号、江西丝苗等曾是南方晚籼生产中的主栽品种。

②斯里兰卡品种 BG90－2　该品种高产、抗病力强，除直接在苏、皖、鄂等省生产种植外，80 年代作为南方中籼育种骨干亲本，育成了扬稻 1～6 号、金陵 57、四喜占等一批高抗白叶枯病兼抗稻瘟病或褐飞虱的品种，广泛种植于苏、浙、皖、湘、鄂、豫等地区（张洪熙，2000）。

③引进的 IRRI 品种　IR20、IR22、IR26、IR28、IR29、IR30、IR36 和 IR54 等品种能抗多种病虫害，都曾作为水稻育种的抗源亲本，衍生了一批生产上大面积种植的抗病品种。如高产、抗白叶枯和稻瘟病的品种二九丰（IR29/原丰早），广泛种植于浙、赣、湘、皖、鄂、闽、蜀等早籼北片稻区，取代了当时感病的当家品种广陆矮 4 号；双抗的湘早籼 3 号（IR36/广解 8 号）、湘早籼 1 号（IR22/红米冬占/广秋矮///ASD7）等品种也都在生产上大面积种植了 10 年以上。IR 品种除了作为对多种病害的抗源外，还对水稻雄性不育基因具有恢复性能，如 IR28、IR30、IR26、IR36、IR54 等品种一直是我国籼型杂交稻育种的主要恢复系和恢复基因源，其覆盖面之广就是不争的事实。

抗性遗传研究表明，由 BG90－2 衍生的金陵 57、南农籼 2 号以及扬稻 1～4 等系列品种和 IR 系统品种衍生的二九丰、83007、三黄占等抗病品种对菲律宾小种 1（PXO61）的抗性均由 *Xa4* 基因控制（沙学延等，1994；施爱农等，1993）。来源于竹印 2 号的华竹矮、青华矮 6 号等品种的抗性基因与 *Xa4* 基因也是等位的。通过基因推导分析上述 3 个抗源及其衍生的大批抗病品种的系谱，其抗白叶枯病基因多没有越出 *Xa4* 的范围（伍尚忠等，1990；章琦等，1990）。地方品种竹占的抗性遗传缺乏明确资料，其抗性与 *Xa4* 基因有什么关系，是否携有其他基因都不清楚。纵观我国籼稻品种（组合）对抗白叶枯病的抗性基本上由 *Xa4* 基因控制，其遗传基础的局限性、狭窄性和脆弱性是显而易见的。*Xa4* 已经在我国南方部分稻区遇到毒性菌株的侵染，其抗性继续下降或全面丧失应是预料中的事。

粳稻品种（组合）的白叶枯病抗源品种如下：

①农垦 58　由它衍生了以农虎、鄂宜、鄂晚、矮粳、城特、秀水、祥湖为编号的一批高产、抗病品种。

②农垦 57　由它衍生了以南粳、盐粳为编号的优质、高抗白叶枯病的品种。

③黎明、福锦、丰锦、秋光、早锦、喜峰等　这些从日本引进的品种都是华北、东北地区 70 年代以来的主栽品种，也是这两个地区育成的新稻 68－11、辽盐 2 号、中百 4 号、红旗 16 等抗病品种以及粳型杂交稻主要不育系黎明 A、秋光 A 等的抗病供体。

抗性遗传研究和系谱分析表明，上述南方稻区的粳稻品种以及华北、东北地区的早粳品种其白叶枯病抗性的来源主要是农垦 57、农垦 58 和源于爱国等的日本品种，所携有的抗白叶枯病基因均与早生爱国 3 号或 Zenith 的成株抗性基因 *Xa3* 是等位的（章琦等，1988；1990；1991；1994；1995）。也就是说，尽管抗源品种较多，其抗性基因只有一个，我国粳稻白叶枯病抗性基因都是 *Xa3* 基因（表 8）。

表 8　携有 *Xa3* 基因的我国粳稻抗白叶枯病品种[b]

Table 8　Chinese japonica rice cultivars carrying *Xa3* gene for BB resistance（Summarized by Zhang，2005）

品种 Cultivar	组合 Cross	抗源 Resistance source	系谱图号 Genealogy	文献[b] reference
辽粳 10 号	BL-6/丰锦	爱国（Ai koku）	—	林世成等，1991
新稻 68-11[a]	山法师系选	爱国（Ai koku）	5-2	同上
中百 4 号[a]	喜峰/南粳 15	农垦 57（良作/爱知旭）	—	章琦等，1990
盐粳 902-2[a]	南粳 11 辐射	农垦 57（良作/爱知旭）	6-3	同上
南粳 15[a]	农垦 57/农垦 51	农垦 57（良作/爱知旭）	6-3	同上
湘虎 25[a]	湘粳 8 号/农虎 6 号	农垦 58	7-5	章琦等，1994
邳早 15[a]	南粳 15 号辐	农垦 58	—	章琦等，1990
蚌珠芒[a]	地方品种	—	—	同上
早生爱国 3 号[a]	引自日本	爱国（Ai koku）	—	章琦等，1991
Zenith[a]	Blue rose	—	—	同上
丹粳 2 号	京引 83 系选	爱国（Ai koku）	5-6	林世成等，1991
南粳 11，南粳 23，盐粳 20	农垦 57 系选	爱国（Ai koku）	6-3	同上
扬粳 201	南粳 11/南粳 33	爱国（Ai koku）	6-3	同上
宿粳 2 号	南粳 15 辐射	爱国（Ai koku）	6-3	同上
南粳 35	农垦 57/桂花黄//IR26///台中育 39	爱国（Ai koku）	6-3	同上
矮粳 23	矮粳 22/桂花黄//农红 73	农垦 58	7-2	同上
鄂晚 5 号	鄂晚 3 号/4243	农垦 58	7-6	同上
鄂宜 105，沪选 19	农垦 58 系选	农垦 58	7-6	同上
农虎 6 号	农垦 58/老虎稻	农垦 58	7-7	同上
嘉湖 6 号	京系 69-408/农虎 6 号	农垦 58	7-7	同上
秀水 48	辐农 709/京引 154//辐农 709	农垦 58	7-7	同上
浙粳 66	矮粳 23/百哥稻＋花培	农垦 58	7-9	同上
黎明 A	BT 台中 65A/黎明	爱国（Ai koku）	8-10	同上
丰锦 A	台中 31A/丰锦	爱国（Ai koku）	8-10	同上

[a] 经抗性遗传分析携有 *Xa 3* 抗性基因（Cultivars genetically analyzed for BB resistance）。

[b] 根据《中国水稻品种及其系谱》中抗源及抗性基因分析资料推测基因（Based on the book *Chinese Rice Varieties and Their Genealogy* and some papers on BB resistance gene analysis from Zhang et al.）。

综上所述，*Xa4* 和 *Xa3* 分别控制了我国籼稻和粳稻品种（组合）的白叶枯病抗性，而大面积长期种植携有单一基因的品种（组合），极易丧失抗病性。伍尚忠等（1983）于 20 世纪 80 年代初在广东发现了对携有 *Xa4* 的 IR26 品种具有毒性的中国白叶枯病Ⅴ型菌之后，该型菌的分布范围迅速扩展至闽、桂、浙、苏、滇等省，发生频率不断上升（徐羡明等，1995；曾列先等，2002），使生产中的一些主栽品种（组合）的抗性逐渐衰退。例如，80 年代中期以来年推广面积最大、一直在 333.3 万 hm^2（5 000 万亩）以上的杂交稻汕优 6 号，到 20 世纪末由于抗性及优势均趋衰落，至 2004 年的种植面积已下降至 29.47 万 hm^2（442 万亩）。李宣铿等（1993）连续鉴定了 1986—1990 年间由全国水稻育种攻关协作组育成的 27 个三系杂交稻对白叶枯病的抗性，只有江优 594 一个组合表现中抗。又据 1991—1993 年全国水稻育种攻关协作组对供试杂交稻组合的抗病性进行的统一评价，在 17 个组合中也仅有江优 594 和博优桂 33 中抗白叶枯病，其余为中感/感病。进入 21 世纪以来，籼型三系杂交稻的恢复系仍然以明恢 63 和 IR36 等为主，其抗性基因都未能超越 *Xa4* 的范围，随时随地面临着中国Ⅴ型菌侵袭暴发流行危害的风险。由抗性遗传基因单一而引起丧失抗性的事例在南方粳稻品种中也有发生，例如浙江省育成的携有 *Xa3* 基因的抗病当家品种秀水 11，1990 年在种植面积最大的宁波地区暴发了白叶枯病，给生产造成严重损失（徐建龙，1996）。

2 日本水稻抗白叶枯病育种

2.1 进展简况

稻作是日本农业的主体，在良种选育、栽培革新、冷害和病虫害治理方面，无论从发展策略、生产技术还是应用基础研究都取得了令人瞩目的进展与成就（闵绍揩等，1989）。稻瘟病是日本水稻生产上最严重的病害。1904 年就开始进行水稻抗稻瘟病育种，最早采用在重病田选择的方法，1910 年开展杂交育种，经历了田间抗性育种、真（垂直）抗性育种，真抗性品种丧失抗性后又回到田间抗性育种三个阶段（Ezuka，1979）。其所形成的广泛收集稻种资源进行抗性鉴定、抗病品种群的划分、品种抗性改良的理念等都为后来的抗白叶枯病育种工作提供了经验，打下了基础。日本在 1923 年开始水稻抗白叶枯病育种，当时在疫区进行自然鉴定，很快在地方品种中筛选出三个主要抗源品种，即滋贺关取 11（Shiga-sekitori 11）、高农 35（Kono 35）和庄兵卫（Shobei），随即开始了抗病杂交育种工作。

爱媛（Ehime）农试场将滋贺关取 11 与小雄（Shoyu）杂交，于 1932 年育成了抗病品种全胜（Zensho）（Ezuka et al.，1969）。20 世纪 60 年代之后，九州农试场又从全胜中系选出抗性更强的全胜 26，并用它育成了有明（Ariake）、丰沃（Hoyoku）、石优（Kokumasari）、不知火（Shiranui）等抗病品种（Soga et al.，1963；Okada et al.，1967）。宫崎（Miyazaki）农试场同时也用全胜 26 及其衍生品种育成了大淀（Oyodo）、丰玉（Toyotama）及南国糯（Nangokumochi）等抗病品种（Ezuka et al.，1978）。

1926 年鹿儿岛农学院在一块感病品种神力（Shinriki）的病田里选出一个抗病株即高农 35。1936 年熊本县（Kumamoto）农试场用高农 35 和旭 1（Asahi 1）杂交，1946 年育成了农林（Norin）27，这是第一个被推荐到生产应用的抗白叶枯病品种（Fujii et al.，1967）。农林 27 曾在白叶枯病流行的日本西南部广泛种植，后来由于感染稻瘟病，面积逐渐下降。九州农试场便将农林 27 与宝（Takara）杂交，育成了抗白叶枯病品种朝风（Asakaze）。

爱知农试场 1926 年用庄兵卫与白千本（Shiro Senbon）杂交，于 1932 年育成了黄玉（Kogyuku）。黄玉不仅广泛用作白叶枯病抗性的研究，也是许多杂交组合的抗源亲本。由它衍生的抗病、优质、高产的优良品种幸风（Sachikaze）和日本晴（Nipponbare）分别于1960和1963年开始在生产上种植。此外还衍生有黄金丸（Koganemaru）和抗性较强的枯叶不知（Hagareshirazu）（Yamaoka et al.，1966）。这些品种均曾在白叶枯病严重的日本西南地区广泛种植，一度有效地控制了该病的危害。

滋贺关取11、高农35和庄兵卫这三个抗源品种，对日本水稻抗白叶枯病育种作出了重要贡献。1930—1972年间，日本育成的农林27、黄玉、全胜26和朝风等28个抗白叶枯病品种，其抗性都是从这三个抗源品种传承下来的。这些被划分归属于黄玉群的品种携有控制日本白叶枯病菌群Ⅰ的抗性基因 *Xa1*。1957年黄玉群的抗病品种“朝风”在白叶枯病危害严重的九州地区遭遇到对 *Xa1* 基因有毒菌株的袭击而感病化（Wasano *et al.*，1987）。对此，日本水稻育种家便开始转向广谱抗性品种的选育，并在20世纪60～70年代开始将育种目标从抗病高产转为优质、抗病、高产，同时启用能抗日本Ⅰ、Ⅱ和Ⅲ三个白叶枯病菌群的早生爱国3号、中新120等携有抗性基因 *Xa3* 的品种（Sakaguchi，1977）作为抗源，育成了70X-46、中国45（Chugoku 45）、BL1和中系314（Chukei 314）等抗病品种（Washio et al.，1966；Ezuka et al.，1978）。随后的抗性遗传研究也证实了这些品种都携有 *Xa3* 基因。*Xa3* 一直是日本抗白叶枯病育种的主要基因源。当发现 *Xa3* 基因能被日本小种Ⅳ和Ⅴ侵染后，育种家又开始考虑应从质量抗性育种转向聚合多个抗性基因和利用水平抗性的质量与数量抗性并举的策略。在利用 *Xa3* 基因的同时又采用对日本Ⅰ～Ⅴ五个小种都具有中等抗性的缅甸品种 Lead rice 作为抗源，与丰沃品种（Hoyoku）杂交和复交，育成了70X-11、70X-37、70X-38、70X-41和70X-42等品系，这些品系高抗日本小种Ⅰ和Ⅴ，中抗小种Ⅱ、Ⅲ和Ⅳ，兼备了质量和数量抗性的特点。这一时期九州农试场与熊本农试场合作，用 Lead rice 为抗源于1973年育成了具有“水平抗性”的日本晴和阿苏稔（Asominori）（Sato et al.，1977）。可以说，进入70年代以后，日本重视聚合不同类型抗病基因源的育种策略是成功的，在大面积上推广的中抗品种基本上控制了白叶枯病的危害。1983年排在全国推广面积前十位的高产优质水稻品种中，丰锦（Toyonishiki）、秋光（Akihikari）、日本晴和西誉（Nishihomare）等都具有较好的抗性。日本晴与阿苏稔等已在日本的中国和九州等地区大面积种植，用它们作抗源品种又育成了西光（Nishihikari）、宫光（Yakaminori）、丰金（Toyokogane）、南海119（Nankai 119）、南海122（Nankai 122）、玉系86（Tamakei 86）等品种（Matsumoto et al.，1977），都对多个日本白叶枯病小种表现中抗（Horise，1996）。此外，从IRRI引进的IR28、IR26、Pelita Ⅰ-1等品种对日本的白叶枯病菌株也有“水平抗性”的表现。通过专化抗性和“水平抗性”的联合应用，至今，水稻白叶枯病在日本广大稻区已得到有效控制，只在零星山区偶有发现。日本植物病理学家野田孝人认为这是20世纪80年代以来生产上注意发展种植具“水平抗性”的优良品种所产生的效果（个人通讯，2003）。

2.2 水稻白叶枯病抗性品种群

日本科学家借鉴本国抗稻瘟病育种的经验，制订出一套白叶枯病抗性评价与利用的方法，确立了日本白叶枯病菌群和抗性品种群的相互关系。抗性品种群的划分就是将供试品种用不同毒性群的代表菌株测试后，把抗性反应型相同的品种归成一类，形成抗性品种群；同样地，将待测菌株用一套鉴别品种进行测试，把毒性反应型相同的菌株归成一类，形成毒性菌株群（简称菌群）。例如，Sakaguchi 等（1968）用 12 个菌株测试金南风、黄玉、兰泰艾玛斯（Rantai Emas）等 20 个品种的反应，结果是：以金南风为代表的一些品种感染所有菌群；以黄玉为代表的一些品种抗菌群Ⅰ、感菌群Ⅱ和Ⅲ；以兰泰艾玛斯为代表的一些品种抗菌群Ⅰ和Ⅱ、感菌群Ⅲ；以早生爱国 3 号为代表的一些品种对Ⅰ、Ⅱ、Ⅲ三个菌群都抗病。据此，待测菌株被划分为Ⅰ、Ⅱ、Ⅲ三个群，待测品种则被划分为金南风、黄玉、兰泰艾玛斯和早生爱国 3 号四个群。后来，当 Yamamoto 等（1977）筛选出能在日本鉴别品种上呈现清晰交叉反应的两个印度尼西亚菌株后，又将它们分别定为日本菌群Ⅳ和Ⅴ。不久将菌群确定为小种。根据早生爱国 3 号抗性品种群的品种只抗小种Ⅰ、Ⅱ和Ⅲ，不抗小种Ⅳ和Ⅴ的情况，又将早生爱国 3 号抗性品种群中抗小种Ⅰ、Ⅱ、Ⅲ和Ⅴ而感染小种Ⅳ的品种划分为 Java 抗性品种群。这样，经 Ezuka 等（1974a）的系统研究，日本白叶枯小种（菌群）由 3 个增加到 5 个，抗性品种群由 4 个增加到 5 个（表 9）。育种家可以参照小种的消长作为制定育种计划的依据，以此来更换使用抗性品种群。

表 9 日本水稻白叶枯病抗性品种群对病菌小种的抗性反应型（Ezuka et al.，1974a）

Table 9 Disease reactions of 5 rice cultivar-groups to 5 races of *Xanthomonas oryzae* pv. *oryzae* in Japan（Ezuka et al.，1974a）

品种群 Cultivar group	代表品种 Representative cultivar	小种[a] Race Ⅰ	小种 Race Ⅱ	小种 Race Ⅲ	小种 Race Ⅳ	小种 Race Ⅴ	抗性基因 *Xa* gene
金南风（Kinmaze）	金南风	S	S	S	S	S	无
黄玉（Kogyoku）	黄玉	R	S	S	S	R	*Xa1 Xa12*
兰泰艾玛斯（Rantai Emas）	特特普（Tetep）	R	R	S	S	R	*Xa1 Xa12*
早生爱国 3 号(Wase Aikoku 3)	早生爱国 3 号	R	R	R	S	S	*Xa3*
爪哇（Java）	Java14	R	R	R	S	R	*Xa1 Xa3 Xa12*

[a] S，感（Susceptible）；R，抗（Resistant）。

3 IRRI 水稻抗白叶枯病育种

3.1 创建“遗传评价与利用研究”

IRRI 建所的目的是改进水稻的产量和品质，为全世界水稻生产服务。其具体研究任务是培育具有高产、优质、抗多种病虫、耐逆境的优良品种及其配套栽培技术，使育成的现代品种符合当代农作制度的要求并在生产上充分发挥其潜力。这样，就必须组织有关学科或专业的科研人员参加，在统一计划下科学分工、通力协作，才能实现上述目标。因此，在 20 世纪 60 年代末他们提出了水稻多学科协作开展“遗传评价与利用研究”这个设想（Concept of multidisciplinary genetic evaluation and utilization），简称 GEU 项目。组织从事品种资源、遗传、育种、植物病理、昆虫、植物生理、土壤化学、栽培等专业的科学家共同参加、互相配合，围绕水稻生产上的需要和问题高效率地完成这项种质评价、遗传改良与育种计划。1973 年正式实施该项目时，投入的科技人员和经费占全所人力财力物力的 40%左右，由一位副所长负责协调总体工作。不仅在所内开展特性鉴定与评价、多抗性筛选、遗传规律、育种方法、亲本选配、杂种后代培育、选择与多点测试、品种适应性鉴定以及示范栽培等研究工作，并组织吸收各国水稻研究机构参加该项目在各地设置的国际水稻遗传评价网（INGER，前称 IRTP）的各项试验圃工作，实现“遗传评价与利用研究”的国际化、全球化。至 21 世纪初，该项目已在 62 个国家设有 15～22 个试验圃。通过参加国之间以及它们与 IRRI 之间水稻种质的自由交换、综合评价，都能各自高效率地选育出一批批高产、优质、多抗、广适的新品种投入生产应用。IRRI 育成的以 IR 编号的优良品种在 20 世纪 80 年代中期编号到 IR77 之后，宣布从那以后提供给各国测试的品种（系），如在生产上表现优良，均由所在国命名，不再用 IRRI 编号。这样，IRRI 的研究成果就可以与全球稻米生产国家共享而得到更充分地利用，从而为实现其建所宗旨和面向全球化更上一层楼。1985 年，时任 IRRI 所长的 M S Swaminathan 在建所 25 周年纪念会上总结该所与各国农业研究机构及国家水稻科研项目的合作，是划时代的、具有强大推动力的全球性大协作，也是共同取得水稻生产创纪录进展的杰出典范，并指出这一成就被其前任所长 N C Brady 称为 IRRI 与合作伙伴们共同创建的一条“应用科学的探索之路（an adventure in applied science）”是十分确切的。它促使全球水稻高产、稳产及多抗性育种工作取得了重大突破。40 多年过去了，笔者认为 GEU 这一科学的研究方式对增强我国作物科学乃至农业科学技术大协作，从而走向大发展仍然有着很大的现实意义。

3.2　抗多个病虫的多抗性育种进展

IRRI 于 1965 年用来自印度的 TKM6、W1264 和印度尼西亚的 Sigadis 等作为抗源品种，开始了水稻白叶枯病抗性改良工作（Khush，1977），相继育成了一批抗病品种（系）。1969 年育成的 IR20（IR262－24－3/TKM6）和 IR22（IR8/Tadukan）是最早在生产上推广应用的两个抗白叶枯病品种，虽然 IR8 和 Tadukan 这两个亲本都感染白叶枯病，但由其杂交育成的 IR22 却是抗病的。之后，在 1973—1975 年间又陆续育成了 IR26、IR28、IR29、IR30 和 IR34 等源自 TKM6 的抗病品种。这些品种曾在亚洲主要稻作地区广泛种植或在各国抗病育种中作为抗源利用。1975—1980 年间作抗源利用的除了上述品种以外，还扩大到 3 个印度尼西亚品种 W1263、Pelita1/1 和 Pelita1/2，以及 Zenith（美国）、BJ1（尼泊尔）、DV85（孟加拉）等 100 多个材料。这一期间，已从这些抗源品种中鉴定出来自日本的 *Xa3*（早生爱国 3 号）、印度的 *Xa4*（TKM6）、孟加拉的 *xa5*（DZ192）、*xa5* 和 *Xa7*（DV85、DV86）及美国的 *xa8*（PI231129）共 5 个抗白叶枯病基因，以及被确认为与 *Xa3* 相同的 *Xa6* 和 $Xa4^b$ 两个基因。携有 *Xa4* 基因的抗源品种在育种中被利用得最为频繁，在 IRRI 的 33 808 个杂交组合中有 18 576 个组合应用了 36 个携有 *Xa4* 基因的品种，占 54.8%；其次，有 3 000 个杂交组合应用了 19 个携有抗谱较广的隐性基因 *xa5* 的品种，占 16.0%；而 *Xa3* 等其他抗性基因的利用率都不到 3%（Mew et al.，1981）。在 IR 系统中，从 IR20 至 IR66，除 IR56 以外的 27 个抗病品种都对菲律宾小种 1 至 6 表现出与 *Xa4* 基因相同的抗谱：即抗小种 1 和 5，中抗小种 4，中感/感染小种 2、3 和 6（Khush，1989；表 10）。这些品种广泛种植于菲律宾各地，70 年代后 IR20、IR26、IR28、IR32、IR34、IR36、IR50 和 IR64 等品种已占全国水稻种植面积的 65%，至 20 世纪 90 年代初，则达到 90%（Mew et al.，1992）。此外，IR 编号的品种也遍布中国、印度、印度尼西亚、孟加拉、斯里兰卡、泰国、尼泊尔等亚洲国家。显然，IRRI 对提高这一广大地区的水稻生产水平和控制包括白叶枯病在内的多种病虫危害发挥了很大作用。

表 10 IR 品种对菲律宾 6 个白叶枯病小种的抗性反应（Khush et al.，1989）

Table 10 Reaction of IR cultivars to 6 Philippine races of *Xanthomonas oryzae* pv. *oryzae* in the Philippines（Khush et al.，1989）

品种 Cultivar	对小种的反应[a] Reaction to race					
	1	2	3	4	5	6
IR5	S	S	S	S	S	S
IR8	S	S	S	S	S	S
IR20	R	MS	S	MR	R	S
IR22	R	MS	S	MR	R	S
IR24	S	S	S	S	S	S
IR26	R	MS	MS	MR	R	S
IR28	R	S	S	MR	R	S
IR29	R	MS	MS	MR	R	S
IR30	R	MS	MS	MR	R	S
IR32	R	MS	S	MR	R	S
IR34	R	MS	S	MR	R	S
IR36	R	MS	S	MR	R	S
IR38	R	MS	MS	MR	R	S
IR40	R	MS	S	MR	R	S
IR42	R	S	S	MR	R	S
IR43	R	MS	S	MR	R	MS
IR44	R	S	S	MS	R	S
IR45	R	MS	S	MR	R	S
IR46	MR	S	S	MS	R	S
IR48	R	R	MR	MR	R	MS
IR50	R	S	S	MR	R	S
IR52	R	S	S	MS	R	S
IR54	R	MR	MR	MR	R	S
IR56	S	S	S	S	S	S
IR58	R	MS	MS	MR	R	S
IR60	R	MS	MS	MR	R	S
IR62	R	S	S	MR	R	S
IR64	R	MS	MS	MR	R	S
IR65	R	S	S	MR	R	S
IR66	R	S	S	MR	R	S

[a] S，感病（Susceptible）；R，抗（Resistant）；MR，中抗（Moderately resistant）；MS，中感（Moderately susceptible）。

IRRI的多抗性育种是以抗病虫为主，兼顾耐低温、干旱和土壤胁迫如耐盐碱等。主要病虫害包括稻瘟病、白叶枯病、通戈洛病、草状矮缩病、褐飞虱、叶蝉、稻螟和稻瘿蚊等4病4虫。上述的IR品种除高产、优质、适应性好外还能抗6～8种病虫害（表11）。

表11 具多抗性基因的IR品种对病虫的抗性反应（IRRI，1985）

Table 11 Disease and insect reactions[a] of IR cultivars with multiple resistance to several pests（IRRI，1985）

品种 Cultivar	稻瘟病 Blast	白叶枯病 Bacterial blight	草状矮缩病 Grassy stunt	通戈洛病 Tungro	褐飞虱[b] Brown planthopper			叶蝉 Green leaf hopper	稻螟 Stem borer	稻瘿蚊 Gall midge
					1	2	3			
IR5	MR[a]	S	S	S	S	S	S	R	MS	S
IR8	S	S	S	S	S	S	S	MR	M	S
IR20	MR	R	S	MR	S	S	S	R	MR	S
IR22	S	R	S	S	S	S	S	S	S	S
IR24	S	S	S	S	S	S	S	R	S	S
IR26	MR	R	S	MR	R	S	R	R	MR	S
IR28	R	R	R	R	R	S	R	R	MR	S
IR29	R	R	R	R	R	S	R	R	MR	S
IR30	MS	R	R	R	R	S	R	R	MR	S
IR32	MR	R	R	R	R	S	R	R	MR	S
IR34	R	R	R	R	R	R	S	R	MR	S
IR36	R	R	R	R	R	R	S	R	MR	R
IR38	R	R	R	R	R	R	S	R	MR	R
IR40	R	R	R	R	R	R	S	R	MR	R

[a] R，抗（Resistant）；MR，中抗（Moderately resistant）；MS，中感（Moderately susceptible）；S，感病（Susceptible）。

[b] 对3个生物型的反应（Reaction to 3 biotypes of brown planthopper）。

IRRI育成的第一个携有*Xa4*基因的抗病品种IR20在20世纪70年代初在菲律宾吕宋岛也遭遇到新毒性菌株"Isabela"的侵袭而丧失了抗性（Kauffman，1974）。这本是预料中的事，因为IRRI已发现IR20的抗病供体TKM6是感染"Isabela"菌株的。*Xa4*的抗谱并不宽，它虽然对东亚和东南亚的白叶枯病优势小种表现抗病，但不抗南亚的某些菌株。表明在推广种植携有*Xa4*的抗病品种时就已潜伏着毒性小种侵袭的隐患。随着携有单一抗病基因的品种扩大种植，诱导病菌小种均势发生变化，促使新兴强毒性菌株的频率迅速上升而暴发为害。当时印度尼西亚、印度、中国和菲律宾等国的许多地区，都是以推广种植携有*Xa4*基因为主的品种来控制水稻白叶枯病的危害，所以无法逃脱抗性丧失的厄运（Mew et al.，1992；2004）。IRRI的育种家随即采用抗谱较广的*xa5*和*Xa7*基因作为抗源相继育成了一批新的优良抗病品种（系）。当他们一发现菲律宾出现小种6后，又从印度引进能抗菲律宾当时所有6个白叶枯病小种的长药野生稻（*O. longistaminata*），将其抗性导入IR24，育成了携有*Xa21*基因、对白叶枯病具广谱高抗的新品系IRBB21，并很快就被全球水稻育种家广泛应用于抗病育种。然而，被广泛利用的 *Xa21* 基因又在韩国、印度、印度尼西亚等国遇到毒性菌株，我国广东省（Ⅳ型、Ⅴ型）、云南致病型（小种）10 也能侵染 *Xa21*。IRRI 育种家 Jena（2004）在韩国的国家作物科学研究所合作培育了粳稻抗白叶枯病品种。育成了 *Xa4*＋*xa5*＋*Xa21* 的三基因粳稻聚合系，该品系能抗所有供试的 16 个韩国菌株，比 *Xa21* 的抗性有了很大改善。由此可见，与病菌作斗争的研究和抗病育种就是这样针锋相对、抢占先机、长期坚持、与时俱进的科学实践。

进入 20 世纪 90 年代至世纪之交，随着分子生物学的迅猛发展，水稻抗病（虫）基因的分子（包括 QTL）定位、标记与克隆技术日臻成熟，IRRI 的抗性育种也逐步转向以常规育种工作为基础的分子标记辅助选择和转基因育种以及分子设计育种的探索。Huang 等（黄宁等，1997）在 IRRI 采用常规育种结合分子标记辅助的选择方法，将 *Xa4*、*xa5*、*xa13* 和 *Xa21* 聚合于 IR24 品种，育成了编号为 5059（*Xa4*＋*xa5*）、5060（*Xa4*＋*xa13*）、5061（*xa5*＋*Xa21*）、5062（*xa5*＋*xa13*）、5063（*xa13*＋*Xa21*）、5064（*Xa4*＋*Xa21*）、5065（*Xa4*＋*xa5*＋*xa13*）和 5066（*Xa4*＋*xa5*＋*Xa21*）等 8 个双基因或三基因聚合系。赵炳宇等（1998，个人通讯）用菲律宾白叶枯病 1～9 小种比较了这套聚合系与 4 个单基因系的抗性，结果表明所有聚合系的抗谱都拓宽了，基因间互补效应十分明显。IRRI 在 Huang 等的研究基础上，最近已把 *Xa4*、*xa5*、*Xa7*、*xa13*、*Xa21* 等抗性基因中的 2～5 个基因累加在一起育成了聚合系 IRBB59～IRBB66（Datta et al，2002），拓宽抗谱、增强抗性，以延长抗性基因的使用寿命。其中 IRBB60（*Xa4*＋*xa5*＋*xa13*＋*Xa21*）、IRBB64（*Xa4*＋*xa5*＋*Xa7*＋*Xa21*）、IRBB65（*Xa4*＋*Xa7*＋*xa13*＋*Xa21*）和 IRBB66（*Xa4*＋*xa5*＋*Xa7*＋*xa13*＋*Xa21*）都高抗菲律宾现有的 10 个 *Xoo* 小种。顺便提一句，上述多基因聚合系中，都有 *Xa4* 基因的参与，因为该基因的残余抗性仍然具有利用价值，在聚合几个抗性基因时，*Xa4* 显示出它是一个优异的搭配基因（Mew，2006）。

3.3 抗病育种方法

IRRI 采用的抗病育种方法仍以杂交育种为主，是在改良株型的基础上进行的。由于涉及的病害较多，每种病害在不同时期都有其流行小种，还要与高产、优质、株型好、稳产等性状逐步聚合，所以用的亲本较多，必须分期分批进行复合杂交，每次复交的子一代都要求有足够大的群体，才有可能使预期的目标基因结合在一起。因此育种规模很大，一般每季杂交计划都在 2 000 个组合以上，其中三交或双交每组合的当代种子至少在 300～500 粒以上，有限回交每次为 100～200 粒，多次回交每次 20～30 粒。对于杂种分离世代则采用系谱法。进行个体选择时，要以各株或株系对有关病害抗性反应的上下代记载为依据，即从最近一次复交的 F_1 开始至 F_4 或 F_5，根据各世代的抗病性反应（或专门检测）、产量潜力和品质指标分株分系观察记载，并联系上下代谱系资料进行选留或淘汰，直至各类目标性状基本上纯合为止（图 1）（Khush，1978）。在转移由单一主基因控制的目标性状时则采用多次回交，如把尼瓦拉野生稻（*O. nivara*）对草状矮缩病的抗性基因转移到农艺性状优异的 IR8、IR20、IR24 等骨干品种时，回交 3～4 次就可达到目的。至于形成纯合品系之后的产量试验和适应性鉴定则与一般育种程序一样。20 世纪 90 年代以后，为了加速选育进程和提高选择目标性状（特别是抗病基因）的准确性，强调在常规育种的基础上尽可能采用分子标记辅助选择育种。

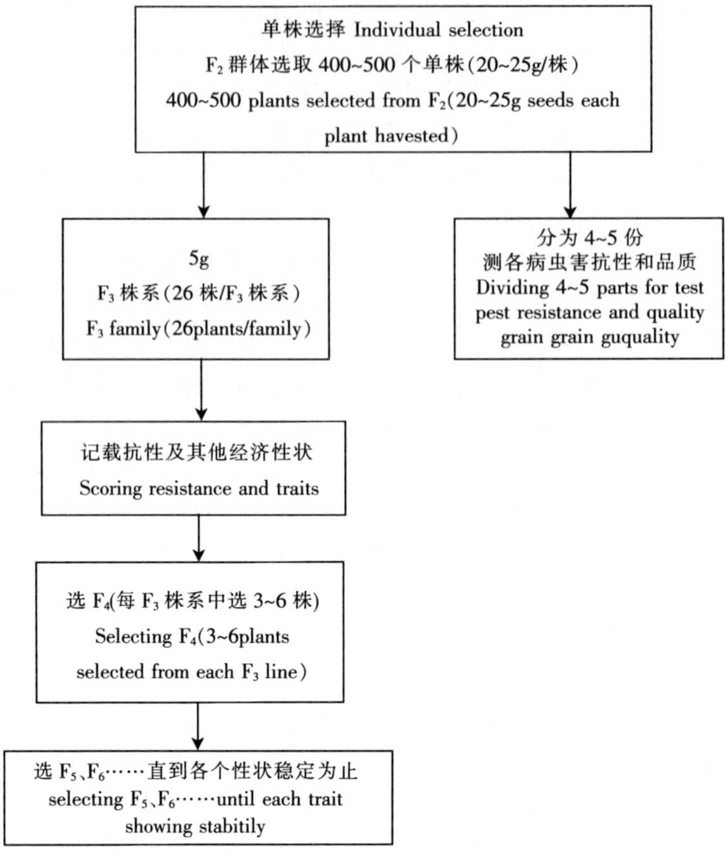

图 1 IRRI 选育病虫多抗性品种的系谱法程序

Fig. 1 Pedigree procedure of breeding for developing rice cultivars with multiple resistance at IRRI (Khush, 1978)

4 亚洲其他国家水稻抗白叶枯病育种

在世界前十位水稻生产国中，除了中国、日本和菲律宾外，印度、印度尼西亚、孟加拉、泰国和越南等亚洲国家都在进行抗白叶枯病育种工作，因手边的资料很少，只能作一般性简介。

印度有 8000 年以上的种稻历史（Oka HI，1988）。65%的人口以稻米为主食，水稻产量占粮食总产量的 52%，病虫害是限制其生产力提高的主要因素之一（Maclean et al.，2002）。1965 年设立“印度全国水稻改良协作项目（AICRTP)”，组织 45 个研究中心和农业大学的科学家开展水稻新品种选育并对育成品种的农艺性状、抗病虫能力、生理特性和所要求的土壤条件等进行全面评价。在 1966—1979 年间育成的 18 个优良水稻品种中，有 4 个能抗白叶枯病并兼抗 1～3 种其他病虫害，即：CR199 - 1 抗白叶枯病（BB)、稻瘟病（BL)、纹枯病（SB）和稻瘿蚊（GM)；RP825 - 24 - 7 - 1 - 1 抗 BB、通戈洛病毒病（RTV)、GM 和褐飞虱（BPH)；RP974 - 56 - 3 - 3 - 1 抗 BB 和 SB；RP974 - 113 - 4 - 15 抗 BB 和 RTV（Pande et al.，1980)。

同一时期，印度与IRRI建立了合作关系，在全国安排35个试验点，承担了包括品种（系）对病虫害小种（生物型）抗谱测定在内的15个国际水稻测试项目（IRTP）并进行种质交换。IRRI鉴定与命名的抗白叶枯病基因*Xa4*的供体品种TKM6就源于印度。印度育种家们推荐TKM6以及引进的IR20、IR36、IR50等数十个品种直接用于生产，并作为抗源育成了一大批以RP编号的抗白叶枯病水稻品种。这些携有*Xa4*的品种曾一度有效地控制了白叶枯病的危害。但在20世纪80年代以后，由于出现一些能侵袭*Xa4*基因的新菌株，上述抗病品种先后丧失了抗性。Goel等（1998）和Marella等（2001）分别在回顾性文章中指出，印度80年代采集的*Xoo*菌株标样中，北部有3/7的菌株和东部6/11的菌株都能侵袭*Xa4*，因此，育种家们改用携有*xa5*、*xa13*和*Xa7*这三个基因的BJ1（印度种质）、DV85、DV86、DZ78和Chinsurah BoroⅡ等品种作为抗源对当时的推广品种进行抗性改良，并广泛发掘和创造新的抗源材料。例如，印度中央水稻研究所（CRRI）的植物病理学家Devadath在1977年曾向IRRI推荐该所保存的编号为12156－1的长药野生稻种质（*O. longistaminata*）能抗御当时印度所有的*Xoo*小种，随后，Khush等（1990）在IRRI用其作为抗源育成了携有抗病新基因*Xa21*的品系IRBB21。由于该品系对白叶枯病的抗谱广、抗性好，便首先在印度广泛推广种植。然而不久发现印度东部三个地区所采集的菌株中有一半能侵袭*Xa21*基因（Goelet al.，1998；Marella et al.，2001），为此，印度育种家又采用由IRRI选育的携有4个白叶枯病抗性基因（*Xa4*＋*xa5*＋*xa13*＋*Xa21*）的聚合系IRBB60作为抗性供体，通过杂交、回交结合分子标记辅助选择技术，对当时的高产品种PR106等进行改良，获得了对白叶枯病抗谱广的新品系IET 17948和IET 17949（Srinivasan et al，2004），有望在生产上取代一些感病的推广品种。

印度尼西亚是世界前十位水稻生产国中的第三位。稻米是印度尼西亚人民的主食，水稻种植面积占粮食作物总面积的40％。病虫害是其高产、稳产的主要限制因素之一（Maclean et al.，2002）。20世纪50年代以来，以印度尼西亚中央农业研究院为主的育种单位育成了许多用于生产的高产、抗病（虫）性较好的水稻品种，在50年代育成的有Sigadis（*Xa4*）、Ramaja和Jelita，60～70年代育成的有Dara、Sintha、Bathara、Dewi Ratha和Pelita－2－2等，70年代开始与IRRI合作，强调在提高产量的基础上，以优质、抗多种病虫、抗逆境和适应性好为主要育种目标，相继育成了诸如Atomita 2、Sadang、Bahbolou、Bogowonto、Citabduy等优良品种，除抗白叶枯病外，还兼抗稻瘟病、通戈洛病毒病、纹枯病和稻飞虱中的1～3种，并有良好的蒸煮品质和食味（李南田，1989）（表12）。同时，他们也注意从IRRI引进许多IR品种作为抗源在育种中利用或直接在生产上种植，例如70年代引进的IR20、IR22等抗病品种以及1971年本国育成的抗白叶枯病高产品种Pelita 1/1都曾作为印度尼西亚水稻育种的主要杂交亲本；IR36就曾在生产上广泛种植了10年之久。后来这些品种在生产上都感染了白叶枯病，就改种能抗御印度尼西亚小种3的Cisadane品种，同时采用Zenith（*Xa3*）、DV85（*xa5*、*Xa7*）、Java14（*Xa1*、*Xa2*、*Xa12*）和携有*xa5*的IR1545－339、Dular等品种作为新抗源以充实和加强抗白叶枯病育种工作（Khush et al.，1989）。80年代以来，印度尼西亚继续与IRRI合作，应用分子标记辅助选择和常规技术，将*xa5*和*Xa7*分别导入IR64后再进行聚合，育成了Angke（*Xa4*＋*xa5*）和Conde（*Xa4*＋*Xa7*）两个抗性基因聚合系的新品种投入生产应用。在2001—2002年，又推广种植了以IR64为遗传背景、聚合了*xa5*和*Xa21*的AR32－19－3－4，在白叶枯病发生较重的湿季，比对照品种

PSB Dc28 和 IR64 显著增产（Tabien et al.，2004）。

表 12　印度尼西亚育成的抗白叶枯病主要水稻品种（选录自李南田，1989）

Table 12　Main BB resistance rice cultivars developed in Indonesia（Li et al.，1989）

品种 Cultivar	年份 Year bred	产量 Yield（kg/hm²）	全生育期/d Days to maturity	食味品质 蒸煮品质 Cooking quality	抗病（虫）性[a] Disease and pest resistance
Sigadis	1953	3 300	145	良	BB
Remaja	1954	4 000	155	良	BB
Jelita	1955	4 000	155	良	BB
Dara	1960	3 000	140	良	BB
Sintha	1963	3 500	145	良	BB
Bathara	1965	4 500	148	良	BB
Dewi Ratih	1969	4 000	140	良	BB
Pelita 1 - 1	1971	5 500	135	良	BB，TG
Pelita2 - 2	1971	5 500	135	良	BB
Atomita2	1983	5 000	125	中	BB
Sadang	1983	6 000	125	良	BB，BPH，WPH，TG
Bahbolou	1983	5 000	125	差	BB，BPH，WPH，TG
Bogowonto	1983	5 000	115	良	BB，BL
Citanduy	1983	5 000	120	良	BB，BL，BPH
Cikapundung	1984	5 000	115	良	BB，BPH

[a] BB，白叶枯病（Bacterial blight）；BL，稻瘟病（Rice blast）；TG，通戈洛病毒病（Tungrovirus）；BPH，褐飞虱（Brown planthopper）；WPH，白背飞虱（Whitebacked planthopper）。

印度尼西亚植物病理学家还根据抗病育种工作需要，将分别携有 10 个抗白叶枯病单基因（*Xa 1*、*2*、*3*、*5*、*7*、*10*、*11*、*13*、*14*、和 *21*）的 IRBB 近等基因系连续三年种植在全国不同农作制（1～4 个耕作季）的各个生态区，考察其自然发病情况，根据其在不同区域、不同季节的发病率差异，提出这些抗病基因在本国合理布局的如下建议：IRBB7（*Xa7*）在不同生态区对强毒性小种都没有严重发病，可以在全国广泛利用；IRBB7（*Xa7*）和 IRBB11（*Xa11*）可用于 Sukamandi 生态区的湿季，IRBB7 和 IRBB10（*Xa10*）可用于该生态区的旱季，以及 Suban 生态区的旱、湿两季（Kadir et al.，2004）。这种有计划地利用抗性单基因近等基因系对国内 *Xoo* 小种进行系统侦察的做法，可为育种家提供详细、具体、实用的科学信息，也是抗病育种计划不可缺少的部分，值得借鉴。

孟加拉的水稻种植面积占农作物总面积的 77%，其产值占农作物总产值的 2/3（Maclean et al.，2002）。白叶枯病是该国水稻生产的主要病害之一。孟加拉国家水稻研究所（BRRI）在 20 世纪 60 年代后期就与 IRRI 建立了以抗病育种为主要内容的合作关系。该所从 IRRI 设置在该国的各类测试圃的品种材料中，选择适于本国种植的品种直接用于生产，同时选用其中的优异种质作为抗源开展抗病（虫）育种工作。1972 年前后，生产上大面积种植的品种主要是 IR20、IR532 - 1 - 176 和 IR26。虽然那时该国也育成了一批诸如 BR817 - 1、BR168 -2B-23、BR171 - 2B - 8、161 - 2B - 25 等抗白叶枯病品种，但其抗性逐年下降。后来，育种家们采用能抗御当地 *Xoo* 优势菌株 BXO9 侵袭的地方品种 Akhnisail、Horibhog、Ghunshi、Fulbaria、Bagdor等以及能抗御另一个优势菌株 BXO5 的 DV85、Bazail 4 和 Nuncha 等品种作为抗源，选育出一批新的优良抗病品种在生产上推广后，才有效地控制了白叶枯病的危害（Khush，1989）。

越南的水稻种植面积占耕地面积的 75%，是仅次于泰国的国际第二大米出口国（Maclean et al.，2002）。白叶枯病是越南水稻生产的一个重要病害。20 世纪 60～70 年代，他们从 IRRI 引进 IR20、IR26、IR28 和 TN73－2（IR1561－228－3）等一批 IR 品种（系），经试验、试种后择优在生产上推广，其种植面积约占当时全国稻田的 40%。这些携有 *Xa4* 的 IR 品种（系）比较抗当地的白叶枯病流行小种，因此也是育种中的主要抗源品种。近年来由于推广密植、增施肥料等措施和引进我国杂交稻种植，北部稻区的白叶枯病危害日趋严重。1998 年越南河内（Ha Noi）农业大学与日本国际合作署合作研究北部稻区水稻品种白叶枯病抗性基因的分布，根据 163 个越南生产种植的品种分别在日本和 IRRI 测试其对 5 个日本小种（Ⅰ～Ⅴ）和 6 个菲律宾小种（1～6）的抗性反应结果，可划分为分别携有 *Xa3*、*Xa4*、*xa5*、*Xa10* 和 *Xa14* 五个抗性品种群，以携有 *Xa10* 基因的品种群为主；除了 *xa5* 品种群外，这些抗性品种群都不抗日本 5 个小种，也不抗越南的菌株（Taura et al.，2004）。因此他们决定先利用 *xa5* 基因来改良越南北部稻区水稻品种对白叶枯病的抗性，等到进一步了解本国白叶枯病小种情况及其分布和本国主要品种的抗性后，再仔细制定更有针对性的抗病育种计划。

泰国的水稻种植面积占耕地面积的 55%。稻米既是泰国人民的主食又是最重要的出口商品，其长粒型优质香米在国际市场上享有盛誉。白叶枯病是泰国水稻生产的主要病害之一，也是影响稻米品质稳定性的重要因素。1967 年日本科学家 Wakimoto（1975）考察了泰国白叶枯病的发生情况，尽管当年该病没有严重流行，但在各个水稻产区几乎都普遍发生，所有当地种植的水稻品种包括深水田的浮稻和新育成的优质品种无一抗病。然而携有 *Xa4* 基因的品种却能抗御当地多数的白叶枯病流行小种。因此他们先后用 IR28、IR32、IR42、IR48 以及 IR50、IR52 等 IRRI 品种作为亲本与推广品种杂交，并在 80 年代后相继育成了几批综合性状好、优质、抗病、丰产的品种，在生产上推广种植。

5 对中国水稻抗白叶枯病基因合理利用的商榷

国内外大量的研究都证实水稻品种与 *Xoo* 菌之间存在着相互定向选择的关系，由于病菌的繁殖速度远快于品种，所以，总是以品种抗性的丧失而告终。我们知道，在基因对基因的病害中，对于寄主中存在着的某一个抗病基因，病原物中都有一个与之相对应的毒性基因。当一个新的抗病基因导入寄主后，寄生在寄主中的病原菌就开始了对新遗传环境的适应过程，这个过程就是寄主与病原菌两者互作的定向选择，因为病原菌具有独特的适应寄主抗性的遗传潜能，其群体遗传结构总是随着品种基因型的变化、发展而变化、发展的。如果携有某个抗性基因的品种其种植面积已经大到足以允许新菌源能以一定数量越冬，那就表明寄生在该品种中的新病菌能够靠自己传宗接代了。定向选择使得新毒性基因在当地病原菌群体中逐渐积累到一定的比例，以致该抗性基因便丧失了原有的功能。在基因对基因的抗病育种中，品种抗性与病菌小种的兴衰循环是极其常见的。抗病基因丧失抗病性能是不可避免的自然法则，只是出现时间或早或迟而已。生产上因为品种抗性基因单一化所造成的严重后果已屡见不鲜。因此，怎样尽可能延长白叶枯病抗性基因的使用年限，怎样拓宽抗性基因的抗谱与功能，怎样合理部署和利用抗性基因才能够抢在病菌小种变异积微见著之前夺取先机、克敌制胜等等，一直是育种家和病理学家最为关切的问题。

但是，目前我国育成的抗白叶枯病的优良品种还不够多，高抗的较少，已知其抗性基因的也不多，而且抗性基因比较单一化，地区分布也不均匀；对病菌小种毒性分化的工作积累还不够，虽然曾经在20世纪80年代末确定了7个致病型，仍有不少新的毒性菌株的归属不清楚，全国统一的鉴别系统也有待完善。在这种情况下，要想在较大范围内统筹兼顾、因地制宜，将各种有效技术措施有步骤地组装上阵、多管齐下，使病菌的侵染、变异、流行、传播始终处在被严密监控、抑制和难以发展的状态是不现实的。应该根据我国的实际情况和工作基础，采取切实可行的技术措施，保护抗病品种的抗性基因尽可能少受侵袭，同时，从长远考虑，把“该补的课”尽快补齐，使各种技术储备日臻丰富成熟，这样就可以标本兼治、克敌制胜。以下是笔者的一些初步设想，愿借此抛砖引玉、兼供商榷。

5.1 抗性基因的合理利用

抗性基因的合理利用是当前和长远控制病菌新毒性菌株发展、肆虐最基本的措施。可根据各地区 *Xoo* 菌优势致病型的分布情况，将携有不同抗病基因的优良品种有计划地在相应地区轮换种植，即所谓的基因轮换，以切断优势菌群的进一步繁衍和扩散，从空间和时间上阻止病菌致病型的定向选择和发展。这是防止新毒性菌群（小种）扎根立足、发展成灾和避免抗病基因迅速丧失抗性的最简易有效的办法。实现基因合理布局，首先要及时掌握菌群时空消长变化的情报乃至其群体遗传结构的信息，这是一项不可或缺的基础工作。

我国水稻白叶枯病致病型的大致分布情况是：华南籼稻区以Ⅳ型为主，对 *Xa4* 有毒性的Ⅴ型菌已从华南扩展至长江流域；在长江下游籼、粳兼种稻区，以Ⅳ型、Ⅱ型为主；北方粳稻区以Ⅱ型和Ⅰ型居多，云南省西南部以及高原粳稻区都分布有毒力强的新致病型。从全国来看，仍以Ⅱ型和Ⅰ型为主，但Ⅳ型和Ⅴ型有所增加（许志刚等，2004）。这仅是个粗略的概况。云南、广东等一些省（市、区）还要根据本地区的具体情况增加若干辅助鉴别品种进行地区性病菌毒性变化监测，才能获得可供本地区应用的基因轮换信息。

5.2 培育抗病基因聚合系

有针对性地将几个抗性基因聚合到同一个优良品种中构筑成抗病基因屏障，是延长品种寿命的一个最有效的途径。随着更多抗白叶枯病基因的染色体定位和克隆，通过常规杂交结合分子标记辅助选择的方法，育成聚有几个主要抗性基因于一体的水稻抗病聚合系已经在国外育种实践中日益普及。近年利用此法以拓宽抗谱、提高抗性水平的许多实例已列举在上文中。我们已经和正在结合开展这方面的工作，引进多基因聚合系，用于改良我国已有的优良抗病品种，可转育到携有良好搭配性能的 *Xa4* 基因的许多品种中。开展抗病聚合系时，一要明确我国现有的主要骨干抗病基因，二要针对其主要缺点“对症下药”，三要分轻重缓急，有计划地、有步骤地将尽可能多的广谱而又互补的抗病基因聚合在一起；后面这个工作可以同时并举，以争取时间。当然，必须采用分子标记辅助选择。

5.3 “水平抗性”品种的利用

携有垂直抗性基因（特别是单基因）的品种其抗病性容易丧失是人所共知的事实。由于寄主的遗传组成单一化而带来病害大流行造成严重减产的问题已屡见不鲜。在这种历史背景下，Van der Plank（1963）提出了“水平抗性”假说。尽管争议不断，但各类大田作物中确实存在着一些品种对某种病害发病晚而轻且发展慢，能较为持久地保持在轻、中感乃至中抗水平的状态。这类品种往往在常年因综合性状优良而增产，在灾（病）年则因耐病而减产轻或少，受到农民的欢迎。我们过去曾过分强调主效抗性基因的选育而忽略了对这类抗性的挖掘与利用，其实二者并不是互相排斥的，可以同时并进、相互配合，主效基因抗性与微效多基因抗性相结合。在疫区，以推广主效基因的品种为主，以控制病害的蔓延；在非疫区可以种植“持久抗性”的品种，因为这类品种往往小种专化性不太明显，可以安全使用。对于水稻白叶枯病来说，无论是生产利用品种还是培育抗病品种，可以考虑采取主效广谱抗病基因为主结合培育具微效多基因性能，能相对持久抗病的优良品种，做到垂直抗性与“水平抗性”齐头并进，可能更有利于生产的持续发展。

5.4 加强基础性研究工作

我国水稻白叶枯病从长远计划考虑，需要大力加强的基础工作有：①尽快建立病菌小种毒性监测网，由南方稻区一个科研单位（包括院校）主其事，至于如何分区协作可通过专门会议讨论确定后立即付诸实施。②在完善全国统一鉴别系统的基础上，完成抗性品种群的划分，并绘制全国品种抗病基因型和病菌毒性群的动态分布图。推广品种的抗性评价、品种抗性基因的鉴定和病原菌毒性群（小种）的监测（短、中、长期）、都是抗病育种和病理研究的常规例行工作，必须长期坚持。③建立永久性的抗病基因资源库和病菌小种库。采用各种技术手段保存、整理、更新携有抗白叶枯病基因的品种（组合）、育种材料（恢复系、不育系或保持系等），以供基因轮换和各种研究之需。病菌小种库主要是供抗性（基因）鉴定及病理研究用。

目前我们对国内现有抗病品种所携有的抗性基因的了解不多，虽知南方稻区多为 *Xa4* 的衍生物，其次为 *Xa3* 的后代；北方粳稻区（面积很小）的抗病品种则来自日本，所携有的基因大多只是推测。需要把我国南北稻区现有生产上表现优良的抗病品种的抗性基因通过国际鉴别系统予以确认，这个摸清家底的工作必须尽快补上，才好考虑以后怎么办。在过渡期间，南方稻区在长江中、下游采用抗Ⅳ型和Ⅱ型菌的主要抗性基因有 *Xa23*、*Xa7*、*Xa21*、*xa5*、*xa13* 和 *xa24t* 等；在华南稻区抗Ⅴ型和Ⅳ型菌的主要抗性基因有 *Xa23*、*Xa7*、*xa5* 和 *xa13* 等；云南高原抗当地主要优势菌的抗性基因有 *Xa22t* 或 *Xa23* 和 *Xa7* 等；北方粳稻白叶枯病疫区采用抗Ⅱ型和Ⅰ型的抗性基因有 *Xa23*、*Xa7*、*Xa21*、*Xa4*、*xa5* 和 *xa24t* 等。

长远计划的实施取得显著成功的实例很多。例如 IRRI 的 GEU 计划为其水稻高产、优质、多抗性新品种的选育、推广应用和基础性研究取得巨大成就与进展打下了扎实基础，就是一个典型。不仅实现了建所的宗旨，而且带动了周边国家农业研究机构中心研究任务的发展。这是主管部门正确决策，不同学科、专业科学家通力协作和科学与生产密切结合三位一体的体现。这确是十分应该吸取的宝贵经验。

水稻对白叶枯病抗性的机理

林汉明[1]　张世宏[2]　章　琦[3]

（1. 香港中文大学生物系；2. 吉林大学；3. 中国农业科学院作物科学研究所）

Mechanism of resistance to bacterial blight

摘要　植物不断受环境中的病原菌挑战，但多数情况都能抗拒它们的侵入。这主要是依靠植物细胞内完备的防御系统。而病原菌也不断改变，以求突破这些防御。所以，植物—病原菌的亲和关系可视作一种协同演化的过程。植物除了既存的防御系统外，还可以通过受伤或感染诱导更强烈的抗病反应。要达到此防御目标，物理屏障、结构改变、具抗菌活性的化学分子及信号传导系统均扮演着重要角色。

植物接触到病原菌后可通过侦察一般微生物的保守结构分子（如 PAMP）开始发现病原菌并引起抗病反应。个别病原菌则突破上述防御措施，将效应分子注入植物细胞使其基本防御失效。这些效应分子包括无毒蛋白，植物又通过抗性蛋白来识别无毒蛋白，由此，植物品种与病原菌小种之间产生了专化性互作关系。

水稻与 *Xoo* 菌之间的关系是上述植物抗病系统的一个典范。水稻对白叶枯病的抗性有既存的和诱导的两类防卫系统。20 世纪 50 年代的研究主要观察了水稻的株型、水孔等形态结构与抗性的关系。Sato 的研究表明水稻固有的多种抗 *Xoo* 物质的活性与其抗性程度是对应的，大多数抗菌物质均具有乙醛和酚类的氧化的木质素。

70～80 年代对白叶枯病的抗性机理研究集中于 *Xoo* 菌侵染诱导寄主代谢变化导致两者的细胞或组织水平之形态和结构变化。Horino 等和 Mew 等的研究明确了水稻- *Xoo* 亲和与不亲和组合在细胞组织形态水平上的特异性互作，包括细菌群体增殖动态、病菌细胞和寄主细胞壁结构发生变化等生理生化过程，抑制细菌的生存以致死亡，增强了寄主的防卫效应。其他的研究进一步揭示了过氧化物酶诱发防卫反应表达的寄主细胞组织及细菌的变化，亲和与非亲和品种- *Xoo* 互作过程中 *PO-CL* 基因表达和同工酶的位置与时间动态，证明了过氧化物酶在参与形成障碍结构、直接抗菌等防卫过程中的贡献，其作用与导管木质素代谢的某些抗菌组分相关。

在分子水平上研究植物对病原物的抗性与调控机制是分子植物病理学研究的中心内容之一，而实质性开展抗性机制研究的理论基础却得益于“基因对基因”假说的建立。

水稻的抗性基因与 *Xoo* 菌的无毒基因的关系是 Flor 提出的“基因对基因”假说的经典模型。并据此确立了水稻抗白叶枯病育种的目标，在 *Xa* 基因的鉴定、克隆和应用中起了理论指导作用。然而，究竟水稻的 Xa 蛋白是否直接识别 *Xoo* 菌的无毒蛋白则未有定论。从拟南芥的研究推出了“防卫理论”，指出有些抗性蛋白的功能是侦察受无毒基因影响或破坏的植物细胞系统组分，从而间接识别无毒蛋白。“防卫理论”可被视作“基因对基因”假说的延伸和补充。目前，有关水稻对白叶枯病的抗性反应途径中所涉及的信息传导的认知仍然有限，相信与蛋白及磷酸化反应以及过氧化衍生物的信号相关。

此外，植物抗性反应中系统获得抗性在广谱及持续抗性中扮演十分重要的角色。这方面的信号传导研究主要以拟南芥为模型。水稻中已发现类似的信号组分基因，如 *NH*1（拟南芥 *NPR*1 的同系物）和 TGA 家族等。但拟南芥内的系统获得抗性受水杨酸信号的调控，而水杨酸在水稻中的角色却不明确，反映不同植物内存在紧密相关但又具独特性的抗病机理。

Abstract Plants are subjected to continuous challenges from pathogens. However, in most cases, plants can successfully prevent the invasion of pathogens, using their sophisticated defense systems. On the other hand, pathogens are undergoing continuous modification to seek ways to break these defense systems. Therefore, the compatibility of plant-pathogen interaction is indeed a process of co-evolution. In plants, in addition to the innate defense system, they also possess defense responses that are triggered by wounding or pathogen invasion. To achieve the purpose of defense, physical barriers, structural modifications, bioactive anti-microbial chemicals, and signal transduction systems all play significant roles.

Plants can detect pathogens by monitoring common and conserved structural molecules (e. g. PAMPs) and induce defense responses. To circumvent this surveillance mechanism, some pathogens evolved effector molecules that can be injected into plant cells to neutralize their basic defense responses. Some of these effectors are also known as Avr proteins. Plants will produce corresponding R proteins to recognize these Avr proteins. This forms the basis for the specific interaction in the incompatibility reactions between corresponding plant cultivars and pathogen races.

The interation between rice and *Xoo* is a typical example of the above. Rice possesses both innate and induced defense systems against *Xoo*. Researches in the 1950s were mainly focused on the correlation between *Xoo* resistance and plant shape and the morphology of hydathodes. On the other hand, Sato' s work suggested that activities of pre-formed anti-*Xoo* chemicals in rice are also positively related to the degree of its *Xoo* resistance. Most anti-microbial chemicals are oxidized derivatives of lignins.

From the 1970s to the 1980s, the studies of *Xoo* resistance mechanism focused on the pathogen-induced metabolic, morphologic, and structural changes at cell and tissue levels. Horino et al and Mew et al demonstrated the specific interactions in the compatible and incompatible groups of rice-*Xoo* combinations, including the dynamic of bacterial population growth, the physiological and biochemical changes in the cell walls of pathogens and host cells, the inhibition of bacterial growth or even the induction of bacterial death. Other studies also revealed the changes in plant cell tissues and pathogens under peroxidase-triggered defense responses. The spatial and temporal expression kinetics of the *PO - CL* gene and the peroxidase isozymes during compatible and incompatible rice-*Xoo* interaction were studied. The results demonstrated the essential role of peroxidases in the formation of structural barriers and anti-microbial activites. Such effects may be due to accumulation of some components of lignin metabolism in vascular tissues.

Molecular analysis of plant resistance to pathogens and the regulation mechanisms is a key issue of molecular plant pathology researches. In fact, these studies were based on the original "gene-for-gene" hypothesis proposed in the 1940s.

The relationship between R genes from rice and *Avr* genes from *Xoo* is a typical model for "gene-for-gene" hypothesis. This helped the indentificaition of targets in *Xoo* resistance breeding programs. It also played an important role in the identification/verification, cloning, and application of *Xa* genes from rice. However, whether the rice Xa proteins will recognize *Xoo* Avr proteins through direct interaction is still unclear. Studies in *Arabidopsis thaliana* led to the proposal of the "guard theory" . This theory stipulates that some R proteins play a role in monitoring cellular system components that are targets of Avr proteins and hence indirectly detect the Avr proteins. The "guard theory" can be considered as an extension and/or a complement to the "gene-for-gene" hypothesis.

Currently, only limited information on the signal transduction pathway related to *Xoo* resistance in rice is available. Phosphorylation cascade and ROS signaling seem to be integral components of this system.

In addition, SAR is important for board-spectrum and durable disease resistance in plants. Investigations of signal transduction pathways leading to SAR have been intensively conducted in the model plant *A. thaliana*. Genes encoding similar signal components have been found in rice, such as *NH*1 (a homolog of *NPR*1 in A. *thaliana*) and members of the *TGA* family. However, while salicylic acid signals can control SAR in *A. thaliana*, its regulatory role in rice is unknown. This suggests that closely related and yet distinct disease resistance mechanisms may exist in different groups of plants.

植物的抗病性是寄主与病原物在协同进化中形成的抵抗病原物侵染的性能。植物针对来自病原物不同方式的侵袭，相应的形成各种不同的抗性机制。植物不同水平的抗病性取决于植物与病原物互作的亲和程度，植物展示的各种抗病过程就是通过各种代谢途径形成了与多种抗病性有关的生理活性物质，其中有植物原来既存的也有接触病原物后诱发的。通过多种代谢步骤、多种酶系统，彼此间或系列串联、或并列、或以不同形式耦联而成一体。从20世纪40年代Flor提出了基因对基因学说，阐明了寄主和病原菌相互关系的遗传基础后，人们就开始了植物抗病性机理的研究。60年代开始发现了寄主对病原的过敏反应，即人们目前普遍认可的寄主主要抗病防卫反应。70年代，随着遗传学和分子生物学的长足发展，人们不仅分析了大量来自病菌的无毒基因与毒性基因，而且确定了来自寄主的一系列抗病及防卫相关基因，以及系统抗病性（SAR）等内容。90年代寄主抗病基因的分离成为全球的研究热点。一些抗病基因，特别是与信号传导有关的基因相继被克隆与鉴定出来，进一步开展了病原菌诱导的寄主多种防卫反应信号传导及交互应答的机理研究，使人们通过揭示植物抗病性的真谛，协调植物与病原物之间的相互关系，使获得寄主持久抗性的愿望可能成为现实。

水稻—白叶枯病菌是典型的基因对基因互作关系的生物模式。早期的研究多半集中在水稻表型抗性因子的观察，逐步进行水稻细胞组织变化的超微结构及生理生化指标的分析。有关抗性基因和无毒基因的功能与结构方面的研究只是近十年的事，但却发展迅速，使人们早期对植物既有的或诱导的、物理或生化的抗性因子的认识更为深入。为便于讨论这些相互关联的复杂机理，本文将从植物既存的抗病性和诱导抗病性两个方面来讨论水稻对白叶枯病抗性的物理、生理生化以及分子机理的研究进展，有关植物与病原物的相互作用的一些经典理论在此也一并作简要回顾。

1 植物与病原物的互作关系

在Flor（1942）提出基因对基因假说后，Vanderplank（1963）把植物抗病性区分为垂直抗性和水平抗性，此理论震动了病理界和育种界。这一假说在不断的争论中得到了充实，并改变了植物病害流行学和抗病育种的方向，这是一个不小的贡献。Vanderplank（1978）提出的“蛋白对蛋白假说”开始触及植物抗病性的实质问题，对于进一步分析植物所形成的各种抗病类型也颇有启迪，因此我们在此重温这些假说。

1.1 基因对基因假说

Flor根据其对亚麻品种与亚麻锈病菌（*Melampsora lini*）小种之间的互作关系研究，提出了基因对基因（gene-for-gene）假说。他是在进行亚麻抗锈病育种时发现抗病品种（不亲和寄主）具有抗病（R）基因，其相应的不亲和病原菌则携有相应的无毒基因（*Avr*）；感病品种（亲和寄主）带有感病（S）基因，亲和病原菌则有毒性基因（*Vir*）。只有当携有无毒基因的病原菌和携有抗病基因的寄主互作时寄主才能表现抗病，即品种具有一个对无毒性小种的抗性基因，其相应的病原菌毒性也受一个基因控制。一些亚麻品种有2、3或4个对无毒性原始小种的抗性基因，病原菌毒性也相应分别受2、3或4个抗性基因控制，就是说对于寄主的每个抗病基因，病菌都有一个相应的无毒基因。随后，许多研究者都证实了在小麦、大麦、燕麦、水稻、玉米、棉花、马铃薯、番茄、向日葵、咖啡等的某些真菌病害中都广泛存在着这种寄主—病原菌间的基因对基因关系。不仅真菌病害如此，细菌、病毒和线虫病乃至高等寄生性植物的病害中也是如此（曾士迈，1995）。基因对基因假说除了有助于寄主—病菌小种鉴别系统的建立、病菌毒性基因和寄主抗病新基因的鉴别外，也是研究寄主—病原物协同进化、寄主抗病与寄生物致病机制的依据。在此理论指导下，研究者得以开展植物抗病机理研究，推论出寄主的抗病过程是通过其抗病基因产物直接或间接与病原物无毒基因的产物相互识别并启动了寄主防卫反应的结果。这个假说在病原真菌、细菌、病毒和寄主关系的研究中已被普遍接受，并在抗病实践中被普遍运用并得到发展。

1.2 蛋白—蛋白共聚假说

在基因对基因假说的基础上，Vanderplank（1978）提出了蛋白—蛋白共聚假说，该假说意指在寄主与病原菌之间存在着基因对基因关系的病害中，两者是通过彼此的蛋白共聚作用（co-polymerization）相互识别的，并以此来解释寄主与病原菌的亲和或不亲和关系。当两者发生共聚后，病原菌消耗了寄主蛋白，寄主则继续合成大量蛋白而促进了代谢，病原菌得以生存，出现了寄主—病原菌亲和性（感病）现象。他以小麦叶锈、棉花角斑病等为例，认为寄主接触到病原菌后随即进行的物质交换具有决定性意义。在亲和组合中寄主细胞核能够经得住细胞受到的侵染，亲和的寄主为其细胞间隙中的大量病菌提供营养物质，这就是范氏所提出的蛋白与蛋白假说的基础。在不亲和组合（寄主抗病）中，经病原菌无毒基因识别所产生的蛋白进入寄主后呈异体蛋白，以一种外来物质的形式游离于寄主细胞中，因而不发生共聚作用，但仍然保持其“酶”的催化作用，从而影响寄主正常的代谢运转，病菌也因缺乏营养而死亡，呈现不亲和（抗病）现象。如棉花角斑病的不亲和寄主品种，在接种后一段时间，病原细菌只限于寄主具有细胞核的薄壁组织中而不进入无细胞核的木质部，病菌对棉花初期侵染的范围也只限于小叶脉之间，将其称为棉花“角斑病”既形象也贴切，这种现象在 Ennanuel 等（2001）等报道的水稻白叶枯病抗性分子实验结果中也有类似的表述，印证了水稻—白叶枯病菌的典型互作关系。

Vanderplank 进一步阐明了蛋白共聚假说的科学依据。他认为只有将寄主和病原菌的蛋白所储存的信息结合在一起，才可能使其在化学接触中产生数百万种不同的潜在方式。我们知道，超过了一定距离（超出化学键之外），分子间就不会发生反应，而分子（包括核酸、蛋白质和复合碳水化合物等三种大分子物质）之间唯有共聚之后，才可能反映寄主和病原菌储存在基因中的遗传信息。蛋白—蛋白的识别可比喻为指纹之间的识别，蛋白的表面有独特的指纹，它们的表面构型和 N、O 及 S 原子的位置各有其特点，使相互补充的蛋白指纹得以识别。Flor 的基因对基因假说中的病原菌对寄主的识别，可以理解为病原菌的无毒因子具有一个指纹，该指纹可以被与寄主中相应的抗病基因产物识别出来而呈现抗病反应。

1.3 水平抗病性与垂直抗病性

这是 Vanderplank 从植物病害流行学的角度根据抗马铃薯晚疫病品种的兴衰循环现象，按流行学的标准所提出的一对术语。

曾士迈（1998）精辟地诠释了 Vanderplank（1963）的水平抗病性与垂直抗病性假说。对水平抗病性的认识，并不是一般误解的所谓“水平抗性”就是“这个品种只要或者必须对病菌不同小种的抗性一致”。Vanderplank 根据当时抗马铃薯晚疫病新品种都很快遭到相应小种的袭击沦为高感品种，而已经在荷兰种植了 30 年的 10 个中抗或中感的品种却一直保持着原有的抗性水平，从植物病害流行学的角度分析了上述现象后，提出了“水平抗病性”。认为水平抗病性的基本特征应该是病原菌的变异与寄主的变异并无因果关系，因而不会因病原菌的适应而丧失抗性，也可以理解为在水平抗病性中不存在寄主—病原菌的分化（特异）性互作。相对而言，垂直抗病性则存在寄主—病原菌的分化（特异）性互作。曾士迈（1995）叙述了 Robinson（1976）采用范氏这一概念并加以发展，将它们应用于植物抗虫性现象以及寄主—病原菌的病害系统中。根据其互作关系分成垂直的抗性系统和水平的抗性系统，垂直抗病性表现具有寄主—病原菌特异性互作（specific interaction）或分化性互作（differential interaction）；不过虽然大多数垂直抗病性表现为抗或感，但事实上在不少病害中也存在着中度抗病的垂直抗病性。如小麦—锈菌就有这种组合，虽然在其抗性分级中高抗（1 级）的比较多，也有中抗（2 级）的。同样的现象也出现在水稻抗白叶枯病基因中，如 *Xa*21 和 *Xa*4 与菲律宾小种 6 和小种 10 的交叉特异性互作，分别表现为“抗”和“中感”，与“感”和“抗”。使我们联想到病原菌与寄主之间这种变异的有机联系，病原菌似乎随着寄主品种的变化也在经历“逐渐”适应的过程。由此寄主的垂直抗病性将因为病原菌的相适、相克，而更易于“丧失”。生产中有些情况是垂直抗病性出现了，但在寄主品种之间或细菌群体之间还没有明显的分界线，这样就可能出现中间类型。这种普遍存在寄主与病原菌之间互作关系的现象已逐渐引起研究者的关注。Li 等（2001）的研究表明携有白叶枯病抗病基因 *Xa4*、*xa5*、*xa13* 和 *Xa21* 的 4 个水稻近等基因系除了对相应的 *Xoo* 小种表现高度抗病的完全抗性外，对其他小种有不同抗性水平的部分抗性，即兼具质量抗性和数量抗性。随着人们对持久抗性的探索，在前人研究的基础上，寄主—病菌抗性与致病性互作的真谛将得到更深入的诠释。

Robinson 提出在垂直体系中，寄主抗病性的差异是垂直抗性的差别，病原菌致病性的差别是“毒性”的差别；在水平体系中，寄主抗病性的差异是水平抗性的差别，病原菌致病性的差别是“侵袭力”的差别。值得注意的是他用“垂直致病性（毒性）”和“水平致病性（侵袭力）”两个新的术语来表达两个体系之间的差别，两者之间是质的区别。

按 Vanderplank（1978）的蛋白对蛋白聚合假说来理解垂直抗病性或感病性是由“聚合作用”决定的，而水平抗病性或感病性则由“催化作用”决定，这一简明的假说与事实是吻合的。由此得出的结论是易于发生的寄主—病原菌相互变异的因果关系是：在寄主方面由感到抗的突变是涉及某一个氨基酸变为疏水性更弱的氨基酸后便减弱了聚合作用；在病原菌方面则由无毒性向毒性的突变涉及某一个氨基酸变为疏水性更强的氨基酸，从而加强了聚合作用。寄主中一个氨基酸的改变可被病原菌的相应改变所抵消，表明寄主所能做的，病原菌也能做到。因此如果给寄主加入一个抗病基因，病原菌则作出相应的回应，这就是垂直抗病性的脆弱性根源，也是作为垂直抗病性基础的“病原菌和寄主的相关变异性”的根源。而在水平抗病性中，病原菌对寄主的部分抗性的反应不能作出上述反应，这可能是其较为稳定的原因。

1.4 非寄主抗病性

非寄主抗病性（nonhost resistance）是植物对大多数潜在病原菌所表现的一种相当稳定的抗病性。植物这种抗病形式的功能运作表现在植物种属的水平上，即某植物种具有的抗病性能针对某种病原菌所有潜在的致病菌。在植物和病原菌的相互关系上，如果该病原菌是寄主植物的非致病菌（nonpathogen），则该植物就是病原菌的非寄主。非寄主抗病性不随病原菌的变异而改变，因此这种抗病性具有稳定、广谱、持久的优点，至今还没有发现病原菌能跨越非寄主抗性的限制而使非寄主变为寄主的记录。这是植物最佳的一种抗病形式，但却受制于其遗传上的复杂性，特别是常规育种难以跨越的种属限制。决定非寄主抗病性的因素极其复杂，有关其抗病机理的几种说法是：①以进化观点提出一种模型，认为寄主和病原菌的基本亲和互作是建立在非寄主抗病性被颠覆后逐代积累进化的基础上的，因此认为既存的抗病机理是构成非寄主抗性的重要部分（Heath，2000）；②强调非寄主抗病性由多基因综合控制，是其主要产物失活而不能被亲和的病原菌识别的结果（Yun et al.，2003）；③因病原菌不能适应植物基本的生理代谢或生物学习性造成（Thordal-Christensen，2003）。上述观点对非寄主抗性进行了某种程度的推论。现今人们更倾向于用多种机制共存的“重叠机理”（overlapping mechanism）来解释非寄主抗性的成因，并且已经了解到在寄主抗性和非寄主抗性中都发生诱导性防卫反应。无疑，这两类防卫反应中存在着类似“搭接”的共有成分可能充当了多种角色。例如，水杨酸/茉莉酸/乙烯（SA/JA/ET）等信号传导系统在抗病基因介导的防卫反应中起着重要作用。有趣的是这些物质在非寄主抗病中也有相似的表现和作用，因此对非寄主抗病性的研究或许也可借鉴研究寄主抗性的方法（赵炳宇，个人通讯，2006）。

针对非寄主抗病性与寄主抗病性在植物表型上的某些共同点（如可诱导防卫反应、基因激活或出现 HR 等），从非寄主中分离这类抗性基因，然后用生物技术导入目标作物也可以获得抗病植株，近年这方面研究已有明显进展。Hulbert 实验室（Zhao et al.，2004）利用他们建立的“水稻细菌性条斑病—玉米”互作系统，首先用筛选的能引起玉米过敏性坏死反应（HR）的水稻细菌性条斑病菌接种玉米，经遗传分析发现 HR 表型与某个遗传位点共分离，由此推测玉米控制水稻细菌性条斑病的抗性是由单个遗传位点控制的。在此基础上，将其定位到玉米第 6 染色体短臂上，并克隆了目标基因 *Rxo1*。最近，他们证明了将玉米 *Rxo1* 基因在水稻中表达能令水稻获得对细菌性条斑病的广谱、稳定抗性（Zhao et al.，2005）。该研究的成果至少提供了两条信息：一是非寄主抗性尽管复杂，但存在一套与寄主抗病性相似的遗传关系；二是非寄主抗病性研究有助于拓宽植物广谱抗病的渠道，这一新发现尚需经过生产实践的验证。这个成功例子也可作为水稻白叶枯病非寄主抗性基因发掘的借鉴，并可通过阐明寄主抗性的效应分子来分析其机理。水稻细菌性条斑病菌能在携有 *Rxo1* 基因的非寄主玉米中引起过敏反应，并且其 AvrRxo1 蛋白具有全核蛋白酶的特征。几乎所有已知的水稻细菌性条斑病小种内均携有 *avrRxo1* 基因，因此携有非寄主玉米的 *Rxo1* 基因的水稻就具有对细菌性条斑病所有小种的广谱抗性，其实质也是符合“基因对基因”假说的，唯一的特殊性仅在于无毒性因子分布在所谓的不同小种中。据此也可进一步推论 *avrRxo1* 基因在对水稻寄主的致病过程中充当了重要角色，仅将玉米的 *Rxo1* 基因导入水稻后能引起对细菌性条斑病的抗性就足以说明这一论点（Zhao et al.，2004；2005）。

2　水稻对白叶枯病的既存抗性

植物有既存的和诱导的两类防卫系统。植物的既存抗病性（preformed resistance）是其第一类防卫系统，即没有受到病原菌侵染之前自身固有的抗病性。既存防卫系统的成分是植物的形态、细胞壁的角质、蜡质、木质素、气孔特殊结构、水孔等。这些成分对病原物的侵入是一个天然屏障，涉及复杂的物理和化学因素。一般来说，植物既有的形态结构和生理生化性状赋予的防御作用发生在抗病过程的早期，但对阻止病原菌的侵入或扩展，赢得时间延缓扩展或引发有效的防卫反应具有一定作用。

2.1 物理的既存抗性因素

水稻白叶枯病菌（*Xoo*）是一种具有潜育特点的寄生性细菌，病原菌在与寄主建立寄生关系之前的侵入阶段，要通过水孔或伤口进入植物体并突破重重理化障碍。这些理化障碍就是物理抗病性和化学抗病性因素。两者的区别在于前者主要是形态学障碍，后者则强调有一定特异性的化学物质，实际上两者有时也难于区分。

水稻抗白叶枯病的形态学包括株型、叶片形态、叶片水孔和排水孔数量等方面。多数抗病品种一般株型紧凑、叶片窄而挺直、开张角度小、叶片茸毛多，感病品种的叶片平展、生长繁茂（Kiryu et al.，1955），植株间的相对湿度较高，叶片相互摩擦的概率加大而增多了接触和造成伤口的机会，有利于病害的散布和扩展，营造了有利的发病条件。

Xoo 主要由分布于叶缘或中脉附近的排水组织向外开口的水孔或伤口进入叶片导管，水孔和排水组织等入侵通道的多寡对品种的抗性可能有一定影响（Ou，1985）。Horino（1984）通过扫描观察了水孔和气孔的物理结构与 *Xoo* 的入侵关系。虽然水稻的水孔与气孔的结构相似，但水孔的大小约有气孔的两倍，而且水孔上部的保卫细胞没有突起，病菌易于侵入，而气孔上部具有口径仅为 0.9μm 的保卫细胞，对于平均有 0.61μm×1.79μm 大小的白叶枯病菌（*Xoo*）未免太小，这可能是其不通过气孔进入的主要原因。寄主形态结构不利于细菌进入的情况在 Tabei（1977）的观察中也有报道。他在相同环境条件下对日本白叶枯病的主要越冬寄主鞘糠草（*Leersia sayanra*）和其亲缘关系较近，但是非越冬寄主苇（*Leersia japonica*）进行了伤口接种，前者呈现出清晰的病斑，后者却没有发病。Horino（1984）观察到苇的保卫细胞外部具有突起，它可能阻挡了细菌的侵入。

2.2 化学的既存抗性因素

植物普遍具有化学的既存抗病性因素。抗病植物可能含有天然抗菌化学物质或抑制病原菌酶活性的物质，它们可能发挥不同的抑制或促进作用，甚至有明显的特异性表现。比如当病原菌侵入亲和品种植株后立即开始增殖，以对数增殖（logarithmic multiplication）速度生长并形成大病斑，而病菌进入非亲和品种后几乎静止生长。亲和与不亲和的病菌—品种组合在生理、生化方面的反应差异与植物体内的营养变化都有密切关系。

植物的细胞壁也是病原物侵入的一个天然屏障。细胞壁的强度与其既有的酚类、木质素、钙和硅等抗菌物质的含量有关。水稻体内可能产生有一定抗病效应的酚类化合物或木质素（lignin）等抗菌物质。

酚类化合物既是许多植物本身固有的抗菌物质，也是植物的次生代谢物质，它们参与寄主—病原菌互作中复杂的生理过程，如氧化还原反应、木质化形成等。由于寄主体内的多酚氧化酶和过氧化物酶的氧化作用转变为具有高毒性的醌类物质而发挥了抗菌作用，并对病原生物的果胶分解酶、纤维素分解酶等致病因子酶产生强烈的抑制作用（章元寿，1996）。在寄主中，酚类物质的含量与病原菌侵染的关系早已引起人们的注意。如洋葱的紫色鳞茎表皮品种比无色表皮品种对洋葱炭疽病（*Colletotrichum circinans*）的抗性强，这是因为紫色洋葱表皮分泌出浓度较高的原儿茶酸和邻苯二酚抑制病菌孢子的萌发，从而产生阻止病菌入侵的效果。方中达等（1963a，1963b）曾分析和比较了水稻对白叶枯病抗、感品种功能叶片的生化成分，发现抗病品种的多元酚和糖含量高于感病品种，高含量的多元酚和糖能阻碍 *Xoo* 侵染，而有利于病菌生长的谷氨酸和天门冬氨酸等游离氨基酸的含量较少。Sato（1979）系统研究了水稻的既有抗菌物质。从 4 个不同抗病水平品种的健康叶片中提取抗菌物质，经纯化后通过柱层析法、生物测定、抗菌组分的分馏以及气相色谱法等步骤，分离出 8 种抗菌组分。其结果可归纳为：①不同抗性水平品种的抗菌物质的活性不同，经 3 次稀释的溶液，在感病对照品种金南风的溶液中仍有少量细菌能够增殖，但病菌在其他三个抗病品种的抗菌溶液中则全部受到抑制；溶液经稀释 8 次后，金南风（感）和黄玉（中感）品种的稀释液便不再抑制细菌增殖，而抗病品种中国 45 和中抗品种 Tetep 的稀释液仍然有 100％的抑制作用。以中国 45 抗菌物质的活性最高，依次为 Tetep、黄玉。②8 种抗菌物质的活性表现不同，其中以转-2-乙烯醛（trans-2-haxenoic acid）的活性最高，在溶液浓度仅为 50μg/ml 时，就能 100％抑制细菌增殖；达到 100％抑制细菌增殖的各抗菌物质的溶液浓度依次是：反-2-已烯酸（trans-2-haxenoic acid）和顺-3-已烯酸（cis-3-haxenoic acid）为100μg /ml，丁香醛（syringaldehyde）和松柏醛（coniferaldehyde）为 200μg/ml，香草醛（vanillin）和对羟基苯（甲）醛（para-hydroxybenzaldehyde）为 400μg/ml；只有乙酰丁香酮（acetosyringone）的活性较低，当溶液浓度 400μg/ml 时，也只有 60％的细菌的增殖受到抑制。③在这些抗菌物质中，只有乙烯酸分布在稻株根部，其他 7 种抗菌物质，均为具有乙醛和酚类的氧化的木质素（lignin）。

木质素是由酚类化合物的聚合物和一些醇酸组成的。在寄主—病原物相互作用中，植物木质素的积累是病原菌入侵后所产生的诱导反应。有些植物病原物并不能分泌分解木质素的酶类，但当植株受伤或受到病菌侵染后，通常在伤口或感染部位产生木质素，木质素和纤维素及其他糖类联结在一起后沉淀于细胞壁上形成木栓化（suberization）而形成病菌入侵的屏障。对于作为典型维管束病害的白叶枯病而言，这种方式似乎作用不大。但有研究指出，水稻内过氧化物酶与木质素代谢产物的作用与 *Xoo* 抗性有关。

3 水稻对白叶枯病的诱导抗性

诱导抗病性是植物的第二类防卫系统。它是指植物受到病原菌侵染后诱发的高度抗病性，也是植物的基本防卫系统，即通常所谓的主动抗病性。它又分为局部的抗性反应和系统抗性反应。局部抗性反应是受到病原菌侵染的寄主局部组织变褐，已感染的细胞及其周围的部分细胞死亡形成枯斑，一般常见于真菌病害。导致产生能远距离输导的诱导性物质使感染区以外的寄主组织产生抗性，即系统抗性反应，其诱导的系统抗性又称为系统获得性抗性（systemic acquired resistance，SAR），具有 SAR 能力的植物组织往往累积病程相关（PR）蛋白。当水稻被病菌侵染后，直接接触细菌的寄主细胞及其周围细胞，迅速诱发多种物理或生化反应，诱导寄主组织形态或结构以及代谢过程的改变以削弱病菌的破坏，其抗病过程是连续的，包括寄主对病原菌的识别及抗病性（不亲和）表达等几个阶段，这些内容将在分子机理部分讨论。

3.1 物理的诱导抗病性因素

病原物侵染引起的植物代谢变化导致细胞或组织水平的形态和结构变化，是物理的诱导抗性因素。多种禾本科植物受到真菌侵染时，在细胞壁与质膜之间，与真菌附着胞相对应的位置上常形成半球形沉淀物，简称乳突（papillae），周围表皮细胞壁也会变色，形成圆型或卵型区域，称为晕圈（halo）（商鸿生，1995）。

水稻白叶枯病菌侵入水稻后，寄主细胞壁结构和病原细胞均发生变化，经历一系列生理生化过程，亲和与不亲和组合的细菌增殖差异很大，增强寄主的防卫效应，能抑制病原细菌的生存以致死亡。

3.1.1　水稻白叶枯病原细菌群体动态（population dynamics）　20世纪70～80年代，病理学家们采用传统病理学方法研究了品种—病菌亲和与不亲和组合接种病菌后间隔不同时间的叶片内 *Xoo* 细菌群体增殖动态。*Xoo* 的入侵对寄主的特异性选择十分明显。例如，Fang 等（1975）观察到 *Xoo* 除了在非寄主植物上的细菌数量稀少之外，在水稻抗、感品种叶片上的水孔趋向反应也不相同；寄主叶片水孔溢出物所含的细菌数量与其亲和性有关，亲和组合比不亲和组合的细菌量要多。Reddy 等（1974）和 Kaku 等（1987）的研究结果也表明，在接种后2～4d的初期的细菌增殖率基本相同，但随着病情发展，亲和组合叶片中细菌的繁殖速度和增殖数量都明显大于不亲和组合。Zhang 等（1985，1987）测试了水稻白叶枯病成株抗性品种在感病和抗病（亲和与不亲和）期不同叶位组织内的细菌增殖动态。采用菲律宾 *Xoo* 小种1测定成株抗性品种 Zenith 在第6、9、10三个叶位（感病时期）和转抗后的第12个叶位中的细菌增殖，比较接种1h（细菌基数）、24h、72h、120h和168h后的单位叶面积细菌（cfu/cm^2）增殖率的结果表明，Zenith 在转抗后的叶片中细菌增殖量明显低于感病期间的叶片，其细菌增殖速度缓慢，细菌量至高峰后便迅速下降，细菌增殖动态与接种后病斑的扩展相对应。抗性特异反应现象实际是两种生物间化学因素互作的直观反映，也是寄主的物理抗性与化学抗性因素交互作用的综合体现。

3.1.2　细胞和组织水平的形态结构变化　Horino（1976，1981）用超微显像技术观察了水稻-*Xoo* 在亲和与不亲和互作过程中寄主细胞结构的变化情况。用日本 *Xoo* 菌群Ⅰ、Ⅱ、Ⅲ接种黄玉和金南风。接种3d后，不亲和组合寄主的导管壁便出现了大量纤维材料（fibrillar material，FM），进入导管的65%的细菌被包封，正常的短棒状细菌开始出现不规则状、生长受到抑制，随之死亡。而亲和性互作的寄主在接种20d后，其导管组织才开始出现FM，并且细菌被相当于由多聚糖荚膜材料组成的电子透明层（electron - transparent layer）包围，这些细菌虽然也受到了FM的轻微抑制，但是其形态正常、生长活跃并且大量增殖。FM的快速出现与白叶枯病的抗性呈对应关系。Horino 的研究均表明FM来源于寄主，并有一定程度的防卫功能。后来，Horino 等（1979）又用 C^{14} 放射性同位素处理白叶枯病菌侵染水稻期间的叶片，放射性自显影观察到在不亲和寄主—病菌组合的导管中充斥了大量FM，叶片组织中 C^{14} 以导管中最多，依次为叶绿体、导管壁，在细菌上的 C^{14} 最少；但在亲和组合中，C^{14} 以叶绿体上最多、依次为导管壁、细菌，在导管中最少。结果表明：①FM与不亲和寄主—病菌组合具有特殊关系，至少有一部分FM是由光合作用产生的 C^{14} 同化物；②在水稻白叶枯病的寄主-*Xoo* 不亲和组合中，FM和导管壁上的 C^{14} 最多，而在细菌上却很少，表明FM与导管壁也存在密切关系。

Mew 等（1984）通过扫描进一步观察了有毒和失毒的 *Xoo* 菌株接种水稻品种后，在水稻叶面水孔上的趋向运行状况。接种24h后，对 TN1 和 Cas 209 两个品种都亲和的毒性菌株 PXO61 的细菌密集于孔口，在水孔外面就开始增殖；失毒菌株 PXO101 在这两个品种的水孔上都不增殖，细菌埋入水孔渗出物中并封住了孔口；又用对 TN1 和 Cas 209 两个品种分别是亲和与不亲和的菌株 PXO86 接种 48h 后，感病品种 TN1 上的细菌密集于孔口周围，随后大量细菌通过水孔进入导管并迅速增殖；在不亲和的 Cas 209 品种上的水孔口的细菌很少，随后细菌便陷于水孔渗出物中失活，少数“侥幸”进入导管组织的细菌不久就形态失常、生长受到抑制以至死亡（图 1）。Feng 等（1975）报道 *Xoo* 细菌经由水孔入侵还有化学物质的辅助作用，水孔分泌物对 *Xoo* 起到正趋化性的作用，亲和与不亲和寄主的分泌物直接影响了细菌进入水孔前的不同趋向，因此水孔不仅是细菌进入导管的入口，也是水稻- *Xoo*的特异性互作以及寄主防卫反应的物质基础。

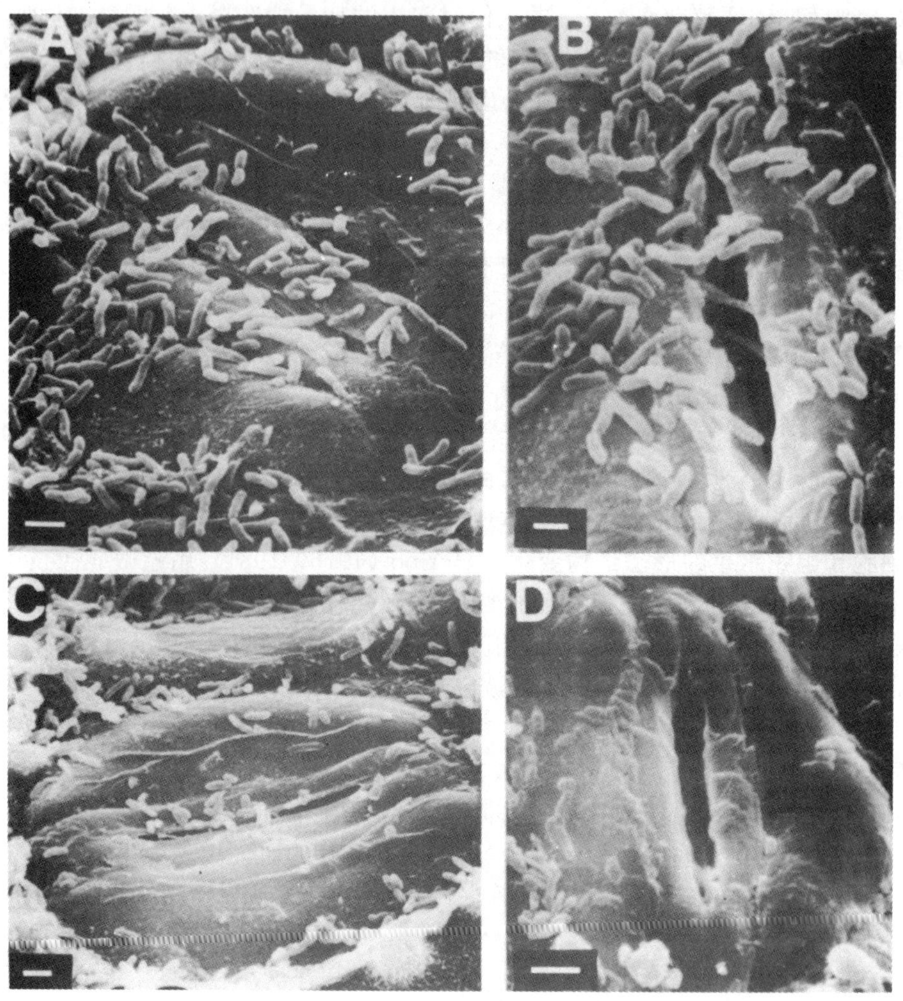

图 1　品种 CAS209 被接种 *Xoo* 毒性菌株、无菌株和微毒菌株后 72 h 的显微扫描图。A、B. 水孔布满了毒性菌株 PXO61 细菌；C. 无毒菌株 PXO86 细菌陷入水孔渗出物；D. 微毒菌株 PXO101 细菌陷入水孔及较少的细菌。标尺＝ 1μm（Mew et al.，1984）

Fig. 1　Scanning electron micrographs of leaf surface of cultivar CAS209 rice 72 h after inoculation with *Xoo*. A and B. Virulent strain PXO61 were fully established at the water pores. Avirulence strain PXO86（C）and weakly virulence strain PXO101（D）showed embedded cells and had lower cell numbers. Bar= 1 μm（Mew et al.，1984）.

3.2 生化的诱导抗病性因素

寄主植物细胞及其周围细胞接触到病原菌后，短期内会诱导发生一系列生化反应来抵抗病原菌的入侵。其结果可产生植保素（phytoalexin）、毒性小分子酚类化合物等抗真菌活性的物质；产生各种病程相关蛋白（PR 蛋白），它们能降解一些真菌细胞壁中的几丁质和葡聚糖，在体外抑制一些真菌病原的生长；此外还产生加固和阻止病原生长的细胞壁成分，如伤口部位胞壁栓质化、病菌入侵的胼胝体（callus）的沉积、富含羟脯氨酸（HPGRP）和富含甘氨酸的糖蛋白（GRP）的聚合以及木质素的聚合和导管中侵填体（ryrose）的生成等；研究较多的是活性氧物质（ROS），它在植物防卫反应中起着极其重要的作用，研究已经证明植物释放的超氧阳离子、过氧化氢、羟自由基等活性氧物质，除了参与毒性脂肪酸和毒性酚类化合物的形成外，其本身对病原菌的杀伤作用也是不可低估的（王钧，1998）。

上述有些情况也可见于水稻抗白叶枯病侵染的生理生化改变及其过敏性反应中。例如，在非亲和性组合中水稻品种的细胞壁能够积累一定的愈创葡萄糖，不仅导致细胞壁的加厚、形成乳突或胼胝体，而且诱导过氧化物酶的产生，这些物质都与植物的抗病性密切相关。

3.2.1 过氧化物酶诱发防卫反应 有研究表明白叶枯病菌侵入不亲和水稻品种接触了维管束的木质部薄壁细胞后，胞内随即积累大量过氧化物酶，加速合成木质素，木质部次生壁加厚，使纹孔直径缩小，缩减了接触病菌的面积。Emmanuel 等（2001）报道了水稻- *Xoo* 亲和与非亲和互作组合寄主细胞新陈代谢活动以及生理生化作用所引起的防卫反应。他们通过扫描成像观察到水稻与 *Xoo* 互作过程中木质部次生壁及薄壁组织细胞的变化，分析了其 *PO-CL* 基因表达和同工酶的位置与动态（图 2、图 3）。他们还比较了亲和与非亲和品种- *Xoo* 组合在接种 12h 和 48h 之间所发生的如下特异性变化：①接种 48h 后，非亲和与亲和品种的木质部导管次生壁分别为 5.8±0.1μm 和 4.8±0.1μm，前者厚于后者，其导管纹孔直径变小，致使细菌进入薄壁组织的空间缩小；②非亲和水稻木质部薄壁组织分泌纤维物质（FM），受其缠绕的病原细菌形态失常，与前人所见相同；这些变化与在木质部薄壁组织和导管壁及中腔内的阳离子过氧化物酶 PO-CL 的累积有关；③接种 12h 内，非亲和水稻品种诱导了防卫特异性的过氧化物酶基因 *PO-CL* 表达，在木质部薄壁组织细胞及其细胞腔中，累积了阳离子的（cationic）过氧化物酶 PO-CL，而在亲和品种中却没有明显的 PO-CL。为了进一步验证寄主与病菌互作中寄主防卫的主动效应，他们又以缺乏抗白叶枯病 *Xa10* 基因的水稻品种，进行其能否被 AvrXa*10* 诱导的反证测试，即采用一个不携有 *Xa10* 基因的品种 IR24，用含有 *avrXa10* 的菌液接种。结果显示，其木质部次生壁没有增厚，表明 AvrXa*10* 未曾识别到 *Xa10* 产物，因为 IR24 根本无 *Xa10* 基因，结果可以初步判断过氧化物酶基因 *PO-CL* 可能具有特异性防卫功能。Reimers 等（1991），Young 等（1995）的研究也能了解 PO-CL 在水稻中所发挥的具体作用，*Xoo* 不亲和水稻品种细胞壁中累积了木质素聚合体（lignin polymers），同时在叶片薄壁组织中新增加了三种过氧化物酶的同工酶（两个阴离子的和一个阳离子的），以功能结构域特异（domain-specific）抗体标定，可以确定 PO-CL 累积了一种由 *Xoo* 病原菌诱导的阳离子过氧化物酶，而且 PO-CL 主要存在于叶肉组织细胞壁和细胞外空间以及导管腔和木质部中。根据 PO-CL 同工酶的活性、时空动态以及所用底物的类型，其在水稻防卫反应中的作用是明显的。Emmanuel 等（2001）的研究也证明过氧化物酶在参与形成障碍结构和直接抗菌作用的防卫过程中具有实际意义，一方面通过形成的障碍结构增强细胞壁和附着在细胞壁上的沉淀作用，加强木质素或木栓层的聚合，促进细胞壁糖蛋白或多聚糖的交叉集结以及抗菌苯酚的二聚化；另一方面，过氧化物酶具有的氧化机能的游离自由基中间产物对病原菌也有直接的毒害作用。

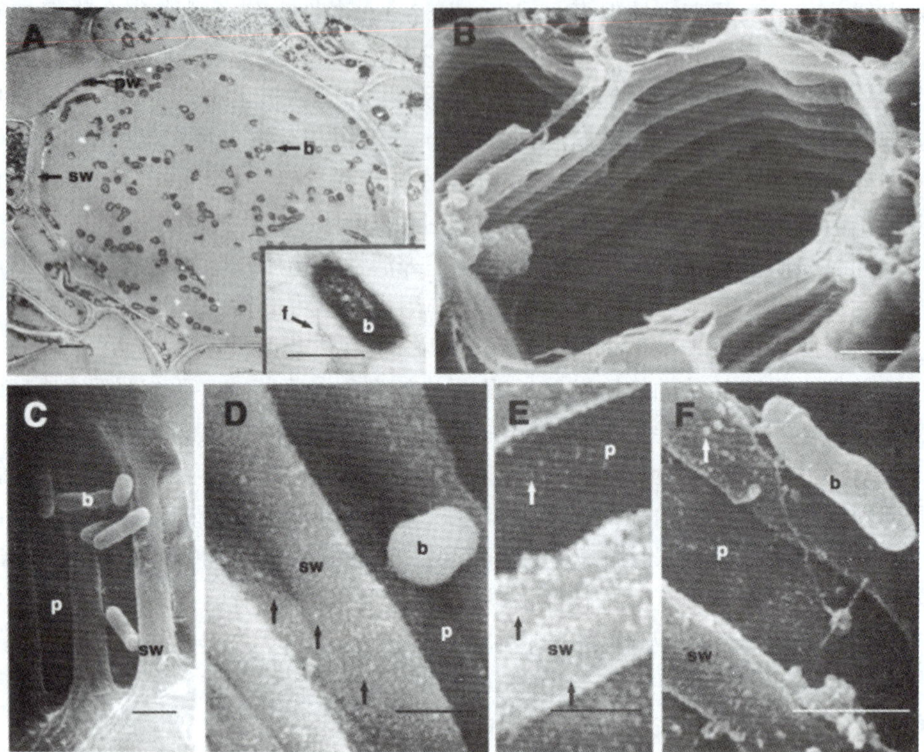

图 2　渗入水稻 IRBB10 品系叶片中的 *Xoo* 无毒菌株的电镜传导和扫描图像（Emmanuel et al.，2001）. A. 转入 pBUavrXa10 后获得无毒功能的 *Xoo* PXO99^{A} 在携有 *Xa10* 基因的水稻品系 IRBB10 的叶片木质部维管束中：伴有纤维状材料的细菌细胞；B. 在未侵染的 12 天苗龄的 IRBB10 之木质部导管伴有与纹孔相连的薄壁组织；C. 与 IRBB10 的不亲和互作的木质部维管束的无毒菌 PXO99^{A}（pBUavrXa10 F_1）细胞；D 和 E. 在不亲和的 *Xoo* 无毒菌 PXO99^{A}（pBUavrXa10 F_1）接种 IRBB10 24 h 后的 PO - CL（箭头）位置；F. PO - CL 与亲和的 *Xoo* 毒性菌株 PXO99^{A}（p HM1）接种 IRBB10 24h 后渗入位置。标尺＝3μm；b，细菌；f，纤维状材料；p，纹孔径；pw，初生壁；SW，次生壁

Fig. 2　Transmission electron microscopy (TEM) images (A) and scanning electron microscopy (SEM) images (B to F) of rice leaf tissue of IRBB10 infiltrated with avirulent *Xoo* strains (Emmanuel et al., 2001)

A. *Xoo* strain PXO99^{A} (pBUavrXa10 F_1) cells within a xylem vessel in IRBB10 (incompatible interaction); inset, a bacterial cell with associated fibrillar material. B. Xylem vessel and associated parenchyma with connecting pits in an uninfected IRBB10 12 - day -old rice leaf. C. *Xoo* strain PXO99^{A} (pBUavrXa10) . F_1 cells in the apertures of pits of a xylem vessel of IRBB10 (incompatible interaction) . D and E. PO - CL location (arrows) afier incompatible *Xoo* stratn PXO99^{A} (pBU avrXa10F_1) interacted with IRBB10 for 24 h. F. PO - CL location (arrows) after compatible. *Xoo* strain PXO99^{A} (pHMI) interacted with IRBB10 for 24 h. Bar=3μm; b, bacteria; f, fibrillar material; p, pit aperture; pw, primary wall; and sw, secondary wall.

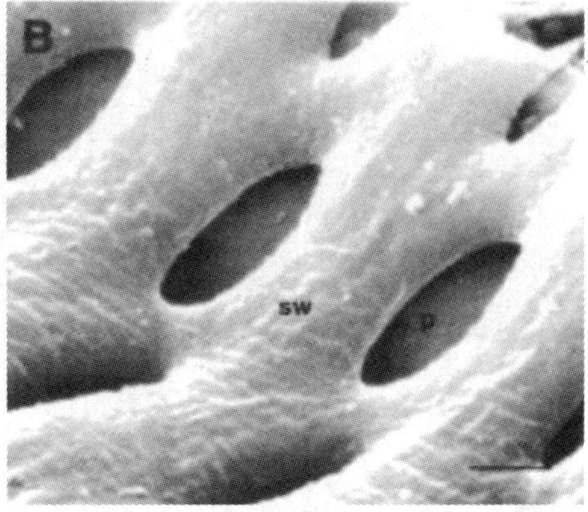

图 3 A. 未接种的 IRBB10 品系 65 天株龄叶片木质部导管扫描图像（Emmanuel et al.，2001）. B. 高倍显示纹孔径的大小（短的直径均值＝2.1μm ±0.5，次生壁厚度均值＝7.4μm ±0.5）. 标尺＝3μm；b，细菌；f，纤维状材料；p，纹孔径；pw，初生壁和次生壁

Fig. 3 A. Longitudinal section of a xylem vessel from a non-inoculated 65－day-old IRBB10 rice leaf viewed by scanning electron microscopy. B. Higher magnification showing the aperture of the pits (average diameter of short axis ＝2.1μm ±0.5) and the thickened secondary walls (average distance between aperture at short axis ＝ 7.4μm ± 0.5). Bar ＝ 3nm; p, pit; and sw, secondary wall (Emmanuel et al., 2001)

3.2.2 过敏性反应（hypersensitive reaction，HR） 这是植物受非亲和性病原菌侵染后在侵染部位产生的抗病反应，属于寄主的局部防卫反应。其结果是经侵染的组织细胞坏死，病菌受寄主的枯死组织抑制并死亡，形成枯斑。过去将过敏反应过程的组织细胞死亡称为坏死（necrosis），随着研究的深入，特别是借鉴了有关动物细胞凋亡的研究成果，目前基本认为这是一种程序性细胞死亡（programmed cell death 或 apoptosis）抗病过程。HR 在植物小种专化抗性和非寄主抗性中占有重要地位，普遍存在于由真菌、细菌、病毒和线虫等多种病原物引起的寄主抗性中。

水稻对白叶枯病的过敏性反应并不经常发生，即使在整个黄单胞菌属（*Xanthomonas*）参与的非亲和互作中也不常见。但是，Kaku 等（1978）的研究却显示个别水稻抗白叶枯病基因与不亲和菌的互作组合能诱导过敏性褐变（browning）反应。此外，他们还观察到 *Xoo* 在番茄、豇豆等非寄主植物上也能诱导出这种反应。后来，Kaku 和 Horino（1977）又用不亲和菌株接种携有某些抗白叶枯病基因的品种，也能诱导出一种不典型的（atypical）褐变反应。如携有白叶枯病 *Xa3* 抗性基因品种接种了不亲和菌株 4d 后诱导了褐变反应，但是接种了亲和菌株 7、8d 后，又呈现典型的白叶枯病的黄白色病斑，表明水稻- *Xoo* 不亲和互作能否诱导过敏性反应取决于品种和小种的互作组合。如图 4（另见彩版 8）所示，在 Kaku 等（1993）用针刺法测试的携有 *Xa1*、*Xa2*、*Xa3* 和 *Xa4* 四个水稻抗白叶枯病基因品种与各自不亲和菌株的反应中，只有携带 *Xa3* 基因品种-*Xoo* 组合表现褐变反应（表 1）。Ogawa 等（1993）就是根据原先被鉴定的携有 $Xa4^b$、*Xa6*、*xa9* 三个白叶枯病成株抗性基因的品种都有这种特殊反应，才重新进行了抗性遗传分析，最后鉴定出表现相同褐变反应的这三个抗性基因与 *Xa3* 实属同一基因座。

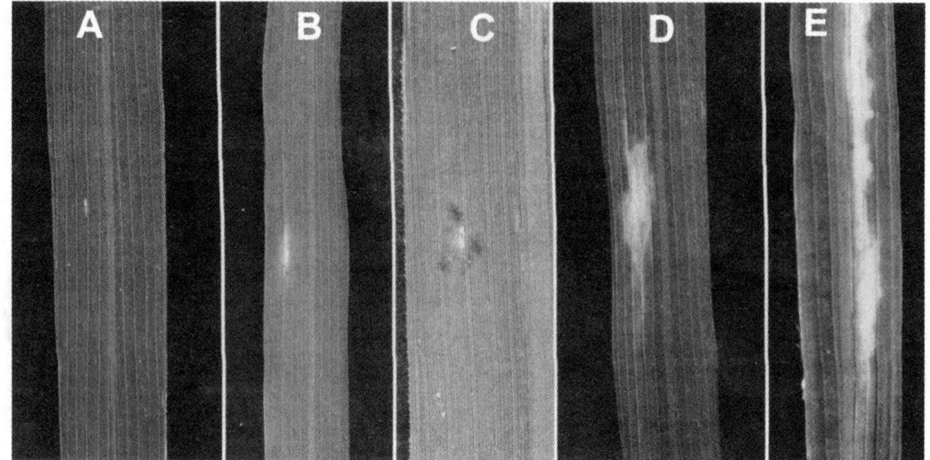

图 4 水稻- *Xoo* 菌不亲和互作组合的侵染反应型（Kaku，1993）。A. *Xa1* 基因控制的无症状反应型；B. *Xa2* 基因控制的小黄斑反应型；C. 携有 *Xa3* 基因品种的褐色病斑型；D. 携有 *Xa4* 基因品种的中小型病斑型；E. 感病反应型

Fig. 4 Infection types in rice - *Xanthomonas oryzae* pv. *oryzae* interaction (Kaku, 1993). A. Symptomless type of reaction controlled by *Xa1*. B. Small yellow lesion type of reaction controlled by *Xa2*. C. Browning type of reaction controlled by *Xa3*. D. Small to middle-sized lesion type of reaction controlled by *Xa4*. E. Susceptible reaction

表 1 已知白叶枯病抗性基因对其非亲和菌株的抗性反应型［据 Kaku（1993）资料整理］

Table 1 Known resistance genes and the reaction to their incompatible strains of *X. oryzae* pv. *oryzae* (Summarized from Kaku, 1993)

抗性基因 Resistance gene	反应型 Reaction type	代表品种 Representative cultivar
Xa1	无症状型（Symptomless）	黄玉、Java 14
Xa2	小黄斑型（Small yellow lesions）	Rantai Emas、Tetep
Xa3	褐变病斑（Browning）	早生爱国 3 号、中国 45
Xa4	小黄斑型	IR20、TKM6
$Xa4^b$	褐变病斑	Semora Mangga
xa5	小黄斑型	IR1545 - 339、DZ192
Xa6	褐变病斑	Zenith
Xa7	小黄斑型	DV85
xa8	小黄斑型	PI231129
xa9	褐变病斑	Sateng
Xa10	无症状型	Cas 209
Xa12	介于无症状型与小黄斑型之间 (Symptomless to Small yellow lesions)	黄玉、Java 14

3.2.3 生物的诱导抗病性 经由诱导因子处理后可以改变植物对病原物反应的局部或系统抗性的现象为诱导抗性（薛应龙等，1998）。植物抗性除了能来自物理和化学因子的诱导处理，还可以经病原物（或其组成部分）等生物性因子的诱导产生。生物因子包括的内容较广，无论哪种植物的病原物还是病原物毒性的强弱，只要能使植物改变抗病性都属于诱导因子的范畴，激发子作为病原物的组成部分也是诱导性因素。植物诱导抗病性和诱导效应的强弱同诱导因子处理时间的长短、前后间隔的距离或接种菌液的浓度等因素都有关系。日本科学家曾研究水稻白叶枯病的诱导抗性。在用非亲和菌株或非病原菌在特定的水稻品种上接种后，再用亲和性菌株接种，其发病就受到明显的抑制。Horino（1976）先用黄玉和枯叶不知（Hagareshirazu）两个品种的亲和菌株接种，然后间隔一定时间和一定的距离再用亲和性菌株作第二次接种，前者的病斑平均长度分别为 6.3cm 和 5.7cm，后者平均病斑长度分别为 1.4cm 和 1.8cm，其病斑扩展受到明显抑制。Horino 还进行了对水稻不致病的非病原物（nonpathogen）的诱导试验。他先在水稻上分别接种了软腐欧氏杆菌（*E. aroideae*）和柑橘溃疡菌（*Xanthomonas citri*），然后接种该水稻品种的 *Xoo* 亲和菌株，结果都表现不同程度的抑制作用，不过其效应较弱，在接种 5d、远于接种点 3cm 处便消失了作用。Watanabe（1976）的类似试验结果表明，前后接种间隔时间以 3d 的抑制效应最大，先后接种点范围在 1～5 cm 之间的距离越短，抑制作用越强，即使间距 5 cm 也有明显作用，间隔 9d 再作第二次接种仍有一定作用。Horino 推测诱发抗病性的机制是由植保素类的抗菌物质引起的。Watanabe（1976）的估计是：①产生了植保素类抑菌物质活化寄主组织抗病反应；②先接种菌的营养竞争阻碍了后接种菌的增殖；③先接种菌生成的抗生性物质造成障碍。但在其实验中发现先接种菌产生的抗生物质量太少，不足以直接抑制先接种菌的生殖。

4 水稻白叶枯病抗性的分子机理

近年来植物抗性的分子机理研究在下列几个方面有了突破性进展，推动了水稻对白叶枯病抗性机理的研究。

4.1 病害特异相关分子（PAMP）介导的抗性

植物有一套针对微生物一般结构特征的固有免疫系统，利用在微生物中特异且结构保守的分子作为识别对象，称为病害特异相关分子（PAMP）（Numberger et al.，2004）。在细菌中常见的 PAMP 包括细胞膜上的脂多糖和细菌鞭毛的鞭毛蛋白（Gomez-Gomez et al.，2002；Ramos et al.，2004）。鞭毛是细菌得以移动的必需工具，鞭毛蛋白的 N 端及 C 端氨基酸序列在细菌界中极度保守，所以鞭毛蛋白被许多植物作为 PAMP 来识别入侵细菌（Felix et al.，1999）。虽然对 *Xoo* 鞭毛蛋白的研究不多，但 *Xanthomonas* 属的鞭毛蛋白有可能是 PAMP（Sun et al.，2006）。此外，鞭毛蛋白这类 PAMP 亦参与非寄主抗病性（Li et al.，2005）。最近的研究显示细菌鞭毛蛋白能与植物表面上的相应受体蛋白直接作用（Chinchilla et al.，2006）。当植物感应到病害的 PAMP 时，就会启动 PAMP 诱发的免疫反应（PTI），并相应阻止病菌在体内生存（Chisholm et al.，2006）。有关 PTI 的研究主要集中在拟南芥系统中。例如 PTI 的启动能够激发有关 MAP 激酶信号传递，诱导抗病相关蛋白（PR proteins）基因表达，其中还产生加固细胞壁的物质 ROS（Nurnberger et al.，2004）。除引起

PTI外，鞭毛蛋白还可能涉及适应小种特异性的免疫反应（Sun et al.，2006）。另一个特别值得关注的现象是在*Xoo*内的一个与鞭毛功能有关的蛋白能与Xa21抗病蛋白互作（Shen et al.，2001），但是目前尚缺少生理作用的相关研究。

4.2 无毒蛋白与抗性的关系

源于植物自身的病害特异相关分子（PAMP）显然不利于病原菌的生存，因此病原菌也形成了一种可将效应分子（effector molecules）释放到植物细胞内来阻止植物细胞内PTI反应的机制（Mudgett.，2005）。研究还发现这种机制不仅影响不亲和互作的抗病过程，并且影响非寄主植物对病原菌的抵抗与识别（Li et al.，2005）。效应分子的释放或分泌方式对植物抗病反应的抑制起着非常重要的作用，其抑制过程也涉及多种抗病途径，如对植物细胞程序化死亡的抑制，对水杨酸或茉莉酸抗病途径的激活抑制，对植物细胞壁重排的抑制，以及对抗性蛋白的激活抑制等。

不同病原菌的效应分子进入植物细胞的分泌方式也不一样（Chisholm et al.，2006；Dangl and McDonell，2006）。细菌的第三类分泌系统具有分泌致病效应分子的功能（Furutani et al.，2003）。已经证明有些*Xoo*的效应分子能以第三类分泌系统进行运输，随着效应分子在植物细胞的分泌，寄主植物也开始产生主动识别功能，并作为抗病信号加以利用。植物中参与识别与启动抗病反应的最典型的物质就是抗病蛋白（R protein），而许多效应分子就是那些与抗病诱导有关的无毒蛋白（Avr proteins）(Buttner et al.，2002；Staskawicz et al.，2001)，因而效应分子参与的抗病过程也符合基因对基因理论。

AvrBs3类蛋白是*Xanthomonas*属中常见的致病因子（效应分子），此类蛋白具有高度保守的C端功能区，在此功能区具有核心定位信号和转录因子同源序列，例如AvrXa10的C端功能区已被证明具有转录激活功能（Zhu et al.，1998），而AvrXa7的C端功能区也被证明具有DNA结合能力。这些研究结果均证明AvrBs3类蛋白的C端具有转录因子的功能(Yang et al.，2000)。迄今发现的*Xoo*无毒蛋白有三个（AvrXa7，AvrXa10及AvrXa27），它们全部属于AvrBs3家族成员。在*Xoo*的基因组中，这是个超基因家族，包含众多成员，因而预示着这些AvrBs3家族成员可能编码更多未被发现的无毒蛋白（Ochiai et al.，2005）。

*Xoo*中的*avrXa7*、*avrXa10*和*avrXa27*基因型分别被水稻抗病基因*Xa7*、*Xa10*、*Xa27*特异识别引起抗性（Hoopkins et al.，1992；Gu et al.，2005）。然而，如果水稻品种未携带相对应的抗病基因（*Xa7*、*Xa10*、*Xa27*），这些效应分子（AvrXa7、AvrXa10和AvrXa27蛋白）将会使相应的*Xoo*菌株具有更强的致病性（毒力）。例如，AvrXa7蛋白就具有此双重功能，携有*avrXa7*的*Xoo*菌株能够在携有*Xa7*基因的水稻中引起抗性（Leach et al.，1996），在未携有*Xa7*基因的水稻植株上，携有*avrXa7*基因的*Xoo*菌株比未携带*avrXa7*基因的菌株表现更强的致病性（Yang et al.，2000；Yang et al.，2004）。如果缺失了病原菌*avrXa7*基因，*Xoo*的致病性就会大打折扣。最近发现AvrXa27能诱导水稻*Xa27*基因的表达并产生抗性（Gu et al.，2006），而转*Xa27*基因水稻也有较广的抗谱，进一步印证了基因对基因关系的合理性。今后研究和解析了更多的AvrXa与Xa组合后，根据抗病机理培育持久抗水稻白叶枯病品种的愿望将成为现实。

效应分子理论对基因对应关系抗性和非寄主抗性的研究均有极为重要的指导价值。如上文所述，水稻细菌性条斑病X．*oryzae* pv. *oryzicola*能在带有*Rxo1*基因的非寄主玉米中引起过敏反应。把 *avrRxo1* 基因转移到 *Xoo* 后，*Xoo* 亦会在带有 *Rxo1* 的非寄主玉米内引起过敏反应，但此反应需依赖 *Xoo* 的第三类运输系统功能（Zhao et al.，2004，2005）。

4.3 抗病基因产物与无毒基因产物之间的作用

在 Flor（1971）的基因对基因关系理论中，抗性基因所产生的抗性蛋白与相对应的无毒基因所产生的无毒蛋白互相识别引起抗性反应。在相当长的一段时间内，人们普遍认为抗性蛋白是直接作用于无毒蛋白的，即所谓配体/受体互作模式。但事实上这种直接互作的例子却为数不多（Jia et al.，2000；Deslandes et al.，2003；Dodds et al.，2006）。另外一种理论则认为抗性蛋白仅具有一种“防卫”功能，即监视细胞内某些重要的蛋白能否正常运作。在相当于无毒蛋白的效应分子干扰或破坏这些作为靶标的蛋白时，抗性蛋白就能感应到这些靶标蛋白结构发生了变化而发挥相应功能，间接与无毒蛋白识别。细菌所产生的效应分子大多具有针对植物蛋白的酶活性（Chisolm et al.，2006）。源于同一个寄主植物的蛋白可能接受多个细菌效应分子的作用。按基因对基因假说，植物细胞针对每一种效应分子都需要产生专一的抗性蛋白，那么抗性蛋白的种类便会多得惊人。如果抗性蛋白针对效应分子（无毒蛋白）对其共享靶标所引起的变化，那么，每个抗性蛋白便可间接识别多个效应分子（无毒蛋白）的酶活性作用（Van der Biezena Jony，1998）。此“防卫”理论较合理地解释了基因对基因假说目前尚不能解释的问题。Chisholm 等（2006）有关拟南芥抗病研究的最新成果也支持了“防卫”理论的合理性。可见基因对基因假说与“防卫”理论就诠释抗病现象的实质是一致的，并且通过互补作用进一步阐明了不同形式的抗病过程（Dangl and McDowell，2006）。针对那些不能改变植物细胞靶标结构的效应分子，抗性蛋白与效应分子（无毒蛋白）的作用是直接的；而较少数的抗性蛋白针对大量的效应分子的作用则可能是间接的，如上文所述。植物在演化过程中，直接作用的抗性基因显示的多态性很大，但防卫作用的抗性基因的多样性则较少。在抗白叶枯病机理的研究中，虽然已找到至少 30 个抗性基因，也从 *Xoo* 基因组序列中分离了大量编码类似无毒蛋白的基因（Ochiai et al.，2005），但能同时成功地分离出相互对应的抗性蛋白与无毒蛋白却很少。因此，它们究竟采用的是直接作用还是间接作用尚需进一步研究。

4.4 抗白叶枯病的信号传导

植物表达抗病性是一个复杂的过程。当抗性蛋白靠识别无毒蛋白辨认到相应的病原信号后，感染的细胞便开始了一连串的变化。如产生大量的 ROS、乙烯及病害相关蛋白等，以加速木质化过程和启动细胞的程序死亡等，从而抑制病菌的扩散（Zipfel et al.，2005；Greenberg et al.，2004；Van Loon et al.，2006）。植物细胞过敏性坏死反应是抗病过程中较普遍的一个现象，而白叶枯病在非亲和寄主内一般不引起典型的过敏反应。但是 Kaku（1993）发现有些抗白叶枯病基因与非亲和小种作用后却呈现出褐斑、小黄斑等症状。*Xoo* 不同小种的毒性差异和抗性基因诱发的特异抗性途径可能起了重要作用。正如拟南芥的 NDR1（Centry et al.，1997）及 EDS1（Falk et al.，1999）信号系统参与的抗病过程。我们知道，拟南芥 R 基因介导的信号传导途径至少包括三条：①依赖 NDRI；②依赖 EDS1；③不依赖 NDRI 和 EDS1。*pbs*1（avrPphB Susceptible 1）、*ndr*1（nonrace-specific disease resistance）、*eds*1（enhanced disease susceptibility）、*pad*4（phytoalexin deficient 4）等基因在参与的信号传导中发挥了以下重要作用：EDS1 基因产物有甘油三酰脂酶的活性，EDS1 可能与信号分子的合成和降解有关；NDR1 编码蛋白中含有两个跨膜区，对促使 R 蛋白向膜的靠近可能起了一定作用；而 PAD4 编码蛋白则与植保素合成有关。

林汉明等（尚未发表数据）利用基因芯片技术对携有抗 *Xoo* 基因 *Xa2*、*Xa4*、*Xa12*、*Xa14*、*Xa21*、*Xa23* 及抗稻瘟病基因 *Pi-b*、*Pi-k*、*Pi-ta* 的水稻近等基因系及其轮回亲本的基因表达谱作了比较分析，获得若干具有分类能力的基因（classifier genes），它们能把白叶枯病及稻瘟病相关的样品区分成两大簇，更可以把各簇内的抗性品种与感病的轮回亲本再分成两组。在研究过程中，还发现了几个与信息传导相关但具体抗病功能尚未清楚的基因，也说明了水稻抗 *Xoo* 的信息传导可能相当复杂。

4.4.1　磷酸化、去磷酸化作用　抗白叶枯病基因引起的抗病反应途径构成了一个复杂的信号网络。如 Xa*21* 基因产物能引起一系列蛋白级联磷酸化反应（Martin et al.，1993；Song et al.，1995），其中研究较为深入的是促分裂原活化蛋白激酶（MAPK）及钙依赖性蛋白激酶（Ligterink et al.，1997；Romeis et al.，1999；2000；Zhang and Klessig，1998）。MAPK 在植物内是一个超家族，它参与生物或非生物胁迫的抗逆过程及植物激素调节的信号传导（Innes，2001），但目前还缺乏这些磷酸化过程与抗病作用的因果关系证据。

4.4.2　ROS 的多重角色和衍生的信号　植物细胞对病原菌侵染的另一个早期反应是迅速产生包括 O^{2-} 和 H_2O_2 等在内的活性氧分子（ROS），这一过程又称为氧突发（oxidative burst）。活性氧在植物抗病防卫反应中的作用主要包括：①直接抑制病原物；②通过氧化交联强化加固细胞壁的作用；③激发植物的细胞过敏性死亡；④诱导合成植保素；⑤调节细胞的氧化还原平衡；⑥激活防卫反应等。

如上所述，植物在制造木质素的过程中能产生多种有毒的酚类化合物，其中多为酚类自由基及活性氧分子（ROS）（Gross，1980），已经证明这些化学物质具有显著的抑菌作用（Horino et al.，1989；Venere，1980）。但植物在抗病过程中活性氧并非单独行使抑菌功能，多半与多种抗性成分串联而发挥作用。例如水稻对 *Xoo* 的过敏反应中，产生植保素等一些常见的抗病化学物质（Nakanishi and Watanabe，1977），植保素的大幅度增加，又促使另一类抗病化学物质 ROS 的增加（Ono et al.，2001），显示了这些化合物的协同作用。

ROS 参与植物抗病过程的事实越来越受到人们的关注（Torres et al.，2006）。研究发现，ROS 在抗病的过程中呈现两阶段作用模式。第一阶段 ROS 累积的时间较短且量少，这与植物通过对 PAMPs 的识别和相应产生基本抗病反应有关；第二阶段可能由抗病蛋白与无毒蛋白的互作引起，ROS 累积的时间持续并且量大，其引起的抗病强度也大。几例研究的报道都表明 ROS 在水稻内还可能参与几种不同的抗 *Xoo* 途径。Chittoor 等（1997）证实水稻经机械损伤或 *Xoo* 感染都可以诱导过氧化物酶基因的表达，过氧化物酶的产物是 ROS，因此 ROS 的作用可能是多元化的，其中与抗病直接相关的就是杀死病菌或受感染的细胞。在 *Xanthomonas* 属中已发现了针对 ROS 的酶系统（Loprasert et al.，1997），用低剂量 ROS 处理后的 *Xoo* 也会产生针对 ROS 的过氧化氢酶，并在细菌内起保护作用（Mongkolsuk et al.，1997）。所以植物可能需要高剂量的 ROS 才能杀死 *Xoo*。

此外，ROS 在植物细胞内还会引起其他作用。除了强化细胞壁外，亦能作为一种信号分子诱导防卫基因的表达及引起受感染植物细胞的程序致死过程。植物 ROS 的产生主要依赖于定位在质膜上的 NADPH 氧化酶复合体。NADPH 氧化酶复合体包括许多成员，其中

gp91^{phox}是复合体中的重要一员。研究发现，羧基端与NADPH氧化酶亚基gp91^{phox}同源的一个拟南芥rbohA在其氨基端可结合两个Ca^{2+}，而Ca^{2+}的浓度对NADPH氧化酶活性可能起着调节作用（Keller et al.，1998）。水稻中也含有编码同源gp91^{phox}亚基的基因（Groom et al.，1996）。Kawasaki等（1999）通过*OsRac*1基因的超量表达和抑制已经证明了水稻中存在该系统，而且他们发现水稻ROS的产生和细胞死亡与*OsRac*1基因表达或抑制之间存在明显的相关性。进一步研究显示*OsRac*1编码一个小GTP蛋白，水稻超量表达*OsRac*1能够引起植株对*Xoo*的抗性（Ono et al.，2001）；反之，如果转基因植株的OsRac1蛋白失活，则导致ROS减少甚至失去由无毒基因引起的过敏反应。不仅如此，*OsRac*1还可能通过调控水稻内的NADPH氧化酸来产生ROS及辅酶A还原酶来制造木质素，最近又有研究显示，*OsRac*1亦能影响由病原引起的MAPK途径，因而显示*OsRac*1在水稻抗性中的重要作用（Lieberherr et al.，2001）。

4.5 诱导性抗病反应

根据诱导方式和抗病特性的不同，诱导性抗病反应大致可分为系统获得抗性（SAR）（Sitcher et al.，1997），诱导性系统抗病反应（ISR）（Pieterse et al.，2002）和伤害诱发抗性（Kesster and Baldwin，2002）三类。这些诱发性抗病反应涉及水杨酸、茉莉酸、乙烯、类固醇、脱落酸等植物激素的参与，并依靠复杂的信号系统来协调和控制（Pieterse et al.，2004）。

系统获得抗性与*Xoo*抗性的关系密切，由局部性的细菌侵染就能引起如同拟南芥发生的SAR（Dong，2004）。启动SAR需要水杨酸的累积和PR蛋白的表达。PR蛋白一般是分泌性或定位在液泡内的，并具有抑菌活性，所以常以PR蛋白或其基因表达来表明SAR的发生情况。水杨酸是在拟南芥SAR中发挥作用的重要信号分子（Fobert and Despres，2005）。例如水杨酸的增加可导致NPR1（Non Expressor of Pathogenesis Related Genes 1）蛋白内半胱氨酸残基还原，从而使NPR1蛋白有可能进入细胞核。此外，水杨酸具有还原TGA家族转录因子的半胱氨酸残基，促进了TGA与NPR1在细胞核内相结合的功能，从而增加TGA转录因子对DNA的结合能力，并相应启动一系列因水杨酸诱发而引起的基因表达。

尽管水杨酸已被认为是一种信号分子，具有把信号从受感染组织传到整株植物的功能，但由于SAR似乎并不需要外源的水杨酸，因而有人认为挥发性较强的乙烯进行信号传递更为合理（Van Loon et al.，2006）。另外，对于水稻而言水杨酸更不具备信号分子的特征，因为水稻内源的水杨酸含量本来极高，而且在病原感染下其水杨酸的变化并不明显（Silverman et al.，1995）。但令人疑惑的是，将拟南芥水杨酸调控基因*NPR*1在水稻中表达，却能引起水稻的抗病表型（Chern et al.，2001），从而推测水稻存在与NPR1调节同源的系统。进一步研究证实水稻中存在NPR1同源蛋白（NH1），它能与TGA型转录因子互作并发挥直接作用（Fitzgerald et al.，2004）。TGA型转录因子是一个较大的基因家族，其成员可能通过与NPR1结合而产生不同效应。在上述实例中，已了解到某些TGA成员的超量表达（over-expression）能够增强水稻对*Xoo*的抗性，因而显示了TGA在抗病过程中所起的正调控效应。然而有趣的是，当水稻TGA2.1基因被敲除（Knockout）后，敲除体对*Xoo*的抗性不但没有降低反而增强了，因而这又显示有些TGA成员作为转录因子对*Xoo*抗性的负调控作用（Pitzgerald et al.，2005）。

水稻 NRR 基因的发现和研究进一步显示 NH1 途径（NPR1 同源途径）在抗 *Xoo* 中的重要作用。水稻 NRR 作为负调控因子存在，它能与 NH1 蛋白直接作用，研究证明水稻超量表达 NRR 基因能减低其基本抗性（Chern et al.，2005）。虽然许多证据支持 NPR1 途径在水稻抗病中的重要作用，但却没有强有力的证据来支持水杨酸的调控功能，这可能反映了单子叶和双子叶植物在抗病进化中的不同。由于拟南芥 NPR1 蛋白会同时影响由水杨酸及茉莉酸引起的信号途径（Pieterse and van Loon，2004），因此茉莉酸是否能通过 NH1 蛋白在水稻中引起 *Xoo* 抗性还需要深入研究。

参 考 文 献

边永高，姚坚，陈仁祥，等．2004. 中熟晚粳秀水 994 的特征特性及栽培要点［J］．浙江农业科学（2）：66－67.

曹立勇，庄杰云，占小登，等．2003. 抗白叶枯病杂交水稻的分子标记辅助育种［J］．中国水稻科学，17（2）：184－186.

陈秀兰，何震天，韩月澎，等．2004. 国家一级优质中粳扬辐粳 7 号的特征特性及栽培技术［J］．江苏农业科学，20（5）：33－34.

陈勇，廖新华，谢岳峰，等．1990. 云南稻种资源对白叶枯病抗性的研究［C］//朱立宏．主要农作物抗病性遗传研究的进展．南京：江苏科学技术出版社：21－30.

陈勇，戴陆园，廖新华．1993. 云南稻种资源［C］//应存山．中国稻种资源．北京：中国农业科技出版社：186－191.

陈寅，向建国，任新国．1984. 水稻白叶枯病菌致病力分化的研究［J］．湖南农学院学报，（4）57－66.

陈利锋，徐敬友．2001. 农业植物病理学［M］．北京：中国农业出版社．

承河元，季伯衡，王杰．1979. 安徽省水稻白叶枯病菌株的鉴别及菌株与水稻品种间相互关系的研究初报［J]. 安徽农学院学报（7）：72－79.

承河元，章健，檀根甲，等．1999. 安徽省水稻白叶枯病菌致病型测定［J］．安徽农业科学，27（5）：484－485.

成国英，苏清实，蔡福民，等．1994. 湖北省水稻白叶枯病菌致病性研究［J］．华中农业大学学报，3（6）：569－575.

陈志良，莫荣传，王春坤．2003. 杂交晚粳稻甬优 2 号制种技术［J］．浙江农业科学（5）：251－253.

程式华，庄杰云，曹立勇，等．2004. 超级杂交稻分子育种研究［J］．中国水稻科学，18（5）：377－383.

程式华．1995. 两系法杂交稻［C］//孙宗修，程式华主编．杂交水稻育种，从三系、两系到一系．北京：中国农业科技出版社：179－181.

褚菊征，邹雪蓉，商玉霞，等．1982. 北京水稻白叶枯病菌噬菌体的形态和血清学的研究［J］．植物病理学报，12（3）：33－40.

邓建平．2005. 2005 年江苏省审定通过的水稻品种（组合）［J］．中国稻米（4）：18－19.

邓建平．2006. 2006 年江苏省审定通过的水稻新品种［J］．中国稻米（3）：20－25.

戴继跃，曾令祥，张再兴．1989. 贵州地方水稻品种抗性研究——抗“两逆两病”种质的频率及其分布［J]. 贵州农业科学（1）：9－14.

丁颖．1983. 广东野生稻不实现象的观察［C］//《丁颖稻作论文选集》编辑组．丁颖稻作论文选集．北京：农业出版社：158－171.

丁伦友．2005. 高抗白叶枯病系列杂交稻的选育［J］。杂交水稻，20（1）：11－14.

董家胜．2005. 2005 年河南省审定通过的水稻新品种［J］．中国稻米（5）：24.

段永嘉，张世光，陆晓东，等．1983. 水稻白叶枯病菌系的研究［J］．云南农业科技（1）：1－4.

段永嘉，张世光．1985. 云南水稻白叶枯病菌系的研究［J］．云南农业科技（3）：3－6.

段永嘉，张世光，陆晓东，等．1987. 水稻白叶枯病菌系研究［J］．云南农业大学学报，2（2）：1－7.

樊颖伦，陈学伟，王春连，等．2006. 水稻抗白叶枯病基因 *Xa23* 的 RFLP 标记定位及其 STS 标记的转

化 [J]. 作物学报，32 (6)：931 - 935.
方中达，刘经芬，朱家玲，等. 1956. 水稻白叶枯病（*Xanthomonas oryzae*）侵染循环的初步研究 [J]. 植物病理学报，9 (2)：173 - 185.
方中达，朱家玲，刘经芬，等. 1957. 水稻品种对白叶枯病抗性测定的方法 [J]. 南京农学院学报 (2)：61 - 68.
方中达，任欣正，陈泰英，等. 1957. 水稻白叶枯病及条斑病和李氏禾条斑病病原菌的比较研究 [J]. 植物病理学报 (3)：99 - 124.
方中达，任欣正. 1959. 李氏禾与水稻白叶枯病在条斑病相互关系的进一步研究 [J]. 植物病理学报 (5)：100 - 110.
方中达，任欣正. 1960. 我国水稻上的一种新的细菌性病害 [J]. 植物病理学报，6 (1)：90 - 92.
方中达，刘经芬，朱家琳，等. 1963. 水稻抗病性研究，Ⅱ. 品种抗病性及其抗性机制的初步研究. Ⅲ. 水稻叶色变化与白叶枯病发展的关系 [J]. 植物病理学报 (6)：107 - 112；113 - 118.
方中达. 1963. 水稻白叶枯病的发生情况 [C] // 方中达. 水稻白叶枯病. 南京：江苏人民出版社：1 - 2.
方中达，许志刚. 1978. 水稻白叶枯病研究中的几个问题 [J]. 中国农业科学 (2)：73 - 77.
方中达，许志刚，过崇俭，等. 1981. 水稻白叶枯病细菌致病性的变异 [J]. 南京农学院学报 (1)：1 - 11.
方中达，许志刚，过崇俭，等. 1990. 中国水稻白叶枯病菌致病型的研究 [J]. 植物病理学报，20 (2)：81 - 88.
方中达. 1998. 植物细菌病害 [C] //方中达. 植病研究方法. 北京：中国农业出版社：156 - 232.
高北平，李桂勇，廖军铖，等. 2003. 两系亚种间杂交稻赣亚 1 号的选育与应用 [J]. 杂交水稻，18 (1)：5 - 7.
高和平，林兴华，余功新，等. 1996. 4 个云南稻种对白叶枯病抗性的遗传研究 [J]. 华中农业大学学报，15 (2)：105 - 109.
高山林，朱立宏. 1982. 水稻白叶枯病抗性遗传研究 [J]. 南京农学院学报 (1)：22 - 35.
广东省农科院植保所. 1977. 关于海南岛杂优水稻急性枯萎型叶枯病鉴定结果的报告 [J]. 广东农业科学 (5)：27 - 30.
广东省农科院水稻研究所，植物保护研究所. 1989. 普通野生稻种资源抗性鉴定 [J]. 广东农业科学 (2)：3 - 6.
过崇俭，王法明，奚文英. 1980. 水稻白叶枯病凋萎型病稻桩中病菌的传病作用 [J]. 植物病理学报 (7)：215 - 219.
谷福林，苏自强，王水方. 2001. 优质杂交粳稻新组合 86 优 8 号 [J]. 杂交水稻，16 (2)：59 - 60.
顾铭洪. 1992. 水稻广亲和性的研究及其进展 [C] //袁隆平. 两系法杂交水稻研究论文集. 北京：中国农业出版社：23 - 29.
顾铭洪，汤述翥. 2001. 两系亚种间杂交水稻育种及其理论基础 [C] //卢兴桂，顾铭洪，李成荃，等. 两系杂交水稻理论与技术. 北京：科学出版社：100 - 132.
郭宗德. 1971. 水稻白叶枯病菌噬菌体 [C] //邱人璋. 水稻病害. 中国农村复兴联合委员会刊印. 123 - 134.
韩玉梅，赵荣杰. 1988. 野生稻对白叶枯病的抗性研究 [J]. 植物保护学报，15 (1)：45 - 48.
华南农学院植保系. 1978. 水稻凋萎型白叶枯病的鉴定研究报告 [J]. 广东农业科学 (2)：30 - 33.
何明，张成婉，罗显芝. 1993. 四川水稻白叶枯病菌致病型研究 [J]. 西南农业学报，6 (增刊)：1 - 5.
何发青，林芳仕，吴松青，等. 2005. 优质早籼三系不育系 H28A 的选育与利用 [J]. 杂交水稻，20 (1)：22 - 25.
何予卿，张宇，孙梅，等. 2001. 利用 ISSR 分子标记研究栽培稻和野生稻亲缘关系 [J]. 农业生物技术学报 (9)：123 - 127.

何震天，陈秀兰，韩月澎，等．2005. 扬辐籼 6 号的选育及特征特性［J］．江苏农业科学，21（3）：49 - 50．

侯小华，李友荣，魏子生，等．2001. 湖南省水稻地方品种资源白叶枯病抗性评价［J］．湖南农业科学（4）：42 - 43.

黄德娟，柳武革，符福鸿，等．2005. 优质抗病高产杂交稻新组合天优 372［J］．广东农业科学，21（5）：24.

黄彗君，符福鸿，李传国，等．2005. 优质高产抗病杂交稻天优 290 的选育［J］．广东农业科学，21（5）：15 - 16.

黄年生，张洪熙，戴正元，等．2003. 优质中籼新品种扬稻 8 号的选育及利用［J］．中国稻米（2）：10 - 11.

黄继武．1989. 贵州地方水稻品种抗性研究［J］．贵州农业科学（1）：9 - 14.

黄斌民．1988. 广东晚稻产量不高不稳的主要原因及对策［J］．广东农业科学（1）：16 - 18.

纪桐云．1989. 水稻病害的发生趋势和防治对策［J］．安徽农业科学（2）：52 - 55.

姬广海，张世光，钱君．2003. 云南水稻白叶枯病菌生理小种初析［J］．植物保护，29（1）：19 - 21.

江苏省农科院稻病组．1981. 水稻白叶枯病菌致病力的变异研究初报［J］．江苏农业科学（4）：44 - 48．

江苏省农科院植保所，粮食所．1980. 稻种资源白叶枯病抗性鉴定研究［J］．江苏农业科学（3）：21 - 28.

江祺祥，刘古春，杨明方，等．2004. 武香粳 14 号特征特性及高产栽培技术［J］．江苏农业科学（2）：26 - 27.

赖传雅．2003. 水稻病害［C］∥赖传雅．农业植物病理学．北京：科学出版社：16 - 17.

李梅芳，等．1983. 水稻花培育种研究［M］．北京：农业出版社：32 - 38.

李爱贞，罗宽，刘毅君．1999. 水稻近等基因系与白叶枯病菌互作的生理指标研究［J］．植物保护学报，26（1）：35 - 39.

李成云．2002. 云南省水稻白叶枯病菌致病力差异的地理分布［J］．西南农业学报，15（4）：71 - 74.

李传国，符福鸿，梁世胡．2000. 杂交稻新组合Ⅱ优 128 的选育与利用［J］．杂交水稻，15（3）：9 - 10.

李道远，陈成斌．1990. 国际野生稻资源利用进展概况［C］∥李道远．广西稻种资源论文选集．南宁：广西农业科学编辑部：14 - 26.

李贵勇，李斌，杨从党，等．2005. 优质、高产、抗病超级稻云超 6 号简介［J］．中国稻米（4）：8.

李宏，周少川，黄道强，等．2005. 核心种质育种理论应用实例——优质稻新品种丰丝占的育成［J］．中国稻米（1）：19 - 20.

李金军，陆金根，张国兴．2003. 优质晚粳稻嘉 991［J］．浙江农业科学（5）：249 - 250.

李柳娟，罗小芬，刘丕庆，等．2005. 广西杂交稻组合抗白叶枯基因 *Xa4* 的分子标记检测［J］．杂交水稻，20（4）：62 - 63.

李南田．1989. 印度尼西亚的水稻［C］∥熊振民，历葆初．世界水稻．杭州：中国水稻研究所：208 - 228.

李民迪，曹丰生，顾满莲，等．2006. 优质杂交晚籼新组合中优 2506［J］．杂交水稻，21（1）：87 - 88.

李金衡．2005. 2004 年云南省审定通过的水稻品种简介［J］．中国稻米（2）：19 - 20.

李金衡．2006. 2005 年云南省审定通过的水稻新品种［J］．中国稻米（2）：28 - 30.

李宣铿，黄河清，柏连阳，等．1993. 杂交稻三系资源及组合抗白叶枯病鉴定［J］．杂交水稻（2）：35 - 36.

李新奇．1987. 广亲和爪哇型品种与籼粳稻的亲和力遗传及杂种优势初步研究［J］．湖南农业科学（3）：4 - 9；21.

李友荣，魏子生，侯小华．1993. 国际水稻种质对白叶枯病的抗性评价与利用研究［J］．湖南农业科学（3）：43 - 44.

李友荣，侯小华，魏子生．1999. 两系杂交稻对白叶枯病的抗性研究［J］．湖南农业科学（2）：36.

李志正，等．1985. 中籼扬稻 1 号及 910 品系对白叶枯病抗性的研究［J］．植物保护学报（1）：29 - 35.

李振歧．1995. 植物病原物的致病性［C］//李振歧，等．植物免疫学．北京：中国农业出版社：15－39.
梁斌，王敬文，明风，等．2003. 云南滇西高原粳稻区白叶枯病菌遗传多样性初探［J］．上海交通大学学报（农业科学版）(21)：38－42.
梁世胡，李传国，符福鸿，等．2004. 高产优质杂交籼稻新组合天优998［J］．杂交水稻，19 (6)：62－63.
林汉明，张世宏，章琦．2007. 水稻对白叶枯病抗性的机理［C］//章琦．水稻白叶枯病的抗性遗传及改良．北京：科学出版社：323－348.
林世成，冯瑞英，康小湖，等．1980. 水稻抗病育种研究 Ⅰ．粳稻的抗瘟性和中丹1号、2号、3号的育成［J］. 中国农业科学 (1)：1－14.
林世成，章琦，冯瑞英，等．1981. 水稻抗病育种研究 Ⅱ．品种（系）对稻瘟病的抗谱鉴定初报［J］．中国农业科学 (3)：1－9.
林世成，闵绍楷．1991. 中国水稻品种及其系谱［M］．上海：上海科学技术出版社：285－371.
林世成，章琦，李道远，等．1992. 普通野生稻对水稻白枯病抗性的评价及遗传研究初报［J］．中国农业科学，6 (4)：155－158.
林世成，阙更生，邢祖颐，等．1993. 广西普通野生稻 RBB16 抗白叶枯病育种初报［J］．广西农业科学 (1)：1－5.
林兴华，余功新，张端品，等．1992. 云南水稻品种冬糯白叶枯病抗性基因定位研究［J］．遗传学报，19 (5)：459－466.
林声和，等．1986. 福建省水稻白叶枯病菌系及其应用研究［J］．福建省农科院学报，1 (2)：48－54.
刘鑫，吴伟．2005. 2004年浙江省审定通过的水稻品种［J］．中国稻米 (3)：21.
刘鑫，吴伟．2005. 2005年浙江省审定通过的水稻品种 (1)［J］．中国稻米 (4)：16－17.
刘鑫，吴伟．2006. 2005年浙江省审定通过的水稻品种 (2)［J］．中国稻米 (2)：27.
刘超，王健康，郭荣良，等．2003. 高产杂交中粳新组合69优8号［J］．杂交水稻，18 (2)：71－72.
刘明放，沈德胜，施去生，等．2004. 早籼糯越糯6号的选育、特征特性和栽培要点［J］．浙江农业科学 (5)：258－259.
陆朝福，李晓兵，朱立煌，等．1996. 用 PCR 技术诊断水稻的白叶枯病抗性［J］．遗传学报（23)：110－116.
罗彦长，王守海，李成荃，等．2003. 应用分子标记辅助选择培育抗稻白叶枯病光敏核不育系 3418S［J］．作物学报，29 (3)：402－409.
罗彦长，吴爽，王守海，等．2005. 聚合抗白叶枯病双基因三系不育系 R106A 的选育研究［J］．中国农业科学，38 (11)：2157－2164.
罗孝和，袁隆平．1989. 水稻广亲和的选育［J］．杂交水稻 (2)：35－38.
卢兴桂，顾铭洪，李成荃，等．2001. 两系杂交水稻理论与技术［J］．北京：科学出版社．
马良勇，李西明，杨长登．2003. 优质早籼中早16的特性特征和栽培技术［J］．中国稻米 (5)：9.
马良勇，杨长登，李西明，等．2006. 超级早稻品种中早22特征特性及栽培技术［J］．中国稻米 (3)：19.
马铮，苏东，徐士库，等．2007. 高产、优质抗病杂交稻新组合Ⅱ优1511［J］．杂交水稻，22 (6)：78－79.
闵绍楷，宋祥甫．1989. 日本的水稻［C］//熊振民，历葆初．世界水稻．杭州：中国水水稻研究所．
莫惠栋．1979. 二项分布在作物遗传育种上的一些应用［J］．遗传，1 (2)：11－16.
彭绍裘，魏子生，等．1981. 野生稻多抗（病）性鉴定［J］．湖南农业科学 (5)：47－48.
潘海军，王春连，赵开军，等．2003. 水稻抗白叶枯病基因 *Xa23* 的 PCR 分子标记定位及辅助选择［J］．作物学报 (29)：931－935.
彭绍裘，魏子生，等．1981. 野生稻多抗（病）性鉴定——高抗白叶枯病的抗源——疣粒野生稻［J］．湖南农业科学（5）：47－48.
彭绍裘，魏子生，毛昌祥，等．1982. 云南省疣粒野生稻、药用野生稻和普通野生稻多抗性鉴定［J］．植物病理学报，12（4）：58－61.
庆容，王辉，兰陆寿，等．2006. 高产抗病杂交籼稻新组合农华优 808［J］．杂交水稻，21（3）：81－82.
钱汉良，陈其志，张翼，等．1997. 江汉平原水稻白叶枯病致病型的变异研究［J］．植物病理学报，27（4）：333－337.
祁祖白，等．1987. 水稻抗白叶枯病基因及其在多代回交情况下的遗传规律的研究［J］．遗传，9（1）：1－4.
全国农业技术推广服务中心．2007. 全国农作物主要品种面积推广情况统计汇编（1982—2006）.
全国农作物品种审定委员会．2005. 2005 年国家审定通过的水稻品种（上）［J］．中国稻米（6）：24－25.
沙学延，朱立宏，吴烈洪．1994. BG－90－2 及其衍生品种抗水稻白叶枯病基因分析［J］．植物病理学报，24（3）：254－257.
邵启明，朱立宏．1992. 籼稻衍生抗源抗白叶枯病的遗传［J］．西南农业大学学报，14（1）：35－38.
沈锦华，李梅芳，陈银全，等．1982. 花药培养在水稻品种改良上的应用［J］．中国农业科学（2）：15－19.
沈锦华，等．1983. 水稻花培育种研究［M］．北京：农业出版社．
盛生兰，龚红冰，周文义，等．2003. 籼型三系恢复系镇恢 084 的选育及利用价值［J］．江苏农业科学，18（6）：19－21.
盛生兰，龚红冰，周文义，等．2005. 优质抗病杂交籼稻新组合协优 084［J］．杂交水稻，20（1）：68－69.
石明松．1981. 晚粳自然两用系的选育及应用初报［J］．湖北农业科学（7）：1－3.
石明松．1985. 对光照长度敏感的隐性雄性不育水稻的发现与初步研究［J］．中国农业科学（2）：44－48.
施爱农，章琦，王春莲，等．1993. 感病品种背景对水稻白叶枯病抗性遗传的影响［J］．北京农学院学报（8）：90－94.
施爱农，章琦，王春连，等．1993. 籼稻品种对三个白叶枯病菌系的抗性遗传分析［J］．华中农业大学学报，12（1）：215－220.
司徒志谋．2003. 2003 年广东省审定通过的水稻品种［J］．中国稻米（4）：24.
孙恢鸿，赖星华，梁跃懋，等．1985. 国外水稻品种白叶枯病抗源筛选研究［J］．广西农业科学（1）：35－37.
孙恢鸿，农秀美，黄福新，等．1987. 广西稻白叶枯病菌致病类型、分布及其变异性研究［J］．广西农业科学（5）：29－33.
孙恢鸿，农秀美，黄福新，等．1992. 广西栽培稻种质资源抗白叶枯病研究［J］．植物保护学报，19（3）：237－241.
孙恢鸿，李清标，黄连江．1993. 广西稻白叶枯菌致病性变异研究再报［J］．广西农业科学（3）：128－129.
孙恢鸿．2004. 我国水稻白叶枯病菌致病力分化研究［J］．植物保护，29（3）：5－8.
谭光轩，任翔，翁清妹，等．2004. 药用野生稻转育后代一个抗白叶枯病新基因的定位［J］．遗传学报，31（7）：724－729.
汤圣祥，余汉勇．2002. INGER 水稻种质的遗传评价和利用［J］．植物遗传资源科学（3）：46－50.
唐红生，孙明法，姚立生，等．2006. 优质中粳稻新品种盐粳稻 9 号的特征及栽培技术［J］．中国稻米（2）：26－27.
万邦惠，等．1980. 水稻白叶枯病抗性遗传的研究［J］．广西农业科学（9）：24－27.
万建民，朱立宏．1985. 水稻对白叶枯病的抗性表达及遗传［D］．南京：南京农业大学．
万建民，朱立宏．1990. 太湖晚粳抗白叶枯病遗传研究［J］．江苏农业学报，6（3）：1－9.

王法明，陈毓苓，邓秀琴，等．1980. 应用反向间接血凝法检测白叶枯病病原［J］．中国农业科学（2）：72－78.
王宝和，胡祝祥，周长海，等．2004. 迟熟中粳糯扬粳糯 1 号的特征特性和栽培技术［J］．江苏农业科学，19（4）：40－41.
王昌明．1986. 水稻对白叶枯病抗性的遗传［J］．植物病理学报，16（3）：161－168.
王春连，章琦，周永力，等．2001. 我国长江以南地区水稻白叶枯病原菌遗传多样性分析［J］．中国水稻科学（15）：131－136.
王春莲，章琦，杨文才．1995. 对水稻品种抗白叶枯病的评价［J］．作物品种资源（2）：26－30.
王春莲，章琦，杨文才，等．2004. 水稻白叶枯病新抗源 Y 238 的鉴定及其近等基因系培育［J］．植物遗传资源学报，5（1）：26－30.
王春莲，戚华雄，潘海军，等．2005. 水稻抗白叶枯病基因 *Xa23* 的 EST 标记及其在分子育种上的利用［J］．中国农业科学，38（10）：1996－2001.
王春莲，章琦，周永力，等．2001. 我国水稻白叶枯病菌的分布及变化［J］．植物病理学报，31（3）：41－45.
王春莲，章琦，周永力，等．2001. 我国长江以南地区水稻白叶枯病原菌遗传多样性分析［J］．中国水稻科学，15（2）：131－136.
王法明，白克勤，孙娥琴，等．1964. 水稻良种农垦 58 在抗白叶枯病性能上的优异性［J］．江苏农学报，3（1）：21－24.
王汉荣，谢关林，余立新，等．1995. 浙江水稻白叶枯病菌系的动态及分布［J］．浙江农业科学（5）：262－263.
王辉．2003. 2002 年安徽省审定的水稻新品种［J］．中国稻米（1）：23.
王辉．2004. 2004 年安徽省审定通过的水稻品种（组合）［J］．中国稻米（5）：23－24.
王辉．2005. 2005 年安徽省审定通过的水稻品种（组合）简介［J］．中国稻米（6）：21－23.
王健龙，唐绍清，胡培松．2003. 超泰型优质香稻中健 2 号［J］．中国稻米（4）：19.
王健康，刘超，郭荣良，等．2003. 杂交中粳 69 优 8 号的特征特性及应用研究［J］．江苏农业科学，18（6）：44－45.
王健康，刘超，郭荣良，等．2006. 优质高产杂交粳稻新组合徐优 201 的选育及应用［J］．杂交水稻，21（2）：22－23.
王建设，朱立宏，张红生，等．2000. 杂交稻抗白叶枯病的遗传机制［J］．作物学报，26（1）：1－8.
王钧．1998. 植物抗病反应的分子机理［C］//余叔文，汤章城．植物生理与分子生物学．2 版．北京：科学出版社：784－807.
王疏，周永力，姚健民，等．1997. 稻曲病白化菌株生物学特性研究［J］．植物病理学报，27（4）：321－327.
王书玉，张桐，薛应征，等．2001. 早熟高产水稻新品种新稻 10 号的选育［J］．中国稻米（5）：16.
王越涛，尹海庆，王生轩，等．2005. 优质粳稻郑花辐 9 号特征特性及其栽培要点［J］．中国稻米（5）：20－21.
王子明，陈风华，吕玉亮，等．2003. 优良籼稻新组合Ⅱ优 118 的选育与应用［J］．江苏农业科学，18（3）：21－22.
王子明，周风明，吕玉亮，等．2005. 优质、高产粳稻新品种华粳 4 号选育与应用［J］．中国稻米（5）：22－23.
王子明，周风明，吕玉亮，等．2006. 优质中熟中粳稻华粳 5 号的选育及高产栽培技术［J］．中国稻米（3）21－22.
吴家道，杨剑波，许传万，等．2001. 水稻抗白叶枯病基因 *Xa21* 转基因水稻及其杂交稻研究［J］．作物学报，27（7）：29－34.

魏景超．1975．水稻病原手册［M］．北京：科学出版社．

吴强，廖兰杰，杨代常，等．1998．野生稻基组随机扩增多态性DNA（RAPD）分析［J］．热带亚热带植物学报（6）：260－266.

吴妙燊，李道远．1990．野生稻遗传资源利用展望［C］//野生稻遗传资源研究论文选编．北京：中国科学技术出版社：97－100.

伍尚忠，徐羡明，陈福坤，等．1979．水稻凋萎型白叶枯病苗期侵染途径病原菌致病力与品种感染性的研究［J］．广东农业科学（4）：43－51.

伍尚忠，等．1980．水稻品种对不同白叶枯病菌株的抗性反应［J］．植物保护学报，7（1）：43－46.

伍尚忠．1983．水稻白叶枯病及其防治［M］．上海：上海科学技术出版社．

伍尚忠，徐羡明，刘景梅．1983．东南亚及华南稻白叶枯病菌株在中国鉴别品种上的致病力比较研究［J］．植物病理学报，13（14）：21－25.

伍尚忠，徐羡明，刘景梅，等．1985．华南及菲律宾稻白叶枯病病原菌致病性比较研究［J］．植物病理学报，15（2）：65－72.

伍尚忠，徐羡明，林壁润，等．1990．华南籼稻抗白叶枯病遗传研究．Ⅰ水稻品种华竹矮及青华矮6号的抗病性遗传．Ⅱ水稻品种玻璃矮的抗病性遗传［C］//朱立宏．主要农作物抗病性遗传研究的进展．南京：江苏科学技术出版社：31－41.

夏怡厚，蔡为艺，陈章艳．1985．水稻品种对白叶枯病的抗性研究［J］．福建农学院学报，14（3）：233－238.

谢崇华，陈永军．2005．高产优质杂交水稻新组合红优527［J］．杂交水稻，20（3）：75－76.

谢岳峰，张端品，余功新，等．1990．15个云南稻品种对白叶枯病抗性遗传研究［C］//朱立宏．主要农作物抗病性遗传研究进展．南京：江苏科学技术出版社：20－33.

谢式坪钰．1971．稻白叶枯病抗性鉴定．稻作病害．台湾省农学会编印，112－113.

谢式坪钰．1978．水稻白叶枯病原细菌 *Xanthomonas oryzae* 之生态［C］//邱人璋．水稻病虫害．中国农村复兴联合委员会刊印，167－183.

谢关林，翁锦屏．1988．浙江省水稻白叶枯病菌致病性变异的观察［J］．浙江农业科学（4）：173－176.

谢关林，翁锦屏．1989．浙江省与菲律宾水稻白叶枯病菌致病性的比较［J］．浙江农业学报（2）：72－77.

肖火根．1989．水稻白叶枯病菌新鉴别品种的筛选及其菌系的研究［J］．云南农业大学学报，4（1）：27－33.

肖思心，陈企屯，茹庆华，等．1988．水稻白叶枯病菌系研究［J］．宁夏农业科技（5）：14－18.

徐卯林，张洪熙，戴正元，等．2004．优质杂交籼稻丰优香占特征特性及栽培技术［J］．中国稻米（3）：15.

徐吉成，朱立煌，陈英，等．1994．用分子标记定位一个未知的抗稻瘟病基因［J］．遗传学报，21（3）：205－214.

徐建龙，林贻滋，张炳林，等．1996．水稻白叶枯病抗性基因 *Xa 21* 的初步利用［J］．浙江农业学报，8（2）：70－73.

徐建龙，林贻滋，翁锦屏，等．1996．水稻白叶枯病抗性基因的聚合及其遗传效应［J］．作物学报，22（2）：129－134.

徐羡明，刘景梅，伍尚忠．1981．广东水稻白叶枯病菌系的研究［J］．广东农业科学（3）：7－9.

徐羡明，刘景梅，伍尚忠，等．1986．水稻品种资源抗白叶枯病的研究［J］．植物保护学报，13（4）：227－233.

徐羡明，刘景梅，伍尚忠．1987．从广东杂交稻的发展看白叶枯病病原菌系的变异［J］．广东农业科学（5）：30－34.

徐羡明，何昆明，伍尚忠，等．1990. 广东地方稻种资源对白叶枯病的抗性鉴定［J］．广东农业科学（5）：31－34.

徐羡明，曾列先，林壁润，等．1994. 广东水稻白叶枯病菌致病型研究［J］．植物保护，20（4）：7－9.

徐羡明，曾列先，伍尚忠，等．1995. 广东珠江三角洲水稻白叶枯病菌的致病型研究［J］．西南农业学报，8（增刊）：115－118.

薛应龙，欧阳光察．1998. 植物抗病的物质代谢基础［C］//余叔文，汤章城．植物生理与分子生物学．2版．北京：科学出版社，770－783.

熊卫，余功新，林新华，等．1994. 五个云南稻种抗白叶枯病基因分析［J］．华中农业大学学报，13（2）：99－110.

许红．1993. 湖北省水稻白叶枯病区域流行特点及其防治对策［J］．华中农业大学学报，12（5）：433－437.

许志刚，邵蔚蓝，方中达，等．1987. 水稻白叶枯病致病型的变异对品种抗性测定的影响［J］．中国农业科学，20（1）：66－70.

许志刚，刘凤权，沈秀萍，等．1998. 水稻白叶枯病和条斑病的流行与预测［J］．西南农业大学学报，20（5）：567－572.

许志刚，孙启明，刘凤权，等．2004. 水稻白叶枯病菌小种分化的检测［J］．中国水稻科学，18（5）：469－472.

许自成，朱军．2000. 杂交稻抗白叶枯病的遗传机制［J］．作物学报，26（1）：1－8.

杨庆文，陈大洲．2004. 中国野生稻研究与利用［M］．北京：气象出版社．

杨仕华，程本义，沈伟峰，等．2004. 我国长江流域籼稻品种选育进展及改良策略［J］．中国水稻科学，18（2）：89－93.

杨仕华，等．2004. 我国南方杂交水稻育种进展［J］．杂交水稻，19（5）：1－5.

杨仕华，等．2005. 1996—2004年国家级审定水稻品种介绍［C］//杨仕华，廖琴．中国水稻品种试验与审定．北京：中国农业科学技术出版社：69－306.

杨仕华，等．2005. 2005年国家审定通过的水稻品种（上）［J］．中国稻米（6）：25－26.

杨仕华，等．2006. 2005年国家审定通过的水稻品种（下）［J］．中国稻米（1）：25－30.

杨伟国，寿建尧，徐兴伟．2008. 杂交粳稻嘉优2号的特征特性与直播栽培配套技术［J］．中国稻米（3）：45－46.

杨文才，章琦，王春莲．1994. 六个籼稻品种对水稻白叶枯病的抗性遗传研究［J］．华中农业大学学报（13）：234－240.

杨文才，章琦，王春连．1995. 对水稻广亲和性品种的抗白叶枯病评价和分析［J］．北京农业科学，13（6）：13－16.

姚立生，顾来顺，孙明法，等．2003. 杂交中籼新组合Ⅱ优559［J］．江苏农业科学，18（4）：26－28.

姚立生，顾来顺，孙明法，等．2003. 籼型三系不育系29A及其新组合29优559的选育［J］．杂交水稻，18（5）：8－9.

姚立生，孙明法，严国红，等．2008. 优质杂交中籼稻新组合天优1120［J］．杂交水稻，25（5）：16－18.

应存山．1989. 世界稻属遗传资源［C］//熊振民，历葆初．世界水稻．杭州：中国水稻研究所：70－78.

应存山．1996. 稻属遗传资源的考察与收集［C］//闵绍楷，申宗坦，熊振民．水稻育种学．北京：中国农业科学技术出版社：93－98.

余功新，张端品，谢岳峰．1990. 水稻品种IR28对我国白叶枯病菌系抗病基因定位研究［J］．作物学报，16（2）：140－152.

余腾琼，叶昌荣，戴陆园，等．2003. 云南高原粳稻区白叶枯病菌的致病性初步鉴定［J］．中国水稻科学，17（3）：255－259.

余腾琼，梁斌，戴陆园，等．2005. 云南高原粳稻白叶枯病菌致病型鉴定及主栽品种抗性反应初析［J］．中国农业科学，38（6）：1148－1155.

袁隆平．2002. 杂交水稻育种［M］．北京：中国农业出版社．

袁世传，应存山，盛景山．1993. 外国稻种资源的引进和利用［C］//应存山．中国稻种资源．北京：中国农业科学技术出版社：133－136.

袁潜华．2001. 主要光、温敏核不育系与两系杂交水稻组合简介［C］//卢兴桂，顾铭洪，李成荃，等．两系杂交水稻理论与技术．北京：科学出版社：225－261.

俞琦英．2003. 2003年浙江省审定通过的水稻品种［J］．中国稻米（3）：19－20.

俞琦英．2004. 2004年浙江省审定通过的水稻品种（组合）［J］．中国稻米（6）：25.

曾春，蔡洲，唐超世，等．2002. 高产优质两系杂交稻组合培杂茂三［J］．杂交水稻，17（3）：54－55.

曾士迈．1995. 植物抗病性的概念和分类．［C］//李振歧．植物免疫学．北京：中国农业出版社：7－14.

曾士迈．1998. 水平抗病性，持久抗病性和抗病性持久化［C］//曾士迈，张树榛．植物抗病育种的流行学研究．北京：科学出版社：67－72.

曾左葵，谢岳峰．1983. 水稻对白叶枯病的抗性遗传研究Ⅰ．水稻对白叶枯病菌系G1701的抗性遗传［J］．华中农学院学报，2（1）：1－9.

曾左葵，谢岳峰．1983. 水稻对白叶枯病的抗性遗传研究Ⅱ．抗病品种对4个白叶枯病菌系的抗性遗传［J］．华中农业大学学报，2（2）：1－10．．

曾列先，黄少华，伍尚忠，等．2002. IRBB21（*Xa21*）对广东稻白叶枯病菌5个小种的抗性反应［J］．植物保护学报，29（2）：97－100.

曾列先，朱小源，杨建源，等．2005. 广东水稻白叶枯病菌新致病型的发现及致病性鉴定［J］．广东农业科学，29（2）：58－59.

翟文学，李小兵，田文忠，等．2000. 由农杆菌介导将白叶枯病抗性基因 *Xa21* 转入我国的5个水稻品种［J］. 中国科学（C辑）（30）：200－206.

翟文学．1999. *Xa21* 的基因工程和 *Xa4* 候选基因的定位克隆［M］．北京：中国科学院遗传研究所．

翟文学，朱立煌．1999. 水稻白叶枯病抗性基因的研究与分子育种［J］．生物工程进展，19（6）：9－15.

张从合，陈金节，蒋家月，等．2007. 两系超级杂交稻新组合两优6号［J］．杂交水稻，22（5）：76－78.

张承妹，陆家安，章振华，等．1993. 水稻花粉株系白叶枯病抗性遗传［J］．遗传学报（20）：88－95.

张端品，谢岳峰．1982. 粳稻白叶枯病的抗性遗传［J］．中国农业科学（5）：17－24.

张端品，谢岳峰，MEW T W. 1984. 水稻品种对白叶枯病成株抗性研究［J］．中国农业科学（1）：40－50.

张洪熙，谭长乐，戴正元，等．2003. 高产优质杂籼新组合K优507选育及栽培技术［J］．江苏农业科学，18（3）：39－40.

张洪熙．2000. 综合育种，全面超越，重点突破［J］．江苏农业科学15（4）：5－7.

张林，张端品，谢岳峰．1988. 水稻亚种间杂交后代白叶枯病的抗性表现［J］．华中农业大学学报，7（2）：123－129.

张效忠．1959. 汉中地区水稻白叶枯病初步调查［J］．陕西农业科学（3）：121－122.

张万霞，杨庆文．2004. 中国野生稻收集、鉴定和保存现状［C］//杨庆文，陈大洲．中国野生稻种研究与利用．北京：气象出版社：19－25.

张耀钢，朱立宏．1991. 衍生抗源籼粳交抗白叶枯病遗传研究［C］//熊振民，梁承业．水稻育种技术基础研究论文集．北京：中国科学技术出版社：39－44.

张再兴，邓光辉．1993. 贵州稻种资源［C］//应存山．中国稻种资源．北京：中国农业科学技术出版社：20.

章琦，林世成，KHUSH G S，MEW T W. 1986. 水稻对白叶枯病（*Xanthomonas campestris* pv. *oryzae*）的抗性遗传研究 Ⅰ．水稻品种对菲律宾白叶枯病菌系PXO 61的抗性遗传［J］．中国农业科学，19

(5)：19－25.

章琦，MEW T W. 1986. 水稻对白叶枯病的成株抗性［J］. 作物学报，12（4）：241－247.

章琦，王春莲，林世成. 1988. 水稻抗病育种研究 Ⅲ. 我国主栽品种对若干白叶枯病菌系的抗性分析［J］. 中国农业科学，21（1）：41－49.

章琦，施爱农，王春莲，等. 1990. 水稻对白叶枯病（*Xanthomonas campestris* pv. *oryzae*）的抗性遗传研究 Ⅱ. 水稻品种对五个菌系的抗性遗传分析［J］. 中国水稻科学，4（1）：1－8.

章琦，施爱农，王春莲，等. 1991. 水稻对白叶枯病（*Xanthomonas campestris* pv. *oryzae*）的抗性遗传研究 Ⅲ. 两个成株抗性基因 *Xa6* 与 *Xa3* 的等位性分析［J］. 作物学报，17（3）：233－237.

章琦. 1991. 我国水稻白叶枯病抗性遗传的评价和利用［J］. 中国农业科学，24（3）：26－36.

章琦，施爱农，王春莲，等. 1994. 9个水稻品种对水稻白叶枯病（*Xanthomonas oryzae* pv. *oryzae*）的抗性遗传研究［J］. 作物学报，20（1）：84－92.

章琦，王春莲，施爱农，等. 1994. 野生稻抗稻白叶枯病性（*Xanthomonas oryzae* pv. *oryzae*）的评价［J］. 中国农业科学，27（5）：1－9.

章琦. 1995. 我国水稻抗白叶枯病基因的利用及策略［J］. 植物保护学报，22（3）：242－246.

章琦，LEACH J E，王春莲，等. 1996. 中国水稻白叶枯病菌系染色体 DNA 的 RFLP 谱型的初析［J］. 作物学报，22（1）：14－18.

章琦，朱立宏，谢岳峰，等. 1996. 我国水稻白叶枯病（*Xanthomonas oryzae* pv. *oryzae*）抗性遗传研究标准化的几点商榷［J］. 中国农业科学，29（4）：85－92.

章琦，施爱农，杨文才，等. 1996. 携有不同主效白叶枯病抗性单基因的6个粳稻近等基因系选育［J］. 作物学报，22（2）：135－141.

章琦. 1996. 水稻主要病害发展趋向及“八五”水稻新品种（组合）的抗病性初析［C］//中国科学技术协会专著论文. 北京：中国科学技术协会：57－60.

章琦，LEACH J E，王春连，等. 1997. 中国水稻白叶枯病菌遗传群体结构多样性及促进水稻育种的研究［C］//中国农业科学论文选萃. 北京：中国农业科学技术出版社：115－121.

章琦，周永力，王春莲，等. 1997. 稻曲病的发生趋势初析——一个值得警惕的稻病［C］//中国科学技术协会专著论文. 北京：中国科学技术协会：84－87.

章琦，施爱农，杨文才等. 1998. 3个粳稻抗白叶枯病近等基因系的构建［J］. 作物学报，24（6）：799－804.

章琦. 1998. 分子标记技术在揭示我国水稻白叶枯病菌群体遗传结构中的作用［J］. 农业生物技术通讯. 4－5.

章琦，赵炳宇，王春莲，等. 2000. 普通野生稻抗水稻白叶枯病（*Xanthomonas oryzae* pv. *oryzae*）新基因 *Xa23*（t）的鉴定和分子标记定位［J］. 作物学报，26（5）：536－542.

章琦. 2005. 水稻白叶枯病抗性基因鉴定进展及其利用［J］. 中国水稻科学，19（5）：453－459.

章琦. 2007. 水稻白叶枯病的抗性遗传及改良［M］. 北京：科学出版社.

章琦. 2009. 中国杂交水稻白叶枯病抗性的遗传改良［J］. 中国水稻科学，23（2）：111－119.

章善庆，唐昌华，曹一平，等. 2008. 优质超级杂交稻中浙优1号示范推广现状与展望［J］. 杂交水稻，23（3）：1－3.

章元寿. 1996. 植物的抗病机制［C］//章元寿，等. 植物病理生理学. 南京：江苏科学技术出版社：296－303.

郑康乐，庄杰云，王汉荣. 1998. 基因聚合提高了水稻对白叶枯病的抗性［J］. 遗传，20（4）：4－6.

邹江石，聂毓琦，满启民，等. 1989. 广亲和选系“02428”在籼、粳稻亚种间杂交的初步利用［J］. 中国农业科学，22（1）：6－14.

周汉钦，潘大建，范芝兰，等. 2006. 优质籼型香稻粤航1号的选育及其应用［J］. 中国稻米（1）：19.

周汉钦，潘大建，范芝兰，等．粳2号的特征特性及保优栽培技术［J］．中国稻米（1）：20.
周明镜，彭淑彬，等．2008. 香型高产杂交水稻新组合冈香707［J］．杂交水稻，23（5）：80－81.
周泰初，董敏玉，王根来．1982. 粳稻品种对白叶枯病的抗性遗传［J］．江苏农业科学，5（4）：1－3.
周毓珍，朱立宏．1985. 水稻抗白叶枯病遗传研究Ⅱ. 几个中间抗源的转育效应［J］．南京农业大学学报，8（3）：10－16.
周毓珍，陆作楣，吕兴泉．1993. 杂交稻恢复系白叶枯病改良研究［J］．南京农业大学学报，16（4）：1－5.
周永力，章琦．1999. 稻曲病菌分离技术的初探［J］．中国水稻科学，13（3）：186－188.
周永力，章琦，王春莲，等．2002. *Xa21* 转基因杂交水稻组合汕优63和汕优559对白叶枯病的抗性［J］．中国水稻科学，16（1）：93－95.
周永力，翟文学，章琦，等．2001. *Xa21* 转基因水稻对白叶枯病的抗性及其遗传［J］．植物病理学报，31（2）：123－129.
周永力，潘雅姣，翟文学，等．2005. 水稻白叶枯病抗性基因 *Xa23* 鉴别菌株突变体库的构建及无毒基因突变体的筛选［J］．生物工程学报（21）：486－488.
周玉萍，袁平荣，杨从党，等．2003. 抗病、优质、高产杂交稻新组合抗优98［J］．中国稻米（2）：1.
朱衡，章琦，王春莲，等．1995. 水稻白叶枯病菌的RAPD分型［J］．生物工程学报，11（3）：281－284.
朱立宏，周毓珍，许志刚．1979. 水稻白叶枯病的抗性遗传的初步研究［J］．南京农学院科学研究简报（4）：1－9.
朱立宏，万建民，邵启明．1990. 籼稻抗白叶枯病衍生抗源及其遗传分析［J］．南京农业大学学报，13（3）：13－18.
朱立宏，万建民，邵启明．1991. 籼稻抗白叶枯病衍生抗源选育及其遗传分析［C］//熊振民，梁承邺．水稻育种技术基础研究论文集．北京：中国科学技术出版社：24－30.
朱友林，吴键胜，王金生．1999. 水稻白叶枯病菌诱导水稻悬浮细胞防卫基因表达的初步研究［J］．植物病理学报（27）：250.
朱永生，陈保棠，余舜武，等．2004. 不对称体细胞杂交转移疣粒野生稻对白叶枯病的抗性［J］．科学通报，49（14）：1395－1398.
朱勇良，陈益海，李余生，等．2003. 优质杂交粳稻新组合86优242［J］．杂交水稻，18（2）：73－74.
赵新平，张端品，谢岳峰．1985. 水稻品种对白叶枯病成株抗性遗传研究［J］．华中农学院学报，4（3）：11－12.
赵新平，张端品，谢岳峰．1986. 水稻对白叶枯病成株抗性基因显性反转的研究［J］．遗传，8（6）：5－9.
赵新平，张端品，谢岳峰．1987. 水稻白叶枯病成株抗性品种病理解剖研究［J］．植物病理学报，17（3）：185－190.
中稻宣．2008. 2007年农业部认定的超级稻品种［J］．中国稻米（1）：25.
中山大学生物系遗传教研室．1978. 水稻白叶枯病抗性遗传的研究［J］．遗传与育种（5）：14.
中国农林科学院．1986. 中国稻作学［M］．北京：农业出版社：271－308.
中国农林科学院．1974. 防治水稻纹枯病的新抗菌素．农业科技通讯（1）．
仲维功，陈志得，杨杰，等．2003. 优质高产粳稻新品种南粳41的选育及栽培技术［J］．江苏农业科学，18（4）：24－25.
钟藏文，肖锋，王长芳，等．1998. 福建省水稻白叶枯病和细条病比较研究［J］．福建农业科技（6）：14.
山田昌雄．1990. 稻瘟病菌致病性的分化［C］//山崎义人，高坂卓尔．林忠专等，译．稻瘟病与抗病育种．北京：农业出版社：356－377.
山崎义人，高坂淖儿．1980. ィネのんっち病と抵抗性育种［M］．博友社．
伊坂实人．1974. 水稻白叶枯病菌在春种作物上的越冬［J］．植物防疫（28）：134－146.

ADHIKARI T B, BASNYAT R C. 1999. Virulence of *Xanthomonas oryzae* pv. *oryzae* on lines containing single resistance genes and gene combinations [J]. Plant Disease, 83 (1): 46-50.

ADHIKARI T B, CRUZ C, ZHANG Q, et al. 1995. Genetic diversity of *Xanthomonas oryzae* pv. *oryzae* in Asia [J]. Appl. Environ. Microbiol., 61: 966-971.

ADHIKARI T B, MEW T W, TENG P S. 1994. Phenotypic diversity of *Xanthomonas oryzae* pv. *oryzae* in Nepal [J]. Plant Disease, 78: 68-72.

ADHIKARI T B, NELSON R J, MEW T W, et al. 1997. Assessment of genetic diversity of *Xanthomonas oryzae* pv. *oryzae* in Nepal. Ⅲ. Proceedings of the Third International Rice Genetics Symposium [C]. 16-20 Oct. 1995. Manila, Philippines: IRRI: 945-949.

AHN S N, BOLLICH C N, TANKSLEY S D, et al. 1992. RFLP tagging of a gene for aroma in rice [J]. Theoretical and Applied Genetics, 84: 825-828.

AHN S N, BOLLICH C N, MCCLUNG A M, et al. 1993. RFLP analysis of genomic regions associated with cooked - kernel elongation in rice [J]. Theoretical and Applied Genetics, 87: 27-32.

AKAGI H, YOKOZEKI Y, INAGAKI A, et al. 1996. Microsatellite DNA markers for rice chromosomes [J]. TAG Theoretical and Applied Genetics, 93: 1071-1077.

ALDRICK S J, BUDDENHAGEN I W, REDDYA P K, et al. 1973. The occurrence of bacterial leaf blight in wild and cultivated rice in Northern Australia [J]. Australia Journal of Agricultural Research, 24: 219-227.

AMANTE-BORDEOS A, SITCH L A, NELSON R, et al. 1992. Transfer of bacterial blight and blast resistance from the tetraploid wild rice *Oryza minuta* to cultivated rice, *Oryza satava* [J]. Theor. APPL. Genet., 84: 345-354.

ARDALES E Y, BARAOIDAN M, YAP I, et al. 1993. Analysis of genetic diversity and population structure of bacterial blight (BB) pathogen in West Java, Indonesia [J]. Int. Rice Research Notes, 18 (4): 28-29.

ARDALES E Y, LEUNG H, VERA - CRUZ C M. et al. 1996. Hierarchical analysis of spatial variation of the rice bacterial blight pathogen across diverse agroecosystems in the Philippines [J]. Phytopathology, 86 (3): 241-248.

AWODEN V A, JOHU V T. 1984. Occurrence of bacterial leaf blight on rice in four Sahelian countries: Senegal, Mali, Niger, and Upper Volta [J]. WARDA Newsl., 5 (1): 36-39.

BECKMAN J S, SOLLER M. 1990. Toward a unified appingroach to genetic mapping of eukaryotes based on sequence tagged microsatellite sites. Nature Biotechnology, 8: 930-932.

BARTOS P, DYCK P L, SAMBORSKI D J. 1969. Adult-plant leaf rust resistance in Thatcher and Marquis wheat: A genetic analysis of the host-parasite interaction [J]. Canadian Journal of Botany, 47: 267-279.

BAW A, MEW T W, 1988. Scoring systems for evaluating varietal resistance to bacterial blight (BB): Lesion size by growth stage [J]. International Rice Research Newsletters, 13: 10-11.

BETH - MUDGETT M. 2005. New insight to the function of phytopathogenic bacterial type III effections in plants [J]. Annual Review of Plant Biology, 56: 509-531.

BHAPKAR D G, CHAVAN V M, KULKARNI N. 1960. Bacterial blight of paddy [J]. Poona. Agr. Coll. Mag., 51: 36-46.

BOKURA U. 1911. Bacterial leaf blight of rice (in Japanese) [J]. Teikoku Nokaiho, 2 (9), 62-66; 2 (10) 54-57; 2 (12) 58-61.

BRAR D S, ELLORAN R, KHUSH G S. 1991. Inter-specific hybrids produced through embryo rescue between cultivated and eight wild species of rice [J]. Rice Genet. Newsl., 8: 91-93.

BRAR D S, DALMACIO R, ELLORAN R, et al. 1996. Gene transfer and molecular characterization of intrigression from wild *Oryza* species into rice [C] // Khush G S. Rice Genetics III Los Banos: IRRI: 477 - 486.

BRAR D S, KHUSH G S. 1997. Alien introgression in rice [J] . Plant Molecular Biology, 35: 35 - 47.

BRINKERHOFF L A. 1963. Variability of *Xanthomonas malvacearum*: the cotton bacterial blight pathogen [R] . Oklahoma State University Technical Bulletin, T - 98: 1 - 96.

BUDDENHAGEN I W, RADDYA P K. 1972. The host, the environment, *Xanthomonas oryza*, and the research pages. 289 - 295. in Rice Breeding . IRRI. P. O. Box 933, Manila, Philippine.

BUTTNER D, BONAS U. 2002. Getting across- bacterial type Ⅲ effector proteins on their way to the plant cell. EMBO J. , 21: 5313 - 5322.

BYUNG-WOOK Y, HELEN A A, CHARLOTTE G, et al. 2003. Loss of actin cytoskeletal function and EDS1 activity in combination, severely compromises non-host resistance in arabidopsis against wheat powdery mildew [J] . The Plant Journal, 34: 768 - 777.

CAUSSE M A, FUITON T M, CHO Y G, et al. 1994. Saturated molecular map of the rice genome based on inter-specific backcross population [J] . Genetics, (138): 1251 - 1274.

CENTURY K S, SHAPIRO A D, REPETTI P P, et al. 1997. NDR1, a pathogen-induced component required for Arabidopsis disease resistance [J] . Science, 278: 1963 - 1965.

CHAMBERLIN D W. 1962. Reaction of resistant and susceptible soybeans to *Xanthomonas phaseoli* var. sojensis [J] . Plant Disease Reporter, 46: 707 - 709.

CHAND J N, WALKER J C. 1964. Relation of age of leaf and varietal resistance to bacterial multiplication in cucumber inoculated with *Pseudomonas lachrymans* [J] . Phytopathology, 54: 49 - 50.

CHATTOPADHYAY S B, MUKHERJII N. 1968. Occurrence in nature of collateral hosts (*Cyperus rotundus* and *C. deformis*) of *Xanthomonas oryzae*, incident of bacterial blight of rice [J] . Curr. Sci. , 15: 441.

CHEN D H, ZEIGLER R S, NELSON R J. 1996. Population structure of *Pyricularia grisea* in two IRRI blast nurseries: implications and applications [C] // KHRSH GS Rice Genetics Ⅲ . Los. Banos: IRRI: 950 - 957.

CHEN H L, WANG S P, ZHANG Q F. 2002. New gene for bacterial blight resistance in rice located on chromosome 12 identified from Minghui 63, an elite restorer line [J] . Phytopathology, 92 (7): 750 - 754.

CHEN H L, CHEN B T, ZHANG D P, et al. 2001. Pathotypes of *Pyricularia grisea* in rice fields of central and southern China [J] . Plant Disease, 85: 843 - 850.

CHEN S, ZHANG Q F. 2000. Improvement of bacterial blight resistance of hybrid rice by molecular marker-assisted selection [J] . Journal of Huazhong Agricultural, 19: 183 - 189.

CHEN W Q, LIANG N, YU J R. 1980. Wild rice in Boluo county [C] // SHEN JH, PANDLE HK, eds. Rice improvement in China and other Asian countries. Los Banos: IRRI; Beijing: CAAS: 75 - 84.

CHEN X, TEMNYKH S, XU Y, et al. 1997. Development of a microsatellite framework map providing genome-wide coverage in rice (*Oryza sativa* L.) [J] . Theoretical and Applied Genetics, 95: 553 - 567.

CHEN Y, HU J, QIAN W, et al. 2003. Characterization and molecular marker screening of a rice bacteria-resistant gene *Xa-min* (t) [J] . Progress in Natural Science, 13: 740 - 745.

CHERN M, FITZGERALD H A, YADAV R C, et al. 2001. Evidence for a disease-resistance pathway in rice similar to the NPR1-mediated signaling pathway in arabidopsis [J] . Plant J. , 27: 101 - 113.

CHERN M, CANLAS P E, FITZGERALD H A, et al. 2005. Rice NRR, a negative regulator of disease resistance, interacts with arabidopsis NPR1 and rice NH1 [J]. Plant J., 43: 623 - 635.

CHERN M, FITZGERALD H A, CANLAS P E, et al. 2005. Overexpression of a rice NPR1 homolog leads to constitutive activation of defense response and hypersensitivity to light. Mol. Plant Microbe Interact., 18: 511 - 520.

CHEUNG M Y, ZENG N Y, LAM H M, et al. 2007. Expression of a RING-HC protein from rice improves resistance to *Pseudomonas syringae* pv. *tomato* DC3000 in transgenic *Arabidopsis thaliana* [J]. J. Exp. Bot., 58: 4147 - 4159.

CHEUNG M Y, ZENG N Y, LAM H M, et al. 2008. Constitutive expression of a rice GTPase-activating protein induces defense responses [J]. New Phytol., 179: 530 - 545.

CHINCHILLA D, BAUER Z, REGENASS M, et al. 2006. The Arabidopsis receptor kinase FLS2 binds flg 22 and determines the specificity of flagellin perception [J]. Plant Cell, 18: 465 - 476.

CHISHOLM S T, COAKER G, DAY B, et al. 2006. Host-microbe interactions: shaping the evolution of the plant immune response [J]. Cell, 124: 803 - 814.

CHITTOOR J M, LEACH J E, WHITE F F. 1997. Differential induction of a peroxidase gene family during infection of rice by *Xanthomonas oryzae* pv. *oryzae* [J]. Mol. Plant Microbe Interact., 10: 861 - 871.

CHOI J E, MATSUYAMA N. 1981. Colony type variants of *Xanthomonas campestris* pv. *oryzae* and their serological properties [J]. Annals of the Phytopathological Society of Japan, 47: 244 - 251.

CHOI Y C, CHO Y S, CHUNG B J. 1978. Kresek disease in Korea. II. Effect of pathotype of pathogens and the use of infected straw on the development of Kresek [J]. Korean Journal of Plant Protection, 17: 23 - 28.

CHU Z, FU B, YANG H, et al. 2006. Targeting *xa13*, a recessive gene for bacterial blight resistance in rice [J]. Theor. Appl. Genet., 112: 455 - 461.

CHU Z, YUAN M, YAO J, et al. 2006. Promoter mutations of an essential gene for pollen development result in disease in rice [J]. Genes Dev., 20: 1250 - 1255.

COOK D, BARLOW E, SEQUERIA L. 1989. Genetic diversity of *Pseudomonas solanacearum*: detection of restriction fragment length polymorphisms with DNA probes that specify virulence and the hypersensitive response [J]. Mol. Plant-Microbe Interact., 2: 113 - 121.

DANGL J L, MCDOWELL J M. 2006. Two modes of pathogen recognition by plants [J]. Proc. Natl. Acad. Sci. U S A, 103: 8575 - 8576.

DATTA K, BAISAKH N, MAUNG T K, et al. 2002. Pyramiding transgenes for multiple resistance in rice against bacterial blight, yellow stem borer and sheath blight [J]. Theoretical and Applied Genetics, 103: 1007 - 1014.

DAVID O, NINO-LIU, PAMELA C, et al. 2006. Pathogen profile *Xanthomonas oryzae* pathovars: model pathogens of a model crop [J]. Molecular Plant Pathology, 7: 303 - 324.

DAVIES D R, GORYSHIN I Y, REZNIKOFF W S, et al. 2000. Three-dimensional structure of the Tn5 synaptic complex transposition intermediate [J]. Science, 289: 77 - 85.

DE - BRUJIN F J. 1992. Use of repetitive (repetitive extragenic palindromic and enterobacterial repetitive intergeneric consensus) sequences and the polymerase chain reaction to fingerprint the genomes of *Rhizobium meliloti* isolates and other soil bacteria [J]. Applied and Environmental Microbiology, 58: 2180 - 2187.

DESLANDES L, OLIVIER J, PEETERS N, et al. 2003. Physical interaction between RRS1-R, a protein conferring resistance to bacterial wilt, and PopP2, a type III effector targeted to the plant nucleus [J]. Proc. Natl. Acad. Sci. U S A, 100: 8024 - 8029.

DEVADATH S, PADMANABHAN S Y. 1969. A preliminary study on the variability of *Xanthomonas* oryzae on some rice varieties [J] . Plant Disease Reporter, 53: 145 - 148.

DEVADATH S, PREMALATHA DATH A. 1970. Mechanism of wilt (kresek phase) in bacterial blight of rice [J] . *Oryza*, 7: 5 - 12.

DEVOS K M, GALE M D. 1992. The use of random amplified polymorphic DNA markers in wheat [J] . Theoretical and Applied Genetics, 84: 567 - 572.

DI Y B, FANG C T, CHU J Z, et al. 1982. Studies on the bacteriophage of bacterial blight of rice (*X. oryzae*) in Beijing, The interrelations between bacteriophage and its host bacteria [J] . Acta Phytophylacica Sinica, 9 (4): 257 - 263.

DODDS P N, LAWRENCE G J, CATANZARITI A M, et al. 2006. Direct protein interaction underlies gene-for-gene specificity and coevolution of the flax resistance genes and flax rust avirulence genes [J] . Proc. Natl. Acad. Sci. U S A, 103: 8888 - 8893.

DONG X. 2004. NPR1, all things considered [J] . Curr. Opin. Plant Biology, 7: 547 - 552.

DURGAPAL J C. 1985. High virulence of *Xanthomonas campestris* pv. *oryzae*, a factor in the 1980 epiphytotic in the non-traditional rice - growing region of northwest India [J] . Ind. J. Agric. Sci., 55 (2): 133 - 135.

EAMCHIT S, MEW T W. 1982. Comparison of virulence of *Xanthomonas campestris* pv. *oryzae* in Thailand and the Philippines [J] . Plant Disease, 66: 556 - 559.

EARDLY B D, MATERON L A, SMITH N H, et al. 1990. Genetic structure of natural populations of the nitrogen - fixing bacterium *Rhizobium meliloti* [J] . Appl. Environ. Microbiol., 56: 187 - 194.

ELSH J, MCCLELLAND M. 1991. Genomic fingerprinting using arbitrarily primed PCR and a matrix of pairwise combinations of primers [J] . Nucl. Acids Res., 19: 5275 - 5279.

ELSTNER E F. 1982. Oxygen activation and oxygen toxicity [J] . Annu. Rev. Plant Physiol., 33: 73 - 96.

ENDO N, OGAWA T. 2004. Breeding. Abstracts of the 1st International Conference on Bacterial Blight of Rice. Dec. 2004. Tsukuba, Japan.

EMMANUEL H, SCOTT A Y, LEACH J E, et al. 2001. Vascular defense responses in rice: perodase accumulation in xylem parenchyma cells and xylem wall thickening [J] . MPMI, 14 (12): 1411 - 1419.

ERCOLANI G L. 1972. Dynamics of infection by *Pseudomonas phaseolicola* in partially resistant population of bean [J] . Phytopathology, 62: 756.

ESAKA S, KONURA T, ITO T, et al. 1969. Historical review of paddy rice breeding work in Aichi prefecture [J] . Agricultural and Forestry Research Council, Aichi prefecture Ext. Mater, 27: 1 - 144.

EXCONDE O R, OPINA O S, PHANOMSAWARA A. 1971. Yield losses due to bacterial leaf blight [J] . Philipp. Agric., 57: 120 - 140.

EZUKA A, HORINO O. 1974. Classification of rice varieties and *Xanthomonas oryzae* strains on the basis of their differential interactions [J] . Bull. Tokai Kinki Natl. Agric. Exp. Stn., 27: 1 - 19.

EZUKA A, HORINO O. 1974. Virulence and aggressiveness in *Xanthomonas oryzae*. Bull. Tokai Kinki Natl. Agric. Exp. Stn; 27: 26 - 34.

EZUKA A, HORINO O, TORIYAMA K, et al. 1975. Inheritance of resistance of rice variety Wase Aikoku 3 to *Xanthomonas oryzae* [J] . Bull. Tokai Kinki Natl. Agric. Exp. Stn., 28: 124 - 130.

EZUKA A, SAKAGUCHI S. 1978. Host-parasite relationship in bacterial blight of rice caused by *Xanthomonas oryzae*. Rev. Plant Proc. Res., 11: 98 - 118.

EZUKA A, WATANABE Y, HORINO O. 1974. Difference in resistance expression to *Xanthomonas oryzae* between seedlings and adults of Wase Aikoku group rice varieties (1) . Bull. Tokai-Kinki Natl. Agric. Exp. Stn., 27: 20 - 25.

EZUKA A. 1979. Breeding for and genetics of blast resistance in Japan [C] // Proceeding of the Rice Blast Workshop. Manila：IRRI：27 - 48.

FALK A，FEYS B J，FROST L N，et al. 1999. EDS1，an essential component of R gene-mediated disease resistance in Arabidopsis has homology to eukaryotic lipases [J] . Proc. Natl. Acad. Sci. U S A，96：3292 - 3297.

FAN Y L，CHEN X W，WANG C L，et al. 2006. Mapping the rice bacterial blight resistance gene *Xa*23 with RFLP markers and converting RFLP to STS marker [J] . Acta Agronomica Sinica，32：931 - 935.

FEINBERG A P，VOGELSTEIN B. 1983. A technique for radiolabeling DNA restriction endonuclease fragments to high specific activity [J] . Analytical Biochemistry，132：6 - 13.

FENG T Y，KUO T T. 1975. Bacterial leaf blight of rice plant. VI. Chemotactic responses of *Xanthomonas oryzae* to water droplet exudated from water pores on the leaf of rice plants [J] . Bot. Bull. Acad. Sin.，16：126 - 136.

FELIX G，DURAN J D，VOLKO S，et al. 1999. Plants have a sensitive perception system for the most conserved domain of bacterial flagellin [J] . Plant J.，18：265 - 276.

FITZGERALD H A，CHERN M S，NAVARRE R，et al. 2004. Overexpression of (At) NPR1 in rice leads to a BTH- and environment-induced lesion-mimic/cell death phenotype [J] . Mol. Plant - Microbe Interact.，17：140 - 151.

FITZGERALD H A，CANLAS P E，CHERN M S，et al. 2005. Alteration of TGA factor activity in rice results in enhanced tolerance to *Xanthomonas oryzae* pv. *oryzae* [J] . Plant J.，43：335 - 347.

FLOR H H. 1942. Inheritance of pathogenicity in Melampsora lini [J] . Phytopathology，32：653 - 669.

FLOR H H. 1971. Current status of the gene-for-gene concept [J] . Annu. Rev. Phytopathol.，9：275 - 296.

FOBERT P R，DESPRES C. 2005. Redox control of systemic acquired resistance [J] . Curr. Opin. Plant Biol.，8：378 - 382.

FRY W E. 1982. Principles of plant disease management [M] . New York：Academic Press：378.

FUJII K，OKADA M. 1967. Progress in breeding of rice varieties for resistance to bacterial leaf blight in Japan [C] // Proceedings of a symposium on tropical agriculture researches. Tokyo：Ministry of Agriculture and Forestry：51 - 61.

FURUTANI A，TSUGE S，OKU T，et al. 2003. Hpa1 secretion via type III secretion system in *Xanthomonas oryzae* pv. *oryzae* [J] . J. Gen. Plant Pathol.，69：271 - 275.

GABRIEL D W，HUNTER J E，KINGSLEY M T，et al. 1988. Clonal population structure of *Xanthomonas campestris* and genetic diversity among citrus canker strains [J] . Mol. Plant - Microbe Interact.，1：59 - 65.

GAO D Y，XU Z G，CHEN Z Y，et al. 2001. Identification of a new gene for resistance to bacterial blight in a somaclonal mutant HX23 (indica) [J] . Rice Genet. Newsl.，18：66 - 68.

GEORG F，JULIANA D D，SIGRID V，et al. 1999. Plants have a sensitive perception system for the most conserved domain of bacterial flagellin [J] . The Plant Journal，18：265 - 276.

GEORGE M L C，BUSTAMAM M，CRUZ W T，et al. 1997. Movement of *Xanthomonas oryzae* pv. *oryzae* in Southeast Asia detected using PCR-based DNA fingerprinting [J] . Phytopathology，87：302 - 309.

GEORGE M L C，CRUI W T，NELSON R J. 1997. DNA fingerprinting of *Xanthomonas oryzae* pv. *oryzae* by ligation-mediated polymerase-chain reaction [J] . Int. Rice Res. Notes，19 (4)：29 - 30.

GOEL R K，KAUR L，SAINI R G. 1998. Effectiveness of different *Xa* genes against *Xanthomonas oryzae* pv. *oryzae* population causing bacterial blight of rice in Punjab (India) [J] . Rice Genetics Newsletter，15：

131 - 133.

GOMES K A, GOMES A A. 1976. Statistical procedures for agricultural research with emphasis on rice [M] . Los Banos: IRRI. 294.

GOMEZ K A, GOMEZ A A. 1984. Statistical procedures for an agricultural research [M] . 2nd ed. New York: John Wiley & Sons, Inc. 97 - 101.

GOMEZ-GOMEZ L, BOLLER T. 2000. FLS2: an LRR receptor-like kinase involved in the perception of the bacterial elicitor flagellin in Arabidopsis [J] . Mol. Cell, 5: 1003 - 1011.

GOMEZ-GOMEZ L, BOLLER T. 2002. Flagellin perception: a paradigm for innate immunity [J] . Trends in Plant Science, 7: 251 - 256.

GONZALEZ C, ONASAYA A, AKATOR K. 2004. Characterization of *Xanthomonas oryzae* pv. *oryzae* strains from Africa. Abstracts of the 1st International Conference on Bacterial Blight of Rice. March 17 - 19, 2004. Tsukuba, Japan: 57.

GORYSHIN I Y, JENDRISAK J, HOFFMAN L M, et al. 2000. Insertional transposon mutagenesis by electroporation of released Tn5 transposition complexes [J] . Nat. Biotech. , 18: 97 - 100.

GOTO K, FUKTATZU R, OKATA K. 1953. Overwintering of the causal bacteria blight of rice in the rice plant and grasses [J] . Agric. Hortic. Jan. , 28: 207 - 208.

GOTO M, OKABE N. 1956. Bacteriophages of *Xanthomonas oryzae*, the pathogen of bacterial leaf blight of rice, collected in the Philippines [J] . Bull. Fac. Agric. Shizuoka Univ. , 15: 31 - 37.

GOTO M. 1964. "Kresek" and pale yellow leaf systemic symptoms of bacterial leaf blight of rice caused by *Xanthomonas oryzae* (Uyeda et Ishiyama) Dowson [J] . Plant Dis. Reptr. , 48: 858 - 861.

GOTO M. 1965. Bacteriophages of *Xanthomonas oryzae*, the pathogen of bacterial leaf blight of rice, collected in the Philippines [J] . Philippine Agric. , 48: 15 - 38.

GOTO M. 1965. Resistance of rice varieties and species of wild rice to bacterial leaf blight and bacterial leaf streak diseases [J] . Philippine Agric. , 48: 329 - 338.

GREENBERG J T, YAO N. 2004. The role and regulation of programmed cell death in plant-pathogen interactions [J] . Cell Microbiol. , 6: 201 - 211.

GROOM Q J, TORRES M A, FORDHAM-SKELTON A P, et al. 1996. *rbohA*, a rice homologue of the mammalian $gp91^{phox}$ respiratory burst oxidase gene [J] . Plant J. , 10: 515 - 522.

GROSS G G. 1980. The biochemistry of lignification [J] . Advanced Botanical Research, 8: 25 - 63.

GU K, TIAN D, YANG F, et al. 2004. High-resolution genetic mapping of *Xa27* (*t*), a new bacterial blight resistance gene in rice, *Oryza sativa* L [J] . Theo. Appl. Genet. , 108: 800 - 807.

GU K, YANG B, TAIN D S, et al. 2005. R gene expression induced by a type-Ⅲ effector triggers disease resistance in rice [J] . Nature, 435 (7045): 1122 - 1125.

WANGGL, HOLS TEN TE, SONA WY, et al. 1995. Construction of a rice bacterial artificial chromosome library and identification of clones linked to the *Xa21* disease resistance locus [J] . Plant J. , 7 (3): 525 - 533.

GUPTA A K, SHARMA S C, SAINI R G. 1986. Variation in pathogenicity of some indian isolates of *Xanthomonas campestris* pv. *oryzae* [J] . Phytopathology, 6: 881 - 883.

HALEY S D, AFANADOR L, KELLY J D. 1994. Selection for monogenic pest resistance traits with coupling- and repulsion-phase RAPD markers [J] . Crop Sci. , 34: 1061 - 1066.

HAMAMATSU C, MIYABAYASHI T, MORISHIMA H. 1992. Variation in bacterial blight resistance within natural populations of wild rice and farmer' s fields [J] . Rice Genetics Newsletters, 9: 88 - 90.

HARTINI R H, NISHIYAMA K, EZUKA A. 1976. Bacteriological characteristics of some isolates of *Xan-*

thomonas oryzae different in their pathogenicity and locality [J] . Contrib. Cent. Res. Inst. Agric. Bogor. , 16: 1 - 18.

HARUSHIMA J, FRITSCH E F, MANIATIS T. 1989. Molecular Cloning: a Laboratory Manual [M] . 2nd ed. New York: Cold Spring Harbor Laboratory Press: 474, 503.

HARUSHIMA Y, YANO M, SHOMURA A, et al. 1998. A high- density rice genetic linkage map with 2275 markers using a single F_2 population [J] . Genetics, 148: 479 - 494.

HAYASHI Y, KYOZUKA J, SHIMAMOTO K. 1988. Hybrids of rice (*Oryza sativa* L.) and wild *Oryza* species obtained by cell fusion [J] . Mol. Gen. Genet. , 214: 6 - 10.

HEAGLE A S, MOORE M B. 1970. Some effects of moderate adult resistance to crown rust of oats (*Avena sativa*) [J] . Phytopathology, 60: 461 - 466.

HEATH M C. 1991. The role of gene-for-gene interactions in the determination of host species specificity [J] . Phytopathology, 81: 127 - 130.

HEATH M C. 2000. Nonhost resistance and nonspecific plant defenses [J] . Curr. Opin. Plant Biol. , 3: 315 - 319.

HEATHER A F, PATRICK E C, MAW - SHENG C, et al. 2005. Alteration of TGA factor activity in rice results in enhanced tolerance to *Xanthomonas oryzae* pv. *oryzae* [J] . Plant J. , 43: 335 - 347.

HIEI Y, OHTA S, KUMASHIRO T, et al. 1994. Efficient transformation of rice (*Oryza sativa* L.) mediated by *Agrobacterium* and sequence analysis of the boundaries of the T-DNA [J] . Plant J. , 6 (2) : 271 - 276.

HILAIRE E, SCOTT A Y, LEACH J E, et al. 2001. Vascular defense responses in rice: Perodase accumulation in xylem parenchyma cells and xylem wall thickening [J] . MPMI, 14 (12): 1411 - 1419.

HO B L, LIM W C. 1976. Effects of bacterial blight on the yield of rice at various stages of growth [J] . MARDI Research Bulletin, 4: 13 - 22.

HOPKINS C M, WHITE F F, CHOI S H. et al. 1992. Identification of a family of avirulence genes from *Xanthomonas oryzae* pv. *oryzae* [J] . Molecular Plant-Microbe Interactions, 5: 451 - 459.

HORINO O, EZUKA A. 1973. A simple screening method for bacterial leaf blight resistance of rice varieties derived from Wase Aikoku 3 [J] . Proc. Assoc. Plant Prot (Hokuriku). , 1: 29 - 32.

HORINO O. 1976. Induction of bacterial leaf blight resistance by incompatible strains of *Xanthomonas oryzae* in rice [C] // Tomiyama K, Daly M, Uritani I, eds. Biochemistry and cytology of plant - parasite interactions. Tokyo: Kodansha, Ltd. : 43 - 55.

HORINO O, HARTINI R H. 1978. Resistance of some rice varieties to bacterial leaf blight and a new pathogenic group of the causal bacterium, *Xanthomonas oryzae* [J] . Centr. Res. Inst. Agric. Bogor. , 44: 1 - 17.

HORINO O. 1978. Distributions of Pathogenic Strains of *Xanthomonas oryzae* (Uyeda et Ishiyama) Dowson in Japan in 1973 and 1975 [J] . Annals of the Phytopathological Society of Japan, 44: 297 - 304.

HORINO O, INABA T. 1979. The distribution of C^{14} in the rice leaves infected with *Xanthomonas campestris* pv. *oryzae* [J] . Ann. Phytopathol. Soc. Jpn. , 45: 557 - 558.

HORINO O, MEW T W, KHUSH G S, et al. 1980. Resistance of Japanese and IRRI differential rice varieties to pathotypes of *Xanthomonas oryzae* in the Philippines and in Japan [J] . IRRI Research Paper Series, 53: 11.

HORINO O, YAMADA T. 1980. Inheritance of resistance of IR26 to bacterial groups I, II, III, IV and V of *Xanthomonas campestris* pv. *oryzae* (Ishiyama 1922) Dye 1978 from Japan [J] . Annals of the Phytopathological Society of Japan, 46: 504 - 509.

HORINO O. 1981. Ultrastructural histopathology of rice leaves infected with *Xanthomonas campestris* pv. *oryzae* on Kogyoku *group rice varieties with different levels of resistance at the seedling stage* [J] . Ann. Phytopathol. Soc. Jpn. , 47: 501 - 509.

HORINO O. 1981. A Survey of geographical distribution of pathogenic groups of *Xanthomonas campestris* pv. *oryzae* from Japan in 1977 and 1979 [J] . Annals of the Phytopathological Society of Japan, 47: 50 - 57.

HORINO O. 1981. Races of *Xanthomonas campestris* pv. *oryzae* in Asia and their virulence against rice cultivars. (2) Indonesia, Bangladesh, Korea, China, India and Nepal [J] . Proc. Assoc. Plant Prot. Hokuriku, 29: 5 - 7.

HORINO O, MEW T W, KHUSH G S, et al. 1981. Comparison of two differential systems for distinguishing pathogenic groups of *Xanthomonas campestris* pv. *oryzae* [J] . Annals of the Phytopathological Society of Japan, 47: 1 - 14.

HORINO O, MEW T W, Yamada T. 1982. The effect of temperature on the development of bacterial leaf blight on rice [J] . Ann. Phytopathol. Soc. Jpn. , 48: 72 - 75.

HORINO O, SIWI B H, MIAH S A. et al. 1983. Virulence of *Xanthomonas campestris* pv. *oryzae* isolated in Indonesia and Bangladesh. Ann. Phytopathol. Soc. Jpn. , 49: 191 - 199.

HORINO O. 1984. Ultrastructure of water pores in *Leersia sayamuka* Makino and *Oryza sativa* L. : its correlation with the resistance to hydathodal infection of *Xanthomonas campestris* pv . *oryzae* [J] . Ann. Phytopathol. Soc. Jpn. , 50: 72 - 76.

HORINO O. 1985. Resistance of high yielding lines of rice and their parents to *Xanthomonas campestris* pv. *oryzae* from Japan [J] . Proc. Assoc. Plant Prot. Hokuriku, 33: 13 - 20.

HORINO O, KAKU H. 1989. Defense mechanisms of rice against bacterial blight caused by *Xanthomonas campestris* pv. *oryzae* [C] // Proceedings of the International Workshop on Bacterial Blight of Rice. Manila: IRRI 135 - 152.

HORISUE N. 1996. Strategy for development of post- Koshikari rice cultivars [C] // Kushibuchi. K. Tasty Rice. vol. 3. Strategies for development of tasty rice looking towards the 21st century. Tokyo: Agriculture, Forestry and Fisheries Technical Information Society: 27 - 56.

HSIEH S P Y, BUDDENHAGEN I W. 1975. Survival of tropical *Xanthomonas oryzae* in relation to substrate, temperature, and humidity [J] . Phytopathology, 65: 513 - 519.

HSU S T. 1966. M. S. thesis/ Dissertation, Nutritional requirements in vitro of *Xanthomonas oryzae* (Uyeda and Ishyama) Dowson and its effect on host resistance [R] . Philippines: University of Philippines, College of Agriculture, 58.

HSU S T. 1967. Nutritional aspect of host-pathogen relationship in bacterial leaf blight of rice [J] . J. Agr & For. , 21: 1 - 22.

HUANG N, ANGELES E R, DOMINGO J, et al. 1997. Pyramiding of bacterial blight resistance genes in rice: marker-assisted selection using RFLP and PCR [J] . Theoretical and Applied Genetics, 95: 313 - 320.

HUANG Z, HE G, SHU L, et al. 2001. Identification and mapping of two brown planthopper resistance genes in rice [J] . Theoretical and Applied Genetics, 102: 929 - 934.

IGOR Y G, JERRY J, LES M H, et al. 2000. Insertional transposon mutagenesis by electroporation of released Tn 5 transposition complexes [J] . Nature Biotechnology, 18: 97 - 100.

IKEDA R, TABIEN R N, KHUSH G S. 1991. Chromosomal location of *Xa21* [J] . Rice Genet. Newsl. , 8: 102 - 103.

IKEDA R, TABIEN R N, OGAWA T. 1995. Linkage analysis of resistance genes to bacterial blight in rice [J]. JARQ, 29: 137 - 142.

IKEHASHI H, ARAKI H. 1984. Varietal screening for compatibility type revealed in afa1 fertility of distant crosses in rice [J] . Jpn. J. Breed, 34: 303 - 312.

INNES R W. 2001. Mapping out the roles of MAP kinases in plant defense [J] . Trends Plant Sci. , 6: 392 - 394.

INOUE Y, GOTO K, OHATA K I. 1957. Overwintering and mode of infection of leaf blight bacteria (*Xanthomonas oryzae*) of rice plant [J] . Bulletin, Division of Plant Breeding and Culture, Tokai-Kinki National Agricultural Experimental Station, 4: 74 - 82.

International Rice Testing Program. 1980. Standard Evaluation System for Rice [M] . Los baños: IRRI, 44.

IRRI. 1985. International Rice Research: 25 Years of Partnership [C] // About the book: an overview, varieties improvement. Los Nanos Laguna, Manila Philippines: IRRI: 1 - 53.

IRRI. Annul Reports for 1966 - 1969.

IRRI. 1990. Crop loss assessment in rice. P. O. Box 933, 1099 Manila, Philippines.

ISHIYAMA S. 1922. Studies on bacterial leaf blight of rice [C] . Tokyo: Report of the Agricultural Experiment Station: 45: 233 - 261.

ISHIYAMS S. 1923. Studies with reference to leaf withering disease of rice [J] . Rept. Dept. Agric. and Commerce, 45 (3): 233 - 261.

IYER A S, MCCOUCH S R. 2004. The rice bacterial blight resistance gene *xa5* encodes a novel form of disease resistance [J] . Mol. Plant - Microbe Interact. , 17: 1348 - 1354.

JABS T, DIETRICH R A, DANGL J L. 1996. Initiation of runaway cell death in an Arabidopsis mutant by extracellular superoxide [J] . Science, 273: 1853 - 1856.

JEAN T G, NAN Y. 2004. The role and regulation of programmed cell death in plant - pathogen interactions [J] . Cellular Microbiology, 6: 201 - 211.

JENA K K, JEUNG J U, SHIN M S, et al. 2004. Pyramiding specific BB resistance genes for improvement of temperate japonica rice in South Korea. Abstracts of the 1st International Conference on Bacterial Blight of Rice. 16.

JENA K K, KHUSH G S. 1990. Introgression of genes from *Oryza officinalis* Well ex Watt to cultivated rice, *O. sativa* L [J] . Theoretical and Applied Genetics, 80: 737 - 745.

JENA K K, KHUSH G S. 1989. Monosomic alien addition lines of rice: production, morphology, cytology, and breeding behavior. Genome, 32: 449 - 455.

JEUNG J U, HEW S G, SHIN M S, et al. 2006. Dynamics of *Xanthomonas oryzae* pv. *oryzae* populations in Korea and their relationship to known bacterial blight resistance genes. Phytopathology, 96 (8): 867 - 875.

JIA Y, MCADAMS S A, BRYAN G T, et al. 2000. Direct interaction of resistance gene and avirulence gene products confers rice blast resistance [J] . EMBO J. , 19: 4004 - 4014.

JIANG C, WANG J Y, LI Q H. 2004. The elite characterizations of wild rice and their utilization in rice breeding [C] // Studies and Applications of Wild Rice in China. Beijing: Qixiang Press: 237 - 241.

KADIR S, SURYADI Y, DARADJAT A A, et al. 2004. Monitoring source of BB resistance by near isogenic lines (NILs) . Abstracts of the 1st International Conference on Bacterial Blight of Rice. 41.

KAJI R, QGAWA T. 1995. Identification of the located chromosome of the resistance gene *Xa7* to bacterial leaf blight in rice [J] . Jpn. J. Breed, 45: 79.

KAKU H, HORINO O. 1977. Browning reaction in rice plant tissues induced by *Xanthomonas oryzae* [J] .

Ann. Phytopathol. Soc. Jpn., 43: 487 - 490.

KAKU H, KIMURA T. 1987. Difference in resistance expression of rice to *Xanthomonas campestris* pv. *oryae* as controlled by resistance genes. 1. Resistance expression controlled by resistance gene *Xa1* [J]. Ann. Phytopathol. Soc. Jpn., 14 - 20.

KAKU H, MEW T W. 1993. Hydathode infection of *Xanthomonas campestris* pv. *oryzicola* [J]. Ann. Phytopathol. Soc. Jpn., 59: 46.

KAKU H, KIMURA T. 1978. Reaction types of rice cultivars to strains of *Xanthomonas oryzae* [J]. Bull. Chukoku Natl. Agric. Exp., 13: 17 - 43.

KARGANILLA A, PARIES-NATURAL M, OU S H. 1973. A comparative study of culture media for *Xanthomonas oryzae* [J]. Philippine Agriculturst,, 57: 141 - 251.

KAUFFMAN H E, PANTULU R. 1972. Virulence patterns and phage sensitivity of Indian isolates of *Xanthomonas oryzae* [J]. Annals of the Phytopathological Society of Japan. 38: 68 - 74.

KAUFFMAN H E, REDDY A P K, HSIEH S P Y, et al. 1973. An improved technique for evaluating resistance of rice varieties to *Xanthomonas oryzae* [J]. Plant Disease Rep., 57: 537 - 541.

KAUFFMAN H E. 1974. The Isabela isolate of *Xanthomonas oryzae* in the Philippines [C] // The paper presented at the International Rice Research Conference. Los Banos, Laguna, Philippines: IRRI.

KAWASAKI T, HENMI K, ONO E, et al. 1999. The small GTP-binding protein rac is a regulator of cell death in plants [J]. Proc. Natl. Acad. Sci. U S A, 96 (19): 10922 - 10926.

KELLER T, DAMUDE H G, WERNER D, et al. 1998. A plant homolog of the neutrophil NADPH oxidase gp^{91phox} subunit gene encodes a plasma membrane protein with Ca^{2+} binding motifs [J]. Plant Cell, 10: 255 - 266.

KESSLER A, BALDWIN I T. 2002. Plant responses to insect herbivory: the emerging molecular analysis [J]. Annu. Rev. Plant Biol., 53: 299 - 328.

KHUSH G S. 1977. Disease and Insect Resistance in Rice [C] // Brady N C. Advances in Agronomy New York: Academic Press. 265 - 341.

KHUSH G S, LIN K Z. et al. 1977. Breeding for resistance to grassy stunt in rice [C] // Proc. 3rd. Int. Congr. SABRAO, Canberra, Australia. Plant Breeding Papers, 1 (4b): 3 - 9.

KHUSH G S. 1977. Virus diseases [C] // Disease and Insect Resistance in Rice. Los Banos: IRRI: 294 - 296.

KHUSH G. S. 1978. Breeding methods and procedures employed at IRRI for developing rice germplasm with multiple resistance to diseases and insects [J]. Trop. Agric. Res. Ser., 11: 69 - 75.

KHUSH G S, MACKILL D J, SIDHU G S. 1989. Breeding rice for resistance to BB [C] // Bacterial Blight of Rice. Manila: IEEI: 207 - 217.

KHUSH G S, BACALANGCO E, OGAWA T. 1990. A new gene for resistance to bacterial blight from *O. longistaminata* [J]. Rice Genetics Newsletter, 7: 121 - 122.

KHUSH G S, ANGELES E R. 1999. A new gene for resistance to race 6 of bacterial blight in rice, *Oryza sativa* L [J]. Rice Genet. Newsl., 16: 92 - 93.

KHUSH G S. 2001. Rice genetics from Mendel to functional genomics [C] // Khush GS, Brar DS, Hardy B, eds. Rice Genetics IV, Proceedings of the Fourth International Rice Genetics Symposium. Los Banos: IRRI: 3 - 25.

KINOSHITA T. 1988. Report of the Committee on gene symbolization, nomenclature and linkage groups [J]. Rice Genetics Newsletter, 5: 3 - 5.

KINOSHITA T. 1993. Report of the committee on gene symbolization and linkage map [J]. RGN

(10): 5 - 10.

KINOSHITA T. 1995. Report of the committee on gene symbolization, nomenclature and linkage groups [J]. Rice Genet. Newsl., 12: 9.

KIRYU T, MIZUTA H. 1955. On the relation between habits of rice and varietal resistance of bacterial leaf blight [J]. Kyushu Agricultural Research. 15: 54 - 56.

KOZAKA T. 1969. Control of rice diseases with resistant varieties [J]. Bull. Tokai-Kinki Natl. Agric. Exp. Stn., 44: 208 - 212.

KUHARA S, KURITA T, TAGAMI Y, et al. 1965. Studies on the strain of *Xanthomonas oryzae* (Uyeda et Ishiyama) Dowson, the pathogen and the bacterial leaf blight of rice, with special reference to its pathogenicity and phage-sensitivity [J]. Bull. Agric. Exp. Stn., 11: 263 - 312.

KUHARA S. 1958. On the pathogen of bacterial leaf blight of rice isolated from severely affected area where resistant variety was widely cultivated [J]. Annals of the Phytopathological Society of Japan, 23: 9.

KULKARNI N R, THOMBRE S B. 1969. Seminar on Problems of rice improvement: Wild rice (*Oryza sativa* L. *spontanea* Rosch), a potential danger for spread of bacterial blight caused by *Xanthomonas oryzae* (Uyeda and Ishiyama) Dowson in Maharashtra. Poona, India: KULKARNI N R, THOMBTRE S B.

KURATA N, NAGARMURA Y, YAMAMOTO K, et al. 1994. A 300 kilobase interval genetic map of rice including 883 expressed sequences [J]. Nat, Genet., 8: 365 - 372.

KUSABA T, WATANABE M, TABEI H, et al. 1958. Varietal difference in resistance to bacterial leaf blight in rice-specialization in pathogenicity (1) [J]. Bull. Agric. Exp. Stn., 23: 9.

KUSABA T, WATANABE M, TABEI H, 1966. Classification of the strains of *Xanthomonas oryzae* (Uyeda et Ishiyama) Dowson on the basis of their virulence against rice plants [J]. Bull. Natl. Inst. Agric. Sci. Ser., 20: 67 - 82.

KOZAKA T. 1969. Control of rice diseases with resistant varieties [J]. Npgyo oyobi (Agric. Hortic)., 44: 208 - 212.

LANDER E S, GREEN P, ABRAHAMSON J, et al. 1987. MAPMAKER: An interactive computer package for constructing primary genetic linkage maps of experimental and natural populations [J]. Genomics, 1: 174 - 181.

LEACH J E, RHOADS M L, WHITE F F, et al. 1988. A repetitive DNA sequence from the *Xanthomonas campestris* pathovars [J]. J. Cell Biochem, 12: 254.

LEACH J E, ROBERT A, GOU A, et al. 1989. Multiplication of *Xanthomonas campestris* pv. *oryzae* in rice leaves [C] // Bacterial Blight of Rice; Proceedings of the International Workshop on Bacterial Blight of Rice. Manila, Philippines: IRRI: 43 - 53.

LEACH J E, KELEMU S, ARDALES E, et al. 1989. The genetics of *Xanthomonas campestris* pv. *oryzae* [C] // Proc. Int. Workshop on Bacterial Blight of Rice. Manila, Philippines: IRRI: 199 - 206.

LEACH J E, WHITE F F, RHOAD M L. 1990. A repetitive DNA sequence differentiates *Xanthomonas campestris* pv. *oryzae* from other pathovars of *Xanthomonas campestris* [J]. Mol. Plant-Microbe Interact., 3: 238 - 246.

LEACH J E, WHITE F F. 1991. Molecular probes for disease diagnosis and monitoring [C] // Khush G S, Toenniessen G H. Rice biotechnology. Oxon: CAB International: 281 - 307.

LEACH J E, RHOADS M L, VERA CRUZ C M, et al. 1992. Assessment of genetic diversity and population structure of *Xanthomonas oryzae* pv. *oryzae* with a repetitive DNA element [J]. Appl. Environ. Microbiol., 58: 2188 - 2195.

LEACH J E, LEUNG H, NELSON R J, et al. 1995. Population biology of *Xanthomonas oryzae* pv. *oryzae* and approaches to its control [J] . Current Opinion in Biotechnology, 6: 298 - 304.

LEACH J E, WHITE F F. 1996. Bacterial avirulence genes [J] . Annu. Rev. Phytopathol. , 34: 153 - 179.

LEE K S, ANGELES E R, KHUSH G S. 2000. Inheritance of resistance to race 6 of bacterial blight in rice [J]. Rice Genet. Newsl. , 17: 73 - 74.

LEE K S, RASABANDITH S, ANGELES E R, et al. 2003. Inheritance of resistance to bacterial blight in 21 cultivars of rice [J] . Phytopathology, 93: 147 - 152.

LEE S W, CHOI S H, HAN S S, et al. 1999. Distribution of *Xanthomonas oryzae* pv. *oryzae* strains virulent to *Xa21* in Korea [J] . Phytopathology, 89: 928 - 933.

LEONARD K J. 1987. The host population as a selective factor [C] // Wolfe M S, Caten C E. Population of Plant Pathogen: Their Dynamics and genetics. Oxford: Blackwell Scientific Publication: 163 - 179.

LEUNG H, NELSON R J, LEACH J E. 1993. Population structure of plant pathogenic fungi and bacteria [J]. Adv. Plant Pathol. , 10: 157 - 205.

LEVY M, ROMAO J, MARCHETTI M A, et al. 1991. DNA fingerprinting with a dispersed repeated sequence resolves pathotype diversity in the rice blast fungus [J] . The Plant Cell, 3: 95 - 102.

LEVY M, CORREA-VICTORIA F J, ZEIGLER R S, et al. 1991. Organization of genetic and pathotype variation in the rice blast fungus at a Colombia 'hot spot' [J] . Phytopathology, 81: 1236 - 1237 .

LI W H, GRAUR D. 1991. Fundamentals of Molecular Evolution [M] . Massachusetts: Sinauer Associates, Sunderlang.

LI L C, QU R D, BEACHY R N, et al. 1993. An improved rice transformation system using the biolistic method [J] . Plant Cell Reports, 12: 250 - 255.

LI C, SUN C Q, MU P, et al. 2001. QTL analysis of anther length and ratio of stigma exsertion, two key traits of classification for cultivated rice (*Oryza sativa* L.) and common wild rice (*O. rufipogon* Griff.) [J] . Acta Genetica Sinica, 28 (8): 746 - 751.

LI X, QIN J. 2005. Modulation of Toll-interleukin 1 receptor mediated signaling [J] . J. Mol. Med. , 83: 258 - 266.

LI Z K, ALMA S, ANGELES E, et al. 2001. Are the dominant plant disease resistance genes similar? A case of rice R genes and *Xanthomonas oryzae* pv. *oryzae* races [J] . Genetics, 159: 757 - 765.

LIANG C Z, GU M H, PAN X B, et al. 2001. RFLP tagging of a new semidwarfing gene in rice [J] . Theoretical and Applied Genetics, 88: 898 - 900.

LIN X H, ZHANG D P, XIE Y F, et al. 1996. Identifying and mapping a new gene for bacterial blight resistance in rice based on RFLP markers [J] . Phytopathology, 86: 1156 - 1159.

LIEBERHERR D, THAO N P, NAKASHIMA A, et al. 2005. A sphingolipid elicitor-inducible mitogen-activated protein kinase is regulated by the small GTPase OsRac1 and Heterotrimeric G-protein in rice [J] . Plant Phys. , 138: 1644 - 1652.

LIBROJO V, KAUFFMAN H E, KHUSH G S. 1976. Genetic analysis of bacterial blight resistance in four varieties of rice [J] . Sabrao J, 8: 105 - 110.

LIGTERINK W, KROJ T, ZUR N U, et al. 1997. Receptor-mediated activation of a MAP kinase in pathogen defense of plants [J] . Science, 276: 2054 - 2507.

LING K C, AGUIERO V M, LEE S H. 1970. A mass screening method for testing resistance to grassy stunt disease of rice. Plant Dis. Rep. , 56: 565 - 569.

LITT M, LUTY J A. 1989. A hypervariable microsatellite revealed by in vitro amplification of a dinucleotide repeat within the cardiac muscle action gene [J] . Am. J. Hum. Genet. , 44: 397 - 401.

LOPRASERT S, ATICHARTPONGKUN S, WHANGSUK W, et al. 1997. Isolation and analysis of the

Xanthomonas alkyl hydroperoxide reductase gene and the peroxide sensor regulator genes *ahpC* and *ahpF-oxyR-orfX* [J]. J Bacteriol., 179: 3944–3949.

LOUWS F J, FULBRIGHT D W, STEPHENS C T, et al. 1994. Specific genomic fingerprints of phytopathogenic *Xanthomonas* and *Pseudomonas* pathovars and strains generated with repetitive sequences and PCR [J]. Appl. Environ. Microbiol., 60: 2286–2295.

LOUWS F J, FULBRIGHT D W, STEPHENS C T, et al. 1995. Differentiation of genomic structure by rep-PCR fingerprinting to rapidly classify *Xanthomonas campestris* pv. *vesicatoria* [J]. Phytopathology, 85: 528–536.

LOZANO J C. 1977. Identification of bacterial leaf blight in rice caused by *Xanthomonas oryzae* in America [J]. Plant Dis. Rep., 61: 644–648.

LU C F. 1998. Towards the map-based cloning of *Xa7*, a resistance gene to bacterial blight disease. In A Phd Thesis. 2–27.

LYER A S, MCCOUCH S R. 2004. The rice bacterial blight resistance gene *xa5* encodes a novel form of disease resistance [J] Mol. Plant-Microbe Interact., 17 (12): 1348–1354.

MACLEAN J L, DAWE D C, HARD Y B, et al. 2002. Rice Almanac, 3rd Edition. IRRI, CIAT. 91.

MANIATIS T, FRITSCH E F, SAMHROOK J. 1982. Molecular cloning: a laboratory manual [M]. New York Cold Spring Harbor Laboratory, Cold Spring Harbor.

MARELLA L S, VERA C M, BERNARDO M A. et al. 2001. Identification of resistance genes effective against rice bacterial blight pathogen in Eastern India [J]. Plant Dis., 85 (5): 506–512.

MARTAIN B, HUMBERT O, CAMARA M, et al. 1992. A highly conserved repeated DNA element located in the chromosome of *Streptococcus pneumoniae* [J]. Nucleic Acids Res., 20: 3479–3483.

MARTIN G B, WILLIAMS J G, TANKSLEY S D. 1991. Rapid identification of markers linked to a *Pseudomonas* resistance gene in tomato by using random primers and near-isogenic lines [C] // Proceedings of the National Academy of Sciences of the United States of America, 88: 2336–2340.

MARTIN G B, BROMMONSCHENKEL S H, CHUNWONGSE J, et al. 1993. Map-based cloning of a protein kinase gene conferring disease resistance in tomato [J]. Science, 262: 1432–1436.

MARUYAMA H. 1908. Disease of rice in Shizuoka prefecture [J]. Shizuoka-ken Nokaiho, 135: 1–16.

MARUYAMA H. 1909. Soybean cake application and leaf blight disease of rice [J]. Shizuoka-ken Nokaiho, 142: 28–29.

MATSUMOTO S, SATO T, WATANABE B, et al. 1977. Studies on the field resistance to bacterial leaf blight in rice [J]. Proc. Assoc. Plant Prot. Kyushu., 23: 170

MATSUO T. 1967. Rice breeding and its prospects [J]. Farming Japan., 1 (1): 36–44.

MAW-SHENG C, HEATHER A F, RAM C Y, et al. 2001. Evidence for a disease-resistance pathway in rice similar to the NPR1-mediated signaling pathway in Arabidopsis [J]. Plant J., 27: 101–113.

MAW-SHENQ C, HANS G E, MAURICIO M B. 1996. The developmentally regulated bZIP factor ROM1 modulates transcription from lectin and storage protein genes in bean embryos [J]. Plant J., 10: 135–148.

MCCOUCH S R, CHEN X, PANAUD O, et al. 1997. Microsatellite marker development, mapping and applications in rice genetics and breeding [J]. Plant Molecular Biology, 35: 89–99.

MCCOUCH S R, KOCHERT G, YU Z H, et al. 1988. Molecular mapping of rice chromosomes [J]. Theoretical and Applied Genetics, 76: 815–819.

MCCOUCH S R, ABENES M L, ANGELES R, et al. 1991. Molecular tagging of a recessive gene, *xa5* for resistance to bacterial blight of rice. RGN, (8): 143–144.

MELCHINGER A E. 1990. Use of molecular markers in breeding for oligogenic disease resistance [J]. Plant Breeding, 104: 1-19.

MEW T W, VERA-CRUZ C M. 1979. Variability of *Xanthomonas oryzae*: specificity in infection of rice differentials [J]. Phytopathology, 69: 152-155.

MEW T W, VERA-CRUZ C M. 1981. Response of rice with different gene resistance to *Xanthomonas campestris* pv. *oryzae* [J]. Proc. Fifth Int. Plant Pathol. Bact. Cali., 323-331.

MEW T W, VERA-CRUZ C M, REYES R C. 1981. Characterization of resistance in rice to bacterial blight. Annals of the Phytopathological Society of Japan, 47: 58-67.

MEW T W, KHUSH G S. 1981. Breeding for bacterial blight resistance in rice [C] // Proc. Fifth Int. Conf. Plant. Bact. Cal. 504-510.

MEW T W, VERA-CRUZ C M, REYES R C. 1982. Pathotypes of *Xanthomonas compestris* pv. *oryzae* in Asia [J]. IRRI Research Paper Series, 75: 7.

MEW T W, VERA-CRUZ C M, REYES R C. 1982. Interaction of *Xanthomonas campestris* pv. *oryzae*. and a resistant rice cultivar [J]. Phytopathology, 72: 786-789.

MEW T W, MEW I P, HUANG J S. 1984. Scanning electron microscopy of virulent and avirulent strains of *Xanthomonas campestris* pv. *oryzae* on rice leaves [J]. Phytopathology, 74: 635-641.

MEW T W, VERA-CRUZ C M. 1985. Virulence of *Xanthomonas compestris* pv. *oryzae* in the Philippines [J]. Phytopathology, 75: 13-16.

MEW T W. 1987. Current Status and future prospects of research on bacterial blight of rice [J]. Annual Review of Phytopathology, 25: 359-382.

MEW T W, BRIDGE J, HIBINO H, et al. 1988. Rice pathogens of quarantine importance [C] // Rice seed health. Los Banos: IRRI: 101-115.

MEW T W, VERA-CRUZ C M, ZHANG Q. 1990. Dose-response of rice cultivars IR20 and Perum Karuppan to virulent and avirulent races of *Xanthomonas campestris* pv. *oryzae* [C] // Plant pathogenic bacteria: Proceedings of the 7th International Conference on Plant Pathogenic Bacteria. 973-983.

MEWT W, VERA-CRUZ C M, MEDALLA E S. 1992. Changes in the race frequency of *Xanthomonas oryzae* pv. *oryzae* in response to rice cultivars planted in the Philippines [J]. Plant Dis., 76: 1029-1032.

MEW T W. 2004. Four decades of bacterial blight of rice research. [C] // Proceedings of the 1st International Conference on Bacterial Blight of Rice. March 17-19, Tsukuba, Japan: [s. n.]: 1

MICHELMORE R W, PARAN I, KESSELI R V. 1991. Identification of markers linked to disease-resistance genes by bulked segregant analysis: a rapid method to detect markers in specific genomic regions by using segregating populations [J]. Proceedings of the National Academy of Sciences of the United States of America, 88: 9828-9832.

MIN S K. 1992. An outline of rice breeding in China [C] // Rice in China. China National Rice Research Institute. 58-67.

MIR G N, KHUSH G S. 1990. Genetics of resistance to bacterial blight in rice cultivar DV86 [J]. Crop Res., 2: 194-198.

MIR G N, KHUSH G S. 1991. Genetics of resistance to bacterial blight in rice (*Oryza sativa* L.) [J]. Indian J. Genet. Plant Breed, 51 (1): 72-78.

MISRO B, RICHHARIA R H, THAKUR R. 1966. Linkage studies in rice (*Oryza sativa* L.) Ⅶ, Identification of linkage groups in indica rice [J]. *Oryza*, 3 (1): 96-105.

MIZUKAMI T. 1961. Studies on the ecological properties of *Xanthomonas oryzae*, the causal organism of bacterial leaf blight of rice plant [J]. Agricultural Bulletin, Saga University. 13: 1-85.

MOHIUDDIN M S, VERMA J P, RAO Y P. 1977. Loss due to bacterial leaf blight of rice [J]. Indian J. Agric. Sci., 47 (5): 221-223.

MOHAN M, NAIR S, BENTUR J S, et al. 1994. RFLP and RAPD mapping of the rice *gm2* gene that confers resistance to biotype 1 of gall midge (*Orseolia oryzae*) [J]. Theoretical and Applied Genetics, 87: 782-788.

MONGKOLSUK S, VATTANAVIBOON P, PRAITUAN W. 1997. Induced adaptive and cross-protection responses against oxidative stress killing in a bacterial phytopathogen, *Xanthomonas oryzae* pv. *oryzae* [J]. FEMS Microbiology Letters, 146: 217-222.

MORINAKA T, KIMURA T, HORI M. 1978. Pathotypes of *Xanthomonas oryzae* isolates collected from Kinki, Chugoku Shigoku districts in Japan [J]. Bull. Chugoku Natl. Agric. Exp. Stn., 14: 1-6.

MUDGETT M B. 2005. New insights to the function of phytopathogenic bacterial type III effectors in plants [J]. Annu. Rev. Plant Biol., 56: 509-531.

MUKOO H, YOSHIDA K. 1951. A needle inoculation method for bacterial blight disease of rice [J]. Ann. of the Phytopathological Society of Japan, 15: 179.

MULTANI D S, JENA K K, BRAR D S, et al. 1994. Development of monosomic alien additional lines and introgression of genes from *Oryza australiensis* to cultivated rice *O. sativa* L [J]. Theor. Appl. Genet., 88: 102-109.

Murashige T, Skoog F. 1962. A revised medium for rapid growth and bioassays with tobacco tissue cultures [J]. Physiol. Plant, 15: 473-497.

MURALIDHARAN K, KHUSH G S, KRISHNAVENI D, et al. 2004. Field performance of genotypes by molecular-assisted breeding to bacterial blight in India [J]. Rice Genetics Newsletter, 21: 29-31.

MURTY V V S, KHUSH G S, JENSES N F. 1973. Inheritance of resistance to bacterial leaf blight, *Xanthomonas oryzae* (Uyeda et Ishiyama) Dowson, in rice [J]. Jpn. J. Breed., 23: 325-328

NAKAI H, GOTO M. 1975. Studies on mutation breeding of rice for bacterial leaf blight resistance. 1. Variations in bacterial leaf blight reactions found in M2 generation [J]. Ibid., 7: 159-170.

NAKAI H, KOBAYASHI M, SAITO M. 1985. Induction and selection of mutations for resistance against bacterial leaf blight in rice [J]. Euphytica, 34: 577-585.

NAKAI H, KUWAHARA M, SAITO M. 1988. Studies of an induced mutant resistant to multiple races of bacterial leaf blight [J]. Rice Genet. Newsl., 5: 101-103.

NAKANISHI K, WATANABE M. 1977. Studies on the mechanism of resistance of rice plant against *Xanthomonas oryzae*. Ⅲ. Relationship between the rate of production of antibacterial substances and of multiplication of pathogenic bacteria in infected leaves of resistant and susceptible varieties. IV. Extraction and partial purification of antibacterial substances from infected leaves [J]. Annuals of the phytopathological Society of Japan, 4: 265-269; 449-454.

NAZARENO N K X, ROELFS A P. 1981. Adult-plant resistance of Thatcher wheat to stem rust [J]. Phytopathology, 71: 181-185.

NEI M, LI W H. 1979. Mathematical model for studying genetic variation in terms of restriction endonucleases [J]. Proc. Natl. Acad. Sci. U S A, 76: 5269-5273.

NEI M, Tajima F. 1981. DNA polymorphism detectable by restriction endonucleases. Genetics, 97: 145-163.

NEI M. 1987. Relative roles of mutation and selection in the maintenance of genetic variability [J]. Phil. Trans. R. Soc. Lond. B 319: 615-629.

NELSON R J，BARAOIDAN M R，VERA - CRUZ C M，et al. 1994. Relationship between phylogeny and pathotype for the bacterial blight pathogen of rice [J]. Appl. Environ. Microbiol.，60：3275 - 3283.

NELSON R J，ARDALES E，BARAOIDAN I，et al. 1996. Exploring the application of molecular markers for improving resistance to bacterial blight [C] // Rice genetics III. Proceedings of the Third International Rice Genetics Symposium. Manila：IRRI：267 - 277.

NISHIDA T. 1909. Bacterial leaf blight of rice [J]. Noji Zappo，127：68 - 75.

NISHIMURA Y. 1961. Studies on the reciprocal translocations in rice and barley [J]. Bull. Natl. Inst. Agric. Sci.，9：171 - 235.

NODA T，MORINAKA T，FUJII K. 1978. Population studies of mixed inoculum of *Xanthomonas oryzae* in susceptible and resistant varieties of rice [J]. Annals of the Phytopathological Society of Japan，44：137 - 141.

NODA T，HORINO O，OHUCHI A. 1987. A survey of geographical distribution of pathogenic races of *Xanthomonas campestris* pv. *oryzae* from Japan [J]. Proc. Assoc. Plant Prot. Hokuriku，35：7 - 13.

NODA T，OHUCHI A. 1989. A new pathogenic race of *Xanthomonas campestris* pv. *oryzae* and inheritance of resistance of differential rice variety，Tetep to it [J]. Ann. Phytopathol. Soc. Jpn.，55：201 - 207.

NODA T，HORINO O，OHUCHI A. 1990. Variability of pathogenicity in races of *Xanthomonas campestris* pv. *oryzae* in Japan [J]. PARQ，23 (3)：182 - 189.

NODA T，YAMAMOTO T，OGAWA T，et al. 1996. Pathogenic Races of *Xanthomonas oryzae* pv. *oryzae* in South and East Asia [J]. JIRCAS Journal，3：9 - 15.

NOH T H，KANG M H，LEE D K，et al. 2004. Variation of race distribution and increase of new race of bacterial leaf blight in Korea [C] // Abstracts of the 1st International Conference on Bacterial Blight of Rice. Ibaraki，Japan：phytopathological Society of Japan，Tsukuba Science City：46.

NURNBERGER T，BRUNNER F，KEMMERLING B，et al. 2004. Innate immunity in plants and animals：striking similarities and obvious differences [J]. Immunol. Rev.，198：249 - 266.

NWIGWE C. 1973. Persistence of *Xanthomonas oryzae* in the infected tissues of rice plants [J]. Plant Disease Reporter，57：227 - 230.

OCHIAI H，INOUE Y，TAKEYA M，et al. 2005. Genome sequence of *Xanthomonas oryzae* pv. *oryzae* suggests contribution of large numbers of effector genes and insertion sequences to its race diversity [J]. JARQ，39：275 - 287.

OCHMAN H，GERBER A S，HARTL D L. 1988. Genetic applications of an inverse polymerase chain reaction [J]. Genetics，120：621 - 623.

OGAWA T，MORINAKA T，FUJII K，et al. 1978. Inheritance of resistance to rice varieties Kogyoku and Java 14 to bacterial group V of *Xanthomonas oryzae* [J]. Ann. Phytopathol. Soc. Jpn.，44：137 - 141.

OGAWA T，FUJII K. 1979. Studies on the breeding of rice varieties resistance to bacterial leaf blight I. Resistance of rice varieties in Japan [J]. Bull. Chugoku Natl. Agr. Exp. Stn.，26：67 - 80.

OGAWA T. 1983. Pathogenic specialization in bacterial blight group I and III of *Xanthomonas campestris* pv. *oryzae* in Japan [J]. Ann. Phytopathol. Soc. Jpn.，49 (1)：69 - 72.

OGAWA T，YAMAMOTO T，KHUSH G S，et al. 1986. The *Xa3* gene for resistance to Philippine races of bacterial blight of rice [J]. Rice Genetics Newsletter，3：77 - 78.

OGAWA T，BUSTO G A，YAMAMOTO T，et al. 1986. Grouping of rice varieties based on reaction to four Philippine races of *Xanthomonas oryzae* pv. *oryzae* [J]. Rice Genetics Newsl.，3：84 - 86.

OGAWA T，YAMAMOTO T. 1986. Inheritance of resistance to bacterial blight in rice [C] // Rice Genetics. Philippines：IRRI：471 - 479.

OGAWA T, YAMAMOTO T, KHUSH G S, et al. 1986. The relationship between genes *Xa3* and *Xa6* for resistance to rice bacterial blight [J]. Rice Genetics Newsletter, 3: 79 - 80.

OGAWA T, LIN L, TABIEN R E, et al. 1987. A new recessive gene for resistance to bacterial blight of rice [J]. Rice Genet. Newsl., 4: 98 - 100.

OGAWA T, YAMAMOTO T. 1987. Reaction of rice cultivars resistance to Japanese and Philippine races of *Xanthomonas* pv. *oryzae* [J]. JARQ, 21: 138 - 145.

OGAWA T, KHUSH G S, MEW T W. 1988. Near-isogenic lines as international differentials for resistance to bacterial blight of rice [J]. Rice Genet. Newsletter, 6: 106 - 107.

OGAWA A T, KHUSH G S. 1989. Major genes for resistance to bacterial blight of rice [C] // Bacterial Blight of Rice. Manila, Philippines: IRRI: 177 - 192.

OGAWA T, YAMAMOTO T. 1989. Resistance gene of rice cultivar, Asominori to bacterial blight of rice [J]. Jpn. J. Breeding, 39: 196 - 197.

OGAWA T, YAMAMOTO T, KHUSH G S, et al. 1991. Breeding near-isogenic lines of rice with single genes for resistance to bacterial blight pathogen (*Xanthomonas campestris* pv. *oryzae*) [J]. Jpn. J. Breed, 41: 523 - 529.

OGAWA T, YAMAMOTO T, KHUSH G S, et al. 1991. Resistance and its inheritance to bacterial blight pathogen (*Xanthomonas campestris* pv. *oryzae*) [J]. Jpn. J. Breed, 41: 211 - 221.

OGAWA T. 1991. Grouping of rice cultivars based on reaction pattern to philippine races of bacterial blight pathogen [J]. Japanese journal of breeding, 41 (1): 109 - 119.

OGAWA T. 1993. Methods and strategy for monitoring race distribution and identification of resistance genes to bacterial leaf blight (*Xanthomonas campestris* pv. *oryzae*) in rice [J]. JARQ, 27 (2): 71 - 80.

OKA H I, Morishima H. 1967. The ancestors of cultivated rice and their evolution. Japan: Applied Genetics, National Institute of Genetics: 145 - 154.

OKABE N, GOTO M. 1963. Bacteriophages of plant pathogens [J]. Ann. Rev. Phytopathol., 1: 397 - 418.

OKADA M, YAMAKAWA Y, FUJII K, et al. 1967. On the new varieties of paddy rice, "Hoyoku, Kokumasari, and Shiranui" and notes on the parental varieties and their origins [J]. Bull. Kyushu Agric. Exp. Stn., 12 (3, 4): 187 - 224.

OLUFOWOTE J O, KHUSH G S, KAUFFMAN H E. 1977. Inheritance of bacterial blight resistance in rice [J]. Phytopathology, 67: 772 - 775

ONO E, WONG H L, KAWASAKI T, et al. 2001. Essential role of the small GTPase Rac in disease resistance of rice [J]. Proc. Natl. Acad. Sci. U S A, 98: 759 - 764.

OU S H, AYAD M R. 1968. Pathogenic races of *Pyricularia oryzae* originating from single lesions and monoconidial cultures [J]. Phytopathology, 58: 179 - 182.

OU S H, Nuque F L, Silva J P. 1971. Pathogenic variations among isolates of *Xanthomonas oryzae* of the Philippines [J]. Plant Disease Reporter, 55: 22 - 26.

OU S H. 1972. Bacterial leaf blight [C] //OU S H Rice Diseases. England: CMI Kew: 54 - 84.

OU S H. 1979. Breeding rice for resistance to blast, A critical review [J] // Proceedings of a Rice Blast Workshop. Manila, Philippines: IRRI: 81 - 137.

OWEN R J, BORMAN P. 1987. A rapid biochemical method for purifying high molecular weight bacterial chromosomal DNA for restriction enzyme analysis [J]. Nucleic Acids Res., 15: 3631.

PANDE H K, SEETHARAMAN R. 1980. Rice research and testing program in India [C] // IRRI, CAAS. Rice Improvement in China and other Asian countries. Manila: Philippines Press: 37 - 49.

PANDEY K R. 1970. Studies on bacterial leaf blight disease of rice [R] . Ph D thesis, Indian Agricultural Research Institute, New Delhi, India.

PARLEVLIET J E. 1976. Evaluation of the concept of horizontal resistance in barley *Puccinia hordei* host-pathogen relationship [J] . Phytopathology, 66: 494 - 497.

PARLEVLIET J E. 1979. Components of resistance that reduce the rate of epidemic development [J] . Annu. Rev. Phytopathol. , 17: 203 - 222.

PARLEVLIET J E. 1981. Disease resistance in plants and its consequences for plant breeding [C] // K J. Plant breeding II. Iowa: Iowa State University Press: 309 - 364.

PENG S Q, WEI Z S, MAO C X, et al. 1982. Identification of multi-resistance of *O. meyeriana*, *O. officinalis*, and *O. sativa* growing in Yunnan province [J] . Acta phytopathologica Sinica, 12 (40): 58 - 61.

PETPISIT V, KHUSH G S, KAUFFMAN H E. 1977. Inheritance of resistance to bacterial blight in rice [J] . Crop Sci. , 17: 551 - 554.

PETERSON B. 1944. Adult-plant resistance of some oat varieties to physiologic races of crown rust [J] . Can. J. Res. , 22: 287 - 289.

PIETERSE C M J, VAN - WEES S C M, TON J, et al. 2002. Signaling in rhizobacteria-induced systemic resistance in *Arabidopsis thaliana* [J] . Plant Biol. , 4: 535 - 544

PIETERSE C M, VAN - LOO L C. 2004. NPR1: the spider in the web of induced resistance signaling pathways [J] . Current Opinion in Plant Biology, 7: 456 - 464.

POTRYKUS I, HARMS C T, LORZ H. 1979. Callus formation from cell culture protoplasts of corn (*Zea mays* L.) [J] . Theor. Appl. Genet. , 54: 209 - 214.

QUENTIN J G, MIGUEL A T, ANTHONY P F. 1996. *rbohA*, a rice homologue of the mammalian *gp91phox* respiratory burst oxidase gene [J] . Plant J. , 10: 515 - 522.

RAFALSK I, ANTONY J, TINGE Y, et al. 1993. Genetic diagnostics in plant breeding: RAPDs, microsatellites and machines [J] . Trends in Genetics, 9 (8): 275 - 280.

RAO P S, KAUFFMAN H E. 1970. A new Indian host of *Xanthomonas oryzae*, incident of bacterial leaf blight of rice [J] . Curr. Sci. , 10: 271 - 272

RAO Y P, MOHAN S K, RADDY O R. 1971. Pathogenic variability in *Xanthomonas oryzae* [J] . Plant Dis. Rep. , 55: 593 - 595.

RASHILD H, YOKOI S, HINATA K, et al. 1996. Transgenic plant production mediated by *Agrobacterium* in indica rice [J] . Plant Cell Reports, 15: 727 - 730.

REDDY P R, NAUAK P. 1974. A new host for bacterial leaf blight pathogen of rice [J] . Curr. Sci. , 43: 116 - 117.

REDDY A P K, KAUFFMAN H E. 1974. Population studies of mixed inoculation of *Xanthomonas oryzae* in susceptible and resistant varieties of rice [J] . Annals of the Phytopathological Society of Japan, 40: 93 - 97.

RADDY O R, OU S H. 1976. Pathogenic variability in *Xanthomonas oryzae* [J] . Phytopathology, 66: 906 - 909.

REDDY A P K. 1989. Bacterial blight: crop loss assessment and disease management [C] //Proceeding of the International Workshop on Bacterial Blight of Rice. Philippines: IRRI: 79 - 88.

RABECCA J N, BARAOIDAN M R, VERA - CRUZ C M, et al. 1994. Relationship between phylogeny pathotype for bacterial blight pathogen of rice [J] . Appl. Environ. Microbiol. , 60: 3275 - 3283.

REINKING O A. 1918. Philippine economic - plant diseases [J] . Philippine J. Sci. , 13, 163 - 274.

REITSMA J, SCHURE P S J. 1950. "Kresek" a bacterial disease of rice [J]. Gontr. Gen. Agr. Res. Sta.

Boger. , 117: 1 - 17.

REIMERS P J, LEACH J E. 1991. Race-specific resistance to *Xanthomonas campestris* pv. *oryzae* conferred by bacterial blight resistance gene *Xa10* in rice (*Oryza sativa*) involves accumulation of a lignin-like substance in host tissues [J] . Physiol. and Mol. Plant Pathol. , 38: 39 - 55.

ROMEIS T, PIEDRAS P, ZHANG S, et al. 1999. Rapid Avr9 - and Cf - 9 - dependent activation of MAP kinases in tobacco cell cultures and leaves: convergence of resistance gene, elicitor, wound, and salicylate responses [J] . Plant Cell, 11: 273 - 287.

ROHLF F J. 1992. NTSYS-pc Numberical Taxonomy and Multivariate Analysis System. Version 1. 80 [CP/DK] . Setauket N. Y. : Exeter Publishing.

RONALD P C, TANKSLEY S D. 1991. Genetic and physical mapping of the bacterial resistance gene *Xa21* [J] . RGN, 8: 142 - 143.

RONALD P C, B ALBANO, R TABIEN, et al. 1992. Genetic and physical analysis of the rice bacterial blight disease resistance locus, *Xa21* [J] . Mol. Gen. Genet. , 236: 113 - 120

SAKAGUSHI S. 1967. Linkage studies on the resistance to bacterial leaf blight *Xanthomonas oryzae* (Uyeda et Ishiyama) Dowson in rice [J] . Bull. Nat. Inst. Agric. Sci. Jpn. , 16: 1 - 18.

SAKAGUSHI S, SUWA T, MURATA N. 1968. Studies on the resistance to bacterial leaf blight , *Xanthomonas oryzae* (Uyeda et Ishiyama) Dowson, in the cultivated and wild rices [J] . Ibid. Ser. D. , 18: 1 - 29.

SAKAGUCHI S. 1977. Breeding of rice strains with highly resistance to bacterial leaf blight [J] . Bull. Natl. Inst. Agric. Sci. , 28: 89 - 115.

SAMBROOK J, MANIATIS T, FRITSCH E F. 1989. Molecular Cloning: a Laboratory Manual [M] . 2 nd ed. Woodbury New York: Cold Spring Harbor Laboratory Pr.

SAS Institute, Inc. 1989. SAS/IML software: usage and reference, version 6 [CP/DK] . 1st ed. Cary, N C: SAS Institute Inc. 501.

SAS Institute, Inc. 1989. SAS/STAT User's Guide. 4th ed. , vol. 1. Cary, N. C: SAS Institute Inc. 943.

SASKAWICZ B J. 2001. Genetics of plant-pathogen interactions specifying plant disease resistance [J] . Plant Physiol. , 125: 73 - 76.

SATO T, CHOI Y S, IWASAKI M, et al. 1976. Distribution of races of *Xanthomonas oryzae* in Kyushu [J] . Ibid. , 42: 357.

SATO T, MATSUMOTO S, WATANABE B, et al. 1977. Studies on the resistance to rice bacterial leaf blight . Field resistance of Asominori [J] . Ann. Phytopathol. Soc. Jpn. , 43: 347.

SATO Z. 1979. Antibacterial substances for bacterial leaf blight pathogen contained in rice [J] . Kagaku to Seibutsu. 17: 83 - 84.

SEEM R C. 1984. Disease incidence and severity relationships [J] . ANNU. Rev. Phytopathol. , 22: 133 - 150.

SHARMA K K, KAUL M L H. 1984. Analysis of yield loss due to bacterial blight of rice [J] . Sci. Oult. , 50 (3): 102 - 104.

SHANTI M L, GEORGE M L C, VERA CRUZE C M, et al. 2001. Identif: cation of resistance genes effective against rice bacterial blight pathogen in Eastern India [J] . Plant Disease, 85 (5): 506 - 512.

SHEN Y, CHERN M, SILVA F G, et al. 2001. Isolation of a *Xanthomonas oryzae* pv. *oryzae* flagellar operon region and molecular characterization of flhF [J] . Mol. Plant-Microbe Interact. , 14: 204 - 213.

SIDHU G S, KHUSH G S. 1978. Dominance reversal of a bacterial blight resistance gene in some rice cultivars [J] . Phytopathology, 68: 461 - 463.

SIDHU G S, KHUSH G S, MEW T W. 1978. Genetic analysis of bacterial blight resistance in seventy-four cultivars of rice, *Oryza sative* L [J] . Theor. Appl. Genet, 53. 105 - 111.

SIDHU G S, KHUSH G S, MEW T W. 1979. Genetic analysis of resistance to bacterial blight in seventy cultivars of rice, *Oryza sativa* L. , from Indonesia [J] . Crop Improv. , 6 (1): 19 - 26.

SIMMONDS N W, SMARTT J. 1999. Principles [M] . [s. l.] Blackwell Science, Ltd.

SIMONS M D. 1955. Adult-plant resistance to crown rust of certain oats selections [J] . Phytopathology, 5: 275 - 278.

SINGH R J, KHUSH G S, MEW T W. 1983. A new gene for resistance to bacterial blight in rice [J] . Crop Sci. , 23: 558 - 560.

SINGH K, VIKAL Y, SINGH S, et al. 2002. Mapping of bacterial blight resistance gene *xa8* using microsatellite markers [J] . Rice Genetics Newsletter, 19: 94 - 96.

SKORN M, PAIBOON V, WIPA P. 1997. Induced adaptive and cross-protection responses against oxidative stress killing in a bacterial phytopathogen, *Xanthomonas oryzae* pv. *oryzae* [J] . FEMS Microbiology Letters, 146: 217 - 222.

SNEATH P H A, SOKAl R R. 1973. Numerical taxonomy: The principles and practice of numerical classification [M] . San Francisco: W H Freeman & Co.

SONG W Y, WANG G L, CHEN L H, et al. 1995. A receptor kinase-like protein encoded [J] . Science, 270: 1804 - 1806.

SONG W Y, PI L Y, CHEN L L, et al. 1997. Evolution of the *Xa21* disease resistance gene family [J] . Plant Cell, 9: 1279 - 1287.

SRINIVASAN B, DAVID PAULRAJ R S, LEENAKUMARI P V, et al. 2004. Improvement of bacterial blight resistance in rice cultivars Jyothi and IR50 by backcross breeding and marker-assisted selection [C] //Abstracts of the 1st International Conference on Bacterial Blight of Rice. Tsukuba. 37.

SRIVASTAVA D N. 1967. Epidemiology and control of bacterial blight disease of rice in India. [C] //Proceeding of a Symposium on Rice Diseases and their Control by Growing Resistant Varieties and other Measures : [s. l.]: [s. n.]: 11 - 18.

STAKMAN E C, PARKER J J, PIEMEISEL F J. 1918. Can biologic forms of stem rust on wheat change rapidly enough to interfere with breeding for rust resistance [J] . Agr. Res. , 14: 111 - 123.

STAKMAN E C, STEWART D M, LOEGERING W Q. 1962. Isolation of physiologic races of *Puccinia graminis* var. *tritici* [J] . US Dcp. Agric. Res. Serv. Bull, 617: 53.

STASKAWICZ B J. 2001. Genetics of plant-pathogen interactions specifying plant disease resistance [J] . Plant Physiol. , 125: 73 - 76.

STICHER L, MAUCH-MANI B, METRAUX P. 1997. Systemic acquired resistance [J] . Annu Rev Phytopathol. , 35: 235 - 270.

SUN W, DUNNING F M, PFUND C, et al. 2006. Within-species flagellin polymorphism in *Xanthomonas campestris* pv. *campestris* and its impact on elicitation of Arabidopsis FLAGELLIN SENSING2 - dependent defenses [J] . Plant Cell, 18: 764 - 779.

SWAMINATHAN M S. 1985. FOREWORD [C] //International Rice Research: 25 Years of Partnership. Los Nanos Laguna: IRRI.

SWINGS J, VAN DEN MOOTER M, VAUTERIN L, et al. 1990. Reclassification of the causal agents of bacterial blight (*Xanthomonas campestris* pv. *oryzae*) and bacterial leaf streak (*Xanthomonas campestris* pv. *oryzicola*) of rice as pathovars of *Xanthomonas* , *oryzae* (ex Ishiyama 1922) sp. nov. , nom. rev. Int [J]. J. Syst. Bacteriol, 40: 309 - 311.

SUN X, CAO Y, YANG Z, et al. 2004. *Xa26*, a gene conferring resistance to *Xanthomonas oryzae* pv. *oryzae in rice*, *encodes an* LRR *receptor kinase-like protein* [J]. Plant J., 37 (4): 517 - 527.

MCCOUCH S R, CHEN X, PANAUD O, et al. 1997. Microsatellite marker development, mapping and applications in rice genetics and breeding [J]. Plant Molecular Biology, 35: 89 - 99.

TABEI H. 1977. Anatomical studies of rice plant effected with bacterial leaf blight, *Xanthomonas oryzae* (Uyeda et Ishiyama) Dowson [J]. Bull. Kyushu. Agric. Exp. Stn., 19: 193 - 257.

TABIEN R E, ABALOS M C, BUSTAMAN M, et al. 2004. Multilines for bacterial blight control in rice [C] //Abstracts of the 1st International Conference on Bacterial Blight of Rice. Tsukuba 51.

TAGAMI Y. 1958. Quantity of bacterial leaf blight disease organism in paddy field water and occurrence of disease [J]. Proc. Assoc. Plant Porot., 4: 63 - 64.

TAGAMI Y, KUHARA S, KURITA T, et al. 1963. Epidemiological studies on the bacterial leaf blight of rice, *Xanthomonas oryzae* (Uyeda et Ishiyama) Dowson. I. The overwintering of the pathogen [J]. Bulletin of the Kyushu Agricultural Experiment station, 9: 89 - 122.

TAGAMI Y, KUHARA S, KURITA T, et al. 1964. Epidemiological studies on the bacterial leaf blight of rice, *Xanthomonas oryzae* (Uyeda et Ishiyama) Dowson. II. Successive change in the population of the pathogen in paddy fields during rice growing period [J]. Bulletin of the Kyushu Agricultural Experiment station, 10: 23 - 50.

TAKAHASHI M. 1964. Linkage groups and gene schemes of some striking morphological characters in Japanese rice [C] // Rice genetics and cytogenetics. [C] //215 - 236.

TAKAISHI M. 1909. Studies on bacterial leaf blight of rice, first report [J]. Dainhon Nokaiho, 340: 53 - 58.

TAKEUCHI H. 1930. On the occurrence of bacterial leaf blight of rice [J]. Bull. Chosen Sotokuhu Agr. Expt. Sta., 5 (1): 63 - 64.

TANAKA K, NAKATSU S, OKU T. 2004. Notes on the occurrence and variation of pathogenic races of *Xanthomonas oryzae* pv. *oryzae* found in Hiroshima Prefecture since 1999 to 2002 [C] //Abstracts of the 1st International Conference on Bacterial Blight of Rice. Tsukuba. 45.

TANKSLEY S, CAUSSE M, FULTON T, et al. 1992. A high density molecular map of the rice genome [J]. Rice Genetics Newsletter, 9: 111 - 115.

TANKSLEY S D, YOUNG N D, PATERSON A H, et al. 1989. RFLP mapping in plant breeding: New tools for an old science [J]. Nat. Biotech., 7: 257 - 264.

TAN Z B, SHEN L S, KUANG H C, et al. 1997. Identification of QTLs for lengths of the top internodes and other traits in rice and analysis of their effects [J]. Chinese J. of Genetics, 24: 15 - 22.

TAURA S, SUGITA Y, KAWAHARA D, et al. 2004. Gene distribution resistant to bacterial blight in Northern Vietnam rice varieties [C] //Abstracts of the 1st International Conference on Bacterial Blight of Rice. Tsukuba: 42

TAURA S, OGAWA T, TABIEN R E, et al. 1987. The specific reaction of Taichung Native 1 to Philippine races of bacterial blight and inheritance of resistance to race 5 (PXO112) [J]. Rice Genet. Newsl., 4: 101 - 102.

TAURA S, OGAWA T, TABIEN R E, et al. 1992. Resistance gene of rice cultivar, Taichung Native 1 to Philippine races of bacterial blight [J]. Japan J. Breed., 42: 195 - 201.

TAURA S, OGAWA T, YOSHIMUR A, et al. 1991. Identification of a recessive resistance gene in induced mutant line XM5 of rice to bacterial blight [J]. Jpn. J. Breeding, 41: 427 - 432.

TAURA S, OGAWA T, YOSHIMURA A, et al. 1992. Identification of a recessive gene to rice bacterial blight of mutant line XM6, *Oryza sativa* L. [J]. Jpn. J. Breeding, 42: 7-13.

THORDAL CHRISTENSEN H. 2003. Fresh insights into processes of nonhost resistance [J]. Curr. Opin. in Plant Biol., 6: 351-357.

THORSTEN N, FREDERIC B, BIRGIT K. 2004. Innate immunity in plants and animals: striking similarities and obvious differences [J]. Immunological Reviews, 198: 249-266.

TOSHIO M, FOLKE S. 1962. A revised medium for rapid growth and bioassays with tobacco tissue cultures [J]. Physiologia Plantarum, 15: 473-497.

TORRES M A, JONES J D, DANGL J L. 2006. Reactive oxygen species signaling in response to pathogens [J]. Plant Physiol., 141: 373-378.

TSUNO K, WKIMOTO S. 1983. Ultrastructural investigation on the host-parasite interaction in the rice leaf parenchymatous tissue infiltrated with bacterial leaf blight and leaf streak pathogens [J]. Ann. Phytopathol. Soc. Jpn., 49: 659-669.

TSUNO K, WKIMOTO S. 1988. Ultrastructural changes in the early stages of interactions between rice leaf parenchymatous tissue and infiltrated bacteria [J]. Ann. Phytopathol. Soc. Jpn., 54: 32-34.

TU J, ONA I, ZHANG Q, et al. 1998. Transgenic rice variety "IR72" with *Xa21* is resistant to bacterial blight [J]. Theoretical and Applied Genetics, 97: 31-36.

TU J, DATTA K, KHUSH G S, et al. 2000. Field performance of *Xa21* transgenic indica rice (*Oryza sativa* L.), 1R72 [J]. Theor. Appl. Genet., 101: 15-20.

UCHIYAMADA H, FUJITA Y, KIMURA K, et al. 1977. Survey of characteristics of domestic and introduced rice varieties (1) [J]. Res. Mater. Hokuriku Agric., 3: 137.

UCHIYAMADA H, FUJITA Y, KIMURA K, et al. 1978. Survey of characteristics of domestic and introduced rice varieties (2) [J]. Res. Mater. Hokuriku Agric., 4: 167.

VAN LOON L C, GERAATS B P J, LINTHORST H J M. 2006. Ethylene as a modulator of disease resistance in plants [J]. Trends in Plant Science, 11: 184-191.

VANDER PLANK J E. 1963. Plant Diseases: Epidemics and control [M]. New York, London: Academic Press,: 349

VANDERPLANK J E. 1978. Genetic and molecular basis of plant pathogenesis [M]. New York: Spring-Verlag Berlin Heidelberg.

VANDERPLANK J E. 1982. Host-pathogen interactions in plant disease [M]. New York: Academic Press: 157.

VAN DER PLANK J E. 1988. Disease resistance in plants [J]. New York: Academic Press: 206.

DER BIEZEN E A, JONES J D. 1998. Plant disease-resistance proteins and the gene-for-gene concept [J]. Trends Biochem Sci., 23: 454-456.

VENERE R J. 1980. Role of peroxidase in cotton resistant to bacterial blight [J]. Plant Sci. Lett., 20: 47-56.

VERA-CRUZ C M, MEW T W. 1986. Phenotypic analysis of the races of *Xanthomonas oryzae* pv. *oryzae* [J]. Phytopathology, 76 (18): 11-17.

VERA-CRUZ, C M NELSON R. LEUNG H, et al. 1992. Reaction of rice cultivars from Ifugao province, Philippines to indigenous strains of the bacterial blight pathogen [J]. IRRI IRRN, 17 (2): 89.

VERA-CRUZ C M, MEW T W. 1989. How variable is *Xanthomonas campestris* pv. *oryzae* [C] //Bacterial Blight of Rice. Manila: IRRI. 153-166.

VERA-CRUZ C M, RAYMUNDO A K, LEACH J E. 1994. Non-radioactive DNA analysis using biotin labeling and chemiluminescent detection [J]. Int. I Rice Res. Notes, 19: 28-29.

VERA-CRUZ C M, ARDALES D Z, SKINNER D Z, et al. 1996. Measurement of haplotypic variation to *Xoo within a single field by rep*-PCR *and* RFLP *analysis* [J]. Phytopathology, 86 (12): 1352 - 1359.

VERSALOVIC J, KOEUTH T, LUPSKI J R. 1991. Distribution of repetitive DNA sequences in eubacteria and application to fingerprinting of bacterial genomes [J]. Nucleic Acids Res., 19: 6823 - 6831.

VOS P, HOGERS R, BLEEKER M, et al. 1995. AFLP: a new technique for DNA fingerprinting [J]. Nucleic Acids Research, 23 (21): 4407 - 4414.

WAKIMOTO S. 1954. The determination of the presence of *Xanthomonas oryzae* by the phage technique [J]. Sci. Bull. Fa Fac. Agric. Kyushu Univ., 14: 495 - 498.

WAKIMOTO S, YOSHII H. 1954. Seasonal change of resistance of rice plants against leaf -blight disease [J]. Sci. Bull. Fac. Agric. Kyushu Univ., 14: 475 - 477.

WAKIMOTO S. 1960. A simple method for comparison of bacterial populations in a large number of samples by phage technique [J]. Annals of the Phytopathological Society of Japan, 22: 159 - 163.

WAKIMOTO S. 1967. Strains of *Xanthomonas oryzae* in Asia and their virulence against rice varieties [C] //Japan: Agric. For. Fish. Res. Coune. Proceedings of the Symposium "Rice Diseases and their Control by Growing Resistant Varieties and other Measures"., Minis. Agric. For. Tokyo. 19 - 24.

WAKIMOTO S. 1975. Bacterial leaf blight disease of rice in a South-East Asia countries [C] //Rice in Asia. Tokyo: the Association of Japanese Agricultural Scientific Societies University of Tokyo Press: 327 - 339.

WANG G L. 1994. RFLP mapping of genes conferring complete and partial resistance to blast in a durably resistant rice cultivar [J]. Genetics, 136: 1421 - 1434.

WANG G L, HOSTEN T E, SONG W Y, et al. 1995. Construction of a rice bacterial artificial chromosome library and identification of clones linked to the *Xa21* disease resistance locus [J]. Plant J., 7: 525 - 533.

WANG G L, SONG W Y, RONALD P, et al. 1996. The cloned gene, *Xa21*, confers resistance to multiple *Xanthomonas oryzae* pv. *oryzae* isolates in transgenic plants [J]. MPMI, 9: 850 - 855.

WANG G L, RUAN D L, SONG W Y, et al. 1998. *Xa21*D encodes a receptor-like molecular with a leucine-rich repeat domain that determines race-specific recognition and is subjective evolution [J]. Plant Cell, 10: 765 - 779.

WASANO K, KASHIMA Y, TOYOKAMA Y, et al. 1987. Varietal resistance to bacterial leaf blight (*Xanthomonas campestris* pv. *oryzae*) of the leading rice varieties cultivated in Kyushu Island of Japan [J]. Bull. Fac. Agric. Saga Univ., 63: 115 - 127.

WASHIO O, KARIYA K, TORIYAMA K. 1966. Studies on breeding rice varieties for resistance to bacterial leaf blight [J]. Bull. Chugoku Agric. Exp. Sci. Ser., 13: 55 - 85.

WATANABE B, KURITA T. 1958. Overwintering of *Xanthomonas oryzae* on weeds [J]. Annals of the Phytopathological Society of Japan, 23: 60.

WATANABE M. 1963. Studies on the nutritional physiology of *Xanthomonas oryzae* (Uyeda et Ishiyama) Dowson. II. Effects of carbon and nitrogen sources on multiplication of bacteria [J]. Annals of the Phytopathological Society of Japan, 28: 201 - 208.

WATANABE M. 1966. Studies on the strains of *Xanthomonas oryzae* (Uyeda et Ishiyama) Dowson, the causal bacterium of the bacterial leaf blight of rice plant [J]. Bulletin of the Faculty of Agriculture, Tokyo University of Agriculture and Technology, 10: 1 - 51.

WATANABE Y. 1975. Ecological studies on kresek phase of bacterial leaf blight of rice [J]. Bull. Tokai-Kinki Natl. Agric. Exp. Stn., 28: 50 - 123.

WATANABE M, NAKANISHI K, KUMAKURA K, et al. 1976. Studies on the mechanisms of resistance in the rice plant infected with *Xanthomonas oryzae*. I. Inhibition of lesion enlargement by the pre-inoculation with the non-pathogenic bacteria [J]. Ann. Phytopathol. Soc. Jpn., 42: 295 - 303.

WATANABE M, NAKANISHI K. 1976. Studies on the mechanisms of resistance in the rice plant infected with *Xanthomonas oryzae*. Ⅱ. Production of an inhibitor in infected leaves [J] . Ann. Phytopathol. Soc. Jpn., 42: 578 - 582.

WATANABE S. 1976. Studies on the breeding of rice varieties resistant to bacterial leaf blight in Sri Lanka [J] . Bull. Tohoku Natl. Agric. Exp. Stn., 54: 1 - 74.

WELSH J, MCCLELLAND M. 1990. Fingerprinting genomes using PCR with arbitrary primers [J] . Nucl. Acids Res., 18: 7213 - 7218.

WELSH J, MCCLELLAND M. 1991. Genomic fingerprints produced by PCR with consensus RNA gene primers [J]. Nucl. Acids Res., 19: 861 - 866.

WILLIAMS J G K, KUBELIK A R, LIVAK K, et al. 1990. DNA polymorphisms amplified by arbitrary primers are useful as genetic markers [J] . Nucl. Acids Res., 18: 6531 - 6535.

WOLFE M S. 1985. The current status and prospect of multiple cultivars and variety mixtures for disease resistance [J]. Annual Review of Phytopathology, 23: 251 - 273.

XINLI S, YINGLONG C, ZHIFEN Y, et al. 2004. *Xa26*, a gene conferring resistance to *Xanthomonas oryzae* pv. *oryzae* in rice, encodes an LRR receptor kinase-like protein [J] . Plant J., 37: 517 - 527.

YAMADAT, HORINO O, SAMOTO S. 1979. Studies on genetics and breeding of resistance to bacterial leaf blight in rice . 1. Discovery of new varietal group on the basis of reaction patterns to five different pathotypes of *Xanthomonas oryzae* (Uyeda et Ishiyama) Dowson in Japan [J] . Annals of the Phytopathological Society of Japan, 45: 240 - 246.

YAMADA T, HORLNO O. 1981. Studies on genetics and breeding of resistance to bacterial leaf blight in rice. V. The multiple alleles resistant to the bacterial groups I and V of *X. c.* pv. *oryzae* of Japan in the varieties, IR28, IR29 and IR30 [J] . Japan. J. Breed., 31 (40): 423 - 431.

YAMAMOTO T, HARTINI R N, MUHAMMAD M, et al. 1977. Variation in pathogenicity of *Xanthomonas oryzae* (Uyeda et Ishiyama) Dowson and resistance of rice varieties to the pathogen [J] . Centr. Res. Inst. Agric. Bogor., 28: 1 - 22.

YAMAMOTO T, MACHMUD M, HIFNI H R. 1977. Variation in pathogenicity of *Xanthomonas oryzae* and classification of rice varieties based on their reaction to the pathogen [J] . Cent. Gen. Agric. Res. Stn., 8: 33 - 45.

YAMAMOTO T, OGAWA T. 1990. Inheritance of resistance in rice cultivars, Toyonishiki, Milyang 23 and IR24 to Myanmar isolates of bacterial blight pathogen [J] . JARQ, 24: 74 - 77.

YAMANUKI S, NAKANURA K, KAYANO M, et al. 1962. First occurrence of bacterial leaf blight of rice in Hokkaido [J] . Ann. Phytopathol. Soc. Jpn., 27: 264.

YAMAOKA M, YOKOI Z, SHIMADA Y, et al. 1966. A new paddy rice variety, "Hagareshirezu" [J] . Bull. Shiga Prefect. Agric. Exp. Stn., 9: 5 - 9.

YANG B, ZHU W, JOHNSON L B, et al. 2000. The virulence factor *AvrXa7* of *Xanthomonas oryzae* pv. *oryzae* is a type III secretion pathway-dependent nuclear-localized double-stranded DNA-binding protein [J] . Proc. Natl. Acad. Sci. U S A, 97: 9807 - 9812.

YANG B, WHITE F F. 2004. Diverse members of the *AvrBs3/PthA* family of type III effectors are major virulence determinants in bacterial blight disease of rice [J] . Mol. Plant-Microbe Interact., 17: 1192 - 120.

YASUYUKI H, JUNKO K, KO S. 1988. Hybrids of rice (*Oryza sativa* L.) and wild *Oryza* species obtained by cell fusion [J] . Mol. Gen. Genet., 214: 6 - 10.

YANG Z F, SUN X L, WANG S P, et al. 2003. Genetic and physical mapping of a new gene for bacterial blight resistance in rice [J] . Theor. Appl. Genet., 106: 1467 - 1472.

YOSHIDA K, MUKOO H. 1961. A method of evaluating resistance of rice to bacterial leaf blight by multi-needle prick inoculation [J]. Shokubutsu Boeki (Plant Prot.), 15: 343 - 346.

YOSHIMURA S, IWATA K. 1965. Studies on evaluation method of varietal resistance to bacterial leaf blight of rice (1). Dip inoculation method and its application (1) [J]. Prot. Assoc. Plant Prot. Hokuriku., 13: 25 - 31.

YOSHIMURA A, MEW T W, KHUSH G S, et al. 1983. Inheritance of resistance to bacterial blight in rice cultivar Cas 209 [J]. Phytopathology, 73. 1409 - 1412.

YOSHIMURA A, MEW T W, KHUSH G S, et al. 1984. Genetics of bacterial blight resistance in IR1545 - 339, a breeding line of rice [J]. Phytopathology, 74: 773 - 777.

YOSHIMURA A, IDETA O, TAKEBAYASHI S, et al. 1992. Integration of RFLP and conventional maps in rice. The location of *GL - 1* AND *XA5* [J]. Jpn. J. Breeding, 42: 124 - 125.

Yoshimura A, Nagato, Y. 2001. Report of the committee on gene symbolization, nomenclature and linkage groups [J]. Rice Genetics Newsletter, 18: 6 - 8.

Yushimura S, Morihashi T, Suzuki Y. 1959. Major weed-hosts for the overwintering of the bacterial leaf blight organism and its distribution and development in Hokuriku District [J]. Annals of the Phytopathological Society of Japan, 24: 6.

YUSHIMURA S. 1963. Diagnostic and ecological studies of rice bacterial leaf blight, caused by *Xanthomonas oryzae* (Uyeda et Ishiyama) Dowson [J]. Bull. Hokuriku Agr. Expt. Stn., 5: 27 - 182.

YOSHIMURA S, NELSON R, YOSHIMURA A, et al. 1992. RFLP mapping of the bacterial blight resistance genes *Xa3* and *Xa4* [J]. RGN, 9: 136 - 138.

YOSHIMURA S, YOSHIMURA A, SAITO A, et al. 1992. RFLP analysis of introgressed chromosomal segments in three near isogenic lines of rice for bacterial blight resistance genes, *Xa1*, *Xa3* and *Xa4* [J]. Jpn. J. Genet., 67: 29 - 37.

YOSHIMURA S, YOSHIMURA A, IWATA N, et al. 1995. Tagging and combining bacterial blight resistance genes in rice using RAPD and RFLP markers [J]. Mol. Breeding, 1: 375 - 387.

YOSHIMURA S, YAMANOUCHI U, KURATA N, et al. 1996. Identification of a YAC clone carrying the *Xa1* allele, a bacterial blight resistance gene in rice [J]. Theor. Appl. Genet., 93: 117 - 122.

YOSHIMURA S, YAMANOUCHI U, SASAKI T, et al. 1998. Expression of *Xa1*, a bacterial blight resistance gene in rice, is induced by bacterial inoculation [J]. Proc. Natl. Acad. USA, 95: 1663 - 1668.

YOUNG S A, GUO A, GUIKEMA J A, et al. 1995. Rice cationic peroxidase accumulates in xylem vessels during incompatible interactions with *Xanthomonas oryzae* pv. *oryzae* [J]. Plant Physiol., 107: 1333 - 1341.

YU Z H, MACKILL D J, BONMAN J M, et al. 1996. Molecular mapping of genes for resistance to rice blast (*Pyricularia grisea* Sacc.) [J]. Theoretical and Applied Genetics, 93: 859 - 863.

YUKOH H, SHOZO O, TOSHIHIKO K, et al. 1994. Efficient transformation of rice (*Oryza sativa* L.) mediated by Agrobacterium and sequence analysis of the boundaries of the T-DNA [J]. Plant J., 6: 271 - 282.

YUN B W, ATKINSON H, GREENLAND A, et al. 2003. Loss of actin cytoskeletal function and EDS1 activity, in combination, severely compromises non-host resistance in Arabidopsis against wheat powdery mildew [J]. Plant J. 34: 768 - 777.

YUN C H. 1991. Ph. D. dissertation: Molecular characterization of a repetitive element of *Xanthomonas oryzae* pv. *oryzae* [R]. Manhattan: Kansas State University.

YUWEI S, PARVEEN S. FRANCISCO G, et al. 2002. The *Xanthomonas oryzae* pv. *oryzae raxP* and *raxQ* genes encode an ATP sulphurylase and adenosine - 5 phosphosulphate kinase that are required for *AvrXa21* avirulence activity [J]. Molecular Microbiology, 44: 37 - 48.

ZADOKS J C, SCHEIN R D. 1979. Epidemiology and plant diseases management [M]. New York: Oxford

Univ. Press: 427.

ZHANG G, ANGELES E R, ABENES M L P, et al. 1996. RAPD and RFLP mapping of the bacterial blight resistance gene *xa13* in rice [J]. Theor. Appl. Genet., 93: 65-70.

ZHANG H S, WANG J F, ZHU L H. 2004. Application of *Xa7*, a resistance gene to the bacterial blight *Xanthomonas oryzae* pv. *oryzae* in the breeding of hybrid rice in China [C] //Abstracts of the 1st International Conference on Bacterial Blight of Rice. Tsukuba, Japan.

ZHANG Q, MEW T W. 1984. Characterizing adult plant resistance to bacterial blight (BB) [J]. International Rice Research Notes, 9: 6.

ZHANG Q, MEW T W. 1985. Adult-plant resistance of rice cultivars to bacterial blight [J]. Plant Disease, 69: 896-898.

ZHANG Q, MEW T W. 1987. The bacteriophage technique for estimating *Xanthomonas campestris* pv. *oryzae* in leaf tissues [J]. Chinese J. Rice Sci., 1 (3): 177-183.

ZHANG Q, MEW T W, VERA-CRUZ C M. 1987. Comparison of virulence of *Xanthomonas campestris* pv. *oryzae* in North China and in the Philippines [C] //Abstracts of the Philippines Conference of Plant Pathology. Manila, Philippines.

ZHANG Q, MEW T W. 1988. Types of resistance in rice to bacterial blight [C] // Bacterial Blight of Rice, Proceedings of the International Workshop on Bacterial Blight of Rice. Manila: IRRI: 125-134.

ZHANG Q, WANG C L, SHI A N. 1990. Evaluation of near-isogenic lines with 8 genes for resistance to strains of bacterial blight in China [J]. Chinese Rice Res. Newsletter, 1: 3-5.

ZHANG Q, LEACH J E, AND WANG C L. 1996. Genetic structure study of bacterial blight for rice breeding in China [C] //Abstracts for East Asia Regional Meeting of the International Program on Rice Biotechnology, Hangzhou.

ZHANG Q, LEACH J E, WANG C L, et al. 1997. Study on the genetic diversity of *Xanthomonas oryzae* pv. *oryzae* to enhance rice breeding in China [C] //Editorial Department of Scientia Agricultura Sinica, Chinese Agriculture Sciences. For the Compliments to the 40th Anniversary of the Founding of the Chinese Academy of Agricultural Sciences. 15-21.

ZHANG Q, LIN S C, ZHAO B Y, et al. 1998. Identifying of a new gene for resistance to bacterial blight from *O. rufipogon* [J]. RGN, 15: 138-142.

ZHANG Q, WANG C L, ZHAO K J, et al. 2001. The effectiveness of advanced rice lines with new resistance gene *Xa23* to rice bacterial blight [J]. Rice Genetics Newsletters, 18: 71-72.

ZHANG Q, WANG C L, ZHAO K J, et al. 2002. Development of near-isogenic line CBB23 with a new resistance gene to bacterial blight in rice and its application [J]. Chinese J. Rice Sci., 16: 206-210.

ZHANG Q, TANG S X. 2004. How to evaluate and identify resistance gene to bacterial blight from wild rice [C] // Proceeding of the 1st International Conferrence on Bacterial Blight of Rice. Tsukuba Science City Ibaraki, Japan.

ZHANG Q, LAM H M. 2007. Studies on genetics and improvement of resistance to bacterial blight in China [C] // 2nd International Rice Bacterial Blight Conference. Nanjing.

ZHANG Q. 2009. Genetics and improvement of resistance to bacterial blight in hybrid rice [J]. Rice Science, 16 (2): 83-92.

ZHANG S, DU H, KLESSIG D F. 1998. Activation of the tobacco SIP kinase by both a cell wall-derived carbohydrate elicitor and purified proteinaceous elicitins from Phytophthora spp. [J]. Plant Cell, 10: 435-450.

ZHANG S P, SONG W Y, Chen L L, et al. 1998. Transgenic elite indica rice varieties, resistant to *Xanthomonas oryzae* pv. *oryzae* [J]. Molecular Breeding, 4: 551-558.

ZHANG Z H. 1982. Application of anther culture techniques to rice breeding [C] // Proc. on Rice Tissue Culture Planning Conference. Manila：IRRI. 55－61.

ZHAO B Y，ZHANG Q，LEUNG H，et al. 1998. Identification and tagging a new rice bacterial blight resistance gene from *Oryzae rufipogon* [C] // Report in Sixth International Symposium on Rice Molecular Biology. Shanghai，& Hangzhou RF.

ZHAO B Y，ARDALES E Y，RAYMUNDO A，et al. 2004. The *avrRxo1* gene from the rice pathogen *Xanthomona oryzae* pv. *oryzicola* confers a nonhost defence reaction on maize with *Rxo1* [J]. Mol. Plant-Microbe Interact.，17：771－779.

ZHAO B Y，ARDALES E Y，BRASSET E，et al. 2005. The *Rxo1/Rba1* locus of maize controls resistance reactions to pathogenic and non-host bacteria [J]. Theor. Appl. Genet.，109：71－79.

ZHAO B Y，LIN X，POLAND J，et al. 2005. A maize resistance gene functions against bacterial streak disease in rice [J]. Proc. Natl. Acad. Sci. U S A，102：15383－15388.

ZHENG K L，Huang N. 1997. Outlook on the application of marker-assisted selection in rice improvement [J]. Hereditas，19 (2)：40－44.

ZHU H，QU F，ZHU L H，et al. 1993. Isolation of genomic DNAs from plants，fungi and bacteria using benzyl chloride [J]. Nucleis Acid Research，21：5279－5280.

ZHU W，YANG B，CHITTOOR J M，et al. 1998. *AvrXa10* contains an acidic transcriptional activation domain in the functionally conserved C terminus [J]. Mol. Plant-Microbe Interact.，11：824－832.

ZIPFEL C，FELIX G. 2005. Plants and animals：a different taste for microbes [J]. Curr. Opin. Plant Biol.，8 (4)：353－360.

图书在版编目（CIP）数据

耕耘：章琦稻病抗性研究选集 / 章琦，林汉明编
. —北京：中国农业出版社，2010. 5
ISBN 978-7-109-14524-5

Ⅰ. ①耕… Ⅱ. ①章… ②林… Ⅲ. ①水稻-病虫害
防治方法-文集 Ⅳ. ①S435. 11 - 53

中国版本图书馆 CIP 数据核字（2010）第 071153 号

中国农业出版社出版
（北京市朝阳区农展馆北路 2 号）
（邮政编码 100125）
责任编辑 张洪光

北京中科印刷有限公司印刷 新华书店北京发行所发行
2010 年 7 月第 1 版 2010 年 7 月北京第 1 次印刷

开本：787mm×1092mm 1/16 印张：43 插页：8
字数：1050 千字 印数：1～1 000 册
定价：110. 00 元